ELECTROMAGNETIC WAVE PROPAGATION

ELECTROMAGNETIC WAVE PROPAGATION

International Conference Sponsored by the Postal and Telecommunications Group of the Brussels Universal Exhibition

EDITED BY

M. DESIRANT
Société Belge de Physique
Loverval
Belgium

J. L. MICHIELS
Laboratoire de Recherches
Physiques, A.C.E.C.,
Charleroi Belgium

1960

ACADEMIC PRESS

LONDON AND NEW YORK

ACADEMIC PRESS INC. (LONDON) LTD.
17 OLD QUEEN STREET
LONDON, S.W.1.

U.S. Edition Published by
ACADEMIC PRESS INC.
111 FIFTH AVENUE
NEW YORK 3, NEW YORK

Library of Congress Catalog Card Number: 59-12033

PRINTED IN GREAT BRITAIN AT
THE UNIVERSITY PRESS
ABERDEEN

FOREWORD

In planning the 1958 Brussels International Exhibition, the organizers envisaged from the outset that one of its most important aspects would be to review scientific progress and its contribution to the welfare of mankind.

Pursuing this idea, Group 38, the Postal and Telecommunications Group of the Exhibition, which included eminent representatives of the Universities, Government Departments and Industry, under the able and energetic chairmanship of Mr. L. Ros, Inspector General of the "Régie des Télégraphes et Téléphones," came to the conclusion that the organization of suitably chosen International Scientific Conferences lying within the field of interest of the Group was one of the most appropriate means of furthering this end.

The papers presented at one such conference form the contents of the present volume. The choice of subject, we feel, needs no particular justification. The importance of a wider and deeper knowledge of electromagnetic wave propagation for the continued progress of modern telecommunication needs no emphasis. Moreover, research on this subject has been particularly active in recent years and has led to advances of major interest. An outstanding illustration of this is the investigation of ionospheric and tropospheric scatter and its application to beyond-the-horizon communication.

The response of the scientists and engineers working in the field to the invitation to participate was most gratifying and in itself an indication of the opportunity of holding the Conference. Some fifty papers were presented by authors from eleven different countries. There was thus every opportunity of a fruitful comparison of the work carried out in different parts of the world.

It is a pleasure to mention our indebtedness to the Minister of Economic Affairs for a stbstantial contribution toward the cost of the Conference and to the Belgian Physical Society for secretarial and other assistance in its organization.

We must also extend our special thanks to the Academic Press Inc. for undetaking the arduous task of publishing the Proceedings of the Conference. In order to avoid unnecessary delay in publication, no attempt has been made to standardize notations and so on, the papers having been prepared according to the customs prevailing in the country of origin.

J. L. MICHIELS
M. DESIRANT

CONTRIBUTORS

L. A. AMES, *Electronic Research Directorate, Air Force Cambridge Research Center, Bedford, Massachusetts.* (pp. 215, and 603)

L. J. ANDERSON, *Smyth Research Associates, San Diego, California.* (p. 209)

E. J. BAGHDADY, *Massachusetts Institute of Technology, Cambridge, Massachusetts.* (p. 183)

B. R. BEAN, *National Bureau of Standards, Boulder, Colorado.* (p. 163)

B. BECKMANN, *Fernmeldetechnisches Zentralamt, Darmstadt, Germany.* (p. 157)

P. BECKMANN, *Institute of Radio Engineering and Electronics, Czechoslovak Academy of Sciences, Prague, Czechoslovakia.* (p. 445)

R. BELLMAN, *The RAND Corporation, Santa Monica, California.* (p. 243)

Å. BLOMQUIST, *Research Institute of National Defence, Stockholm, Sweden.* (p. 127)

H. BREMMER, *Philips' Research Laboratories, N.V. Philips' Gloeilampenfabrieken, Eindhoven, Netherlands.* (p. 253)

F. CARASSA, *Laboratorio Centrale Radio della Fabbrica Italiana Magneti Marelli, Milano, Italy.* (p. 471)

G. CARLSON, *Research Institute of National Defence, Stockholm, Sweden.* (p. 459)

F. du CASTEL, *Centre National d'Etudes des Télécommunications, Issy-les-Moulineaux (Seine), France.* (pp. 591 and 671)

A. M. CONDA, *National Bureau of Standards, Boulder, Colorado.* (pp. 103 and 661).

P. W. COUCH, *Wright Air Development Center, Wright-Patterson Air Force Base, Ohio.* (p. 479)

E. DEWAN, *Electronics Research Directorate, Air Force Cambridge Research Center, Bedford, Massachusetts.* (p. 575)

W. DIEMINGER, *Max-Planck-Institut für Aeronomie, Lindau (Harz), Germany.* (p. 699)

P. R. DROUILHET, *Jr., Wright Air Development Center, Wright-Patterson Air Force Base, Ohio.* (p. 149)

W. J. FAY, *Smyth Research Associates, San Diego, California.* (p. 317)

B. FRIEDMAN, *University of California, Berkeley, California.* (p. 261)

M. W. GOUGH, *Research Division, Marconi's Wireless Telegraph Co., Great Baddow, Essex.* (p. 557)

J. S. GREENHOW, *University of Manchester, Jodrell Bank Experimental Station, Cheshire.* (p. 493)

S. GRUBER, *Air Force Cambridge Research Center, Bedford, Massachusetts.* (p. 299)

O. HACHENBERG, *Heinrich-Hertz-Institut der Deutschen Akademie der Wissenschaften, Berlin, Germany.* (p. 143)

E. HARNISCHMACHER, *Ionosphären Institut Breisach im Fernmeldetechnischen Zentralamt der Deutschen Bundespost, Breisach (Rhein), Germany.* (p. 527)

J. L. HERITAGE, *Smyth Research Associates, San Diego, California.* (p. 317)

G. A. ISTED, *Research Division, Marconi's Wireless Telegraph Co., Great Baddow, Essex.* (pp. 515 and 541)

C. JAUQUET, *Laboratoire de Recherches Physiques, Ateliers de Constructions Electriques de Charleroi, Charleroi, Belgium.* (p. 271)

R. KALABA, *The RAND Corporation, Santa Monica, California.* (p. 243)

T. J. KEARY, *U.S. Navy Electronics Laboratory, San Diego, California.* (p. 277)

J. M. KELSO, *Space Technology Laboratories, Los Angeles, California.* (p. 291)

U. KÜHN, *Betriebslaboratorium für Rundfunk und Fernsehen, Berlin-Adlershof, Germany.* (p. 615)

A. W. LADD, *Communications Division, Hughes Aircraft Company, Los Angeles, California.* (p. 357)

E. A. LAUTER, *Observatorium für Ionosphärenforschung, Kühlungsborn, Germany.* (p. 505)

E. A. LEWIS, *Electronics Research Directorate, Air Force Cambridge Research Center, Bedford, Massachusetts.* (p. 335)

F. A. LOSEE, *Communications Division, Hughes Aircraft Company, Los Angeles, California.* (p. 357)

S. G. LUTZ, *Communications Division, Hughes Aircraft Company, Los Angeles, California.* (p. 357)

C. L. MACKEY, *Rome Air Development Center, Griffiss Air Force Base, Rome, New York.* (p. 685)

F. MARIANI, *Istituto di Fisica dell' Università and Istituto Nazionale di Geofisica, Roma, Italy.* (p. 451)

H. S. MARSH, *Air Force Cambridge Research Center, Bedford, Massachusetts.* (p. 719)

E. J. MARTIN, *Electronics Research Directorate, Air Force Cambridge Research Center, Bedford, Massachusetts.* (pp. 215 and 603)

P. MISME, *Centre National d'Etudes des Télécommunications, Issy-les-Moulineaux (Seine), France.* (pp. 591 and 671)

H.-G. MÖLLER, *Max-Planck-Institut für Aeronomie, Lindau (Harz), Germany.* (p. 699)

E. L. NEUFELD, *University of Manchester, Jodrell Bank Experimental Station, Cheshire.* (p. 493)

K. A. NORTON, *National Bureau of Standards, Boulder, Colorado.* (p. 375)

A. OMHOLT, *The Auroral Observatory, Tromsö, and The Institute of Theoretical Astrophysics, University of Oslo, Norway.* (p. 75)

K. W. OTTEN, *Wright Air Development Center, Wright-Patterson Air Force Base, Ohio.* (pp. 63 and 149)

G. B. PARRENT, *Jr., Air Force Cambridge Research Center, Bedford, Massachusetts.* (p. 55)

L. PENNINCKX, *Régie des Télégraphes et des Téléphones, Bruxelles, Belgium.* (p. 49)

P. QUARTA, *Laboratorio Centrale Radio della Fabbrica Italiana Magneti Marelli, Milano, Italy.* (p. 471)

I. RANZI, *Centro Radioelettrico Sperimentale "G. Marconi", Roma, Italy.* (p. 29)

K. RAWER, *Ionosphären Institut Breisach im Fernmeldetechnischen Zentralamt der Deutschen Bundespost, Breisach (Rhein), Germany.* (p. 647)

R. Rawhouser, *Wright Air Development Center, Wright-Patterson Air Force Base, Dayton, Ohio.* (p. 227)

G. C. Rider, *Research Division, Marconi's Wireless Telegraph Co., Great Baddow, Essex.* (p. 631)

T. F. Rogers, *Electronic Research Directorate, Air Force Cambridge Research Center, Bedford, Massachusetts.* (pp. 215 and 603)

G. Rose, *Max-Planck-Institut für Aeronomie, Lindau (Harz), Germany.* (p. 699)

R. Schünemann, *Heinrich-Hertz-Institut der Deutschen Akademie der Wissenschaften, Berlin, Germany.* (p. 15)

P. Springer, *Aerial Reconnaissance Laboratory, Wright Air Development Center, Dayton, Ohio.* (p. 227)

K. Toman, *Geophysics Research Directorate, Boston, Massachusetts.* (p. 1)

L. G. Trolese, *Smyth Research Associates, San Diego, California.* (p. 209)

J. Voge, *Centre National d'Etudes des Télécommunications, Issy-les-Moulineaux (Seine), France.* (pp. 591 and 671)

K. Vogt, *Fernmeldetechnisches Zentralamt, Darmstadt, Germany.* (p. 157)

H. Volland, *Heinrich-Hertz-Institut der Deutschen Akademie der Wissenschaften, Berlin, Germany.* (p. 143)

J. R. Wait, *National Bureau of Standards, Boulder, Colorado.* (pp. 87, 103 and 661)

A. T. Waterman, *Jr., Stanford Electronics Laboratories, Stanford University, Stanford, California.* (p. 111)

S. Weisbrod, *Smyth Research Associates, San Diego, California.* (pp. 209 and 317)

H. J. Wirth, *U.S. Navy Electronics Laboratory, San Diego, California.* (p. 277)

E. Wolf, *The Physical Laboratories, University of Manchester, Lancashire.* (p. 119)

M. S. Wong, *Electronics Research Directorate, Air Force Cambridge Research Center, Bedford, Massachusetts.* (p. 37)

CONTENTS

Ray-geometry Considerations for Highly Elevated Antennas

K. TOMAN

Geophysics Research Directorate, Boston, Mass.

A minimum property of the circle * and a maximum property of two concentric circles is stated. Based on these properties geometrical conditions are described as they apply to a spherical earth and ionosphere for highly elevated antennas. Under simplifying assumptions the ray paths are obtained and intersected radially at great angular distances from the antenna. For given parameters and because of the requirement that each point along the radial orbit corresponds to only one ray path it is found that optimum conditions exist for the largest possible range of positive and negative elevation angles. The importance of the minimum property regarding the MUF-concept is pointed out.

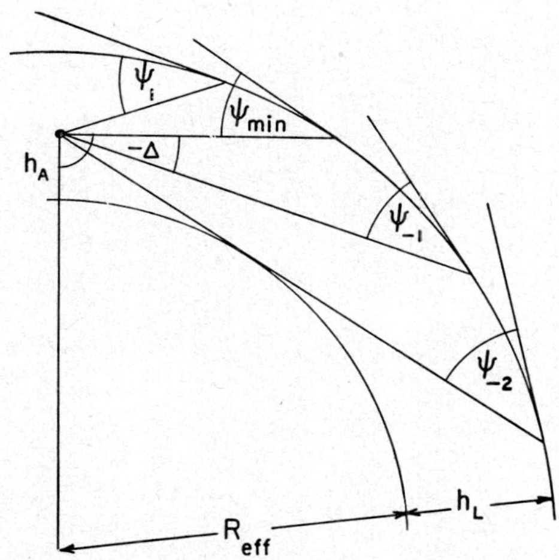

FIG. 1. A minimum property of the circle.

I. THE MINIMUM PROPERTY OF THE CIRCLE [4]

Without including the geometrical proof which was reported elsewhere [5] the general formulation of the minimum property is as follows: If one chooses an arbitrary point P inside of a circle other than its center, a normal s upon the diameter, both through P, intersects the circle with ψ, which is the angle between the normal and the tangent to the circle at the intersection point. Rotating s on P in either sense increases ψ. For the conditions stated above ψ must be a minimum. This is illustrated in Fig. 1. $R_{\rm eff}$ is the symbol for the modified earth

* The minimum property was reported at the Spring URSI-meeting, 23–26 April 1958 in Washington, D.C.

radius. h_A represents the height above ground of an antenna. h_L is the height of the reflecting layer. The normal upon the radius through the point at h_A intersects the circle with ψ_{min}. If positive and negative deviations from the right angle are of equal magnitude the angles of intersection ψ_1 and ψ_{-1} are equal.

The minimum property is of importance as it pertains to the reflection of radio waves from an ionospheric layer. For frequencies above about 30 kc/s the reflection coefficient of the ionosphere depends upon the angle of incidence at the layer. Consequently, the minimum property of the angle ψ corresponds to a maximum property for the reflection coefficient.

For increasingly negative elevation angles, which become feasible for highly elevated antennas, ψ increases and the reflection coefficient decreases until the bulge of the earth obstructs the radio beam. This is obtained when the beam is tangential to the earth. Accounting for uniform refraction the actual radius of the earth is modified by a factor of 1·25. This factor has been adopted by the National Bureau of Standards for ionospheric radio propagation [1].

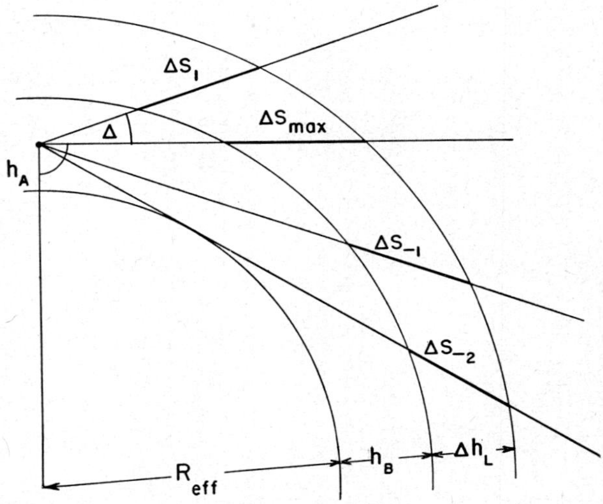

FIG. 2. A maximum property of two concentric circles.

II. THE MAXIMUM PROPERTY OF TWO CONCENTRIC CIRCLES

The geometrical proof was reported elsewhere [6]. The general formulation of the maximum property is as follows: If one chooses an arbitrary point P inside of two concentric circles a normal s upon the diameter, both through P, forms two secants. The difference in length between the secants of both circles diminishes if s is rotated on P in either sense. For the conditions stated above this difference is a maximum. This is illustrated in Fig. 2. h_B represents the height of the bottom of a layer of thickness Δh_L. The normal upon the radius through the point at h_A intersects both circles and forms a segment Δs_{max}. If positive and negative deviations from the right angle are of equal magnitude the segments Δs_1 and Δs_{-1} are equal.

The maximum property is of importance as it pertains to the absorption of a radio wave passing through an ionospheric layer of finite thickness. The absorption depends on the absorption coefficient integrated over the length of the ray path in the absorbing medium [2]. In the following analysis only the geometrical behavior of ray paths is described. Moreover it is assumed that ray paths are not refracted inside the absorbing layer. This assumption conflicts less with actual physical conditions the higher the frequency of the radio wave.

III. The Earth Geometry

The above mentioned geometrical properties are generally valid. Applied to a spherical earth and ionosphere and selecting specific values for various parameters the following geometrical conditions were obtained. As shown in Fig. 3

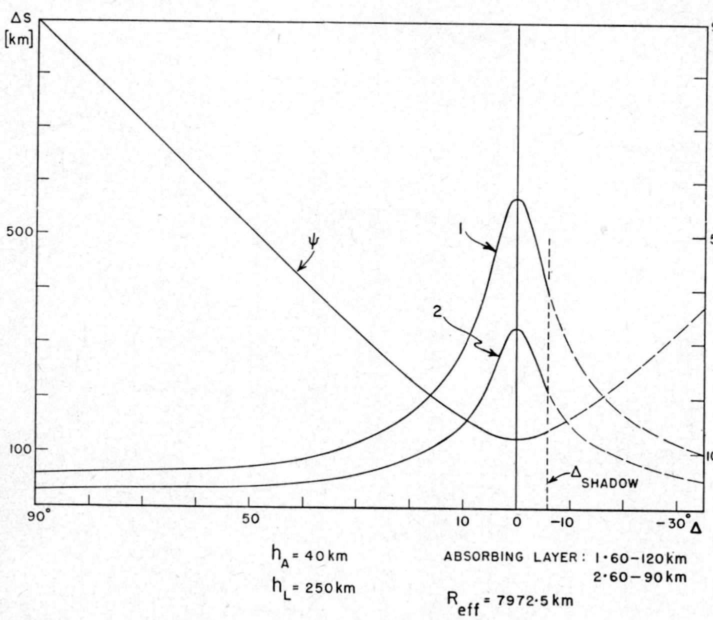

FIG. 3. Reflection angle ψ and absorption path length Δs as a function of the elevation angle Δ.

a radio source is assumed to be at a height $h_A = 40$ km above the surface of the earth. The effective earth radius is 7972·5 km. The elevation angle Δ of a linear ray is progressively moved from 90° to about $-35°$. The behavior of the absorption path length Δs as a function of Δ for two absorbing layers, one extending from 60–120 km (1) and the other from 60–90 km (2), reveals a sharp maximum at $\Delta \equiv 0$. Δs is a minimum for $\Delta = 90°$. In either case the ratio of $\Delta s_{max}/\Delta s_{min}$

is approximately equal to 10. The beam, originating at h_A, grazes the earth for a negative elevation angle marked Δ_{shadow}. In this case $\Delta_{shadow} = -5{\cdot}7°$. For a reflecting layer height of $h_L = 250$ km, ψ is displayed as a function of Δ. ψ is a minimum for $\Delta \equiv 0$. Like the distribution of Δs, the ψ-distribution is symmetrical with respect to $\Delta = 0$. As pointed out before, this property holds for any point inside of the circle; thus it is true for any antenna height below the absorbing layer. It can also be seen that for elevated antennas these optimum properties for ψ and Δs are within the observation range until the bulge of the earth commences its obstruction at Δ_{shadow}.

Somewhat different results are obtained when the antenna touches the bottom of the absorbing layer. This is illustrated in Fig. 4. The absorbing layer extends

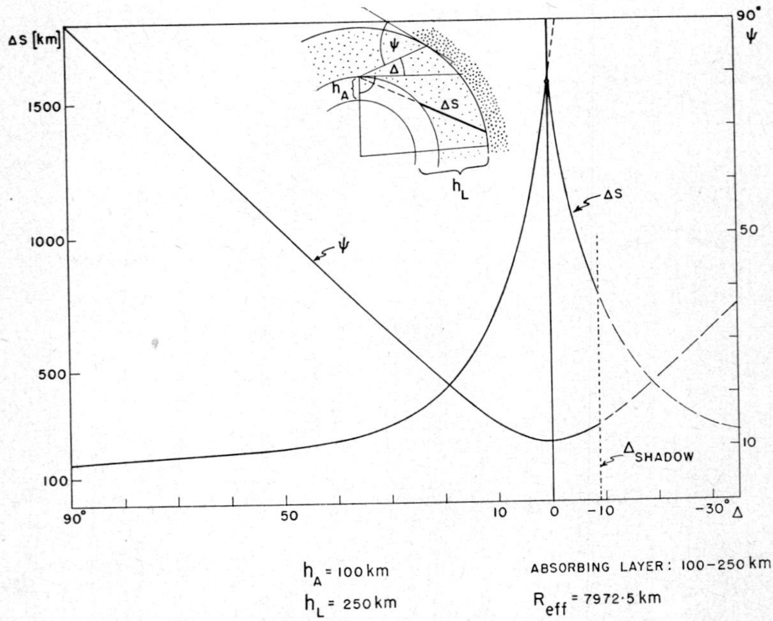

FIG. 4. Reflection angle Ψ and absorption path length Δs as a function of the elevation angle Δ.

from 100–250 km. The reflection takes place at 250 km. Again, the maximum of Δs is obtained at $\Delta = 0$. Although the general shape of the curve is different from the one in Fig. 3, Δs is still symmetrical relative to $\Delta = 0$. The same is true for ψ as a function of Δ. In this example the shadow angle is about 9°.

Fig. 5 displays the dynamics of Δs relative to Δ for values of h_A which increase from below to above the absorbing layer. The layer extends from 60–120 km. For an antenna height of 40 km Δs is a continuous function of Δ as was also shown in curve 1 of Fig. 3. For an antenna height of 60 km Δs has the shape similar to the display of Fig. 4. In both cases Δs_{max} is obtained for zero elevation angle Δ. For an antenna height of 90 km, which places the ray source inside the absorbing region, the maximum of Δs does no longer occur at $\Delta = 0$. The Δs-distribution also ceases to be symmetrical. As the antenna height is progressively increased the maximum of Δs increases while advancing towards more negative values of Δ. Once h_A lies above the absorbing region, Δs_{max} remains constant, although the maximum moves to the right as the antenna height continues to rise. The dotted line represents the locus of the shadow angles for the selected antenna heights.

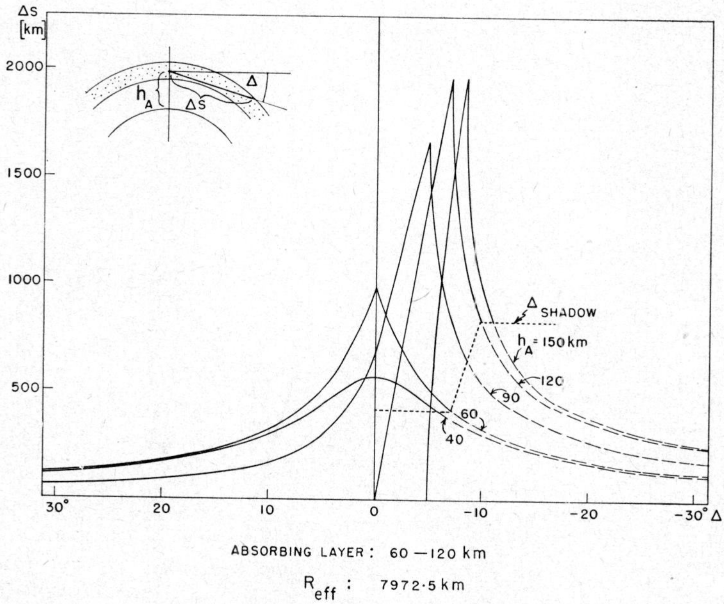

FIG. 5. Absorption path length Δs as a function of the elevation angle Δ for various antenna heights h_A.

It is now of interest to conceive an experiment in which the optimum properties of the circle are utilized for a radio transmission experiment. These optimum properties are perhaps observable if an orbiting satellite is monitored from a highly elevated receiver. In this case the satellite itself may be transmitting or it may act as a passive reflector. In either case the geometry is simple. A more complicated geometry exists when these optimum properties are to be observed by intersecting the ray paths along a radial orbit at great angular distances from the elevated source. This case will now be treated in more detail. Fig. 6 illustrates the spherical geometry involved. The ray path length from the source at h_A to the reflecting layer is x. The distance from the reflection point to the orbit is y. The shadow line of the direct ray is obtained when the ray is tangent to the earth (Δ_{shadow}). y_s is the length of the reflected ray where it intersects the shadow line. The latter is considered an upper limit for the radial orbits. It is convenient to also introduce a limiting condition when the reflected ray grazes the earth. This occurs for an elevation angle which is equal in magnitude to Δ_{shadow}. $K = \infty$ marks one point of infinite convergence factor.

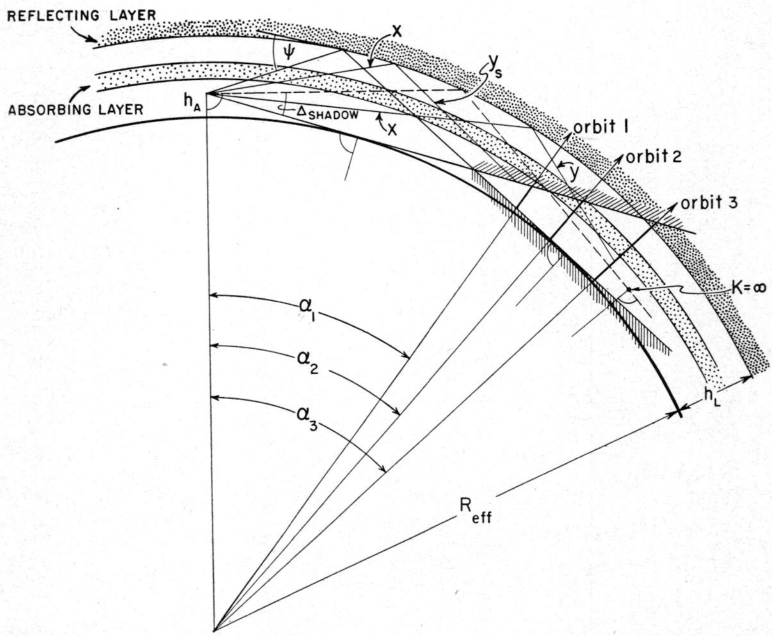

FIG. 6. Ray geometry for a spherical earth.

Fig. 7 illustrates the total absorption path length Δs_{total} as a function of the elevation angle Δ. The absorbing layer extends from 60-120 km. The reflecting layer is at 250 km. The antenna height $h_A = 40$ km. The distribution displays a transition from $2\Delta s$ to Δs as the orbits intersect the ray paths inside the absorbing region. For example, for $\alpha = 24°$ the curve for $2\Delta s$ passes through the maximum at $\Delta \equiv 0$. For increasing values of $-\Delta$ the absorption path length diminishes uniformly. At $\Delta \doteq 2°$ intersection commences inside the absorbing region. For $\Delta \doteq 5°$ the radial orbit intersects the ray path above the absorbing region and the Δs-distribution takes over.

IV. THE CONVERGENCE FACTOR K

The convergence factor is a measure of the focusing effect. This effect is important in the vicinity of $\Delta = 0$. K is defined [3] as follows:

$$K = \left(1 - \frac{2xy}{r_0(x+y)\sin\psi}\right)^{-1}$$

where x, y, and ψ were already identified. r_0 is the radius of curvature of the smooth reflecting layer ($r_0 = R_{eff} + h_L$; Fig. 6). Advancing from the source at h_A (Fig. 6) along a ray path of elevation angle Δ, the convergence factor K remains identically equal to unity. After reflection at a distance x from the source K becomes greater than unity as y increases. Eventually it reaches infinity. This is the point where initially diverging rays intersect each other after reflection. A further increase of y leads to diminishing values of K of opposite sign.

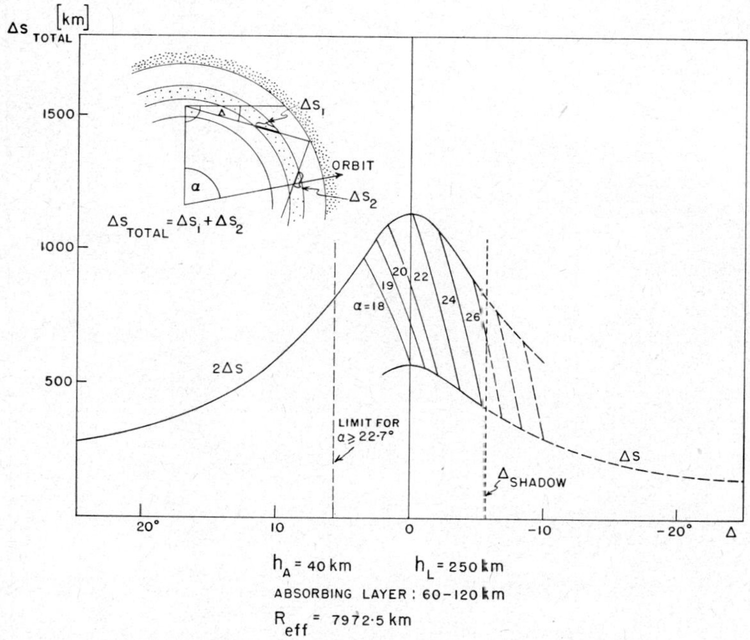

FIG. 7. Absorption path length Δs encountered for radial orbits at various angular distances α.

The behavior of $x + y$ as a function of Δ for K as a parameter is shown in Fig. 8. In this case the antenna height $h_A = 120$ km and the layer height $h_L = 250$ km. The curve for x ($K = 1$) represents the locus of the length of the ray paths between source and reflection point. $(x + y)_{shadow}$ is the distance from the source along the ray path to the interception with the shadow line (Δ_{shadow}). An observation point which is to remain below the line of sight must lie above the locus of $(x + y)_{shadow}$.

For $K = \infty$ the locus of $x + y$ as a function of Δ reaches a minimum for $\Delta \equiv 0$. Thus, $\Delta = 0$ represents an optimum condition not only with respect to ψ_{min} and Δs_{max} but also with respect to $K = \infty$. For values of $K \neq \infty$ this minimum is displaced from zero.

For simplicity it is desirable to restrict an intersecting orbit to positive values of K. Together with the requirement of remaining below the shadow line, i.e.

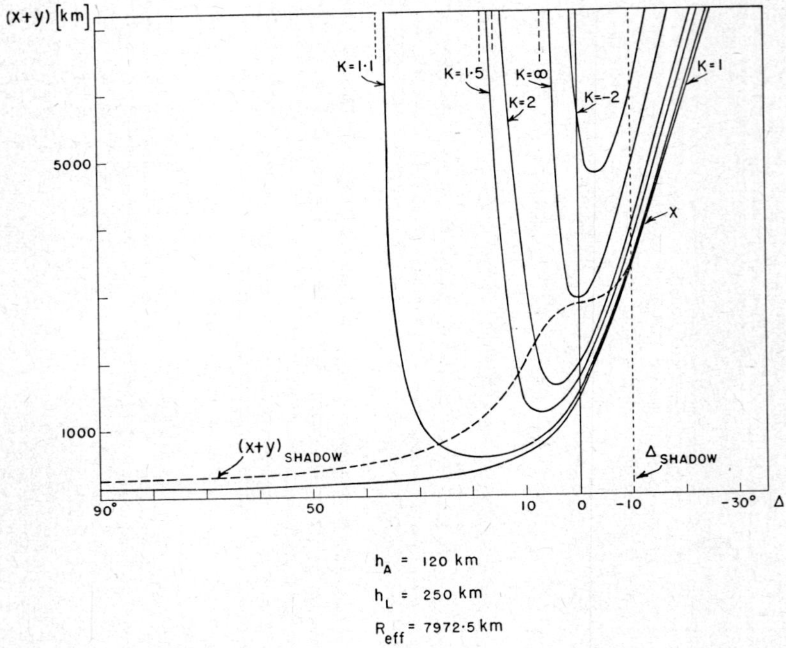

FIG. 8. Convergence factor K ($h_A = 120$ km).

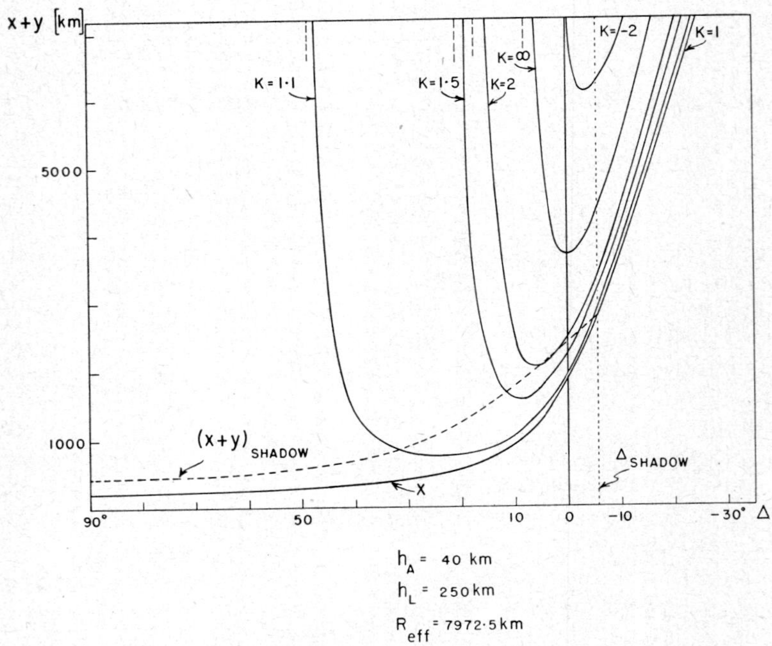

FIG. 9. Convergence factor K ($h_A = 40$ km).

$(x + y) > (x + y)_{\text{shadow}}$, this restriction shows that there remains only a small range between the locus of $(x + y)_{\text{shadow}}$ and $K = \infty$. This range can be increased if a smaller antenna height is used. Fig. 9 displays similar conditions as before except that the loci of $x + y$ were obtained for $h_A = 40$ km. The reduced antenna height corresponds now to a smaller shadow angle but the range between the curves of $(x + y)_{\text{shadow}}$ and $K = \infty$ is much wider.

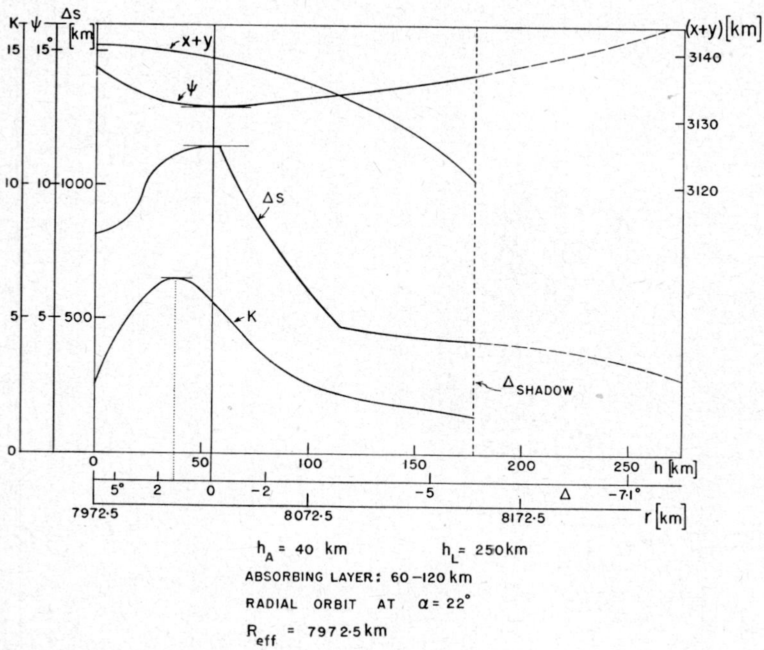

FIG. 10. Dynamic behavior of geometrical parameters for a radial orbit at $\alpha = 22°$.

The minimum property of ψ, which determines the behavior of the reflection coefficient and thus the behavior of the received signal strength as a function of Δ, may be obscured by the focusing effect in the vicinity of $\Delta = 0$. It is of interest to study the dynamics of $x + y$, ψ, Δs, and K as a function of Δ while advancing along a radial orbit. For an angular distance $\alpha = 22°$ the maximum of K is displaced from $\Delta = 0$ as it is illustrated in Fig. 10. On the abscissa the height h along the radial orbit is plotted on a linear scale. The ordinates carry the scales for various parameters. h_A equals 40 km. Δs reaches its maximum at $\Delta = 0$. Its nonuniform distribution is attributed to the radial orbit intersecting the ray paths inside of the absorbing region which, in this case, extends from 60–120 km. ψ is a minimum strictly at $\Delta = 0$. The maximum value of K occurs at $\Delta = +1{\cdot}5°$. The total length of the ray paths $(x + y)$ can be seen to vary only slightly.

V. Optimum Conditions

Fig. 11 displays the caustic of the ray paths. The rays are intersected radially at various angular distances α from the source. r is the distance along the orbit measured from the center of the earth. The antenna height $h_A = 40$ km, the layer height $h_L = 250$ km. For an angular distance $\alpha = 19°$, for example, a ray originating at h_A for an elevation angle $\Delta = 5°$ intersects the radial orbit at

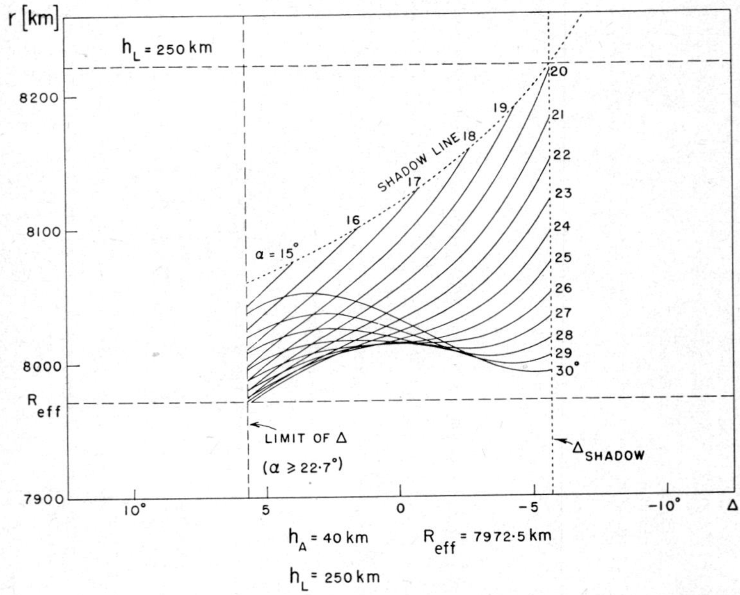

Fig. 11. Caustic of ray paths intersected radially at various angular distances from the source ($h_A = 40$ km).

approximately 25 km above ground. This height, measured along the radial orbit, increases monotonically as Δ decreases. The total range of Δ extends from the limit of $\Delta = +5{\cdot}7°$, where the reflected ray grazes the earth, to the locus of the shadow line beyond which the source at h_A is in line of sight from the orbit. For $\alpha = 22°$ the lower limit is given by $\Delta_{\text{shadow}} = -5{\cdot}7°$. For $\alpha = 19°$ the slope of the curve is positive. For large values of α ($\alpha > 26°$) the curves inflect and there are also negative slopes. When this happens r ceases to be a single-valued function of Δ. Consequently, more than one ray intersect one point on the orbit. In this analysis it is desirable to have available the largest possible range of Δ. It is also required for r to remain single-valued ($dr/d\Delta > 0$). This is fulfilled for

$$\left.\frac{dr}{d\Delta}\right|_{\Delta=0} > 0.$$

Fig. 12 is similar to Fig. 11 but h_A is now 80 km. The limiting range of Δ is larger. The inflection of $r(\Delta)$ commences for a radial orbit at an angular distance

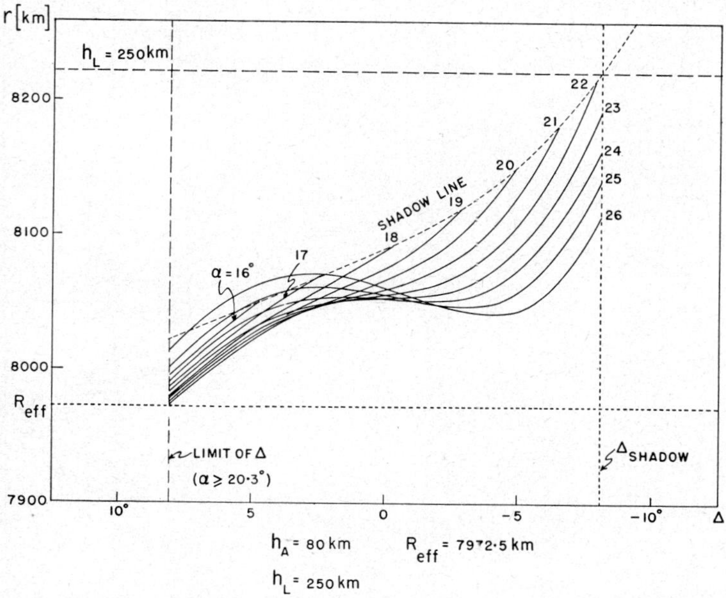

FIG. 12. Caustic of ray paths intersected radially at various angular distances from the source ($h_A = 80$ km).

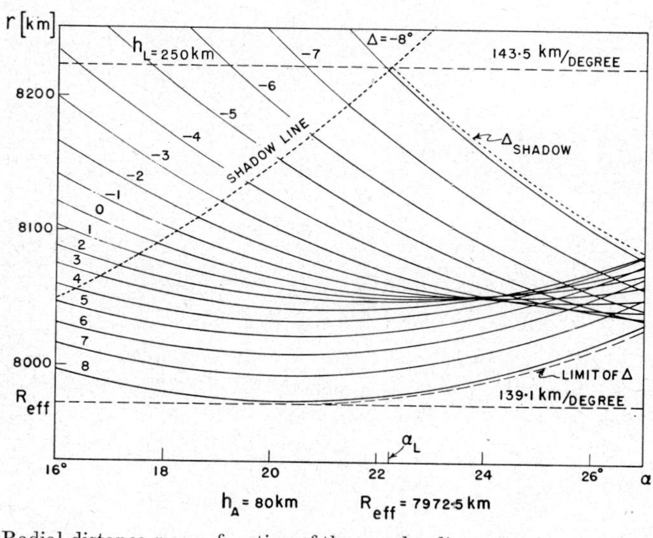

FIG. 13. Radial distance r as a function of the angular distance α for various values of the elevation angle Δ ($h_A = 80$ km).

$\alpha \doteq 24°$, which is less than 26° obtained for $h_A = 40$ km. Thus, an increase of h_A is associated with diminishing angular distances beyond which $r(\Delta)$ ceases to be single-valued. Although the increase of h_A increases the Δ-range it reduces the upper limit of the angular distances for the radial orbits which in turn reduces the range of Δ. This property yields optimum conditions.

Figure 13 is derived from Fig. 12. It displays r as a function of α. Δ is the

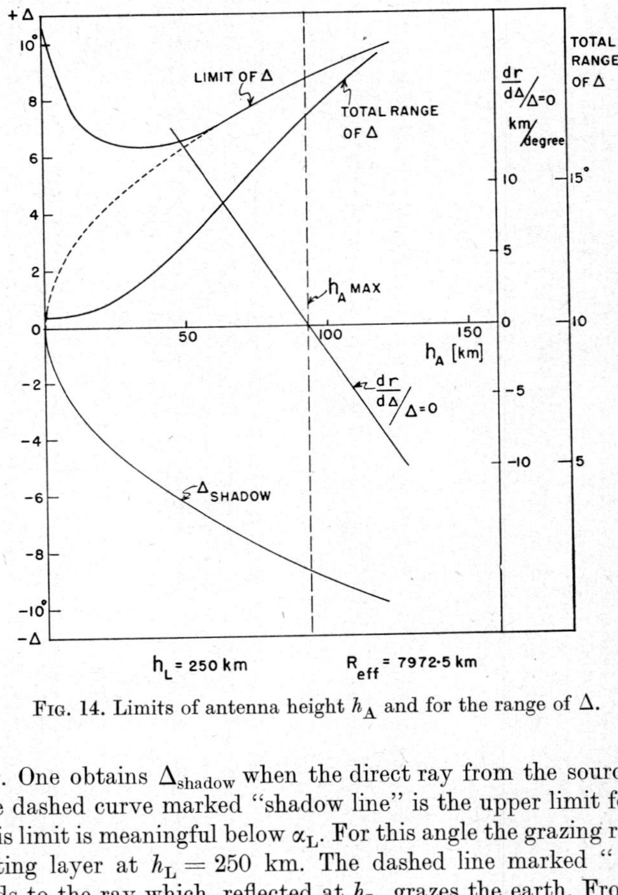

FIG. 14. Limits of antenna height h_A and for the range of Δ.

parameter. One obtains Δ_{shadow} when the direct ray from the source grazes the earth. The dashed curve marked "shadow line" is the upper limit for the radial orbits. This limit is meaningful below α_L. For this angle the grazing ray intersects the reflecting layer at $h_L = 250$ km. The dashed line marked "limit of Δ" corresponds to the ray which, reflected at h_L, grazes the earth. From Fig. 13 it can be seen that a radial orbit ($\alpha =$ constant) does not necessarily yield the largest range of Δ. An orbit which begins at $\alpha = 19°$ and curves away from the shadow line and reaches Δ_{shadow} at about 23° comprises a larger range of Δ than a radial orbit at, e.g., $\alpha = 22°$.

Figure 14 shows the distribution of the limit of Δ for $\Delta > 0$ as well as Δ_{shadow}, both as a function of antenna height h_A. An increase of h_A increases the total range of Δ. For an antenna height of, e.g., $h_A = 70$ km the range of Δ is 15°. The previously mentioned requirement for $r(\Delta)$ to remain single-valued is met by using the intersection of $\left.\frac{dr}{d\Delta}\right|_{\Delta=0}$, for convenience obtained for α_L, with the abscissa. This determines the upper limit of h_A. Any antenna height below $h_{A\max}$ can be used at an angular distance α_L from the radial orbit.

VI. The MUF-concept

For zero antenna height a radio beam moved progressively from the zenith to the horizon, encounters diminishing values of ψ while the transmission distance on the surface increases. As ψ decreases higher frequencies can be used. For highly elevated antennas, however, a radio beam encounters also increasing values of ψ for negative elevation angles Δ in the vicinity of the grazing angle thus reducing the MUF for extreme ranges. It was already shown that $\psi(\Delta)$ is symmetrical with respect to $\Delta = 0$. It may be desirable to waive the restriction of receiving at the earth's surface and to extend the MUF-concept to the case where transmitter and receiver are both elevated. In this case the minimum property of ψ may become useful.

VII. Conclusion

It is best to conclude with the words of Einstein, who said: "As far as the laws of mathematics refer to reality they are not certain and as far as they are certain they do not refer to reality." The geometry presented in this paper seems rather certain. Whether it refers to reality is a question of the future.

Acknowledgment

I wish to thank Raymond Robertson for his assistance in this work.

References

1. Kirby, R. C. (1955). *Rep. nat. Bur. Stand.* No. 3554.
2. Little, C. G. & Leinbach, H. (1958). *Proc. Inst. Radio Engrs, N.Y.* **46**, 334.
3. "Radio Wave Propagation," p. 193; consolidated summary technical report of the Committee on Propagation of the National Defense Research Committee. Chas. R. Burrows, chairman, Stephen S. Attwood, editor. Academic Press Inc., N.Y. (1949).
4. Toman, K. (1958). *Proc. Inst. Radio Engrs, N.Y.* **46**, 495.
5. Toman, K. (1959). *Math. Gaz.* **43**, No. 343, 46–48.
6. Toman, K. A maximum property of two concentric circles. (Accepted for publication by the *Math. Gaz.*)

Über die Fluktuation des Atmosphärischen Brechungsindex in Bodennähe. Ergebnisse einiger Messungen mit einem Mikrowellenrefraktometer

R. Schünemann

Heinrich-Hertz-Institut der Deutschen Akademie der Wissenschaften, Berlin

Die Ausbreitung ultrakurzer elektromagnetischer Wellen weit hinter dem Horizont wird durch die Fluktuationen des atmosphärischen Brechungsindex bestimmt. Eine größere Anzahl von Registrierungen mit einem Mikrowellen-Refraktometer bei sonnigem Wetter, Regen, Gewitter und Übergangslagen zeigt die Abhängigkeit des Fluktuationscharakters von der Witterung. Die Messungen wurden in Bodennähe durchgeführt; weitere Messungen in größeren Höhen sind vorgesehen.

The fluctuations of the refractive index control the propagation of ultra-short radio waves far beyond the horizon. A large number of registrations with a micro-wave-refractometer on sunny days, in rain, thunderstorm and change of weather, shows the dependence on the meteorological conditions. The measurements were taken near the ground. Other measurements, at greater heights, will soon be carried out.

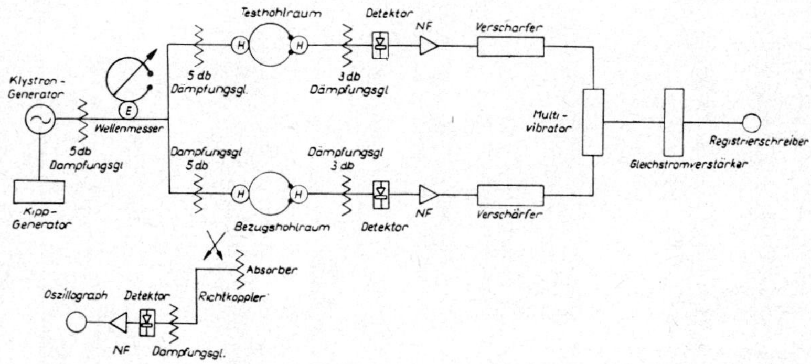

Abb. 1.—Prinzipschaltung des Refraktometers.

Die experimentellen Ergebnisse von Ausbreitungsmessungen im Ultrakurzwellenbereich über größere Entfernungen und ihre theoretischen Deutungen haben ein erhebliches Interesse auf die Untersuchung der atmosphärischen Feinstruktur gelenkt. Für die Registrierung der Mikrofluktuationen des Brechungsindex wurden deshalb in den U.S.A. Refraktometer entwickelt, mit denen Änderugen in der Größenordnung von 1×10^{-7} des mittleren Wertes noch aufgezeichnet werden können. Eine große Zahl von Messungen, insbesondere der horizontalen und vertikalen Struktur mit Hilfe von Flugzeugen und Luftschiffen wurden bereits dort durchgeführt. Ein Teil der Ergebnisse ist durch die Literatur bekannt.

Wegen der etwas andersartigen klimatischen Verhältnisse in Europa hielten wir es für erforderlich, Feinstrukturmessungen auch hier durchzuführen und insbesondere zunächst die Abhängigkeit des Fluktuationscharakters von der Wetterlage zu untersuchen.

Wir haben im Heinrich-Hertz-Institut der Deutschen Akademie der Wissenschaften in Berlin ein Refraktometer gebaut, das im Prinzip der Entwicklung des "National Bureau of Standards" in Boulder gleicht.

Ein Oszillator mit einer Wellenlänge von 3·2 cm wird mit einem Sägezahn frequenzmoduliert. Das Signal wird gleichzeitig zwei Hohlraumresonatoren gleicher Dimension zugeführt. Diese Hohlräume haben keinen Einfluß auf die

Refraktometer-Registrierung bei geschlossenen Hohlräumen mit einer Zeitkonstante um 1 sec, Vorschub 120 mm/h, Empfindichkeit 20 mA/MHz = 0·08 N/mm.

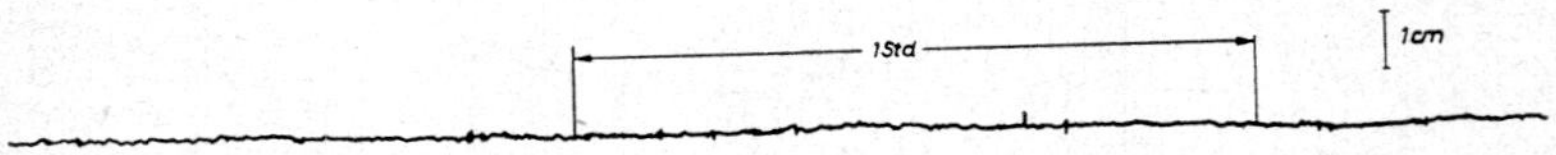

Refraktometer-Registrierung bei geschlossenen Hohlräumen ohne Zeitkonstante Vorschub 120 mm/h. Empfindlichkeit 20 mA/MHz = 0·08 N/mm.

ABB. 2.

Oszillatorfrequenz selbst. Unterscheiden sich die Dielektrizitätskonstanten der Medien im Innern der Resonatoren und damit die Resonanzfrequenzen, dann werden bei dem Durchlaufen des Oszillatorfrequenzbereiches die Resonanzspitzen zu verschiedenen Zeiten auftreten. Diese Zeitdifferenz ist ein direktes Maß für den Unterschied der Brechungsindices. Die Impulse werden gleichgerichtet und lösen in Sperrschwingerschaltungen Nadelimpulse aus. Der zeitlich erste öffnet einen Multivibrator, der von dem folgenden zweiten wieder gesperrt wird. Am Ausgang des Multivibrators erhält man einen neuen Impuls, dessen Breite ein Maß für die Frequenzverstimmung des Meßhohlraums gegenüber dem Bezugshohlraum ist. Die aufeinanderfolgenden Impulse werden durch eine Diode in einen Gleichstrom umgewandelt, der zur Registrierung benutzt wird.

Mit der entwickelten Apparatur wurde eine Empfindlichkeit erreicht, die es gestattet, Fluktuationen des Brechungsindex von 1×10^{-7} des mittleren Wertes noch meßbar aufzuzeichnen. Durch einen sorgfältigen Aufbau wurde eine ausreichende elektrische Konstanz der Apparatur erreicht, wie die Registrierungen bei geschlossenen Hohlräumen in der Abbildung 2 zeigen.

Mit einer Geräte-Zeitkonstanten von 1 Sekunde ergab sich das obere Bild, der untere Verlauf wurde bei einer Konstanten von 30 Millisekunden aufgezeichnet. In beiden Fällen betrug die Empfindlichkeit der Apparatur 0·08 N-Einheiten für 1 mm Schreiberausschlag.

Das gesamte Meßgestell zeigt die Abb. 3.

Die beiden unteren Einschübe enthalten die Stromversorgungsgeräte, darüber befindet sich der Phasenmesser und oben die Kontrolleinrichtung. Direkt an der Frontplatte des Phasenmessers ist der Bezugsresonator befestigt. Er kann mit dem Skalenknopf einer Mikrometerschraube an der Unterseite abgestimmt werden. An der linken Seite des Gestells liegt der Anschluß für den Wellenleiter zum Meßhohlraum.

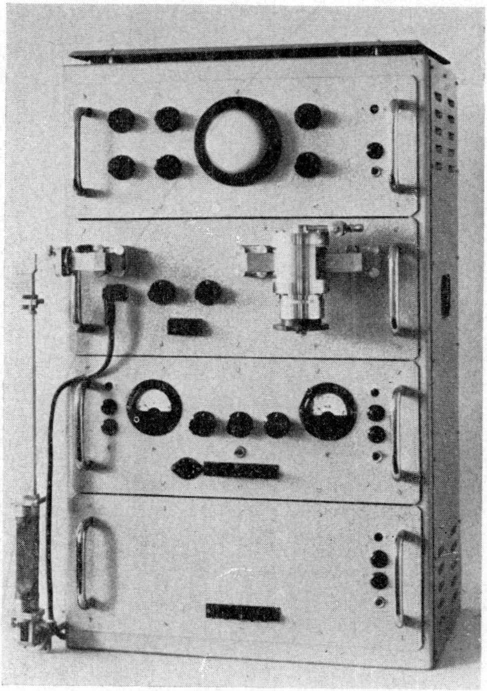

Abb. 3. Meßgestell.

Die Endplatten des Meßhohlraums sind mit einer Anzahl von Löchern versehen, die ein Durchfluten mit dem zu untersuchenden Medium gestatten. Bei Windstille sorgt ein kleiner Ventilator für die erforderliche Durchströmung.

Die folgenden Abbildungen zeigen einen Teil der bisher durchgeführten Registrierungen, die wir bei verschiedenen Wetterlagen erhielten. Die ersten Messungen wurden auf einem 20 m hohen Turm durchgeführt. Anschliessend wurde an einem 80 m-Gitterturm des aerologischen Observatoriums in Lindenberg in verschiedenen Höhen registriert. Hier war es auch möglich, mit Hilfe eines Seilaufzuges die vertikale Struktur des Brechungsindex bis in 70 m Höhe aufzuzeichnen.

Die nächsten Abbildungen zeigen zunächst Registrierungen bei Hochdruck-Wetterlagen.

Am 16.6.58 (Abb. 4) erstreckte sich ein umfangreiches Hoch von der Barentsee bis zum Mittelmeer. Bei nur ganz leichter Schleierbewölkung lagen die Temperaturen bei 20°C. Schwache südliche Winde hatten eine Geschwindigkeit von weniger als 2 m in der Sekunde.

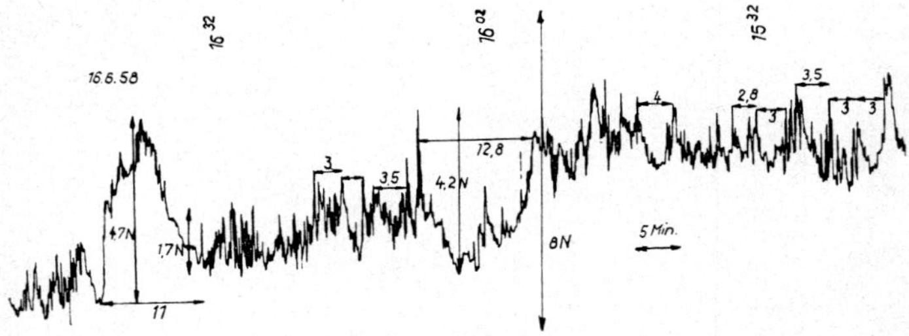

Abb. 4.

Typisch für solche Wetterlagen scheint der stark unruhige Verlauf des Brechungsindex zu sein. Die Fluktuationen lagen zwischen 0·1 N-Einheiten und etwa 2 N-Einheiten. Neben den schnellen Fluktuationen, etwa 3 bis 5 in der Minute, wurden auch größere Perioden von 3 bis 4 Minuten, weniger häufig zwischen 11 und 12 Minuten Dauer mit Schwankungen von 4 bis 5 N-Einheiten beobachtet. Bei einer Windgeschwindigkeit von 1·5 m/sek errechnet sich für diese eine horizontale Ausdehnung von rund 1000 m.

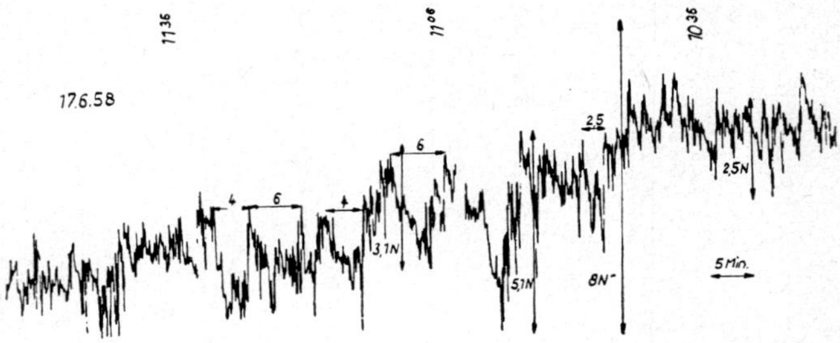

Abb. 5.

Ein ähnliches Bild ergab sich auch am 17.6.58 bei gleicher Wetterlage, nur die Windgeschwindigkeit war mit 6·5 m/sek im Mittel etwas größer (Abb. 5).

Die schnellen Fluktuationen erreichten an diesem Tag bis zu 2·5 N-Einheiten. Für die Schwankungen mit größeren Perioden von 2·5 bis 6 Minuten ergaben sich horizontale Ausdehnungen zwischen 900 m und 2000 m.

Auch am 24.6.58 zeigte die Registrierung bei teils heiterem, teils wolkigem Wetter eine größere Turbulenz (Abb. 6).

Bei einer Windgeschwindigkeit von 1 bis 2·5 m/sek traten schnelle Schwankungen bis zu 5 N-Einheiten auf.

Am 11.7.58 (Abb. 7) hatte sich das Hochdruckgebiet abgeschwächt, mehrere kleine Hochdruklagen über Mitteleuropa gestalteten das Wetter aber noch freundlich. Die Fluktuationen erreichten noch 3·6 N-Einheiten.

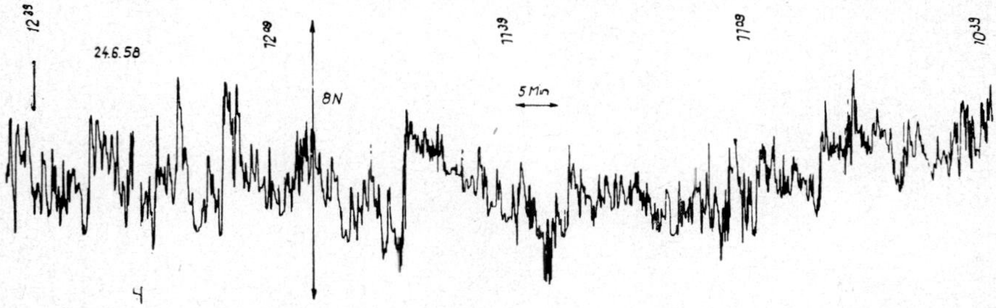

ABB. 6.

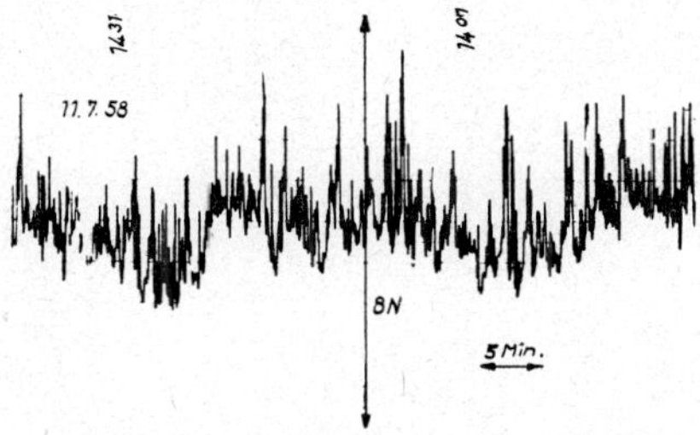

ABB. 7.

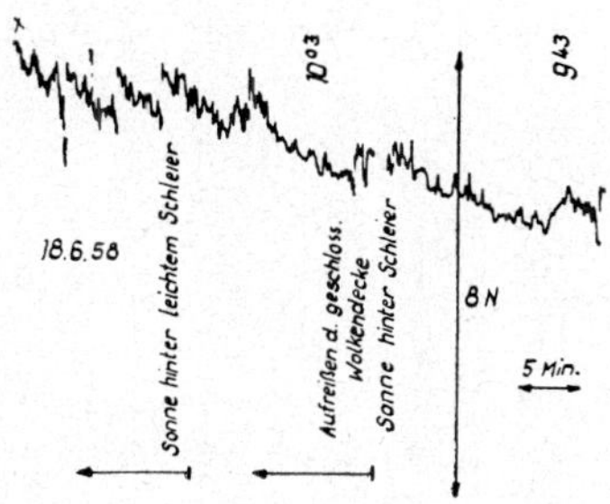

ABB. 8.

Demgegenüber wurde bei kühlerem, aber trocknem Wetter mit geschlossener Wolkendecke immer wieder ein recht ruhiger Verlauf festgestellt, wie ihn die Registrierung vom 18.6.58 (Abb. 8) zeigt.

Über Nacht war Meeresluft aus dem Westen eingeflossen und am Vormittag überquerte eine Kaltfront das Beobachtungsgebiet. Die Fluktuationen hatten

Amplituden zwischen 0·1 N-Einheiten und 0·8 N-Einheiten. Mit dem Aufreißen der Wolkendecke kurz vor 10 Uhr wurde der Verlauf unruhiger.

Die Einwirkung von Regen bei stärkerem, böigen Wind von 7 bis 8 m/sek Geschwindigkeit zeigen die nächsten Registrierungen. Bei Einsetzen des Regens am frühen Nachmittag des 18.6.58 (Abb. 9) mit einzelnen Tropfen war der Fluktuationsverlauf zunächst noch verhältnismäßig ruhig.

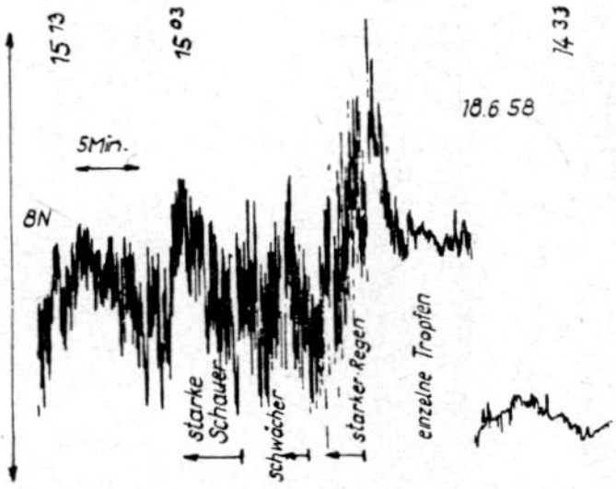

ABB. 9.

Schnelle und große Schwankungen traten aber sofort mit Beginn des Schauers auf.

Die Registrierung vom 20.6.58 zeigt noch einmal einen Verlauf bei Regenwetter mig leichtem Wind zwischen 2 und 4 m/sek (Abb. 10).

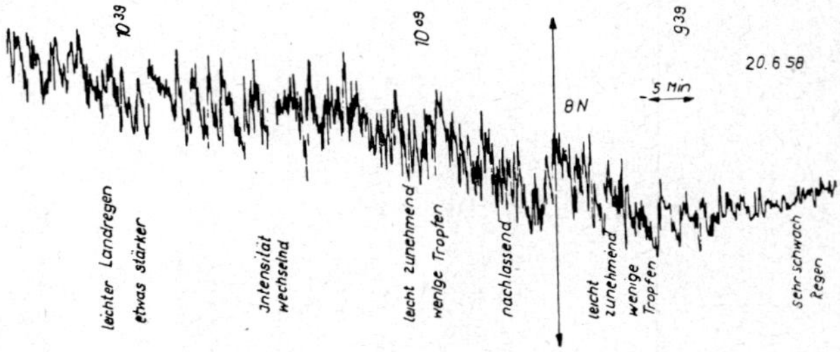

ABB. 10.

Bei Beginn gegen 9^{20} Uhr, als die ersten Tropfen fielen, war der Gang mit Amplituden bis zu 0·7 N-Einheiten noch ruhig. Das änderte sich jedoch bald. Bei leichtem Regen wechselnder Intensität verstärkten sich die Amplituden bis zu 2 N-Einheiten. Ab 11^{30} Uhr ging der Niederschlag in einen gleichmäßigen leichten Landregen über. Dabei ergab sich, wie die Abb. 11 zeigt, eine merkliche Beruhigung im Brechungsindex-Verlauf.

Auch an anderen Tagen mit ähnlicher Witterung wurde beobachtet, dass mit Einsetzen des Regens zunächst stärkere Schwankungen auftraten, die nach einiger Zeit bei gleichmäßigem Regen wesentlich zurückgingen. Weitere Beobachtungen sind aber erforderlich, um festzustellen, ob ein solches Verhalten kennzeichnend für diese Witterung ist.

Die Abb. 12 zeigt die Registrierung bei einem leichten Gewitter, das in wenigen Kilometern Entfernung vom Beobachtungsort vorüberzog.

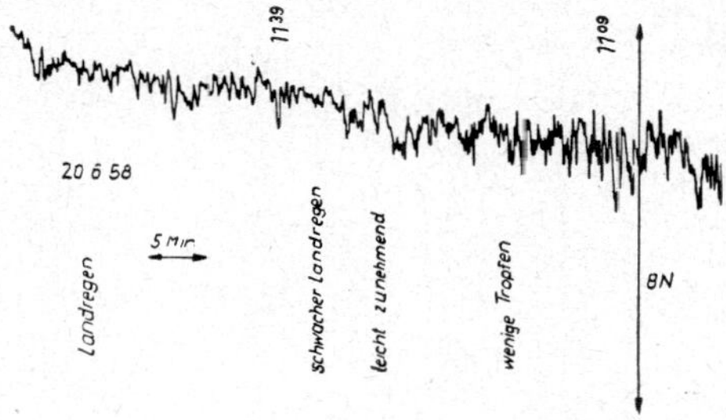

ABB. 11.

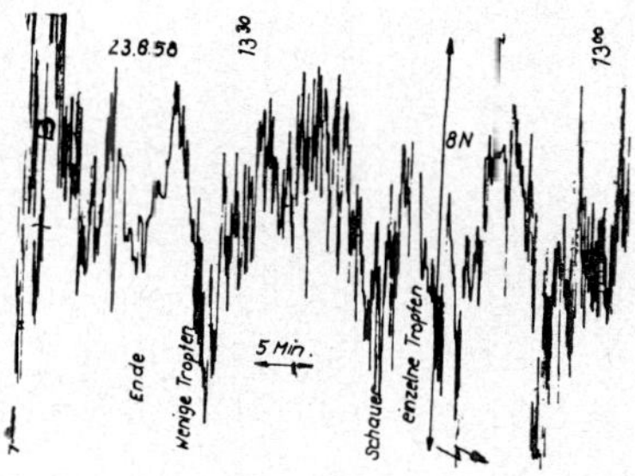

ABB. 12.

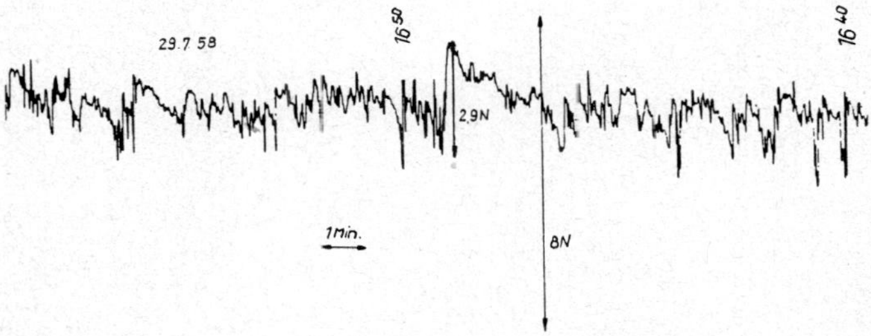

ABB. 13.

Ein ausgedehntes Tief mit Zustrom kälterer Luftmassen lag am 23.6.58 über Europa. Während der Registrierung flaute der Wind von etwa 4 m/sek bis zur Windstille ab und es traten kleine Schauer auf. Die schnellen Fluktuationen hatten Amplituden bis zu 5 N-Einheiten, die sich etwas später auf 7 N-Einheiten verstärkten.

Die nächsten Abbildungen zeigen Ausschnitte aus einer Registrierung, die vom 29.7.58 nachmittags bis zum Vormittag des 30.7.58 kontinuierlich durchgeführt wurde. In der Zeit von 16^{40} Uhr bis 17^{00} Uhr (Abb. 13) herrschte bei heiterem Wetter mit einzelnen Cumuluswolken ein böiger Wind mit Geschwindigkeiten zwischen 6·8 und 9·5 m/sek. Der Brechungs-index-Verlauf war unruhig mit Fluktuationen zwischen 0·2 N-Einheiten bis 2·9 N-Einheiten.

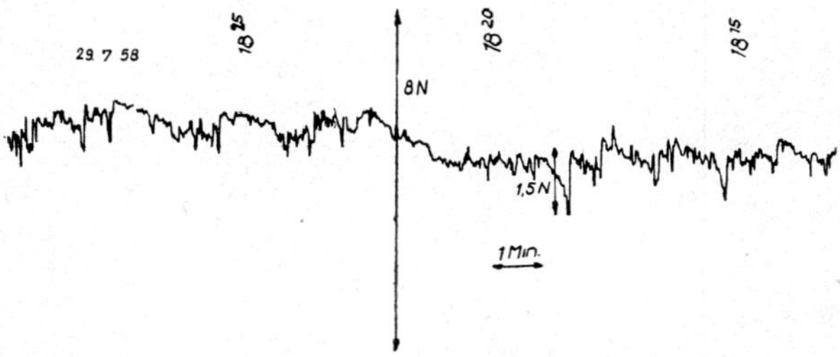

Abb. 14.

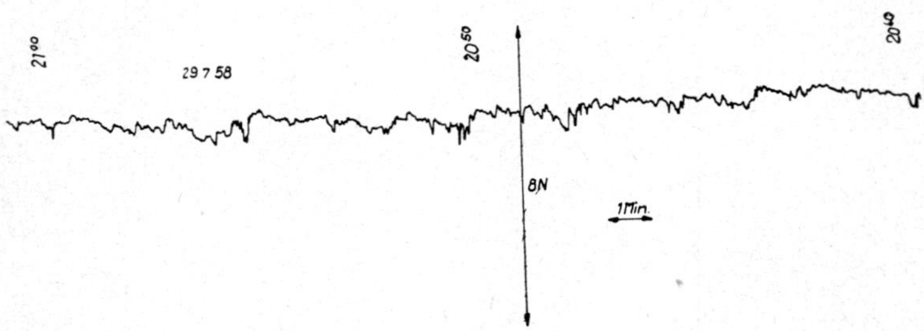

Abb. 15.

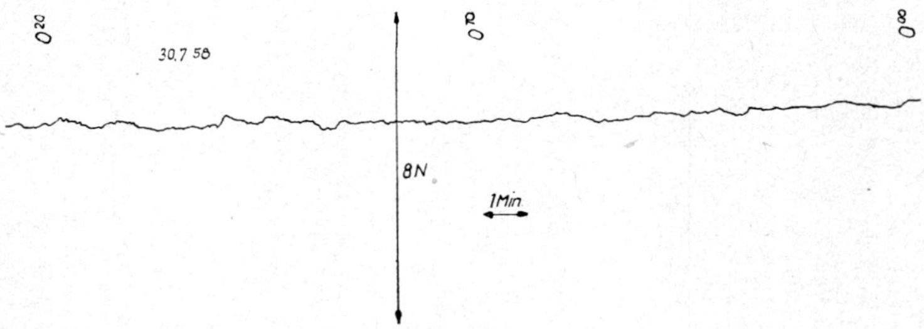

Abb. 16.

Es wurde hierbei ein 5 mal schnellerer Papiervorschub als bisher (10 mm in der Sekunde) benutzt.

Zwischen 18^{15} Uhr und 18^{30} Uhr (Abb. 14) war der Wind zwar noch böig, seine Geschwindigkeit aber auf 5 bis 6·8 m/sek zurückgegangen.

Das Wetter war weiterhin heiter. Die Registrierung zeigte zu dieser Zeit einen etwas ruhigeren Verlauf mit Fluktuationen zwischen 0·1 bis 0·7 N-Einheiten. Nur einzelne Schwankungen erreichten nocn 1·5 N-Einheiten.

Zwischen 20^{40} Uhr und 21 Uhr (Abb. 15) waren die Fluktuationen weiter stark zurückgegangen, es wurden 0·1 bis 0·5 N-Einheiten beobachtet. Die Windgeschwindigkeit lag zwischen 3·6 und 4·7 m/sek bei geringer Bedeckung.

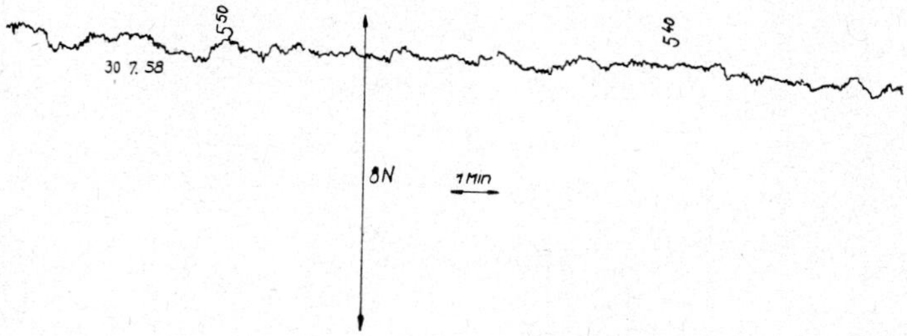

ABB. 17.

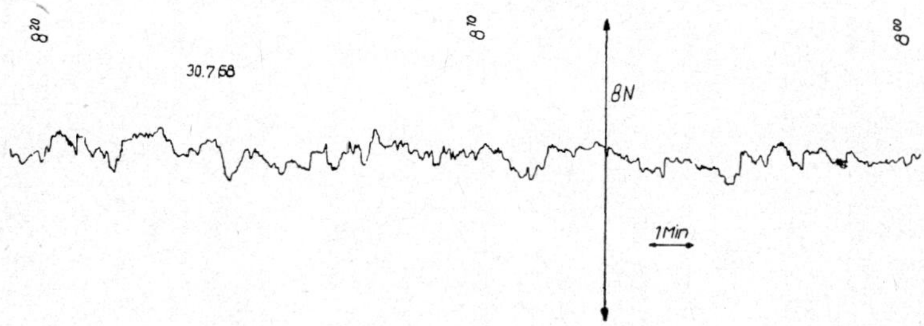

ABB. 18.

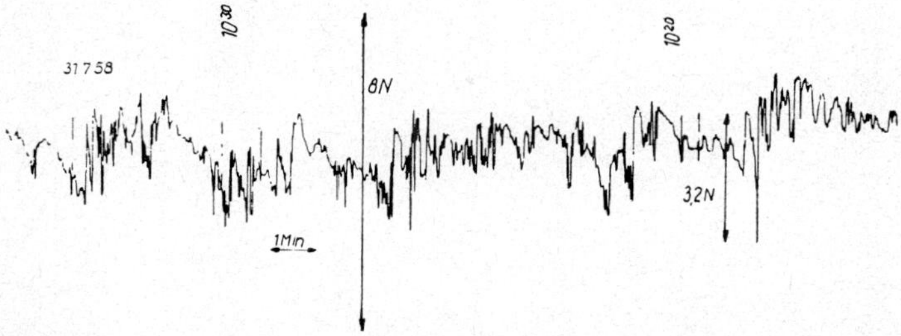

ABB. 19.

Um Mitternacht zwischen 0^{00} Uhr und 0^{20} Uhr (Abb. 16) wurden nur noch sehr geringe Fluktuationen beobachtet.

Es war sternklar mit einzelnen hohen Wolken und einer Windgeschwindigkeit von 1 bis 2 m/sek.

Der ruhige Charakter blieb auch noch bei Sonnenaufgang kurz nach 4 Uhr erhalten. Erst nach 5^{30} Uhr (Abb. 17) wird die Unruhe wieder etwas größer. Die Windgeschwindigkeit hatte sich mit 1 bis 2 m/sek nicht verändert.

Die Unruhe wuchs dann mit forschreitender Tageszeit, wie die Registrierung zwischen 8^{00} Uhr und 8^{20} Uhr (Abb. 18) zeigt. Es traten wieder Fluktuationen bis zu 1 N-Einheit bei Windgeschwindigkeiten zwischen 2 und 3 m/sek auf. Die Wolkendecke hatte sich zu dieser Zeit geschlossen. Es kann vermutet werden,

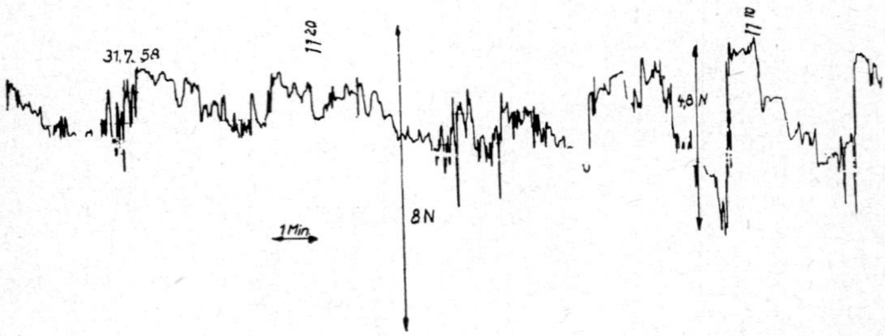

ABB. 20.

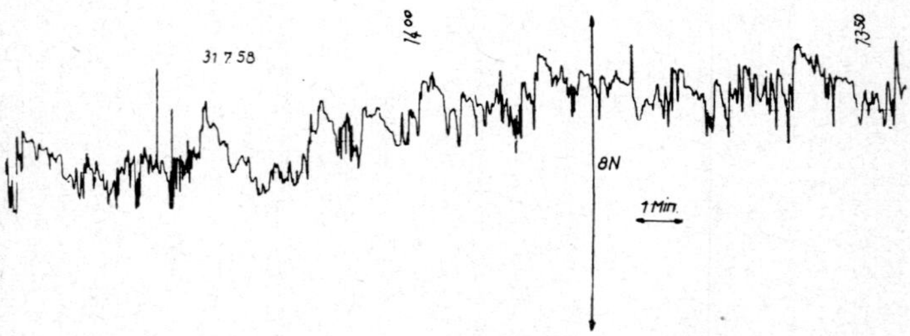

ABB. 21.

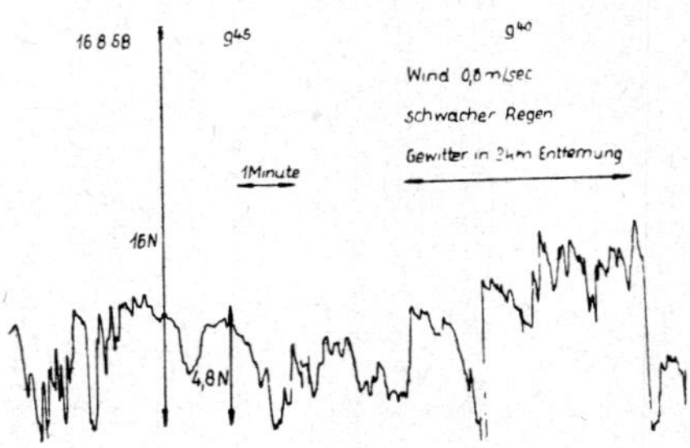

ABB. 22.

dass eine Bodeninversion in den Nachtstunden die Turbulenz in dieser Schicht stark reduziert hatte. Mit wachsender Tageszeit heben sich solche Inversionen vom Boden ab, es entstehen freie Inversionen, unter denen sich wieder Turbulenzen ausbilden können.

Bei der Registrierung am 31.7.58 zwischen 10^{15} Uhr und 10^{35} Uhr (Abb. 19) war das Wetter teils heiter, teils wolkig mit Windgeschwindigkeit von 6 m/sek.

Es zeigte sich wieder der bei solchem Wetter meist beobachtete stark unruhige Verlauf mit Fluktuationen bis zu 3 N-Einheiten. Diese Unruhe wuchs noch und erreichte nach 11 Uhr (Abb. 20) bis zu 4·8 N-Einheiten.

Der Witterungscharakter blieb an diesem Tag erhalten und das Fluktuationsbild änderte sich nur wenig, wie die Registrierung um 14 Uhr (Abb. 21) zeigt.

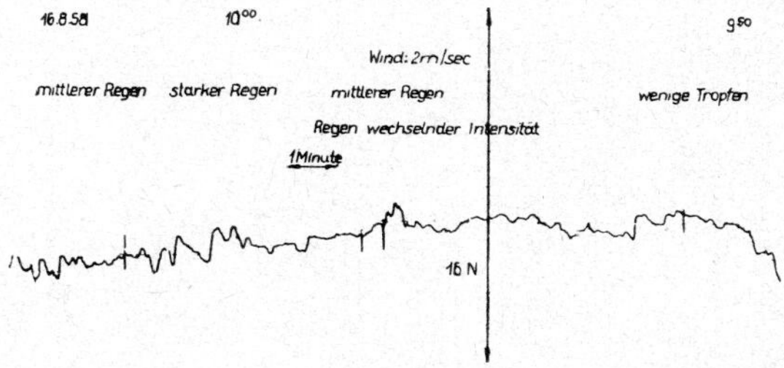

ABB. 23.

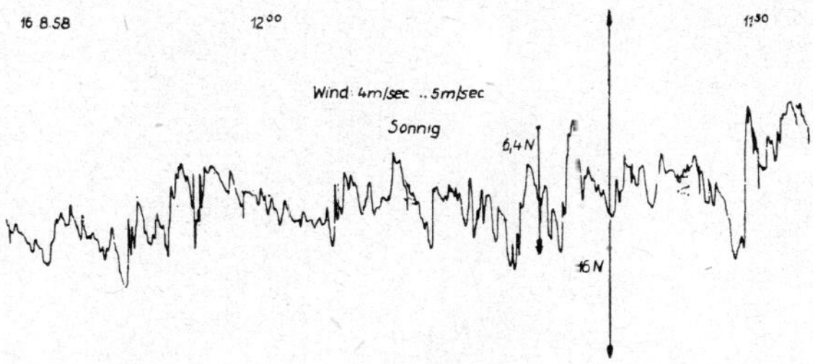

ABB. 24.

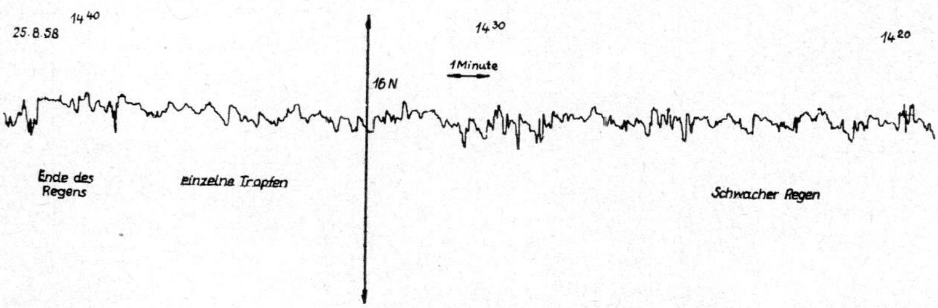

ABB. 25.

Die nächste Abbildung (Abb. 22) zeigt noch einmal eine Registrierung bei einem Gewitter, das in 2 km Entfernung vom Meßort vorüberzog. Am 16.8.58 lag ein Tief über der Nordsee und Norddänemark und eine Störungsfront überquerte Mitteldeutschland. Bei dieser Aufnahme war die Empfindlichkeit der Apparatur von bisher 8 N-Einheiten für den Vollausschlag auf 16 N-Einheiten verringert worden.

Bei nur schwachem Regen und geringer Windgeschwindigkeit wurden stärkere Schwankungen bis zu 5 N-Einheiten beobachtet, die nach Abklingen der Gewittertätigkeit noch etwa 7 Minuten anhielten. Kurz danach nahm der Regen

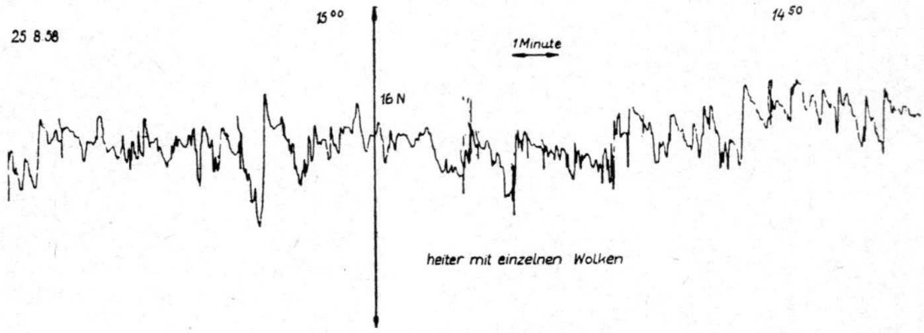

Abb. 26.

Abb. 27.

erheblich zu und es trat eine auffallende Beruhigung im Brechungsindex-Verlauf ein (Abb. 23).

Man kann annehmen, dass die vorhergehenden starken Schwankungen durch die Gewittertätigkeit ausgelöst wurden, während der darauf folgende Regen zum Abklingen der Fluktuationen beitrug.

Wenig später klärte das Wetter auf und es traten bei etwas auffrischendem Wind von 4 bis 5 m/sek wieder größere Fluktuationsamplituden auf, wie sie bei sonnigem Wetter meist beobachtet wurden (Abb. 24).

Die Registrierungen vom 25.8.58 (Abb. 25 und 26) zeigen noch einmal den Fluktuationsverlauf bei Übergang von regnerischem in sonniges Wetter. Kurz nach Aufhören des Regens wurden die Fluktuationen des Brechungsindex wieder wesentlich größer.

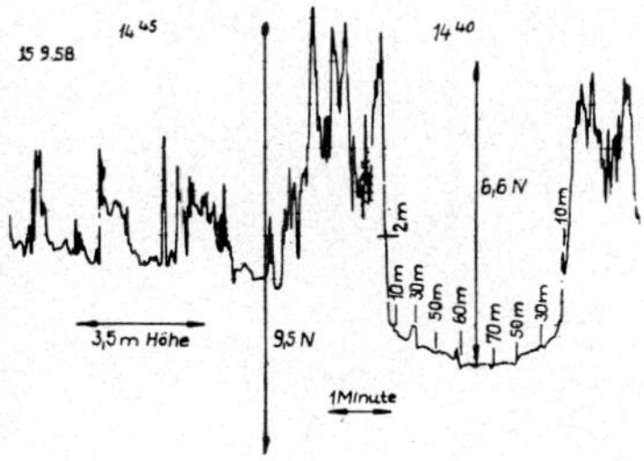

Abb. 28.

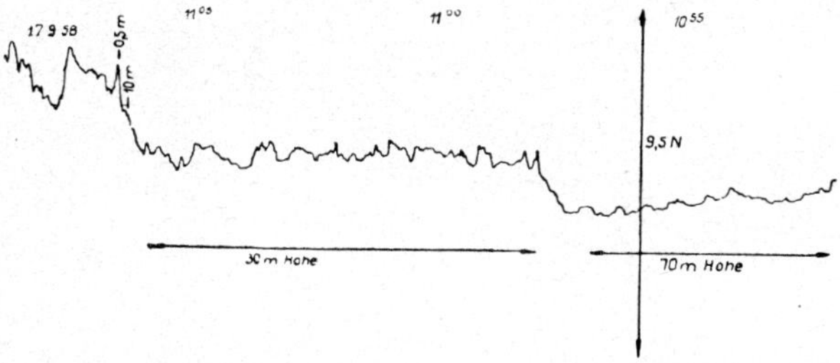

Abb. 29.

Es folgen jetzt noch 2 Registrierungen, bei denen das Refraktometer an einem Gitterturm des aerologischen Observatoriums Lindenberg kontinuierlich bis in 70 m Höhe hochgezogen wurde, wie die Abbildung 27 zeigt.

Bei bedecktem Himmel mit leichtem Nieselregen wurde am 15.9.58 (Abb. 28) vom Erdboden bis zur Höhe von 70 m eine Brechungsindex-Änderung von 6 bis 7 N-Einheiten festgestellt, wobei der größte Anteil, etwa 5 N-Einheiten, in dem untersten Bereich unter 10 m Höhe eintrat.

Während am Erdboden die Fluktuationsamplitude noch 1 bis 1·5 N-Einheiten betrug, wurden über 10 m nur noch geringe Schwankungen beobachtet.

Im Anschluß an diesen Aufzug wurde noch einige Minuten in 3·5 m Höhe registriert. Bei ähnlichen Schwankungen wie am Erdboden wurde hier schon der größte Teil der Änderung des mittleren Brechungsindex-Wertes zwischen 0 m und 70 m erreicht. Es ergab sich also an diesem Tage eine starke Abnahme des Brechungsindex über dem Erdboden mit größeren Schwankungen nur in diesem kleinen Höhenintervall.

Die nächste Abbildung (Abb. 29) zeigt die Registrierung bei einem Abstieg des Refraktometers aus 70 m. Höhe am 17.9.58, wobei in 70 m und 30 m einige Minuten lang registriert wurde.

Bei ähnlichem Wetter wie am 15.9.58 war in 70 m Höhe der Verlauf recht ruhig, bei 30 m zeigte sich schon eine leichte Zunahme der Unruhe, die dann in 0·5 m Höhe wieder stark wurde.

Registrierungen an anderen Tagen ergaben aber auch bei einem ruhigen Verlauf des Brechungsindex in Bodennähe eine erhebliche Fluktuation in 70 m Höhe. Hier ist anzunehmen, dass eine Inversionsschicht unterhalb von 70 m Turbulenzen in Bodennähe weitgehend ausschloß.

Bisher wurde erst eine verhältnismäßig geringe Anzahl solcher Registrierungen durchgeführt. Zahlreiche Messungen in allen Jahreszeiten werden noch erforderlich sein, um gültige Aussagen über Zuordnung von Wetter und Fluktuationscharakter zu gewinnen.

Researches on Backscatter of Radiowaves

I. Ranzi

Centro Radioelettrico Sperimentale "G. Marconi," Viale Trastevere, Roma

Experiments on backscatter sounding with a rotating Yagi, which are now being performed near Rome, are described.

Comparison is made between backscatter data and skip distances deduced from vertical sounding at the mid-point of a 1500 km path: a disagreement is found, probably due to a lack of the method currently employed for scaling vertical sounding data, which gives values of MUF appreciably lower than the actual ones.

These results were confirmed by experiments of backscatter sounding on 22·3 Mc/s, with a simultaneous determination of the time of the evening failure of 21·64 Mc/s waves received from Daventry, and with vertical sounding at the mid-point.

I. Introduction

The main purpose of this work is to test the accuracy attainable in the determination of skip distances from backscatter data, in view of the practical applications.

In a first part of the work, we made a comparison between the backscatter ranges deduced from records taken on 18·6 Mc/s and the skip distances deduced from vertical sounding near the point of reflection of backscattered waves.

In a second part, the same comparison, made on 22·3 Mc/s, was completed by the simultaneous determination of the time of evening failure of 21·64 Mc/s waves received from the Daventry broadcast station; the reflection of these waves in the F2 layer occurred near the zone of vertical sounding.

II. Experimental Equipment

The experimental station is placed at Torrechiaruccia (lat. 42·03° N., long. 11·84° E.), at 55 km from Rome. The apparatus is equipped with a rotating horizontal four elements Yagi aerial, tuned to 18·6 Mc/s. The elements were adjusted so as to obtain a very high back-to-front ratio: for common transmit receive working, a value of 66-dB was attained. For the same conditions, the azimuthal beam width for a 3-dB drop in response resulted of 56°. The speed of rotation of the aerial is 2 r.p.m.

The transmitter has a peak pulse power of about 2 kW and a pulse length of 1 msec; the repetition rate is 25 s^{-1}.

The output of the receiver is applied to the intensity modulating electrode of the recording cathode ray oscillograph.

The 35-mm recording film moves with uniform speed during the uniform rotation of the aerial: a directional mark (for S) is transmitted to the oscillograph by a suitable switch connected to the aerial.

III. Comparison with Vertical Sounding

As has been said, we compared our data on backscatter distance with the skip distance deduced from vertical sounding near the point of reflection of backscattered waves in the F2 layer.

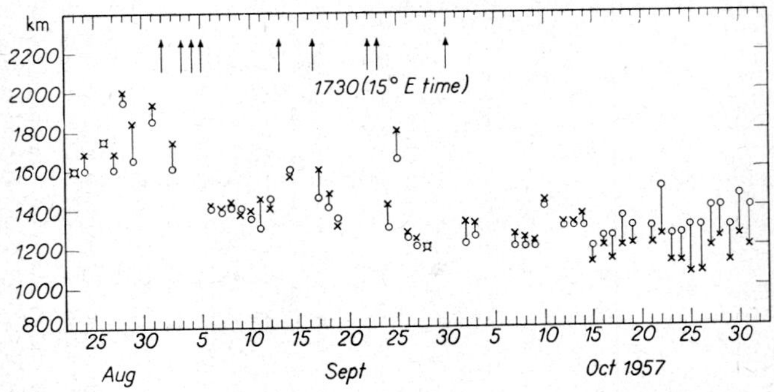

Fig. 1. Comparison between equivalent backscatter oblique paths (circles) and skip distances deduced from vertical sounding at mid-point (crosses), at 1230 (15° E. Time).

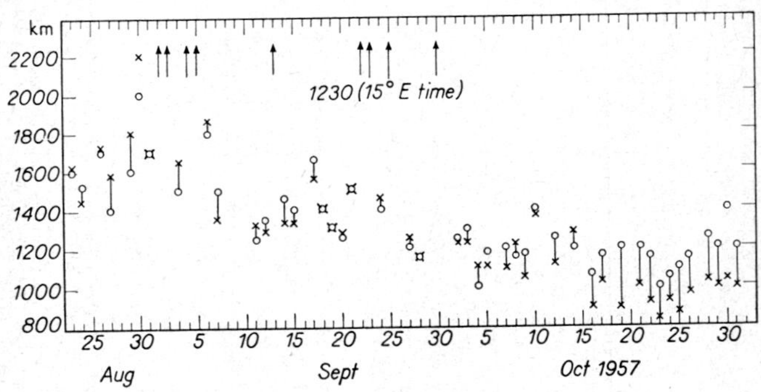

Fig. 2. Comparison between equivalent backscatter oblique paths (circles), and skip distances deduced from vertical sounding at mid-point (crosses), at 1730 (15° E. Time).

For this comparison we chose the vertical sounding data obtained at the ionospheric station of Breisach (48·1° N., 7·6° E.), at 740 km from our station, in the direction 24° W. from N.: the skip distances for 18·6 Mc/s were found by using the nomogram of the Central Radio Propagation Laboratory.

In Figs. 1 and 2 we report some results: the circles represent the equivalent backscatter oblique path lengths, and the crosses indicate the skip distances deduced from vertical sounding data of Breisach.

For a skip distance greater than 1400 km, theory shows that the difference between the equivalent oblique path length and the actual skip distance is of the order of 200 km (when the virtual height of reflection of the waves in the ionosphere is about 300 km).

Analysis of our results shows that there is a striking disagreement between the deduced skip distances and the theoretical predictions for ground backscatter. In many cases, the equivalent path length is less than the skip distance, as deduced from vertical sounding: only during the second half of October, the equivalent path becomes longer than the calculated skip distance: however, it is evident that this effect depends on the reduction of the skip distance, since, in accordance with theoretical deductions, the difference between equivalent and actual backscatter path length rapidly increases when the skip distance becomes less than 1000 km.

Such a discrepancy may be explained only by assuming that the method currently employed for calculating the skip distance from vertical sounding data is not correct.

A number of authors have reported experimental values of MUF higher than those predicted by the present theory on propagation mode.

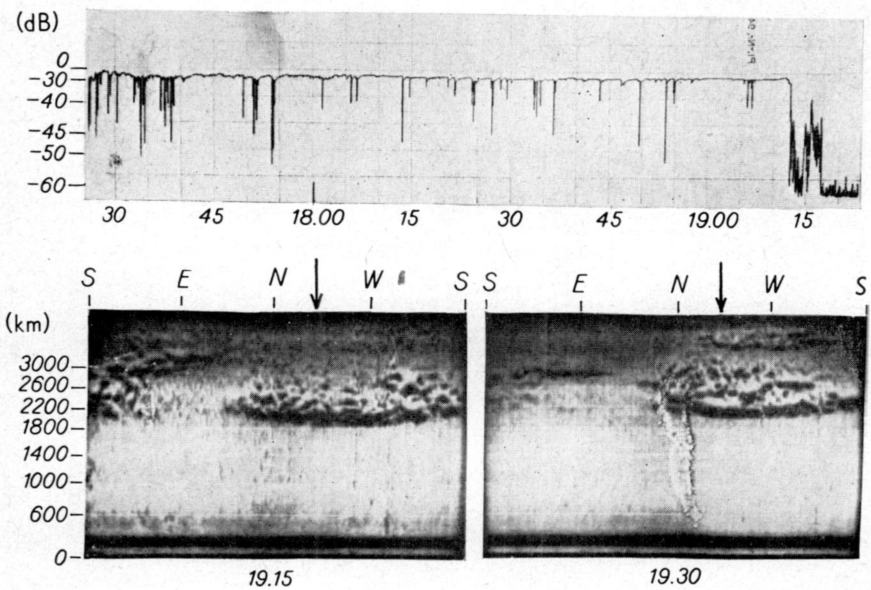

Fig. 3. 7 March, 1958. Above: Record of the field received from Daventry (21·64 Mc/s). Below: Records of backscattered echoes on 22·3 Mc/s, at 1915 and 1930 (15° E. Time): the arrow indicates the direction of Daventry.

IV. Simultaneous Propagation Observations*

For a better understanding of the causes underlying the said discrepancy, we proposed to carry out a comparison between the data obtained from backscatter and the ones from vertical sounding, at the time when the field received at

* In collaboration with A. Porreca.

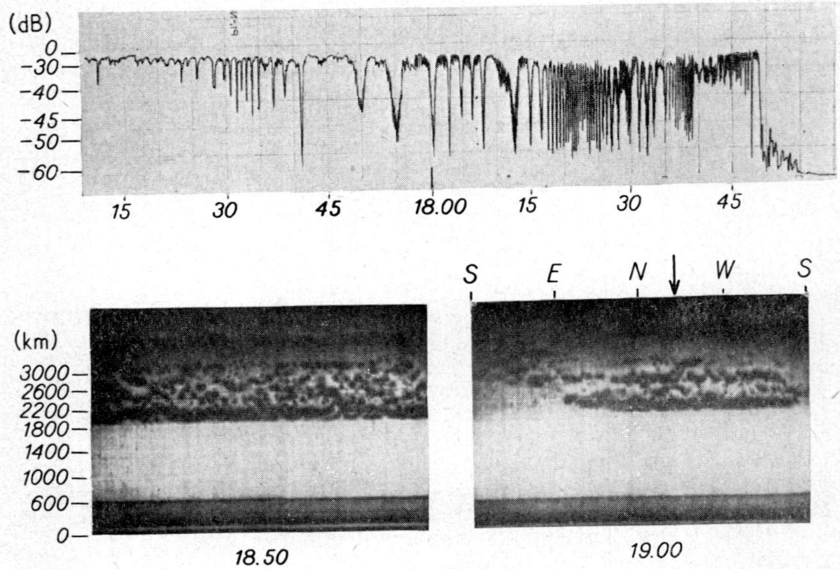

FIG. 4. 10 March, 1958. Above: Record of the field received from Daventry (21·64 Mc/s). Below: Records of backscattered echoes on 22·3 Mc/s; at 1850 hours; the aerial is kept fixed in the direction of Daventry (the duration of the record is 30 sec).

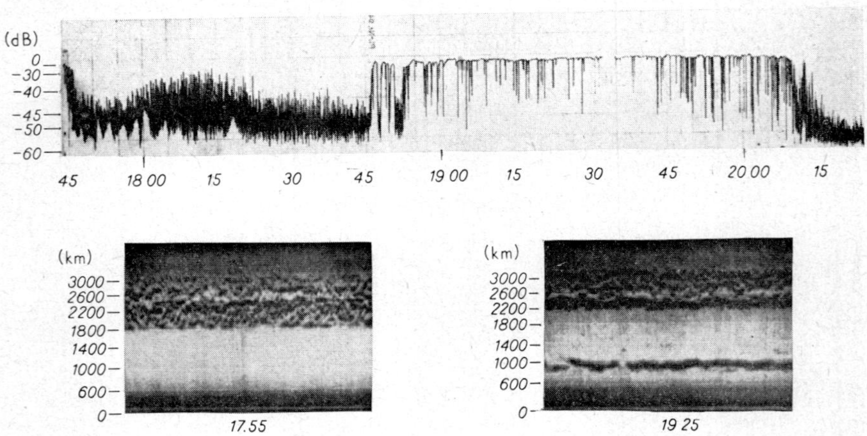

FIG. 5. 13 March, 1958. Above: Record of the field received from Daventry (21·64 Mc/s). Below: Records of the backscattered echoes from the direction of Daventry, at 1755 and 1925 hours (the duration of each record is 30 sec).

Rome from the broadcast station of Daventry, on about 21·5 Mc/s, presents the evening failure.

The location of the said broadcast station appeared to be particularly favourable for our purpose: in fact, roughly mid-way between Daventry (52° 15′ N.,

01° 08′ W.) and our station for field recording (located at Rome, in the Istituto Superiore delle Poste e Telecomunicazioni, at 41° 52′ N., 12° 27′ E.) is situated the ionospheric station of Breisach (48° 03′ N., 07° 35′ E.). The data from Breisach could thus be used for evaluating, by standard method, the MUF corresponding to the distance separating Daventry from Rome (1540 km) at the time of evening failure of the signals from Daventry, i.e., at the time when this distance becomes the minimum skip distance for the frequency radiated from Daventry.

The frequency of 22·3 Mc/s used for backscatter sounding during these experiments is very close to the Daventry broadcast frequencies of 21·47, 21·64, 21·55 Mc/s.

The experiments under consideration were performed during the period from 10 February to 15 March 1958.

Figures 3, 4 and 5 show some typical results obtained in our records of field intensity and backscattered echoes. It is readily apparent that the time of evening failure of the Daventry signals could be determined with great accuracy: only when the sporadic E layer is present, this time is no longer sharply defined; this event occurred only once (13 March) and, as we shall see, confirmed the practical advantages afforded by the backscatter sounding method.

In the following Table we report the results of our observations.

TABLE I

		Reception from Daventry		Vertical Sounding		Deviation %	Backscatter equiv. path km
		Frequency Mc/s	Time of failure 15° E.M.T.	foF2 Mc/s	MUF 1540 Mc/s		
1958							
Feb.	12	21·470	1855	8·9	17·1	– 20	1800
	14	21·470	1900	9·0	17·5	– 18	1800
	15	21·550	1918	9·9	19·2	– 11	1800
	19	21·470	1850	9·2	17·7	– 17	1800
	22	21·640	1930	9·2	17·6	– 18	—
March	4	21·640	1837	10·1	18·9	– 12	1850
	5	21·640	1955	9·8	18·5	– 14	—
	6	21·640	1904	10·4	19·1	– 12	1800
	7	21·640	1918	10·5	18·7	– 13	1850
	10	21·640	1850	11·1	20·5	– 5	1850
	11	21·640	1912	10·8	19·6	– 9	1800
	12	21·640	1922	10·5	19·3	– 11	1800
	13	21·640	1747	10·9	19·3	– 11	1800
	14	21·640	2020	10·2	19·5	– 10	—

We note that the MUF corresponding to the distance between Daventry and Rome (1540 km) was evaluated by means of the C.R.P.L. nomogram, on the basis of the hourly observations of foF2 and MUF 3000 made at Breisach. By interpolation between these hourly data, we obtained the ones relevant to the time of the field failure, taking in due account the difference of longitude between Breisach and the mid-point along the Daventry-Rome path.

The percentage deviations of the computed MUF values from the actual ones are given in the sixth column of the Table.

In the last column we report the values of the backscatter minimum equivalent oblique path, recorded at Torrechiaruccia on the frequency of 22·3 Mc/s, in the direction of Daventry, at the time at which the reception of the signals from Daventry stopped in Rome. The slight distance separating the mid-points of the Rome-Daventry and of the Torrechiaruccia-Daventry paths may be regarded as utterly negligible for our present purpose.

Inspection of the above Table shows marked deviations of the computed MUF values from the corresponding actual values: these discrepancies appear to increase when the foF2 decreases.

Particularly noteworthy is the excellent agreement between the values of the backscatter paths observed at the time of the evening failure of the signals from Daventry, though the time at which the failure occurs, may vary to an appreciable extent from day to day.

The median value of the minimum equivalent path is 1800 km: only in three cases out of ten we found the different value of 1850 km. With the help of the C.R.P.L. nomogram, it is possible to evaluate, with sufficient accuracy, the skip distance for 22·3 Mc/s, corresponding to a skip distance of 1540 km for 21·64 Mc/s (the frequency of Daventry), and to values of foF2 comprised between 10 and 11 Mc/s (range of values more frequently observed at Breisach during the experiments).

The skip distance for 22·3 Mc/s is so found to be between 1600 and 1620 km. This result is in full agreement with theoretical predictions: in fact the difference between the backscatter minimum equivalent path and the actual skip distance is of the order of 200 km, when the virtual height of reflection of the waves in the ionosphere is about 350 km.

In Fig. 3, below the record of the field received from Daventry on 21·64 Mc/s, we report the records of backscattered echoes obtained at 1915 and 1930 (15° E.M.T.) of the same day (7 March). The arrow marks the Daventry direction. The increase of the backscatter equivalent path from 1800 to 2000 km, which occurs during the time interval under consideration, is clearly detectable; the reception from Daventry stops at 1918.

Records shown in Fig. 4 refer to 10 March. The backscattered echoes were recorded at 1850 hours, while the Yagi aerial was kept fixed in the direction of Daventry.

A case of particular interest occurred on 13 March and is shown in Fig. 5. At 1747 hours we observe a first failure of the field, which must be attributed to the increase of the skip distance for the waves reflected from the F2 layer. After a period of deep fading, we note an increase of the field strength, which continues up to about 2010: this increase is due to the formation of the sporadic E layer, as it is clearly shown by inspection of the records of backscattered echoes from the direction of Daventry. In the record of 1920 hours (Fig. 5), echoes from an equivalent distance of 900–1000 km are due to sporadic E, while the equivalent path of backscattered echoes reflected in the F2 layer has increased to 2100 km.

The period we chose for our experiments was particularly favourable: in effect, during summer the pattern of the backscattered echoes is often complicated: the interpretation needs skilled personnel.

We may conclude that the method of backscatter ionospheric sounding may lead, when properly applied, to values of skip distances more reliable than those obtained from vertical sounding, and provides a satisfactory means of monitoring propagation conditions along a given path.

Ionospheric Ray Tracing with Analogue Computer

MING S. WONG

Electronics Research Directorate, Air Force Cambridge Research Center, Bedford, Mass.

Computed ray patterns portraying the spatial distribution of rays emitted from a radio transmitter are shown for isotropic media where the refractive index varies with height, and with height and distance. These patterns illustrate the types of propagation anomalies—extremely far-reaching rays; trapping of rays in ducts between the ionosphere and earth's surface, without as well as with skip distances; regions of reduced radio signal intensity; regions with enhanced, and dense spatial fadings of, signal—which appear prominently when radio transmitters are close to ionospheric layers as in the satellites.

An account is given of current work on solution of the differential equations, given by Hamiltonian optics, for ray paths in non-isotropic, three-dimensionally inhomogeneous media; arbitrary angles between the rays and the geomagnetic field are included.

I. INTRODUCTION

In ionospheric propagation this simplified picture is often used: oblique transmission is viewed as occurring via a ray which, originating from a transmitter, travels along an upward-slanting straight line to the virtual height of an ionospheric layer, and which is thence reflected along a downward-slanting line to a distant receiver. The ionosphere is thus represented as a specularly reflecting mirror. The condition for reflection to occur, derived from Snell's law, is that the electron density must attain with increasing height a value of $1{\cdot}24 \times 10^4 f^2 \cos^2 I$ electrons/cm^3 approximately, where f is frequency in Mc/s and I is angle of incidence of the ray upon entering the ionosphere.

The above simplified picture—incorporating as it does a curious mixture of refraction phenomena (in the use of Snell's law) and of mirror-like reflection—has a defect in that it omits any account of the spatial gradient of the electron density N. It should be borne in mind that what is customarily referred to as ionospheric reflection, say in oblique transmission, is actually a phenomenon of refraction (gradual bending of the direction of electromagnetic energy transport). In so far as refraction is concerned it is well established, from experience in air-to-air tropospheric propagation, that the gradient of N would play a crucial role, in producing surprising propagation anomalies, when the radio transmitter is anywhere above, or within, or not too far below a layer where N varies steeply with height. Thus with increasing use of radio transmitters in satellites near the ionosphere, it becomes important to give realistic account of the gradient of N. Moreover these satellites will yield increasing data on detail structures of the ionosphere—data with which significant effects of the ionosphere on propagation can be assessed with the aid of ray tracings (ray-path calculations).

Ray tracings—to be done rapidly, and for realistic descriptions of the ionosphere—are made possible by three facilities at our disposal: (1) Hamiltonian optics, which gives the desired ray-tracing equations for non-homogeneous, non-isotropic media, (2) electronic computers, and (3) the Appleton-Hartree formula for phase refractive index.

Section II below sets down the ray equation first for isotropic media, then for non-isotropic media. The electron density may vary arbitrarily with three spatial coordinates, and the earth's magnetic field may be included.

In Section III we show representative ray patterns in isotropic media, taken mostly from anomalies of atmospheric refraction in tropospheric propagation. Radio holes, antiholes, trapping of radio waves in ducts, and dense spatial fadings of radio signal—these constitute the known types of tropospheric, refraction anomalies [6]. These anomalies of atmospheric refraction occur in the domain of primary illumination or visibility of the radio transmitter (this domain is defined by the space reached by appreciable density of rays from the transmitter). Some of the anomalies have appeared in certain features observed in the radio signals received from satellites. The other anomalies, not yet observed in satellite experiments, are clearly predicted from experience in air-to-air tropospheric propagation.

Section IV describes work in progress on ray-tracing techniques using an analogue computer (electronic differential analyzer) for non-isotropic media.

II. Ray-tracing Equations

The starting point of ray-tracing theory is Fermat's principle of stationarity in the time taken by a ray for travelling from one point to any other point in the medium. By a ray we mean the path followed by the transport of electromagnetic wave energy between the transmitter and a radio field-intensity point in space.

A. Isotropic Media

With Fermat's principle, the calculus of variations quickly yields

$$\frac{d^2h}{dx^2} = \left\{\left(1+\frac{h}{a}\right)^2 (1+y'^2) + h'^2\right\}\frac{1}{\mu}\frac{\partial\mu}{\partial h} + \left\{\left(1+\frac{h}{a}\right)^2 (1+y'^2) + 2h'^2\right\}\frac{1}{a+h} - \left\{1+y'^2+\frac{h'^2}{(1+h/a)^2}\right\}\frac{h'}{\mu}\frac{\partial\mu}{\partial x} \tag{1}$$

$$\frac{d^2y}{dx^2} = \left\{1+y'^2+\frac{h'^2}{(1+h/a)^2}\right\}\left\{\frac{1}{\mu}\frac{\partial\mu}{\partial y} - \frac{y'}{\mu}\frac{\partial\mu}{\partial x}\right\} \tag{2}$$

for determining ray paths. Here h is height radially above a reference level at or anywhere above the earth's surface; x is arc distance measured along the geocentric sphere passing through the reference level, and along the initial direction of propagation of the ray; y gives the small deviation, in lateral distance, of the ray from its initial plane of propagation; μ is the phase refractive index given by

Equation 9 (with $f_H = 0$); a is earth radius ($1/a = 0{\cdot}048 \times 10^{-6}$/ft $= 1{\cdot}57 \times 10^{-4}$/km); the super-primes (′) denote differentiation with respect to x.

Equations 1 and 2 are convenient to use when h' is not too large compared with unity. Applicable for μ variable with h, x, and y, they are a generalization of an equation given by Hartree [2] *et al.* for μ varying only with h.

B. Non-isotropic Media

In this case we have to distinguish from the group velocity (of electromagnetic energy transport) two phase velocities, one along a ray (along which energy is transported) and the second along the associated wave normal. Applying Fermat's principle to this complicated situation, Hamiltonian optics leads to Hamilton's equation [5] for ray paths; these equations are formally identical with Hamilton's equations in classical mechanics. From them Haselgrove [3] derived the following for ray paths in spherical coordinates:

$$\dot{r} = \left(y_1 - c^2\mu \frac{\partial \mu}{\partial y_1}\right)\frac{1}{\mu^2} \tag{3}$$

$$\dot{\theta} = \left(y_2 - c^2\mu \frac{\partial \mu}{\partial y_2}\right)\frac{1}{\mu^2 r} \tag{4}$$

$$\dot{\phi} = \left(y_3 - c^2\mu \frac{\partial \mu}{\partial y_3}\right)\frac{1}{\mu^2 r \sin\theta} \tag{5}$$

$$\dot{y}_1 = \frac{c^2}{\mu}\frac{\partial \mu}{\partial r} + \dot{\theta} y_2 + \dot{\phi} y_3 \sin\theta \tag{6}$$

$$\dot{y}_2 = \left(\frac{c^2}{\mu}\frac{\partial \mu}{\partial \theta} - \dot{r} y_2 + \dot{\phi} y_3 r \cos\theta\right)\frac{1}{r} \tag{7}$$

$$\dot{y}_3 = \left(\frac{c^2}{\mu}\frac{\partial \mu}{\partial \phi} - r y_3 \sin\theta - \dot{\theta} y_3 r \cos\theta\right)\frac{1}{r \sin\theta} \tag{8}$$

Here super-dots (·) denote differentiation with respect to time t; r, θ, and ϕ are the spherical coordinates, with origin at the earth's center, of position along a ray; y_1, y_2, and y_3 are spherical components of a vector parallel to the associated wave normal; c is the velocity of light in vacuum.

In deriving Equations 3–8 Haselgrove used an identity of Cartan. By putting Hamilton's equations in covariant form, and using the property that these equations are invariant with respect to coordinate transformations, we can alternatively derive (3–8) by elementary procedure.

The Appleton-Hartree formula for the phase refractive index μ along a wave normal can be put in the form

$$\mu^2 = 1 - \frac{N(r, \theta, \phi)/1{\cdot}24 \times 10^4 f^2}{1 - M \pm \sqrt{(M^2 + (f_H \cos\psi/f))^2}} \tag{9}$$

$$M \equiv \frac{(f_H \sin\psi/f)^2}{2(1 - N/1{\cdot}24 \times 10^4 f^2)} \tag{10}$$

In Equations (9, 10) the frequency f is in Mc/s, N is the number of electrons per cm^3, f_H is geomagnetic gyro-frequency in Mc/s, ψ is the angle between the ray and the geomagnetic field. If this field is taken to be that of a magnetic dipole, and the polar axis of the spherical coordinates taken to coincide with the dipole axis, we have

$$\cos\psi = \frac{y_1 \cos\theta + y_2 \sin\theta/2}{\sqrt{(y_1^2 + y_2^2 + y_3^2)}\,\sqrt{(\cos^2\theta + \sin^2\theta/4)}}$$

$$f_H = 0{\cdot}9\sqrt{(1 + 3\cos^2\theta)}\ \text{Mc/s}$$

It is noteworthy that the partial derivatives of μ appear explicitly in Equations (1, 2) and (3–8), in accord with the fact that the gradients of μ play crucial roles in creating anomalous effects as mentioned earlier.

Also, the Appleton-Hartree formula (9, 10) is notable in that it presents μ, which actually is a function of six variables, in the form containing purely algebraic operations on factors or terms each of which is a function of at most three variables—of either (r, θ, ϕ) or (y_1, y_2, y_3). This reduces the representation of μ to being a problem accessible to foreseeable computer techniques.

III. Ray Patterns in Isotropic Media

Figures 1–3 show ray patterns for μ depending only on height h; Fig. 4 is for μ depending on h and x (round-earth distance). For these figures the deviation y, in lateral distance, of the rays from the initial propagation plane is null.

In each ray pattern the radio transmitter is pictured as emitting rays in various directions from a fixed point. The $1/\mu\, d\mu/dh$ versus h profile, and the position of the transmitter relative to the peaks of this profile, are important. For each ray pattern the rays were obtained by solving Equation (1), with $h = h(x; h_0, A_0)$ as solutions, for a fixed initial value on h (corresponding to the transmitter height h_0) and successively for various initial values on dh/dx (corresponding to initial angles A_0 of the rays). An electronic differential analyzer was used to compute and plot the rays. All the terms in h/a, h' and y' were neglected from Equation (1) in computing Figs. 1–4.

Figure 1. The transmitter is at $h = 180$ km. The $1/\mu\, d\mu/dh$ versus h profile has a peak between 100 and 160 km; above this height $1/\mu\, d\mu/dh$ is constant at -2×10^{-4}/km, chosen to be nearly the negative of $1/a = 1{\cdot}57 \times 10^{-4}$/km ($a$ is earth radius.) The corresponding profile of electron density N versus h is for the frequency of 20 Mc/s. N rises to a value of 10^6 electrons/cm^3 at 250 km (what N does above this height is immaterial to the ensuing discussion). The numbers on the rays, falling along an arc, denote the initial slope of the rays; the numbers just above the x-axis denote the slope of the rays upon their arrival at the earth's surface at $h = 0$.

The particular value fo $1/\mu\, d\mu/dh$ in the height interval around the transmitter's height is crucial in Fig. 1. In that interval, where $1/\mu\, d\mu/dh$ is near $-1/a$, the rays all have small curvatures; this follows from Equation (1). Thus the rays emitted

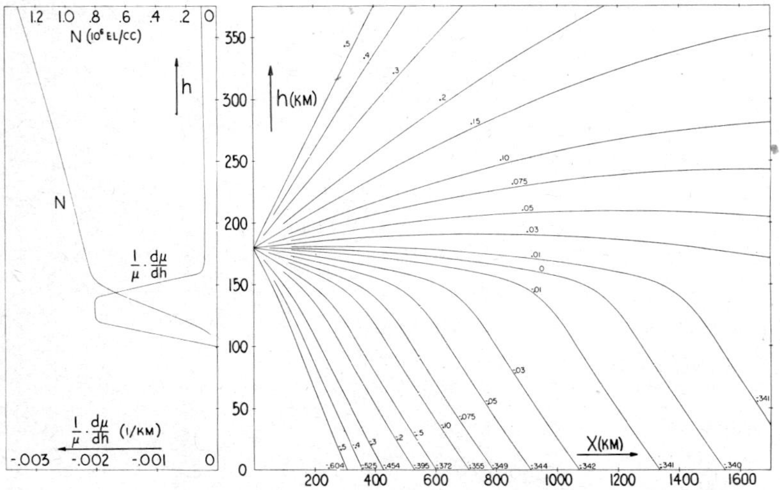

FIG. 1. Ray pattern showing requirments on $\frac{1}{\mu}\frac{d\mu}{dh}$ and on N to produce (a) far-reaching rays, (b) trapping of these rays. Skip distances do not occur in figure or in subsequent hops of trapped rays beyond figure.

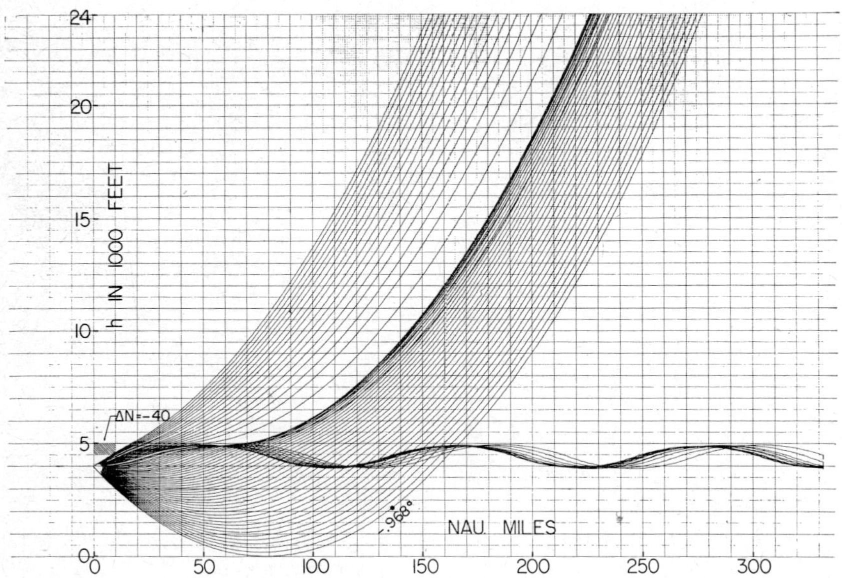

FIG. 2. Pattern with trapping of rays in a duct, plus a pair of ray-scarce and ray-crossing regions. Skip distances occur in duct.

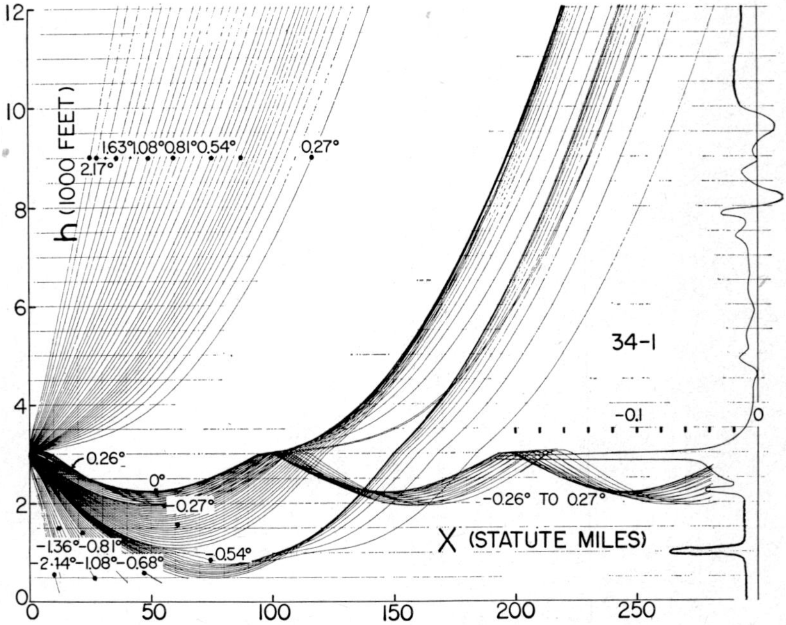

FIG. 3. Pattern with one duct of trapped rays, plus two pairs of ray-scarce and ray-crossing regions. Skip distances occur in duct.

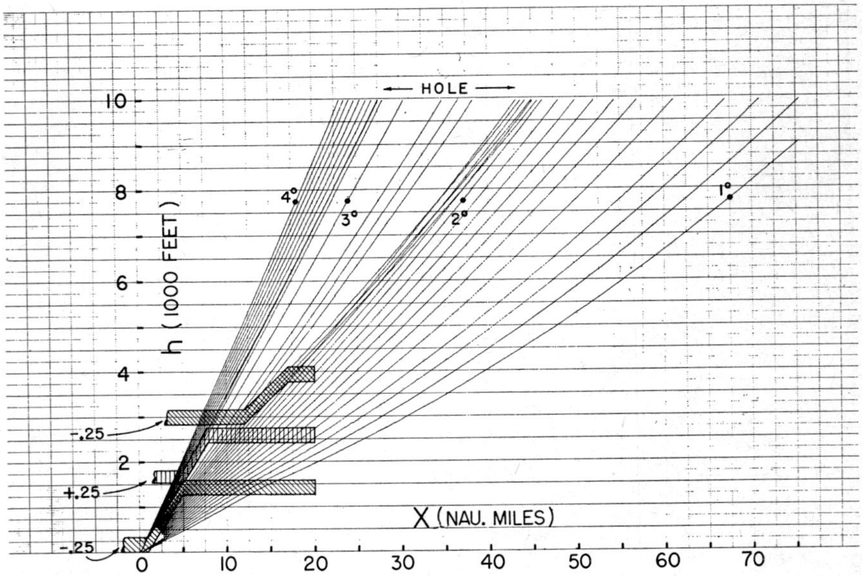

FIG. 4. Ray pattern with ray-scarce region caused by slanting, and properly deployed, layer segments.

at initial angles within several degrees of the horizontal direction travel unusually large distances before turning downward toward earth. These same rays will, after reflection at the earth's surface, travel upwards to the ionosphere, thence downward to earth again. The trapping of these rays between the ionosphere and earth's surface—analogous to trapping of rays in tropospheric radio ducts—will continue as far as the electron density N, determined by the area under the $1/\mu \, d\mu/dh$ versus h profile, remains sufficiently high with increasing distance. Provided that $1/\mu \, d\mu/dh$ continues to have values near $-1/a$ at the downward turning regions of the trapped rays, these rays will reach distances of many thousand kilometers after few hops between the ionosphere and earth. As there is favorable density of trapped rays at the earth's surface for distances up to 1600 km traversed by the first hops of these rays, the surface ray density would remain considerable in subsequent hops. Thus the trapping of the rays (those with initial angles within several degrees of the horizontal) in Fig. 1 would not give rise to any skip distance, such as familiar in ionospheric propagation, but instead will give a more or less continuum of rays, with at least appreciable ray density everywhere for a very large distance along the earth's surface.

These results of Fig. 1 serve as a possible explanation for an interesting case of radio observations on Sputnik III: in this case the Sputnik's 20 Mc/s transmission was received at Bedford, Mass., with at least appreciable signal intensity continuously for more than 7500 km of satellite travel.

Figures 2, 3 and 4 show ray patterns taken from tropospheric propagation; the height and distance intervals in them are smaller than for the ionosphere. However, all the considerations relating to production of the anomalies (radio holes, antiholes, trapping of rays in ducts, and dense spatial fadings of radio signal) apply for the ionosphere.

Figure 2. The shaded rectangle near the transmitter represents a layer between 4500 and 5000 ft; the transmitter is at $h = 4000$ ft. The change in μ (proportional to change in N for the ionosphere) through the layer is 40×10^{-6}. $1/\mu \, d\mu/dh$ has a constant value of $-0{\cdot}08 \times 10^{-6}$/ft within the layer, and of $-0{\cdot}012 \times 10^{-6}$/ft outside the layer.

Figure 2 shows trapped rays in an elevated horizontal duct, and concomitantly a narrow pair of hole and antihole above the duct. The hole is the area with reduced density of rays, and thus implies reduced radio signal intensity. The antihole is the area with crossings of two interfering groups of rays. A caustic locus—formed by intersections of consecutive rays and made visible by heavy concentration of rays—demarcate the antihole from the hole. An antihole implies enhanced, and spatial fadings of, radio signal intensity.

The trapped rays in Fig. 2 also has a caustic locus, and all pass through periodically spaced foci on successive hops in the duct. (How far these trapped rays reach downward, from the top of the layer, is set by the change in μ through the layer.) Unlike Fig. 1, the trapped rays in Fig. 2 give high rays densities—say along a horizontal line within the duct—only at narrow distance intervals, with wider gaps or skip distances in between.

Thus in Fig. 2, for distances beyond the region reached by the non-trapped rays, the radio signal intensity along a horizontal line in the duct would rise sharply on crossing the caustic locus at the end of a skip distance, then would decay gradually, then would become very low in the next skip distance, and so on repetitively for subsequent hops. Features such as these are contained in Dewan's [1] report of a radio observation at Bedford, Mass., on the 20 Mc/s transmission from Sputnik I.

Figure 3. The transmitter is at $h = 3000$ ft, at a sharp peak in the $1/\mu\, d\mu/dh$ versus h profile shown at the right; the $1/\mu\, d\mu/dh$ scale is in units of 10^{-6}/ft.

The $1/\mu\, d\mu/dh$ peak at the transmitter produces trapping of rays in the horizontal duct (which has skip zones like Fig. 2, and unlike Fig. 1); it also produces concomitantly a wide hole followed by an antihole above the duct. The lowest $1/\mu\, d\mu/dh$ peak produces the second pair of hole and antihole. The $1/\mu\, d\mu/dh$ peaks above the transmitter do not produce any anomaly; to cause anomalies, the peaks above the transmitter must have sufficient areas under them; the peaks below the transmitter always have discernible effects.

Figure 4. The transmitter is at bottom of the first of three stair-shaped layers; as shown by the shaded areas, each layer has a slanting segment adjoining two horizontal segments. $1/\mu\, d\mu/dh$ has a constant value of $-0{\cdot}25$, $0{\cdot}25$ and $-0{\cdot}25 \times 10^{-6}$/ft in the bottom, center, and top layer. The numbers on the rays denote their initial elevation angles.

The slanting segments of the two negative layers are properly oriented and deployed, and given a sufficiently negative value of $1/\mu\, d\mu/dh$, in order to produce an extensive radio hole beginning at an elevation angle of 3° (relatively high for tropospheric propagation) in a ground-to-air propagation geometry. The positive layer in Fig. 4 neither aids nor handicaps the creation of the hole; it is included to conform to experimental data that the algebraic change in μ through the portion of the troposphere concerned should not exceed 10^{-4}. It is emphasized that if the two negative layers were not properly oriented and deployed—for example if they were horizontal, without the slanting segments—the hole shown would not be possible.

Applied to the ionosphere and radio astronomy, the slanting segments of the layers in Fig. 4 may be aligned along geomagnetic lines at ionospheric heights. Electrons may on occasions (say at times of auroral activity) concentrate around the magnetic lines, giving rise to field-aligned columns with intense negative and positive values of $1/\mu\, d\mu/dh$ as comtemplated in Fig. 4. Possible radio holes caused by magnetic-field-aligned ionization has been suggested by Rush and Colins [4] to explain some anomalies in radio-astronomical observations.

IV. Ray Tracing Technique for Non-isotropic Media

Except for one major exception, all the complications of Equations (3–8) are within easy scope of electronic computers, analogue or digital. This exception consists of the large and exacting requirements on function generating equipments used for inputs to the computation. Seven function generating systems are required—to accommodate μ, and the partial derivatives of μ with respect to r, θ, ϕ, y_1, y_2, y_1. Each system involves lengthy algebraic operations on several individual function generators; each function generator accommodates an arbitrary function of up to three variables.

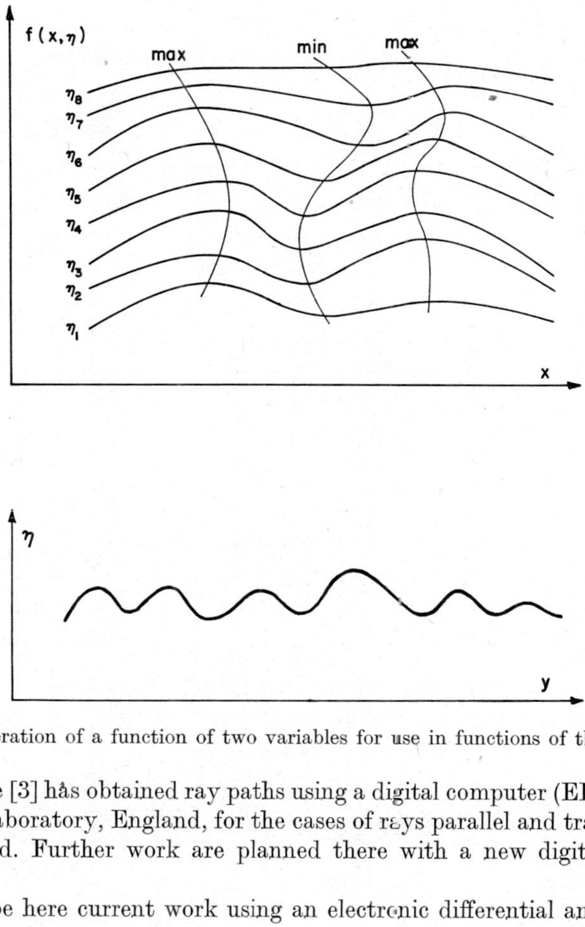

FIG. 5. Generation of a function of two variables for use in functions of three variables.

Haselgrove [3] has obtained ray paths using a digital computer (EDSAC) at the Cavendish Laboratory, England, for the cases of rays parallel and transverse to a magnetic field. Further work are planned there with a new digital computer (EDSACII).

We describe here current work using an electronic differential analyzer (analogue computer).

To generate a function of three variables, we proceed as follows: Take an arbitrary function of h, as the $1/\mu\ d\mu/dh$ versus h profile in Fig. 3, for example. The most significant effects of possible variations with x and y of this profile arise from variations in the peaks of this profile. Thus we focus attention on the amplitude P, the height H, and the height interval W of each peak in a profile function $f(h, x_0, y_0)$ of h at a fixed point (x_0, y_0) in the x-y plane.

We let P, H and W to be each a function of x and y as shown schematically in Fig. 5. A diode function-generating circuit, for a fixed biasing voltage η_1 and variable input voltage x gives an output voltage $f(x, \eta_1)$ for approximating a desired function of x through a number of (usually 10 to 20) sloping-line segments. By varying the bias η as a parameter, we obtain the family of curves for $f(x, \eta)$ in the upper half of Fig. 5. By generating a output voltage η, again using a diode function-generating circuit, as an arbitrary (suitably chosen) function of y, serving as input voltage in the lower half of Fig. 5—and by using this output η as a continuously variable parameter in the upper half of the figure—we obtain a desired function of three variables, $f(h, x, y)$. The function obtained this way is of a type which is significant for portraying ionospheric effects, and which is accessible to current computer art.

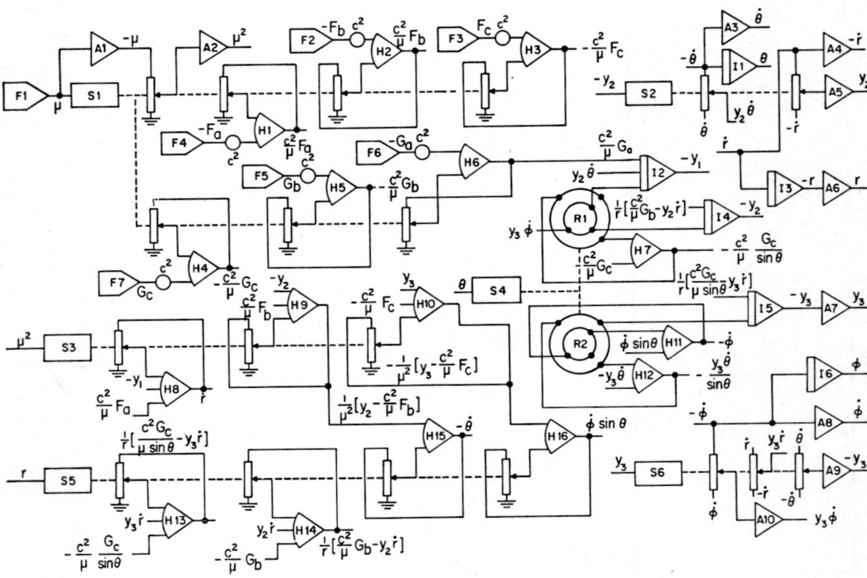

FIG. 6. Schematic diagram showing connections of computing elements for solving equations (3-8).

Figure 6 shows schematically the connections of computing elements in an electronic differential analyzer for solving Equations (3–8).

In the figure, A stands for summing and/or inverting amplifier, I for integrating (and summing) amplifier, H for high gain (and summing) amplifier, S for servo (which converts a computing voltage into a shaft rotation), R for resolver (a combination of sine and cosine potentiometers used to obtain trigonometric functions).

F_a, F_b and F_c are abbreviations for μ^2 times the partial derivatives of μ with respect to y_1, y_2 and y_3. G_a, G_b and G_c are abbreviations of the partial derivatives of μ with respect to r, θ and ϕ. The letter F followed by an Arabic numeral stands for function-generating system: $F1$ accommodates μ; the others, $F2$ through $F7$, accommodate F_a, F_b, F_c, G_a, G_b and G_c (not in this order, however).

The small vertical rectangles in the figure stand for linear potentiometers driven by servos as indicated. The small circles stand for scale-factor potentiometers, used to multiply voltages by constant factors. The output of the amplifiers A is equal to the negative sum of the inputs. The output of the high-gain amplifiers H is such that the sum of the inputs becomes zero.

For performing a multiplication, in Fig. 6: A servo (for example $S6$) drives a linear potentiometer to a position proportional to its input (y_3 in this example); a voltage proportional to another variable ($-\dot{\phi}$ in example) is applied to the ends of this potentiometer; the output at the arm of the potentiometer is then proportional to the product of the input to the servo and the input to the potentiometer ($-y_3\dot{\phi}$ in example).

For performing a division: One input to a high-gain amplifier H is the dividend; the "feedback" circuit around the amplifier is closed through a servo-driven potentiometer positioned proportionally to the divisor; the output of the high-gain amplifier is equal to minus the quotient. Example: In amplifier $H1$, the dividend is $-c^2F_a$; the divisor is μ the quotient is $c^2/\mu\, F_a$.

The amount of computing elements indicated in Fig. 6, exclusive of the function-generating systems $F1$ through $F7$, are available within a commercially made computer such as the REAC 400 (made by Reeves Instrument Co., New York). The function-generating systems required would likely double the amount of computing equipments.

V. Remarks

The isotropic equations (1, 2) are within easy scope of current analogue computers, even when the ionospheric electron density varies three dimensionally with height and horizontal directions. These equations give ray patterns which serve well to explain anomalous propagation effects in recent satellite experiments, and will likely be successful for the portrayal, analysis and prediction of propagation effects in future satellite experiments, say at frequencies above 20 Mc/s. Some of the effects so far observed in radio signals from satellites suggest the essential roles of two-and three-dimensional variations in the ionospheric electron density.

It is not clear whether the more general, non-isotropic equations (3–8) will give rise to any new anomaly in addition to the radio holes, antiholes, trappings of rays in ducts, and dense spatial fadings of radio signal which are now known. It is probable that the Equations (3–8) will disclose significant new results of detail, if not of kind, principally for propagation experiments at frequencies close to the geomagnetic gyro-frequency (near 1 Mc/s).

The writer is indebted to Dr. P. Newman for support, to P. Nawrocki and Professor I. Kay for exposition of Hamiltonian optics, to Dr. L. Bauer and staff of Project Cyclone for exploring techniques of three-dimensional, input-function generation.

References

1. Dewan, E. M. (1958). An Interesting Propagation Effect of Sputnik I. Paper at this Conference.
2. Hartree, D. R., Michel, J. G. L. & Nicolson, P. (1946). In report of Conference on " Meteorological Factors in Radio Propagation ", p. 161. London, The Physical Society.
3. Haselgrove, J. (1955). In report of Conference on "Physics of the Ionosphere", p. 362. London, The Physical Society, (1957) *Proc. phys. Soc. Lond. B*, LXX, 653.
4. Rush, S. & Colin, L., (1958). *Proc. Inst. Radio Engrs, N.Y.* **46**, 356.
5. Synge, J. L. "Hamilton's Method in Geometrical Optics" chapter 2. Univ. of Maryland: Inst. for Fluid Dynamics and appl. Math., Lecture Series No. 9 (1951). "Geometrical Mechanics and de Broglie Waves," chapter 2. Cambridge University Press, London (1954).
6. Wong, M. S. (1958). *Proc. Inst. Radio Engrs, N.Y.* **46**, 1628.

Choix de l'Emplacement d'une Station de Base ou de Radiodiffusion en Ondes Métriques

L. PENNINCKX

Régie des Télégraphes et des Téléphones, Bruxelles

Pour trouver les emplacements favorables pour les stations de base ou de radiodiffusion en ondes métriques, on utilise en Belgique une *carte en relief* du pays, appliquée sur une surface sphérique dont le rayon équivaut aux 4/3 du rayon terrestre. On l'éclaire par une petite ampoule placée au point et à la distance de la surface qui correspondent à l'emplacement envisagé. Des essais systématiques ont montré que la photo de la partie éclairée de la carte donnait une excellente idée de la couverture obtenae en réalité.

Il est très important de placer les stations de base ou de radiodiffusion en ondes métriques en des points tels que la couverture de la région environnante soit aussi complète et, généralement, aussi étendue que possible, avec une hauteur d'antenne minimum, de façon à réduire le prix des pylônes et le danger qu'ils présentent pour les avions.

On utilise en Belgique une *carte en relief* du pays, qu'on éclaire par une petite ampoule électrique placée au point et à la distance de la surface qui correspondent à l'emplacement envisagé pour l'antenne. On photographie la partie éclairée; on compare l'image obtenue à la région effectivement couverte par la station.

I. ETUDE DE LABORATOIRE

La carte de Belgique en plâtre (moulage d'une carte composée par des couches d'égale épaisseur) est au 1/160.000; l'échelle des hauteurs est huit fois plus grande: 1/20.000. Pour que la portée optique sur la carte soit semblable à la portée réelle, il faut multiplier par huit la courbure de la terre. Le rayon terrestre fictif, égal aux 4/3 du rayon terrestre réel, est représenté par une longueur de:

$$\frac{40.000.000 \times 4}{2\pi \times 160.000 \times 8 \times 3} = 6{,}65 \text{ m.}$$

La carte en relief a donc été appliquée sur une surface sphérique de 6,65 m de rayon.

On emploie une ampoule de lampe de poche de type plat. Le filament est à 2 mm environ du sommet du verre: la hauteur équivalente minimum est de l'ordre de 35 à 40 m. Pour des hauteurs plus grandes, il suffit d'éloigner quelque peu l'ampoule de la surface de la carte; si des hauteurs moindres devaient être représentées, on prendrait une ampoule de type moins courant, formée d'un filament noyé dans une petite boule de verre.

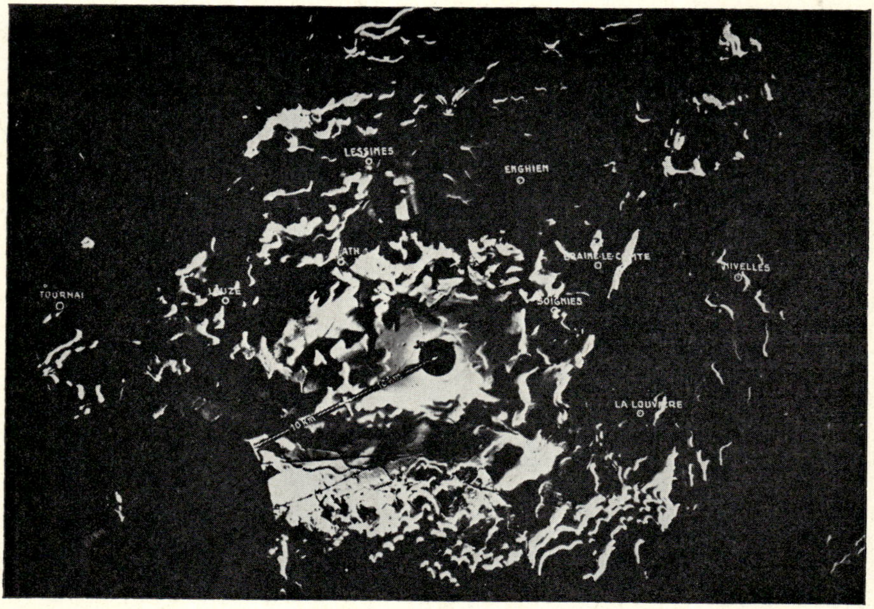

FIG. 1. Jurbise 65 m.

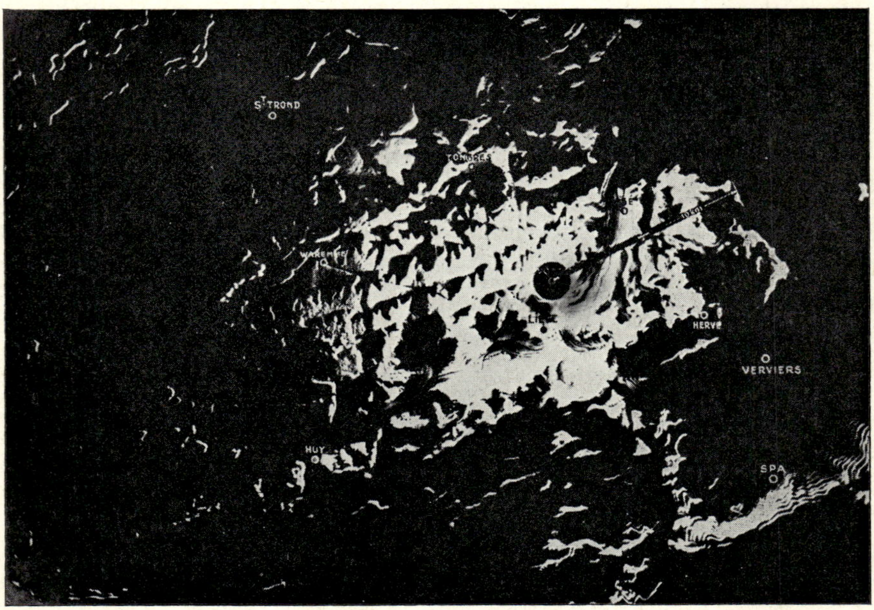

FIG. 2. Vottem 35 m.

A titre d'exemple, on a considéré deux stations de base destinées à couvrir les régions de Mons et de Liège par un réseau public de radiotéléphonie mobile. Les antennes en sont omnidirectionnelles.

Près de Mons, on a choisi la station de réception de Jurbise, où un pylône de 65 m était disponible (Fig. 1).

Près de Liège, au contraire, on a recherché sur la carte en relief un des meilleurs emplacements: le village de Vottem (Fig. 2).

II. Etude sur le Terrain

Les essais eurent lieu dans les conditions suivantes:

Station de base:

fréquence: 159,42 Mc/s;
puissance émise: 250 W;
antenne: dipôle demi-onde vertical.

Station mobile (camion):

fréquence: 164,46 Mc/s;
puissance émise: 10 W;
antenne: quart d'onde vertical sans contrepoids, à 2 m du sol.

Modulation de phase (déviation: $\pm$ 10 kc/s; précaccentuation: 6 dB par octave). Travail en alternat. Récepteurs étalonnés pour fournir une mesure approximative de champ (courant de grille du premier limiteur converti en tension à l'entrée et champ reçu).

Station de base de Jurbise (d'octobre 1957 à janvier 1958):

altitude du sol: 75 m;
hauteur de l'antenne au-dessus du sol: 65 m;
altitude de la région couverte: 20 à 130 m.

Station de base de Vottem (de février à avril 1958):

altitude du sol: 120 m;
hauteur de l'antenne au-dessus du sol: 12 m;
(la hauteur équivalente de l'ampoule est de 35 m; cette différence est peu importante, parce que les dénivellations du sol sont d'un ordre de grandeur bien supérieur)
altitude de la région couverte: 30 à 600 m.

A chaque point de mesure (environ 1000 points pour chaque station de base), on recherchait le maximum du champ reçu en déplaçant lentement le camion sur une distance de l'ordre du mètre (demi-onde); on notait le champ et la qualité de la parole aux deux extrémités de la liaison.

L'étude des résultats obtenus amène les constatations suivantes:

1° Les points éclairés sur la carte en relief donnent sur le terrain une bonne qualité de transmission dans les deux sens. Les contours de la partie éclairée correspondent, à une précision de quelques centaines de mètres, à la limite des communications commerciales.

Cependant, la qualité est parfois encore satisfaisante dans certaines parties sombres. Les points où ce phénomène s'observe semblent en général séparés de la station de base par une crête peu accentuée: un effet de diffraction doit probablement intervenir (cas de la route Leuze—Ath—Enghien, par exemple, Fig. 1).

2° A chaque valeur de la qualité subjective de la transmission, cotée sur 5, correspond une grande dispersion des intensités de champ mesurées.

3° Comme la station de base est 25 fois plus puissante que la station mobile, les champs reçus aux deux extrémités de la liaison, toutes choses égales par ailleurs, devraient être dans un rapport de 5.

Or, on a trouvé pour ce rapport une valeur médiane de 18 à Jurbise, 25 à Vottem. La zone de bonne réception de la station de base est plus étendue que celle dans laquelle le véhicule est bien entendu à cette station; elle dépasse nettement le contour de la partie éclairée de la carte en relief.

Une grande hauteur au-dessus de la région environnante est donc plus avantageuse à l'émission qu'à la réception (cf. Egli: Radio Propagation Above 40 Mc Over Irregular Terrain. *Proc. Inst. Radio Engrs, N.Y.* (1957); Fig. 2, p. 1385).

4° L'influence de la structure fine du relief et de toutes les irrégularités locales du terrain est telle, qu'il est pratiquement impossible de dessiner des courbes équichamps valables, même avec l'aide de la photo de la carte en relief.

5° L'influence des conditions météorologiques est peu apparente (le temps était le plus souvent couvert).

6° Le champ reçu est généralement fort inférieur à celui déduit des courbes de propagation de l'Atlas du C.C.I.R., qui sont basées, notamment, sur l'hypothèse que la surface de la terre est une sphère régulière. Des valeurs supérieures à celles données par ces courbes s'observent cependant en quelques points élevés et dégagés, atteints en visibilité directe.

III. Conclusions

1° On a proposé plusieurs méthodes pour le choix de l'emplacement et de la hauteur d'antenne d'une station en ondes métriques destinée à couvrir une région déterminée. Celle de l'éclairage d'une carte en relief par une ampoule déplaçable est de loin la plus rapide et la plus facile à appliquer. La correspondance entre la partie éclairée de la carte et la zone réellement couverte sur le terrain est tellement bonne que l'étude sur carte peut suffire à elle seule à déterminer les régions atteintes. On la complétera seulement par quelques essais sur le terrain dans les zones intéressantes et apparemment "sombres."

2° La méthode ne tient pas compte des obstacles locaux, tels que les bâtiments et les arbres (cf. C.C.I.R., Question n° 138, *Commission d'Etudes* n° 5), dont l'influence est considérable à moyenne distance; mais il est certain qu'aucune méthode globale d'étude de la propagation ne peut donner de précision à ce sujet.

3° L'échelle du 1/160.000 convient pour la carte de Belgique. Pour les grands pays, il faudrait sans doute faire plusieurs cartes, avec certains recouvrements. L'échelle pourrait être réduite, à condition de conserver la même finesse de détail.

4° Il est difficile d'établir une correspondance entre la puissance des ampoules et celle des émetteurs: la diffraction de la lumière sur les arêtes n'est pas la même que celle des ondes sur les crêtes du terrain; il faudrait normaliser le temps de pose, la sensibilité et le développement de la pellicule, etc.

5° On pourrait représenter une antenne directive par une ampoule partiellement masquée, pour que le flux lumineux ne s'en échappe que dans un certain angle.

6° Le procédé convient remarquablement pour l'étude d'un service mobile terrestre. On ne peut garantir des communications commerciales que sur des routes couvertes avec une certaine continuité: c'est le cas dans toute la partie éclairée.

Pour les services de télévision ou de radiodiffusion sonore en ondes métriques, on n'obtient évidemment qu'une image générale de la région couverte. Beaucoup d'auditeurs ou de téléspectateurs habitant dans les zones sombres pourront avoir encore une excellente réception avec une antenne de gain approprié placée à un endroit et une hauteur déterminés empiriquement sur place. En sens inverse, en pleine zone éclairée, des obstacles locaux, qui n'apparaissent pas sur la carte en relief, pourront gêner la réception à tel ou tel groupe de maisons. Néanmoins, même pour la télévision, le procédé de la carte en relief a rendu en Belgique d'inestimables services pour le choix des emplacements des stations et l'évaluation rapide des populations desservies.

On the Propagation of Correlation in Wave Fields*

George B. Parrent Jr.

Air Force Cambridge Research Center, Bedford, Mass.†

Using the formulation of coherence theory introduced by Wolf [4] a general solution for the mutual coherence function in a field illuminated by a plane polychromatic source is given. The limiting forms of the mutual coherence function for coherent and incoherent quasi-monochromatic fields are obtained; and using these forms the corresponding limits of the general solution referred to above are examined in detail. In particular it is shown that in a coherent field the mutual coherence function is, apart from a simple periodic factor, expressable as the product of a wave function with its complex conjugate, each factor depending on the co-ordinates of one point only. It is also shown that an incoherent field cannot exist in free space.

In several fields employing electromagnetic theory, visible and microwave optics, radio astronomy and scatter communications, many of the current problems reduce ultimately to the determination of the correlation between the electromagnetic disturbances at two points in the field. Historically workers in visible optics were the first to become concerned with this type of problem in an effort to bridge the gap between coherent and incoherent fields, to treat partially coherent fields. In surveying the optical literature on the subject of partial coherence one cannot fail to notice that there are as many definitions of the basic quantities and concepts as there are papers on the subject. However, the formulation of the problem introduced by Wolf [4] is the most general and in fact it can be shown [3] that the quantities introduced by the earlier authors can be deduced from a knowledge of Wolf's mutual coherence function.

The fundamental quantities in the study of partial coherence are: (1) the mutual coherence function, $\Gamma_{12}(\tau)$, defined as the complex cross-correlation of the analytic signals associated with the disturbances at the two points; (2) the complex degree of coherence, $\gamma_{12}(\tau)$, defined as the mutual coherence function normalized by the square root of the product of the intensities at the two points; and (3) the mutual intensity, $\Gamma_{12}(0)$, defined as the zero ordinate of the mutual coherence function. Mathematically these definitions may be expressed respectively as follows:

$$\Gamma_{12}(\tau) = \langle V_1(t+\tau)V_2^*(t)\rangle, \tag{1}$$

where the sharp brackets denote time average,

$$\gamma_{12}(\tau) = \frac{\Gamma_{12}(\tau)}{\sqrt{(\Gamma_{11}(0)\Gamma_{22}(0))}}, \tag{2}$$

and

$$\Gamma_{12}(0) = \langle V_1(t)V_2^*(t)\rangle. \tag{3}$$

*A fuller treatment of the subject matter of this paper has been accepted for publication in the *Journal of the Optical Society of America.*

†The research discussed herein was conducted during the author's stay at the Physical Laboratories, The University, Manchester.

From the Schwarz inequality it is readily seen that the modulus of the degree of coherence is bounded, by zero and unity, i.e., that

$$0 \leqslant |\gamma_{12}(\tau)| \leqslant 1. \tag{4}$$

The extremes of 0 and 1 characterize incoherence and coherence respectively.

In this paper the general solution for the mutual coherence function in a field illuminated by a plane polychromatic source will be given. For the more practical case of nearly monochromatic illumination a more detailed analysis will be given showing the implications of coherence and incoherence on the form of $\Gamma_{12}(\tau)$; and the corresponding limits of the general solution mentioned above are also investigated.

The definitions reviewed above were formulated in terms of Gabor's analytic signal [2] and to establish the notation we shall review briefly the formation of the analytic signal, $V(t)$, associated with the real signal $V^r(t)$. Let $V^r(t)$ represent a real scalar function (e.g., a Cartesian component of the electric vector) possessing a Fourier representation,

$$V^r(t) = \int_0^\infty a(\nu) \cos [\phi(\nu) - 2\pi\nu t] d\nu. \tag{5}$$

We associate with $V^r(t)$ another real function $V^i(t)$, formed by changing the phase of each spectral component by $\pi/2$, i.e.,

$$V^i(t) = \int_0^\infty a(\nu) \sin [\phi(\nu) - 2\pi\nu t] d\nu. \tag{6}$$

Then the analytic signal is defined as

$$V(t) = V^r(t) + iV^i(t); \tag{7}$$

and it can be shown that $V^r(t)$ and $V^i(t)$ are Hilbert transforms, i.e.,

$$V^i(t) = \frac{P}{\pi} \int_{-\infty}^{\infty} \frac{V^r(t')}{t'-t} dt' \quad \text{and} \quad V^r(t) = -\frac{P}{\pi} \int_{-\infty}^{\infty} \frac{V^i(t')}{t'-t} dt'. \tag{8}$$

P denotes Cauchy's principal value at $t' = t$. From the theory of Hilbert transforms we require the theorem relating the derivatives of transform pairs, i.e.,

$$\frac{d}{dt} V^i(t) = \frac{P}{\pi} \int_0^\infty \frac{(d/dt') V^r(t')}{t'-t} dt'. \tag{9}$$

Using this theorem and the fact that $V^r(t)$ in vacuum satisfies the scalar wave equation, it follows immediately that $V(t)$ satisfies the wave equation

$$\nabla^2 V(t) = \frac{1}{c^2} \frac{\partial^2 V(t)}{\partial t^2}, \tag{10}$$

where c is the velocity of light in vacuum.

If Equation (1) is differentiated first in the coordinates of P_1 and then P_2 and the order of averaging and differentiating are interchanged we obtain

$$\nabla_s^2 \Gamma_{12}(\tau) = \frac{1}{c^2} \frac{\partial^2 \Gamma_{12}}{\partial \tau^2}(\tau) \quad (s = 1, 2). \tag{11}$$

Thus in vacuum the mutual coherence function satisfies the two wave equations (11). This result was first obtained by Wolf [4] by different arguments. To obtain the solution to Equation (11) for a field illuminated by a plane polychromatic source we introduce the coordinate system illustrated in Fig. 1, where σ is the

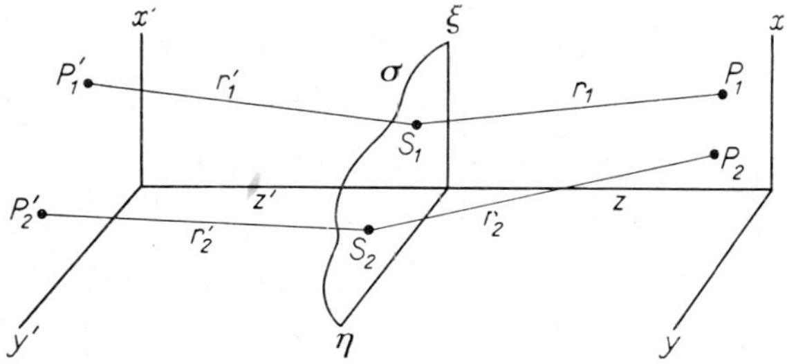

FIG. 1.

plane of the source with coordinates (ξ, η); S_1 and S_2 are points in the source. P_1 and P_2 are points in the illuminated field; and P_1' and P_2' are the mirror images of P_1 and P_2 with respect to the plane σ.

Since $\Gamma_{12}(\tau)$ is a cross-correlation of analytic signals it has a Fourier spectrum of the form

$$\Gamma_{12}(\tau) = \int_0^\infty G_{12}(\nu)e^{-2\pi i\nu\tau}\,d\nu. \tag{12}$$

If Equation (12) is substituted into Equation (11) and the terms rearranged, we obtain the two Helmholtz equations for $G_{12}(\nu)$,

$$[\nabla_s^2 + k^2(\nu)]G_{12}(\nu) = 0, \quad (s = 1, 2), \tag{13}$$

with $k(\nu) = \dfrac{2\pi\nu}{c}$. These equations can be solved formally for an arbitrary source distribution using Green's theorem to give

$$G(P_1, P_2, \nu) = \frac{1}{(4\pi)^2}\int_{\sigma_1}\int_{\sigma_2} G(S_1, S_2, \nu)\frac{\partial \mathscr{G}_1}{\partial n}\frac{\partial \mathscr{G}_2}{\partial n}\,d\sigma_1 d\sigma_2, \tag{14}$$

where $\mathscr{G}_1$ and $\mathscr{G}_2$ are Green's functions satisfying Equation (13) and the boundary condition that they vanish over the plane of the source. Imposing Sommerfeld's radiation condition on $V_1^r(t)$ and $V_2^r(t)$ gives finally

$$G(P_1, P_2, \nu) = \frac{1}{(2\pi)^2}\int_{\sigma_1}\int_{\sigma_2} \cos\theta_1 \cos\theta_2(1-ikr_1)(1+ikr_2)G(S_1, S_2, \nu)\frac{e^{ik(r_2-r_1)}}{r_1 r_2}\,d\sigma_1, d\sigma_2, \tag{15}$$

where

$$\cos\theta_s = \frac{z_s}{r_s}.$$

The mutual coherence is obtained by taking the Fourier transform of both sides of (15) which gives formally

$$\Gamma_{12}(\tau) = \frac{1}{(2\pi)^2}\int_{\sigma_1}\int_{\sigma_2} \cos\theta_1 \cos\theta_2 \mathscr{F}\left[\Gamma\left(S_1, S_2, \tau - \frac{r_1 - r_2}{c}\right)\right] d\sigma_1 d\sigma_2 \tag{16}$$

where $\mathscr{F}$ is the differential operator,

$$\mathscr{F} = 1 + \frac{r_1 - r_2}{c}\frac{\partial}{\partial\tau} - \frac{r_1 r_2}{c^2}\frac{\partial^2}{\partial\tau^2}. \tag{17}$$

Equation (16) gives the mutual coherence for a field illuminated by a plane polychromatic source (with an arbitrarily prescribed distribution of coherence) in terms of the mutual coherence on the source 'delayed' by $(r_1 - r_2)/c$. In most practical problems, however, one deals with quasi-monochromatic illumination. In such a field $\Gamma_{12}(\tau)$ may be written as

$$\Gamma_{12}(\tau) = \Gamma_{12}(0)e^{-2\pi i\bar{\nu}\tau} \qquad \left(|\tau| \ll \frac{1}{\Delta\nu}\right), \tag{18}$$

where $\bar{\nu}$ is the mean frequency, and $\Delta\nu$ is the spectral range of the radiation. $\Gamma_{12}(0)$ is the mutual intensity function of Zernike. An approximate expression for $\Gamma_{12}(0)$ for points on a plane illuminated by an incoherent source was given independently by Van Cittert [1] and Zernike [5]. We will show shortly the conditions under which this law is valid. First, however, we must digress to determine the form of $\Gamma_{12}(0)$ in coherent and incoherent fields. Substitution of (18) into (11) gives

$$[\nabla_s^2 + \bar{k}^2]\Gamma_{12}(0) = 0 \qquad s = 1, 2 \qquad \left(|\tau| \ll \frac{1}{\Delta\nu}\right). \tag{19}$$

Thus the mutual intensity in a quasi-monochromatic field is propagated according to the two Helmholtz equations of (19). A coherent field is characterized by

$$|\gamma_{12}(0)| = 1, \tag{20}$$

which implies that

$$\Gamma_{12}(0) = A_1 A_2 e^{i\phi_{12}}, \tag{21}$$

where $A_s = \sqrt{(\Gamma_{ss}(0)}$. If (21) is substituted into (19) we obtain after equating the imaginary parts to zero

$$(\nabla_1\phi_{12})^2 = \left[\frac{\nabla_1^2 A_1}{A_1} + \bar{k}^2\right] = f^2(x_1, y_1, z_1); \tag{22}$$

and of course a similar equation involving ∇_2. Equation (22) is of the form of the eikonal equation of optics or the Hamilton-Jacobi equation of dynamics and can be solved by standard variational methods to give

$$\phi_{12} = \alpha(P_1) + \beta(P_2). \tag{23}$$

From the definition of $\Gamma_{12}(\tau)$ and the assumption of stationary it follows that

$$\Gamma_{12}^*(\tau) = \Gamma_{21}(-\tau), \tag{24}$$

which in turn implies that

$$\alpha(P_1) = -\beta(P_1). \tag{25}$$

and

$$A_1(P_1) = A_2(P_1).$$

Hence

$$\Gamma_{12}(\tau) = e^{-2\pi i\bar{\nu}\tau} U(P_1)U^*(P_2) \qquad \left(|\tau| \ll \frac{1}{\Delta\nu}\right). \tag{26}$$

Thus in a coherent quasi-monochromatic field the mutual coherence function can be expressed, apart from a periodic factor, as the product of a wave function, U, with its complex conjugate and each factor is a function of one point only. A coherent source may be defined as one for which $\Gamma_{12}(0)$ is of the form given by (26) for all pairs of points in the source.

To examine the implications of incoherence we recall that incoherence is by definition characterized by

$$|\gamma_{12}(0)| = 0, \tag{27}$$

which implies

$$\Gamma_{12}(0) = \begin{cases} 0 & P_1 \not\equiv P_2 \\ I(P_2) & P_1 \equiv P_2 \end{cases}. \tag{28}$$

We shall now show that an incoherent quasi-monochromatic field cannot exist in free space, i.e. $\Gamma(P_1P_2, 0)$ cannot have the form indicated by (28) in any finite volume. For consider a finite volume, V, in which the field is assumed to be incoherent ((28) is satisfied throughout V). Let $\sum$ be any closed surface contained in V. Then the mutual intensity for any two points, P_1 and P_2, contained in $\sum$ is given by

$$\Gamma(P_1, P_2, 0) = \int_\Sigma \int_\Sigma \Gamma(S_1, S_2, 0) F(\overline{P_1 - S_1}, \overline{P_2 - S_2}) dS_1 dS_2, \tag{29}$$

where S_1 and S_2 are points on Σ; $\overline{P_i - S_i}$ is the distance from P_i to S_i;

$$F = \left.\frac{\partial \mathscr{G}_1}{\partial n} \frac{\partial \mathscr{G}_2}{\partial n}\right|_\Sigma ;$$

and $\mathscr{G}_1$ and $\mathscr{G}_2$ are Green's functions satisfying the double Helmholtz equations and vanishing on $\sum$. Since $\sum$ is contained in V, $\Gamma(S_1, S_2, 0)$ must be of the form given by (28). Therefore either: (a) the integral in (29) vanishes, or (b) $\Gamma(S_1, S_2, 0)$ is sufficiently singular at $S_1 \equiv S_2$ to preserve the contribution from this surface. Since (29) holds for all points within $\sum$ it must hold at $P_1 = P_2$; and therefore the possibility (a) above which implies $\Gamma(P_1, P_2, 0) \equiv 0$ is contrary to hypothesis (28). If in accordance with (b) the contribution from $S_1 = S_2$ is preserved (29) can be written as a single integral of the form

$$\Gamma(P_1, P_2, 0) = \int_\Sigma \Gamma(S_1, S_1, 0) F(\overline{P_1 - S_1}, \overline{P_2 - S_1}) dS_1, \tag{30}$$

since the double integral is identically zero when $S_1 \neq S_2$; but $\Gamma(S_1, S_2, 0) = I(S_1)$. Therefore

$$\Gamma(P_1, P_2, 0) = \int_{\Sigma} I(S_1) F(\overline{P_1 - S_1}, \overline{P_2 - S_1}) dS_1. \tag{31}$$

Comparison of Equations (29) and (31) with the aid of Equation (28) shows that

$$\Gamma(S_1, S_2, 0) = I(S_2)\delta(S_2 - S_1), \tag{32}$$

where $\delta(S_2 - S_1)$ is the three dimensional Dirac delta function. Since the above argument can be applied to any other surface in $\sum'$ it follows that (32) must be satisfied by all points P_1 and P_2 in V.

The expression on the right of (32), however, is not a solution to the two Helmholtz equations and furthermore is again contrary to the hypothesis (28). Therefore an incoherent field cannot exist in free space. However, an incoherent *source* may be defined as a source for which Equation (32) is satisfied by all pairs of points in it. It will be shown that consistent with the above argument, such a source gives rise in general to a partially coherent field.

We are now in a position to examine the limiting forms of the general (16). For a quasi-monochromatic field Equation (16) reduces to

$$\Gamma_{12}(\tau) =$$

$$\frac{e^{-2\pi i\bar{\nu}\tau}}{(2\pi)^2} \int_{\sigma_1} \int_{\sigma_2} (1 - ikr_1)(1 + ikr_2) \cos\theta_1 \cos\theta_2 \Gamma(S_1, S_2, \theta) d\sigma_1 d\sigma_2 \left(|\tau| \ll \frac{1}{\Delta\nu} \right), \tag{33}$$

which for a coherent source gives (after substituting form (26))

$$\Gamma_{12}(\tau) = e^{-2\pi i\bar{\nu}\tau} U(P_1) U^*(P_2) \qquad \left(|\tau| \ll \frac{1}{\Delta\nu} \right). \tag{34}$$

Thus, a coherent source must always give a coherent field.

The incoherent limit obtained by substituting (32) and integrating over S_2 gives

$$\Gamma_{12}(\tau) =$$

$$\frac{e^{-2\pi i\bar{\nu}\tau}}{(2\pi)^2} \int_{\sigma} I(S)(1 - ikr_1)(1 + ikr_2) \cos\theta_1 \cos\theta_2 \frac{e^{ik(r_1 - r_2)}}{r_1 r_2} d\sigma. \left(|\tau| \ll \frac{1}{\Delta\nu} \right) \tag{35}$$

Formula (35) is a generalization of the van Cittert-Zernike theorem and shows that the field from an incoherent source is in general partially coherent with a degree of coherence given by an integral over the intensity distribution of the source. This result is consistent with our earlier demonstration that the incoherent field cannot propagate.

Conclusion

In this paper is given the mutual coherence function for a field illuminated by a plane polychromatic source with an arbitrary but known distribution of mutual coherence. In the special case of quasi-monochromatic fields it is shown that: (1)

a coherent field is characterized by a mutual coherence function which splits into the product of a wave function with its complex conjugate, each factor depending on coordinates of one point only, multiplying a periodic function of τ, and (2) that an incoherent field cannot exist in finite volume. In the narrow spectral width approximation the limiting cases of the general solution are examined and it is shown that: (1) a coherent source always gives a coherent field, and (2) that an incoherent source gives in general a partially coherent field. The van Cittert-Zernike theorem is found to represent an approximate form of the incoherent limit of the quasi-monochromatic solution.

Acknowledgment

The author wishes to express his gratitude to Dr. E. Wolf for suggesting the problems treated in this paper and for helpful discussions as the work progressed.

References

1. Cittert, P. H. van (1934). *Physica 's Grav.* **1**, 201.
2. Gabor, D. (1946). *Proc. Inst. elect. Engng, N.Y.* **93**, 429.
3. Parrent, G. B. In preparation.
4. Wolf, E. (1955). *Proc. roy. Soc. A* **230**, 246.
5. Zernike, F. (1938). *Physica 's Grav.* **5**, 785.

Radio Wave Propagation Simulator

Klaus W. Otten

Communication and Navigation Laboratory, Wright Air Development Center, Wright-Patterson Air Force Base, Dayton, Ohio

The design considerations and applications of a Radio Wave Propagation Simulator are described. It is a device similar to an analogue computer which is intended to simulate the time varying properties of a communication channel utilizing radio wave propagation. Possible instrumentations of a dynamic model of the radio propagation channel in terms of a bandpass filter analogue are discussed. Design data for a simulator for a 30 kc-bandwidth communication channel in the 3 to 30 Mc frequency range is given.

I. Introduction

During the development phase of radio communication systems extensive testing is conducted to determine the effect of the propagation medium on the performance of the system. Usually the performance of the system under development is compared with that of a similar operational system. Since the systems cannot be tested under exactly the same conditions, the comparison tests must be analyzed on a statistical rather than instantaneous basis. Statistical evaluation requires long test periods in order to gain sufficient information during the periods of critical propagation condition.

Usually the percentage of time during which a critical propagation condition exists is very low; moreover, the time of occurrence of this condition is difficult to predict. This makes extended tests very inefficient and extremely costly, especially for air-to-ground and air-to-air systems tested over long distances. Little information is gained from these tests as it is difficult to correlate the cause of various types of distortion with the effect on the instantaneous operation of the system.

A radio wave propagation simulator can replace the costly and time consuming tests of communication systems conducted in the field under almost unknown conditions by short time laboratory tests under programmed conditions. A simulator will allow the instantaneous correlation of every effect with the factors which cause it. (Fig. 1.)

II. The Model of a Radio Wave Propagation Channel

A radio communication system consists of a transmitter, a transmitting antenna, a receiving antenna, a receiver and the propagating medium linking both antennas. The last of these is defined as a communication channel in the following discussion. It is assumed that only one antenna is in use at each terminal.

A general model of a communication channel between two stations should be capable of describing its characteristics for any geographic location of the stations and any distance between them. For vehicular stations, the change of distance has to be included. The relative speed between the terminals may be as much as several thousand kilometres per hour. The general form or structure of the model should be independent of the carrier frequency. However, any specific model (specific set of parameters) may be valid only for one such frequency.

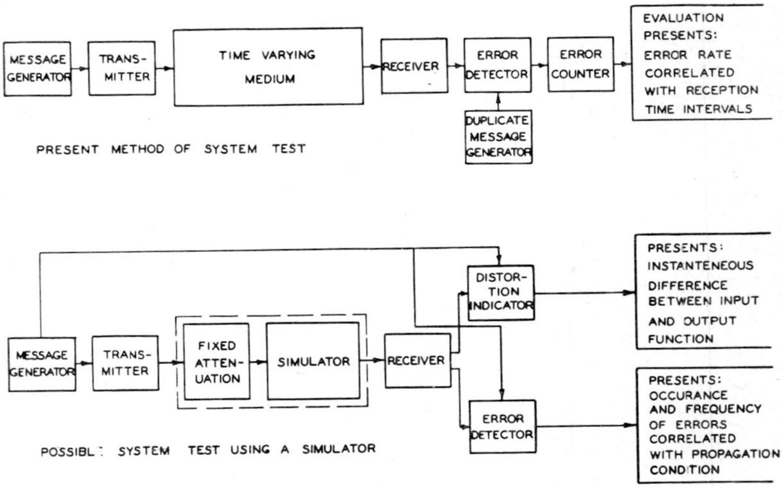

FIG. 1. Test of a communication system for digital data.

To derive a general model of a communication channel one has to know qualitatively and quantitatively how the propagating medium affects a signal passing through it. For a first simplified description of the channel properties it is assumed that only one path exists between the two terminals, but that the path lengths may change during the simulated test. The absolute minimum path length defines the minimum values of the parameters.

For the minimum distance between two communicating stations during a test period (for a point-to-point communication system it is the shortest possible path using the lowest possible reflecting layer) there exists a lower bound on the attenuation. This is the free space attenuation corrected by the antenna gain figures at both terminals. If the channel deviates from the ideal free-space condition the attenuation is higher. The added attenuation is introduced by several factors. The primary one is the constantly changing distribution of absorbing or reflecting particles within the medium. This changing distribution causes time varying absorption losses and—if the signal is bent or reflected—time varying reflection losses. The changing height of the reflecting layers causes a variation of path length and therefore of free space attentuation. This latter attenuation can be different for different frequencies because of the non-homogeneous ion density in the layers. If one or both of the stations are vehicular the minimum free space attenuation is exceeded due to the increased distance. In addition, the changes in effective antenna gain due to the antenna pattern and the variation of antenna orientation introduce additional attenuation. Considering all effects the total attenuation of a single path can be described as a fixed minimum attenuation with a time-varying positive component superimposed on it.

A fixed time is required for the signal to travel through the minimum path length. This fixed time delay is a function of the path length and is equivalent to a phase shift which, for a given path length, is a linear function of the frequency.

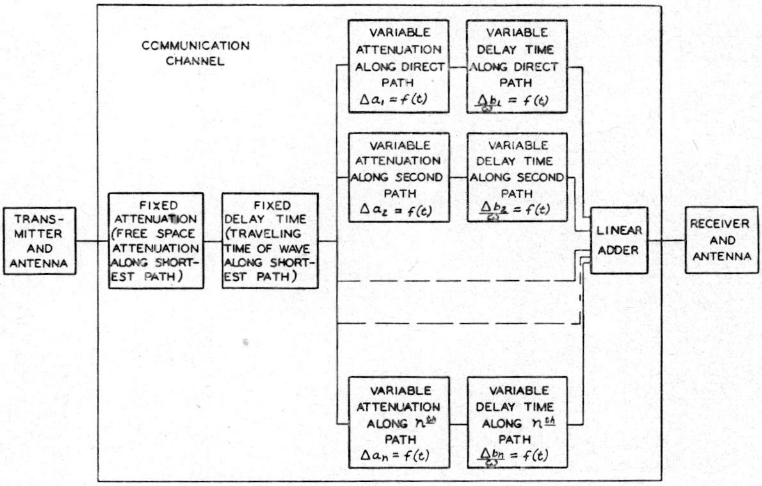

FIG. 2. Model of a noise-free radio communication channel with multipath.

The same factors that give rise to the variable attenuation cause variable time delay. This variable time delay fully describes the doppler effect resulting from a relative speed between the two terminals. This effect is encountered whenever the path length is changing; either by the movement of one station relative to the other or by the movement of a reflecting layer.

The effects of the various single path propagation phenomena on the radio signal can be summarized:

1. The signal is attenuated and delayed (or shifted in phase) by a fixed minimum amount. This attenuation and delay depend only on the shortest path encountered during the test period and the highest obtainable system gain.
2. Additional attenuation and delay are time varying functions which result from uncontrollable factors modifying the properties of the channel. No direct correlation between attenuation and delay time exists.

A multipath channel is a parallel arrangement of different single-path channels. These channels differ in fixed attenuation and delay as well as in the functions describing the variations of the additional attenuation and delay. The linear addition of outputs of these channels generates the signal with the typical multipath distortion. The relative phase of the individual signals together with their amplitudes determine the instantaneous amplitude of the received signal. Selective fading results from destructive interference between at least two paths. (Fig. 2.)

A device simulating this simplified model of a communication channel can now be envisioned as an electrical analogue simulating fixed and variable attenuation and delay time independently for each path of a channel. For a single path, it is a frequency independent variable attenuator in cascade with a variable time delay or phase shifting device. The latter must have a linear phase-versus-frequency relationship within a pass band which is wider than the bandwidth of the signal.

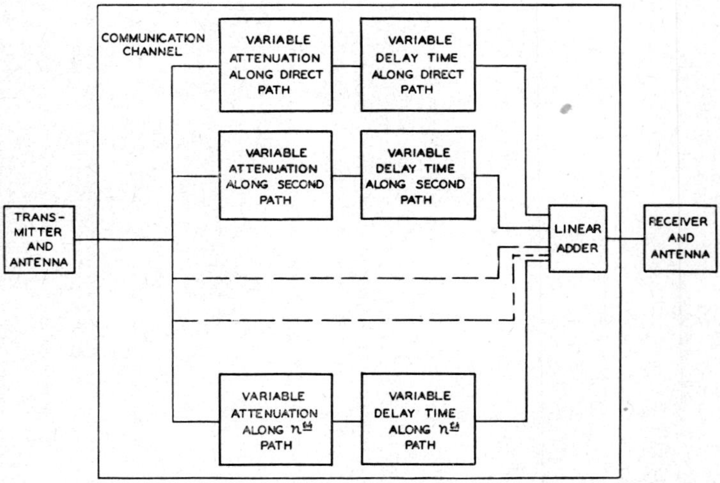

FIG. 3. Simplified model of a noise-free radio communication channel with multipath.

III. SIMPLIFYING ASSUMPTIONS MADE FOR THE DESIGN OF A SIMULATOR

For the design of a practical simulator several simplifying assumptions can be made:

1. The constant minimum attenuation and delay is of no interest and therefore need not be simulated.
2. The propagation phenomena can be simulated at a carrier frequency lower than the original carrier frequency as long as this new frequency is high in comparison to the signal bandwidth.
3. The signal is band-limited.

With these three assumptions in mind it will be possible to construct a device which can realistically simulate the properties of a radio communications channel.

Due to the first assumption, only the changes of the communication channel during a test condition need to be simulated. Instead of the wide range of absolute values of the parameters which may be expected for all possible test conditions only the relatively small range of variation of these parameters must be handled. This range is described by the maximum deviations of the parameters from their minimum values as expected during a typical operation. It may be relatively short for an assumed point-to-point link, and can become larger with increasing distance between the two stations, as it is encountered in air-to-ground or air-to-air communications. It can become very large for the simulation of channels between missiles and their ground stations.

The assumption that the simulation can be performed at a frequency different from the actual carrier frequency permits operation of the simulator at a standardized carrier frequency. With a single simulator laboratory tests of systems operating over a wide range of carrier frequencies can be conducted, provided that the signal bandwidth is small compared to the simulator carrier frequency.

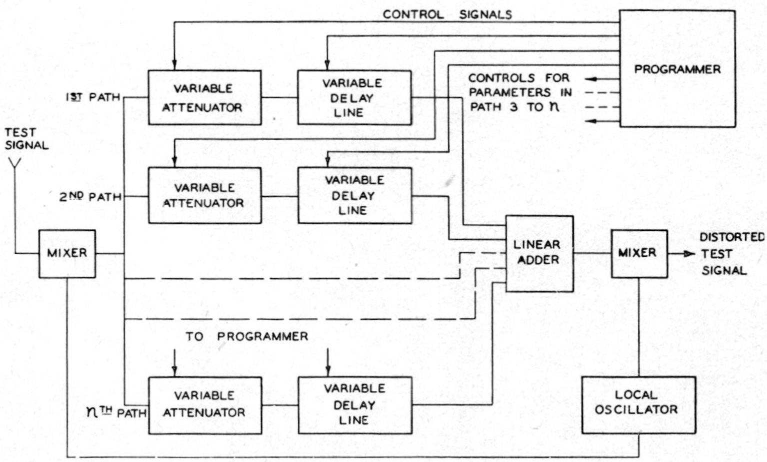

FIG. 4. Basic simulator with frequency conversion for noise-free channel with multipath.

IV. THE BASIC SIMULATOR

From the foregoing the following components, interconnected as shown in Fig. 4, are necessary for the simulation of a noise-free channel with multipath:

1. Variable delay devices in cascade with variable attenuators to form the individual paths.
2. A linear adder to combine the channel outputs.
3. Frequency converters at the input and output to shift the frequency band of the test signal into the passband of the simulator.
4. A programming unit to control the variable parameters. This unit must be capable of programming any possible sequence of propagation conditions which can be simulated by this device.

V. The Complete Simulator

The more complete simulator of Fig. 5 will permit the simulation of a noisy radio channel with multipath. It consists of the basic simulator with added components to introduce noise and interfering signals of all required forms into the communication channel. These additional components are:

1. A white-noise generator.
2. Radio-signal generators with audio- and pulse-modulated outputs.
3. Additional linear adders to superimpose the noise on the signal.

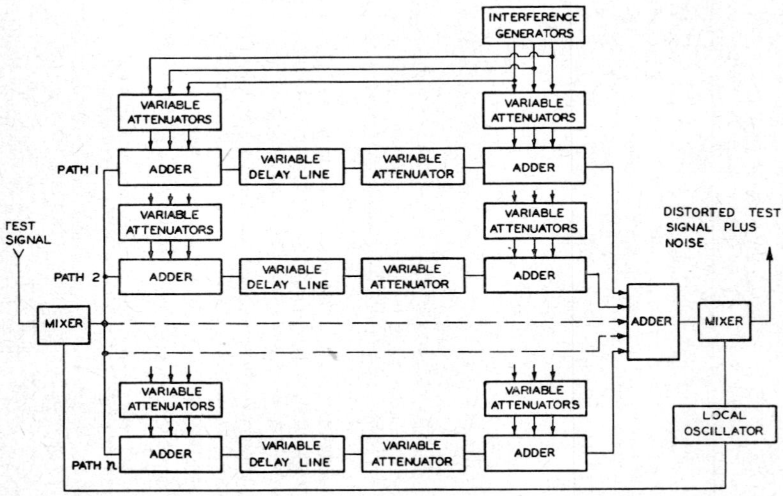

Fig. 5. Complete simulator with frequency conversion for test of systems with a noisy multipath channel.

VI. The Technical Realization of the Components

The design and construction of most of the components is well within the state-of-the-art. However, the requirements for the variable delay device which is the nucleus of the whole simulator, cannot be satisfied by current techniques.

VII. Simulation of Varying Time Delay

The required delay lines must be essentially continuously variable. Maximum delay times of 5 msec, corresponding to a 1500 km change or difference in path length, are considered sufficient for most simulation purposes. In the following discussion the bandwidth of the signal is assumed to be less than 40 kc, permitting the simulation of most communication channels with carrier frequencies below 50 Mc. Various types of delay lines have been investigated as to their suitability for this application.

Distributed-constant delay lines, such as helical windings on a ferromagnetic core, have very high attenuation and relatively small delay times per unit physical length. Step-wise variable delay lines of this type have been manufactured, but the delay times are too small for this application. If cascaded, too many elements are required. The multitude of interconnections with imperfect matching would introduce high distortion.

A lumped-constant delay structure would require an excessive number of elements. If each section of the line contributes a delay equal to one period of the carrier frequency, the number of sections per millisecond delay per 100 kc of carrier frequency is 100. For a delay of several milliseconds and useable carrier frequencies thousands of sections are required.

A continuously variable delay could be obtained by fabricating the delay line out of electrically variable inductances and capacitors. The complexity and the stability problems involved make the approach impractical and too costly.

Ultrasonic delay lines with a solid medium are suitable for very short delay times and for wideband signals. However, no practical means are known for varying this delay time continuously over a wide range.

Ultrasonic delay lines having a liquid medium (i.e. water or mercury) with varying delay times are feasible in principle, and may be realizable after intensive study and development. The water delay line appears promising. The speed of sound in water is approximately 1/200,000 that of light, consequently practical delay lines become short enough for laboratory use. For a 5-msec delay time the length of an unfolded water delay line is approximately 7·5 m. By addition of a solvent to the water, a liquid medium with a zero thermal coefficient over a narrow temperature range for the velocity of sound can be obtained. With accurate temperature control sufficient stability of the line can be expected. Quartz crystals may serve as transducers for this application as they can be designed with a high relative bandwidth (up to 50 % of the carrier frequency) and for high carrier frequencies (in the megacycle range). One crystal would have a fixed position at one end of the tank, the other one would be movable over the full length of the tank. An accurate servo system with a fast response time could be used to control its position. Alternately the water delay line can be folded, with both receiver and transmitter fixed at one end of the tank and a movable corner reflector to determine the path length. The physical length of the delay line could thus be reduced to approximately 4 m for a 5 msec delay time; however, the increased likelihood of spurious reflections may make the folded line impractical.

Liquid delay lines are especially suitable for simulation of wideband signals because a high carrier frequency can be used. For carrier frequencies of the order of a few megacycles no frequency conversion is necessary. Wide band signals at higher carrier frequencies can be accommodated by frequency conversion.

The disadvantages of liquid delay lines are the difficulty of eliminating undesired reflections and the compensation required for the varying attenuation which increases linearly with the delay time. A calibrated automatic gain control amplifier covering a wide amplification range must be controlled by a signal that is proportional to the attenuation of the control signal in the delay line. Extensive research will have to precede the design of an operational delay line because of the lack of experience in this area.

Magnetic storage devices offer the most promising realization of the required variable delay line for signals whose bandwidth does not exceed a few tens of thousands of cycles. Magnetic tape mechanisms, magnetic discs and magnetic drums are the storage devices to be considered for this application. All require the capability of varying the relative position of recording and playback heads. The reproduced signal is an exact replica of the recorded signal attenuated by a fixed amount and delayed by the time required for the transporation of the magnetic carrier from one head to the other. The delay time is varied by changing the distance between the two heads. The usable signal bandwidth is determined primarily by the speed of the magnetic carrier, secondarily by the construction of the head, the spacing between head and magnetic layer, and the properties of the magnetic layer. The requirement for reproducible simulation can only be met if the transfer function from head to head is constant. Short-time speed fluctuations cause small random variations in the time delay; changes in the head to drum spacing cause variation in the output signal amplitude. Reproducibility requires accurate speed control of the magnetic carrier and gain control of the compensating amplifier. A frequency stabilized power source and a large mechanical inertia is required to keep short time speed fluctuations small. Variations in the attenuation, including those caused by inhomogeneities in the magnetic material, can be partially compensated by the introduction of a feedback loop.

A magnetic tape mechanism with a tape loop permits very long delay times; however, difficulties are encountered in maintaining the head-to-head transfer function constant and in obtaining the required high speed without jitter. The stresses which occur in a tape moving at high speed makes extended trouble-free operation questionable.

Magnetic disks and magnetic drums have limited delay-bandwidth products. The mechanical properties of the drum or disk material set an upper limit on the circumferential speed. An air gap between head and magnetic layer is necessary to avoid excessive wear at the high speeds. Small variations in the spacing due to the unavoidable eccentricities, especially if the head position is to be changed, result in output signal amplitude variations. An AGC circuit with a time constant which is small compared to the time of one full rotation can partially compensate this effect.

The magnetic disk has different frequency responses for the different tracks due to the different circumferential speeds. This effect may be negligible if only a few channels are used so that the ratio of the largest to smallest track radius is close to unity or if the bandwidth of each channel is limited to that of the channel with the narrowest bandwidth.

A magnetic drum offers the advantages of small compact construction and equal characteristics for all channels. At present it is considered as the most suitable approach to the parallel arrangement of variable delay lines needed for the multipath simulation. The design of magnetic drums for computer applications has been mastered and enough experience is available in the industry to modify such a design for the development of the specific delay line drum arrangement required for a simulator. Figure 6 shows a continuously variable time delay device using a magnetic drum with automatic attenuation control.

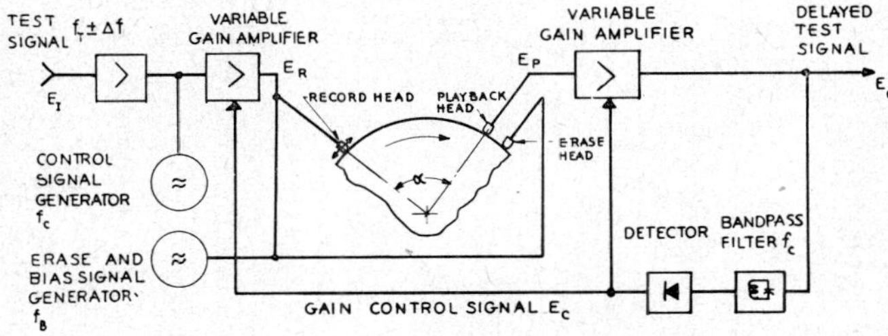

FIG. 6. Variable time delay device using a magnetic drum.

$$\text{Delay Time } \Delta T \sim \alpha$$

$$E_O = E_I$$

$$f_C < f_T - \Delta f$$

$$E_C \sim \frac{E_P}{E_R}$$

$$f_B \gg f_T$$

VIII. Simulation of Variable Ionospheric Attenuation

The variations of attenuation due to changes in the absorption, etc., are very slow. Differences in attenuation between several paths contributing to a received signal are assumed to be less than 40 db; this is within the signal-to-noise ratio obtainable for the delay lines.

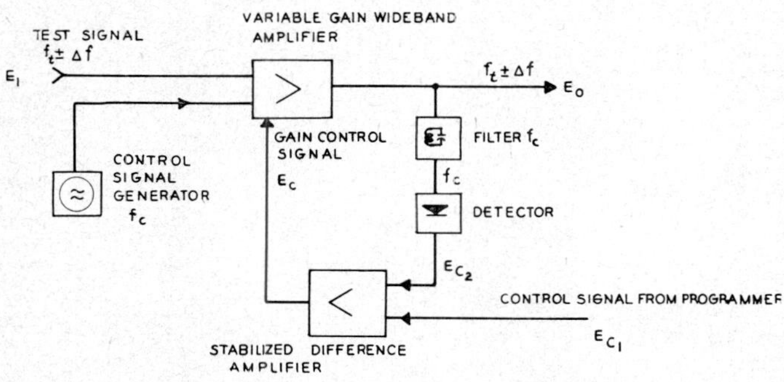

FIG. 7. Electronically controlled attenuator.

Servo driven potentiometers can be cascaded, but the desired degree of reproducibility requires a precise electromechanical servo system which would be costly and require considerable maintenance.

Fully electronically controlled devices are preferable if a high stability can be guaranteed. Variable gain amplifiers do not have adequate long term stability of the control parameters. Periodic adjustment of the gain versus control voltage function is required. A more reliable and more complex solution is the inclusion of a feedback loop in the control circuit. The simulated signal together with a pilot signal of lower frequency is amplified by the wideband amplifiers. The pilot signal is filtered out, detected, and together with the control voltage fed into a difference amplifier, the output of which is a dc voltage for the control of the variable gain amplifiers. The accuracy of the attenuator will then be independent of the characteristics of these amplifiers. An electronically controlled variable attenuator is illustrated in Fig. 7.

IX. Programming Unit

The programming unit permits the composition of a sequence of propagation conditions which are of interest for a particular test, and the control of all parameters in accordance with this program. A program to simulate a desired sequence of propagation conditions can be synthesized manually. By the proper correlation of the programming signals in the various channels selective fading, doppler shift, absorption fading, etc., can be simulated and jamming signals introduced.

The nucleus of the programming unit is a tape recorder with the required read-in and play-back circuitry for multichannel recording. Since the variation of all control parameters is relatively slow a channel bandwidth in the order of cycles is sufficient. This permits the recording of the varying dc control signals either in a frequency-shared or a time-shared manner on one channel. The first could be done in form of a parallel recording of various FM signals with carrier frequencies distributed over the bandwidth of the recorder. The latter requires commutators sampling the control signals and encoding the samples in either pulse duration, pulse amplitude, pulse position, or in a binary code. Conversion and recording techniques well known in telemetry can be applied directly. The encoding and decoding equipment and the recording devices which can serve as programming units for simulators with 20 or more channels are readily commercially available.

X. A High Frequency Channel Simulator

To illustrate the complexity and the required stability of a practical simulator the proposed design and operation of a high frequency channel simulator is described. It is intended to simulate a noisy multipath channel within the frequency range from 3 to 30 Mc. The bandwidth of the signal is limited to 30 kc.

A magnetic drum is used as the time delay device, since the small signal bandwidth permits a low carrier frequency. This requires a conversion of the test signal into a frequency within the passband of the drum. An I.F. frequency of 200 kc permits enough resolution for observation of multipath effects even for the highest frequency component in the signal (8 I.F. cycles per signal cycle).

The output signal from the transmitter of the system under test is attenuated by an external device to the input power level of the simulator (for example, $\frac{1}{10}$ w).

In a mixer the test signal is heterodyned with a signal of proper frequency from the local oscillator. The 200 kc I.F. signal is filtered and amplified to the nominal input level of the recording head. From here on the signal is processed in six separate channels.

Six recording tracks on a magnetic drum provide the variable time delay for each path. In each track one servo motor controls the positioning of the recording head. A count of the revolutions of the servo-motor reference position indicates the location of the movable head. For operation at 60 c/s the direct drive of the drum with a 3600 r.p.m. synchronous motor is convenient. The drum rotates once every 16·67 msec. For the 30 kc bandwidth, plus a safety margin, and a head-to-drum spacing of 0·03 mm the circumferential speed of the drum is selected to be 30 m/s. The delay time in each channel varies from 1 msec, as given by the minimum spacing between the recording and the playback head, to 6 msec. To cover this 5 msec delay time range one head has to move 108°. One full cycle of the carrier frequency corresponding to 5 μsec delay time takes a length of only 0·57 mm on the recording track. If a reproducible resolution of the total phase shift of only $\pm$ 30° is required, the servo motor head drive system must position the head around the drum to 0·047 mm. If the phase jitter resulting from short time speed variations is limited to the same $\pm$ 30°, the speed must be controlled to $\pm$ 0·05 % or 5 parts in 100,000. This is within the state of the art.

The attenuation between the two heads is estimated to be in the order of 70 db, and will vary with the head-to-drum distance variations. The gain of the input and output amplifiers must be controlled automatically to compensate for this changing attenuation. A control signal of constant amplitude with a frequency of 100 kc is passed through the channel simultaneously with the simulation signal. At the output of the playback amplifier the control signal is filtered out; an AGC loop maintains the control signal output constant, thus stabilizing the combined record-playback gain. A response time of the control loop of 0·2 msec (corresponding to 4·3° of arc) is considered sufficient to compensate for small eccentricities.

Each channel has a variable attenuator of 50 db attenuation range. A two-stage amplifier with variable μ-tubes can handle this range at the required signal levels without introducing noticeable distortion. The gain is set by the control voltage from the programmer and is electronically stabilized as described above. The same 100 kc control signal which serves for the control of the head-to-head transfer function is used as control signal in the closed loop system for setting the variable attenuators.

A simple resistive network serves as a linear adder to combine the six independently attenuated and delayed signals. In addition it adds the outputs of a noise generator and three interference generators, each of which can be independently attenuated.

In the second mixer the distorted I.F.-signal is heterodyned to its original frequency and fed into the receiver.

The programming unit delivers the control signals for all variable parameters. Six recording heads, six channel attenuators, and four interference attenuators must be independently controlled. In addition, timing and reference signals are supplied by the tape recorder.

XI. Application of the Radio Channel Simulator

The principal applications are as a laboratory tool in communication system design and as a substitute for field tests in later system evaluation.

Comparison tests of devices which are designed to have similar performance characteristics may show advantages and disadvantages of the various devices for operation under specific propagation conditions. The results may lead to the design of new devices incorporating the advantages and avoiding the disadvantages of the tested equipment. Receivers and demodulators can be tested at threshold levels under certain critical propagation conditions, thereby aiding in the design of new modulation and demodulation techniques intended to overcome these conditions. In testing the effect of any design changes on the performance of a system reproducible conditions are available for determining directly the degree of improvement obtained. Tests with different standardized recorded programs emphasizing certain propagation conditions as the control signal source permit the assignment of performance ratings for various applications to the tested items.

Performance tests of new techniques, intended to be used at higher carrier frequencies and for wider information bandwidths than that of the simulator, can be performed on the simulator if all functions are scaled down into its frequency range. Although this does not permit the test of actual hardware. useful information on the limitations of the proposed technique can be obtained.

If proper signal generators are available as interference signal sources, the sensitivity of systems to unintelligent and intelligent jamming can be tested for ideal as well as for critical propagation conditions.

Ionization by Auroral Particles

A. Omholt

The Auroral Observatory, Tromsö and The Institute of Theoretical Astrophysics, University of Oslo

Arguments for the assumption that almost all the auroral excitation and ionization is due to fast primary electrons are reviewed. Accepting this assumption, the rate of ionization in homogeneous aurorae has been computed from the absolute luminosity and compared with simultaneous observations of the auroral luminosity and radio echoes from E-layer heights. The results yield an apparent recombination coefficient of about 10^{-6} cm^3 sec^{-1}. It seems difficult to explain this high value without assuming a high value of λ, the negative ion to electron ratio. The ionization effect in the lowest ionosphere by X-rays emitted as bremsstrahlung from the primary electrons has been considered. It is concluded that below an aurora with international brightness coefficient III the rate of ionization between 65 and 85 km is of the order of magnitude 10 ion pairs cm^{-3} sec^{-1}.

I. Introduction

During the last decade there has grown up a rapidly increasing interest in the ionospheric conditions over the polar caps. This field includes the disturbances associated with geomagnetic storms and aurorae, which are rather important to the propagation of radio waves. It seems natural to divide these disturbances of the normal polar ionosphere into two groups: (a) increase in the electron densities owing to an increased rate of ionization, (b) variations owing to the effect of the geomagnetic storm on the electron content in the ionosphere, such as redistribution of the electrons due to dynamic or electromagnetic effects, increased recombination and other effects.

In this paper we shall confine ourselves to the first type of disturbances, particularly the increase in ionization associated with the appearance of visual aurorae. We shall then first review the evidence regarding the nature of the fast primary particles which presumably cause the aurorae, and then discuss the ionization, direct or indirect, which such particles may cause.

II. The Nature of the Primary Particles

Since Meinel [21] verified Vegard's [32] detection of the hydrogen lines in the auroral spectrum and measured the doppler profile from aurorae in the zenith, there has been some tendency to assume that the primary particles which presumably excite the aurora may be protons or, less likely, other positive ions. The Størmer-Birkeland theory has been reconsidered by Bennett and Hulbert [5] (cf. also [4]) on the assumption that a self-focusing stream of fast protons and slow electrons hits the earth. While there is no doubt that fast protons do hit the atmosphere during auroræ, the available evidence do not support the view that protons alone are able to maintain any aurora. The faintness and great variation in the intensity of the hydrogen lines rather suggest that the presence of fast protons in the stream of primary particles is a less important phenomenon. While the hydrogen lines are always observed in auroral arcs at low latitudes ([9], [26]), they are usually not detectable in arcs at high latitudes (cf. [28]). Further, from the intensity of the hydrogen lines it is possible to obtain an estimate of the upper limit of the total energy carried by the protons and the amount of ionization due to protons. From the brightness of the N_2^+ bands in the auroral spectrum it is possible to estimate the total rate of ionization, and this turns out to exceed the rate of ionization that can be due to protons by a factor 10 or more [27, 28].

It has also been thought that some auroral forms, particularly rays, may be due to electric discharges in the atmosphere. Extensive photoelectric measurements on the intensity ratio between the [OI] line λ5577 and the First Negative nitrogen band λ4709 do not support these views (cf. [29]). These two emissions have widely different excitation potentials and their relative intensity should therefore be very sensitive to the excitation conditions, which should vary very much along rays if these were due to discharges [6]. The measurements show that this intensity ratio is about 20 and almost constant (within $\pm$ 25 %) in all types of aurorae.

Since we have no evidence of other positive ions entering the atmosphere during aurorae, it may be permitted to assume that fast electrons are the main cause of the auroral excitation. This is also supported by the recent detection of X-ray emission from aurorae which is discussed in section IV.

III. Ionization and Recombination During Aurorae

1. In an earlier paper [28] it was shown that the total rate of ionization in an aurora is probably about 500 times the rate of emission of quanta in the band λ4709 of the First Negative system of N_2^+, or about 20 times the total rate of emission of quanta in this band system. Adopting this ratio the values given in Table I have been computed. The rate of ionization at the height of maximum emission has been computed, assuming that the equivalent thickness l of the aurora is about 30 km, in accordance with measurements of homogeneous arcs by Harang [11] (i.e., $l = \int L(h)dh/L_{\max}$ where $L(h)$ is the emission per cm^3 as a function of the height h).

2. Radio echoes from aurorae may be divided into two groups. The first group consists of echoes received at very high frequencies, above about 20 Mc/s. This type of echoes has been much discussed during the recent years and it appears likely that these echoes are due to reflections or scatter from irregularities in the ionization [18]. The second group consists of echoes received in the ordinary high frequency range (1–20 Mc/s) by conventional equipment. These are frequently designated E_s-echoes and attributed to extensive ionization in the E-layer associated with magnetic storms and aurorae.

A qualitative study by Meek [19, 20] suggested that the latter type of echoes is closely correlated with the appearance of aurorae, and indeed a quantitative

TABLE I

International brightness coefficient of aurora in zenith*	I	II	III	IV
Emission of First Negative bands in KR*	1	10	10^2	10^3
Rate of ionization in ion pairs cm^{-2} (column)$^{-1}$ sec^{-1}	2×10^{10}	2×10^{11}	2×10^{12}	2×10^{13}
Rate of ionization in ion pairs cm^{-3} sec^{-1} at the height of maximum emission	7×10^3	7×10^4	7×10^5	7×10^6

*Cf. [15] 1 KR = 10^9 quanta cm^{-2} (column)$^{-1}$ sec^{-1}. The total emission of the First Negative bands is almost equal to that of the green oxygen line λ5577.

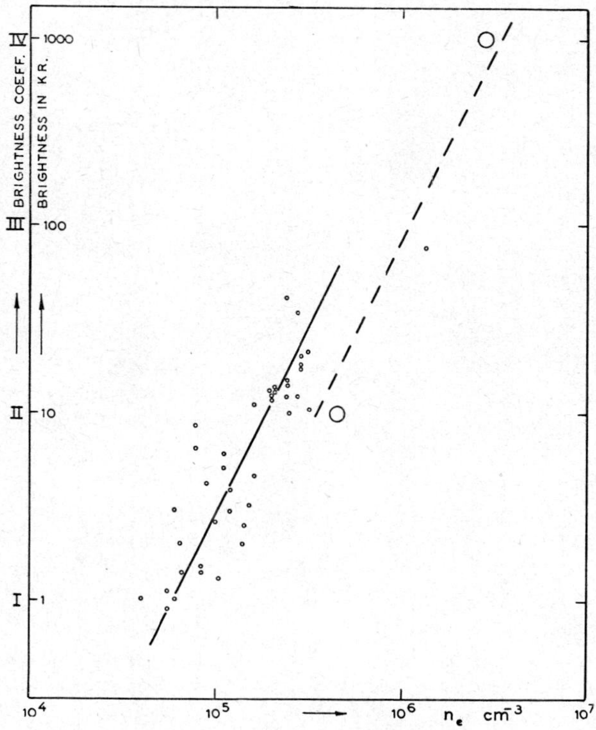

FIG. 1. Brightness of aurora in the zenith versus maximum electron density of the associated E_s-layer.

correlation between the electron densities deduced from inospheric records and the absolute luminosity of homogeneous aurorae in zenith was found [23]. These results are plotted in Fig. 1, where the absolute intensity of the First Negative band system of N_2^+ emitted from the aurorae in zenith is plotted against the maximum electron density in the E-layer (virtual height 95 to 130 km) found from ionospheric records. If the rate of ionization at the height of maximum emission (and presumable maximum electron density n_e) is proportional to the brightness of the aurora, as assumed in Table I, then the brightness should be proportional to n_e^2 if the ordinary recombination law is obeyed. The correlation coefficient between the brightness and n_e^2 is 0·81. With a recombination coefficient of $\alpha = 2 \times 10^{-6}$ cm³ sec⁻¹ the straight line drawn in Fig. 1 is derived theoretically, using the values given in Table I. We may expect that there is a variation in the thickness l of the luminous layer, the ratio between the total rate of ionization and the rate of emission of the First Negative bands, as well as the recombination coefficient. The scatter of the measured points is therefore not unexpected.

The results displayed in Fig. 1 are also supported by measurements by Knecht [16], who measured the maximum electron density in the E-layer during aurorae and found a good correlation with visually estimated intensities of inactive aurorae.

In Fig. 1 an attempt has been made to compare the two sets of observations. Knecht never observed echoes with critical frequencies above 14 Mc/s, and from his results it appears that 15 Mc/s ($n_e = 2{\cdot}8 \times 10^6$ cm^{-3}) should correspond to the very brightest aurorae, which we will assume corresponds to international brightness IV. This point is marked by the upper large circle in Fig. 1. The lowest critical frequencies observed were about 6 Mc/s ($n_e = 4{\cdot}5 \times 10^5$ cm^{-3}), corresponding to "very faint" aurora. If we assume that this is equal to brightness II the point marked by the lower of the large circles in Fig. 1 is obtained. Aurorae of brightness I are hardly visible even to a trained observer with a well dark-adapted eye. The dashed line in Fig. 1 is derived from the values given in Table I, adopting the usual recombination law and a recombination coefficient of $5{\cdot}8 \times 10^{-7}$ cm^3 sec^{-1}.

Also Stoffregen reports (private communication) that with his very sensitive equipment he never or seldom records critical frequencies above 15 Mc/s. If this critical frequency corresponds to brightness IV the recombination coefficient is 9×10^{-7} cm^3 sec^{-1}, again adopting the values given in Table 1.

It may seem peculiar that Omholt [23] seldom observed echoes from bright aurorae, at frequencies above 5 Mc/s, whereas Knecht [16] only observed echoes above 6 Mc/s, although the frequency ranges of the two recorders were almost the same (1–20 and 1–25 Mc/s). This may, however, be due to a different sensitivity versus frequency.

The observations reported here appear to be consistent with the following assumptions:

(a) The maximum rate of ionization within a homogeneous aurora is approximately proportional to the luminosity of the aurora.

(b) The square of the electron density is proportional to the rate of ionization.

(c) The recombination coefficient is about 10^{-6} cm^3 sec^{-1}.

Since it appears that the height in the atmosphere of an aurora varies systematically with the brightness [12] it may well be that the thickness l of the luminous layer as well as the recombination coefficient vary somewhat with the brightness, but the data available are too scarce to yield reliable informations.

However, Fig. 1 may indicate that bright (i.e., low) aurorae are associated either with a lower recombination coefficient than the average or with a thinner luminous layer. This is also indicated by data published by Omholt [23], which include the virtual height of the maximum electron density. Harang's [11] results do really show that low aurorae have a less vertical extension than high aurorae.

3. The recombination coefficient which was obtained in the preceding section is appreciably higher than the effective recombination coefficient deduced from time variations of the electron concentration in the normal E-layer. The average value of this recombination coefficient is about 3×10^{-8} cm^3 sec^{-1} [22]. As pointed out by Bates [3], however, the apparent recombination coefficient depends strongly on what process we consider. If two species of ions are formed, and the one recombines sufficiently faster than the other, the relative concentration of the fastest recombining ion is low, and the time variations in the electron density is essentially determined by the recombination coefficient of the slowest recombining ions.

If we, on the other hand, consider the equilibrium case, and consider only direct recombination between the two species of ions and the electrons, then

$$Q = \alpha_{av} n_e^2, \tag{1}$$

with

$$\frac{1}{\alpha_{av}} = \frac{c_1}{\alpha_1} + \frac{c_2}{\alpha_2}, \tag{2}$$

where Q is the total rate of ionization, $c_1 Q$ and $c_2 Q$ the rates of formation of the two species of ions and α_1 and α_2 their recombination coefficients. α_{av} is by Bates [3] called the average recombination coefficient.

The positive ions formed in the E-layer during aurorae are probably N_2^+, O^+, O_2^+ and a small amount of N^+. Bates [3] advances arguments for assuming that charge transfer (or ion-atom exchange) from O^+ and N^+ takes place so rapidly that we may adopt the view that effectively only N_2^+ and O_2^+ ions are formed. These ions may recombine directly with the electrons through dissociative recombination.

The most favourable assumption for explaining the high recombination coefficient during aurorae, without invoking negative ions, is that N_2^+ recombines so much more rapidly than O_2^+ that the effective recombination coefficient of 3×10^{-8} cm^3 sec^{-1} is essentially that of O_2^+, and that only the term for O_2^+ needs to be considered in Equation (2) for the case of aurorae. Then

$$\alpha_{av} = \frac{1}{c_1} \times 3 \times 10^{-8} \text{ cm}^3 \text{ sec}^{-1}. \tag{3}$$

If we neglect the formation of N^+ it seems reasonable to assume that $c_1 = 0{\cdot}2$, the fraction of oxygen in air. This yields $\alpha_{av} = 1{\cdot}5 \times 10^{-7}$ cm³ sec⁻¹, a value which is still appreciably lower than the one deduced from the auroral measurements. Even with a liberal allowance for errors in the adopted numerical values it seems difficult to justify theoretically a higher value of α_{av} than 3×10^{-7} cm³ sec⁻¹, which is still three to four times too low.

Since at night radiative detatchment of electrons from negative ions can be ignored, it may be tempting to try to involve the formation of negative ions. One can show that in this case one gets for equilibrium:

$$Q = \alpha'_{av}(1 + \lambda)n_e^2, \tag{4}$$

with

$$\frac{1}{\alpha'_{av}} = \frac{c_1}{\alpha_1 + \lambda\alpha_{i1}} + \frac{c_2}{\alpha_2 + \lambda\alpha_{i2}}, \tag{5}$$

and

$$\lambda = \frac{\eta n}{\kappa n + \alpha_{i1}n_1^+ + \alpha_{i2}n_2^+}, \tag{6}$$

where λ is the negative ion/electron ratio, α_{i1} and α_{i2} the effective coefficients of recombination between the two species of positive ions (with concentrations n_1^+ and n_2^+) and the negative ions, ηn the probability of attachment for an electron and κn the probability of collisional detachment.

For small electron concentrations (small n_1^+ and n_2^+) λ is equal to n/κ, this must also be the case if the recombination coefficient and thus λ shall be independent of n_e (and thus of n_1^+ and n_2^+) [24]. Again adopting the view that N_2^+ recombines very much faster than O_2^+, we then obtain

$$Q = \frac{1}{c_1}\left(\alpha_1 + \frac{\eta}{\kappa}\alpha_{i1}\right)(1 + \eta/\kappa)n_e^2. \tag{7}$$

If either $\eta/\kappa > 1$ or $\eta\alpha_{i1}/\kappa > \alpha_1$, the formation of negative ions may contribute appreciably to the apparent recombination coefficient. Presently our knowledge about the various coefficients is not sufficient, but it may not be unlikely that any of these conditions are fulfilled. Further, the condition that λ should be approximately independent of n_e requires that $\alpha_i n^+ \ll \kappa n$ for electron densities up to nearly 10^6 cm⁻³.

It may be added that Sayers's [30] measurements on dissociative recombination of N_2^+ suggest that the coefficient for this process is only about 10^{-7} cm³ sec⁻¹. Although this value was obtained for an electron temperature of 3200°K, it may cast doubt on our assumption that N_2^+ recombines very fast and that these ions can be neglected. This emphasizes the need for a high value of λ. D. K. Bailey [1] has also through his analysis of the ionospheric disturbances following the solar flare of 23 February 1956 found evidence for the presence of large amounts of negative ions during the night.

It is also fair to point out that the rate of ionization adopted in Section III. 2 may be somewhat too high, so that the value of α derived from the observations is too high. Also, dynamic effects in the upper atmosphere may make the conditions more complicated and contribute to a lowering of the maximum electron density. Presently the state of affairs is not quite satisfactory, but the disagreement between the observations and the theoretical results is not excessive.

Also, we shall again emphasize that these results are valid for homogeneous aurorae only and do not apply to ray-structured forms which have a complicated geometry and frequently lie appreciably higher in the atmosphere.

IV. Ionization by X-rays from Aurorae

1. The detection of X-rays in the upper atmosphere in the auroral zone [31] and of an enhanced intensity of X-rays during aurorae over Minnesota [33, 34] suggests that emission of X-rays from aurorae may cause important ionization in the deeper layers of the ionosphere. Winckler *et al.* interpreted the X-rays as bremsstrahlung from electrons with energies of about 50 keV. They concluded from their measurements that the flux of primary electrons with this energy must be about $10^7 - 10^8$ cm^{-2} (column)$^{-1}$ sec^{-1}. If we take into account that each 50 keV electron produces less than 2000 ion pairs along its path, it is seen from Table I that the required electron flux is not undue.

The occurrence of polar black-outs during aurorae suggests that ionization takes place also below the aurora itself, and there is experimental evidence that the absorption effect must be produced at least partly below 85 km [17]. Bates [2] has suggested that the Lyman α radiation emitted from neutralized primary protons may contribute to this ionization, but it seems not likely that this is the case [25]. It is evident, however, that if aurorae are caused by fast incident electrons, X-rays must be emitted as bremsstrahlung, since the electrons are stopped in the atmosphere. X-rays in the energy range in question (1–100 keV) may penetrate quite deep into the atmosphere and cause important ionization in the D-region. This has been pointed out by Chapman & Little [7].

For electrons in air the fraction f_0 of the initial electron energy which is converted into X-rays is approximately

$$f_0 = 7 \times 10^{-9} E_0 \tag{8}$$

where E_0 is the initial electron energy in eV. The fraction which is converted into X-rays with energies between E and $E + dE$ is

$$df = 13 \times 10^{-9} \left(1 - \frac{E}{E_0}\right) dE; \; (E \leqslant E_0) \tag{9}$$

(cf. [10]). Since the ionization efficiency of X-rays is about the same as that for fast electrons, f_0 will also be the ratio between the total rate of ionization due to X-rays and that due to electrons, and df/f_0 is the fraction of the X-ray ionization which is performed by X-rays within the interval E to $E + dE$.

If we adopt the values given in Table I, the values given in Table II are obtained from Equation (8).

TABLE II

International brightness coefficient of aurorae in zenith		I	II	III	IV
The total rate of X-ray ionization due to the aurorae in ion pairs cm^{-2} (column of the aurora)$^{-1}$ sec^{-1} for various assumed electron energies	3 keV	4×10^5	4×10^6	4×10^7	4×10^8
	10 keV	13×10^5	13×10^6	13×10^7	13×10^8
	30 keV	4×10^6	4×10^7	4×10^8	4×10^9
	100 keV	13×10^6	13×10^7	13×10^8	13×10^9

2. The next question which arises is where in the atmosphere this X-ray ionization takes place. Since the X-rays are emitted in various directions a proper solution of this problem is very difficult and depends on the geometry of the aurora. A certain idea about the distribution of the ionization may be obtained by a crude approximation which we shall outline in this section.

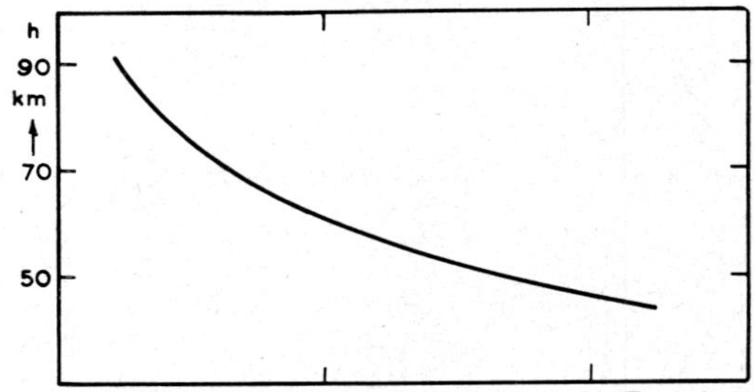

FIG. 2. Height of maximum rate of ionization versus energy of the X-rays.

A parallel beam of monochromatic X-rays which penetrates into the atmosphere will form the well known Chapman layer, the variation of the rate of ionization along the path being given by

$$F(x) = \exp(x - \exp x), \tag{10}$$

where

$$x = -s/H,$$

H being the scale height and s the height in the atmosphere above the point of maximum rate of ionization. The maximum rate of ionization occurs where

$$\xi(\mu/\rho) \sec \chi = 1 \tag{11}$$

μ/ρ being the mass absorption coefficient, ξ the depth in the atmosphere in unit mass/area, and χ the inclination of the beam to the vertical. Equation (10) shows that most of the ionization occurs within about $\pm H$ from the height where the rate of ionization has its maximum. We will call this height h_{max}.

In our first approximation we shall assume that the X-rays penetrate vertically downwards. Adopting the Rocket Panel model atmosphere the variation of h_{max}

with E, the energy of the X-rays, has been computed from Equation (11) for this case. μ/ρ was taken from "Handbook of Chemistry and Physics." The result is shown in Fig. 2, from which it is evident that most of the X-ray ionization above 40 km is performed by X-rays with energies less than 30 keV.

The fraction of the electron energy which is converted to X-rays with $h_{\max}$ between h and $h + dh$ is (cf. eq. (9))

$$df = 13 \times 10^{-9} \left(1 - \frac{E}{E_0}\right) \left(\frac{-dE}{dh_{\max}}\right) dh \tag{12}$$

where $dE/dh_{\max}$ and $E(h_{\max})$ may be derived from Fig. 2. df/dh has been computed for various heights as a function of the electron energy E_0. The result is displayed in Fig. 3.

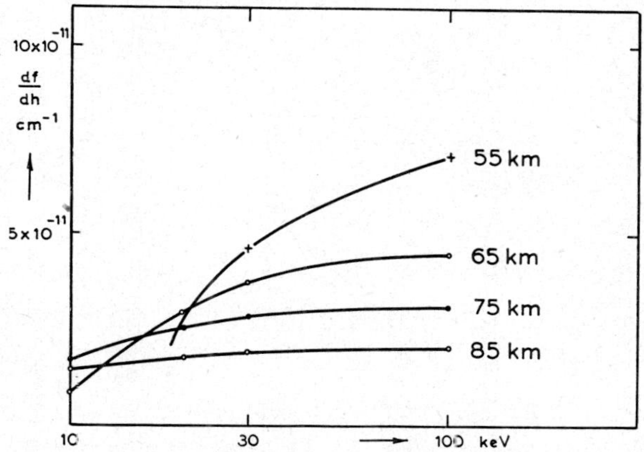

FIG. 3. df/dh versus energy of the primary electrons (cf. the text).

It is most likely that the majority of the auroral electrons have energies around 10 to 20 keV. Electrons with an energy of 20 keV are stopped at a height of about 105 km (Rocket Panel model atmosphere, 95 km if the atmosphere of Horowitz & La Gow [13] is adopted). For these energies it is seen from Fig. 3 that df/dh is about 2×10^{-11} cm^{-1} (within a factor 2) between 65 and 85 km. Since for a particular X-ray energy the ionization is concentrated around $h_{\max}$, this means that for each ion pair produced per cm^2 column in the aurora there is produced approximately 2×10^{-11} ion pair per cm^3 between 65 and 85 km, provided all the X-rays are emitted vertically downwards. We may also infer from Fig. 3 that for electron energies around 20 keV the rate of ionization vary very little with height. For an aurora of brightness III the rate of ionization should then be about 40 ion pairs cm^{-3} sec^{-1} (cf. Table I). Higher primary energies, as may seem more consistent with the results obtained by Winckler *et al.* [33, 34] would yield a higher rate of ionization for the same auroral brightness, particularly at low heights.

The model treated here is of course very crude, and we must expect that the rate of ionization will be somewhat lower because the X-rays are not emitted vertically downwards, but in all directions. The polar diagram will vary with the electron energy. For energies below 10 keV a rather large fraction will be emitted backwards, i.e., upwards, whereas for energies around 60 keV most will be emitted forwards at angles between 30° and 60° with the direction of the electrons. Further complications arise because the electrons may follow spiralling paths around the magnetic lines of force and they may show a dispersion in energy.

The fact that the directions of the X-rays are inclined to the vertical will increase the value of h_{max} a few kilometres. It will also dilute the radiation, particularly when the X-rays are spread out from a distinct auroral form. Because of this a weak aurora which covers most of the sky may be as effective in creating polar blackouts as a strong aurora which covers only a small fraction of the sky. This seems also sometimes to be the case (W. Stoffregen, private communication).

It is likely that the maintenance of the day-time electron density around 75 km ($\approx 10^3$ cm^{-3}) requires a rate of ionization which is less than 1 cm^{-3} sec^{-1}. Below this level the rate of ionization decreases rapidly [14]. From the crude model treated in this paper one should expect that the rate of ionization during aurorae may exceed that during the day, particularly at lower heights, where the electron absorption efficiency is high. Chapman & Little [7] suggest that the recombination coefficient is higher at night than in the day because at night negative ions may be formed (cf. Section III. 3). Presently it is not possible to make any reliable estimate of the recombination coefficient in the lowest ionosphere at night.

Altogether we may say that the arguments advanced in this paper make it possible that the ionization by X-rays plays an important role in creation of polar black-out conditions, but much more experimental and theoretical evidence is urgently needed.

V. Radio Noise from Aurora

Ellis [8] has pointed out that radio noise may be emitted from aurorae as Cerenkov radiation from the primary particles. We shall only point out here that the data given in Table I strengthen his arguments. Since each ion pair requires an energy of about 30 eV, the production of an aurora of brightness III requires a flux of 3×10^9 electrons if the electron energy is 20 keV (velocity about 0·3 c). The density in the beam will then be about 0·4 electrons/cm^3, which is 400 times the density adopted by Ellis for a particle velocity of 0·3 c.

Acknowledgment

The research reported in this paper has been sponsored in part by the Geophysics Research Directorate of the U.S. Air Force, Cambridge Research Center, Air Research and Development Command, under contract AF 61 (514)–1123, through the European Office, Brussels.

REFERENCES

1. Bailey, D. K. (1958). Private communication.
2. Bates, D. R. (1955). *Ann. Geophys.* **11,** 253.
3. Bates, D. R. "Solar Eclipses and the Ionosphere." Ed. W. J. G. Beynon & G. M. Brown. Pergamon Press (1956).
4. Bennett, W. H. (1958). *Astrophys. J.* **127,** 731.
5. Bennett, W. H. & Hulbert, E. O. (1954). *Phys. Rev.* **95,** 315.
6. Chamberlain, J. W. "The Airglow and the Aurorae." Ed. E. B. Armstrong & A. Dalgarno Pergamon Press (1956).
7. Chapman, S. & Little C. G. (1956). *J. atmos. terr. Phys.* **10,** 20.
8. Ellis, G. R. (1957). *J. atmos. terr. Phys.* **10,** 302.
9. Fan, C. Y. & Schulte, D. H. (1954). *Astrophys. J.* **120,** 563.
10. "Handbuch der Physik," vol. 30. Ed. S. Flügge. Springer Verlag (1957).
11. Harang, L. (1946). *Geofys. Publ.* **16,** No. 6.
12. Harang, L. "The Aurorae." Chapman & Hall Ltd. (1951).
13. Horowitz, R. & La Gow, H. E. (1957). *J. geophys. Res.* **62,** 57.
14. Houston, R. E. Jr. (1958). *J. atmos. terr. Phys.* **12,** 225.
15. Hunten, D. M., Roach, F. E. & Chamberlain, J. W. (1956). *J. atmos. terr. Phys.* **8,** 345.
16. Knecht, R. W. (1956). *J. geophys. Res.* **61,** 59.
17. Landmark, B. (1958). *J. atmos. terr. Phys.* **12,** 79.
18. Lowell A. C. B., Forsyth, P. A. & Harang, L. (1957). *Ann. Geophys. Year* **3,** 337. Pergamon Press.
19. Meek, J. H. (1953). *J. geophys. Res.* **58,** 445.
20. Meek, J. H. (1954). *J. geophys. Res.* **59,** 87.
21. Meinel, A. B. (1951). *Astrophys. J.* **113,** 50.
22. Mitra, A. P. & Jones, R. E. (1954). *J. geophys. Res.* **59,** 391.
23. Omholt, A. (1955a). *J. atmos. terr. Phys.* **7,** 73.
24. Omholt, A. (1955b). *J. atmos. terr. Phys.* **7,** 345.
25. Omholt, A. (1956). *J. atmos. terr. Phys.* **9,** 28.
26. Omholt, A. (1957). *Astrophys, J.* **126,** 461.
27. Omholt, A. (1958). Proc. from conference on "Ionospheric Physics and Radio Astronomy." Essen, 1957. Ausschuss für Funkortung. In press.
28. Omholt, A. (1959a). *Geofys. Publ.* **20,** 11.
29. Omholt, A. (1959b). *Geofys. Publ.* **21,** 1.
30. Sayers, J. "Solar Eclipses and the Ionosphere." Ed. W. J. C. Beynon & G. M. Brown. Pergamon Press (1956).
31. Van Allen, J. A. (1957). *Proc. nat. Acad. Sci., Wash.* **43,** 57.
32. Vegard, L. (1939). *Nature, Lond.* **144,** 1089.
33. Winckler, J. R. & Peterson, L. (1957). *Phys. Rev.* **108,** 903.
34. Winckler, J. R., Peterson, L., Arnoldy, R. & Hoffman, R. (1958). *Phys. Rev.* In press.

Diffractive Corrections to the Geometrical Optics of Low Frequency Propagation

JAMES R. WAIT

National Bureau of Standards, Boulder, Colo.

The influence of caustics in low frequency radio propagation is studied. Correction factors to be applied to the geometrical optical representations are presented in the form of graphs. Particular attention is paid to the treatment of the caustic which occurs at the geometrical horizon of the first hop sky wave. It is shown that the apparent infinity in the convergence coefficient can be removed when the wave nature of the problem is considered. Antipodal phenomena are also discussed.

I. INTRODUCTION

The computation of field strengths at low and very low radio frequencies are usually based on the (wave-guide) mode series or the geometrical optics (hop theory). The latter method has certain limitations which are not always evident in the standard references. Difficulty is first encountered when the first hop sky wave leaves and returns to the earth tangentially. In this situation, the geometrically optically computed convergence coefficient becomes infinite. This is a caustic of the system. In this case, it is shown to be convenient to introduce a correction factor which is a function of frequency (or of time in pulse problems) to remove the objectionable infinity. A similar phenomenon occurs at the antipode of the transmitter where there is a focus or axial caustic. The behavior near the antipode for a symmetrical earth and concentric ionospheric reflecting layer is also considered by examining the mode series representation of the field. Finally, the complete geometrical series with appropriate diffractive corrections for the total field is discussed.

II. THE CONVERGENCE COEFFICIENT

It has proven convenient in many geometrical optical calculations to represent the focusing action of the spherical ionosphere by a convergence coefficient. The geometrical meaning of this quantity is seen easily from the following simple derivation. The model chosen consists of a smooth spherical earth of radius a and a smooth concentric reflecting shell of radius $a + h$; the height of this shell above the earth is h. For the moment, the refractive index of the material between the concentric shells is taken as unity.

The convergence coefficient α is now given by $(q'/q)^{\frac{1}{2}}$ where q represents the actual cross section at P of a narrow pencil of rays leaving the transmitter at Q, and q' the cross-section which would exist in the case of undisturbed linear propagation. Some consideration shows that $q' = \hat{R} \sin \phi d\phi d\delta$ where δ is the angle of an individual ray with the plane of the paper in Fig. 1 and $\hat{R}$ is the length of the shortest ray that can be drawn from Q to P with a reflection point at the ionosphere. Now the actual cross-section at P is equal to $q = q_r \cos \theta_r$ where

$$q_r = a^2 \sin \theta \, d\theta \, d\delta$$

is the cross section as is viewed from the center of the earth. It is now easily shown that

$$\alpha = \left(\frac{q'}{q}\right)^{\frac{1}{2}} = \frac{\hat{R}}{a}\left[\frac{\tan \phi \mid \partial\phi/\partial\theta \mid}{\sin \theta}\right]^{\frac{1}{2}} \tag{1}$$

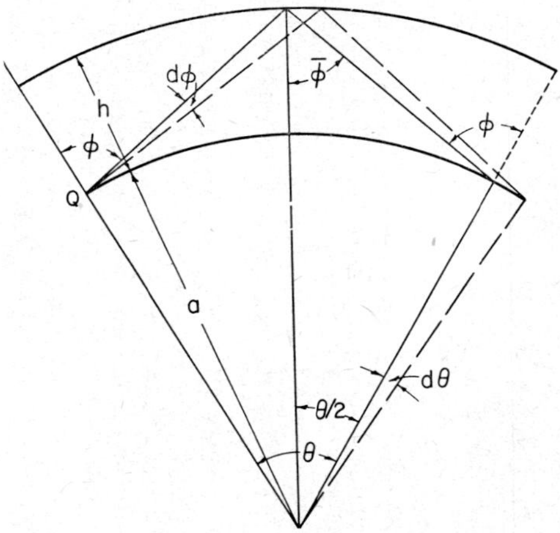

FIG. 1.

which was derived by Lord Rayleigh [10] in equivalent form. When this is applied to the single hop ray between Q and P shown in Fig. 1 it follows that

$$\alpha = \left(\frac{a+h}{a}\right)\left[\frac{2 \sin (\theta/2)}{\sin \theta}\right]^{\frac{1}{2}}\left[\frac{(h+a) - a \cos (\theta/2)}{(h+a) \cos (\theta/2) - a}\right]^{\frac{1}{2}} \tag{2}$$

since

$$\tan \phi = \frac{(h+a) \sin (\theta/2)}{(h+a) \cos (\theta/2) - a} \tag{3}$$

The above expression for the convergence coefficient can be readily generalized to m hops and is expressed by [1]

$$\alpha_m = \left(\frac{a+h}{a}\right)\left[\frac{2m \sin (\theta/2m)}{\sin \theta}\right]^{\frac{1}{2}}\left[\frac{(h+a) - a \cos (\theta/2m)}{(h+a) \cos (\theta/2m) - a}\right]^{\frac{1}{2}} \tag{4}$$

which is identical in form to the single hop formula except that the quantity $\theta/2$ in the expression for α is replaced by $\theta/2m$ except in the factor

$$\sqrt{(\sin \theta)}$$

in the denominator. Most of the following discussion will be restricted to the single hop case (i.e., $m = 1$) but the arguments are easily generalized to the multiple hop case as will be indicated.

The quantity α increases from

$$\frac{a + h}{a}$$

for θ equal to zero to infinity when θ approaches θ_c which is a zero of the denominator of α . θ_c is obtained from

$$\cos \frac{\theta_c}{2} = \frac{a}{a + h} \tag{5}$$

where also $\phi = \pi/2$. In the language of optics, this point is called a caustic. In calculations of field strength by geometrical-optical methods, this infinity at the caustic is usually counteracted by the form of the antenna patterns at the grazing angles $\phi = \pi/2$. This comes about because the vertical patterns, assuming vertical electric dipoles at Q and P, are each $\sin \phi\, [1 + R_v(\phi)]$ where $R_v(\phi)$ is the Fresnel reflection coefficient at angle of incidence ϕ on the imperfectly conducting ground. At grazing incidence $R_v(\phi)$ approaches -1 and since $[1 + R_v(\phi)]$ occurs twice, the resulting zero is of higher order than the zero in the denominator of the convergence coefficient. Thus Bremmer [1] shows calculations of the sky wave field intensity of the various hops (for a finitely conducting spherical earth and concentric ionosphere) which approach zero near the caustic and become zero at the caustic and remain zero thereafter. Now if the proper values for the antenna pattern are employed [11] there is no zero in the direction $\phi = \pi/2$ corresponding to grazing incidence and consequently, the infinity at the caustic is no longer cancelled. It is thus necessary to examine the behavior at the caustic more closely.

There have been numerous discussions of the evaluation of the fields at caustic points. The procedure proposed originally by Debye [3] and later by Picht [8] was to construct appropriate solutions of the wave equation which reduce to the known geometrical optical solution as the wave number k approach infinity. The problem has been discussed more recently by Luneberg [6] and Kay and Keller [5] who made some useful extensions. On the caustic, the field is quite generally proportional to

$$(k)^{\frac{1}{2} - \frac{1}{n}}$$

where $k = 2\pi/\text{wavelength}$ and where n is an integer greater than 2. Sufficiently far away from the caustic the field is given by geometrical optics and is independent of k. At a perfect focus such as at the antipode in the sphere problem $n = \infty$ and the field is thus proportional to $k^{\frac{1}{2}}$. (More will be said about this case later on.) As it turns out here, the caustic occurring in the convergence coefficient is of the lowest order (namely, $n = 3$) and consequently the field is proportional to $k^{\frac{1}{6}}$. This is demonstrated as follows.

In a rigorous formulation of the problem, the field of a downcoming wave from a spherical smooth reflector must satisfy the wave equation and for infinite frequency, transform to the geometrical optical solution. Some consideration shows that in spherical coordinates it must be of the form

$$(kr)^{-\frac{1}{2}} H_{\nu+\frac{1}{2}}(kr) P_\nu(-\cos\theta)$$

where $H_{\nu+\frac{1}{2}}$ is a Hankel function and P_ν is a Legendre function. Since the asymptotic form of this expression is to be identified with a ray subtending an angle ϕ with the radius vector, it follows that for r near a, $\nu + \frac{1}{2} \sim ka \sin\phi$. It should be noted that if $kr \gg 1$ and $\sin\phi$ is somewhat less than unity, the so-called second order Debye expansion is valid. That is

$$(kr)^{-\frac{1}{2}} H_{\nu+\frac{1}{2}}(kr) \cong \frac{1}{kr} S_\nu(kr)$$

where

$$S_\nu(z) = \left(1 - \frac{n^2}{z^2}\right)^{-\frac{1}{4}} e^{\pm i\left[\int_n^z \left(1-\frac{n^2}{x^2}\right)^{\frac{1}{2}} dx\right]} e^{\mp i\pi/4} \tag{6}$$

and where $n = \nu + \frac{1}{2}$. This is the basic representation from which the geometrical optics representations are derived (see for example Bremmer [1]). Now when $\nu + \frac{1}{2}$ becomes near ka (or n near z), corresponding to ϕ approaching $\pi/2$, it is necessary to use the third order or Hankel approximation. In this case

$$(kr)^{\frac{1}{2}} H_{\nu+\frac{1}{2}}(kr) \cong \left(\frac{\pi z}{6}\right)^{\frac{1}{3}} \frac{z}{n} \left\{1 - \frac{n^2}{z^2}\right\}^{\frac{3}{4}} H_{\frac{1}{3}}(\rho)\, e^{\pm i(5\pi/12-\rho)} S_\nu(z) \tag{7}$$

where $H_{\frac{1}{3}}(\rho)$ is the Hankel function of order one third with argument ρ given by

$$\rho = \frac{n}{3}\left[\frac{z^2}{n^2} - 1\right]^{3/2} \tag{8}$$

and as before $n = \nu + \frac{1}{2}$ and $z = ka$. When z is not near n such that $\rho \gg 1$ the above third order expression for H_n transforms to the second order representation on replacing $H_{\frac{1}{3}}(\rho)$ by its leading term in its asymptotic expansion. This indicates that the field at and near the caustic is simply obtained by multiplying the geometrical optical field by a factor $R(\rho)$ which is the ratio of the third order to the second order representation of the function $H_n(z)$. This is simply

$$R(\rho) \cong \left(\frac{\pi\rho}{2}\right)^{\frac{1}{2}} H_{\frac{1}{3}}(\rho)\, e^{-i(5\pi/12-\rho)} \tag{9}$$

which approaches unity for $\rho \gg 1$. In terms of the angle ϕ,

$$\rho \cong \frac{ka\cos^3\phi}{3\sin^2\phi}. \tag{10}$$

Near the caustic, ρ is small compared to unity and since

$$H_{\frac{1}{3}}(\rho) \sim \frac{+\, i2^{\frac{1}{3}}\Gamma(1/3)}{\pi\rho^{\frac{1}{3}}} \tag{11}$$

it follows that

$$R(\rho) \sim 1{\cdot}3466\rho^{\frac{1}{6}}e^{+i\pi/12} \qquad (12)$$

$$\sim 1{\cdot}3466\left(\frac{ka}{3}\right)^{\frac{1}{6}}\frac{(\cos\phi)^{\frac{1}{2}}}{(\sin\phi)^{\frac{1}{3}}}e^{+i\pi/12}.$$

In terms of the geometry in Fig. 1.

$$(\cos\phi)^{\frac{1}{2}} \cong \left[\frac{(h+a)\cos(\theta/2) - a}{(h+a)\sin(\theta/2)}\right]^{\frac{1}{2}} \qquad (13)$$

which approaches zero at the caustic in just a way to cancel the infinity in the geometrically optically computed convergence coefficient α.* For example on taking the limit

$$(\alpha R)_c = \alpha R_{\phi\to\pi/2} = \left(\frac{2}{\sin\theta_c}\right)^{\frac{1}{2}}\left[\left(\frac{a+h}{a}\right)^2 - 1\right]^{\frac{1}{2}} 1{\cdot}3466\left(\frac{ka}{3}\right)^{\frac{1}{6}} e^{+i\pi/12} \qquad (14)$$

This quantity which is proportional to the field at the caustic varies as the one sixth power of the frequency. It is in agreement with a formula obtained by Norton [7] who used a somewhat different method.

It is now clear that the convergence coefficient α in the usual geometrical optical formulae should be replaced by αR which enables the concept to be utilized as described above right up to the caustic. The question now arises as to what happens beyond the caustic. It might be reasoned that the field should decrease beyond this point and the convergence should therefore not increase. Closer inspection indicates however that the convergence as defined does increase. This can be seen by noting that

$$\left|\tan\phi\frac{\partial\phi}{\partial\theta}\right|$$

and R are essentially constant beyond the caustic so that αR varies as

$$\hat{R}(\sin\theta)^{-\frac{1}{2}}.$$

Therefore, beyond the caustic

$$\frac{\alpha R}{(\alpha R)_c} \sim \frac{\hat{R}}{\hat{R}_c}\left(\frac{\sin\theta_c}{\sin\theta}\right)^{\frac{1}{2}}$$

$$\cong \left[\frac{2[(1+h/a)^2-1]^{\frac{1}{2}} + \theta - \theta_c}{2[(1+h/a)^2-1]^{\frac{1}{2}}}\right]\left(\frac{\sin\theta_c}{\sin\theta}\right)^{\frac{1}{2}} \qquad (16)$$

or

$$\alpha R \cong \left(\frac{2}{\sin\theta}\right)^{\frac{1}{2}}\left\{\left[\left(\frac{a+h}{a}\right)^2 - 1\right]^{\frac{1}{2}} + \left(\frac{\theta-\theta_c}{2}\right)\right\} 1{\cdot}3466\left(\frac{ka}{3}\right)^{\frac{1}{6}} e^{+i\pi/12} \qquad (17)$$

*In the case of m hops.

$$(\cos\phi)^{\frac{1}{2}} \cong \left[\frac{(h+a)\cos(\theta/2m) - a}{(h+a)\sin(\theta/2m)}\right]^{\frac{1}{2}} \qquad (15)$$

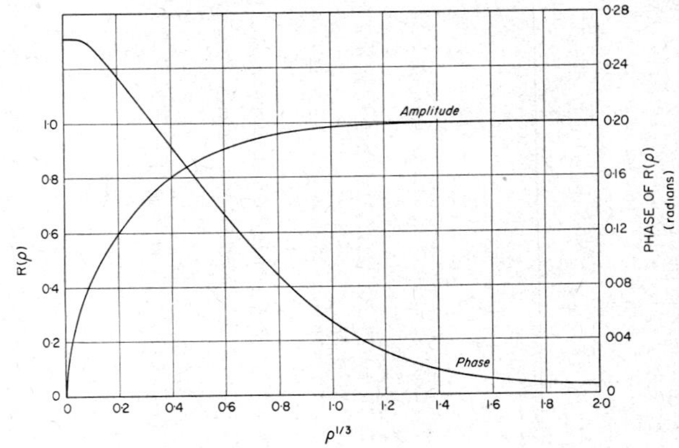

FIG. 2.

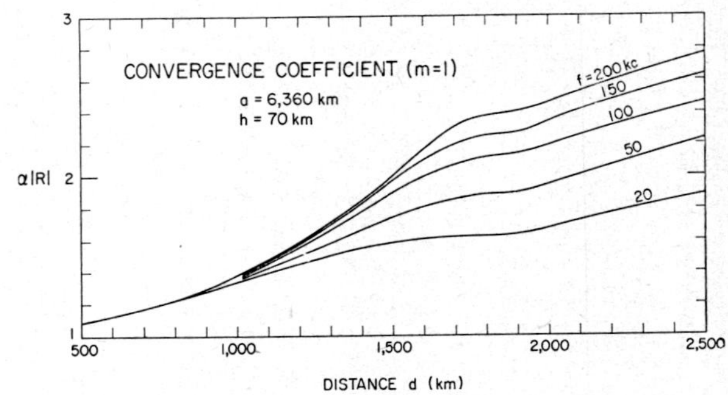

FIG. 3A.

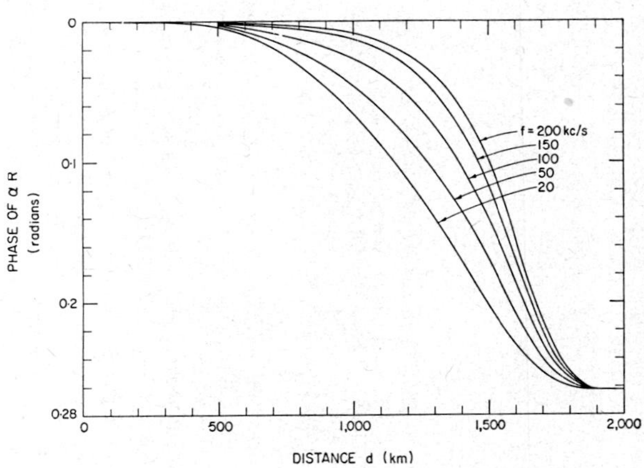

FIG. 3B.

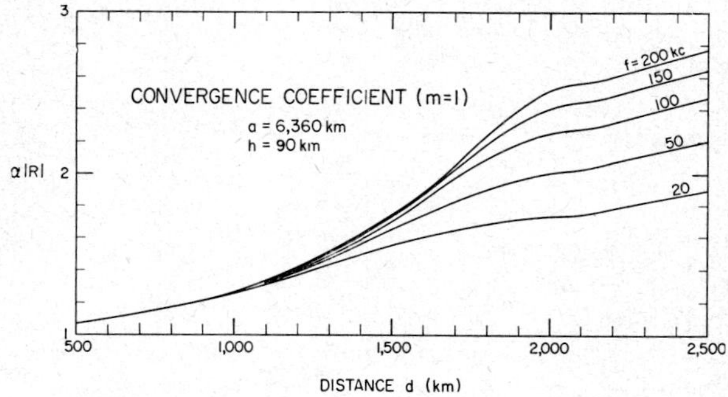

FIG. 4A.

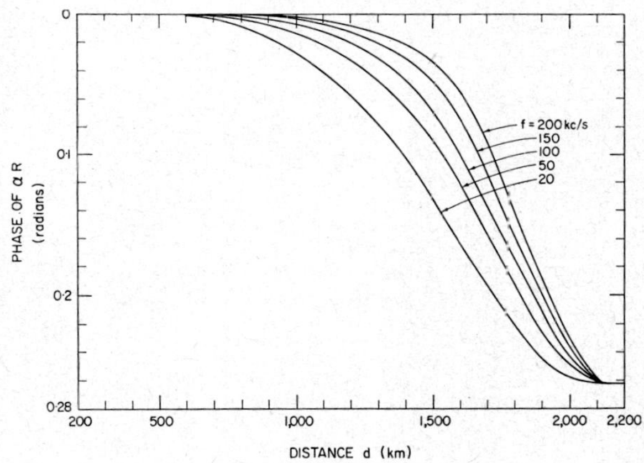

FIG. 4B.

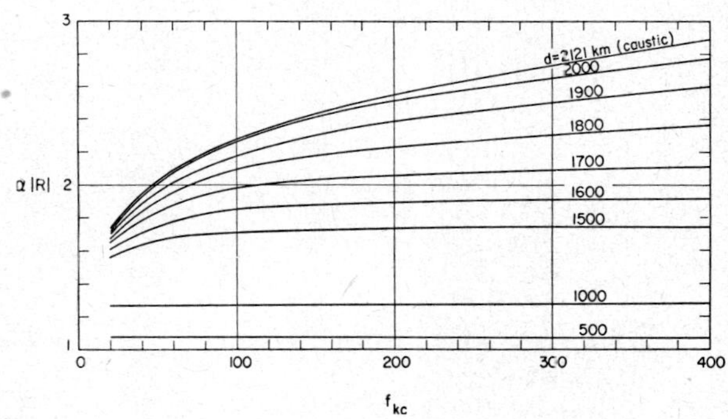

FIG. 5.

To illustrate the significance of the preceding results some quantities have been plotted. The amplitude and phase of the quantity $R(\rho)$ is shown in Fig. 2 as a function of $\rho^{\frac{1}{3}}$. The amplitude and phase of the resultant convergence factor αR is illustrated for various frequencies in Fig. 3a and 3b, respectively for a reflecting layer height h of 70 km. The corresponding curves are shown in Fig. 4a and 4b for a height of 90 km.

It can be observed from Fig. 5 that the convergence is generally decreased as the frequency is lowered. In fact, near the caustic the convergence is varying as the one-sixth power of the frequency.

Caution should be exercised in drawing too many conclusions from the shape and behavior of these curves because of the extreme idealizations of the model. The ionospheric reflecting layers are not smooth and the full amount of the convergence would not be realized except possibly at the lowest frequency shown. The influence of roughness has been discussed recently by Norton and the reader is referred to his treatment. Another factor which plays some role is the variation of the refractive index of the atmosphere between the ground and the lower edge of the ionospheric reflecting layer. Methods based on geometrical optics for including this effect has also been developed by Norton. Unfortunately, at low frequencies the variation of the refractive index with height is comparable to the wavelength thus partially invalidating ray tracing.

III. Convergence for Pulse Excitation

In the above discussion the time variation is harmonic, corresponding to an excitation of a continuous wave with a single frequency. In certain instances the incident field at the reflected layer is a pulse. Within the rather limited confines of geometrical optics the dispersion would be expected to result solely from the imperfect conductivity of the reflecting layer. This effect has been discussed previously, both for an isotropic and an anisotropic sharply bounded ionosphere [15]. To illustrate the dispersion resulting from the frequency dependence of the convergence coefficient, it is desirable to examine a step-function incident pulse. To this end, the caustic correction factor can be expressed as a function of ω as follows

$$R(\omega) = \frac{H_{\frac{1}{3}}^{(2)}(\omega\beta)}{(2/\pi\omega\beta)^{\frac{1}{2}}} e^{i\omega\beta} e^{-i5\pi/12} \tag{18}$$

where

$$\beta = \frac{a \cos^3 \phi}{3c \sin^2 \phi}.$$

This can be rewritten in terms of the modified Hankel function of order one third, $K_{\frac{1}{3}}$, as follows

$$R(s) = K_{\frac{1}{3}}(\beta s) e^{\beta s} \left(\frac{2\beta s}{\pi}\right)^{\frac{1}{2}} \tag{19}$$

where $s = i\omega$. For a step function incident field at $t = 0$, the form of the reflected field (assuming an ideal reflector) is then the inverse Laplace transform,

$r(t)$, of $R(s)/s$. Making use of Erdelyi [4] *et. al's.* transform pair No. 515–(23) with $\mu = -\frac{1}{2}$ and $\nu = \frac{1}{2}$, it readily follows that

$$\begin{aligned} r(t) &= P_{-\frac{1}{6}}(1 + t/\beta) \text{ for } t > 0 \\ &= 0 \text{ for } t < 0 \end{aligned} \tag{20}$$

which is the Legendre function of order $(-1/6)$ and argument $(1 + t/\beta)$. As t/β approaches zero (i.e., for small time) $r(t)$ becomes unity. Therefore, away from the caustic the leading edge of the step is preserved. Near the caustic however β is small and

$$r(t) \simeq \frac{\Gamma(1/3)}{(\pi)^{\frac{1}{2}}\Gamma(5/6)} \left(\frac{\beta}{2t}\right)^{\frac{1}{6}} \text{ for } t \gg \beta \tag{21}$$

which is varying as the inverse one sixth power of time. The latter result was obtained from the asymptotic form of the Legendre function $P_{\frac{1}{6}}$ for large arguments. It can also be checked by noting that its Laplace transform is

$$1{\cdot}3466\ (\beta s)^{\frac{1}{6}}/s$$

which is consistent with Equation (12).

IV. Antipodal Focusing

As indicated in the beginning of the paper, the field of an individual mode or hop depends on the angular distance θ by the Legendre function $P_\nu(-\cos\theta)$ where $\nu + \frac{1}{2} \simeq ka \sin\phi$ and ϕ is the angle of incidence at the ground of the ray in question.

Now the asymptotic expansion of the Legendre function can be written in the form

$$P_\nu(-\cos\theta) \simeq \frac{e^{i[ka \sin\phi(\pi-\theta)-\pi/4]} + e^{-i[ka\sin\phi(\pi-\theta)-\pi/4]}}{(2\pi\ ka \sin\phi \sin\theta)^{\frac{1}{2}}}, \tag{22}$$

which is valid if θ is not near π (i.e., away from the antipode). Each of the exponential terms can be identified as an azimuthal wave travelling toward and away from, respectively, the antipode. The amplitude of these components are varying as $(\sin\theta)^{\frac{1}{2}}$ or $[\sin(\pi - \theta)]^{\frac{1}{2}}$. Near the antipode it is possible to write

$$P_\nu(-\cos\theta) \simeq J_0(ka \sin\phi(\pi - \phi))$$

where J_0 is the Bessel function of the first kind of order zero. It can be noted that the first term of the asymptotic expansion of J_0 is the same as that of $P_\nu(-\cos\theta)$. It is now of interest to indicate the way in which the geometrically optically computed field may be corrected to give the correct behavior at or near the antipode. The term $\exp i[ka \sin\phi(\pi - \theta) - \pi/4]$ is a geometrical optical ray incident at an angle ϕ and its direction of propagation is towards the antipode (i.e., away from the source). As it propagates through the antipode, it changes its phase by $\pi/2$ and emerges at the other side as a wave of the form $\exp i[ka \sin\phi(\theta - \pi) + \pi/4]$. In analogy to the correction factor applied at the caustic a factor R_a is now defined as the ratio of the field in the antipodal region to the incident geometrical optical field. Clearly, for $\theta < \pi$, this is given by

$$R_a \simeq \frac{\dfrac{J_0[ka \sin \phi(\pi - \theta)]}{e^{ika(\pi-\theta)}\, e^{-i\pi/4}}}{[2\pi ka \sin \phi(\pi - \theta)]^{\frac{1}{2}}} \tag{23}$$

Near the caustic such that $ka \sin \alpha(\pi - \theta) \ll 1$ it is seen that

$$R_a \simeq [2\pi ka \sin \phi(\pi - \theta)]^{\frac{1}{2}} e^{i\pi/4} \tag{24}$$

which varies as the square root of frequency. This is the expected behavior of the field at a focus or an axial caustic.

V. The Complete Geometrical Series with Certain Diffractive Corrections

It is now desirable to actually apply the previous results to the computation of the first hop sky wave. The situation is illustrated in Fig. 6. The surface of the

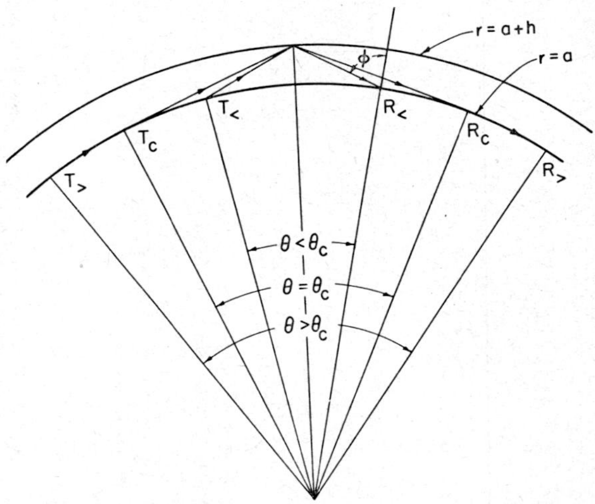

Fig. 6.

earth is defined by $r = a$ and a smooth reflecting layer is located at $r = a + h$. It is convenient to distinguish three cases. In the first, the receiving antenna located at $R_<$ is illuminated by the one reflected sky wave for the transmitting antenna located at $T_<$. In this situation $\phi < \pi/2$. In the second, the two antennas located at T_c and R_c are located at the critical points where ϕ just becomes $\pi/2$. In the third case, the antenna located at $T_>$ and $R_>$ beyond the critical points such that the first hop sky wave propagates in the diffraction zone or shadow.

The vertical electric field at R for a vertical electric dipole source at T can now be written in the form [14]

$$E = E_0 e^{-i\Phi} + 80{\cdot}5 \sum_m \frac{(\alpha R)_m \sin^2 \phi_m (F)^2}{d_{km} + 300 t_m^{ms}} Q_m \, e^{-i2\pi f_{kc} t_m^{ms}} \tag{25}$$

In the above, E_0 and Φ are the amplitude and phase (lag) of the ground wave normalized such that E_0 is 100 mv/m at one mile (1·609 km); these quantities have been computed and are available in convenient graphical form [13]. $(\alpha R)_m$ is the convergence coefficient defined above for the mth hop sky wave. d_{km} is the distance in kilometres between transmitter and receiver measured along the ground. f_{kc} is the frequency in kilocycles. t_m^{ms} is the time delay in m-secs between the direct (mth hop) sky wave and the direct wave along the ground (assuming free space velocity in both cases). Clearly, for the mth hop sky wave

$$300 t_m^s = 2m \left[(a + h)^2 + a^2 - 2a(a + h) \cos \frac{\theta}{2m} \right]^{\frac{1}{2}} - a\theta \,.\, km$$

which is valid if $\theta < \theta_c$ where θ_c is given by $\cos(\theta_c/2) = \dfrac{a}{a + h}$. The same expression also holds for $\theta \geqq \theta_c$ if $m \geqq 2$ and $\theta < \theta_c'$ where θ_c' is given by $\cos(\theta_c'/4) = a/a + h$.† In this case, $300 t_1^s = a[2 \tan(\theta_c/2) - \theta_c] km$.

The factor F, in the above expression, is the cut-back factor discussed extensively in another paper [12]. It accounts for the influence of finite conductivity and earth curvature on the resultant pattern of the antennas. When θ is less than but near θ_c, F can be replaced by the factor $[1 + R_v(\phi_m)]/2$ where $R_v(\phi_m)$ is the Fresnel reflection coefficient for a vertical polarized plane wave incidence at angle ϕ_m on the earth. As stated before, ϕ_m is given by

$$\cos \phi_m = \frac{(h + a) \cos(\theta/2m) - a}{(h + a) \sin(\theta/2m)}.$$

When θ is near θ_c, the more accurate value for the amplitude and phase of F must be employed. These are plotted for various frequencies in Figs. 7 and 8 as a function of $(\theta - \theta_c)$ expressed in degrees. When $\theta - \theta_c$ is sufficiently negative (i.e., the left hand portion of the figures) the quantity F becomes identical to the Fresnel value. The quantity Q_m involves the reflection coefficient of the ground and the reflection and conversion coefficients at the ionosphere. It's form has been discussed elsewhere. In the case of the first hop sky wave it becomes simply the reflection coefficient ${}_vR_v(\bar{\phi}_1)$ for a wave incident at angle $\bar{\phi}_1$ at the ionosphere. It might be noted that $(a + h) \sin \bar{\phi}_1 \simeq a \sin \phi_1$. In the case $m = 2$ and 3,

$$Q_2 = [{}_vR_v(\bar{\phi}_2)]^2 R_v(\phi_2) + {}_vR_h(\bar{\phi}_2) R_h(\phi_2)\, {}_hR_v(\bar{\phi}_2) \tag{26}$$

and

$$\begin{aligned} Q_3 = {} & [{}_vR_v(\bar{\phi}_3)]^3 [R_v(\phi_3)]^2 \\ & + 2\, {}_vR_h(\bar{\phi}_3) R_v(\phi_3)\, {}_vR_h(\phi_3) R_h(\phi_3)\, {}_hR_v(\phi_3) \\ & + {}_vR_h(\bar{\phi}_3) R_h(\phi_3)\, {}_hR_h(\bar{\phi}_3) R_h(\phi_3)\, {}_hR_v(\bar{\phi}_3) \end{aligned} \tag{27}$$

† $\theta = \theta_c'$ corresponds to the situation when the second hop sky wave leaves and arrives at the earth tangentially. Without further modification, the theory is not valid there.

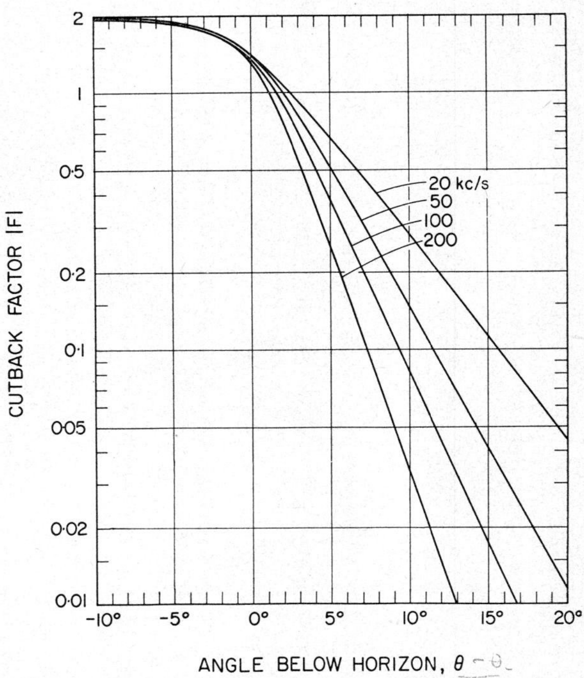
2
1
0·5
0·2
0·1
0·05
0·02
0·01
CUTBACK FACTOR |F|
20 kc/s
50
100
200
-10°
-5°
0°
5°
10°
15°
20°
ANGLE BELOW HORIZON, θ

FIG. 7A.

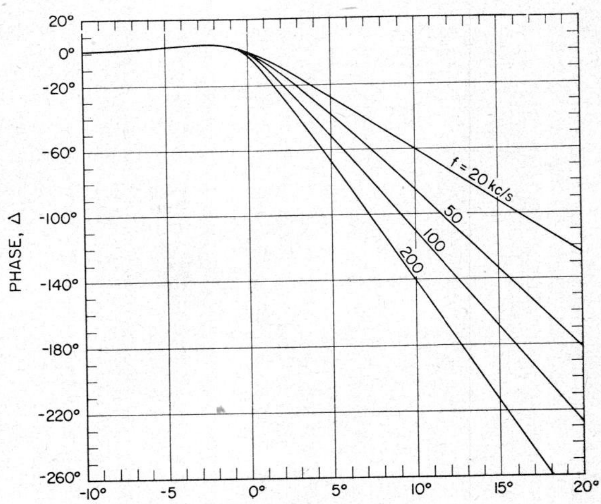
20°
0°
-20°
-60°
-100°
-140°
-180°
-220°
-260°
PHASE, Δ
f = 20 kc/s
50
100
200
-10°
-5
0°
5°
10°
15°
20°
ANGLE BELOW HORIZON, θ

FIG. 7B.

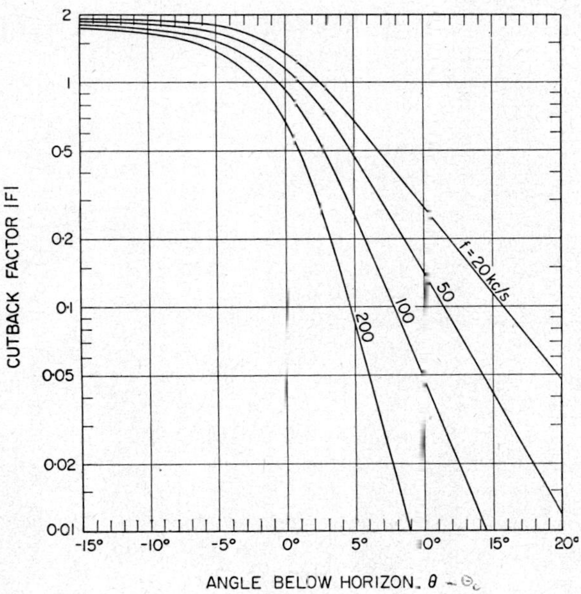
2
1
0·5
0·2
0·1
0·05
0·02
0·01
CUTBACK FACTOR |F|
-15°
-10°
-5°
0°
5°
10°
15°
20°
ANGLE BELOW HORIZON, θ
f = 20 kc/s
50
100
200

FIG. 8A.

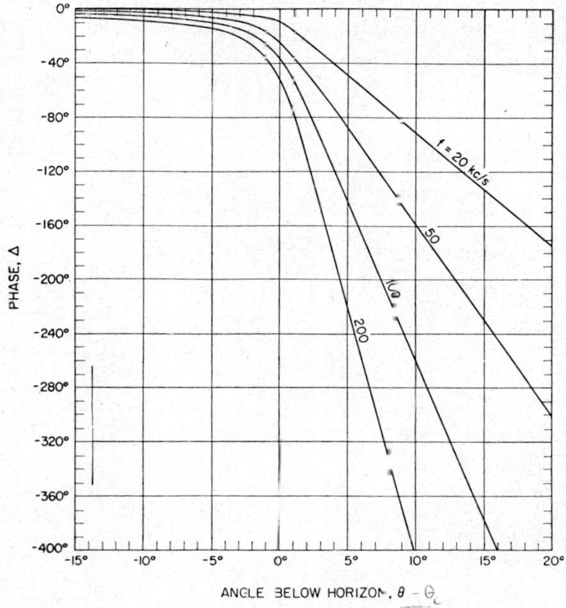
0°
-40°
-80°
-120°
-160°
-200°
-240°
-280°
-320°
-360°
-400°
PHASE, Δ
-15°
-10°
-5°
0°
5°
10°
15°
20°
ANGLE BELOW HORIZON, θ
f = 20 kc/s
50
100
200

FIG. 8B.

where $(a + h) \sin \bar{\phi}_m \simeq a \sin \phi_m$. The coefficients $R_v(\phi)$ and $R_h(\phi)$ are Fresnel coefficients at the ground for the electric vector parallel and perpendicular to the plane of incidence, respectively. These coefficients are defined by

$$R_v(\phi) = \frac{N^2 \cos \phi - (N^2 - \sin^2 \phi)^{\frac{1}{2}}}{N^2 \cos \phi + (N^2 - \sin^2 \phi)^{\frac{1}{2}}} \tag{28}$$

and

$$R_h(\phi) = \frac{\cos \phi - (N^2 - \sin^2 \phi)^{\frac{1}{2}}}{\cos \phi + (N^2 - \sin^2 \phi)^{\frac{1}{2}}} \tag{29}$$

for a homogeneous smooth ground whose complex refractive index is N. The coefficients ${}_vR_v$, ${}_vR_h$, ${}_hR_v$ and ${}_hR_h$ define the complex ratio of a specified electric field component in the wave after reflection to a specified electric field component before reflection. The first subscript denotes whether the electric field in the incident wave is parallel (v) or perpendicular (h) to the plane of incidence and the second subscript refers in a like fashion to the reflected wave.‡ ${}_vR_v$ and ${}_hR_h$ are reflection coefficients and ${}_vR_h$ and ${}_hR_v$ are often described as conversion coefficients. The form of these ionospheric coefficients is rather complicated even for the idealization of a sharply bounded ionosphere with the superposition of a uniform earth's magnetic field [2] [15]. Some numerical results for these quantities are available where Booker's quasi-longitudinal approximation is invoked [15]. Recently, more exact calculations have been carried out by Yabroff [16] who still, however, retains the idealization of a sharply-bounded ionosphere. As Yabroff points out, the quasi-longitudinal approximation is sufficiently valid for the computation of reflection coefficients at VLF but is not valid for the transmission coefficients (which do not enter the present problem). In any case, it would be desirable to employ Yabroff's formulas for any future calculations.

Acknowledgments

I would like to thank Mr. K. A. Norton and A. D. Watt for stimulating discussions and Mrs. Alyce M. Conda for performing the computations.

References

1. Bremmer, H. "Terrestrial Radio Waves." Elsevier Publishing Company (1949). See also vol. xvi, "Encylopedia of Physics," Springer (1957).
2. Budden, K. G. (1951). *Phil. Mag.* **42**, 833–850.
3. Debye, P. (1909). *Ann. Phys., Lpz.* **30**, 755.
4. Erdelyi, Magnus, Oberhettinger & Tricomi, "Tables of Integral Transforms," vol. 1. McGraw-Hill, New York (1954).
5. Kay, I. & Keller, J. B. (1953). *N.Y. Univ., Inst. Math. Sci. Rep.* No. EM-55.

‡In the terminology of radio engineering the polarizations indicated by (v) and (h) are described as vertical and horizontal, respectively.

6. Luneberg, R. K. (1949). *Inst. Math. Sci. Rep.* No. EM-15.
7. Norton, K. A. (See this volume).
8. Picht, J. (1925). *Ann. Phys., Lpz.* **77,** 685.
9. Poeverlein, H. (1958). *J. atmos. terr. Phys.* **12,** 126–139.
10. Rayleigh, Lord "Theory of Sound" (1896).
11. Wait, J. R. & Conda, A. M. (1958). *Trans. Inst. Radio Engrs, N.Y. AP*-**6,** 348–360.
12. Wait, J. R. & Conda, A. M. (1958). *Trans. Inst. Radio Engrs, N.Y. AP*-**6,** 348–360.
13. Wait, J. R. & Howe, H. H. (1956), *Nat. Bur. Stand. Circ.* No. 574.
14. Wait, J. R. & Murphy, A. (1956). *Geofis. pur. Appl.* **35,** 61–72.
15. Wait, J. R. & Perry, L. B. (1957). *J. geophys. Res.* **62,** 43–56. (Also recent unpublished results of J. R. Johler.)
16. Yabroff, I. W. (1957). *Proc. Inst. Radio Engrs, N.Y.* **45,** 750–753.

Radiation from a Slot on a Large Corrugated Cylinder

JAMES R. WAIT and ALYCE M. CONDA

National Bureau of Standards, Boulder, Colo.

The radiation fields are computed for an axial slot on a cylinder whose surface is regularly corrugated. The mathematical idealization is a magnetic line source on a cylindrical surface whose boundary impedance is specified. Special attention is given to surfaces whose radii of curvature are large compared to the wave-length.

I. INTRODUCTION

Considerable interest has been shown recently in the radiation characteristics of slots or similar apertures on convex metallic surfaces. Such radiators have extensive application to high speed aircraft or missiles. It has been proposed from time to time that the radiation characteristics could be controlled to some extent by modifying the surface on which the slots are cut. One suggestion is to coat the surface with a dielectric film which may also serve as a protective covering. Another scheme, which is somewhat equivalent, is to corrugate the surface in a periodic fashion. If the spacing between the grooves is small compared to the wavelength, the surface can be approximately characterized by an inductive surface reactance. For example, if the grooves are rectangular in shape, with depth h, width d and periodicity s, the average surface impedance Z for propagation transverse to the grooves is given by

$$Z \simeq \frac{i\eta_0}{N}\frac{d}{s}\tan Nkh$$

with $\eta_0 = 120\pi$, $k = 2\pi/\text{wavelength}$ and where N is the refractive index of the material filling the grooves. The above result is valid if $d \ll h$ and $s \ll \lambda$ [5, 8]. In the present case $Z = iX_s$ where X_s is real and is called the surface reactance.

The propagation and excitation of surface waves on *flat* reactive surfaces has been discussed extensively in the literature. Reference is made to several recent papers on the subject [1, 6, 9, 12]. There have been several investigations on surface wave propagation on curved reactive surfaces but none of these have really considered the excitation or the radiation aspects of the problem [2, 3, 7].

It is the purpose of the present paper to consider the radiation from an axial slot or an axial magnetic line source on a circular cylinder whose surface has a finite inductive reactance X_s. The prime objective is to obtain the dependence of the radiation pattern on the reactance X_s.

II. FIELD REPRESENTATIONS

The model chosen is an infinite circular cylinder of radius a. The excitation is considered to be a magnetic line source of finite length which is on the cylindrical surface and axially directed. This line source may be either an axial slot of infinitesimal width or a small axial loop. Choosing a cylindrical coordinate system (ρ, θ, z) co-axial with the cylinder, the surface is defined by $\rho = a$ and the source is located on the surface. θ is the azimuthal angle measured such that the observer is at $\theta = -\pi/2$. (See Fig. 1.) The azimuthal variation of the field

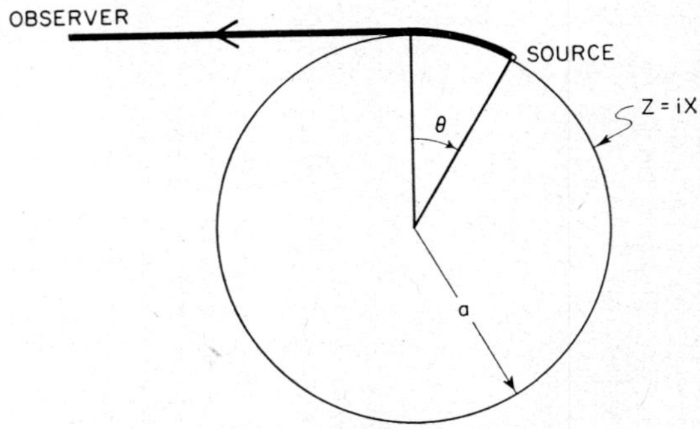

FIG. 1. Illustrating radiation from a magnetic line source on a circular cylinder with a reactive boundary impedance.

at large distances from the cylinder is by definition the E-plane radiation pattern, $F(\theta)$. When the surface of the cylinder possesses a surface impedance Z it was shown previously that [10]

$$F(\theta) = \frac{2}{\pi i k a} \sum_{n=0}^{\infty} \frac{\epsilon_n e^{in\pi/2} \cos n(\pi/2 + \theta)}{H_n^{(2)\prime}(ka) + G H_n^{(2)}(ka)} \tag{1}$$

where $G = -iZ/\eta_0$ with $\eta_0 \simeq 120\pi$. In the above $\epsilon_0 = 1$, $\epsilon_n = 2$ $(n \neq 0)$. $k = 2\pi/\lambda$, $H_n^{(2)}(ka)$ is the Hankel function of the second kind of order n and argument ka, and

$$H_n^{(2)\prime}(ka) = [dH_n^{(2)}(x)/dx]_{x=ka}.$$

The above representation for $F(\theta)$ is of a type which is often called the "harmonic series." It is very poorly convergent and is difficult to use for numerical computation unless ka is small. An alternative representation which is valid for large values of ka is in terms of a complex integral. It was derived previously [11] as an extension of the basic work of Fock [4]. Therefore

$$F(\theta) \simeq e^{-ika\theta} \quad g(X)$$

with

$$g(X) \simeq \frac{e^{-ika\theta}}{\sqrt{\pi}} \int_{\infty e^{-i2\pi/3}}^{\infty} \frac{e^{-iXt}}{W_1'(t) - qW_1(t)} dt \tag{2}$$

and where $X = \left(\frac{k}{2}\right)^{\frac{1}{3}} \theta,$

$$q = -i\left(\frac{ka}{2}\right)^{\frac{1}{3}}\frac{Z}{\eta_0},$$

$$W_1(t) = e^{-2\pi i/3}\left(\frac{\pi}{3}\right)^{\frac{1}{2}}(-t)^{\frac{1}{2}}H^{(2)}_{\frac{1}{3}}\left[\left(\frac{2}{3}\right)\left(-t\right)^{3/2}\right],$$

$$W_1'(t) = dW_1(t)/dt.$$

As indicated above, the integration contour is along a line making an angle $2\pi/3$ with the real axis and the real axis itself. This representation for $F(\theta)$ is valid for $(ka/2)^{\frac{1}{3}}\pi \gg 1$ and $|\theta|^3 \ll 1$. The function $g(X)$ was investigated by Fock for the case $q = 0$ corresponding to a perfectly conducting boundary and more recently, by the present authors for imperfectly conducting boundaries where

$$q \simeq |q| e^{-i3\pi/4}.$$

In the present situation, the interest is in a reactive boundary such that $Z = iX_s$ where the real quantity X_s is the surface reactance.* When $X_s > 0$, the surface is inductive and when $X_s < 0$, the surface is capacitive. Of special interest is the inductive case since this is the expected condition for surface wave trapping.

When X is a *large negative number*, the integral $g(X)$ can be evaluated by a saddle point method and leads to

$$g(X) \cong \frac{2X}{X - iq}\, e^{iX^3/3}. \tag{3}$$

When $X > 0$ it is possible to express $g(X)$ as a sum of residues of the poles of the integrand. Then [4, 11]

$$g(X) \cong -i2\sqrt{\pi}\sum_{s=1}^{\infty}\frac{e^{-iXt_s}}{(t_s - q^2)W_1(t_s)} \tag{4}$$

where t_s are the solutions of

$$W_1'(t) - qW_1(t) = 0.$$

When $|q|$ is less than about unity, the following series is useful

$$t_s = t_s^\circ + \frac{1}{t_s^\circ}q - \frac{1}{2(t_s^\circ)^3}q^2 + \left[\frac{1}{3(t_s^\circ)^2} + \frac{1}{2(t_s^\circ)^5}\right]q^3 - \left[\frac{7}{12(t_s^\circ)^4} + \frac{5}{8(t_s^\circ)^7}\right]q^4 \ldots \tag{5}$$

where t_s° is a solution of $W_1'(t) = 0$. The convergence of the residue series is poor unless X is greater than about 1·0. It does not converge at all if $X < 0$. Furthermore, unless q is small, the series formula for t_s is not adequate. Consequently, for many cases of practical interest it is necessary to evaluate the integral for $g(X)$ or F by numerical means. This has been done for the following values of q

$$q = 0,\ 0{\cdot}161,\ 0{\cdot}336 \text{ and } 0{\cdot}830$$

* X_s is not to be confused with X.

corresponding to an inductive surface reactance X_s which is related to q by

$$q = \left(\frac{ka}{2}\right)^{\frac{1}{3}} \frac{X_s}{\eta_0}.$$

The quantities plotted in Fig. 2a and 2b, as the amplitude $|F|$ and phase Δ_1 are defined by

$$\begin{aligned} F &= e^{-ika\theta} g(X) \qquad (6) \\ &= |F| \, e^{-ika\theta} \, e^{iX^3/3} \, e^{i\Delta} \text{ for } \theta \geqq 0 \\ &= |F| \, e^{-ika\theta} \, e^{i\Delta} \text{ for } \theta \leqq 0. \end{aligned}$$

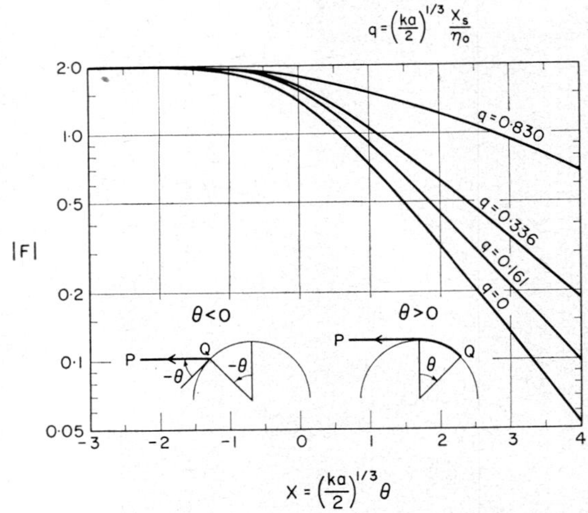

FIG. 2A. Space factor for an inductive curved boundary.

The case $q = 0$ corresponds, of course, to the perfectly conducting cylinder and was treated originally by Fock [4]. When q increases, the inductive reactance of the surface becomes finite and it appears that $|F|$ does not diminish as rapidly in the shadow (i.e., $X > 0$) as for a perfectly conducting surface. In this sense, the pattern is being broadened into the shadow and diffraction effects are consequently lessened. On examining the phase curves, it is seen that Δ is becoming increasingly negative as q increases. This means that the phase velocity of the surface wave is decreasing with increasing q in accord with the well known behavior for plane surfaces.

To extend the range of validity of Equation (6) to larger values of θ, it is necessary to replace Equation (6) by

$$F \simeq e^{-ika\theta} g(X) + e^{-ika(\pi-\theta)} g(X') \qquad (7)$$

where

$$X = (ka/2)^{\frac{1}{3}}\theta$$

and

$$X' = (ka/2)^{\frac{1}{3}}(\pi - \theta).$$

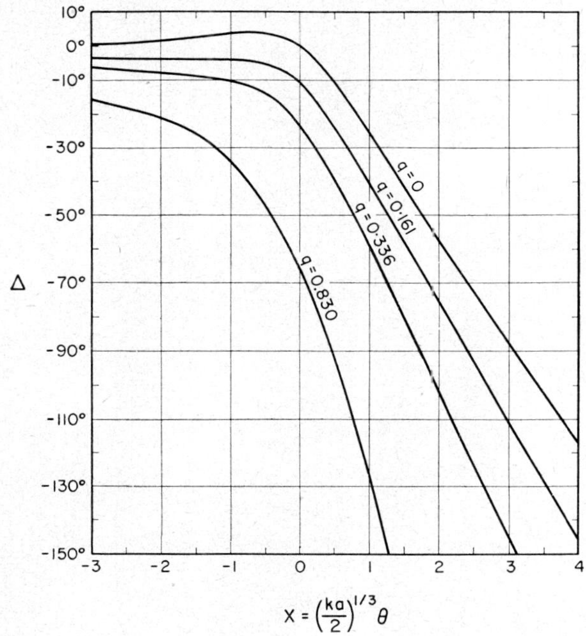

Fig. 2B. Phase of space factor for an inductive boundary.

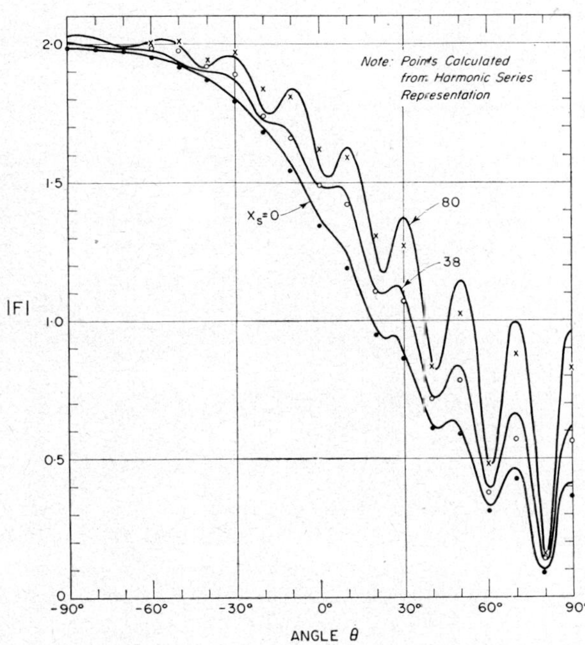

Fig. 3A. Space factor for a circular cylinder with an inductive boundary (ka = 8).

The second term of Equation (7) corresponds to a wave which has "crept" around the rear of the cylinder by the "long route." It can be neglected when

$$(ka/2)^{\frac{1}{3}}(\pi - \theta) \gg 1.$$

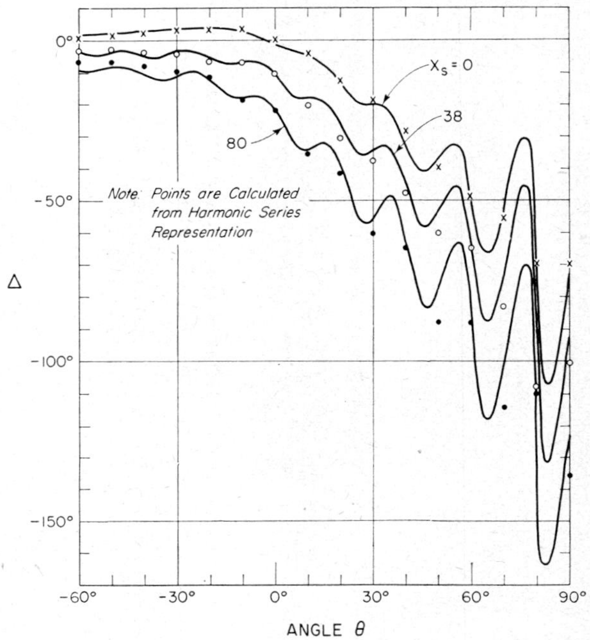

FIG. 3B. Phase of space factor for a circular cylinder with an inductive boundary (Ka = 8).

As an illustration of the parametric curves in Fig. 2a and 2b, the pattern is computed from Equation (7) for a cylinder whose circumference is 8 wavelengths (i.e., $ka = 8$). The values of q on the curves corresponds to 0·0, 0·161, and 0·336. Physically, these are equivalent to a cylinder with a corrugated surface whose inductive surface reactance X_s is 0, 38 and 80 ohms. Employing the harmonic series representation, F is computed for every ten degrees from $\theta = -90°$ to $+90°$. These points are also indicated in Figs. 3a and 3b. The agreement between the curves and the points is reasonable in view of the fact that the complex integral representation is valid only if ka is large.

III. Conclusion

The results presented here show that an inductive surface may well enhance the radiation from a slotted cylinder antenna. The patterns have been computed from the complex integral representation by numerical integration, saddle-point evaluation and summing of the residues. The validity of this approach is verified, at least in part, by comparing the results with the numerical data from the exact harmonic series. It is difficult to compare these results with those of Elliott [2] for azimuthal surface waves on corrugated cylinders since he did not consider the excitation problem. It is possible, however, to compare the attenuation factor (i.e., the imaginary part of t_s) of the function $F(\theta)$ deep in the shadow (i.e., $X \gg 1$) with the corresponding attenuation factor of Elliott's. There appears to be good agreement for the slope of the curve for $q = 0{\cdot}833$ in Fig. 2a which is the only one that can be directly compared.

The research reported in this paper was supported by the Electronics Research Directorate of the U.S. Air Force Cambridge Research Center.

References

1. Barlow, H. E. M. & Cullen, A. L. (1953). *Proc. Instn elect. Engrs* **100,** 321.
2. Elliott, R. S. (1955). *J. appl. Phys.* **27,** 368–376.
3. Elliott, R. S. (1956). *Trans. Inst. Radio Engrs, N.Y. AP*-**4,** 422–427.
4. Fock, V. A. (1946). *J. Phys., Moscow,* **10,** 130–140 and 399–409.
5. Hurd, R. A. (1954). *Canad. J. Phys.* **32,** 727–734.
6. Miller, M. A. (1955). *J. tech. Phys., Moscow* **25,** 1810. (1952). *C.R. Acad. Sci., U.R.S.S.* **87,** 571.
7. Miller, M. A. & Talanov, Val. (1956). *J. tech. Phys., Moscow* **12,** 2665–2673.
8. Rotman, W. (1951). *Proc. Inst. Radio Engrs, N.Y.* **39,** 952.
9. Wait, J. R. (1957). *J. Res. nat. Bur. Stand.* **59,** 365–377.
10. Wait, J. R. & Conda, A. M. (1957). *J. Res. nat. Bur. Stand.* **59,** 507–516.
11. Wait, J. R. & Conda, A. M. (1958). *Trans. Inst. Radio Engrs,* N.Y. *AP*-**6,** 348-360.
12. Zucker, F. J. (1954). *Proc. Symp. Mod. Advances Microwave Techniques,* p. 403 (Polytechnic Inst. Brooklyn). (Many references are given in this comprehensive paper.)

A Rapid Beam-swinging Experiment in Transhorizon Propagation*

A. T. WATERMAN, JR.

Stanford Electronics Laboratories, Stanford University, Stanford, Calif.

By using a broadside phased array for an antenna, a narrow beam can be swung rapidly and in quick succession through a limited sector by fast control of the phasing, rather than by movement of the entire antenna structure. This technique is used at the receiving end of a 101-mile beyond-the-horizon transmission path in order to probe the portion of the troposphere through which the signal is propagated. At the frequency employed of 3·12 kMc, a 0·49° beam is swung in azimuth through a 4·2° sector each tenth of a second.

A variety of phenomena are observed with this technique which have not been directly apparent in slower beam-swinging experiments. The beam-broadening effect attributed to atmospheric scattering is not always evident on any one sector scan. However, the change from scan to scan is frequently rapid enough so that a time average would show the broadening. At times the scan-to-scan changes are systematic and show a continuity indicative of a motion of the scattering or reflecting regions; in some cases this motion is too rapid to be accounted for by transport of air, thus implying a wave motion rippling through the atmosphere. At other times the atmospheric structure is too fine to be resolved by the beam width employed, and the time variations are too rapid to reveal a continuity from one scan to the next.

Narrow beam antennas have been used very effectively as devices for probing the atmosphere in transhorizon propagation experiments [1, 2, 3, 4, 6, 8, 9]. The unmanageable nature of the large physical structure required for a narrow beam is largely circumvented if a phased array is used and a mechanism provided for rapid variation of the phase of each radiating element.

The experiment described here utilizes such an array for the receiving antenna. A 3000 Mc circularly polarized signal is transmitted, with sufficiently wide beam width (about 5°) to illuminate the pertinent portion of the atmosphere. The receiving antenna [7] is an 8-element broadside array (see Fig. 1). Each element in the array consists of an offset parabolic reflector and helical feed. The helix in each feed is connected to a syncro drive so that the desired phase shifting can be accomplished by mechanical rotation of the helix about its axis. The syncros are then driven at multiple rates from a centrally located gear train. Thus the overall rate of beam swinging is controlled by the speed of the master motor driving the gear train.

The horizontal beam width of the entire antenna depends on the lateral length of the overall array. At 3000 Mc, a 30-ft long array has a beam width of about $\frac{1}{2}$°. In the array which has been constructed each element (consisting of reflector and feed) is mounted along a pipe supported on an A-frame structure, so that the element can be slid laterally along the pipe, thus affording an adjustment of the length of the array and consequently of the width of the antenna

*The research reported in this document was made possible through the U.S. Army Signal Corps under Contract DA36(039)SC-73151.

pattern. The interval between successive principal maxima in the array pattern depends on the spacing between adjacent elements in the array, or, for an array of fixed length, on the number of elements. This angular interval is also the sector over which the array can scan without ambiguity. The sector of interest in these transhorizon propagation experiments covers only 4° or 5° in azimuth. Consequently only 8 elements are needed in the array and these may then be placed at a 4-ft spacing. Utilization of a parabolic reflector in each element does not affect the array factor over this limited sector but does increase the capture area of the overall antenna. It also establishes the vertical beam width, in this case about 5°.

FIG. 1. Phased array used as the receiving antenna in rapid beam-swinging experiment.

For operation in this frequency range, all the component parts are of convenient physical size. The helices [5] consist of 3·4 turns of No. 12 copper wire and are 7·8 cm long. The half-power beam width for each helical feed is 60°, a convenient value for illuminating the offset section of paraboloidal reflector. This offset places the feed itself and associated syncro drive out of the path of the reflected radiation and in a conveniently accessible position.

To provide a continuous azimuthal scan, the helical feeds are rotated at multiple rates—or rather, the right 4 helices are rotated at odd multiple rates in one sense and the left 4 at corresponding rates in the opposite sense. With this procedure, a rotation rate of 2100 r.p.m. for the outermost helices corresponds to a scan rate in which one principal maximum moves on and the next moves into its place in one-tenth second (i.e., 10 sector scans per sec). It is to be noted that this beam-swinging is not a back-and-forth sector scan; the principal maxima in the array factor rotate uniformly like the spokes of a wheel and move one after another through the sector being scanned.

At the frequency employed, 3·12 kMc, the horizontal beam width of the array and the azimuthal sector scanned, as measured on a line-of-sight test range, are 0·49° and 4·2° respectively.

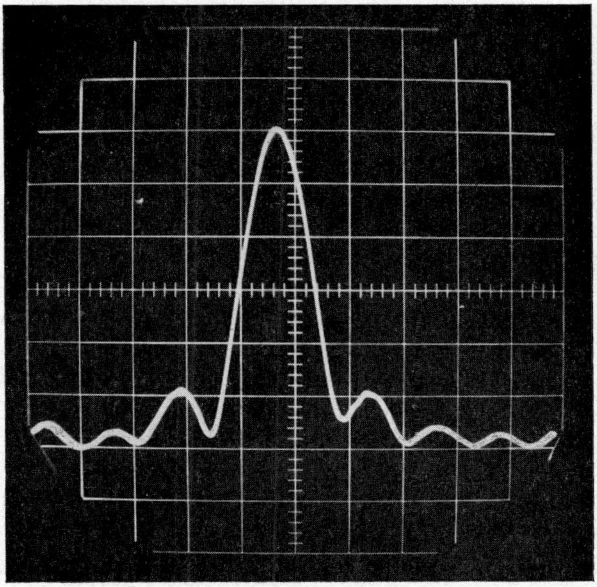

FIG. 2. Measured antenna-pattern of the array over a 4·2-degree sector.

This equipment has been employed in propagation measurements which, though preliminary, are nevertheless startling enough in their implications to warrant reporting promptly. A c-w transmitter was located 101 miles distant from the receiving array, well beyond line of sight, and the system response was observed as the receiving antenna beam scanned in azimuth through the great-circle bearing to the transmitter.

In the accompanying figures, which are pictures of an oscilloscope face, signal level is displayed as a vertical deflection and the azimuth in which the beam is pointing is synchronized with the horizontal sweep. Figure 2 shows the response of this antenna system to a low-power test transmitter within line-of-sight of the receiving location. It serves to illustrate the beam-width (0·49°) and sidelobe level (better than 10 db down) of the array. It also provides an azimuth reference for the sector scan. The 10 divisions of horizontal displacement in the figures correspond to the 4·2° sector. On this scale, the great-circle bearing to the 101-mile distant transmitter (not to the test source in Fig. 2) is 7·2 divisions from the left-hand edge.

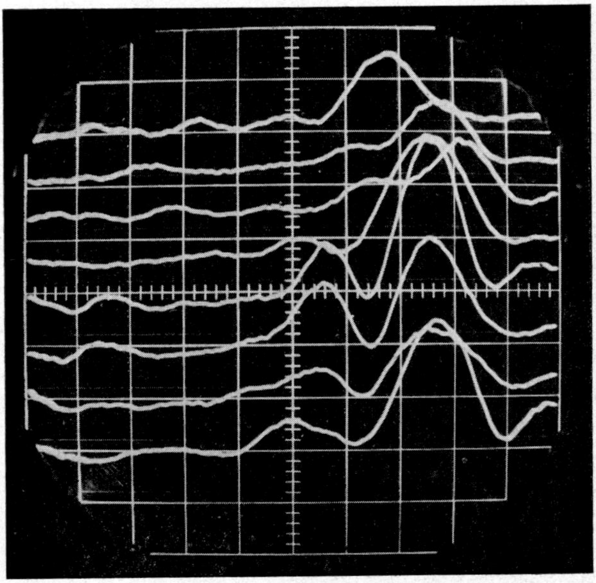

FIG. 3.

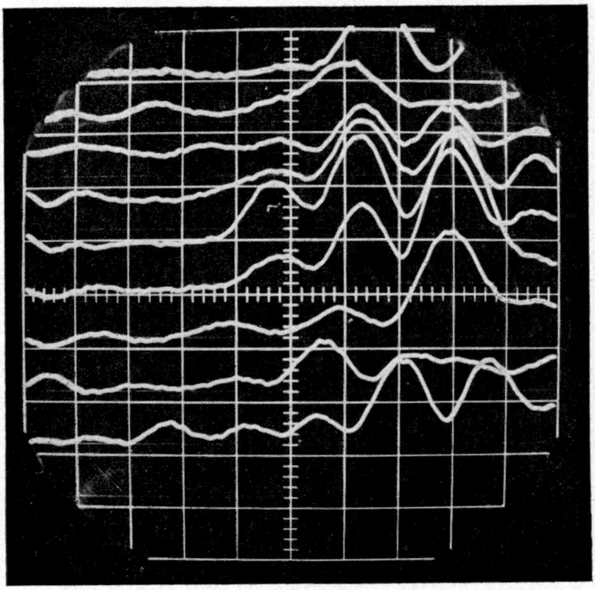

FIG. 4.

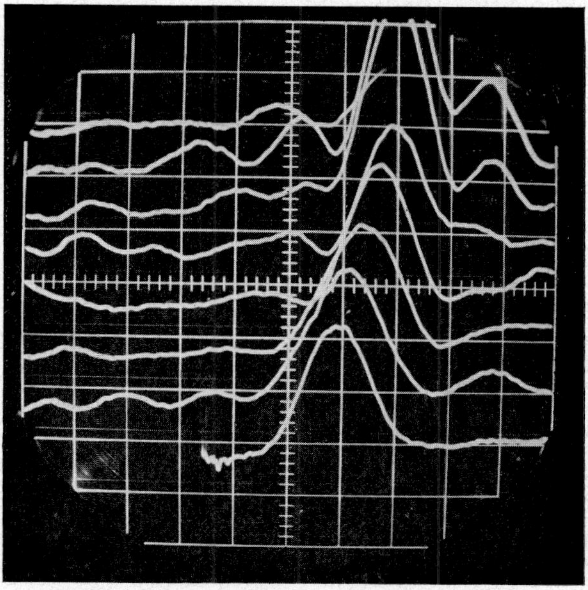

Fig. 5.

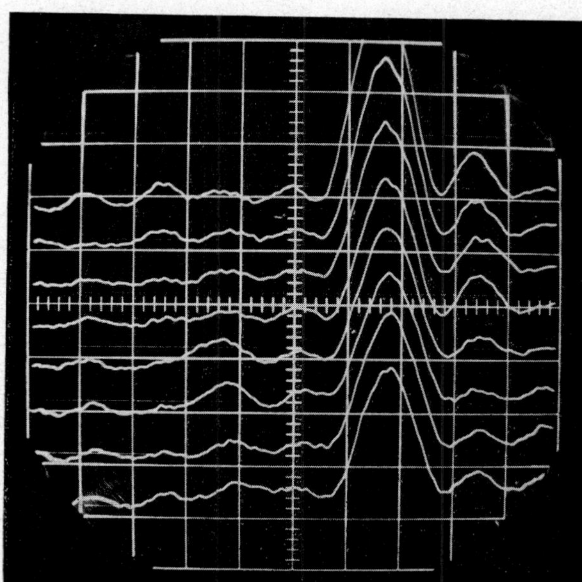

Fig. 6.

Figure 3 is an example of receiving-system response to a signal propagated over the 101-mile path. There the display is the same, except that a raster containing alternate antenna sector-scans is employed. Thus, each scan consumes $\frac{1}{10}$ of a second, but there is $\frac{1}{5}$ of a second between sucessive scans in the raster. It will be noticed that the effective antenna pattern obtained via the atmospherically scattered signal varies rapidly from scan to scan. The beam-broadening effect attributed to random scattering [2, 9] is not very marked on any one scan, but if a succession of scans is averaged, the shift in angle of arrival plus the apparently discrete off-path contributions (which show up here looking like exaggerated side lobes) result in considerable beam broadening.

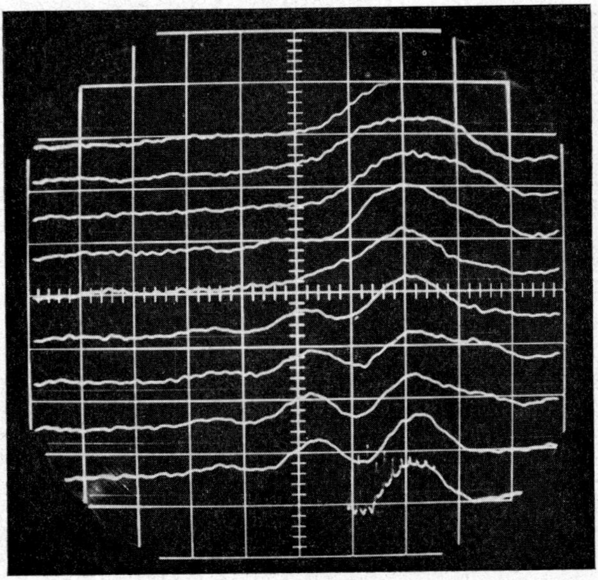

Fig. 7.

Figs. 3-7, samples of received signal; raster display of receiving system response to 101-mile-distant transmitter; each azimuth scan covers 4·2° (10 dimensions in the figures) in 0·1 seconds; successive scans displayed are 0·2 seconds apart; great-circle bearing to transmitter is 7·2 divisions from left.

An example in which there appears to be a small number of discrete scatterers which change quite rapidly is shown in Fig. 4. A slower beam-swinging experiment would not reveal this fine structure; only an average broadening would be observed.

The predominant characteristic evident in Fig. 5 is the rapid and continuous motion of one discrete scatterer. Such a motion, of course, does not necessarily imply a physical translation at this speed (which would correspond roughly to 600 or 800 m.p.h.), but might be caused by a wave motion in a reflecting layer.

By way of contrast, Fig. 6 shows a constant pattern with very little beam broadening. This constancy, when observed, occurred at night, in contrast to the more wildly fluctuating patterns seen during daytime.

A final example has been selected for Fig. 7 in which several of the sweeps show an instantaneous beam broadening. This type of pattern, which so far has not been observed frequently, might be what one would expect if there were a large number of small incoherent scatterers.

The above figures are samples of observations which have been made recently and have been repeated sufficiently to establish their reality. A detailed analysis of the results has not yet been undertaken. Nevertheless, some pertinent remarks can be made.

There are times when the antenna beam, though broad in elevation and only moderately narrow in azimuth, appears to have sufficient resolution to distinguish discrete scatterers. (The term "scatterer" is used here in its broadest sense; it might well refer to an appropriately oriented portion of a partially reflecting layer.) These scatterers, in such cases, are small in number (1 to 4); their resolvability implies a spatial separation of at least 500 to 1000 ft. At other times, the scatterers are not resolved, and few conclusions can be drawn at the moment.

Frequently, particularly in the daytime, the atmosphere changes structure in $\frac{1}{5}$ sec to such an extent that little continuity can be found in the changes. In such cases one cannot distinguish a random type of turbulent process from a systematic wave motion. At other times the $\frac{1}{5}$ sec interval between scans is sufficiently short to establish a continuity of change from one situation to another. In these cases, there are many instances in which the apparent motions are more rapid than can reasonably be accounted for in terms of tropospheric winds; a wave motion of some sort is implied, such as ripples along an interface.

Much of the beam broadening does not occur instantaneously but shows up in a time average owing to the rapid changes in the scattering or reflecting centers.

Acknowledgment

The author would like to express his gratitude to Mr. R. E. Miller for his competent direction of the construction and operation of the array as well as other equipment, both transmitting and receiving. In addition, thanks are due other members of the Stanford Electronics Laboratories who took an active part in this work: W. H. Huntley, Jr., G. K. Durfey, D. B. Coates, C. L. Liston and P. Brunetti.

References

1. Ament, W. S. & Katzin, Martin (1956). *Trans. Inst. Radio Engrs, N.Y. CS*-**4**, 118–112.
2. Chisholm, J. H., Portmann, P. A., deBettencourt, J. T. & Roche, J. F. (1955). *Proc. Inst. Radio Engrs, N.Y.* **43**, 1317–1335.

3. Chisholm, J. H., Roche, J. F. & Jones, W. J. Experimental investigations of the angular scattering and multipath delays for transmission beyond the horizon. Paper presented at URSI meeting in Washington, D.C., 22–25 May, 1957.
4. Doherty, L. H. A 216-mile 2700 Mc scatter link. Paper presented at URSI meeting in Washington, D.C., 22–25 May, 1957.
5. Huntley, W. H. Jr., *Stanf. Elect. Lab. Tech. Rep.* No. 461–3, 20 September 1957. (Signal Corps Contract DA-36 (039)SC-73151.)
6. Kummer, W. H. & Hogg, D. C. Characteristics of signals received on a large aperture antenna in propagation beyond the horizon. Paper presented at URSI meeting in Washington, D.C., 22–25 May, 1957.
7. Miller, R. E., Durfey, G. K. & Huntley, W. H. Jr. *Stanf. Elect. Lab. Tech. Rep.* No. 461–5. (Signal Corps Contract DA-36(039)SC-73151.)
8. Waterman, A. T. Jr. (1953). *Stanf. Elect. Lab. Tech. Rep.* No. 8. (Contract Nonr 225–10).
9. Waterman, A. T. Jr., Bryant, N. H. & Miller, R. E. (1957). *Trans. Inst. Radio Engrs, N.Y. AP*–**5**, 260–266. (This also appeared as *Stanf. Elect. Lab. Tech. Rep.* No. 461–2, 1957, under Signal Corps Contract DA–36(039)SC–63189.)

Some Aspects of a Rigorous Scalar Treatment of Electromagnetic Wave Propagation *

E. Wolf †

The Physical Laboratories, University of Manchester

The work which will be described in this paper arose from an attempt to understand the full connection between Maxwell's electromagnetic theory and the scalar methods which are used in optics to analyse the structure of optical images. This attempt led some years ago to the introduction, by H. S. Green and the present speaker [4] of a rigorous scalar theory of electromagnetic wave propagation. In the present paper it will be shown that this theory suggests an interesting new model for energy propagation. Before doing so it will be useful to summarize briefly the main results obtained in the earlier investigation.

I. Summary of the Earlier Investigation

An electromagnetic field in vacuum can always be represented in terms of a magnetic vector potential **A** and an electric scalar potential ϕ in terms of which the electric field **E** and the magnetic field **H** are given by

$$\mathbf{E} = -\frac{1}{c}\dot{\mathbf{A}} - \operatorname{grad}\phi, \quad \mathbf{H} = \operatorname{curl}\mathbf{A}, \tag{1}$$

the dot denoting differentiation with respect to time. The equations which the potentials themselves satisfy take a particularly simple form if the potentials are chosen so as to be coupled by the *Lorentz' condition*

$$\operatorname{div}\mathbf{A} + \frac{1}{c}\dot{\phi} = 0. \tag{2}$$

Even when subjected to this condition the potentials are not unique, the field being invariant under the *gauge transformation*

$$\mathbf{A} \to \mathbf{A} + \operatorname{grad}\chi, \qquad \phi \to \phi - \frac{1}{c}\dot{\chi}, \tag{3}$$

where χ is arbitrary. Moreover, if χ is a solution of the homogeneous wave equation, the new potentials will again be coupled by the Lorentz' condition. As is well known in any region which does not contain currents and charges, the gauge may be so chosen that $\phi = 0$. The field may then be derived from a

*The research described in this paper was begun in 1957, when the author was a guest at the Institute of Mathematical Sciences, Division of Electromagnetic Research, New York University. It was sponsored in part by the Air Force Cambridge Research Center, Air Research and Development Command, under Contract No. AF 19(604)1717. The research is being continued under the sponsorship of the same Agency, through its European Office, under Contract No. AF 61(052)–169.

† Now at the Institute of Optics, University of Rochester, Rochester, N.Y., U.S.A.

vector potential alone, which, according to equation (2) satisfies the subsidiary condition

$$\operatorname{div} \mathbf{A} = 0. \tag{4}$$

Thus in such a region, the field can be represented by three scalar waves, the Cartesian rectangular components of **A**, which are coupled by one relation, namely (4). We shall now show that this coupling is strong enough to make possible a reduction by one in the number of the scalar wave functions needed to represent the field.

We express **A** in the form of a Fourier integral

$$\mathbf{A}(\mathbf{x}, t) = \int \{\mathbf{a}(\mathbf{k}, t) \cos \mathbf{k}\,.\,\mathbf{x} + \mathbf{b}(\mathbf{k}, t) \sin \mathbf{k}\,.\,\mathbf{x}\} d\mathbf{k}, \tag{5}$$

where (since **A** is real) the Fourier coefficients **a** and **b** are real and the integration is taken throughout half of the **k**-space, e.g., that part throughout which the component k_z of **k** is non-negative. The condition (4) implies the *transversality* of the plane waves on the right of (5), i.e.,

$$\mathbf{k}\,.\,\mathbf{a} = \mathbf{k}\,.\,\mathbf{b} = 0. \tag{6}$$

To utilize fully the conditions (6), we associate with each **k** two unit base vectors $\mathbf{l}_1(\mathbf{k})$ and $\mathbf{l}_2(\mathbf{k})$ which are orthogonal to **k** and to each other. This may be done in many ways. A particularly convenient choice of $\mathbf{l}_1$ and $\mathbf{l}_2$ is as follows: Let **n** be any constant vector and take

$$\mathbf{l}_1(\mathbf{k}) = \frac{\mathbf{n} \wedge \mathbf{k}}{|\,\mathbf{n} \wedge \mathbf{k}\,|}, \quad \mathbf{l}_2(\mathbf{k}) = \frac{\mathbf{k} \wedge \mathbf{l}_1(\mathbf{k})}{|\,\mathbf{k} \wedge \mathbf{l}_1(\mathbf{k})\,|}. \tag{7}$$

We now resolve each of the Fourier coefficients along these base vectors:

$$\left.\begin{aligned} \mathbf{a}(\mathbf{k}, t) &= a_1(\mathbf{k}, t)\mathbf{l}_1(\mathbf{k}) + a_2(\mathbf{k}, t)\mathbf{l}_2(\mathbf{k}), \\ \mathbf{b}(\mathbf{k}, t) &= b_1(\mathbf{k}, t)\mathbf{l}_1(\mathbf{k}) + b_2(\mathbf{k}, t)\mathbf{l}_2(\mathbf{k}). \end{aligned}\right\} \tag{8}$$

Next we form the complex combinations

$$\left.\begin{aligned} \alpha(\mathbf{k}, t) &= a_1(\mathbf{k}, t) + i a_2(\mathbf{k}, t), \\ \beta(\mathbf{k}, t) &= b_1(\mathbf{k}, t) + i b_2(\mathbf{k}, t), \end{aligned}\right\} \tag{9}$$

and regard α and β as the Fourier coefficients of a new function V:

$$V(\mathbf{x}, t) = \int \{\alpha(\mathbf{k}, t) \cos \mathbf{k}\,.\,\mathbf{x} + \beta(\mathbf{k}, t) \sin \mathbf{k}\,.\,\mathbf{x}\} d\mathbf{k}, \tag{10}$$

where the integration is again taken over half of the **k**-space, viz. $k_z \geqslant 0$. We call V *the complex scalar potential of the field.*

We have now replaced the real vector potential **A** by the complex scalar potential V. Conversely it is easy to see that from the complex potential V, the magnetic vector potential **A** and hence the field vectors may be derived. For in the first place, from the knowledge of V, the quantities α and β and consequently a_1, a_2, b_1 and b_2 may be derived by Fourier inversion. Then **a** and **b** are given by the relations (8), and **A** is obtained by the Fourier synthesis in accordance with (5). Thus our complex potential completely specifies the field.

Moreover, because **A** satisfies the wave equation

$$\nabla^2 \mathbf{A} = \frac{1}{c^2} \ddot{\mathbf{A}}, \tag{11}$$

it may be shown (cf. reference [4]) that

$$\nabla^2 V = \frac{1}{c^2} \ddot{V}, \tag{12}$$

i.e., V represents a wave.

For the purpose of later discussion, the following results, also derived in the paper already referred to, are needed. In terms of V, *the energy density* w of the field may be expressed in the form

$$w = \frac{1}{8\pi}\left(\frac{1}{c^2} \dot{V}\dot{V}^* + \nabla V \,.\, \nabla V^*\right) \tag{13}$$

and *the energy flow* **S** is given by

$$\mathbf{S} = -\frac{1}{8\pi}\left(\dot{V}^* \nabla V + \dot{V} \nabla V^*\right) \tag{14}$$

where the asterisk denote the complex conjugate. These expressions, which satisfy the usual conservation law

$$\frac{dw}{dt} + \operatorname{div} \mathbf{S} = 0, \tag{15}$$

are formally identical with the quantum mechanical expressions for the probability density and the probability current.

These are the main results derived in the paper already referred to. It may be mentioned that the analysis was extended by P. Roman [6] to regions which include currents and charges and the method was used by him and also by K. Nagy [5] in quantum mechanical investigations. This scalar representation was recently employed by J. Focke [3] in the analysis of an optical diffraction problem.

II. A New Model for the Energy Transport in a Monochromatic Field

For the purpose of our present discussion it will be convenient to express the transformation $\mathbf{A} \to V$ and the inverse transformation $V \to \mathbf{A}$ in a somewhat more compact form. Also, instead of using a Fourier integral representation involving sines and cosines we shall find it more convenient to use a Fourier representation involving exponentials, i.e., we represent the vector potential in the form

$$\mathbf{A}(\mathbf{x}, t) = \int \mathbf{c}(\mathbf{k}, t) e^{i\mathbf{k}\cdot\mathbf{x}} d\mathbf{k}, \tag{16}$$

where the integration is now taken over the *whole* **k**-space. Since **A** is real, we have the subsidiary condition

$$\mathbf{c}(-\mathbf{k}, t) = \mathbf{c}^*(\mathbf{k}, t). \tag{17}$$

We now set

$$\left.\begin{aligned} \mathbf{L}(\mathbf{k}) &= \mathbf{l}_1(\mathbf{k}) + i\mathbf{l}_2(\mathbf{k}) \quad \text{when } k_z \geqslant 0, \\ \mathbf{L}(-\mathbf{k}) &= \mathbf{L}^*(\mathbf{k}), \end{aligned}\right\} \tag{18}$$

where $\mathbf{l}_1$ and $\mathbf{l}_2$ are the same base vectors as before. A straightforward calculation shows that the complex scalar potential $V(\mathbf{x}, t)$ may be expressed in the form*

$$V(\mathbf{x}, t) = \int \mathbf{L}(\mathbf{k}) \,.\, \mathbf{c}(\mathbf{k}, t) e^{i\mathbf{k}\cdot\mathbf{x}} d\mathbf{k}. \tag{19}$$

With the help of the Fourier inversion formula, we readily find that the transition from **A** to V and its inverse may be expressed as the following linear transforms:

$$V(\mathbf{x}, t) = \int \mathbf{A}(\mathbf{x}', t) \,.\, \mathbf{M}(\mathbf{x}' - \mathbf{x}) d\mathbf{x}', \tag{20a}$$

$$\mathbf{A}(\mathbf{x}, t) = \mathscr{R} \int V(\mathbf{x}', t) \mathbf{M}^*(\mathbf{x}' - \mathbf{x}) d\mathbf{x}', \tag{20b}$$

where the kernel **M** is the Fourier transforms of **L**, i.e.,

$$\mathbf{M}(\mathbf{x}) = \frac{1}{(2\pi)^3} \int \mathbf{L}(\mathbf{k}) e^{i\mathbf{k}\cdot\mathbf{x}} d\mathbf{k}, \tag{21}$$

and the symbol $\mathscr{R}$ in (20b) denotes the real part.

Let us now consider a monochromatic wave of angular frequency ω. Under fairly general conditions (cf. [2], pp. 41–42), and subject to Sommerfeld's radiation condition in the half-space $z > 0$, the wave may be expressed in the form of an angular spectrum of plane waves:

$$\mathbf{A}(\mathbf{x}, t) = e^{-i\omega t} \int_{-\infty}^{+\infty} \int_{-\infty}^{+\infty} \mathbf{a}(k_x, k_y) e^{i\mathbf{k}\cdot\mathbf{x}} dk_x, dk_y + e^{i\omega t} \int_{-\infty}^{+\infty} \int_{-\infty}^{+\infty} \mathbf{a}^*(k_x, k_y) e^{i-\mathbf{k}\cdot\mathbf{x}} dk_x dk_y, \tag{22}$$

where **k** is the vector with components

$$k_x, k_y, k_z = + \sqrt{\left(\frac{\omega}{c}\right)^2 - k_x^2 - k_y^2}. \tag{23}$$

If we now apply to (22) the transformation (20*a*) we see immediately that the complex scalar potential associated with the monochromatic wave, (22), is of the form

$$V(\mathbf{x}, t) = U_-(\mathbf{x}) e^{-i\omega t} + U_+(x) e^{i\omega t} \tag{24}$$

*The domains of integration in the expressions (19) and (21) are the same as in (16), i.e., the whole **k**-space.

A simple calculation gives the following expressions for U_- and U_+:

$$\left.\begin{aligned} U_-(\mathbf{x}) &= \int_{-\infty}^{+\infty}\int_{-\infty}^{+\infty} \mathbf{L}(\mathbf{k}) \,.\, \mathbf{a}(k_x, k_y) e^{i\mathbf{k}\cdot\mathbf{x}} dk_x\, dk_y, \\ U_+(\mathbf{x}) &= \int_{-\infty}^{+\infty}\int_{-\infty}^{+\infty} \mathbf{L}(\mathbf{k}) \,.\, \mathbf{a}^*(k_x, k_y) e^{-i\mathbf{k}\cdot\mathbf{x}} dk_x\, dk_y. \end{aligned}\right\} \tag{25}$$

It is to be noted that in general, U_- and U_+ are *not* complex conjugates of each other.

If we substitute from (24) into (12) we immediately find that U_- and U_+ each satisfy the Helmholtz equation, i.e.,

$$\left.\begin{aligned} \nabla^2 U_- + k^2 U_- = 0, \\ \nabla^2 U_+ + k^2 U_+ = 0, \end{aligned}\right\} \tag{26}$$

where $k = \omega/c$ is the vacuum wave number; hence each of the complex potentials

$$V_-(\mathbf{x}, t) = U_-(\mathbf{x}) e^{-i\omega t}, \qquad V_+(\mathbf{x}, t) = U_+(\mathbf{x}) e^{i\omega t}, \tag{27}$$

is propagated as a wave.

Before examining the physical significance of these two waves we note some very simple properties which they possess. Let w_- and w_+ denote the energy densities and $\mathbf{S}_-$ and $\mathbf{S}_+$ the energy flow vectors of the two waves. On substituting from (27) into (13) and (14) we obtain

$$w_- = \frac{k^2}{8\pi}\left(U_- U_-^* + \frac{1}{k^2} \nabla U_- \,.\, \nabla U_-^*\right), \tag{28}$$

$$\mathbf{S}_- = -\frac{ikc}{8\pi}\left(U_-^* \nabla U_- - U_- \nabla U_-^*\right), \tag{29}$$

and

$$w_+ = \frac{k^2}{8\pi}\left(U_+ U_+^* + \frac{1}{k^2} \nabla U_+ \,.\, \nabla U_+^*\right), \tag{30}$$

$$\mathbf{S}_+ = \frac{ikc}{8\pi}\left(U_+^* \nabla U_+ - U_+ \nabla U_+^*\right). \tag{31}$$

These formulae show that the energy density and the energy flow vector of each of these two waves are *time-independent*.

Let a_- and a_+ represent the amplitudes and ϕ_- and ϕ_+ the phases of U_- and U_+, i.e.,

$$U_- = a_- e^{i\phi_-}, \qquad U_+ = a_+ e^{i\phi_+}, \tag{32}$$

where the a's and ϕ's are real. On substituting from (32) into (29) and (31) we find that

$$\mathbf{S}_- = \frac{kc}{4\pi} a_-^2 \nabla\phi_- , \tag{33}$$

and

$$\mathbf{S}_+ = -\frac{kc}{4\pi} a_+^2 \nabla\phi_+ , \tag{34}$$

i.e., *the energy flow vector of each of the two waves is at right angles to its surface of constant phase.*

In the usual representation, the orthogonality between the energy flow and the surface of constant phase is, in general, valid only in the limit of the geometrical optics approximation, i.e., in the high frequency limit $\omega = kc \to \infty$ (cf. [1], pp. 112–113). In the present representation this orthogonality is seen to hold rigorously for each of the two "partial waves" V_- and V_+, irrespective of the frequency ω.

Now according to (24), the complex scalar potential of the monochromatic field (22) is represented by the sum of the waves V_- and V_+, and we must therefore consider their combined effect. On substituting from (24) into the formulae (13) and (14) and taking a time average (denoted by angular brackets) over a time interval which is large compared to the period $T = 2\pi/\omega$, we obtain the following expressions for the time averaged values of the energy density w and of the energy flow vector $\mathbf{S}$ of the monochromatic field represented by (22):

$$\langle w \rangle = w_- + w_+, \tag{35}$$

$$\langle \mathbf{S} \rangle = \mathbf{S}_- + \mathbf{S}_+. \tag{36}$$

Since there are no mixed terms on the right hand sides of these formulae it follows that in the time average, the waves V_- and V_+ do not give rise to interference effects. We have thus established the following result: *Any monochromatic wave field in free space may be represented in terms of two complex, time-harmonic scalar waves. These waves are mutually incoherent in the sense that the time averaged energy density and the time averaged energy flow of the field are equal to the sum of the energy densities and of the energy flow respectively of these two waves. Moreover, the energy flow of each of these partial waves is in the direction of the normal to its surface of constant phase.*

It remains to consider the physical significance of the two partial waves. This is best seen by considering a typical plane wave component of each. According to (25) the component is

$$\mathbf{L}(\mathbf{k}) \,.\, \mathbf{a}(k_x, k_y) e^{i(\mathbf{k}\,.\,\mathbf{x} - \omega t)} \quad \text{for} \quad V_-, \tag{37}$$

and

$$\mathbf{L}(\mathbf{k}) \,.\, \mathbf{a}^*(k_x, k_y) e^{-i(\mathbf{k}\,.\,\mathbf{x} - \omega t)} \quad \text{for} \quad V_+. \tag{38}$$

To find what vector potentials correspond to the scalar potentials (37) and (38) we apply the transformation (20b) to each. We then find

$$\mathbf{A}_- = \mathscr{R}\{[\mathbf{L}(\mathbf{k}) \,.\, \mathbf{a}(k_x, k_y)]\mathbf{L}^*(k) e^{i(\mathbf{k}\,.\,\mathbf{x} - \omega t)}\}, \tag{39}$$

$$\mathbf{A}_+ = \mathscr{R}\{[\mathbf{L}(\mathbf{k}) \,.\, \mathbf{a}^*(k_x, k_y)]\mathbf{L}^*(\mathbf{k}) e^{-i(\mathbf{k}\,.\,\mathbf{x} - \omega t)}\}, \tag{40}$$

where $\mathbf{A}_-$ and $\mathbf{A}_+$ are the vector potentials corresponding to (37) and (38) respectively. If we set

$$\mathbf{L}(\mathbf{k}) \,.\, \mathbf{a}(k_x, k_y) = g e^{i\varphi}, \tag{41}$$

$$\mathbf{L}(\mathbf{k}) \,.\, \mathbf{a}^*(k_x, k_y) = h e^{i\chi}, \tag{42}$$

where g, h, ψ and χ are real, and remember that $\mathbf{L} = \mathbf{l}_1 + i\mathbf{l}_2$, $\mathbf{L}^* = \mathbf{l}_1 - i\mathbf{l}_2$, the expressions (39) and (40) become

$$\mathbf{A}_- = g\{\mathbf{l}_1(\mathbf{k}) \cos (\mathbf{k} \cdot \mathbf{x} + \psi - \omega t) + \mathbf{l}_2(\mathbf{k}) \sin (\mathbf{k} \cdot \mathbf{x} + \psi - \omega t)\}, \tag{43}$$

$$\mathbf{A}_+ = h\{\mathbf{l}_1(\mathbf{k}) \cos (\mathbf{k} \cdot \mathbf{x} + \chi - \omega t) - \mathbf{l}_2(\mathbf{k}) \sin (\mathbf{k} \cdot \mathbf{x} + \chi - \omega t)\}. \tag{44}$$

We see that $\mathbf{A}_-$ and $\mathbf{A}_+$ each represent a *circularly polarized plane wave*, one being *left handed*, the other *right handed*. Hence *if a monochromatic electromagnetic vector potential is decomposed into an angular spectrum of plane waves, and then each plane wave is decomposed into two circularly polarized components, one left handed, the other right handed, then* V_- *is the complex scalar potential formed by all the circularly polarized components of one kind, and* V_+ *is the complex scalar potential formed by all the circularly polarized components of the other kind.*

This result has an important bearing on the representation of an *unpolarized* wave, e.g., in connection with images of visual and radio stars. On physical grounds it is obvious that for an unpolarized quasi-monochromatic wave, the amount of energy carried by the two partial waves (V_+ and V_-) must be the same; hence, for such a wave we must have

$$\langle w \rangle = \langle 2w_- \rangle = 2\langle w_+ \rangle, \tag{45}$$

$$\langle \mathbf{S} \rangle = 2\langle \mathbf{S}_- \rangle = 2\langle \mathbf{S}_+ \rangle. \tag{46}$$

Thus *the time averaged energy density and the time averaged energy flow of an unpolarized quasi-monochromatic electromagnetic wave may be rigorously derived from a single complex wave function, by means of the relations* (28), (29) [*or* (30), (31)].

This last result is of interest for the theory of image formation in optical systems of wide angular aperture (cf. [3],) for which the usual optical scalar theory is inadequate. Work is continuing along these lines and it is hoped to publish the results, as well as a fuller analysis of the subject matter of this paper at a later date.*

References

1. Born, M. & Wolf, E. "Principles of Optics." Pergamon Press, London (1959).
2. Bouwkamp, C. J. (1954). *Rep. Progr. Phys.* **17**, 42.
3. Focke, J. (1957). *Optica Acta*, **4**, 124.
4. Green, H. S. & Wolf, E. (1953). *Proc. phys. Soc. Lond. A.* **66**, 1129.
5. Nagy, K. (1955). *Acta. phys. hung.* **4**, 327.
6. Roman, P. (1955). *Acta. phys. hung.* **4**, 209.

* *Footnote added in proofs:* A paper giving a fuller analysis as well as a generalization to non-monochromatic fields of some of the results described here, will be published in *Proc. phys. Soc. Lond.*

Local Ground Wave Field Strength Variations in the Frequency Range 30-1000 MHz

Åke Blomquist

Research Institute of National Defence, Stockholm

Results from a number of field strength recordings of the ground wave will be presented. The measurements have been made over different types of terrain and at different seasons both at vertical and horizontal polarization.

Local field strength variations due to vegetation and ground moisture conditions with low antennas are discussed and field strength distribution curves are given.

I. General

The most predominant factors when determining the characteristics of VHF and UHF ground wave propagation are the terrain irregularities and vegetation of the ground, and with very low antennas, as used in mobile communication, also the ground constants. In the frequency range 30–1000 MHz the electrical properties of the ground have such values that the ground can be regarded as dielectric at propagation over land and fresh water. The influence due to variations in the refractive index of the atmosphere is not dealt with here.

From theoretical formulas field strength curves can be calculated. But these curves give only average values for an arbitrary path and do not take into account the seasonal and local terrain variations in the field strength of the path concerned. This paper will deal with the field strength deviation around these average values based on a general investigation during the last five years in terrain typical for middle Sweden.

II. Ground Wave Formula for Smooth Ground

In the following discussion the propagation factor F will be used defined as the ratio between the actual field strength and the free space level at the same distance. This F includes all the factors that govern the propagation due to the influence of the ground and atmosphere. As only standard atmosphere conditions will be considered and as the antenna heights are small F is less than 1.

Dealing with ground wave propagation in the Sommerfeld region over smooth ground with low vertically polarized antennas we can express F as follows,

$$F=\frac{\lambda}{d\,.\,\pi}\cdot\frac{\epsilon_1\epsilon_2}{\sqrt{((\epsilon_1-1)(\epsilon_2-1))}}\sqrt{\left(1+4\pi^2\left(\frac{h_1}{\lambda}\right)^2\cdot\frac{\epsilon_1-1}{\epsilon_1^2}\right)}\sqrt{\left(1+4\pi^2\left(\frac{h_2}{\lambda}\right)^2\cdot\frac{\epsilon_2-1}{\epsilon_2^2}\right)} \tag{1}$$

where d = distance in meters; λ = wavelength in meters; ϵ = relative dielectric constant; h_1 and h_2 are the antenna heights in meters over ground of different dielectric constant ϵ_1 and ϵ_2 respectively.

For the deduction of this equation see [1] where a more theoretical treatment of the ground wave problem is given.

As shown in this reference a similar formula can be derived for horizontal polarization. But in the case of horizontally polarized waves the effect of ϵ is very small even near the ground for all frequencies concerned.

From Equation (1) it is seen that the 1st derivate of F with regard to the wavelength λ has a zero, $\lambda = \lambda_0$. Further the 2nd derivate is positive, so F has a minimum value for $\lambda = \lambda_0$,

$$F_{\min} = \frac{2 \, . \, \sqrt{(\epsilon_1 \epsilon_2)}}{d[(\epsilon_1 - 1)(\epsilon_2 - 1)]^{\frac{1}{4}}} \, . \, (h_1 + h_2) \tag{2}$$

$$\lambda_0 = 2\pi \, . \, \frac{[(\epsilon_1 - 1)(\epsilon_2 - 1)]^{\frac{1}{4}}}{\sqrt{(\epsilon_1 \epsilon_2)}} \, . \, \sqrt{(h_1 h_2)} \tag{3}$$

Having definite antenna heights h_1 and h_2 it is found that for each value of ϵ there is one wavelength which is more unfavourable than the others out of propagation aspect. With such low antennas as are common in mobile communication sets this minimum value occurs in the frequency range concerned.

This is shown in Fig. 1, where the quantity 20 log $(F \, . \, d) - F$ is multiplied by d to make the curves independent of distance—is given as a function of frequency for some common values of ϵ. The antenna heights of both the transmitter and the receiver are 2·5 or 10 m. The curves and Equation (1) show that the influence of the dielectric constant decreases with increasing frequency, as does this influence when the antenna height is increased.

How do these curves agree with the measured field strengths?

III. Field Strength Recordings

A number of field strength recordings of the ground wave have been made over different types of terrain and at different seasons at vertical as well as horizontal polarization. The frequencies 44, (66), 88, 182, 450, 950 and in some cases also 35, 55 and 169 MHz have been used. When these tests were carried out the receiving and recording apparatus were placed at a fixed position while the transmitter was movable.

An analysis of such field strength recordings measured over approximately plain ground in open fields and woods has been made. The transmitting and receiving antenna heights were 4 m and 2 m respectively. It is found that the field strength median value over a 100 m distance seems to obey an inverse distance squared law except at the highest frequencies. In Fig. 2 the exponent n in the Equation $E = \text{const} \, . \, d^{-n}$, where E is the field strength and d the distance, is shown as a function of the frequency at four different places with rather typical terrain conditions.

At the places 1 and 2 there are mostly open fields, the country being slightly more rough at place 2. Woods are dominating at the places 3 and 4 with the smoothest terrain at place 3.

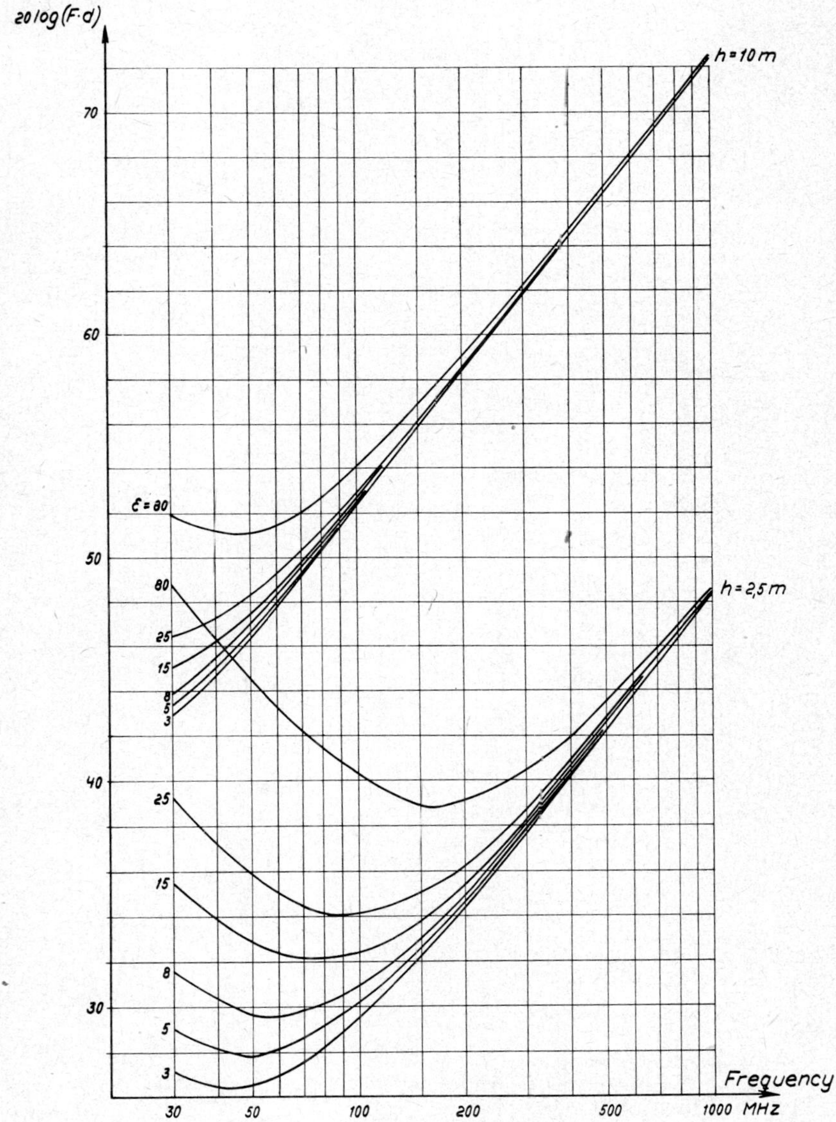

Fig. 1. Frequency dependence of propagation factor over smooth ground, vertical polarization calculated values, both antennas at height shown.

The curves in Fig. 2 are not valid for very short distances, viz., $d < \epsilon . \lambda$, which is a very approximate expression but yields in practice if $\epsilon \geq 10$.

In open fields at 950 MHz there is a tendency to an inverse distance relation which is most pronounced at horizontal polarization, while at the same frequency in woods with a little rough terrain the field strength seems to decrease slightly more than with the squared distance.

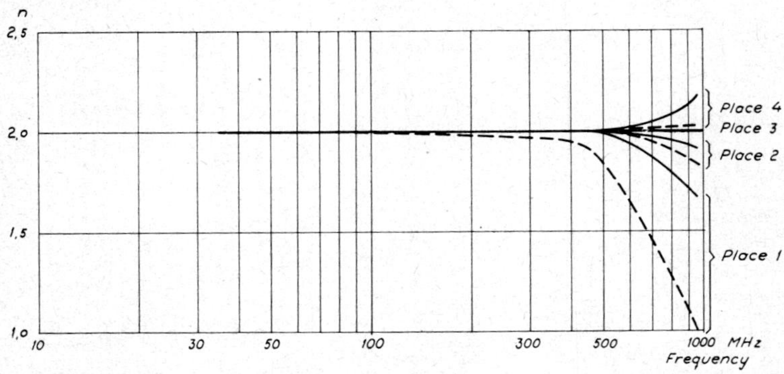

FIG. 2. Exponent n in E = const d^{-n} as a function of frequency at the different place.
——— Vertical polarization - - - - Horizontal polarization.

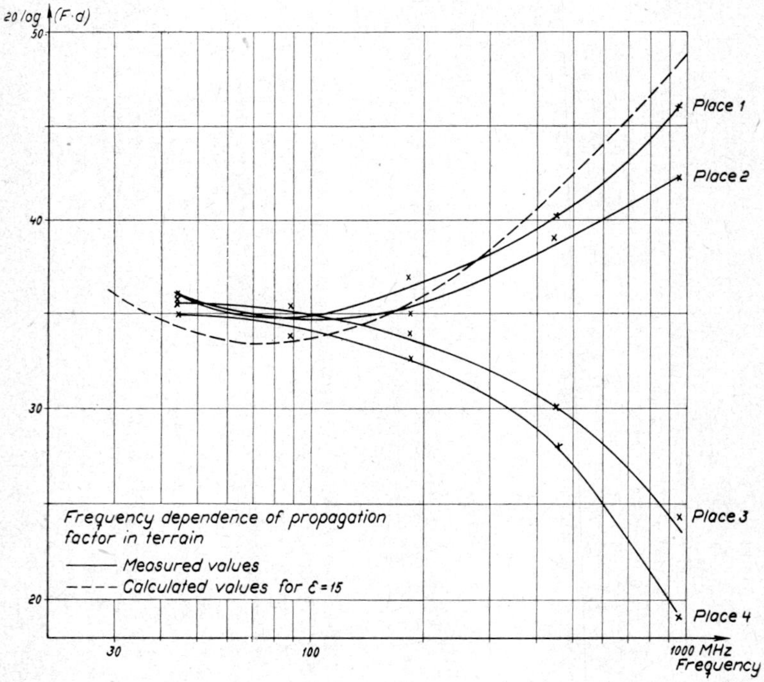

FIG. 3.

If the median field strength curves from open fields and woods are compared, the ground having the same dielectric constant in both cases, it is found that the curves for the wooded ground lie below those for the open field. This is shown in

Fig. 3 for the places discussed above. The deviation is very small at the lowest frequencies but increases with the frequency with different amount due to the roughness of the terrain. The terrain is smoothest at place 1 and roughest at place 4.

But as the curves for both woods and open fields on the whole obey the inverse distance squared law this deviation must be caused by a ground loss near the antennas. This problem will be discussed more thoroughly later on.

The measured curves in Fig. 3 are compared with a curve calculated for $\epsilon = 15$, a common average value of the dielectric constant. The influence of the ground

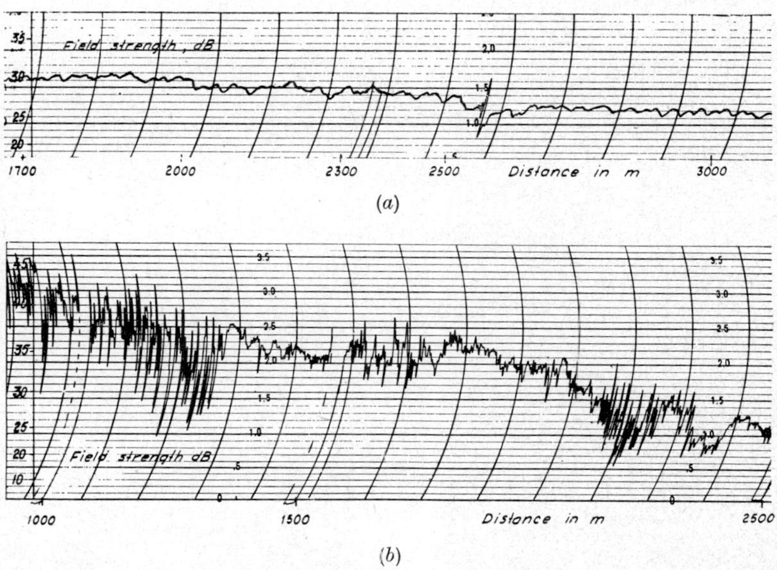

FIG. 4. Typical field strength variations (*a*) in an open field, (*b*) in a forest. Frequency 44 MHz, vertical polarization.

constants is of great importance up to about 100 MHz. The influence due to the terrain is significant all over the frequency range but as the frequency increases the number of objects possible to cause multipath propagation from scattered and reflected wave energy also increases and thus making the terrain frequency dependence considerable.

IV. Analysis of the Variations in Open Fields and Woods

There is a characteristic difference between the field strength recordings from the open field and the wood as seen in Fig. 4. An analysis of such curves shows that the variations in open fields occur at random and are due to discrete local irregularities. The occurrence of the fluctuations in woods shows an obvious frequency dependence.

In the lower picture of Fig. 4 at about 1300 m distance the amplitude starts getting very small. Some days before the measurement was made a violent storm had torn up nearly all the trees by their roots—there were only single trees left. This shows that the amplitude of the signal variations is dependent on the number of trees and it is also an indication that only the terrain near the antenna affects the wave energy.

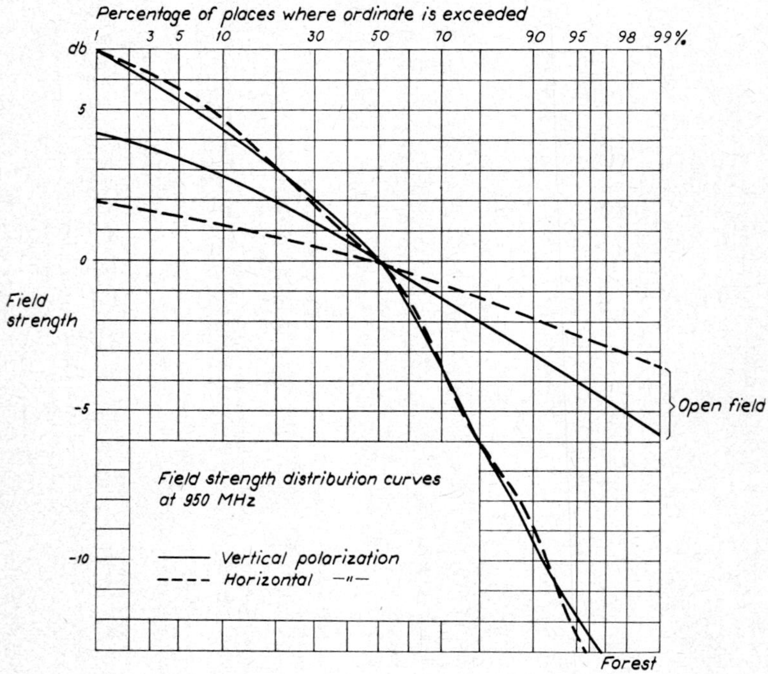

Fig. 5.

Field strength distribution curves for an open field and a forest for both types of polarizations at 950 MHz are given in Fig. 5. The curves are referred to ± 50 m distance from a fixed point and are drawn on a log-normal probability paper.

The curves for the forest tend to give a Rayleigh distribution while the curves for the open field are more of the log-normal type.

At a lower frequency, 35 MHz, the curves for both the open field and the forest are found to be log-normally distributed (Fig. 6).

At increasing frequency the total field is composed of a still greater number of field vectors which give an increasing tendency towards a Rayleigh distribution. Naturally this effect is most pronounced at rough and wooded terrain.

Though it would be possible to speak about the standard deviation the 10 % and 90 % levels are more severely defined. The difference between these levels has been determined as a function of frequency. As seen in Fig. 7 this difference is much greater with vertical than with horizontal polarization but at the highest frequencies in the range concerned even horizontally polarized waves are much affected by the vegetation.

In the forest the number of objects possible to affect horizontally polarized waves increases with the frequency to a greater extent than is the case with

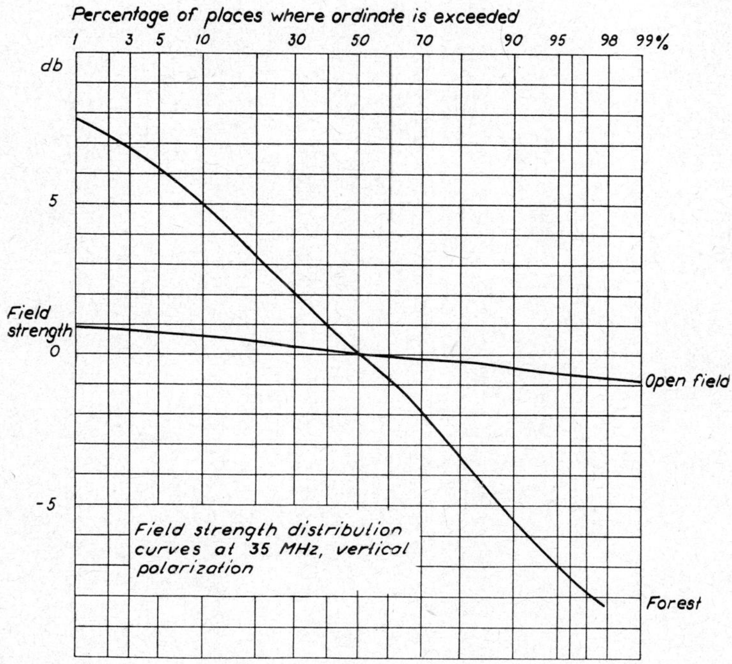

FIG. 6.

vertical polarization. In the open field, however, the signal is only influenced by the ground itself which seems to affect vertically polarized waves more than horizontally.

Even the field strength variations around a single tree are very great which is shown in [1].

In the parts of Sweden where the measurements were made there are mostly pine-forests. So no difference in field strength variations between seasons has been observed when the trees are and are not in leaf.

As snow flattens the inhomogeneities of the ground and thus variations from ground covered with snow might be expected to be less important than from bared ground, no such evidence has been clearly found. Therefore the fluctuations seem to derive from the trees and the underwood, the branches of which when covered with snow may form surfaces enabling the energy to be reflected and distributed in various directions or—which is rather probable at the lowest frequencies—the penetration depth is much larger than the thickness of the snow so the coverage of snow has negligible influence.

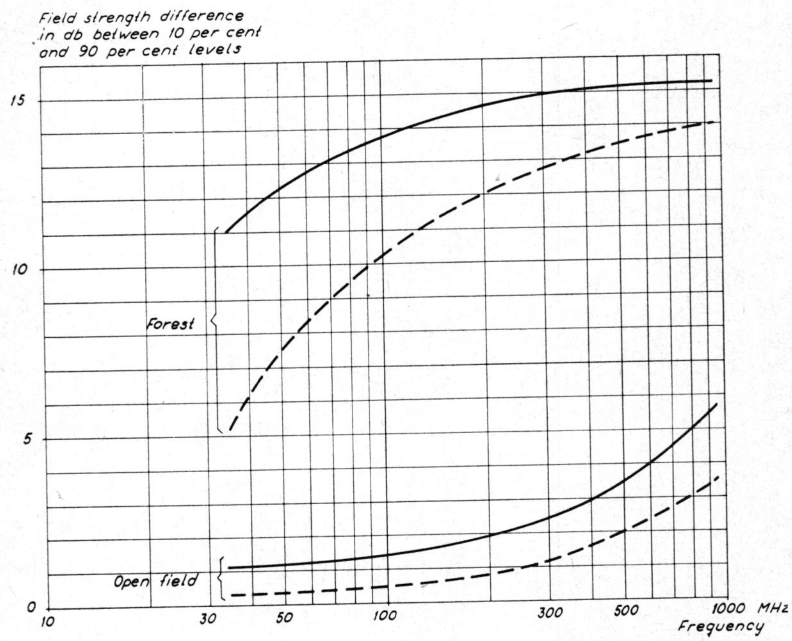

FIG. 7. Frequency dependence of local field variations in an open field and in a forest. Vertical polarization ———, horizontal polarization - - - -.

As mentioned before most of the measurements have been made in pine-forests where the trees were up to 20 m high. In such forests with little underwood the exponential distance attenuation factor over short paths is very small for frequencies below 1000 MHz, but in a forest with dense leafy underwood attenuation factors as high as 0·02 db/m at 30 MHz and 0·09 db/m at 1000 MHz at vertical polarization have been measured at very small heights and over short paths. The curve of the attenuation factor as a function of frequency on logarithmic scales seems to be rather linear. With raised antennas or for longer paths but still within the horizon the attenuation factor decreases.

V. Terrain Variations with one Antenna Elevated

In all the measurements earlier described both the transmitting and receiving antenna heights have been small. Some tests have been made to determine the height-gain curves for vertical polarization and 44 MHz at different distances up to 40 km and at heights up to 2000 m. These measurements were carried out with the transmitting equipment borne by a helicopter and the signal was received and recorded simultaneously in seven places chosen within an area of about 100 × 200 m. Tests were made with the transmitter both above land and above sea. In both cases the receiving equipment was placed on land, but near the shore in the tests when the transmitter was above sea. The receiving antenna heights were 1·5–2·5 m.

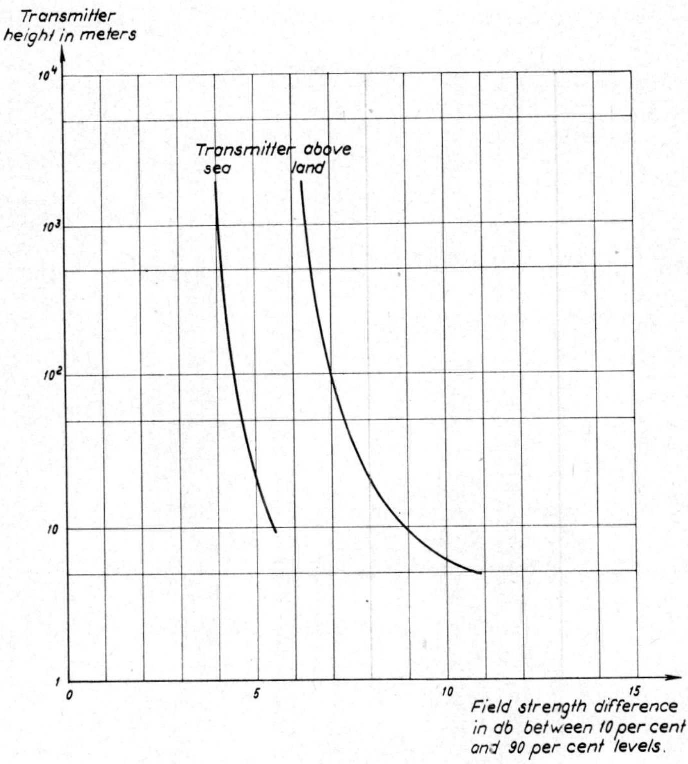

FIG. 8. Field strength terrain variations with transmitting antenna elevated at 44 MHz vertical polarization. Antenna heights of receivers 1·5–2·5 m.

Due to the terrain the recorded field strength curves at the different places show dispersion. It depends on the transmitting antenna height but does not seem to alter very much with the distance. The field strength difference in db between the 10 % and the 90 % levels of this dispersion is shown in Fig. 8 for both types of transmission paths as a function of transmitter height. This difference is much larger for the all over land path and the value for the lowest antenna height quite agrees with the values obtained in Fig. 7.

Over the land path the wave energy is transmitted both directly and by ground reflections through the vegetation and over the irregularities of the ground. At most heights of the transmitter the ground reflections occur near the receivers. Concerning the sea path the vegetation has less importance as the receivers were placed near the shore.

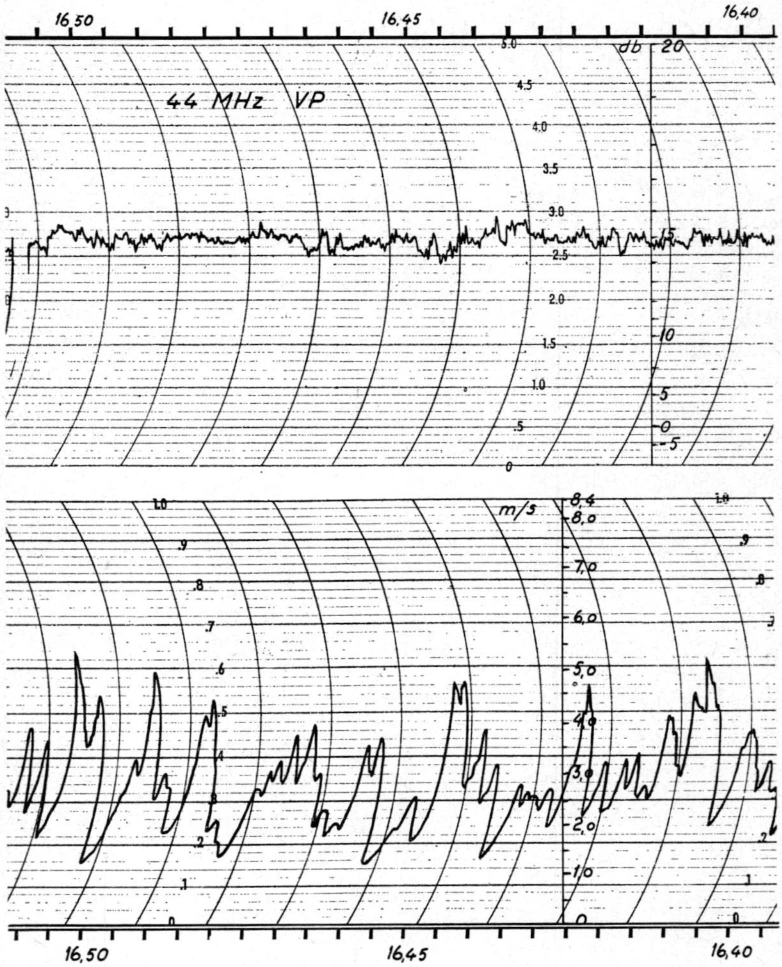

FIG. 9. Field strength variations due to wind velocity.

VI. SIGNAL VARIATIONS DUE TO THE WIND VELOCITY

Up to now only space variations in the received field strength appearing when the antennas are moved have been considered.

Between two fixed terminals with low antennas in a forest a type of signal variation due to the motion of the trees has been measured. The combination of direct and reflected waves is affected by the wind as the phase difference between the waves varies when the trees move. This fading phenomenon often amounts to

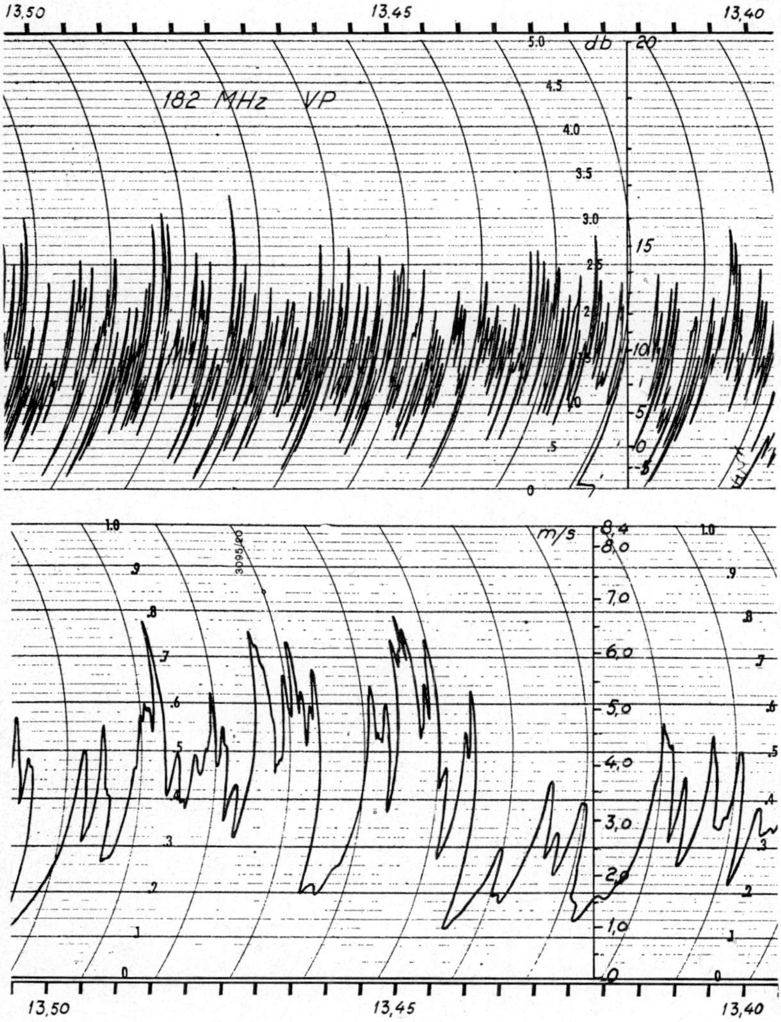

FIG. 10. Field strength variations due to wind velocity.

a considerable amount in amplitude at quite moderate winds and is most important at the highest frequencies in the range.

In Figs. 9, 10 and 11 typical recordings of these signal variations are shown. In each figure the upper curve gives the field strength and the lower curve the wind velocity at the same time interval. The scales for the field strength on one hand and the wind velocity on the other are the same in all the three figures. There is an interval of 30 sec between two consecutive time markings. Figure 9 shows the conditions at 44 MHz vertical polarization. The

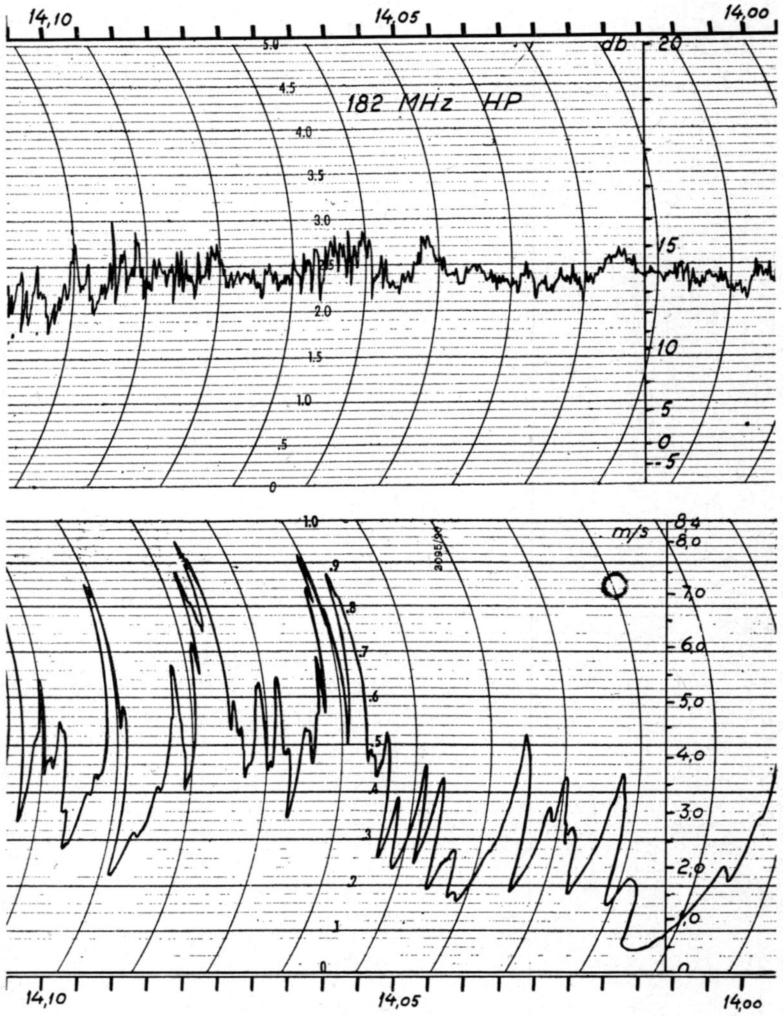

FIG. 11. Field strength variations due to wind velocity.

amplitude of the field strength variations is about 1·5 db at an average wind velocity of about 3 m/s. At a higher frequency, 182 MHz, vertical polarization, (Fig. 10) the variations are much more evident, and the amplitude amounts to about 15 db at wind velocities slightly above those in Fig. 9. Keeping the frequency at 182 MHz but changing to horizontal polarization (Fig. 11) there is a remarkable difference in the signal variation amplitude, which here only amounts to 3 db although the wind velocity sometimes reaches almost 8 m/s. In this figure at 14·06 (i.e., 6 min past 2) the effect of a change in wind velocity on the field strength record is seen.

Most of these measurements have not yet been thoroughly analysed. So for instance nothing about the type of distribution and the influence of the direction of wind compared to the path direction can be mentioned here.

It should be noticed that field strength curves measured in similar ways in open fields—and, at very calm weather, also in forests—have been found to be straight lines. As seen the effect of the wind is very small at 44 MHz, but even at 88 MHz it is apparent. The fading rate of the signal variations shows frequency dependence.

VII. Local Influence of Moisture in the Ground

VHF field strength recordings at vertical polarization over smooth ground have shown variations up to 14 db over the same paths at different seasons. The measured field strengths correspond to ϵ-values from about 3 to 30, as calculated from Equation (1).

Investigations have shown that the value of the effective dielectric constant is determined mainly by the water content of the ground and is comparatively independent of the type of ground.

In most parts of Sweden the ground moisture has an average annual variation with maxima in autumn and in spring and a low minimum in summer.

The influence of the ground moisture and thus the ϵ-value has been found to decrease with increasing frequency if the antenna heights are kept constant. At the lowest frequencies the range of communication sets with low antennas varies seasonally up to a ratio of more than 2 : 1, as the field strength obeys the inverse distance squared law.

In many cases it is required to know how the propagation conditions within a certain area have changed from one time to another. Then a method has been used based on the relation between w, the water content, and ϵ, the dielectric constant, of the ground. It is, however, difficult to determine the absolute value of w, but the following expression has been found to give a useful measure of the variations of the ground moisture content

$$w = C + \Sigma (p - e) \tag{4}$$

where p is the precipitation (mm); e is the evaporation (mm); C is the water content at the start of observation.

The quantities p and e are relatively easy to determine and the summation is made over a suitable time interval, e.g., every 24 h.

If at a certain time a field strength is measured the knowledge of the variation of ground water content makes it possible to predict the field strength at some time afterwards. The method has proved to give good results in time intervals of a few months.

A more thorough treatment of these problems is given in Reference [1]. This also applies to the influence of temperature and snow which will be discussed very briefly here.

The temperature coefficient of the dielectric constant of the ground is very small, so no direct influence of temperature can be expected. An analysis of field strength measurements from different seasons has shown no correlation with ground or air temperature. The influence of temperature is small even with the ground frozen but a covering of snow may cause a great decrease in the dielectric constant.

VIII. The Terrain Loss near the Antenna

In the preceding discussion an indication has appeared that only the terrain very near the antennas affects the field strength. This is similar to the influence of the dielectric constant of the ground.

The decrease in field strength caused by the vegetation and the irregularities of the ground can be regarded as a distribution of the wave energy. It is thus concentrated to a small area around the antenna and determined by the terrain there and the frequency. Further this loss is independent of distance.

This type of field strength reduction is the consequence of a scattering phenomenon, the energy is spread at random in different directions by trees and terrain irregularities. In spite of the similarity with the ground constant influence this ground wave loss does not increase the energy absorbed by the ground.

Some words might be said about the probable dimensions of this small area within which the terrain loss arises. To deal with this problem theoretically is difficult. But with a sufficient material of measurements it would be possible to estimate the size.

Measurements performed with a movable antenna along paths where sudden changes in vegetation (cf. Fig. 4) or ground constants (mixed paths) have occurred give a measure how far from the antenna the ground, wooded or not, has an influence on the received signal or in other words the antenna efficiency. The height of the trees and the type of terrain must be regarded in comparison with the wavelength. The ground constants as well as the vegetation are not of the same importance in all directions from the antenna.

It has been found that the electrical properties of the ground have the greatest importance in the direction of propagation from the antenna up to a certain distance Δd depending on the wavelength and the value of the dielectric constant. This distance

$$\Delta d = \lambda \cdot \frac{\epsilon_1 \cdot \epsilon_2}{\sqrt{(\epsilon_1 - 1)(\epsilon_2 - 1)}} \tag{5}$$

is recognized as the proportionality factor in Equation (1). The ground close to the antenna in the opposite direction does not seem to be of the same importance.

On the whole the measurements indicate that this area may have the shape of an ellipse, the major axis of which is coincident with the transmission direction. The antenna is then assumed to be in the focus farthest from the other terminal.

Let the Equation of the ellipse have the form

$$r = \frac{p}{1 - e \cos v} \tag{6}$$

where r = radius vector; $p = b^2/a$; $2a$ = major axis; $2b$ = minor axis; e = eccentricity; v = angle between major axis and radius vector.

The center of the vector r is considered to be the site of the antenna and when the angle v passes all values from 0 to 2π the area swept out by r is regarded as being equal to the area in question.

The value of the ellipse eccentricity must be due to the special conditions of the actual place, at least to some extent. If only the effect of the dielectric constant is considered r becomes equal to Δd in Equation (5) for $v = 0$.

The value of r when $v = \pi$ does not seem to depend much on the dielectric constant but is estimated to be a multiple n of the wavelength λ. Then the eccentricity is

$$e = \frac{\Delta d - n\,.\,\lambda}{\Delta d + n\,.\,\lambda} \qquad \Delta d > n\,.\,\lambda \tag{7}$$

which shows that e is independent of the wavelength. In most cases $\epsilon_1 = \epsilon_2 \gg 1$, then Equation (7) reads

$$e = \frac{\epsilon - n}{\epsilon + n} \qquad \epsilon > n \tag{8}$$

The value of n seems to be about 5 and a common average value of ϵ is 15, then $e = \frac{1}{2}$, $a = 10\lambda$ and $b = 5\sqrt{3}\lambda$. The area of the ellipse is $\pi ab = 50\pi\sqrt{3}\lambda^2$.

Although Equation (5) does not take into account the influence of the vegetation the calculated values seem to yield a fairly good measure useful for practical purposes within this frequency range.

IX. Concluding Remarks

In the preceding paragraphs each of the different conditions causing ground wave losses has been discussed separately. This is a trial to combine the effect of all these factors into a level diagram. It is not a common one, however, regarding the whole transmission path but only the characteristic energy losses that the wave is subjected to near the antenna. The field within a certain small area is built up from the following contributions:

(1) The median field strength is determined by:
- (a) the dielectric constant of the ground;
- (b) the terrain loss.

(2) The variations around the median value are determined by:
- (a) the local space variations within a small area due to the trees and underwood and the irregularities of the ground;
- (b) the wind velocity, a time depending factor superposed on the actual field strength value.

The diagram is shown in Fig. 12 where an example is given to point out the different local losses and variations to be considered every time when a signal is to be received by a low antenna placed in the terrain. The vertical axes may be regarded as having db-scales showing 20 log (F . d). As an example typical conditions at 182 MHz have been chosen. Within a certain small area, assume the dielectric constant to have an average value which inserted in Equation (1) gives 20 log (F . d) = 36. Then a terrain loss of 2·5 db is assumed. In this case this applies to vertical and horizontal polarization as well, and then the field strength median value has been determined. The local space variations are different for

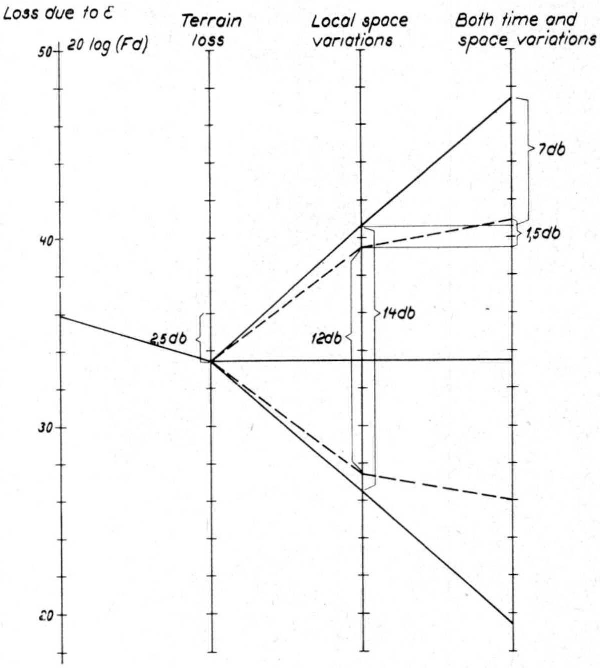

FIG. 12. Level diagram showing the characteristic ground wave features within a small area. Example: 182 MHz, vertical polarization ——— horizontal polarization – – – –.

vertical and horizontal polarization, ± 7 db and ± 6 db respectively around the median value. At any point within the considered area a time depending factor due to the wind velocity is superposed on the field strength value. In the example these time variations are much more emphasized for vertical than for horizontal polarization as shown above in Figs. 10 and 11. So the fourth axis of the diagram gives the actual field with time as well as space variations around the median value.

If a period of several months is considered also the median value varies especially at the lowest frequencies due to seasonal changes in the moisture content of the ground.

REFERENCE

1. Josephson, B. & Blomquist, Å. (1958). *Trans. Inst. Radio Engrs, N.Y. AP*–**6**, no. 2.

Comparison between Radio Wave Burst Emission on 3·2 cm and Contemporary Absorption in the Ionospheric D-region

O. HACHENBERG and H. VOLLAND

Heinrich-Hertz-Institut der Deutschen Akademie der Wissenschaften, Berlin

The time variation of the excessive absorption in the ionospheric D-layer during a sudden ionospheric disturbance is compared with the corresponding solar radio burst on 3·2 cm. wavelength. Using the continuity equation of the ionospheric electron density, the intensity variation of the ionizing radiation can be determined. It is found that this intensity variation is in qualitative agreement with the intensity variation of 3·2 cm. emission.

The critical electron density in the cm-range is $1{\cdot}1\ 0{\cdot}10^{12}$ electrons/cm^3 for 3·2 cm wavelength and $11{\cdot}10^{12}$ electrons/cm^3 for 1 cm wavelength. Comparing these data with the electron density within the solar atmosphere—e.g., the model of Boehm-Wittense [2]—we notice that the origin of emission of these wavelengths is probably the lower chromosphere.

On measuring solar bursts within the frequency interval between 1 cm and 3·2 cm, we consequently can see down into the gaseous volume causing the flare. Therefore the range of cm wavelengths is very well qualified for the comparison of the solar radio wave emission with optical observations or with the ionospheric absorption caused by UV radiation.

Presuming this, we have made a comparison between the burst measurements at 3·2 cm and the measurements of excessive absorption in the ionospheric D-region during a sudden ionospheric disturbance (S.I.D.).

The exact determination of the absorption in the ionospheric D-region during a S.I.D. is difficult because of the fadings and the relative high amount of absorption. Thus the non-deviative absorption was determined simultaneously from three methods:

(1) Measurements with a pulse transmitter on 3·86 Mc/s.
(2) From the f-min of ionospheric sweep frequency soundings (both these measurements were made at Juliusruh/Rügen).
(3) From signal strength measurements at oblique incidence (transmitting station at Stuttgart—receiving station at Neustrelitz; frequency 6·03 Mc/s).

In Figs. 1 and 2 two examples show the course of 3·2 cm radio burst emission and of additional absorption derived with the different methods described above. (Measurements of radio emission of the sun are made in Berlin-Adlershof.)

A median curve is fitted to the observed points of excessive absorption during the S.I.D.

From comparing the results, we can state the following characteristics:

(1) Both curves are starting at the same time.
(2) The ascent of the absorption curve is slower than the ascent of the radio burst, the maximum of absorption being 5 to 10 min later than the burst maximum, which has been reached few minutes after burst begin.

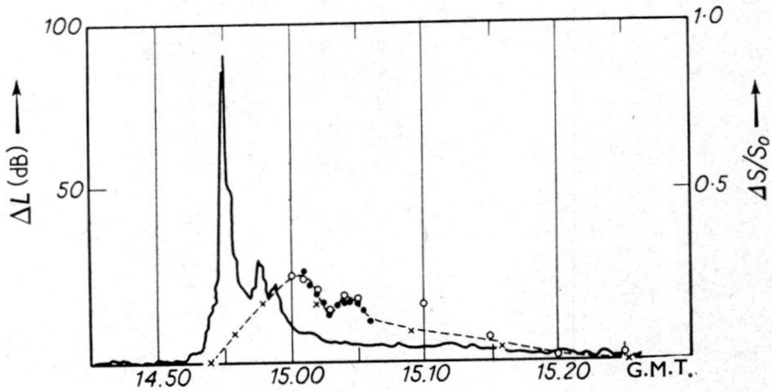

FIG. 1.

• pulse reflection method 3·86 Mc/s.
○ f-min. method.
× oblique reflection (Stuttgart broadcasting station 6·03 Mc/s).
1958, March 20.
—— burst $\lambda = 3{\cdot}2$ cm.
– – – additional absorption in the D-region, related to 3·86 Mc/s.

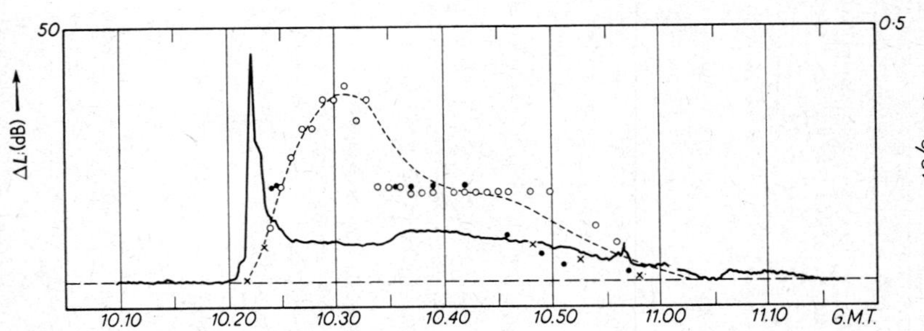

FIG. 2.

• pulse reflection method 3·86 Mc/s.
○ f-min. method.
× oblique reflection (Stuttgart broadcasting station 6·03 Mc/s).
1958, March 21.
—— burst $\lambda = 3{\cdot}2$ cm.
– – – additional absorption in the D-region, related to 3·86 Mc/s.

(3) After reaching the maximum both curves generally descent monotonously, the absorption descending generally slower than the burst, except in the case of a " post burst increase," which is not always present.

(4) The end of excessive absorption is later than the end of the burst.

Only in the case of a "post burst increase" the end of the burst and the end of the excessive absorption are almost at the same time.

An analysis of the form of burst curves on 3·2 cm, as published elsewhere [3], has shown that the short emission peaks are a typical element of the burst curve, while the "post burst increase" is not always seen.

The ionospheric equation of recombination

$$\frac{dN}{dt} = q - \alpha N^2$$

was solved with plausible data for the case of a simple eruption. Figure 3 shows a theoretical increase of ion production q and the calculated increase of electron

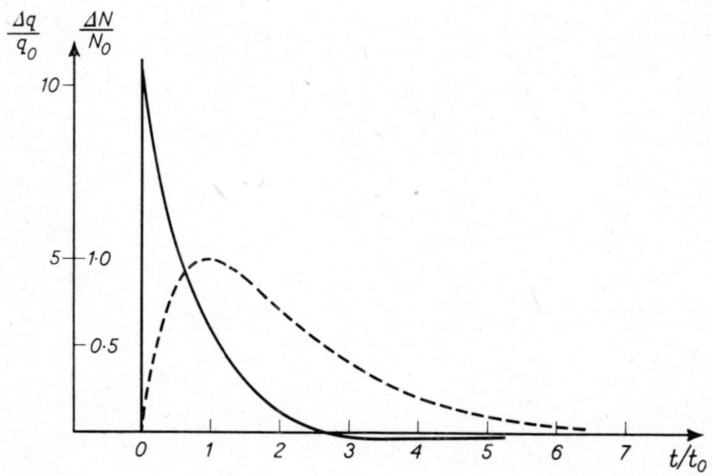

FIG. 3. Excessive ion production during a solar flare Δq (full line) and theoretical increase of electron density in the ionospheric D-layer ΔN (broken line). (q_0 and N_0 are the normal values before the begin of solar flare.)

density in the maximum of the D-layer N. If we assume a simple model of a Chapman-layer, the electron density N is proportional to the non-deviative absorption L and the ion production q is proportional to the causing solar emission S. The curves in Fig. 3 are therefore equivalent to the eruptive UV-emission and the contemporary excessive absorption.

The comparison of the calculated curves with the curves in Figs. 1 and 2 shows that the observed initial phase qualitatively corresponds with the theoretical example in Fig. 3, while the apparent parallel decay of absorption and eruption may be caused by the "post burst increase" which seems to be an additional effect to the initial eruption.

Theoretically there is

$$\mu_{max} = \frac{e \, . \, n_{max}}{\alpha \, . \, N_0 \, . \, t_0}$$

with

$\mu_{max} = \left(\frac{\Delta S}{S_0}\right)_{max}$ = maximal increase of UV-emission.

$n_{max} = \left(\frac{\Delta L}{L_0}\right)_{max}$ = maximal increase of excessive absorption.

t_0 = time difference between the maximum of the UV-emission and the maximum of absorption.

α = recombination coefficient.

N_0 = electron density in the maximum of the normal undisturbed Chapman-layer representing the D-region.

e = the base of the natural logarithmus.

Figure 4 gives (μ_{max}) 3·2 cm versus $\frac{n_{max}}{t_0}$.

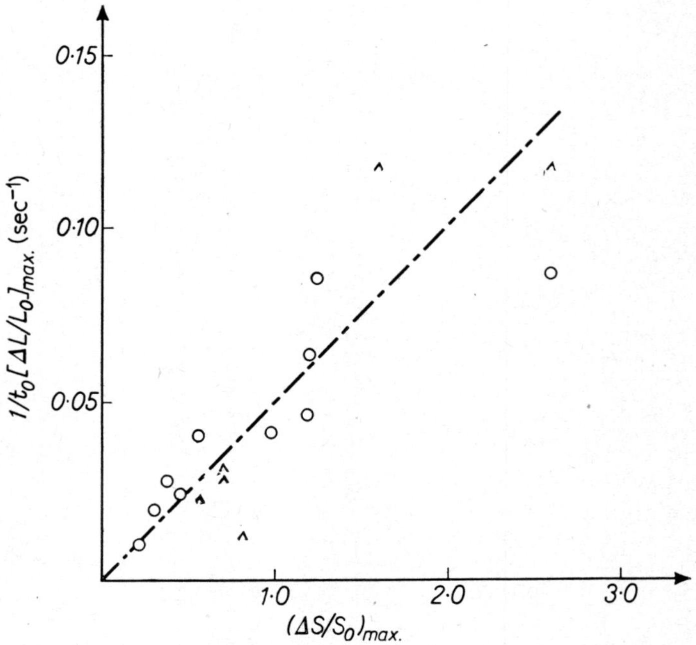

FIG. 4. Maximal 3·2 cm radio burst emission $(\Delta s/s_0)_{max}$ versus comtemporary maximal excessive absorption in the D-region $(\Delta L/L_0)_{max}$ divided by t_0 · (t_0 is the time difference between burst maximum and absorption maximum.)

It shows some linearity for a number of investigated flare effects. From the given value of $\alpha N_0 = 5 \cdot 10^{-4}$ sec^{-1} [after Appleton [1]] and the slope of the curve in Fig. 4 we may conclude that the maximum emission in the UV-range is approximately 30 times greater than the maximum at 3·2 cm radio wave emission.

The examples presented here permit the conclusion that the qualitative course of bursts at 3·2 cm wavelength is similar to the course of emission on those wavelengths which are responsible for the ionisation of the D-region and therefore for the excessive absorption during a S.I.D.

The complete paper of this work will be published elsewhere [4].

References

1. Appleton, E. V. (1953). *J. atmos. terr. Phys.* **3,** 282.
2. Boehm, Wittense (1955). *Z. Astrophys.* **36,** 145.
3. Hachenberg, O. (1958). *Z. Astrophys.* **46** 67.
4. Hachenberg, O. & Volland, H. (1959). *Z. Astrophys.* **47,** 69.

A Simple Air/Ground Meteor Burst Communication System

PAUL R. DROUILHET, JR. and KLAUS W. OTTEN

Communication and Navigation Laboratory, Wright Air Development Center, Wright-Patterson Air Force Base, Dayton, Ohio

A meteoric scatter communication system has been designed to provide reliable low data rate communications between a fixed station and a mobile or airborne station. Considerable average power economies are achieved, particularly at the mobile station, by utilizing a technique for probing for the existence of a useable meteor trail basically different from that of any existing operational or experimental meteoric scatter link.

I. INTRODUCTION

Up to the present, the only means available for two-way air/ground communication beyond line-of-sight ranges has depended on the reflection of high-frequency radio waves by the ionosphere. Unfortunately, the various forms of ionospheric disturbances cause this propagation mode to be undependable; this situation is particularly acute in arctic areas, where the relatively high incidence of propagation disturbances plus the lack of a network of UHF or VHF ground facilities make it doubly important that some alternative form of communications be available.

Rather than attempt to obtain the desired 100 % reliability for the total communications requirement of the air/ground link, it is worthwhile to consider a system which will provide a minimum of communications capability, but will provide this minimum with extreme dependability. Such a system would be used to transmit coded high-priority messages, e.g., position reports, commands, etc., when the normal higher capacity channels are not useable. If necessary in the system design, long term channel capacity would be sacrificed in order to increase the reliability of transmission of a short message in a minimum time.

Because of its limited capacity such a system would supplement rather than replace the normal communications facilities of an aircraft. It is, therefore, doubly important that its size, weight, and power consumption be kept to a minimum so that it can be installed in the modern high-performance aircraft which, by the nature of their mission, need it most.

II. METEOR BURST PROPAGATION

To provide the required highly-reliable back-up communications, a form of propagation must be used which is (relatively) insensitive to ionospheric disturbances. This, in combination with the requirement for long range, limits the choice to low frequency or VHF scatter. Although with large ground installations reliable ground-to-air low frequency communications can be attained, the high power and large transmitting antenna required make LF impractical for the air-to-ground link.

An ionospheric scatter link likewise requires higher power and larger antennas than are practical for general airborne application. However, the scattering of signals from individual trails of "larger" meteors produces short periods during which the path attenuation falls well below that of the normal ionospheric scatter path, thus providing an intermittent link requiring only moderate transmitter powers and antenna gains. Because the communications path is open only for short intervals, of the order of 0·1 to 1·0 sec. at a time, special techniques are necessary to transmit information efficiently over this path.

At the present time, several point-to-point communications links are in operation using meteor burst propagation for the transmission of digital data and teletype. Because of the intermittent nature of the communications path, means must be provided to sense when the path is open and key the system accordingly. In the existing links both stations transmit continuously on slightly different frequencies; the reception at either station of the other's probe signal initiates transmission of the message. Message transmission continues until the received signal fades out, at which time it ceases and the system resumes probing.

III. The Air/Ground Meteor Burst System

Several factors make this type of operation undesirable for an air/ground link. Continuous transmission from the aircraft expends much more power than is actually required for the communication process, thus creating a severe cooling problem and requiring a larger transmitter. More important, the continuous radiation of a signal on a preassigned frequency is not consistent with the security requirements of many military aircraft operations. Simultaneous transmission and reception on two closely spaced frequencies cannot be easily accomplished in an aircraft; in ground installations sufficient isolation may be obtained by physically separated transmitting and receiving antennas, or by bulky coaxial line filters in conjunction with a single antenna. Neither of these solutions are practical for an airborne station.

In order to overcome these difficulties, a different mode of operation has been devised for the Air/Ground Meteor Burst System. Link probing is done exclusively by the ground station; the airborne station transmits only when a communication path exists and a message is to be sent. As an added benefit in some applications, this type of operation allows both ground and airborne station to operate on the same frequency.

Figure 1 shows the nature of the probe signal transmitted by the ground station; each interrogation pulse consists of a synchronizing signal, selective call, and a link command code. During the interval between interrogation pulses the ground station listens for a reply from an airborne station. This interval must be equal to at least twice the maximum expected link propagation time plus the time necessary for recognizing the reply signal. Because of the limited maximum range of meteor burst propagation, the maximum expected link propagation time may be determined quite accurately.

At the instant of occurrence of a meteor trail which produces a communications path between the two stations, the system may be in one of four states depending on which, if either, of the stations desires to transmit a message. The sequence of link operations for each initial state can be best analyzed by considering each state separately.

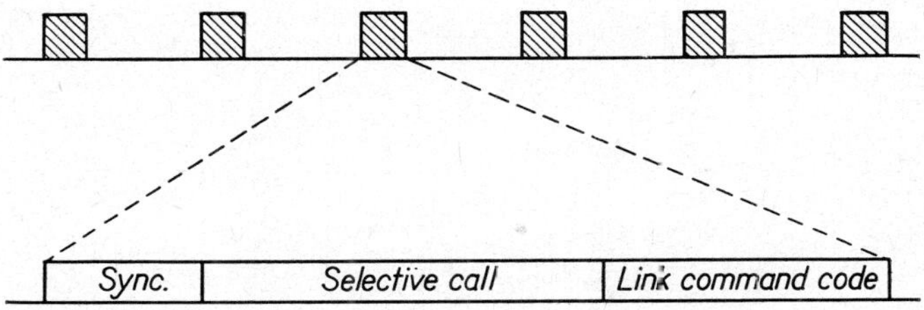

FIG. 1. Probe signal.

1. *Neither station desires to transmit a message.* The probe signal transmitted by the ground station contains a general call sign plus the link command code "Ready-to-Receive, Ground" (RTRG). The airborne station does not reply.

2. *The airborne station desires to transmit a message.* See Fig. 2. Again the probe signal contains the general call sign and the link command code RTRG. Immediately on receipt of this signal, the airborne station transmits a preamble consisting of the synchronizing signal, the aircraft call sign, and the link command

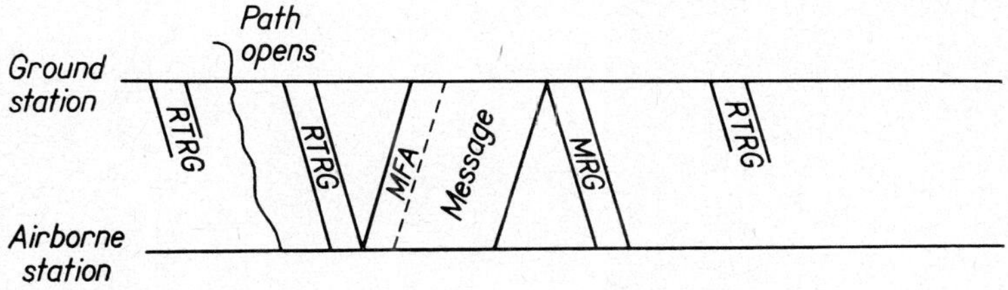

FIG. 2. Air to ground message transmission.

code "Message Follows, Aircraft" (MFA); this preamble is followed by the message of 150 bits (100 information bits plus 50 added bits to provide error detection). If the message is received completely and with no detected errors the ground station transmits a confirmation signal consisting of sync, aircraft call sign, and the link command code "Message Received, Ground" (MRG).

If the aircraft receives the confirmation signal, it transmits the next message if it has one or, if not, makes no further reply. If the confirmation signal is not received, the airborne station assumes that the ground station did not receive the message. The same message is then retransmitted on next burst.

When the ground station receives no reply to one of its transmissions, it assumes that the link is broken and resumes probing.

3. *The ground station desires to transmit a message.* See Fig. 3. The ground station now transmits the call sign of the aircraft to be contacted followed by the link command code "Ready-to-Transmit, Ground" (RTTG). The aircraft replies with its call sign and "Ready-to-Receive, Aircraft" (RTRA) on receipt of which the ground station transmits a preamble including "Message Follows, Ground" (MFG) followed by the message. After confirmation from the aircraft, "Message Received, Aircraft" (MRA), the ground station transmits another preamble and message to the same aircraft or resumes probing.

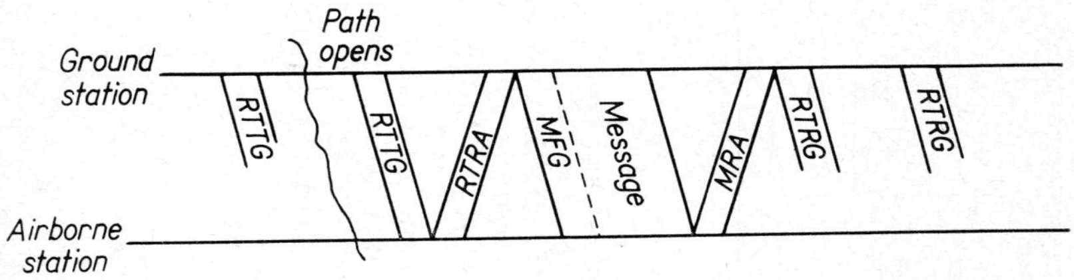

FIG. 3. Ground to air message transmission.

4. *Both stations desire to transmit messages.* In this case priority is arbitrarily assigned to the airborne station, which replies to the RTTG probe with MFA and its message. If this is received correctly by the ground station, the latter transmits MFG and its message, implying confirmation; the exchange continues until all messages are transmitted or the link is broken.

Whenever a received message is incomplete or contains errors, the link is considered broken and the ground station resumes probing.

IV. System Design Considerations

In selecting certain of the system parameters for optimum system performance it is most important to keep in mind the primary design objective: to provide a limited back-up communications capability with very high short-term reliability. This consideration must take precedence over maximization of long term message handling capacity.

A. Operating Frequency Range

It is well known that as the operating frequency is increased, the duty cycle of a meteor burst link decreases; to maintain a specified transmission rate, the transmitter power must be increased with frequency. Conversely the effect of the ionosphere, and particularly the severity of ionospheric disturbances, decreases markedly with increasing frequency. The optimum operating frequency will be one high enough to avoid in as much as possible increased absorption during disturbed periods, but low enough to provide useful information rates with reasonable transmitter power.

Initial system tests will be conducted in the 50 Mc region. Concurrently results obtained from test programs being conducted by other agencies will be used to determine whether this or some other frequency range will be most satisfactory for operational use.

B. Transmission Data Rate

For a specified bit error rate at the detector output the signal power required at the receiver input increases in proportion to the data rate (in bits per second). However, if the data rate is decreased to permit the use of lower signal levels the burst length required to complete a message transmission increases. Figure 4

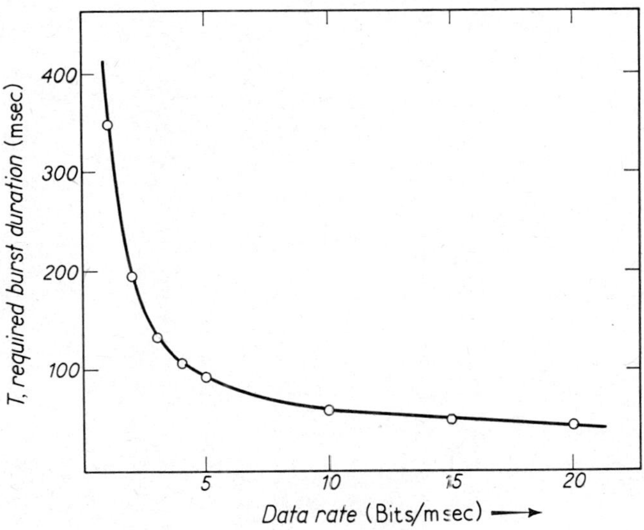

FIG. 4.

shows minimum burst length versus bit rate for a ground-to-air message transmission. A path delay of 7 msec is assumed corresponding to the approximate maximum delay encountered in a meteor burst link. Each combination of synchronizing signal, call sign, and link command code requires 43 bits; the message itself requires 150 bits. To complete the message transfer requires four transmissions, with their attendant path delays, comprising a total of 322 bits. Thus, the time T in milliseconds for a ground-to-air message transmission is

$$T = 28 + \frac{322}{BR}$$

where BR is the data rate in bits per millisecond.

The curve shown in Fig. 4 must be combined with data on Time above Threshold versus Threshold Setting obtained from long term meteor burst propagation studies; the result will be a determination of the data rate which will permit utilization of the maximum number of bursts, with emphasis on the periods of disturbed ionospheric conditions. Sufficiently instrumented propagation studies are presently in progress to provide the data needed for this determination.

C. Transmitter Power and Duty Cycle

In the airborne components of this system minimization of size, weight, and cooling requirements is of prime importance. The most critical item in this regard is the transmitter power amplifier and its associated power supplies. In order to attain the desired communication capability it is expected that a transmitter power output of the order of 5 kw will be required. A transmitter designed to supply such power continuously would be prohibitive in size and weight; however, by taking full advantage of the low transmitter duty cycle a much smaller unit can give satisfactory performance.

The considerations in designing for low duty cycle are very different than in the case of radar transmitters where megawatt pulses may be obtained from kilowatt transmitters. Although the long-time average duty cycle is low, the meteor burst transmitter must be capable of operating for periods of several hundred milliseconds at a time. Contrary to the radar case it is not feasible to store the energy required for such a transmission in electrical form; rather the power supply must be capable of supplying the full peak power with good regulation. Thus the peak loads are placed directly on the aircraft generators.

Probably the most critical component in designing for transmissions of this length is the final amplifier tube. The passive circuit elements will in general have relatively long thermal time constants; internal tube temperatures, however, will very rapidly rise to values exceeding their allowable maximums. In addition peak cathode current limitations of most tubes prevent taking full advantage of the low duty cycle.

The optimum tube for this application would utilize special cathode structures permitting high peak currents with relatively low heater power, and have elements designed to withstand the heating which occurs during relatively long pulses. With such a tube peak to average power ratios of 10 or 20 : 1 should be obtainable. Using currently available tubes average power ratings must be 20–50 % of the desired peak powers; with such tubes the power efficiency will be low, the filament power alone exceeding the average power output at the low duty cycles involved.

D. Modulation Technique

In order to reduce to a minimum the required transmitter power output full advantage must be taken of improved modulation and detection techniques to lower the receiver threshold. First system tests will employ frequency shift keying with "matched filter" (i.e., synchronously gated integrator) detection. However, additional development will attempt to exploit the advantages of phase modulation, using bit-to-bit phase comparison for detection [1], to further reduce the receiver threshold.

E. Antennas

The operational problem of properly orienting a directional antenna plus the obvious size and shape limitations at the frequency range of interest preclude the use of directional antennas on most aircraft. About the best which can be hoped for is an essentially omnidirectional airborne antenna, with a minimum of deep nulls in directions of interest.

The lack of gain in the airborne antenna may be made up to only a limited extent by increasing the gain of the ground antenna. Aside from the operational problem of maintaining an aircraft in the "field of view" of a high gain antenna, narrowing the beamwidth (i.e., increasing the gain) beyond a certain point means that some otherwise useable meteor trails will now not be illuminated. An optimum compromise will maintain maximum communication capability without making unduly difficult the problem of orienting the antenna. It is presently thought that an antenna with approximately a 20° horizontal beam width and 20 db gain will provide close to an optimum pattern.

V. Summary

The proposed technique for air/ground meteor burst communications overcomes most of the drawbacks of the more conventional systems. As this system is still in the early development stage, it is not possible to quote performance figures or even final system parameters; however, it is felt that an optimization of these parameters will result in a system filling a serious gap in existing air/ground communications.

Reference

1. Lutz, S. G., Losee, F. A. & Ladd, A. W. Pulse Phase-Change Signaling in the Presence of Ionospheric Multipath Distortion. Presented at the International Conference on Electromagnetic Wave Propagation, Liège, Belgium. October 1958.

Backscatter-Beobachtungen an Telegraphie-Signalen

B. Beckmann und K. Vogt

Fernmeldetechnisches Zentralamt, Darmstadt

Die typischen Merkmale eines Tagesganges der Rückstreu-Feldstärke von Kurzwellen-Telegraphiesignalen werden mit Hilfe eines Filmes erläutert. Für die Darstellung der Rückstreu-Amplituden auf dem Schirm einer Braunschen Röhre wird die Ringmethode gewählt. Aus der zeitlichen Änderung der Zahl der Ringe und ihrer Durchmesser können Rückschlüsse auf die jeweili gen Ausbreitungsbedingungen gezogen werden.

Gestatten Sie mir, kurz zu berichten über Backscatter-Beobachtungen an Telegraphie-Signalen. Die Deutsche Bundespost führte Versuche durch, um diese Beobachtungen für eine Abschätzung der Ausbreitungsbedingungen während des kommerziellen Betriebes auf oder in der Nähe der Sendestation zu benutzen.

Telegraphie-Signale haben eine längere Dauer als die Laufzeit der Backscatter-Echos. Deshalb ist es notwendig, die Backscatter-Amplitude der Randkurve des in die Zwischenfrequenz transponierten Signals zu entnehmen. Durch Interferenz zwichschen dem Bodensignal und dem Backscatter-Signal entstehen Stufen in der Randkurve. Wegen der unregelmäßigen Folge und Zeitdauer der Telegraphie-Signale ist es nicht möglich, eine Synchronisation in der Amplituden-Zeit-Darstellung auf dem Schirm des Kathodenstrahl-Oszillographen zu erreichen. Aus diesem Grunde benutzten wir eine neue "Ringmethode" für die Beobachtung der Backscatter-Amplitude von Telegraphie-Signalen, die keine Synchronisation benötigt. Wenn man die Zwischen-frequenz des Empfängers mit 90° Phasenverschiebung auf die vertikalen und horizontalen Platten des Kathodenstrahl-Oszillographen legt, werden die Amplituden des Signals als Kreise auf dem Schirm abgebildet.

In der linken Spalte auf Bild 1 sehen Sie Backscatter-Messungen mit kurzen Impulsen. 1a z.B. zeigt einen einfachen Backscatter-Sprung über die F-Schicht. Links sehen Sie das Bodensignal, rechts mit einer bestimmten Laufzeitverzögerung das Backscatter-Paket. 1b zeigt ein Beispiel für einen Mehrfach-Sprung, hier zweifach. 1c zeigt den Fall ohne Backscatter. Wenn der Sender mit normalen Telegraphie-Signalen getastet ist, ergeben sich in der mittleren Spalte die entsprechenden Bilder d-f. Unten (Bild f) ist ein Telegraphie-Signal ohne Backscatter gezeigt. Oben (Bild d) und in der Mitte (Bild e) können Sie die erwähnten Sprünge in den Randkurven sehen. Die Amplitude am Ende der Signale stellt Backscatter dar. Die entsprechenden Ringdarstellungen sind in der rechten Spalte g-h gezeigt. Der Durchmesser des inneren Ringes zeigt die Backscatter-Amplitude.

Wir versuchen, die ionosphärischen Ausbreitungsbedingungen durch die Backscatter-Amplituden des ersten Sprunges auf der Seite der Sendestation

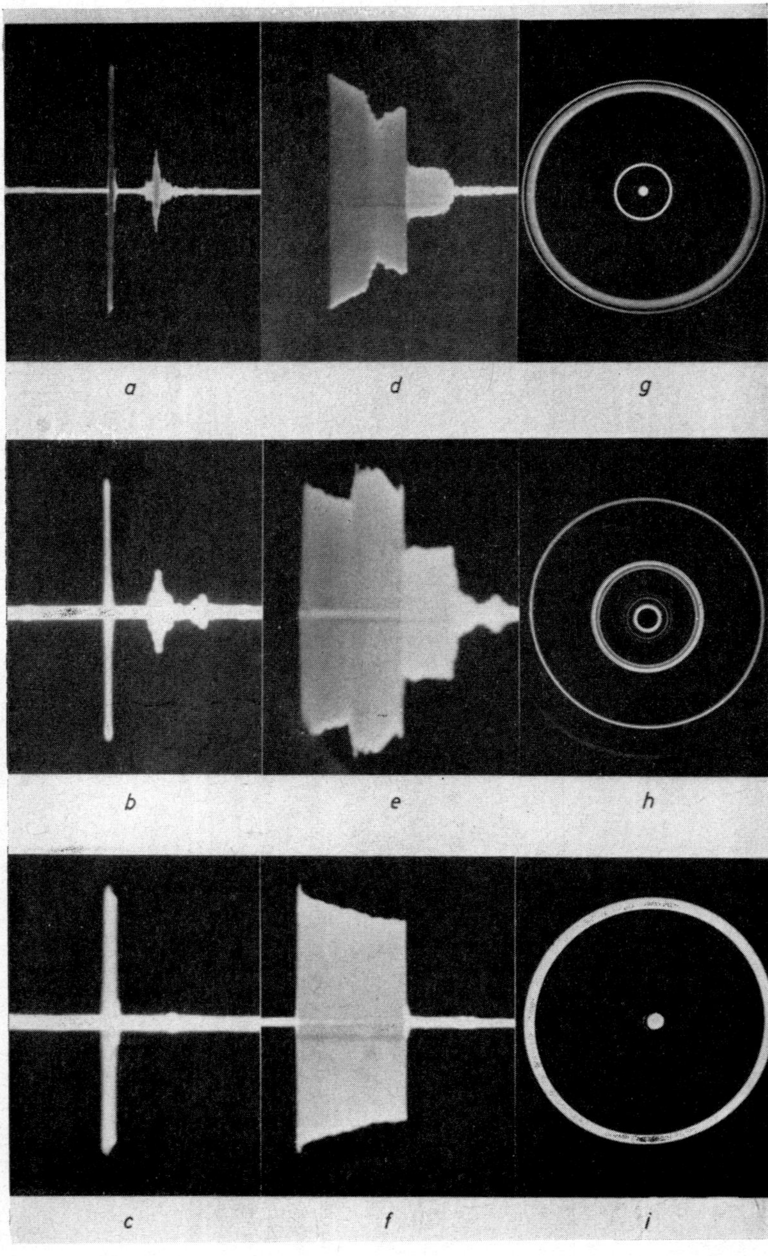

FIG. 1.

abzuschätzen, die wir als Ringdurchmesser beobachten. Die Versuche zeigen, daß normalerweise die Amplitude mit einer Abnahme der Sprungentfernung und der ionosphärischen Absorption wächst. Da beide Daten die Ausbreitungsbeding-

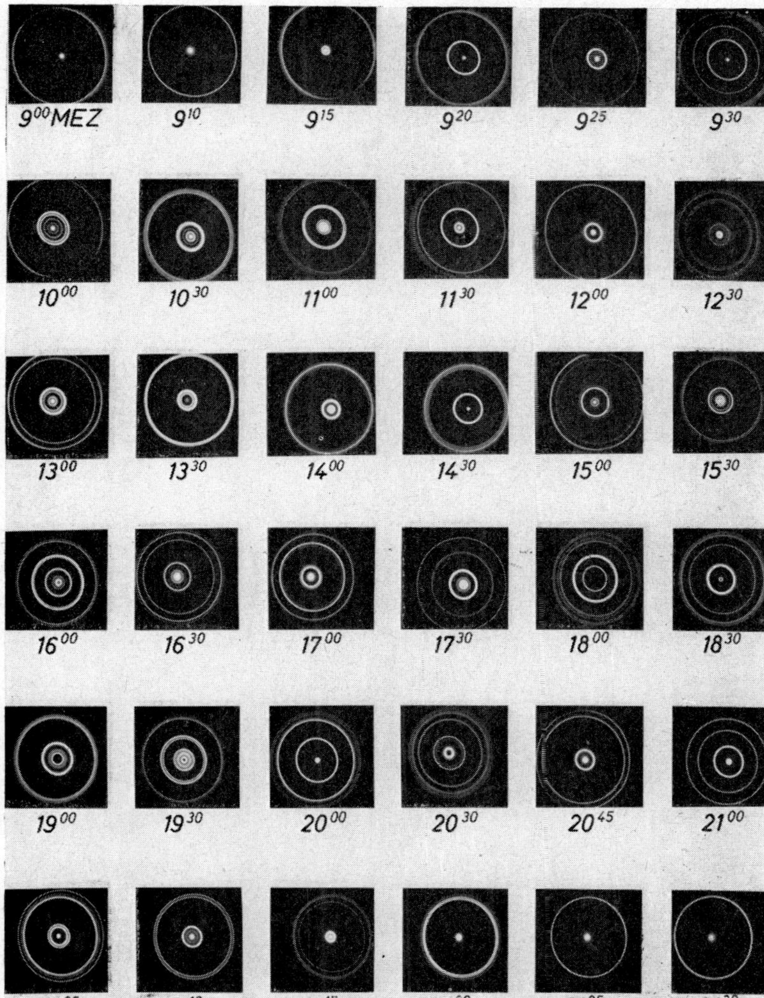

FIG. 2.

ungen bestimmen, können wir die Bedingungen als Funktion der Betriebsfrequenz und der Zeit abschätzen.

Die tägliche Änderung kann folgendermaßen gekennzeichnet werden: Wenn die kritische Frequenz der F-Schicht in den Morgenstunden wächst, ist das erste Ansteigen der schwachen Backscatter-Amplitude zu bemerken, zu einer Zeit, wo die Betriebsfrequenz ungefähr der MUF für etwa 3000 km Entfernung entspricht. Wenn die kritische Frequenz weiter anwächst, tritt zunächst gleichzeitig eine Zunahme der Backscatter-Amplitude ein, und es werden abhängig von der Übertragungsrichtung auch Mehrfachsprünge sichtbar. Gegen Mittag macht sich die ionosphärische Absorption mehr oder weniger stark bemerkbar. In den Nachmittagsstunden, wenn die Absorption nachläßt, erscheinen wieder Mehrfachsprünge, die um Mittag wegen zu hoher Absorption nicht vorhanden waren. Nähert sich in den Abendstunden die Betriebsfrequenz der MUF für etwa 3000 km Entfernung, verschwindet die Backscatter-Amplitude. Möglicherweise ist noch ein Restwert vorhanden, hervorgerufen durch Backscatter-Effekte im E-Gebiet.

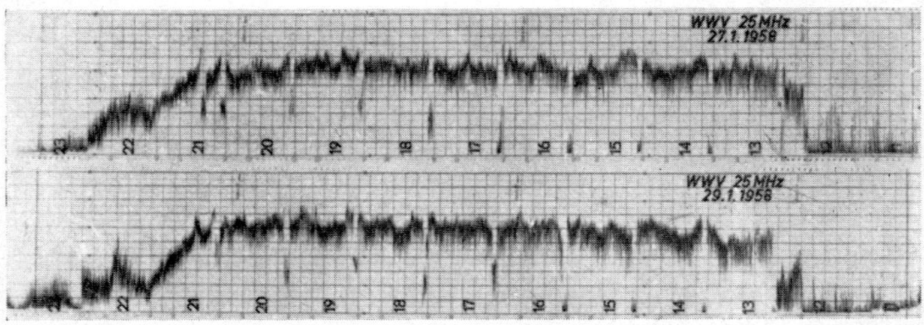

FIG. 3.

Im folgenden Film sehen Sie ein Beispiel für diese zeitliche Veränderung der Backscatter-Amplitude. Ein Sender unserer Betriebsstation Bonames in der Nähe von Frankfurt/Main wurde auf der Frequenz 25·4 MHz mit Telegraphie-Signalen getastet. Es wurde eine Rhombusantenne für die Richtung Nordamerika benutzt (Azimut 294°). Die Backscatter-Beobachtungen wurden auf der Empfangsstation Eschborn durchgeführt in 10 km Entfernung von der Sendestation, wobei eine Rhombusantenne mit demselben Azimut verwendet wurde.

Der erste Teil des Films wurde am Morgen des 29 Januar aufgenommen und der zweite Teil, der die Entwicklung am Nachmittag zeigt, am 27 Januar 1958. Alle ein bis zwei Minuten wurde ein Ringbild aufgenommen. Der Film zeigt die zeitliche Veränderung der Ringbilder nicht kontinuierlich, sondern sprungweise in Zeitabständen von einer bis zwei Minuten. Jedes Bild dauert etwa eine halbe Sekunde, der Film hat so eine Dauer von etwa sieben Minuten für den Zeitraum vom Morgen bis zum Abend. Fig. 2 zeigt Bildausschnitte.

Die drei bemerkenswerten Phasen des Backscatter, die Sie in Bild 1 sahen, sind in den entsprechenden Teilen des Films eingefügt. Das erste Auftreten von Backscatter zeigt sich um 10·15 MEZ, etwa zwei Stunden vor dem Aufgang der Frequenz 25 MHz für die Verbindung Frankfurt/Main-Nordamerika (WWV), was etwa der Zeitdifferenz zwischen den beiden Reflexionspunkten entspricht. Nach dem Anwachsen der Backscatter-Amplitude, d.h. des inneren Ringdurchmessers im Film, als Folge der wachsenden Grenzfrequenz sehen Sie den Einfluß der Mittagsdämpfung. Die Backscatter-Ringe werden etwa um 12·30 MEZ wieder etwas kleiner. Am Nachmittag nach 16·00 MEZ erscheinen als Folge der abnehmenden Dämpfung Doppelsprünge. Sie sehen gleichzeitig zwei Backscatter-Ringe mit wachsendem Durchmesser. In den Abendstunden verschwindet die Backscatter-Amplitude um 21·15 MEZ gleichzeitig mit den Signalen des oben erwähnten Senders WWV 25 MHz, verursacht durch den Sonnenuntergang an dem uns nächstgelegenen Reflexionspunkt. Die äußeren Ringe, die durch die Bodenwelle und durch die Interferenz zwischen Bodenwelle und Backscatter entstehen, werden zeitweilig größer als der Schirm des Kathodenstrahl-Oszillographen und sind deshalb in einigen Fällen im Film nicht sichtbar. Vor Beginn des Films möchte ich Ihnen noch die erwähnten Registrierungen von WWV in Bild 3 zeigen.

Die hier erläuterte einfache Backscatter-Beobachtungsmethode kann auf kommerziellen Stationen für die Abschätzung der Ausbreitungsbedingungen und für die Wahl der optimalen Betriebsfrequenz von Nutzen sein.

Atmospheric Bending of Radio Waves

B. R. Bean

National Bureau of Standards, Boulder, Colo.

Meteorological data from many diverse climates and seasons indicate that the refractive index gradient over the first kilometer above the earth's surface is highly correlated with the surface value of the refractive index. Further, 60 % of the bending of radio rays passing completely through the earth's atmosphere is accomplished in the first kilometer above the earth's surface. These observations have led to the development of several models of the height distribution of the refractive index that yield more realistic values of radio ray bending than the effective earth's radius approach and have the further advantage of being adjustable for any season or location for which standard surface weather observations are available.

I. Introduction

The purpose of the present paper is to present several models of the height distribution of atmospheric refractive index that should be of use in radio propagation problems. As a result of considering the average refractive index structure of the atmosphere as well as results of recent statistical studies of radio ray bending, it has been found that these models may be expressed in terms of the value of the refractive index at the earth's surface.

II. Background

The effects of atmospheric refraction of radio waves may be expressed in many ways. A convenient method, often used in classical astronomy, is to specify the atmospheric refraction in terms of the angle through which a radio ray turns as it passes through the atmosphere. This angle, called the bending and designated as τ, is simply that angle formed by the intersection of the tangents to the ray at the two points being considered. The geometry of radio ray bending is shown on Fig. 1.

The bending, $\tau_{1,2}$ of a radio ray passing between levels in the atmosphere at which the refractivities are N_1 and N_2, can be given in radians by the approximate expression [9, 10]:

$$\tau_{1,2} = - \int_{N_1}^{N_2} \cot \theta dN \,.\, 10^{-6} \tag{1}$$

where the local elevation angle, θ, at any point along the path is the acute angle between the ray and the tangent to an imaginary sphere concentric with the earth. Equation (1) is derived from ray theory and implies the limitation that the medium must change slowly in distance equal to λ/θ where λ is the wavelength of the radio frequency being considered and θ is the elevation angle.

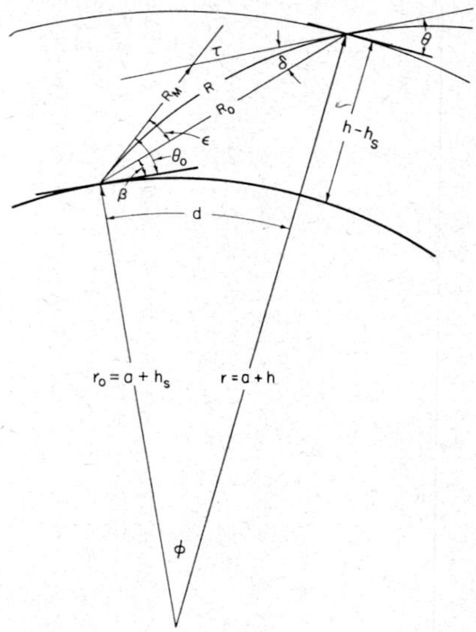

FIG. 1. Geometry of the refraction of radio waves.

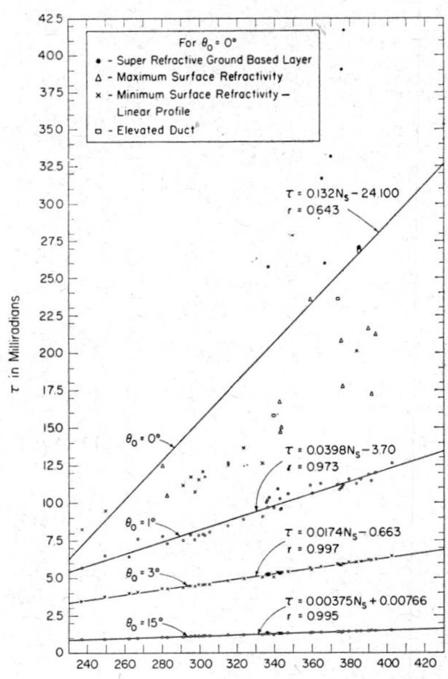

FIG. 2. Regression lines of total bending, τ, on the surface refractivity, N_s.

The refractivity, N, is given by, [11]

$$N = (n - 1)10^6 = \frac{77 \cdot 6}{T}\left(P + \frac{4810 e_s RH}{T} \right) \quad (2)$$

where n is the usual refractive index, P is the station atmospheric pressure in millibars, RH is the percent of the saturation vapor pressure, e_s, in millibars at the absolute temperature, T, in degrees Kelvin.

The refractivity usually changes more rapidly for a given vertical distance in the earth's atmosphere than for the same distance in the horizontal direction. Thus it is customary to assume that the atmosphere is horizontally stratified and that a single refractivity profile may be used to represent the atmosphere for the purpose of calculating the bending of radio waves as they pass through the earth's atmosphere. We shall see that this assumption is a reasonable one for rays with initial elevation angles greater than about 2°. As a result of this assumption the gradient of N enters directly into the integral expression for τ and the bending of the ray is a sensitive function of the distribution of temperature and humidity in the atmosphere. This is especially true when the local elevation angle, θ, is less than about 1°.

The total bending of a radio ray passing from the earth's surface, where $N = N_s$, completely through the atmosphere to where $N = 0$ may be obtained by integrating Equation (1) by parts:

$$\tau = -\int_{N_s}^{0} \cot \theta dN \,.\, 10^{-6} = N_s \,.\, 10^{-6} \cot \theta_0 - \int_{(\cot \theta)_{N=0}}^{\cot \theta_0} N d(\cot \theta) \,.\, 10^{-6} \quad (3)$$

where θ_0 is the initial elevation angle at the earth's surface.

The second term of Equation (3) contributes less than 3·5 % of the total for $\theta_0 = 10°$ and becomes negligibly small as θ_0 increases. Thus the first term of Equation (3) asymptotically approaches τ with increasing θ_0, and the total bending can be specified by the initial elevation angle and the surface value of N.

The author has recently examined the variation of the total bending of radio rays as a function of N_s [2]. The results of this study are summarized on Fig. 2 for various initial elevation angles and emphasizes a very pertinent characteristic of the total bending of radio rays, namely, the sensitivity of the bending to profile type for small values of θ_0. Although this was based upon an extreme range of N profiles at a number of climatically diverse locations, the total bending may be well represented by a linear function of N_s, regardless of profile type for initial elevation angles as small as 10 mrad. Note, however, that when $\theta_0 = 0$, it was necessary to distinguish between the total bendings calculated for the various profile types. It is important to remember that in the subsequent development of a model atmosphere that this model is intended to represent average conditions in temperate regions and will be misleading when applied at $\theta_0 < 10$ mrad. in the presence of extreme refractivity profile conditions such as shown on Fig. 2. However, at elevation angles of 10 mrad. or more, where the associated correlation coefficients are at least 0·9, it was concluded that the total bending could be well represented by surface weather conditions alone.

This rather surprising result has led to further investigations of these data to determine the relative bending imparted by various portions of the atmosphere. This may be accomplished by expressing the bending of radio rays from the earth's surface to any arbitrary height as a percentage of the total bending as in Fig. 3. It is seen that for minimal refraction conditions (i.e., designated as linear

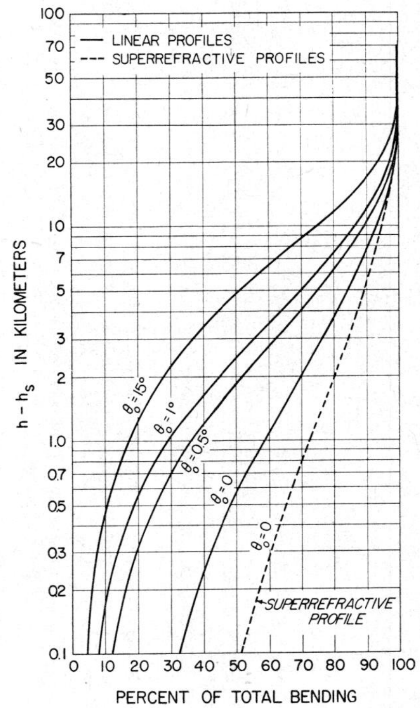

FIG. 3. Percentage of total bending as a function of height in the earth's atmosphere.

profiles) at $\theta_0 = 0$ that at least 30 % of the total bending is accomplished in the first 100 m, 60 % in the first kilometer and 95 % in the first 10 km above the earth's surface for a ray departing with an initial elevation angle of 0°. The superrefractive profiles (those with initial gradients between — 0·120 N units per meter and — 0·150 N units per meter as compared with the normal — 0·0392 N units per meter) produce 52, 72, and 95 % of the total bending at these same heights, indicating an even stronger dependence upon the lower atmosphere. It seems quite reasonable, then, that the initial value of refractivity would be intimately related to the average N structure in the lowest layers of the atmosphere and thus might lead one to expect the observed correlation between the bending and N_s.

III. Average Refractive Index Structure of the Atmosphere

An examination of long term climatic data of N_s and of ΔN, where ΔN is the difference between N_s and the value of N at 1 km, $-\Delta N = N_s - N$ (1 km), reveals that a correlation between the gradient and the surface value does exist. The results of this study, shown on Fig. 4, indicate that the gradient may be expressed as:

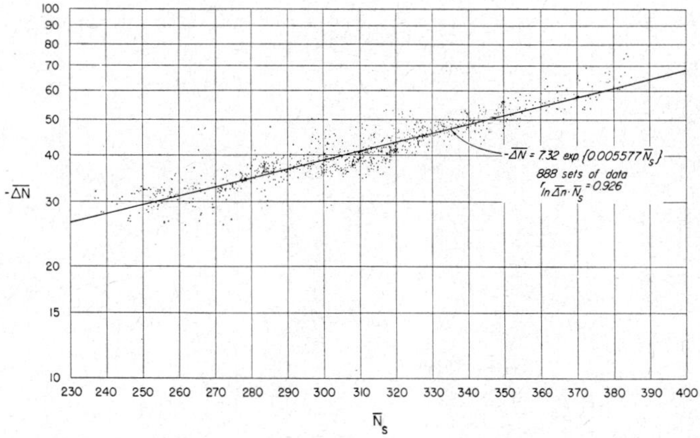

FIG. 4. Eight year average N_s versus 6-year average ΔN at 03·00 and 15·00 GMT.

$$-\overline{\Delta N} = 7{\cdot}32 \exp\{0{\cdot}005577\, \overline{N}_s\} \qquad (4)$$

Equation (4) is based upon a least squares determination from 888 sets of monthly mean values of $\overline{\Delta N}$ and $\overline{N}_s$ from 45 weather stations scattered about the United States and representing many diverse climates. This relationship between ΔN and N_s is expected to represent the best estimate of a majority of individual N profiles and certainly will closely agree with average conditions for the United States with one notable exception, southern California in the summer. These data were eliminated from Fig. 4 and the least squares determination because of their singularly large range of ΔN compared to their small range of N_s which resulted in a marked singularity of points rising from the main body of the data. This obvious departure of data points (24 points out of a total of 912) plus the well known unique nature of the southern California summer climate were taken as sufficient justification for ignoring these points, although, as we shall see, the ray bendings based upon the ΔN obtained from Equation (4) are in rather good agreement with the values calculated from the actual profile, even in southern California.

The relationship between ΔN and N_s as expressed by Equation (4) suggests that a model of the atmospheric refractivity structure in the first kilometer above the earth's surface may be based upon knowledge of N_s alone. A consideration of many refractivity profiles indicates that the refractivity generally decreases exponentially in the height increment of 1 km above the surface to the tropopause. Above the tropopause the refractivity usually decreases more rapidly but still in an exponential fashion. An examination of many refractivity profiles, summarized in Table I, indicates a minimum in the range of refractivity in the neighbourhood of 9 km above sea level.

The refractivity has a mean value of about 105 at 9 km above sea level, and corresponds to the altitude, as Humphreys points out [4], where the density of the atmosphere is approximately the same in all parts of the world and all seasons of the year. Since the first term in the expression for refractivity is proportional to the air density, and the water vapor term is negligible at an altitude of 9 km, it seems quite reasonable to expect that the refractivity would also tend to be constant at this altitude. This observation was further examined by Stickland [12], who has shown that the refractivity does indeed tend to be constant at 9 km with a value of about 105.

TABLE I

REFRACTIVITY STATISTICS AS A FUNCTION OF ALTITUDE ABOVE SEA LEVEL AS DERIVED FROM INDIVIDUAL RADIOSONDE OBSERVATIONS

Altitude in kilometers	$\overline{N}$	Maximum N	Minimum N	Range*
4	197·1	209·5	186·5	23·0
5	172·3	184·0	165·0	19·0
6	151·4	161·0	146·0	15·0
7	134·0	139·5	129·5	10·0
8	118·4	121·5	113·3	8·2
9	104·8	108·0	100·0	8·0
10	92·4	97·0	86·0	11·0
11	81·2	86·0	70·0	16·0
12	70·7	76·0	60·5	15·5
14	53·2	60·0	44·5	15·5

*Range = maximum N-minimum N.

IV. DEVELOPMENT OF THE REFERENCE ATMOSPHERES

The present study has been concerned with characteristics of atmospheric refraction of radio waves as well as the average refractivity structure of the atmosphere. At this point a model atmosphere may be developed in a variety of ways:

(1) A mathematical model of the atmosphere may be developed that is in good agreement with average refraction conditions but is not necessarily a good representation of the refractivity structure of the atmosphere.

(2) Specify an atmosphere that is in good agreement with the refractivity structure of the atmosphere and derive estimates of average refraction from this atmosphere.

Actually, the preceding analysis allows development of model atmospheres along both lines of attack. The central point to both methods is the relation between ΔN and N_s as specified by Equation (4). Any structure function for the refractivity decay within the first kilometer above the earth's surface may be evaluated from Equation (4) since this equation only specifies the difference in the value of N at the earth's surface and 1 km above the earth's surface. One very simple approach is to assume a single exponential distribution of refractivity:

$$N = N_s \exp\{- c_e(h - h_s)\} \tag{5}$$

where

$$c_e = \ln \frac{N_s}{N(1\ \text{km})} = \ln \frac{N_s}{N_s + \Delta N}$$

and these equations are used to determine N at all heights. This model of the atmospheric refractivity has the advantage of being a close representation of the refractivity structure within the first kilometer. Further, the single exponential model is a continuous function and may be used for theoretical considerations. This model of the atmosphere has been adopted for use within the National Bureau of Standards by adoption of specific values of the constants in Equation (5). These constants are given in Table II and specify the CRPL Exponential Reference Atmospheres, 1958.

TABLE II

TABLE OF THE CONSTANT c_e FOR THE CRPL EXPONENTIAL RADIO REFRACTIVITY ATMOSPHERES*

$$N = N_s \exp[- c_e(h - h_s)]$$

ΔN	N_s	c_e per km
0	0	0
22·3318	200·0	0·118400
29·5124	250·0	0·125625
30·0000	252·9	0·126255
39·2320	301·0	0·139632
41·9388	313·0	0·143859
50·0000	344·5	0·156805
51·5530	350·0	0·159336
60·0000	377·2	0·173233
68·1295	400·0	0·186720
70·0000	404·9	0·189829
90·0406	450·0	0·223256

*The International Radio Consultative Committee of the International Telecommunication Union meeting in Geneva during July, 1958, recommended the adoption of the basic reference atmosphere:

$$N(h) = 300 \exp\{- 0{\cdot}139h\}$$

where h is the height above the ground in kilometers.

An alternative model of the atmosphere may be obtained by assuming that N decreases linearly in the first kilometer above the surface:

$$N = N_s + \Delta N\{h - h_s\}, \quad h_s \leqslant h \leqslant h_s + 1 \tag{6}$$

where ΔN is obtained from Equation (4). The assumption of a linear gradient produces a great simplification in radio propagation problems since it has been shown [8] that the behavior of radio waves in such an atmosphere is nearly the same as that expected with no atmosphere for an earth with effective radius ka' where a' denotes the actual earth's radius and k is defined by

$$\frac{1}{k} = 1 + \frac{a'\Delta N}{n}$$

It may be assumed that $N = 105$ at 9 km above sea level and that the refractivity decreases exponentially [1] in the interval of 1 km above the earth's surface, $h_s + 1$, to the value of 105 at 9 km. This assumption means that N may be expressed by

$$N = N_1 \exp\{- c(h - h_s) - 1\}, \quad h_s + 1 \leqslant h \leqslant 9 \tag{7}$$

where

$$c = \frac{1}{8 - h_s} \ln \frac{105}{N_1}$$

and N_1 is the value of N at $h_s + 1$. Note that the only two variables in Equation (7) are the altitude of the earth's surface above sea level and the value of N_1 which, in turn, is a function of N_s.

Above 9 km the refractivity may be described in terms of the well known exponential decay of atmospheric density. The expression:

$$N = 105 \exp\{- 0{\cdot}1424(h - 9)\}, \quad h \geqslant 9 \text{ km} \tag{8}$$

was determined by a least squares analysis of the Rocket Panel [7] data. This expression is also in agreement with the ARDC Model Atmosphere—1956 [5] and Dubin's [3] conclusion that a standard density distribution may be used at elevations in excess of 20,000 ft. The reader may question the arbitrary adoption of a single value of N at 9 km and a single exponential decrease of N above that height. This choice was motivated by the knowledge that the bending of radio rays emanating from the earth's surface is relatively insensitive to variations of N at 9 km and above. Although this model of the atmosphere has the advantages of the effective earth's radius approach within the first kilometer, it does have the disadvantage, for theoretical considerations, of having discontinuities in the profile at 1 and 9 km.

This three part model of the atmosphere has also been adopted for use at the National Bureau of Standards. The constants adopted for use are given in Table III and specify the CRPL Reference Refractivity Atmosphere, 1958.

The station elevations corresponding to given combinations of N_s and ΔN were chosen to correspond with an average decay of N with station elevation. Although the error in neglecting this height dependence has been estimated to be no more than a few per cent it could be important in such high precision applications as radar tracking of earth satellites. It should be remembered in subsequent applications that a unique feature of these reference atmospheres is the height dependence of N_s. This feature was built in so that the reference atmospheres would be completely specified by the single parameter N_s.

The CRPL Reference Refractivity Atmosphere is illustrated on Fig. 5 for

TABLE III

CONSTANTS FOR THE CRPL REFERENCE ATMOSPHERE, 1958

N_s	h_s ft	a' miles	$-\Delta N$	k	a_e miles	c per km
0	0	3960·0000	0	1·00000	3960·00	0
200	10,000	3961·8939	22·3318	1·16599	4619·53	0·106211
250	5,000	3960·9470	29·5124	1·23165	4878·50	0·114559
301	1,000	3960·1894	39·2320	1·33327	5280·00	0·118710
313	700	3960·1324	41·9388	1·36479	5403·88	0·121796
350	0	3960·0000	51·5530	1·48905	5896·66	0·130579
400	0	3960·0000	68·1295	1·76684	6996·67	0·143848
450	0	3960·0000	90·0406	2·34506	9286·44	0·154004

Note: a_e is the effective earth's radius and is equal to $a'k$.

$a' = a + h_s$, where h_s is the altitude of the earth's surface above sea level and $a \equiv 3960$ miles.

$$c = \frac{1}{8 - h_s} \ln \frac{N_1}{105}$$

several values of N_s. Also, for comparison, a single exponential atmosphere is given as well as the N distribution assumed for the 4/3 earth radius concept with $N_s = 301$. It can be seen that the 4/3 earth assumption agrees with the reference atmosphere in the first kilometer, which is to be expected since $N_s = 301$ is the value of N_s required to yield the 4/3 earth's gradient from Equation (4). Above 1 km, however, the 4/3 earth atmospheric model decreases considerably more rapidly with altitude than do the reference atmospheres. It is notable that both of the adopted reference atmospheres are in closer agreement with the Rocket Panel and ARDC N distribution than is the 4/3 earth model and, significantly, it is at high heights that the difference in bending of the radio rays in the two models of the atmosphere is most pronounced. This point is illustrated in Fig. 6 where the ray paths in the reference atmosphere are compared with those in the 4/3 earth atmosphere. It is interesting to note that the 4/3 earth ray at $\theta_0 = 0$ is in very good agreement with the values from the reference atmospheres for distances out to 200 miles and heights up to 14,000 ft. For ranges of 600 miles, where the ray reaches heights of about 200,000 ft, the

4/3 ray departs radically from the model atmospheres. Because of its greater refractivity gradient and resultant increased bending at high elevations the 4/3

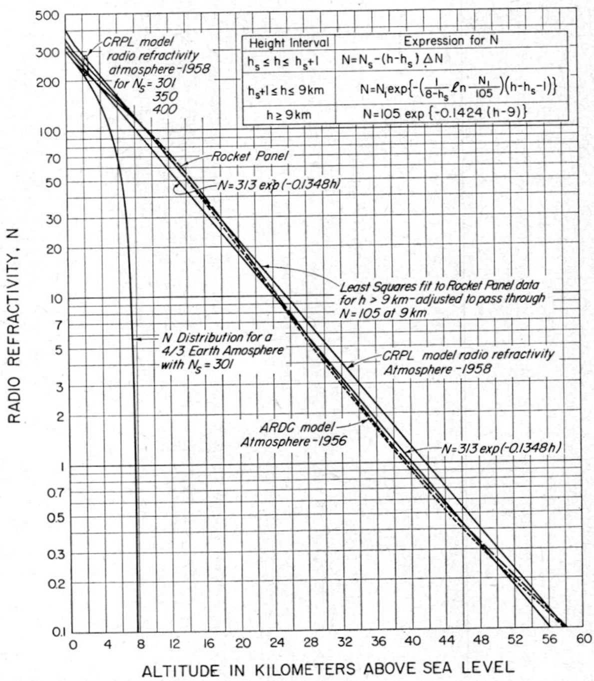

Fig. 5. CRPL model radio refractivity atmosphere, 1958.

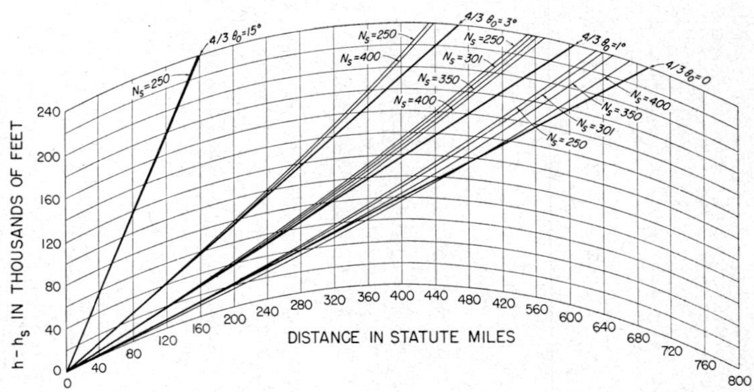

Fig. 6. Comparison of rays in the CRPL Reference Atmosphere, 1958 and the 4/3 earth atmosphere.

ray is about 9000 ft lower than the $N_s = 400$ reference atmosphere and 36,000 ft lower than the $N_s = 250$ reference atmospheres.

V. Comparison of Reference Atmospheres with Long Term Mean Data

Radio ray bendings were calculated for the reference atmospheres and compared to similar values obtained from 5 year mean N profiles from a variety of U.S. radiosonde stations. Mean profiles were obtained from both summer and winter for a variety of climates as represented by the states of Florida, Texas, Maine, Illinois, Nevada, California, North Dakota, Washington, Nebraska, Wyoming and the District of Columbia.

Comparisons of the long term mean bendings with the reference atmospheres are shown in Figs. 7a and 7b. These figures were selected to illustrate the range of agreement between the reference atmospheres and the long term means. Figure 7a gives a comparison for a small initial elevation angle, $\theta_0 = 0$ and a small height increment, $h - h_s = 3$ km, and shows that both reference atmospheres tend to set a lower limit to the bendings. In this case, the exponential reference atmospheres appear to be in better agreement with the long term mean bendings than the reference atmosphere. The numbered data points for Washington, D.C., Omaha, Neb., and Santa Maria, California are of special interest. Washington and Omaha have the only long term mean N profiles with initial N gradients (i.e., $-0{\cdot}112$/m and $0{\cdot}106$/m respectively) that are significantly greater than would be expected from the ΔN versus N_s relationship. Both of these stations have an unusually large humidity decrease near the ground. The third point, Santa Maria, California, is of interest since it is in relatively good agreement with the reference atmospheres even though it represents the southern California summer climate which was excluded from the original ΔN versus N_s relationship. This agreement is attributed to the reference atmospheres being in good agreement with the N distribution below the California elevated inversion and that a majority of the bending is accomplished below the elevated inversion height of about 500 m. Figure 7b shows a similar comparison for a high initial elevation angle, $\theta_0 = 15°$, and a large height increment, $h - h_s = 70$ km. This comparison shows that both of the reference atmospheres are in rather close agreement with the long term mean bendings.

In considering the comparisons of Figs. 7a and 7b one might ask if they reflected the form of the basic equation for bending; namely, at low angles is τ determined by the N gradient throughout the N profile, and at high angles is τ essentially a function of the value of N at both ends of the N profile, i.e., the limits of integration? Thus one might expect the deviations to be smaller if the comparisons were made on the basis of a function of the N gradient such as ΔN. Such a comparison is given by Figs. 7c and 7d for the same initial elevation angles and height increments as before. It is seen that the ΔN-specified reference atmospheres improve the agreement for the low angle case but decidedly decrease the agreement for the high angle case.

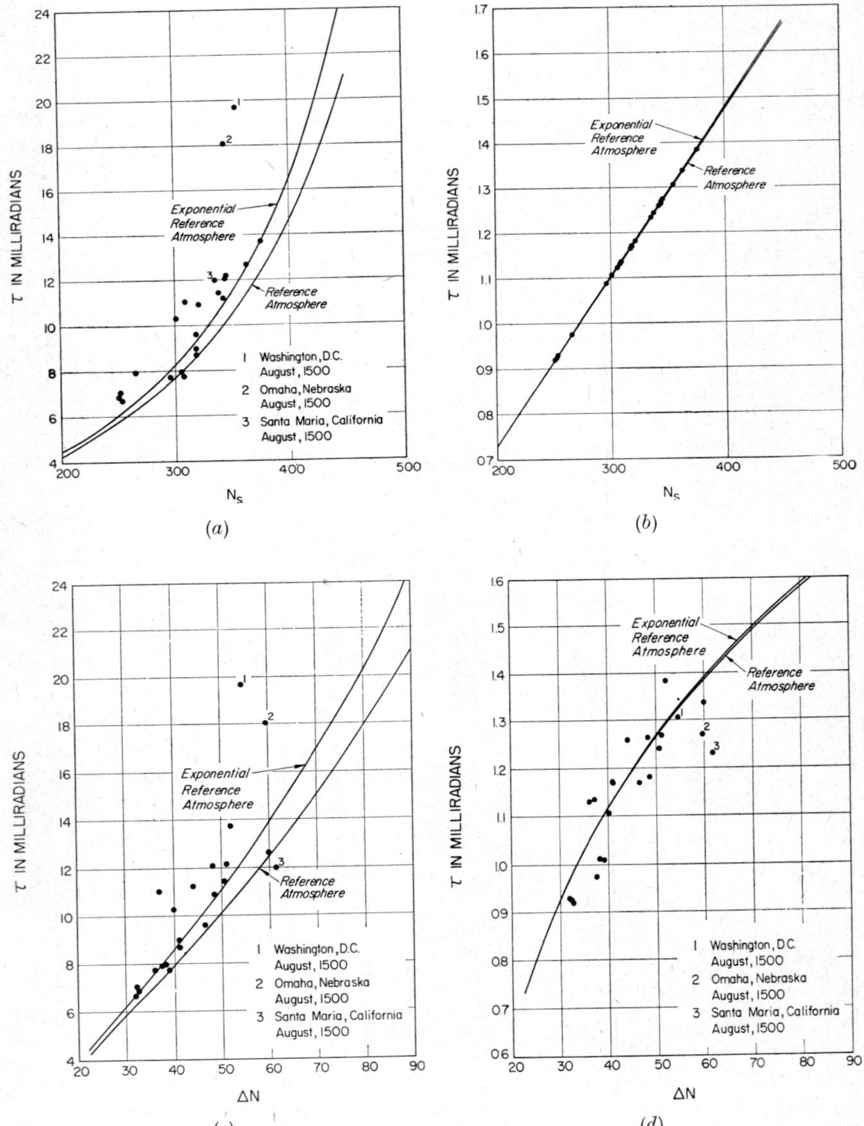

FIG. 7. Comparison of τ vs. N_s and ΔN as obtained from the CRPL Reference Atmosphere 1958 and 5-year mean radiosonde data.

(a) $\theta_0 = 0$, $h - h_s = 3$ km.

(b) $\theta_0 = 15°$, $h - h_s = 70$ km.

(c) $\theta_0 = 0$, $h - h_s = 3$ km.

(d) $\theta_0 = 15°$, $h - h_s = 70$ km.

A numerical evaluation of the root mean square (r.m.s.) deviation of the long term mean bendings from both the reference atmospheres determined both as a

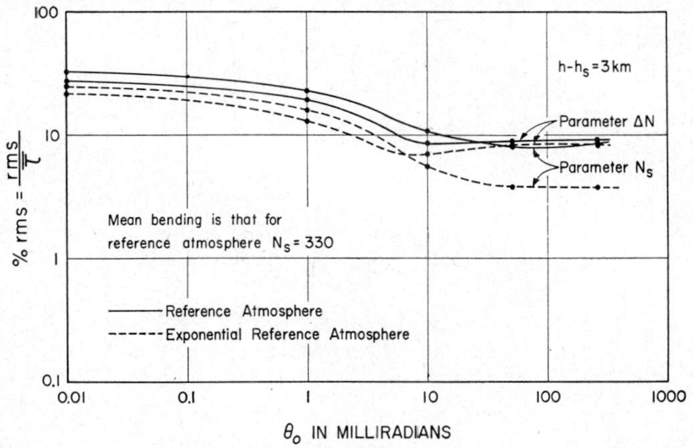

FIG. 8. Comparison of per cent. r.m.s. deviations of 5-year mean profile bendings about CRPL reference atmosphere using two parameters, N_s and ΔN.
$h - h_s = 3$ km.

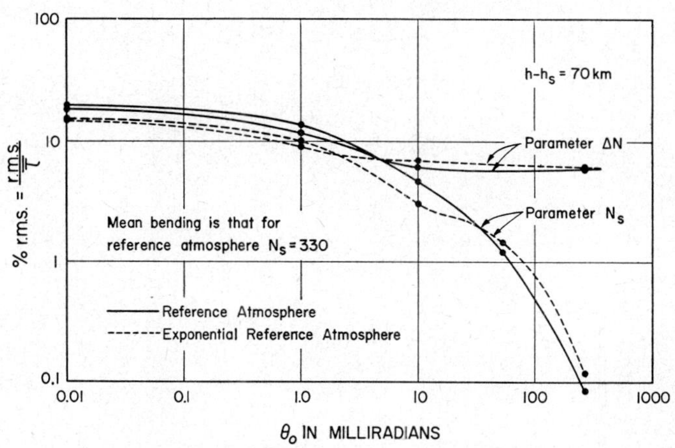

FIG. 9. Comparison of per cent. r.m.s. deviations of 5-year mean profile bendings about the CRPL reference atmospheres using two parameters, N_s and ΔN.
$h - h_s = 70$ km.

function of ΔN and N_s was made for a variety of initial elevation angles for the height increments 3 and 70 km. Figure 8 summarizes the r.m.s. deviations for the $h - h_s = 3$ km case. It is seen that for $\theta_0 < 10$ mrad. (about 0·5°) the ΔN-specified reference atmospheres have the smaller r.m.s. deviations. Also, the exponential reference atmospheres, whether specified by ΔN or N_s, have smaller r.m.s. deviations than the reference atmosphere.

It is seen for the 70 km case, Fig. 9, that the N_s-specified reference atmospheres have a significantly smaller r.m.s. deviation than the ΔN-specified atmospheres for $\theta_0 > 5$ mrad. Again it is seen that the exponential reference atmosphere generally has the smaller r.m.s. deviation. However, the slightly smaller r.m.s. deviations associated with the reference atmosphere for $\theta_0 > 10$ mrad. is felt to be a reflection of that model's closer agreement with the actual N structure of the atmosphere at high heights.

VI. Limitations of the Reference Atmospheres

The reference atmospheres are based upon the observed correlation of ΔN and N_s and consequently are only applicable under the assumption that this relationship is valid. There are many individual instances of subrefraction, superrefraction and ducting for which Equation (4) is not applicable. However, Equation (4) is expected to be useful in the consideration of long term mean conditions. Even in this case, however, it has been seen that there are locations where superrefractive layers near the earth's surface produce significantly greater bending than one would expect for the observed value of N_s.

The concept of radio ray bending is not a useful one under the condition of ducting or trapping of radio waves. One may obtain an estimate of the frequency of occurrence of ducting from a consideration of radiosonde observations using Snell's law expressed in cylindrical coordinates as:

$$nr \cos \theta = n_0 r_0 \cos \theta_0 \tag{9}$$

where r is the radial distance from the center of the earth to the point under consideration and the other variables are as previously defined. The zero subscript refers to the value of the variable at the earth's surface. For the present study a duct is defined to occur when the refraction is sufficient to direct a ray either back to earth or in a circular path at a constant height above the earth, i.e.,

$$\frac{n_0 r_0 \cos \theta_0}{nr} \geqslant 1 \tag{10}$$

The frequency of occurrence of ground based ducts was determined by applying Equation (10) to 5 years of the individual radiosonde observations at Fairbanks, Alaska, Washington, D.C., and Swan Island (Caribbean Sea) for the months of February, May, August and November. The results of this analysis, shown on Fig. 10, indicate that ducts do not occur more than about 15 % of the time. It is interesting to note the opposite annual trends of the arctic station, Fairbanks, and the tropical station, Swan Island. The arctic station has a winter-time maximum of about 9 % associated with strong cooling of the surface layer of the atmosphere but does not have over a 2 % incidence of ducts in the other months analysed. Swan Island has a summer-time maximum of about 14 % associated with the strong humidity gradient near the surface. Washington, D.C., has a well-defined annual cycle with a maximum of about 5 % associated with summertime radiation inversions.

Although the percentage occurrences of ducts given above are perhaps too small due to the slow response of the radiosonde and its resultant inability to report layers near the surface, these percentages are felt to indicate the relative seasonal and geographical distribution of the incidence of ducts.

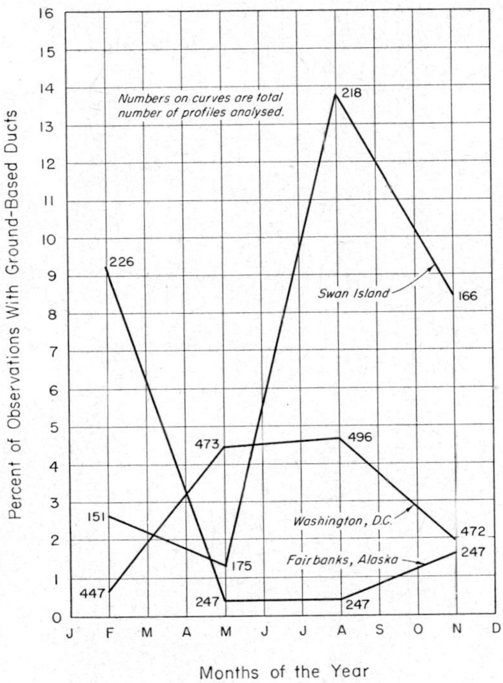

FIG. 10. Frequency of occurrence of ground based ducts.

The bendings given by the reference atmospheres were calculated under the assumption that the atmosphere was horizontally stratified and that the refractive index did not change with horizontal distance. Although this is a difficult assumption to check under a wide variety of meteorological conditions several experiments have been made in certain locations where detailed measurements have been made of both the vertical and horizontal N structure. One of the experiments, The Canterbury Project of New Zealand [6] offers a convenient method of examining the effect of horizontal changes in N profile upon ray bending. The radio ray paths were determined for the particular case with marked superrefraction near the surface and change of N profile with distance that is shown on Fig. 11. The value of bending obtained by detailed ray tracings through this atmosphere was compared with the values predicted under the assumption of horizontal homogeneity of the N profile by using the same regression technique as illustrated by Fig. 2. A comparison of these values, shown on Fig. 12, indicates rather good agreement

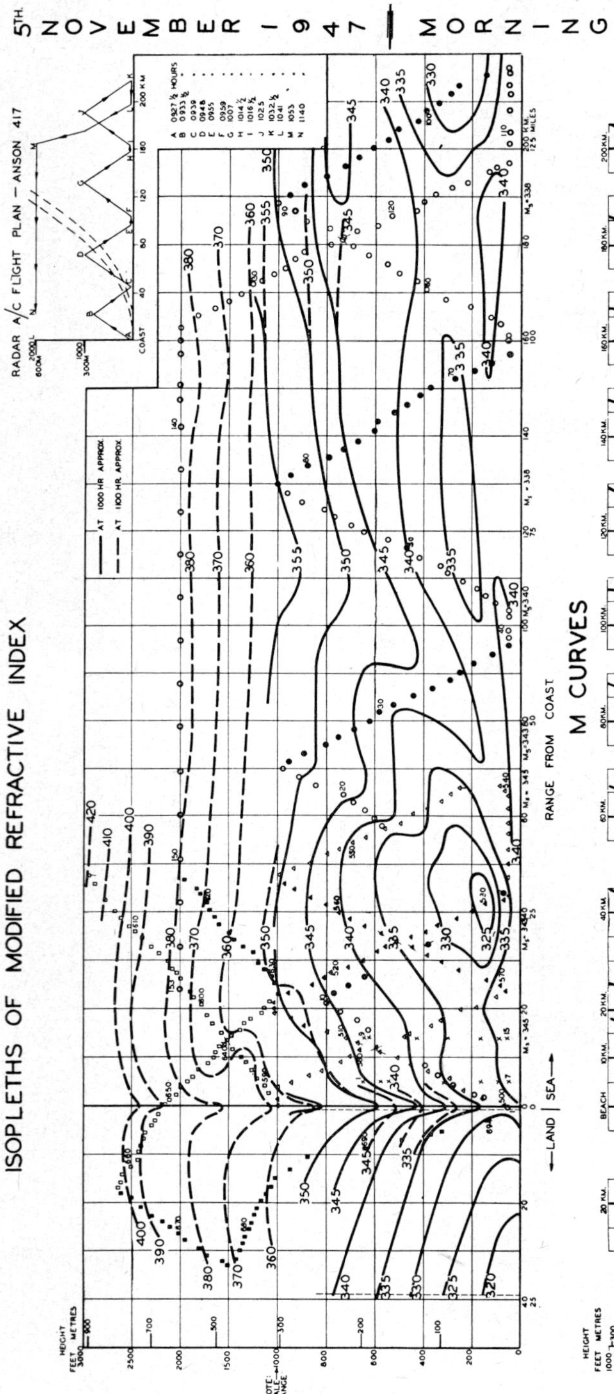

FIG. 11. Isopleths of modified refractive index—Canterbury Project—5 November 1947. (Reproduced with the kind permission of the New Zealand Department of Scientific and Industrial Research.)

between the predicted and calculated ray bendings for initial elevation angles at least as large as 3°. Further calculations are currently in progress and indicate that similar agreement will be obtained for $\theta_0 = 2°$ but that the effect of the surface duct produces progressively more disagreement as the initial elevation angle becomes smaller. This preliminary study indicates that, even in the worst case of ducting, the effect of horizontal changes of N profile are perhaps negligible for initial elevation angles of at least 3°. The results of this preliminary study are felt to be fortuitous and consequently a second case is currently under study.

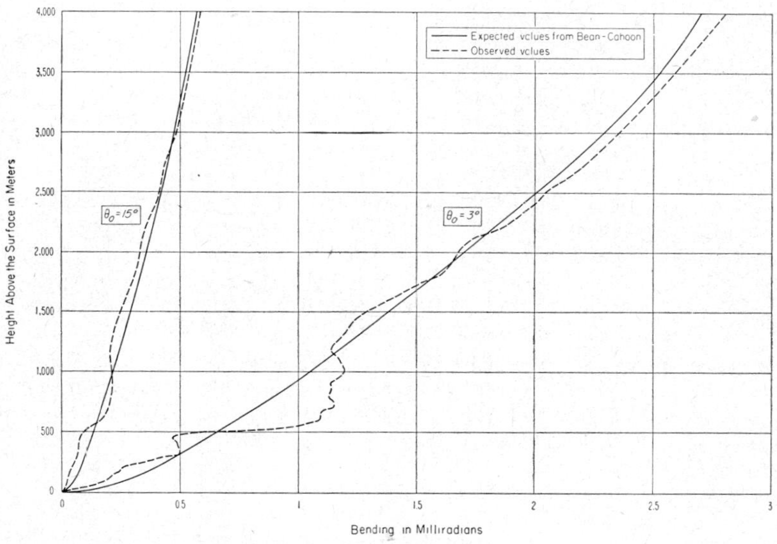

FIG. 12. Comparison of calculated and expected values of τ for Canterbury Project, 5 November 1947.

VII. Conclusions

Perhaps the most pertinent conclusion of the present study is its starting point, namely: 50 to 60 % of the total bending of radio rays occurs in the first kilometer above the surface for small initial elevation angles. This observation was taken to mean that a model atmosphere must particularly describe the initial N structure of the atmosphere. This observation also explains why the 4/3 earth model of bending gives such good results over moderate ranges and heights, i.e., the 4/3 atmosphere has reasonably realistic N gradients within the first kilometer. Other salient points of this study are that a rather well-defined exponential relationship exists between N_s and ΔN and that the refractivity is nearly constant at an altitude of 9 km. By utilizing this information, model atmospheres may be constructed. The bending of radio rays derived from these model atmospheres are in good agreement with the bending obtained from mean radiosonde data although they yield values noticeably lower than those obtained from super-refractive profiles. This latter point led to a critical examination of whether ΔN or N_s is the better predictor. The general conclusion of this examination is that N_s is the better predictor for $\theta_0 > 10$ mrad. and that ΔN is the better predictor for $\theta_0 < 10$ mrad., although neither N_s norΔN are in good agreement with the bending from all profiles at $\theta_0 = 0$. This latter conclusion suggests that it would be most pertinent for future research to investigate more thoroughly the N structure in the first kilometer.

Another result is that a single exponential decay of refractivity with height yields values of bending that are more in agreement with many observations than the reference atmospheres. This is somewhat surprising since the exponential reference atmospheres are in marked disagreement with the actual atmosphere at high elevations. The better agreement of the exponential reference atmospheres evidently reflects that an exponential decay of refractivity is more representative of the N structure in the lower atmosphere. However, this atmosphere does not agree with all profiles at $\theta_0 = 0$, thus also indicating the need of further investigation of the N structure near the surface.

Further tentative conclusions are that the calculation of atmospheric bending assuming horizontal homogeneity of the refractivity profile is valid for initial elevation angles in excess of about 2°. It should be emphasized that this conclusion is based on the examination of a single day's observations in the presence of a surface based duct. The assumption of horizontal homogeneity is probably valid to even smaller values of the initial elevation angle for the majority of meteorological conditions. This latter speculation is based upon the conclusion that ducts are not indicated by radiosonde measurements to occur more than 15 % of the time.

The models of atmospheric refractivity given above were derived by assuming various height distributions of N. Other models of the height distribution could as easily have been adopted. As an example, an exponential N distribution could have been adopted in place of the linear N distribution of the reference atmosphere. Although this would have resulted in better agreement with the long term mean bendings, it was felt that this advantage was outweighed by the resultant loss in simplicity of the effective earth's radius concept for many applications.

Acknowledgments

The author wishes to express his gratitude to his colleagues G. D. Thayer and B. A. Cahoon for their aid in many aspects of this work. The material summarized above represents but a fragment of the total information available from the author in the form of informal laboratory reports.

References

1. Bean, B. R. (1953). *Proc. Inst. Radio Engrs, N.Y.* **41**, 549–550.
2. Bean, B. R. & Cahoon, B. A. (1957). *Proc. Inst. Radio Engrs, N.Y.* **45**, 1545–1546.
3. Dubin, M. (1954). *J. geophys. Res.* **59**, no. 3, 339–344.

4. Humphreys, W. J. "Physics of the Air," p. 82. McGraw-Hill Book Co., Inc., New York (1940).
5. Minzer, R. A. & Ripley, W. S. (1956). *Air Force Surveys in Geophysics*, no. 86.
6. Report of Factual Data from the Canterbury Project. Dep. Sci. Industr. Res., Wellington, N.Z. (1951).
7. Rocket Panel, The (1952). *Phys. Rev.* **88**, 1027–1032.
8. Schelling, J. C., Burrows, C. R. & Ferrell, E. B. (1933). *Proc. Inst. Radio Engrs, N.Y.* **21**, 427–463.
9. Schulkin, M. (1952). *Proc. Inst. Radio Engrs, N.Y.* **40**, 554–561.
10. Smart, W. M. "Spherical Astronomy," Chapter II. Cambridge University Press, London (1931).
11. Smith, E. K. & Weintraub, S. (1953). *Proc. Inst. Radio Engrs, N.Y.* **41**, 1035–1037
12. Stickland, (Miss) A. C. "Meteorological Factors in Radio-Wave Propagation," pp. 253–267. The Physical Society, London (1946).

Signal-Cancellation Techniques for Capturing the Weaker of Two Cochannel FM Signals

Elie J. Baghdady

Massachusetts Institute of Technology, Cambridge, Mass.

A conventional FM receiver captures the stronger of two cochannel FM signals and suppresses irretrievably the weaker signal. But this is only a limitation of conventional FM demodulators. In this paper, two general, laboratory-tested, signal-cancellation techniques—*dynamic trapping* and *feedforward*—are described to show how the message carried by the weaker of two cochannel FM signals can be reproduced with negligible distortion even when its amplitude is much smaller than that of the stronger signal. The application of these methods to the realization of a two-carrier, single-channel, "amplitude-discrimination" FM multiplex system is also discussed.

I. Introduction

The process of frequency modulation offers important advantages in low-distortion transmission, noise suppression, and rejection of interference, but not without certain attendant limitations that will at times impair the usefulness of the communication link. However, it is becoming increasingly apparent that some of these limitations are really limitations of the conventional demodulation technique rather than of the fundamental modulation process. For example, the operations that the sum of two cochannel signals encounters in going through a well-designed conventional FM receiver will usually deliver a substantially undistorted replica of the message that is carried by the stronger signal, and exclude almost completely the effect of the presence of the weaker signal. This stronger-signal capture effect is a great asset of a communication system, as long as there is some assurance that the desired signal will be the stronger of the two competing signals within the same channel. But in situations in which the desired signal is just as likely to be the weaker of the two as it is to be the stronger, the desired signal will often be suppressed irretrievably.

The development of signal-processing techniques that would make it possible to capture the weaker signal, when this is desirable, would not only facilitate more reliable communication, but would also give great promise of providing more efficient use of the crowded frequency spectrum.

In this paper, we shall show how the message of the weaker of two cochannel FM signals can be reproduced with negligible distortion even when this signal is of much smaller amplitude than the stronger one. Two laboratory-tested techniques—*dynamic trapping* and *feedforward*—will be discussed. These methods aim fundamentally at switching the roles of "weaker" and "stronger" in the two signals before they reach the final FM demodulation stage. The performance of these systems in the realization of a two-carrier, single-channel, "amplitude-discrimination" FM multiplex system will also be described.

II. Dynamic-trapping Techniques

In dynamic-trapping systems, some means is first devised for extracting the frequency modulation of the stronger signal with reasonable accuracy. This information is then used to guide a trap system, in order to attenuate the stronger signal. The attenuation, or cancellation, of the stronger signal is achieved either by making a *dynamic trap* track the signal, or by "freezing" the instantaneous frequency of the signal so that it always equals the resonant frequency of a *fixed trap*. Thus, if the modulation of the stronger signal is applied to change the value of a simulated reactance that forms a variable tuning element in a trap circuit, the attenuation band of the trap can be made to follow the instantaneous-frequency position of the stronger signal. Alternatively, the stronger-signal modulation can be made to change the frequency of a local oscillator so that it always differs from the instantaneous frequency of the stronger signal by an amount that equals the resonant frequency of a stationary trap.

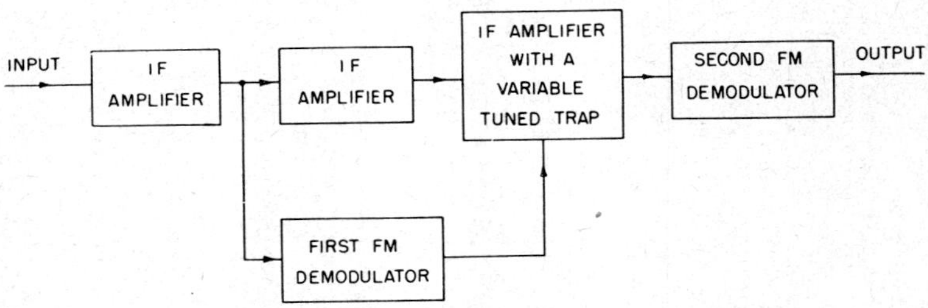

Fig. 1. Block diagram of variable-tuned trap system.

A. Variable-tuned Trap System

Figure 1 shows a block diagram of a dynamic-trap receiver. In this system, a well-designed conventional FM demodulator is used to determine the instantaneous frequency behavior of the stronger signal from the resultant of the two input signals. The output of this demodulator, after proper filtering, will specify the frequency behavior of the stronger signal. This information is then used electronically to change the center frequency of the variable-tuned trap so that it follows the instantaneous position of the stronger signal on the frequency scale. In this way, the trap is made to attenuate the undesired signal below the level of the desired, but initially weaker, signal. With the desired signal thus transformed into the stronger of the two competing signals, a second conventional FM demodulator is used (after the trap system) to extract the desired message.

In order to appreciate the potentialities of the variable-trap system, as well as its limitations, it is important to examine the basic mechanism of its operation. First of all, note that the reasoning that has led us to devise this system depends upon our ability to visualize the stronger signal as a sinusoid whose position on the instantaneous-frequency scale varies, essentially, in accordance with the modulation of this signal. The justification for this attitude, as far as the contribution of the stronger signal to the response of the I.F. amplifier that drives the trap is concerned, depends upon the assumption that the I.F. filter is able to follow, within reasonable bounds, the variable-frequency excitation of the stronger signal in a quasi-stationary manner. The precise conditions under which this assumption can, in general, be made have been worked out and published by the author [1, 2], and need not be stated here.

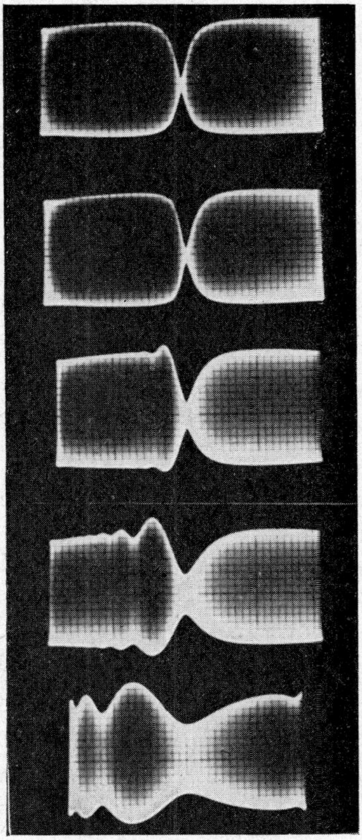

FIG. 2. Envelope of a sinusoidally modulated FM signal as it sweeps at different rates over the passband of an I.F. amplifier whose response is modified by the introduction of a fixed trap. The frequency of the modulating tone is f_m c/s.

Next, note that the trap filter will be swept in frequency by the desired signal at a relative rate that equals the sum of the instantaneous-frequency modulations of the two signals. If the trap filter is too sluggish with respect to this sweep rate, distortion will be introduced into the modulation of the desired signal by the transients that will be set up in the response of the trap filter whenever it is swept by the signal. Figures 2 and 3 illustrate the nature of the AM and FM transients that may be experienced in the envelope and frequency of the signal.

But, it turns out that there usually is no need for an unusually sharp trap attenuation band, and therefore this type of distortion will not be of a serious nature.

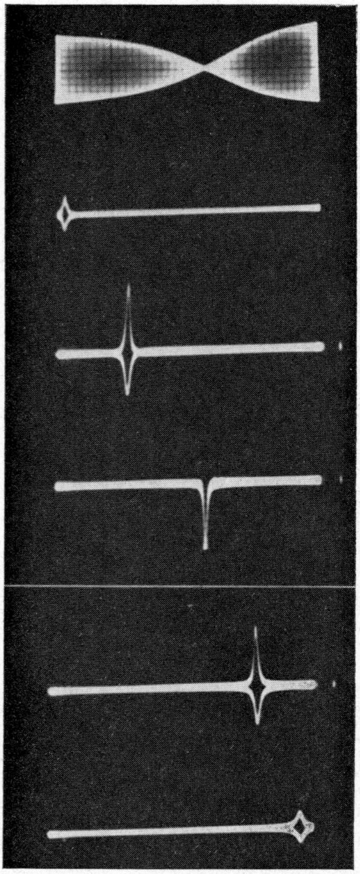

FIG. 3. FM transient produced by a variable trap that sweeps sinusoidally through a stationary carrier. The frequency position of the carrier is changed from picture to picture to illustrate its effect upon the FM transient. Response of I.F. amplifier as modified by stationary trap is shown in the top oscillogram.

Finally, the determination of the limits within which the simulated reactance can be varied in order to vary the resonant frequency of the trap filter in the desired manner is a problem in the theory of the reactance modulator. This theory has been explored in numerous publications on basic FM theory and time-variant systems. The proper phasing of the signal that must modulate the simulated reactance in the necessary manner is vitally important for the successful tracking of the stronger-signal frequency by the resonance frequency of the trap.

The derivation of this modulating signal depends upon the ability of the first FM demodulator to extract a close replica of the modulation of the stronger signal from the resultant of the two input signals. A conventional FM receiver will usually perform this task for all values of the input weaker-to-stronger signal amplitude ratio that are bounded by a certain value, $\rho < 1$. This value, ρ, is a function of the receiver design and of the interference conditions, and it will serve as an index of the stronger-signal capture capabilities of the receiver when it is measured under the most adverse condition of interference (two unmodulated signals separated by one I.F. bandwidth). The quantity ρ will be referred to as the stronger-signal *capture ratio* of the FM receiver.

B. Performance of the Experimental Variable-trap System

A system that operates on these principles has been built and tested [6]. This system was designed with sufficient flexibility to make it possible to study the

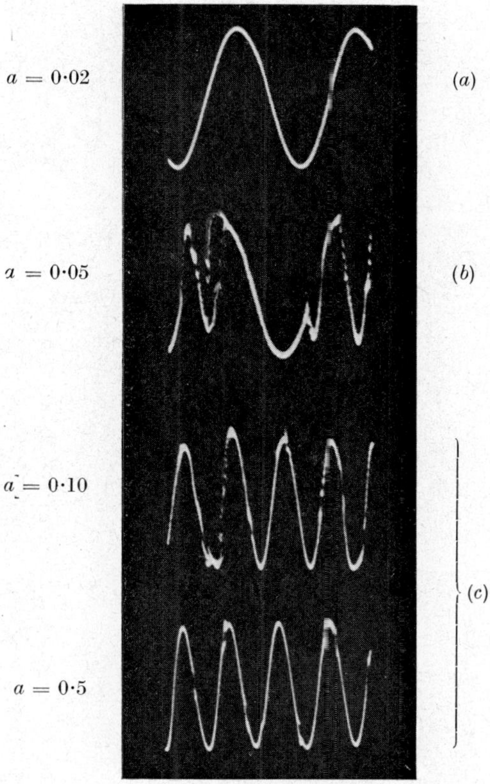

FIG. 4. Oscillograms of retrieved weaker-signal modulation in the output of the experimental variable-tuned trap receiver. $\delta \simeq 0{\cdot}05$. (*a*) Detected modulation of stronger signal (400 c/s). (*b*) Received waveform at center of capture-transition range. (*c*) Retrieved weaker-signal modulation (1000 c/s).

effect (upon the performance of the receiver) of varying the important characteristics of the trap, such as its maximum attenuation, and the width of its attenuation band. The detailed circuitry has also been described [5]. This experimental system can be adjusted to extract the message of the weaker signal when its amplitude is less than 0·04 of the amplitude of an interfering signal whose spectrum overlaps the spectrum of the desired signal completely.

The oscillograms in Fig. 4 show some informative samples of the observed performance. Although the stronger signal is suppressed completely by this system, a certain amount of inherent distortion remains in the retrieved message. This distortion results from the fact that the frequency modulation causes both

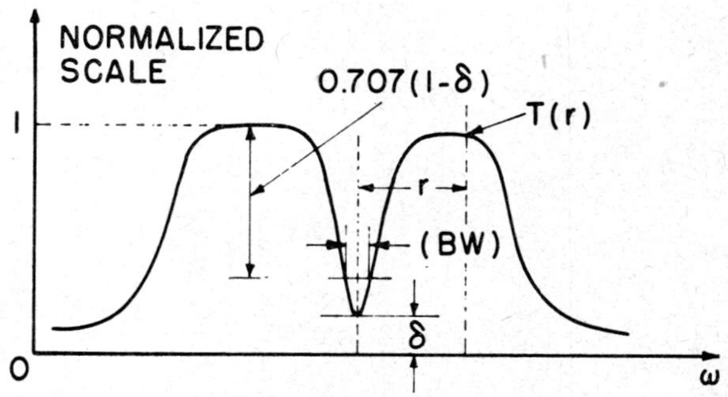

FIG. 5. Definition of trap bandwidth (BW), and trap attenuation factor, δ. $T(r)$ denotes the normalized over-all I.F. response as modified by the introduction of the trap. The variable r denotes the frequency deviation (in rad./s) from the center frequency of the trap. $T(0) = \delta$.

signals to have substantially the same instantaneous-frequency values during certain short time intervals. During these intervals, the desired weaker signal will also find itself in the strong-attenuation band of the trap, and a short burst of distortion will result because the second FM demodulator will be unable to differentiate between the two signals. To a lesser extent, some of this distortion is also caused by the sluggishness of the trap as is shown in Fig. 3.

In order to evaluate the capture performance of an FM receiver in a meaningful, quantitative, and reproducible manner, we have devised a test procedure that simulates the cochannel interference by the superposition of two carriers whose modulations can be easily identified and separated. A suitable choice of modulation that simplifies the necessary measurements is a sinusoid of known frequency. The measured capture performance of the receiver can then be presented in terms of a plot of the fundamental component of stronger- (or weaker-) signal modulation frequency as a function of the weaker-to-stronger signal amplitude ratio, a. Generally speaking, a more complete presentation would involve both weaker- and stronger-signal capture characteristics plotted on the same coordinates, as well as a plot of the total distortion as a function of a.

The capture characteristics of a dynamic-trap receiver are determined largely by the two trap parameters that are defined in Fig. 5. The more important of the two is the maximum trap attenuation factor δ. In essence, δ represents the factor by which the amplitude of the undesired (but initially stronger) signal is ideally multiplied. Consequently, while the transition in the capture from one signal to the other occurs in a conventional receiver over a range of a values that is centered about $a = 1$, the transition range for a dynamic-trap receiver should be centered about $a = \delta$. If ρ_1 denotes the capture ratio of the FM demodulator that guides the trap, and ρ_2 denotes that of the FM demodulator that is driven by the trap system, the dynamic-trap receiver will capture the weaker signal with a minimum of distortion for all values of weaker-to-stronger signal amplitude ratio a in the range $\delta/\rho_2 \leqslant a \leqslant \rho_1$.

Some indications of how the shape of the trap attenuation characteristic might affect the shape of the capture characteristics can be arrived at in the following manner: Let the input signals have amplitudes E_s at p rad./s and aEs at $p + r$ rad./s, where $a < 1$ and $r \ll p$. Also, with reference to Fig. 5, let $T(r)$ denote the normalized I.F. transmission as modified by the presence of the trap when the trap center frequency coincides with the frequency of the stronger signal. Note that within the desired I.F. passband, $T(r)$ can have a minimum value of $T(0) = \delta$, and a maximum value of unity when the effect of the trap upon the I.F. response is substantial only within a small frequency band. Thus, the signals at the output of the trap system will be described by a sinusoid of amplitude $T(0)E_s = \delta E_s$ at p rad./s and a sinusoid of amplitude $aE_sT(r)$ at $p + r$ rad./s. For very small values of a, that are such that $a\rho_2 < \delta$, where ρ_2 represents the capture ratio of the FM demodulator that follows the trap system, the initially stronger signal (at p rad./s) will continue to predominate by a sufficient amount to dictate the level of the average voltage at the output of the receiver, irrespective of the value of r. This means that when $a\rho_2 < \delta$, the FM demodulator that follows the trap system will capture the initially stronger signal over the whole modulation cycle (all values of r), and the stronger-signal capture characteristic as a function of a will be flat at its maximum value.

As a is increased so that aE_s becomes greater than δE_s, the signal amplitudes at the output of the trap will become equal at some frequency difference $r = r_c$ which is determined by $aT(r_c) = \delta$. The initially weaker signal at $p + r$ rad./s will be the stronger signal at the output of the trap for all $r > r_c$, and the smaller of the two for $r < r_c$. Consequently, if ρ_2 is close to unity, the initially stronger signal will be captured only over that part of the modulation cycle for which $r < r_c$. The fraction of the time during which the initially stronger signal will be captured will decrease rapidly with increasing values of a when $T(r_c) \simeq 1$, because the points (on the transmission plateaux on both sides of the trap dip) at which $aT(r_c) = \delta$ will then converge rapidly toward the center frequency of the trap attenuation band.

When a reaches values for which $aT(r_c) = \delta$ on the steep flanks of the trap dip, the motion of the points of signal-amplitude equality will be mostly downward with a small component (small decrease in r_c) of horizontal convergence toward the center frequency of the dip. Consequently, the fraction of the time during which the initially stronger signal will be captured will now decrease slowly with increasing a, and the slope of the falling portion of the stronger-signal capture characteristic will taper off (to the point of reaching almost a zero value).

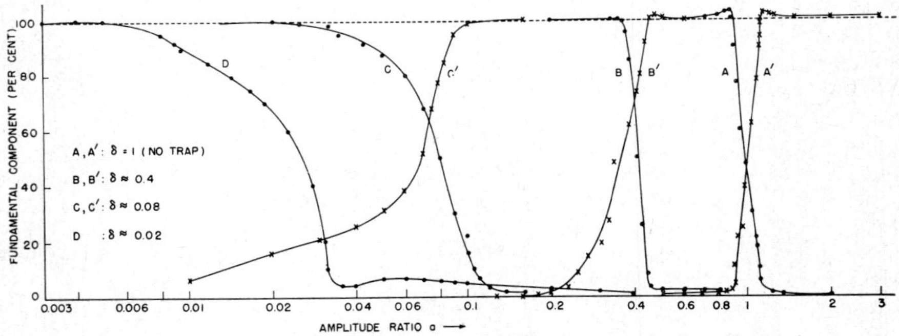

FIG. 6. Capture characteristics of variable-trap receiver for four values of δ; bandwidth is fixed at 15 kc. Curves A, A′, B, B′, C, C′: E_s is modulated ± 30 kc at 400 c/s; E_w is modulated ± 30 kc at 1 kc. Both signals are centered in the passband. Curve D: E_s is modulated ± 35 kc at 400c/s and centered in passband; E_w is unmodulated and at edge of passband. (Prime denotes weaker-signal capture characteristics.) Experimental receiver operates at a frequency of 3·2 Mc, and with an I.F. bandwidth of 120 kc.

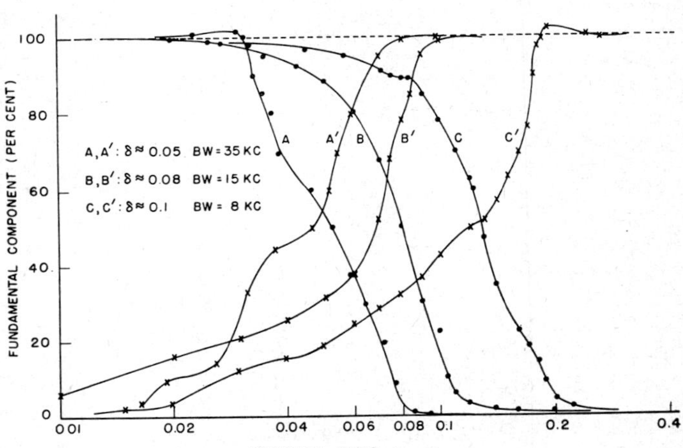

FIG. 7. Capture characteristics of variable-trap receiver for three values of bandwidth. E_s is modulated ± 30 kc at 400 c/s. E_w is modulated ± 30 kc at 1 kc. Both signals centered in passband.

Two sets of measured capture characteristics are presented in Figs. 6 and 7 to illustrate the effect of each of the trap parameters. The signals used in the determination of each curve are described in the figure captions. In Fig. 6, curve A shows the measured capture characteristics of the receiver without the trap. It represents the performance of the I.F. amplifier and the second FM demodulator in conventional FM receiver operation. The capture-transition range is relatively narrow and is centered at $a = 1$. Curves B, C, and D demonstrate the effect of introducing a variable-tuned trap of fixed bandwidth for three different values of δ. In each case, the capture-transition range is shifted from the neighborhood of $a = 1$ to the neighborhood of $a = \delta$.

In Fig. 7, curves A, B, and C represent measured capture characteristics for different values of trap bandwidth. In each measurement, the value of δ was adjusted to the minimum value attainable with the indicated values of bandwidth and within the limitations of the general experimental arrangement. The trap bandwidth is thus seen to affect only the abruptness and the detailed shape of the capture curves within the capture-transition range. Otherwise, the effect of the trap attenuation bandwidth on the over-all capture performance characteristics is not pronounced.

C. Fixed-tuned Trap System

Results that are similar to those just described can also be achieved by using a fixed trap. With reference to Fig. 8, the message of the stronger signal is

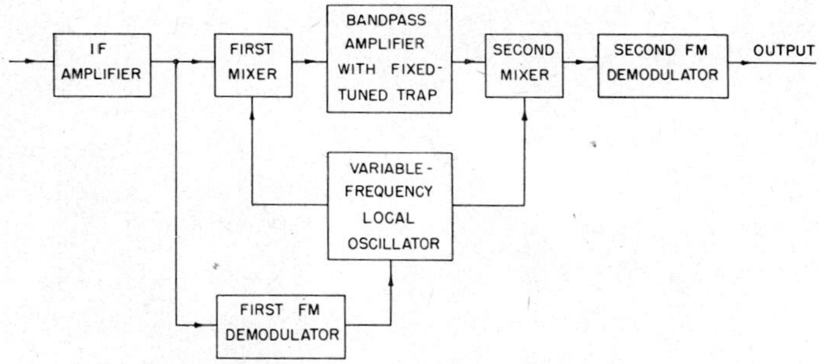

FIG. 8. Block diagram of fixed-tuned trap system.

again derived at the first FM demodulator, and it is used to control a reactance-simulator arrangement in order to deviate the local-oscillator frequency. The purpose of this deviation is to keep the difference in frequency between the local-oscillator voltage and the input stronger signal equal to the constant center frequency of the trap attenuation band. This amounts to the regeneration of only the stronger signal at a second intermediate frequency that differs from the first by an amount that equals the constant center frequency of the fixed trap. Other systems are now being studied that will achieve a substantial approximation to this result without having to extract the modulation of the stronger signal for use in modulating a local oscillator.

In Fig. 8, we also note that the same local oscillator is made to drive a second stage of frequency conversion whose signal source is the output of the fixed trap. Since the output of the fixed trap is made up of a fixed-frequency residual signal at the center of the band plus a predominant modified signal that carries the algebraic sum of the modulations of the two input signals, the second mixing operation will restore the original frequency modulations to the modified signals,

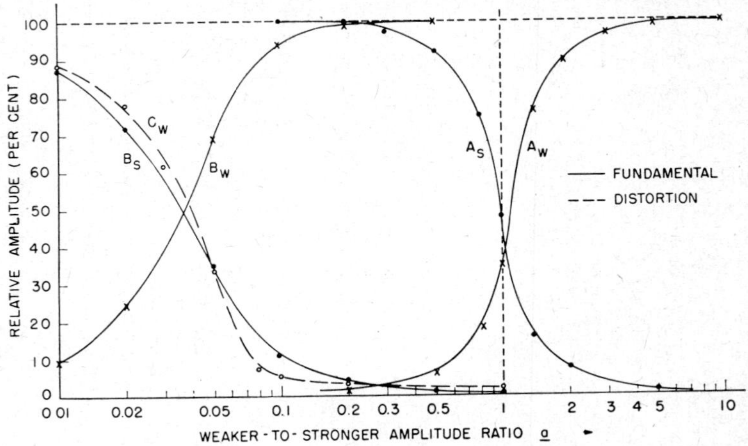

FIG. 9. Capture characteristics of experimental receiver with (curves B_s and B_w) and without (curves A_s and A_w) the fixed-tuned trap system. The receiver operates at a frequency of 10·7 Mc. and with an I.F. bandwidth of 200 kc. Signals are centered within the passband. E_s is modulated $\pm$ 75 kc at 400 c/s; E_w is modulated $\pm$ 75 kc at 1000 c/s.

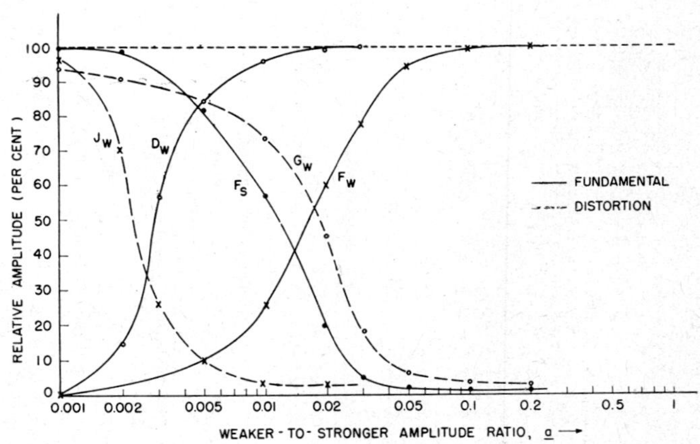

FIG. 10. Capture and total-distortion characteristics of experimental fixed-tuned trap receiver. J_w and G_w are curves of the total distortion that accompanies the message of the weaker signal.

J_w and D_w: stronger signal unmodulated; weaker signal modulated $\pm$ 75 kc at 1000 c/s.

F_s, G_w, F_w: stronger signal modulated $\pm$ 35 kc at 400 c/s; weaker signal modulated, $\pm$ 35kc at 1000 kc.

but the desired message will now be carried by the stronger of the two. A second conventional FM demodulator following the second mixer operates on the modified signals to deliver the message of the desired (originally weaker) signal.

The theoretical considerations that determine the scope of usefulness and the limitations of the present system are similar to those that were pointed out in connection with the variable trap. But a significant advantage of the fixed trap is that the value of δ must be controlled at one frequency only, whereas, in the variable trap, δ must be controlled over a wide frequency band.

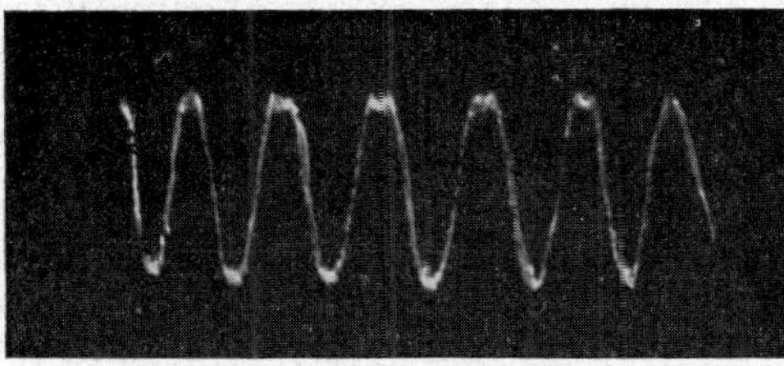

(a) $a = 0{\cdot}005$

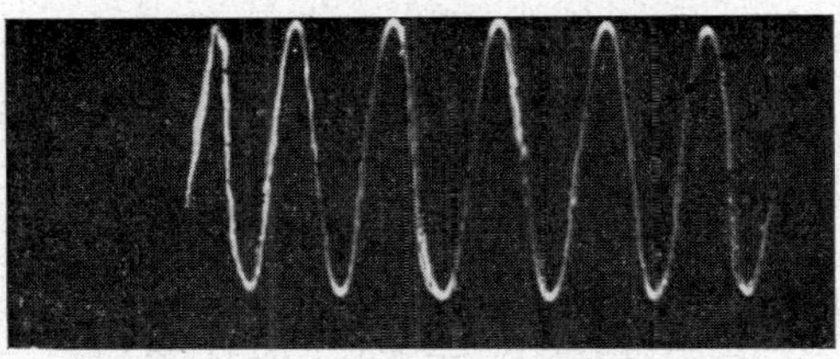

(b) $a = 0{\cdot}01$

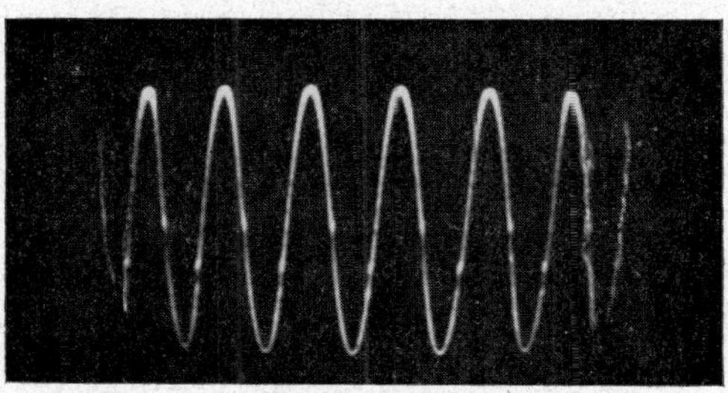

(c) $a = 0{\cdot}05$

FIG. 11. Oscillograms of retrieved weaker-signal modulation in the output of the experimental fixed-trap receiver. Stronger signal unmodulated.

D. Performance of an Experimental Fixed-trap System

An experimental system that embodies these principles has been built and tested [7]. The trap circuit used here was similar to the one used in the variable-tuned trap system. Representative experimental results are summarized in Figs. 9, 10, and 11.

In Fig. 9, the curves A_w and A_s are the weaker- and stronger-signal capture characteristics of the I.F. amplifier and second FM demodulator of the present system, operated as a conventional receiver. The introduction of the trap system results in the characteristics B_w and B_s. The trap system thus shifts the center of the capture transition range from $a = 1$ to $a \simeq 0{\cdot}04$. The total distortion that is superimposed upon the fundamental modulation frequency of the weaker signal is plotted as curve C_w.

The best performance of the present experimental system is illustrated by the capture characteristics of Fig. 10 and the oscillograms of Fig. 11.

III. Feedforward Technique

In feedforward systems, the resultant of the two input signals is channeled through two independent unilateral signal paths whose outputs are recombined at a later stage in the receiver. In this way, the signals can be processed so that their relative amplitudes at the output of one path are significantly different from their relative amplitudes at the output of the other path. Appropriate superposition of the resulting receiver-path outputs would then lead to the suppression of whichever of the two signals is undesired. Specifically, in one of the signal paths, a cascade of narrow-band limiters (or some other means of enhancing the predominance of the stronger over the weaker signal) might be used. In the other path, the primary object would be to provide a phase-shift characteristic that is identical with that of the first path within the desired passband. If the outputs of these paths are combined subtractively, the resulting signal cancellations can be made to suppress whichever of the two signals is undesired. The parameter that decides which of the two signals is to predominate is the ratio of path outputs. Control of this ratio enables the receiver to switch from capture of the weaker to capture of the stronger signal.

A. Analysis of a Simple Feedforward System

An interesting example of a feedforward system is shown in the block diagram of Fig. 12a. The limiter and amplifier in the lower signal path are, in effect, equivalent to one idealized narrow-band limiter. The upper signal path contains two narrow-band limiters. Each idealized narrow-band limiter is assumed to incorporate an ideal filter of one I.F. bandwidth and to deliver an output sinusoid whose amplitude is k volts in response to an input sinusoid of amplitude $E_{\text{in}} > 0$. Limiters are inserted in each of the two signal paths in order to ensure that the properties of the output signal will be independent of the input signal level, as long as the resultant input signal amplitude exceeds a certain threshold value.

A system that is identical in its effect with the one shown in Fig. 12a, but which requires one less limiter, is shown in Fig. 12b. The arrangement in Fig. 12b is the more desirable for practical realization. It is easy to show that if the first limiter in Fig. 12b is not narrow-band, the combination of the second-limiter and amplifier outputs in phase opposition cannot effect the type of cancellation that is necessary for achieving beneficial changes in the interference conditions at the output.

The interference is considered to arise from the simultaneous presence of two signals of amplitudes E_s and aE_s $(a < 1)$ and frequencies p and $p + r$ rad./s $(r \ll p)$ within the passband of the I.F. amplifier. To simplify the analysis, these two carriers will be assumed to have constant amplitudes; their frequency modulations are assumed to be so slow that, for the time being, the frequencies can be

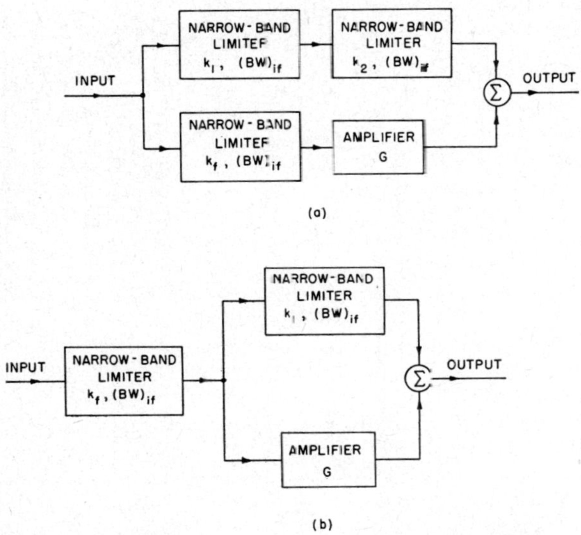

FIG. 12. (a) Simple feedforward system. (b) Simplified form of the system of (a).

considered stationary. Thus, the resultant signal at the output of the I.F. amplifier can be expressed as

$$e(t) = E_s \cos pt + aE_s \cos (p + r)t \text{ with } a < 1 \text{ and } r \ll p. \tag{1}$$

The ideal limiter is, by definition, a device that will operate upon the resultant of the two carriers and deliver an output signal given by

$$e_l(t) = k \cos [pt + \theta(t)], \tag{2}$$

where k is a constant, and

$$\theta(t) = \tan^{-1} \frac{a \sin rt}{1 + a \cos rt}. \tag{3}$$

By a direct Fourier analysis [3]

$$e_l(t) = k \sum_{n=-\infty}^{\infty} A_n(a) \cos (p - nr)t. \tag{4}$$

In the analysis of the equivalent feedforward systems of Fig. 12a and b, we note that for a given weaker-to-stronger signal amplitude ratio a at the input, the conditions of interference that arise with the frequency differences $r = 0$ and

$r = (BW)_{if}$ are of special interest. The importance of these two limiting conditions stems from the fact that the amount of reduction in the equivalent weaker-to-stronger signal amplitude ratio that will be effected by a chain of narrow-band limiters decreases with a decrease in the value of the frequency difference, r, between the two signals, relative to the limiter bandwidth [4]. The effect of narrow-band limiting upon the FM disturbance caused by interference between

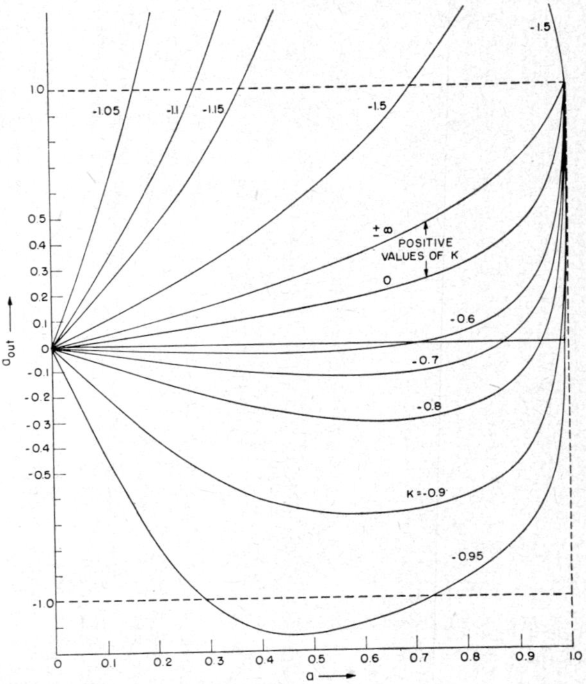

FIG. 13. Plots of the ratio of weaker-to-stronger signal amplitude, a_{out}, at the output of the system of Fig. 12 as a function of the input ratio a, when the frequency difference exceeds one-half the I.F. bandwidth.

two signals will increase with an increase in the degree of frequency-band limitation suffered by the amplitude-limited resultant of the two signals when they go through the limiter filter. For $r = 0$, the amplitude-limited resultant of the two input signals will experience no frequency-band limitation in going through the limiter filters, whereas the greatest possible band limitation will be experienced with $r = (BW)_{if}$. When the frequency difference r lies between $r = 0$ and $r = (BW)_{if}$ the effect of passing the resultant signal through the system will be intermediate between the extremes indicated for $r = 0$ and $r = (BW)_{if}$.

Thus, for frequency differences that exceed one-half of the I.F. bandwidth, the spectrum at the output of the idealized narrow-band limiter will consist of only the components whose frequencies are p and $p + r$ rad./s. From Equation (4), we find that these components have amplitudes given by $k_1A_0(a)$ and $k_1A_{-1}(a)$ if they are observed at the output of the first-encountered narrow-band limiter, and $k_2A_0(a')$ and $k_2A_{-1}(a')$, $a' \equiv A_{-1}(a)/A_0(a)$, if they are observed at the output of the second-encountered limiter, and so on. If the two signal paths are assumed to have identical phase characteristics, an additive combination of the path outputs results in two signals at p and $p + r$ rad./s with new relative amplitudes. The quantity

$$a_{\text{out}} = \frac{A_{-1}(a') + KA_{-1}(a)}{A_0(a') + KA_0(a)}, \tag{5}$$

with k_f, k_2, and G as defined in Fig. 12, and

$$K \equiv (k_f/k_2)G, \tag{6}$$

constitutes the ratio of the amplitude of the signal at $p + r$ rad./s (which corresponds to the originally weaker signal) to that of the signal at p rad./s.

A family of curves for a_{out} versus a is shown in Fig. 13, with K as a running parameter. Note that K is the ratio of the output of the lower signal path to the output of the upper signal path when the input excitation is a single unmodulated carrier. Negative values of K can be interpreted as indicating the condition in which the phase characteristics of the two paths differ by a constant value of π (or by an odd multiple of it) but are otherwise identical functions of frequency. From its definition in Equation (6) the parameter K depends only upon certain design constants of which the gain of the amplifier is an easily controllable factor. Evidently, K would not be a constant that is subject to *a priori* adjustment for all usable signal levels at the input if the lower signal path in Fig. 12a did not contain the limiter that is shown, or if the first limiter in Fig. 12b were absent. That the parameter K should be substantially independent of the input signal levels is an important requirement for achieving interference-suppression performance that is independent of the input stronger-signal level.

The curves of Fig. 13 reveal some interesting possibilities in relation to the capture of the weaker or of the stronger of the two input signals under the conditions of this analysis. For example, for values of K that lie in the range $-0{\cdot}8 < K < 0$, the system will depress the ratio of weaker-to-stronger signal amplitude to a value that is smaller than 0·4 for all input values of this ratio that are below 0·9. This represents substantial enhancement of the predominance of the originally stronger signal. The curve for $K = -0{\cdot}6$ shows that the value of a_{out} is less than 0·05 for all a less than 0·8.

Values of a_{out} that exceed unity correspond to the situation in which the combination of the signal-path outputs enables the originally weaker signal to emerge as the stronger of the two. Thus, for $K = -1{\cdot}05$, a_{out} is greater than one for all a in the range $0{\cdot}16 < a < 1$. This potentiality for capturing the weaker signal is even more clearly brought out by the curves of $1/a_{\text{out}}$ versus a shown in Fig. 14. It is evident that as $-K$ approaches 1, from the right, the range of a values in which the originally weaker signal will emerge as the stronger widens and its lower limit approaches $a = 0$. But values of K that are centered about -1 indicate signal cancellation that becomes more and more complete as K approaches -1. It is clear, therefore, that a limit on how closely K can approach -1 is placed by considerations that relate, first, to the random noise level at the output of the system; second, to the signal-level requirements and sensitivity of the stages driven by it; and last, but not least, to the role that the presence of other sideband components will play in deciding the character of the resultant output signal when the frequency difference r becomes smaller than one-half of $(BW)_{if}$.

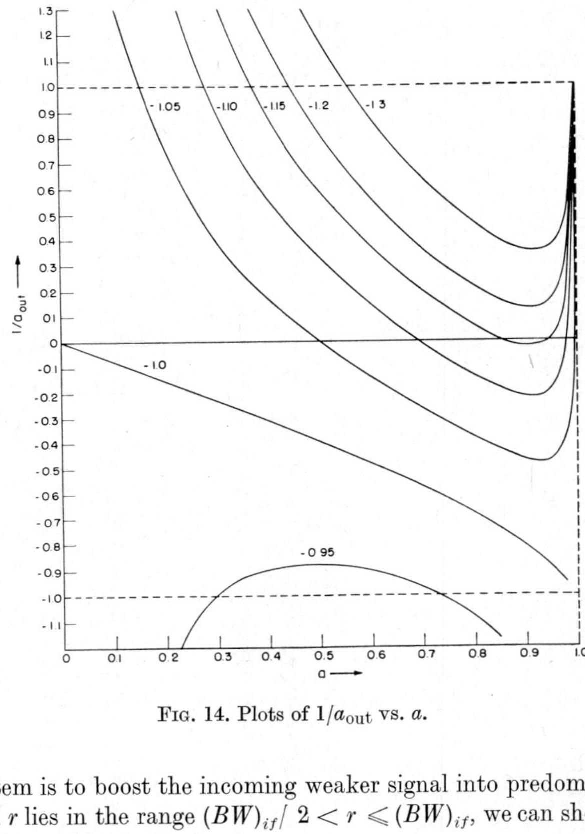

FIG. 14. Plots of $1/a_{\text{out}}$ vs. a.

If the system is to boost the incoming weaker signal into predominance at the output when r lies in the range $(BW)_{if}/\,2 < r \leqslant (BW)_{if}$, we can show that $-K$ must be chosen within the range

$$\rho_{UW} > -K > \rho_{LW}$$

with ρ_{UW} and ρ_{LW} as plotted in Fig. 15. If however, we wish to employ this system in order to improve the predominance of the incoming stronger signal, then K must lie in the range $K > -\rho_s$, with ρ_s as plotted in Fig. 15. Figure 15 also shows a plot of K_∞, the value of K for which the cancellation at the output of the system will reduce the amplitude of the component at p rad./s to zero, and a plot of K_0, the value of K that will lead to the nullification of the amplitude of the signal at $p + r$ rad./s If K is chosen equal to K_∞, so that the component at p rad./s is completely cancelled out at the output for a certain value of a, then a sudden removal of the stronger signal at the input will cause the amplitude of the desired signal at the output of the system to drop to a value that equals $(3/4)a$ of its value in the presence of the stronger signal at the input. In the trap system, the sudden removal of the stronger signal at the input would throw the desired signal to the center frequency of the trap.

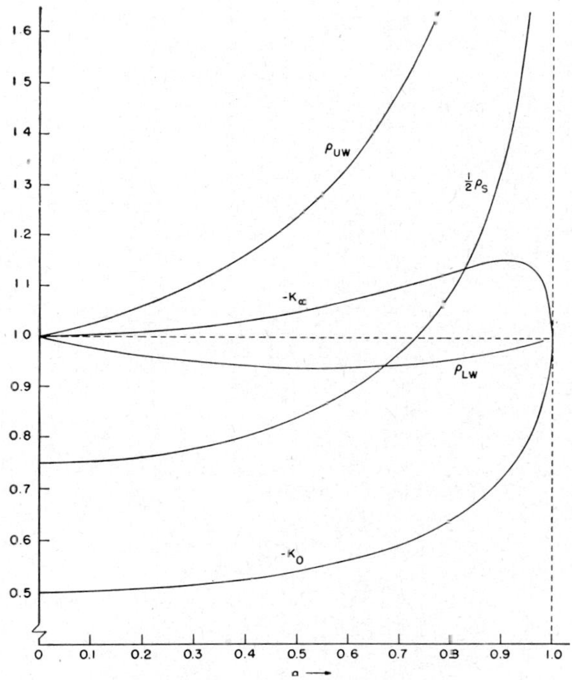

FIG. 15. Plots of the functions ρ_{UW}, ρ_{LW}, ρ_s, K_0, and K_∞ that are used in the text.

Consider, next, the situation in which r approaches zero. For values of r that are less than

$$r_{\min} = \beta \frac{1-a}{1+a}(BW)_{if} \tag{8}$$

where $\beta \simeq 0{\cdot}2$ for an ideal filter, the signal at the output of each limiter filter will approach the amplitude-limited resultant of the two input signals more and more closely. Consequently, the signal at the output of the system will approach

$$e_{\text{out}}(t) = k_2(1 + K) \cos [pt + \theta(t)] \tag{9}$$

with $\theta(t)$ as defined in Equation (3). This shows that for values of r that make up a small fraction of the limiter-filter bandwidth, the average frequency of the output signal will always equal the frequency of the stronger of the two input signals. This means that if the chosen value of K enables the system to deliver an output signal whose average frequency equals the frequency of the weaker signal when r falls in the range $(BW)_{if}/2 < r < (BW)_{if}$, this condition for the capture of the weaker signal will not subsist as r takes on values that are small fractions of the I.F. bandwidth. Consequently, with a given value of input weaker-to-stronger signal amplitude ratio, the capture at the output of the system will shift from the weaker to the stronger signal as r is decreased, and back to the weaker signal as r is increased again. The transition in the capture will take place within a frequency-difference range centered about a value of r that is intermediate between $r_{\min}$ (as given by Equation (8)) and $(BW)_{if}/2$. While r is going through this range of values, the reception at the output of the succeeding FM demodulator will be marred by the severe distortion that is usually experienced with conventional FM receivers when a exceeds the capture ratio of the receiver and approaches unity. Evidently, if either of the systems of Fig. 12a and b is used to facilitate the capture of the weaker signal, it should be followed by an FM demodulator of high stronger-signal capture capability in order to minimize the duration of the severe distortion that accompanies the capture transition from one signal to the other. However, if the system is used to suppress the weaker signal, no capture transitions will arise in the course of a modulation cycle. Hence, the system should always perform better as a weaker-signal suppressor when the frequency difference between the two input sinusoids can be expected to go to zero during every modulation cycle.

Detailed numerical analysis indicates that as the frequency difference between the two signals decreases, the amount by which the relative amplitudes of the spectral components at the output of the upper signal path will differ from the relative amplitudes of the corresponding components in the lower signal path decreases also. For values of the frequency difference that constitute small fractions of $(BW)_{if}$, there is no excess narrow-band limiting effect in the upper signal path, and the relative amplitudes of corresponding components in each path are the same. This means that the ability of this system to enhance the capture of either the stronger or the weaker of the two signals also decreases with a decrease in the value of r relative to $(BW)_{if}$. If the system is adjusted to boost the weaker signal, then the most unfavorable instantaneous-frequency position for this signal within the passband is the one in which the weaker signal falls nearer one of the band edges than does the stronger signal while the frequency difference r equals some critical value, $r_{\text{crit}} < (BW)_{if}/2$. This value of r is a very complicated function of K and a. The most detrimental distortion in the reception of the weaker signal is actually the distortion that arises while the average frequency at the output is in transition from the value dictated by the frequency of one of the signals to the value dictated by the frequency of the other. The difference between these average values is small when r is small.

The capture-transition distortion, which appears whenever the two signals approach a condition of frequency crossover, constitutes a performance limitation on the simpler (and more practical) forms of the feedforward technique in applications that require close reproduction of the weaker-signal message. If the two signals are cochannel, they may often pass through zero frequency difference, with consequent severe distortion in the weaker-signal reception at the receiver output. However, if the center frequencies of the signals are separated so that their instantaneous frequencies seldom, or never, coincide, the distortion will not be present. In a way, the feedforward system of Fig. 12 is an extremely simple realization of a simulated ideal bandpass filter with extremely sharp cutoff characteristics for FM signals that can be expected to sweep over non-overlapping (or only slightly overlapping) frequency channels.

The weaker- (or stronger-) signal capture-enhancement performance of the feedforward scheme can be improved significantly over the performance indicated in the results of the preceding analysis of the system of Fig. 12, at the expense of increased complication of the system. One way to do this is to increase the number of narrow-band limiters in the upper signal paths shown in Fig. 12. Another way is to use feedback around the limiter, but these possibilities will not be considered here. Although small differences in the phase shifts of the two paths may not affect the over-all performance materially, the difficulty in achieving proper phasing of the combined outputs within tolerable limits is a decided disadvantage of the more complicated systems.

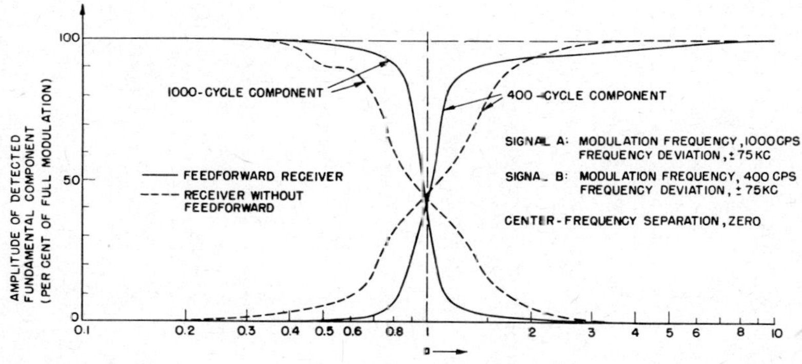

FIG. 16. Stronger-signal capture characteristics of experimental receiver, with and without feedforward.

B. Performance of an Experimental Feedforward System

An experimental feedforward system based on the block diagram of Fig. 12b was built and tested [8]. The measured capture performance of this system is illustrated by the curves in Figs. 16, 17, and 18, and by the oscillograms of Figs. 19 and 20. The signals used in the measurements are described in the figures. The curves of Fig. 16 indicate a substantial improvement in the stronger-signal capture ratio of the receiver by the introduction of the feedforward circuit with the appropriate adjustment.

The curves of Fig. 17 show that the circuit is able to alter the relative signal strengths so that the weaker of two cochannel signals predominates in the approximate range $0{\cdot}6 < a < 0{\cdot}8$, as well as over the reciprocal range. The appearance of the detected waveform at the output is shown in the oscillogram of Fig. 19c for $a = 0{\cdot}7$. The curves were measured with the system adjusted for

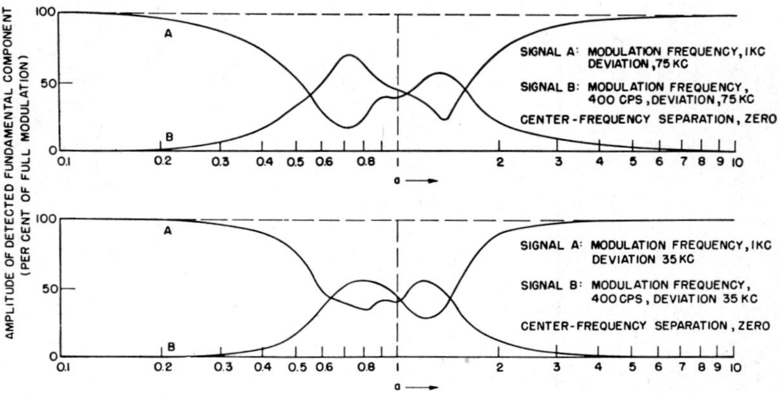

FIG. 17. Weaker-signal capture characteristics of the feedforward receiver: cochannel signals.

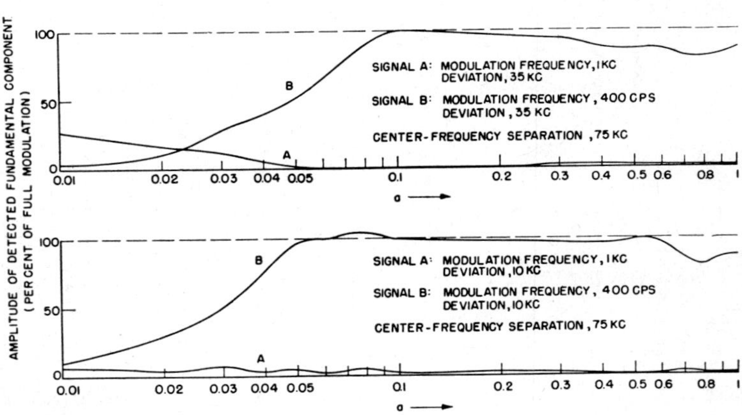

FIG. 18. Weaker-signal capture characteristics of the feedforward receiver; optimum performance with half-channel signals.

best weaker-signal capture at $a = 0{\cdot}7$. Similar unrecorded results, with almost the same quality of output waveform, were also observed when the adjustment was made for best weaker-signal capture at a values of less than $0{\cdot}1$. The adjustment of the signal combination ratio at the output of the feedforward system for best performance changes slightly with the value of a. The adjustment in Fig. 19b was also made for best capture at $a = 0{\cdot}7$, but the stronger signal was unmodulated. Notice that the stronger signal is captured momentarily when the frequencies of the two signals become approximately equal. The intensity of the capture-transition distortion has been smoothed out by the audio filter.

The quality of weaker-signal capture performance improves rapidly as the unmodulated carrier frequencies are separated. The performance that was achieved, with an optimum setting for each value of a, is illustrated by the

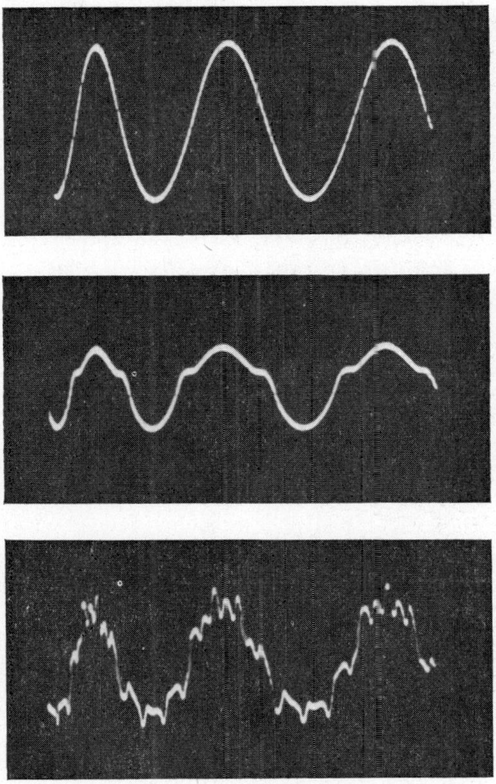

(*a*)

(*b*)

(*c*)

Fig. 19. Capture performance of experimental feedforward receiver when the signals have the same center frequency.

(a) Weaker-signal modulation (400 c/s). (b) Detected weaker-signal modulation with capture-transition distortion from unmodulated stronger signal. (c) Detected weaker-signal modulation with capture-transition distortion from a stronger signal with 1000-c/s tone modulation.

capture curves of Fig. 18 for the situation in which the signal frequencies sweep contiguous, but separate, half-channels. The oscillograms of Fig. 20 illustrate the quality of the received waveforms. The experimental receiver that incorporated the feedforward system was designed so that the weaker- and the stronger-signal messages were simultaneously available at two independent outputs, with the stronger- and weaker-signal "channels" sharing all circuits up to and including the amplifier path of the feedforward system. At that point, two feedforward arrangements were made, one to capture the stronger signal, and the other to capture the weaker signal. The signals were then fed to two separate FM demodulators that delivered the waveforms shown in Fig. 20.

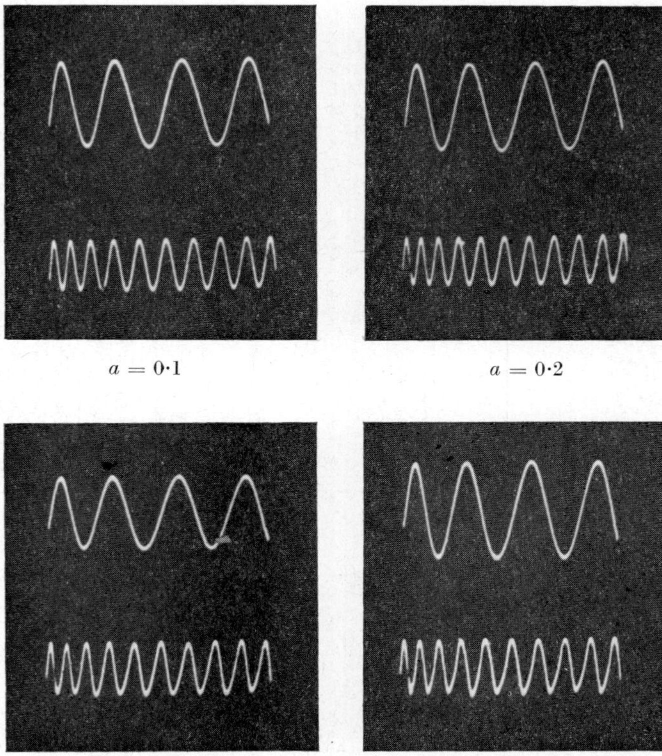

$a = 0{\cdot}1$ $a = 0{\cdot}2$

$a = 0{\cdot}5$ $a = 0{\cdot}9$

FIG. 20. Simultaneous outputs of the weaker-signal channel (upper trace) and the stronger-signal channel (lower trace) of experimental feedforward receiver designed for simultaneous duplex reception. The 150-kc I.F. channel (at 10·7 Mc) is divided into two 75-kc bands. Each carrier is centered in one band and is swept with a deviation of $\pm$ 35 kc. The circuits were adjusted for best capture in each case.

IV. APPLICATION TO FM MULTIPLEXING

We may interpret the fundamental principle of multiplexing as being the deliberate creation of fundamental differences, either between the carriers or between the modulations impressed on the same carrier, which will enable circuits that can distinguish these differences to separate the desired messages from the undesired in a multitude of simultaneously transmitted messages. The successful separation of the transmitted messages involves, fundamentally, the solution of an interference problem. Carriers can be made easily distinguishable, and hence separable, on the basis of carrier frequency or phase, or on the basis of the time of the occurrence of an event that characterizes the carrier. Modulations can be made distinguishable either through the use of subcarriers, if all of the messages are to modulate the same parameter of the ultimate carrier, or through the modulation of separate distinguishable parameters of the carrier by separate messages, or by a combination of both.

The possibility of achieving an adequate capture of the weaker signal with economical modifications in a conventional FM receiver suggests that a new system of "amplitude-discrimination" duplexing could be used to double the number of messages that are transmitted on a given FM channel. In this system, a given transmitter would radiate two cochannel radio-frequency carriers that need generally differ only in amplitude. In a commercial broadcast application of this system, the compatibility of the system to existing conventional FM receivers can be assured by a proper choice of the amplitude ratio of the two carriers. FM receivers that are intended to receive the duplex transmission could, if desired, be equipped with two outputs—one for each program. In this way, binaural and other programs could be transmitted on a single FM channel.

In some applications, this possible technique for doubling the number of messages transmitted over a single channel might be considered to augment, rather than to compete with, other existing techniques. But, in any case, we shall not attempt to debate the merits of various systems here. Our purpose is to point out some interesting results that were obtained with the experimental models of the systems discussed in the preceding sections. These results are summarized in Figs. 17, 18, 19, 20, 21, and 22. Figure 21 shows experimental curves that illustrate the manner in which the total distortion that appears on the retrieved message of the weaker signal varies with the frequency separation of the unmodulated carriers, when the weaker signal is captured with the experimental variable-tuned trap system. Figure 22 shows total-distortion curves and capture characteristics that were obtained with the experimental fixed-trap system when the signal frequencies were swept over non-overlapping, but contiguous, frequency ranges.

These results indicate, in general, that in applications in which stringent requirements on the quality of reproduced messages at the output are imposed, the assigned channel must be divided into two contiguous subchannels in order to minimize the unavoidable distortion that will result from frequency crossovers between the two carriers. Such applications might, for example, be concerned with the transmission of two independent high-quality programs or with binaural transmission of one such program. The techniques that have been described for the separation of the carriers amount, in these applications, to simple, effective, and economical realizations of a close approximation to the ideal filter with the extremely sharp transition from passband to rejection band. In other applications, in which the emphasis is placed on the usefulness rather than on the high quality of the received replicas of the transmitted message, each carrier could, if it is so

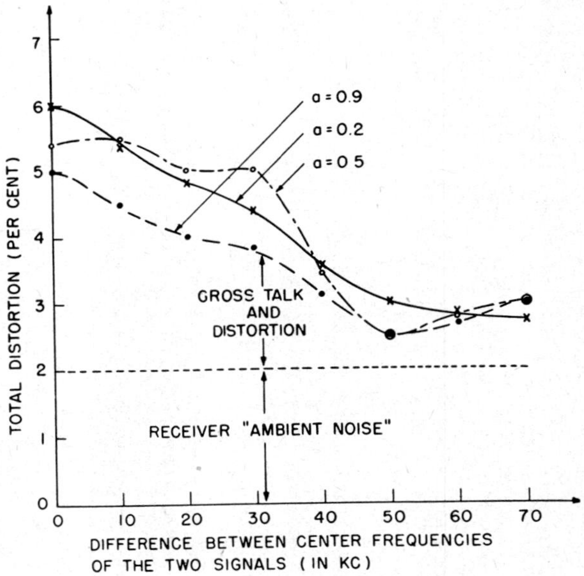

FIG. 21. Variation of the total distortion that is superimposed upon the retrieved message of the weaker signal, with the center-frequency separation between the two signals, when the weaker signal is captured with the experimental variable-tuned trap system. The "ambient noise" of the system itself is approximately 2 %; the remainder is largely cross talk from the stronger signal. Stronger signal modulated ± 20 kc at 1000 c/s. Weaker signal modulated ± 20 kc at 400 c/s.

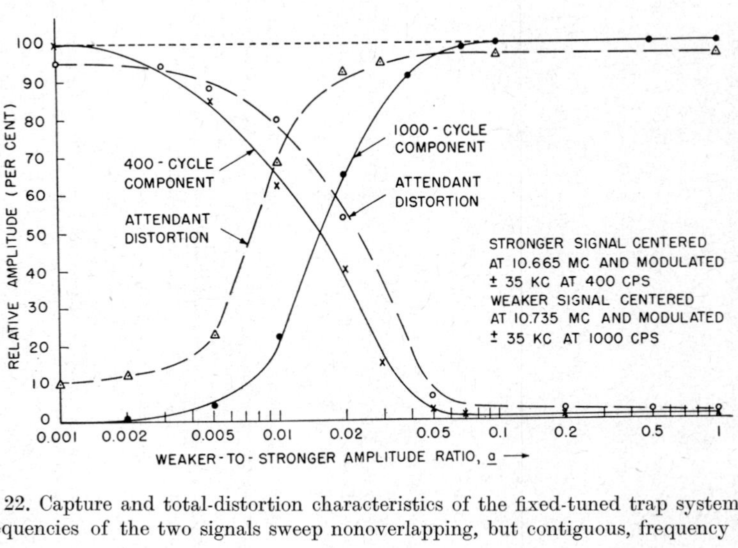

FIG. 22. Capture and total-distortion characteristics of the fixed-tuned trap system when the frequencies of the two signals sweep nonoverlapping, but contiguous, frequency bands.

desired, be swept in frequency over the entire assigned channel. Note that only the received message of the weaker carrier will then show unavoidable distortion as a result of this deliberate interference. The stronger signal is still capturable within stringent requirements on quality.

Acknowledgment

The author wishes to acknowledge the contributions of his students, G. J. Rubissow, R. H. Small, and F. I. Sheftman, who tirelessly carried out the experimental work and recorded the experimental results that are described in this paper while they were working on their Master's theses under the author's supervision.

References

1. Baghdady, E. J. (1957). *Tech. Rep.* 332, Res. Lab. Electron., Mass. Inst. Technol.
2. Baghdady, E. J. (1958). *Trans. Inst. Radio Engrs, N.Y. CT*-5, no. 3, 202–214.
3. Baghdady, E. J. (1956). *Tech. Rep.* 252, Res. Lab. Electron., Mass. Inst. Technol.
4. Baghdady, E. J. (1958). *Proc. Inst. Radio Engrs, N.Y.* **46**, 728–738.
5. Baghdady, E. J. & Rubissow, G. J. (1959) *Electronics* **32**, 64–66.
6. Rubissow, G. J. (1958). S.M. Thesis, Dep. Elect. Engng, Mass. Inst. Technol.
7. Sheftman, F. I. (1958). S.M. Thesis, Dep. Elect. Engng, Mass. Inst. Technol.
8. Small, R. H. (1958). S.M. Thesis, Dep. Elect. Engng, Mass. Inst. Technol.

Simplified Method for Computing Knife-edge Diffraction in the Shadow Region

L. J. ANDERSON, L. G. TROLESE and S. WEISBROD

Smyth Research Associates, San Diego, Calif.

A simplified method of computing knife-edge diffraction in the shadow region is presented which is applicable to most "obstacle gain" paths, that is, paths with a dominant mountain obstacle. The method is applicable to the 4-ray model, where specular ground reflection takes place on both sides of the obstacle. The summation of four field components with differing amplitude and phase coefficients is replaced by a product of three terms, which can be evaluated conveniently from graphs. The method can also be used for 2-ray model—reflection on one side of the obstacle only, and the single ray model—no reflection. The area of applicability and accuracy of the method are discussed.

On a propagation path dominated by a prominent mountain obstacle, the received field strength is usually computed by use of the classical knife-edge diffraction theory of optics. In most cases of practical interest in radio propagation, one must consider not only the direct ray from transmitter to receiver via the knife-edge, but also the rays which are reflected from the ground on either or both sides of the knife-edge [2, 3, 4, 5]. In cases where such reflections occur all rays must be added in their proper phase and amplitude, in order to determine the resultant field:

$$E = |E_0| \sum_j F_j \exp i\psi_j \tag{1}$$

where E_0 is the free space field (no knife-edge present), F_j is the amplitude and ψ_j the phase of the jth ray. The phase ψ includes four components, the phase β due to diffraction over the knife-edge, the phase of the free space path length differences, Δ_D, associated with each ray, the phase change caused by diffraction path length differences δ, and the phase change due to reflection from the ground. Figure 1 shows the geometry. T–R is the free space path length associated with ray 0 (T–O–R), while T–R' and T'–R are the free space paths for the singly reflected ray 1 (T–O–R'), and ray 2 (T'–O–R). The doubly reflected ray 3 (T'–O–R') is similarly referred to T'–R' which has the same path length as T–R.

The amplitude F and phase β due to knife-edge diffraction are given by

$$F = (S + \tfrac{1}{2})\sqrt{2}\sin(\beta + \pi/4) \tag{2}$$

$$\tan(\beta + \pi/4) = (S + \tfrac{1}{2})/(C + \tfrac{1}{2}) \tag{3}$$

where S and C are the Fresnel integrals

$$C = \int_0^v \cos \tfrac{1}{2}\pi v^2\, dv$$

$$S = \int_0^v \sin \tfrac{1}{2}\pi v^2\, dv$$

v is the dimensionless parameter which expresses the diffraction geometry given by

$$v = h(2D/\lambda d_1 d_2)^{\frac{1}{2}} \qquad D = d_1 + d_2 \tag{4}$$

where λ is the wavelength and other quantities are as shown in Fig. 1. The height

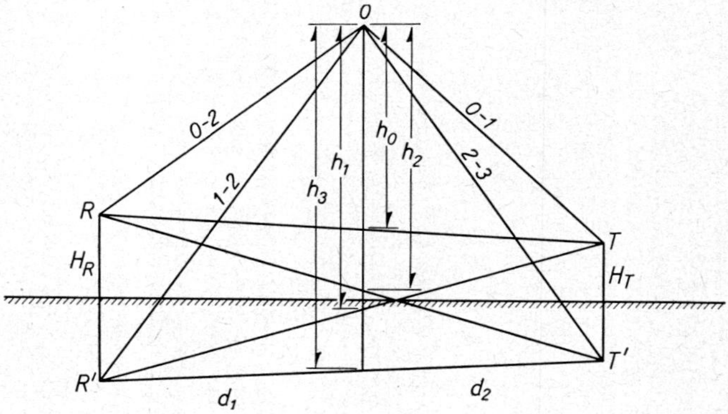

FIG. 1.

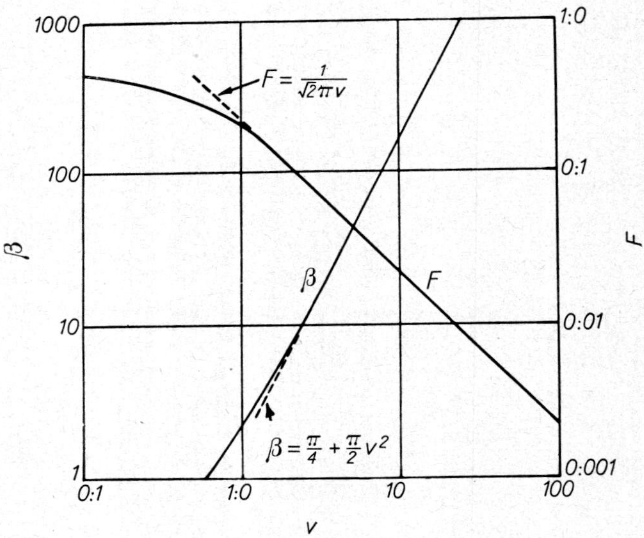

FIG. 2. Amplitude and phase *vs* V_0 in the shadow region.

of the knife-edge h_j takes on the different values h_0, h_1, h_2, and h_3, for the different rays.

Computation of the field by evaluation of the proper amplitude and phase terms and substitution in Equation (1) is a tedious procedure, particularly if the field strength is desired for a number of terminal heights so as to obtain a height-gain curve. A greatly simplified computation method can be developed by use of the asymptotic approximations for Equations (2) and (3)

$$F_j = 1/\sqrt{2\pi v_j} \quad \text{and} \quad \beta_j = \pi/4 + \tfrac{1}{2}\pi v_j^2 \tag{5}$$

Which are valid for $v > 1$ as shown in Fig. 2. If we define

$$\begin{aligned} v_0 &= h_0(2D/\lambda d_1 d_2)^{\frac{1}{2}} \\ v_1 &= v_0 + \Delta v_1 \\ v_2 &= v_0 + \Delta v_2 \\ v_3 &= v_0 + \Delta v_1 + \Delta v_2 \end{aligned} \tag{6}$$

and

$$\begin{aligned} \beta_1 - \beta_0 &= k\delta_1' \\ \beta_2 - \beta_0 &= k\delta_2' \\ \beta_3 - \beta_0 &= k\delta_3' \\ k &= 2\pi/\lambda. \end{aligned} \tag{7}$$

Then substitution in Equation (1) gives

$$\frac{E}{|E_0|} = \frac{1}{\sqrt{2\pi v_0}}\left[1 + \frac{\rho_1 \exp i[k(\delta_1' + \Delta D_1) + \phi_1]}{1 + \Delta v_1/v_0} + \frac{\rho_2 \exp i[k(\delta_2' + \Delta D_2) + \phi_2]}{1 + \Delta v_2/v_0} + \frac{\rho_1\rho_2 \exp i[k(\delta_3' + \Delta D_3) + \phi_1 + \phi_2]}{1 + \Delta v_1/v_0 + \Delta v_2/v_0}\right] \tag{8}$$

where $\rho_1 \exp i\phi_1$ and $\rho_2 \exp i\phi_2$ are the ground reflection coefficients on the receiver and transmitter sides, respectively.

In order to determine the diffraction path length differences δ_1 and δ_2, as well as the free space (no obstacle) path differences, ΔD_1, ΔD_2, and ΔD_3, we obtain from Fig. 1,

$$\begin{aligned} \delta_1 &= T\text{-}O\text{-}R' - T\text{-}O\text{-}R = H_R(h_0 + h_3)/d_1 \\ \delta_2 &= T'\text{-}O\text{-}R - T\text{-}O\text{-}R = H_T(h_0 + h_3)/d_2 \\ \Delta D_1 &= T\text{-}R' - T\text{-}R = [D^2 + (H_R + H_T)^2]^{\frac{1}{2}} - [D^2 + (H_R - H_T)^2]^{\frac{1}{2}} \\ &= 2H_R H_T/D. \\ \Delta D_2 &= T'\text{-}R - T\text{-}R = \Delta D_1 = 2H_R H_T/D \\ \Delta D_3 &= T'\text{-}R' - T\text{-}R = 0 \end{aligned} \tag{9}$$

$$\begin{aligned} h_1 &= h_0 + 2H_R d_1/D = h_3 - 2H_T d_2/D \\ h_2 &= h_0 + 2H_T d_2/D = h_3 - 2H_R d_1/D \\ h_3 - h_0 &= 2H_R d_1/D + 2H_T d_2/D. \end{aligned} \tag{10}$$

By use of Equations (4), (5), and (7), δ_1' can be expressed as

$$\begin{aligned} \delta_1' &= (D/2d_1d_2)(h_1 - h_0) = (D/2d_1d_2)(h_1 - h_0)[(h_0 + h_3) - (h_3 - h_1)] = \delta_1 - \Delta D_1 \\ \delta_2' &= \delta_2 - \Delta D_2 \\ \delta_3' &= (D/2d_1d_2)(h_3 - h_0)^2 = (D/2d_1d_2)(h_3 - h_0)(h_3 + h_0) = \delta_1 + \delta_2. \end{aligned} \tag{11}$$

Using results of Equation (11) together with the approximation that

$$1 + \Delta v_1/v_0 + \Delta v_2/v_0 \approx (1 + \Delta v_1/v_0)(1 + \Delta v/v_0) \tag{12}$$

Equation (8) can be factored and reduced to

$$\left|\frac{E}{E_0}\right|^2 = \frac{1}{2\pi^2 v_0^2}\left[1 + \left(\frac{\rho_1}{1 + \Delta v_1/v_0}\right)^2 + \frac{2\rho_1 \cos(k\delta_1 + \phi_1)}{1 + \Delta v_1/v_0}\right] \times \left[1 + \left(\frac{\rho_2}{1 + \Delta v_2/v_0}\right)^2 + \frac{2\rho_2 \cos(k\delta_2 + \phi_2)}{1 + \Delta v_2/v_0}\right] \tag{13}$$

Using Equations (5), (6), (7) and (11)

$$\beta_{1,2} - \beta_0 = \pi \Delta v_{1,2}(v_0 + \Delta v_{1,2}) = k\delta'_{1,2}.$$

From which we get

$$1 + \frac{\Delta v_{1,2}}{v_0} = (1 + 2k\delta'_{1,2}/\pi v_0^2)^{\frac{1}{2}} \tag{14}$$

Using Equation (14) and the half angle formulas, Equation (13) can be put in a little more convenient form.

$$\left|\frac{E}{E_0}\right|^2 = {}^2\left[\frac{1}{2\pi^2 v_0^2}\right] \times \left[\frac{4\rho_1 \cos^2 \frac{1}{2}(k\delta_1 + \phi_1)}{(1 + 2k\delta_1/\pi v_0'^2)^{\frac{1}{2}}} + \frac{[(1 + 2k\delta_1/\pi v_0'^2)^{\frac{1}{2}} - \rho_1]^2}{1 + 2k\delta_1/\pi v_0'^2}\right] \times \left[\frac{4\rho_2 \cos^2 \frac{1}{2}(k\delta_2 + \phi_2)}{(1 + 2k\delta_2/\pi v_0'^2)^{\frac{1}{2}}} + \frac{[(1 + 2k\delta_2/\pi v_0'^2)^{\frac{1}{2}} - \rho_2]^2}{1 + 2k\delta_2/\pi v_0'^2}\right] \tag{15}$$

where

$$v_0' = v_0/(1 + \Delta D/\delta_1) \tag{16}$$

If the three terms on the right hand side of Equation (15) are designated by A, B_1 and B_2, respectively, Equation (15) can be expressed as

$$10 \log \left|\frac{E}{E_0}\right|^2 = 10 \log A + 10 \log B_1 + 10 \log B_2 \tag{17}$$

Figure 3 is a plot of 10 log A versus v_0, and Fig. 4 is a plot of 10 log B versus $2\delta/\lambda$ with v_0' as a parameter for the special case of $\rho = 1$ and $\phi = \pi$.

Using Figs. 3 and 4, calculation of field strength of the knife-edge diffraction in the shadow region becomes extremely simple once the various geometric parameters have been determined with the aid of Equations (6) and (9).

If either ρ_1 or ρ_2 are zero, Equations (13) and (15) reduce to the "two ray" case where reflection occurs only on one side. If both ρ_1 and ρ_2 are zero, Equations (13) and (15) reduce to the familiar "one ray" case.

I. Accuracy of Simplified Method

To check the accuracy of the above method, comparisons were made with the much more laborious rigorous method. For $v_0 = 1{\cdot}26$ and $\Delta v_1/v_0 = \Delta v_2/v_0 = 0{\cdot}45$, the error was less than 0·1 db. For higher values of v_0 and smaller $\Delta v/v_0$ which

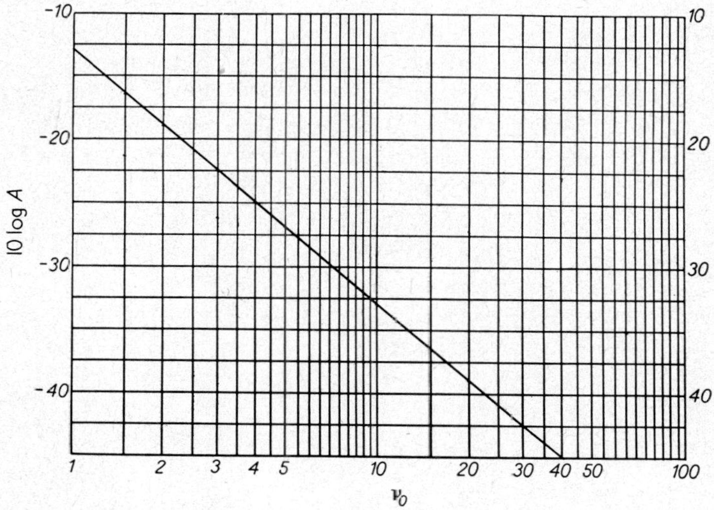

FIG. 3. 10 log A *vs* V_0.

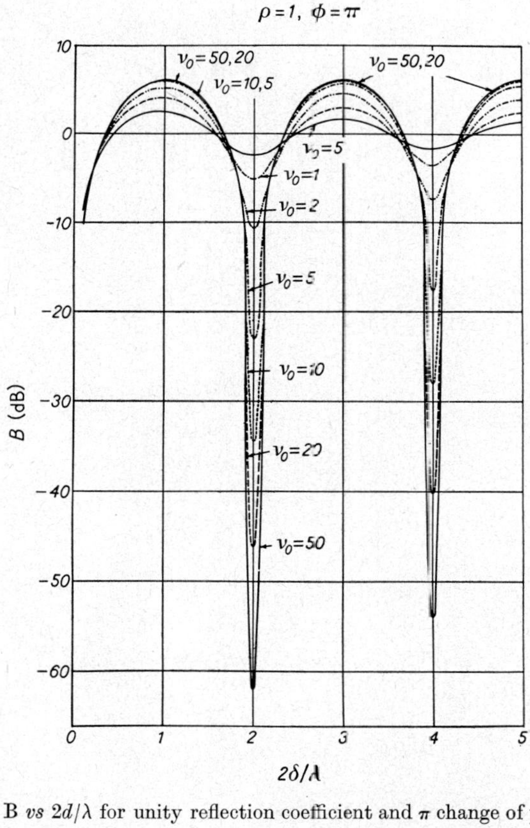

FIG. 4. 10 log B *vs* $2d/\lambda$ for unity reflection coefficient and π change of phase on reflection.

are the usual cases, the errors are at most only a few hundredths of a db. Even for the extreme cases of $v_0 = 1$, $\Delta v_1/v_0 = \Delta v_2/v_0 = 1$, the error was about $\frac{3}{4}$ of a db. For $v_0 = 0{\cdot}7$ and $\Delta v_1/v_0 = \Delta v_2/v_0 = 0{\cdot}5$, the error was less than 3 db. Thus, it appears that for all cases of practical interest in mountain obstacle paths, the error introduced by the simplified method is negligible.

II. Extension to Non-flat Earth

Up to this point the flat Earth geometry of Fig. 1 was used. The method can be readily extended to the much more general case of arbitrarily tilted reflection planes provided that δ_1 and δ_2 are taken as differences in path computed from the actual profile in question. The proof is very simple and can be readily obtained from examination of Fig. 1. Imagine the lines R–R' and T–T' rotated by a small angle about their respective centers. Physically these angles are very small, so that the cosines are very nearly equal to one. The values of h_0, h_1, h_2 and h_3, and therefore the differences between β's in Equation (7), remain unchanged. The free space path length and the ray path length over the obstacle change by very nearly equal amounts and thus Equation (15) remains unchanged, provided, of course, that δ_1 and δ_2 are computed as path differences based on the actual profile.

Acknowledgments

The authors are indebted to Mr. J. H. Crysdale and Dr. James H. Wait, who made valuable suggestions which helped to clarify some points which were not adequately treated in a former paper by the authors [1].

The work contained in this paper was partially supported by the Signal Communications Department, Army Electronic Proving Ground, Fort Huachuca, Arizona.

References

1. Anderson, L. J. & Trolese, L. G. (1958). *Trans. Inst. Radio Engrs, N.Y. AP*–**6**, No. 3, 281–286.
2. Crysdale, J. H. (1955). *Proc. Inst. Radio Engrs, N.Y.* **43**, 627.
3. Dickson, F. H., Egli, J. J., Herbstreit, J. W. & Wickizer, G. S. (1953). *Proc. Inst. Radio Engrs, N.Y.* **41**, 967–969.
4. Matsuo, S. (1950). Report of the Electrical Communication Laboratory, 621.39.001.6(047.3), Ministry of Telecommunications, Tokyo.
5. Schelleng, J. C., Burrows, C. R. & Ferrell, E. B. (1933). *Proc. Inst. Radio Engrs, N.Y.* **21**, 427–463.

The Airborne Measurement of 1·36 m Fields to Ranges in Excess of 900 Miles and at Altitudes from the Surface to 40,000 Ft.

L. A. Ames, E. J. Martin and T. F. Rogers

Electronics Research Directorate, Air Force Cambridge Research Center, Bedford, Mass.

A survey of recent results of an extensive airborne field strength measurement program is presented. The dependence of 1·36 meter field strength upon altitude and distance is examined in some detail. Of additional interest, continuous ionospheric propagation at 800–1,000 miles has been observed at this wavelength.

I. Introduction

In the years following World War II, intensive investigations were initiated, particularly in the United States, of the propagation of VHF/UHF radio waves to distances beyond the line-of-sight, the expected limit of useful communications at VHF and higher frequencies. These studies, which were originally inspired by the unexpected long radar ranges occasionally encountered during the war, became of paramount importance not only to problems such as cochannel television interference in this country; but to the necessity of obtaining more economical use of the radio frequency spectrum. Once it had been established that these fields were omni-present at many times the line-of-sight range and with magnitudes much greater than earlier theories had predicted, the exploitation for communications began. This was particularly true of the U.S. Air Force in its requirement for highly reliable long distance communication systems to link our northern defense networks. Normal long range communications in the already overcrowded HF radio bands were found to be unreliable and at times impossible in the arctic. Tropospheric "scatter," as extended-range UHF became popularly called, offered great promise for the transmission of wide-band information over long hops without requiring repeater stations. Initially it was felt that the utility of this communication mode was limited to point-to-point ground circuits of the order of 100 to 200 miles in length.

In 1953 AFCRC initiated a program to determine the extent, strength and characteristics of 1·36 m fields as functions of receiver altitude and range. Here the use of an airborne receiving terminal seemed most desirable not only since this permitted measurements at various altitudes but also because it was possible by this method to determine the distance decay along a path holding time essentially constant. This technique, however, is not readily adaptable to the accumulation of long term fading statistics for a given point in space. Only through very many flights over the path during a seasonal cycle can such estimates be made.

It was soon evident that these fields were not simply surface phenomena but that they were present at all altitudes up to 15,000 ft out to distances beyond 400 statute miles. Sufficient data were accumulated by the end of 1954 to indicate that the distance dependence of the fields well beyond the radio horizon was a monotonically decreasing function fairly well represented by either an exponential or inverse power of the distance law. The effect of receiver altitude could not be evaluated too well at that time but there was some evidence of height gain.

In order to describe completely the tropospheric field as a function of distance and altitude, it was necessary to make measurements out to the greatest possible

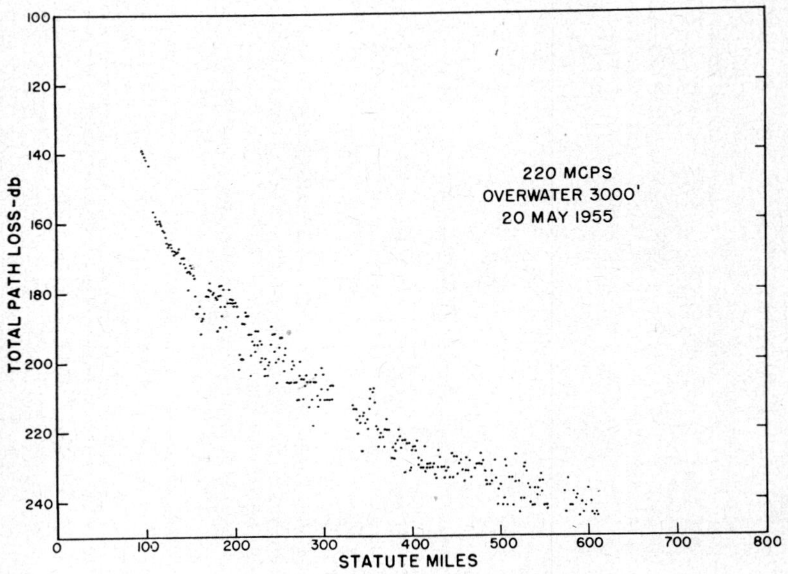

FIG. 1.

distance and up to the highest possible altitude. The distance requirement was partially met in 1955 when the C-54 employed in the earlier measurements was used to extend our low altitude measurements to beyond 600 miles (Fig. 1). These flights graphically demonstrated the decreasing rate of decay with distance deep in the "scatter field." Within the past few months new transmitting and receiving equipment has permitted an additional extension of our measurement range yielding some surprising results out to 1000 miles.

In order to extend our altitude capability a B-29 was instrumented in the winter of 1955–56, and several flights were conducted at 30,000 ft. During one of these, the C-54 was flown simultaneously along the path at low altitude providing the basis for height-gain predictions. Some concern was felt for Starkey's hypothesis that much beyond-the-horizon propagation might be explained by specular reflection from the tropopause and layers in the vicinity of the tropopause. Since all of the B-29 measurements were made beneath the tropopause, a B-47 with an altitude capability of 40,000 ft was obtained and instrumented in the following winter. This permitted consistent measurements above the winter tropopause.

II. Experimental Parameters

For all measurements prior to this year the signal source was an experimental radar feeding about 500 kW peak power to a 30 × 70 ft dipole array having a power gain of close to 500. The B-29 and B-47 were both outfitted with 5 element yagi-type antennas having a measured gain of 10. The antennas used

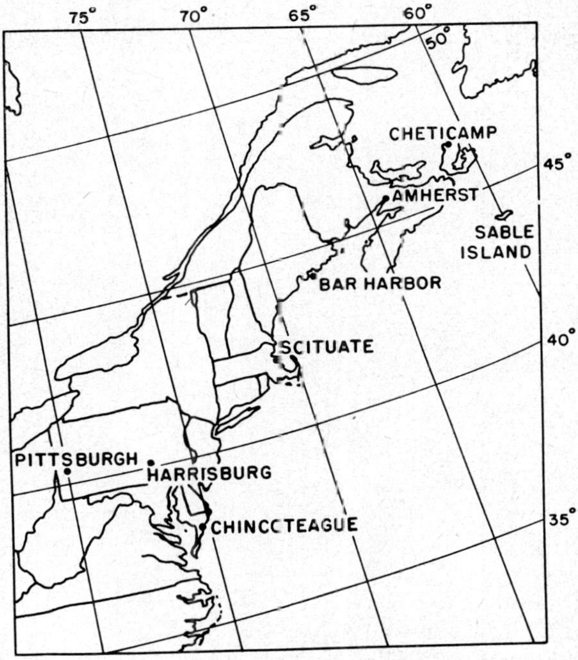

Fig. 2.

on the C-54 were also yagi-type arrays. The receivers employed in the early measurements had a meter sensitivity of about 10^{-15} W. The new measurement system includes a 40 kW CW transmitter and a receiver with a sensitivity of about 10^{-19} W.

The flight path (Fig. 2) was over the North Atlantic, originating on the Massachusetts coast south of Boston and terminating southeast of Sable Island. The radar beam which can be steered electro-mechanically was swept through an arc of $\pm 5°$. Even with this latitude for position error, expert navigation was required both to maintain the aircraft within this 10° arc and hold the drift within limits which permitted maximum utilization of receiving antenna gains. Each time the radar beam illuminated the aircraft, the peak received signal was recorded as one data point.

III. Measurements

In ascertaining some basis for predicting the effect of increased altitude on the field strengths, some consideration was given to the current "radio-scattering" theories which indicated that the source of maximum power ought to be at the receiver's radio horizon. The implication for height-gain is shown in Fig. 3.

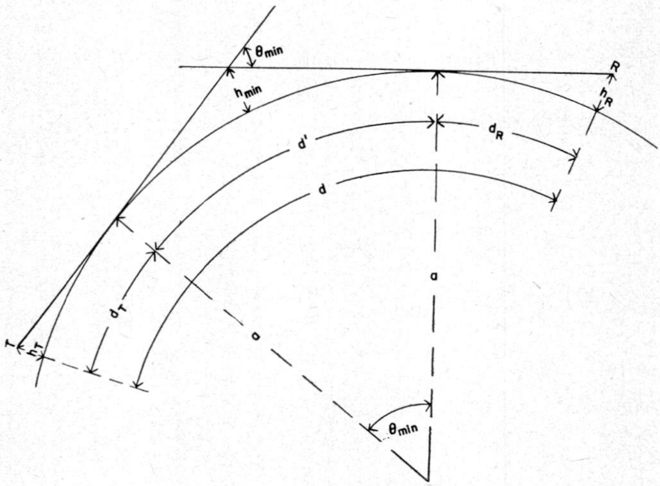

Fig. 3. Geometry for a smooth earth path.

Here we see that an increase in receiver height, h_R, results principally in a reduction in the minimum scattering angle, $\theta_{\min}$, or equivalently a reduction in the distance between radio horizons (indicated as d' in the figure). This immediately suggests choosing the distance between radio horizons as the principal parameter to be used in describing the smooth earth tropospheric field. We have called this distance "effective distance" and have assumed that the magnitude and characteristics of the fields are dependent upon this parameter. Norton's group at the U.S. National Bureau of Standards have expressed field strength dependence in terms of an "angular distance" for the general propagation path. This reduces essentially to our " effective distance " for the smooth earth path. Using this simplified concept, the smooth earth height gain function for great heights can be expressed in terms of the reduction in "effective distance" obtained in elevating a receiver (or transmitter) to a given height.

To test the above assumption, simultaneous measurements were taken at 3000 ft with the C-54 and at 30,000 ft with the B-29 in the winter of 1956. The resultant data are plotted in Fig. 4 as total transmission loss versus the actual surface path length. No relationship between the two measurements is readily apparent in this form. When these data are plotted in Fig. 5 as path loss relative to free space versus "effective distance," they are found to coincide almost exactly. Here the agreement is found to be good not only in absolute magnitude,

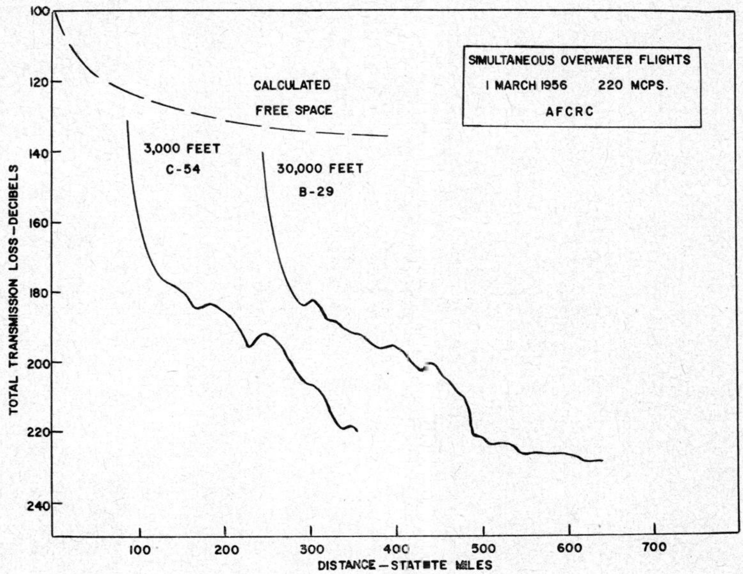
SIMULTANEOUS OVERWATER FLIGHTS
1 MARCH 1956 220 MCPS.
AFCRC
CALCULATED
FREE SPACE
3,000 FEET
C-54
30,000 FEET
B-29
TOTAL TRANSMISSION LOSS—DECIBELS
100
120
140
160
180
200
220
240
100
200
300
400
500
600
700
DISTANCE—STATUTE MILES

FIG. 4.

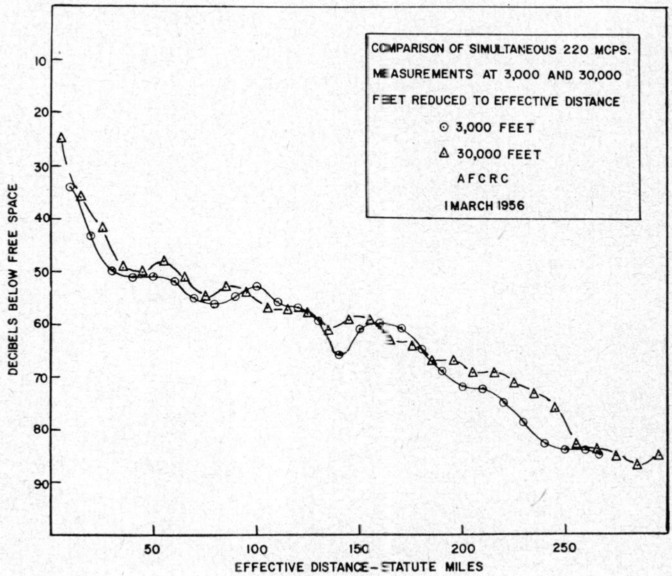
COMPARISON OF SIMULTANEOUS 220 MCPS.
MEASUREMENTS AT 3,000 AND 30,000
FEET REDUCED TO EFFECTIVE DISTANCE
⊙ 3,000 FEET
△ 30,000 FEET
AFCRC
1 MARCH 1956
DECIBELS BELOW FREE SPACE
10
20
30
40
50
60
70
80
90
50
100
150
200
250
EFFECTIVE DISTANCE—STATUTE MILES

FIG. 5.

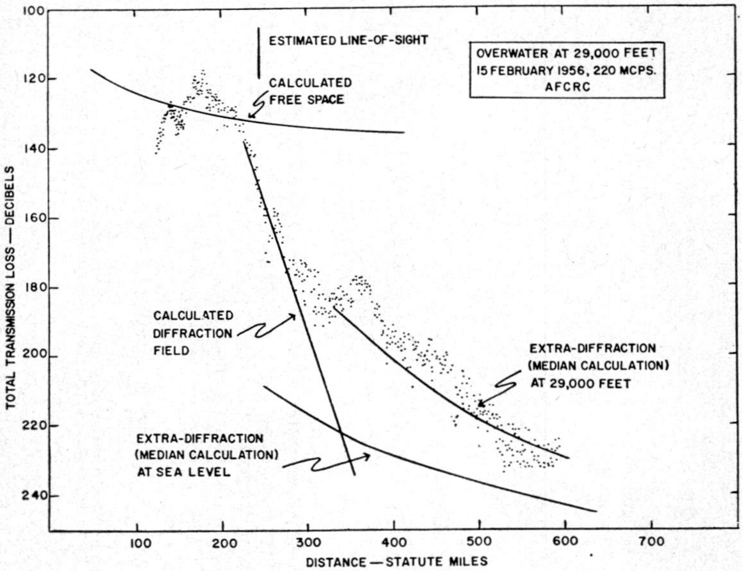
ESTIMATED LINE-OF-SIGHT
OVERWATER AT 29,000 FEET
15 FEBRUARY 1956, 220 MCPS.
AFCRC
CALCULATED
FREE SPACE
CALCULATED
DIFFRACTION
FIELD
EXTRA-DIFFRACTION
(MEDIAN CALCULATION)
AT 29,000 FEET
EXTRA-DIFFRACTION
(MEDIAN CALCULATION)
AT SEA LEVEL
TOTAL TRANSMISSION LOSS—DECIBELS
DISTANCE—STATUTE MILES
100
120
140
160
180
200
220
240
100
200
300
400
500
600
700

FIG. 6.

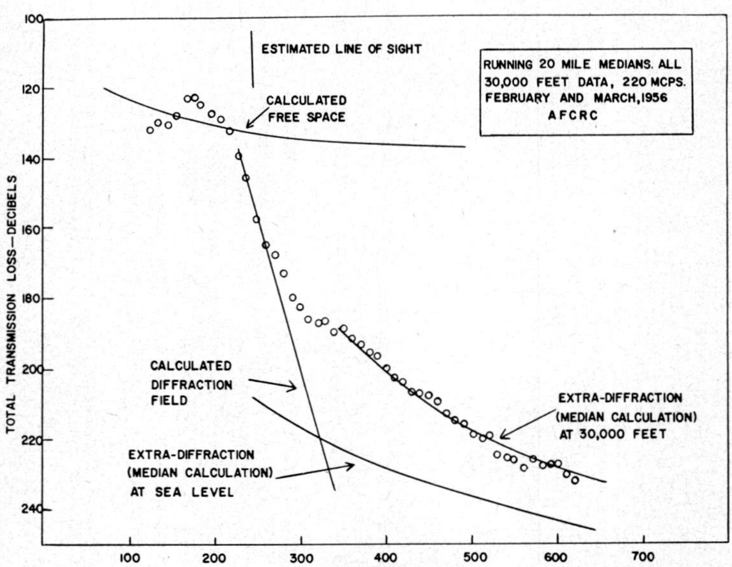
ESTIMATED LINE OF SIGHT
RUNNING 20 MILE MEDIANS. ALL
30,000 FEET DATA, 220 MCPS.
FEBRUARY AND MARCH, 1956
AFCRC
CALCULATED
FREE SPACE
CALCULATED
DIFFRACTION
FIELD
EXTRA-DIFFRACTION
(MEDIAN CALCULATION)
AT 30,000 FEET
EXTRA-DIFFRACTION
(MEDIAN CALCULATION)
AT SEA LEVEL
TOTAL TRANSMISSION LOSS-DECIBELS
DISTANCE-STATUTE MILES
100
120
140
160
180
200
220
240
100
200
300
400
500
600
700

FIG. 7.

but in the slopes of the decay curves and observed spatial variations as well. This supports the validity of choosing "effective" distance as the key parameter in calculating path loss values over a smooth earth.

On the basis of this evidence, field strength estimates drawn from surface observations were extended to high altitudes. A typical flight at 29,000 ft is compared in Fig. 6 to the resultant calculated median curves for that altitude and the surface. In addition, all the data obtained in five flights with the B-29 were reduced to running 10 mile medians and compared with the calculated curves for 30,000 ft and the surface. This is shown in Fig. 7. It is readily seen

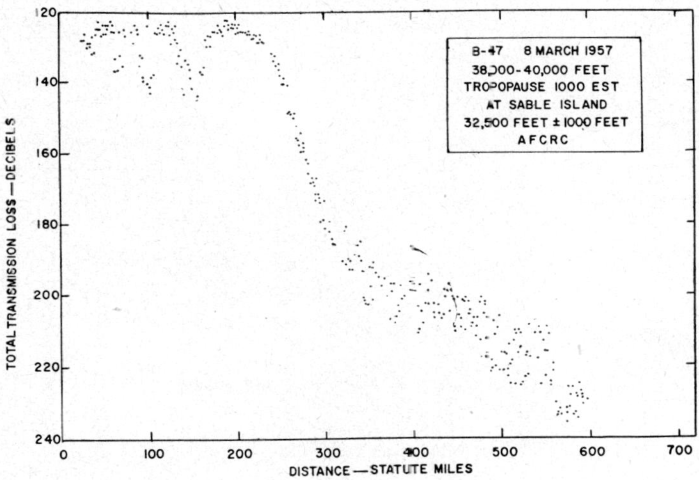

FIG. 8.

that considerable height-gain is indicated by these measurements and that the amount of height-gain decreases with increasing distance. It must be noted here that the calculated curves were for temperate latitude yearly median values and the data were primarily winter data. Since ground statistics have indicated that in winter the fields are generally at their lowest, it was expected that the measured values would be low compared to the yearly medians instead of being in the observed close agreement.

For the B-29 measurements, the tropopause along the path was from 1000-5000 ft above the receiving terminal. In order to validate the smooth earth "effective distance" concept for heights above the trosopause, a B-47 similarly instrumented was flown the following winter at altitudes between 38,000 and 40,000 ft with essentially the same experimental conditions as before. In this case, for fuel considerations an altitude of 38,000 ft was maintained from over the transmitting site out to about 400 miles at which point a climb up to 40,000 ft was begun. Seven flights were successfully completed of which 6 were at altitudes of from 4,000 to 11,000 ft above the tropopause, and the seventh very near to the tropopause. The first flight was made on 8 March 1957 with a tropopause height of around 32,000 ft. This flight is shown in Fig. 8 displaying the rather regular and anticipated characteristics at the various ranges. The last flight was completed on 1 April 1957 with the tropopause above 36,000 ft. The raw data from this flight are shown in Fig. 9. The latter flight in which the tropopause was probably penetrated at least once, shows an as yet unexplained oscillatory characteristic. The distance range obtained in these measurements was limited principally by interference. When the signal level dropped to within 10–15 db of noise toward the end of the flights, high-powered transmitters within line-of-sight of the aircraft made it difficult to determine exactly the level of the desired signal. In Fig. 10 a summary of the B-47 data is plotted in the form of medians

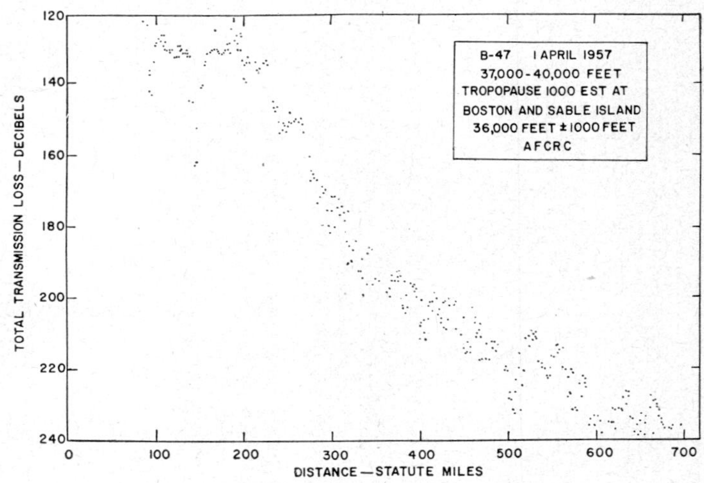

FIG. 9.

over 10 mile intervals for all flights with the range of values indicated. A comparison was made as before with the calculated yearly median values at this altitude and at the surface. A winter curve for this altitude was calculated as well and the data seems to be more representative of these expected winter values.

In addition, a height-gain curve was calculated for 400 miles on the basis of the "effective" distance decay rate of median path loss observed on surface measurements. All valid airborne measurements taken by the Air Force Cambridge Research Center at 400 miles up to 38,000 ft were compared in Fig. 11 with these predicted values. Considering the limited statistics available for a fixed distance, the agreement was believed to be quite good. Large height gains may be expected using the "effective" distance calculation, but greater fading depths should also be expected at altitude in keeping with the observed decrease in fading range with "effective distance."

As a continuation of our airborne investigations of long-distance 1·36 m field strengths, we have developed the equipment mentioned above as well as experimental techniques which permit measuring a path attenuation of about 280 db both on the ground and in the air. Prior to undertaking airborne measurements a surface experiment was conducted during February 1958. A receiving site was

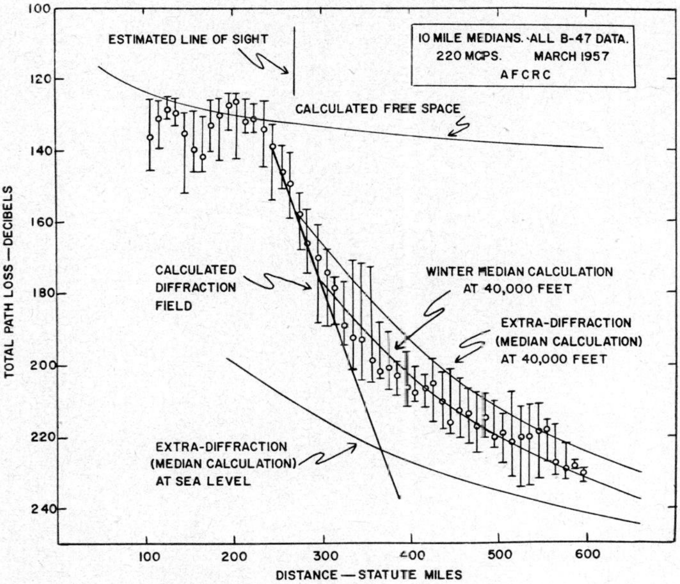

FIG. 10.

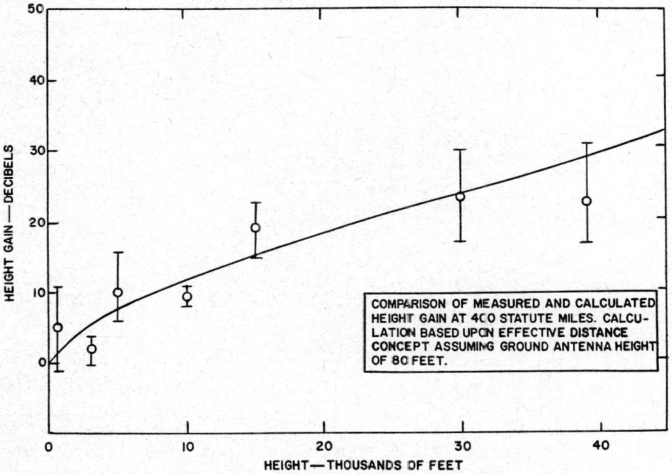

FIG. 11.

established near Stephenville, Newfoundland, about 735 statute miles from the transmitter. The measured median path attenuation was 266 $\pm$ 5 db. With this promising evidence in hand, the first airborne measurement was made on 18 March 1958 at an altitude of 10,000 ft. In the region beyond 600 miles, the rate

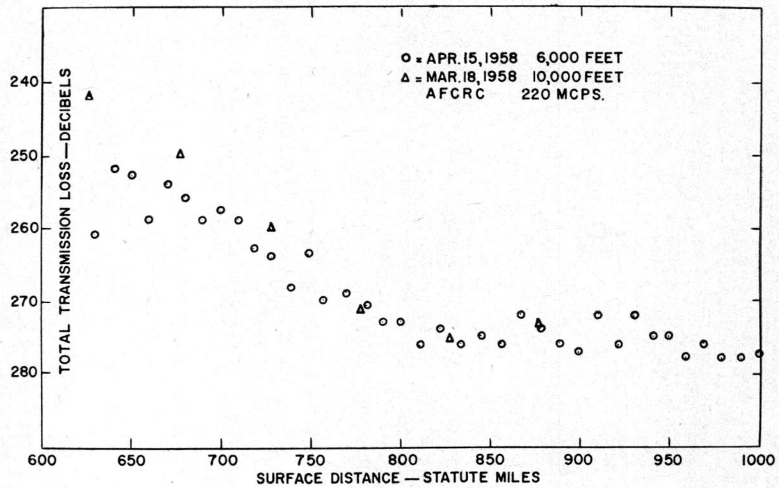

FIG. 12.

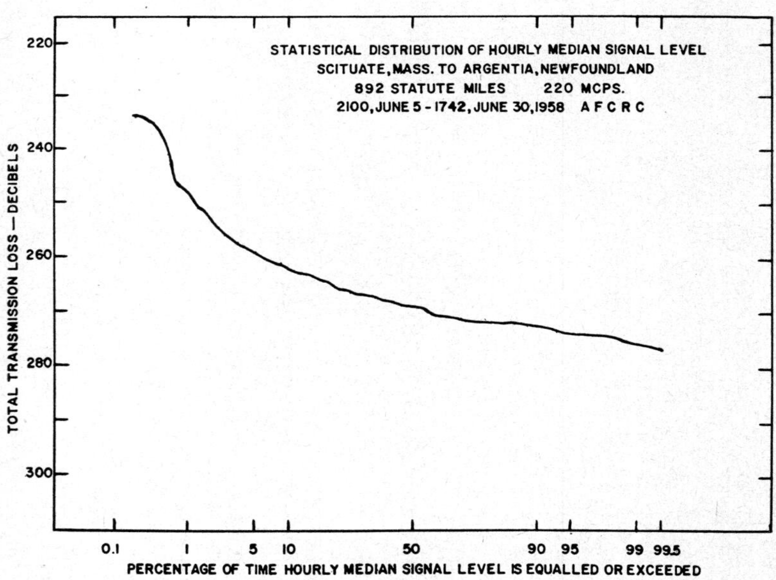

FIG. 13.

of decay increased markedly as expected from measurements at a shorter wavelength. However, at a distance of 750–800 miles, and out to the predetermined aircraft turn-around at 850 miles, the fields began a much less rapid decay. In addition, the signal trace changed in this region, exhibiting a spiky, rapid fading which has been characteristic of signals received on 6 m. To determine whether these fields were ionospheric in origin, a second airborne measurement was conducted on 15 April; this time the scheduled turn-around was at 1000 miles. The results of the two flights are presented in Fig. 12 and it can be seen that the median transmission loss throughout the 800–1000 mile region is fairly constant at about 275 db $\pm$ 5 db. Both the variation with distance and the fading rate offered strong evidence that we had measured ionospheric scatter fields at 1·36 m.

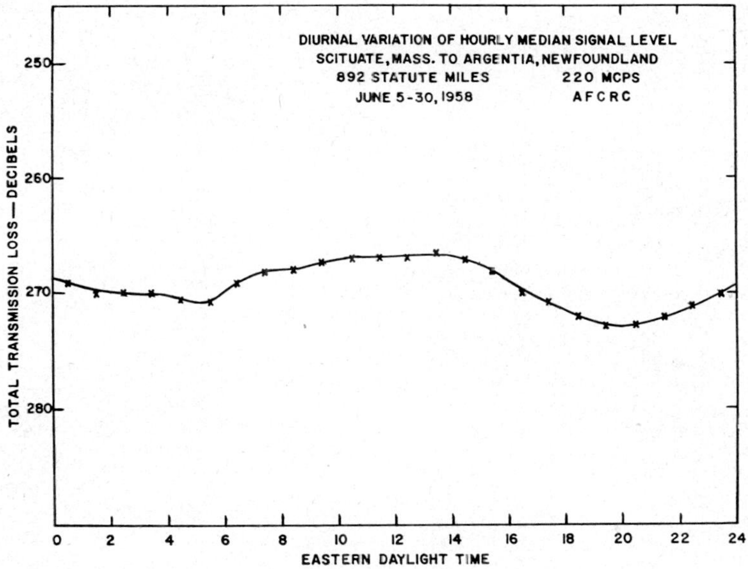

FIG. 14.

An additional surface experiment was conducted during June using a receiving site near Argentia, Newfoundland, at a range of about 900 miles. There is no doubt about the persistence of the signal level since it remained consistently detectable during the entire measurement period of nearly 600 hours. A cumulative distribution of the hourly median path losses is shown in Fig. 13. In addition, Fig. 14 displays the diurnal variation of the hourly median field intensities. It can be seen that this variation, although slight, is quite similar to that observed at 6 m with broad-beam antennas in the ionospheric scatter region.

IV. CONCLUSIONS

The decay of tropospheric fields well beyond the radio horizon at 1·36 m is a monotonically decreasing function of distance out to about 800 statute miles.

The "effective" distance concept has been shown to be quite accurate and useful for calculating path losses over a smooth earth in the distance region of 100–600 miles.

Large height gains are predicted in the tropospheric "scatter" field which decrease with increasing distance. This is precisely what has been observed in our airborne measurements.

"Scatter" fields exist above the tropopause and their magnitude and characteristics have not differed to any great extent from predictions.

The ionosphere persistently manifests its influence at a wavelength much shorter than has heretofore been demonstrated to exhibit characteristics of continuous ionospheric propagation.

Acknowledgments

The authors wish to express their appreciation to R. Barrett, J. Frazier, T. Willson, N. Conger, R. Miner, A. Orange, R. Howell, J. Reegan and E. Hoell, who were responsible for the actual experimental measurements; to the personnel of the Antenna Laboratory at AFRC who designed the various airborne antennas; the Airborne Engineering people at L. G. Hanscom Field who performed the detailed engineering and aircraft modifications, and to the various Air Force flying personnel without whose skill and cooperation, these measurements could not have been made.

Instantaneous Electronic Ray Tracing Computer for the Solution of Electromagnetic Propagation Problems

P. Springer and R. Rawhouser

Aerial Reconnaissance Laboratory, Wright Air Development Center, Wright-Patterson Air Force Base, Dayton, Ohio

A new analog computer capable of providing instantaneous ray tracing solutions to the Wave Propagation Equation has been developed for the Wright Air Development Center, United States Air Force. With this computer and current refractive index profiles it is possible to determine tropospheric propagation conditions while they exist. Excellent correlation between ray tracing predictions and measured propagation data is obtained.

Ray tracing techniques have been used for a number of years in determining the effects of atmospheric refraction on radio propagation. D. F. Hartree, J. G. Michel and P. Nicolson utilized a differential analyser at Manchester, England, as early as 1942 in computing solutions to the rigorous equation for radio propagation.

Almost six years later Ming S. Wong of the Wright Air Development Center, USAF became interested in Hartree's work and set about to utilize the computational facilities available at the Wright Air Development Center in an attempt to explain some of the propagation anomalies which were adversely affecting the performance of Air Force radar and communication systems.

Wong used the following generalization of the propagation equation given by Hartree, *et al.*, for the case of n depending only on h

$$\frac{d^2h}{dx^2} = h'\left\{\frac{1}{n}\frac{\partial n}{\partial h} + \frac{2}{a+h}\right\} - h'\,\frac{1}{n}\frac{\partial n}{\partial x}\left\{1 + y'^2 + \frac{h'^2}{(1+h/a)^2}\right\}$$
$$+ \left\{\frac{1}{n}\frac{\partial n}{\partial h} + \frac{1}{a+h}\right\}\left(1+\frac{h}{a}\right)^2(1+y'^2)$$

$$\frac{d^2y}{dx^2} = \left\{\frac{1}{n}\frac{\partial n}{dy} - y'\,\frac{1}{n}\frac{\partial n}{\partial x}\right\}\left\{1 + y'^2 + \frac{h'^2}{(1+h/a)^2}\right\}$$

$$(h' = dh/dx \quad y' = dy/dx)$$

where $h =$ height;

$x =$ distance;

$n =$ refractive index;

$a =$ radius of the earth

By considering only the rays falling within 6° of the central ray, the equation was simplified to yield the following equations

$$\frac{d^2h}{dx^2} \simeq \frac{\partial n}{\partial h} + \frac{1}{a}$$

$$\frac{d^2y}{dx^2} \simeq \frac{\partial n}{\partial y}.$$

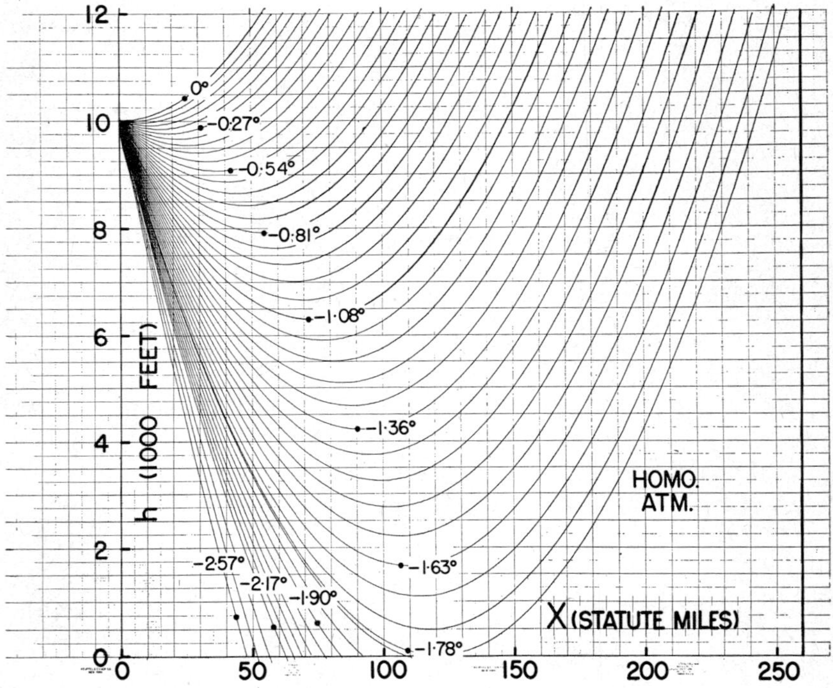

FIG. 1.

To utilize the simplified equation Wong obtained refractive index profile data from meteorological and microwave refractometer soundings which he then plotted on coordinate paper. By breaking up the refractive index profile into small segments he obtained the gradient curve by graphical differentiation. A wire was then cemented onto the paper, to conform to the contours of the gradient curve. This was fitted to a drum on the input function generator and essentially served as a variable contact of a potentiometer to generate the input voltage. The information was then introduced into the REAC (Reeves Electronic Analog Computer) or the GEDAC (Goodyear Analog Computer) in order to compute the ray family. Later more refined techniques made it possible to eliminate the mechanical slide wire and substitute a diode and relay switching system, whereby the curve was approximated by a series of sloping line segments.

Solutions to the equation for the given initial conditions, were computed and plotted on paper by the analog computer.

Figure 1 shows a typical ray tracing presentation. Here the function for a uniform homogeneous atmosphere was fed into the machine. The rays emanating from the transmitter actually follow straight lines but the computer must take into account the curvature of the earth. Since the computer does not work with curve coordinants a linear coordinate system must be used. In this system the rays are shown curving away from the earth. It must be noted that the distance scale utilized is quite compressed as compared to the altitude scale. The ray

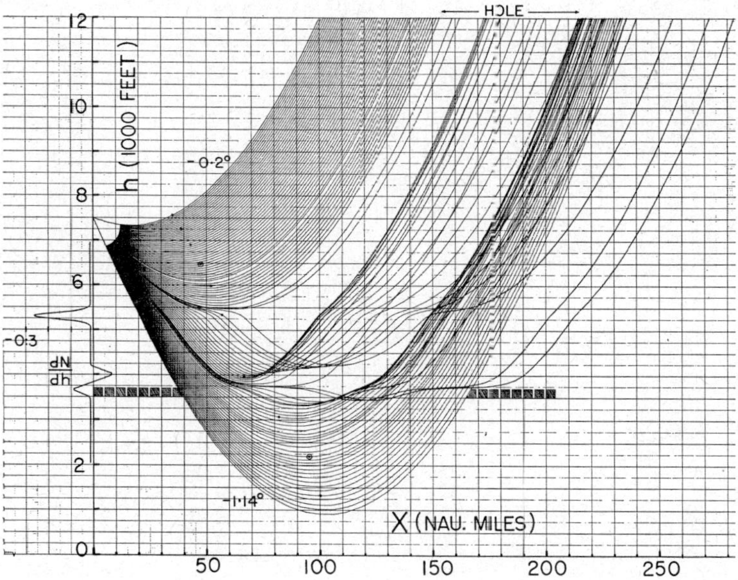

FIG. 2.

angles are therefore greatly exaggerated and a complete family of rays covers only $\pm 6°$. A better understanding of the value of these tracings can be gained from Fig. 2 which is a ray tracing picture of propagation through an atmosphere commonly existing in the central part of the United States in the summer. The absence of rays in a given area on the tracing indicates that the energy has been bent away by the changing refractive index gradient.

Over a period of years the experimental unit of the Wave Propagation Section conducted studies of air-to-air radio propagation in many parts of the world while gathering simultaneous microwave refractometer data and meteorological data. From the refractive index data Wong and co-workers laboriously compiled hundreds of ray tracing families on the GEDA computer. The radio propagation data was then used as a basis to assess the value of ray tracing techniques as later discussion will indicate. The results were quite gratifying. They were so successful, that the need for a specialized computer capable of providing an immediate picture of radio propagation for a given set of atmospheric conditions rather than a computer capable of a post-mortem analysis as to what had previously caused the radio anomaly, became evident. The need for an instantaneous ray tracer dictated the requirements for a computer having a cathode ray scope presentation. A rapid method for synthesizing the refractive index profile and generating the proper input function, together with convenient methods for setting in the initial conditions of transmitter height and angular coverage was also required. Since the Sperry Piedmont Division of Sperry-Rand Corp. has acquired considerable experience in the design of repetitive analog computers, they were awarded a contract for the development of the instantaneous Ray Tracer for Propagation Studies.

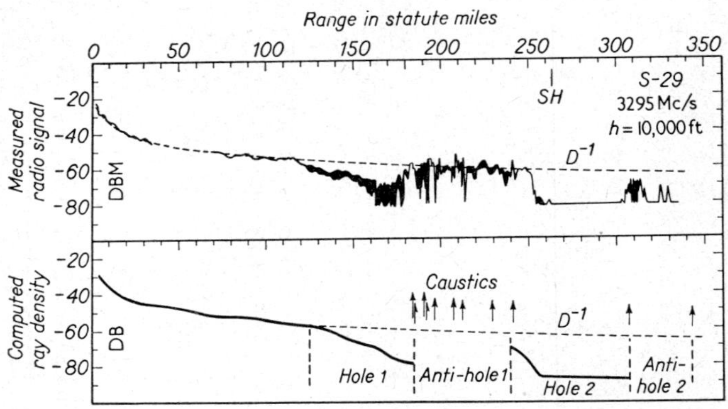

FIG. 3.

Let us now consider Fig. 3 which represents a simplified block diagram of the computer required for the solution of the equation:

$$\frac{d^2h}{dx^2} \simeq \frac{dn}{dh} + \frac{1}{a}$$

where h = height of transmitter; a = radius of the earth; x = horizontal distance; n = refraction index gradient.

The complete ray tracer consists of a function generator and memory unit, a computer unit, and the necessary display unit. Since an electronic analog computer must use time as the independent variable the distance x must be made a linear function of time, i.e., $x = kt$ and our equation becomes

$$\frac{1}{k}\frac{d^2h}{dt^2} \simeq \frac{dn}{dh} + \frac{1}{a}.$$

Assume we have a repetitive function generator which will generate a function.

$$f(x) = \frac{dn}{dh} + \frac{1}{a}.$$

In keeping with good engineering practice for computer design, differentiation operations will be performed by means of integrators. Since we have a second order differential equation it will be necessary to perform two integrations. Our block diagram becoming Fig. 4. Here we have a function generator, a first integrator and a buffer, which provides isolation, and reverses the sign of the h. At this point the initial angle information is introduced into the circuit. A second integrator and buffer provides height as an output. At this point the initial height information is fed into the circuit. The output of the second integrator is then fed back into the input to the function generator to close the circuit loop. It also provides height information, to the vertical plates of a cathode ray oscilloscope. A sync generator synchronizes the operation of the integrators, and provides synchronization voltages to the sweep generator.

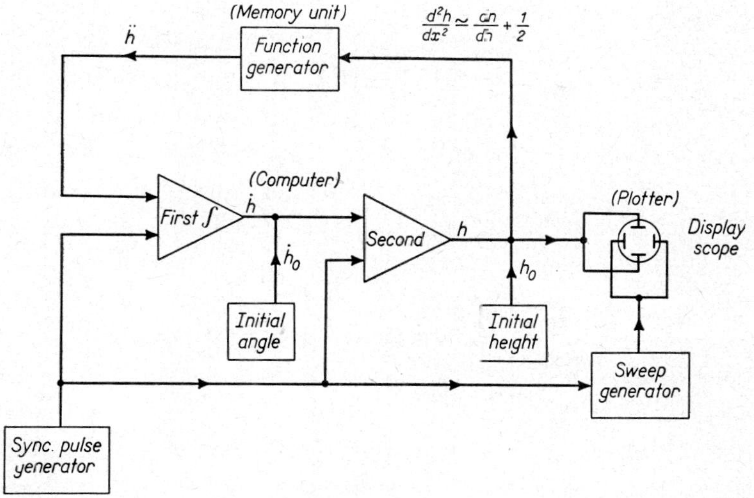

FIG. 4.

While, as propagation specialists our interests lie mainly in the application of the ray tracer to propagation problems, it may be of interest to examine a few of the more important basic circuits employed in the instantaneous computer.

I. FUNCTION GENERATOR

The Electronic Ray Tracer utilizes a function generator based on switching circuits. These are used to advantage in this application since the refraction gradient profile is represented by straight-line segments over the working altitude The derivatives of straight-line segments give step functions which are easily synthesized by switching circuits. Quite naturally, the greater the number of straight-line segments used, the greater the accuracy between the computer's simulated refraction gradient profile and the actual profile. Consequently, a function generator with 20 step circuits is provided.

All inputs of the function generator are common and all outputs are tied together and connected directly to the first integrator input. The step circuits are all identical and interchangeable, facilitated by etched wiring boards. The input level representing altitude is compared with a reference voltage also representing altitude, adjustable through the altitude potentiometer. Reference voltages vary between 0 volts and + 20 volts. Referring to Fig. 5, a schematic of one of the step circuits, when the input signal exceeds the reference voltage, transistor Q101 is turned on and it conducts heavily. This difference signal is transmitted through the high gain section of the step circuit consisting of Q101

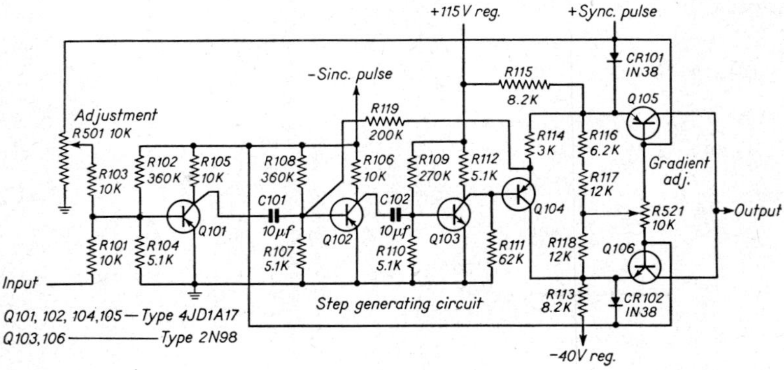

Fig. 5.

through Q103. Transistor Q104 is normally saturated, holding the output stage Q105 and Q106 at cut off. With the application of a signal, Q104 is cut off followed by Q105 and Q106 turning on, and conducting to an amount determined by R521 (the refraction gradient potentiometer adjustment. With R521 set at the center, Q105 and Q106 conduct equally and no output occurs. Adjusting the refraction gradient potentiometer adjustment to the left or right, makes either Q105 and Q106 conduct more than the other, and the difference current flows in the output load.

II. First Integrator

The output of the function generator is connected directly to the input of the first integrator. For zero input transistor 305 and 306 conduct the same amount thus neither adding nor subtracting any current flowing through Q307 and Q308. A signal to the input of the integrator will cause Q305 or Q306 to conduct more or less depending upon the polarity of the inputs. The change is coupled to Q307 and Q308 which previously conducted equal amounts of current. One of the two will now conduct more current the other will conduct less, also depending upon input signal polarity. The difference current will flow into the integrating capacitor C303 or C304 charging it plus or minus. At the end of the computing period C303 or C304 is left with a charge on it which is discharged to 0 V through one of the collector diodes of Q307 or Q308.

III. Second Integrator

The second integrator is basically identical to the first integrator. The design philosophy utilized is the same as that used for the first integrator. The only changes required were those necessary to obtain a larger dynamic range, thus necessitating more current in the output stage. The second integrator is shown schematically in Fig. 6. The slope of the output of the first integrator and initial angle are introduced at the second integrator inputs. The second integrator, like the first, works into emitter followers.

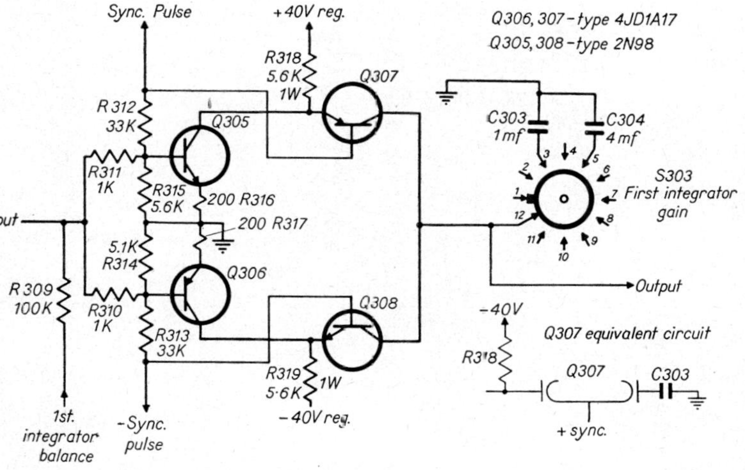

FIG. 6. First integrator schematic.

IV. Second Integrator Emitter Follows

Two cascaded stages of grounded-collector circuits are used after the first integrator, to prevent loading non-linearities. Transistors Q307 through Q320 comprise these circuits. Since the output of the second integrator is ray altitude versus time (actually the ray pattern), initial altitude is introduced after the first emitter follower.

V. Initial Angle

A motor-driven potentiometer, R301, provides automatic initial angle, and a manually adjusted potentiometer, R303, is utilized for manual *initial angle* as shown in Fig. 7. So that sector scanning may be accomplished, the manual initial angle is selected through S304 which turns on the synchronous motor, B301, and connects the automatic *initial angle* potentiometer to the second integrator input. Adjusting R303 to a desired angle, and turning S304 to AUTO will give automatic sweeping about the angle selected by R303.

The drive motor for the automatic initial angle potentiometer is a 30 r.p.m. synchronous motor. The combination of 30 r.p.m. and a computing rate of

60 rays per second results in a family of 120 rays for a complete sweep of automatic.

A problem is set into the computer merely by adjusting the "height" and "gradient" potentiometers of as many function cards as are needed for the desired gradient. A precision "calibrate" 10 turn potentiometer is incorporated in the computer as a comparitor to set the proper voltages. Limits for the variables, range and altitude, are selected and proportional parts are interpolated between the limits.

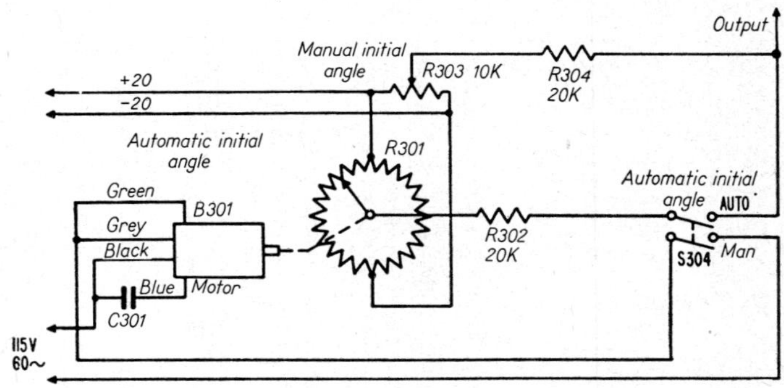

FIG. 7. Initial angle schematic.

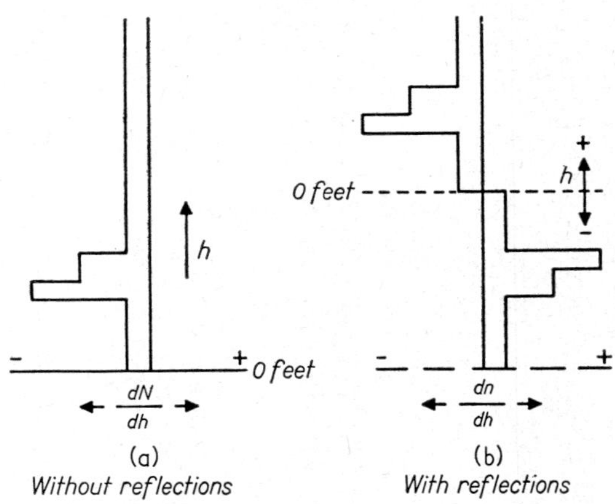

FIG. 8. Set-up showing reflections.

The process has been made easier by means of a switch on the cathode ray display which changes the picture to the gradient (rather than the ray pattern). The effect of height and gradient adjustments is seen on the scope and the magnitudes set by means of known calibration points.

The ray tracer can be made to operate in a way which exhibits the reflection of rays from the earth. In this mode of operation switching circuit in the computer reverses the direction of the ray whenever it goes to ground level. In order to accomplish this object, the computer must be set up in a special way. The function $dn/dh + 1/a$ must be inserted twice with an axis of symmetry about the zero altitude as shown in Fig. 8. In other words a mirror image must be produced. This is necessary since the input to the integrators must reverse sign when ground level is reached. Therefore, twice as many step circuits must be used in setting in the function. The important point is that the zero altitude reference has been moved to mid-range of the scale; it is now at a dial setting of 500 on the calibrate potentiometer. Proportional changes must be made for other altitudes.

Integration of the ray path proceeds on either side of the zero height axis. However, the afore-mentioned switching circuit reverses the direction of travel of h (i.e., the sign of its voltage) every time it passes through zero. The display is thus all on one side of a reference line and thus exhibits the result of reflections.

Figure 9 shows the ray tracing computer on the extreme right. The large screen on the left is used to provide more detail, and better visibility. When used with the camera and hood, together with an illuminated plexiglass screen, it provides a means of recording, a calibrated ray tracing. The unit in the center of the picture is an additional function generator, which makes it possible to consider the propagation of a signal through an atmosphere having refractive index variation in both height and distance which I will explain later. Figure 10 is self-explanatory, showing the appearance of the ray tracer with the case removed. It will be noted that transistors are used throughout. Etched circuit boards are used for the step circuits and in other parts of the computer where space is at a premium.

I will now present a few figures which are direct photographs of the screen of the cathode ray scope of the electronic ray tracer, taken by means of Polaroid Lantern Slide Film in a Polaroid camera. These can be taken in a matter of seconds, while the GEDAC requires several hours to compute a ray family. Changing the initial altitude, that is, the position of the airborne transmitter can be simulated on our ray tracer by a simple turn of the altitude knob. It is thus possible to observe immediately whether changing the altitude of the aircraft will improve or degradate the performance of the radar or communications system insofar as the propagation link is concerned.

Figure 11 shows a ray family for a homogeneous atmosphere. Unfortunately the cathode ray presentation of the ray tracer was not photogenic when these pictures were taken. The high angle rays appear as they should, but the low angle rays are somewhat confused by the appearance of extra ray traces. These are due to the fact that the automatic initial angle potentiometer is a continuous 360° potentiometer with fixed stops at zero and 180°. Rays are swept from high to low angle during the first half turn of the potentiometer, and then repeated from low angle to high angle during the second half revolution. Any slight misalignment of turns in the second half of the potentiometer as compared to the first half will cause the rays to not exactly coincide. This defect could easily be remedied by blanking out the voltages generated during the second half revolution.

FIG. 9.

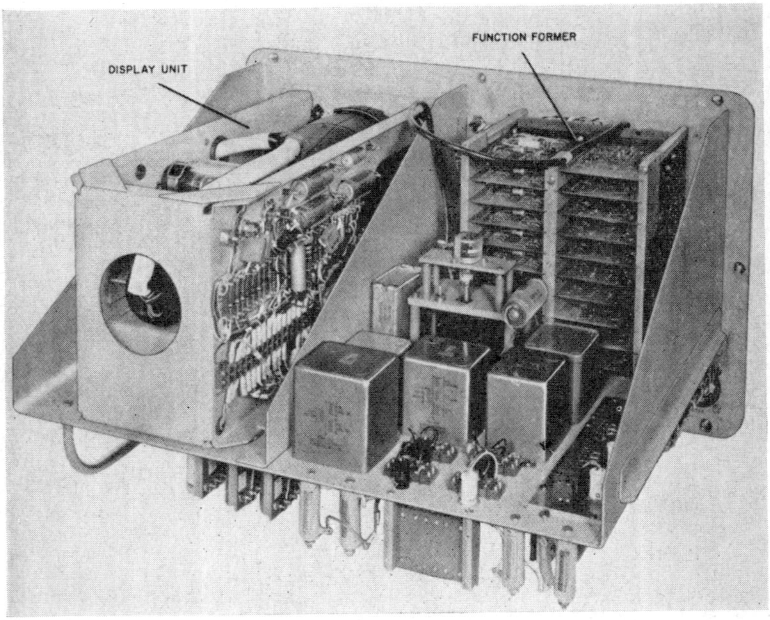

FIG. 10. Rear view of electronic ray tracer removed from case.

By taking into consideration the fact that the double trace rays indicated by arrows are superfluous it can be seen that the rays are uniformly spaced. It will be noted that they occupy only about half of the horizontal sweep of the screen.

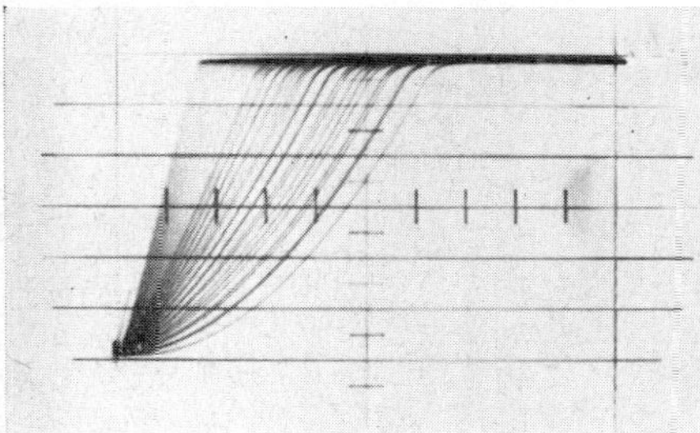

FIG. 11.

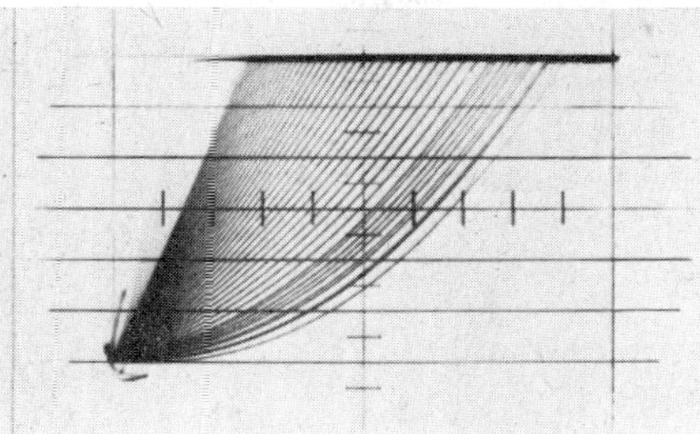

FIG. 12.

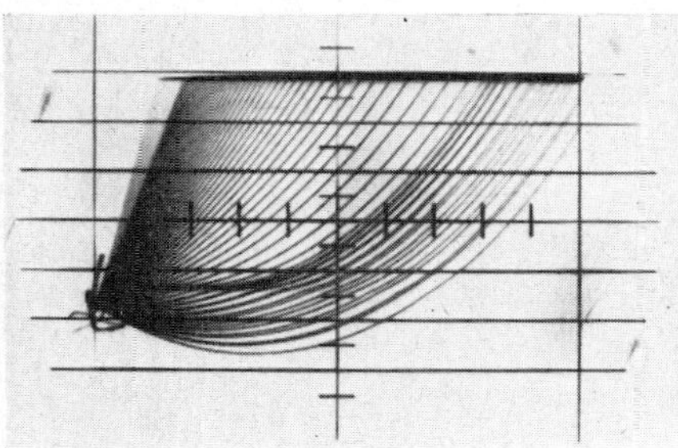

FIG. 13.

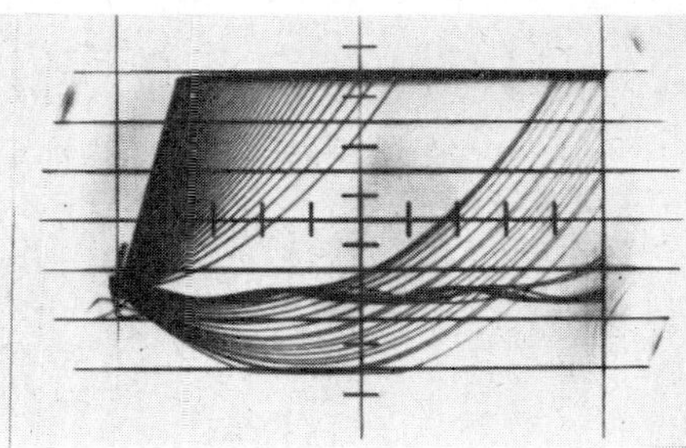

FIG. 14.

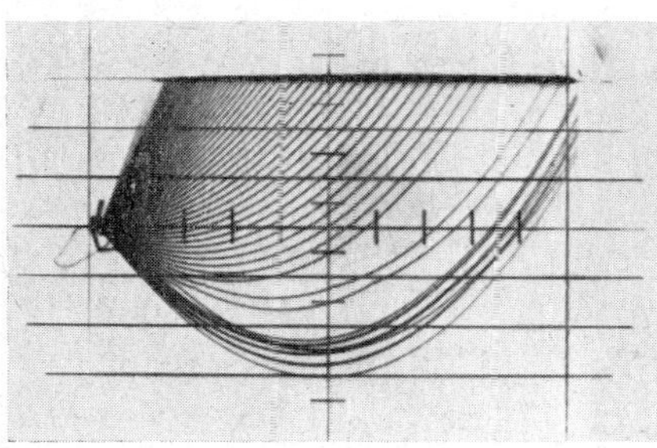

FIG. 15.

The ordinate scale being 4000 ft altitude per division and abscissae being a total distance of 320 miles. When we now introduce a uniform gradient into the ray tracer Fig. 12 roughly corresponding to a so called standard atmosphere, it will be observed that the rays are refracted downward, thus increasing the distance to radio horizon. Figure 13 shows the same conditions as before, with a smooth uniform duct placed at between 7000 and 7500 ft, above the transmitter, this does not effect radio propagation, except for a small offset in the ray path, which may not be very noticeable for the relatively thin duct used on the cathode ray scope. Moving the transmitter into the layer, by turning the altitude potentiometer produces the ducting condition shown in Fig. 14 causing the scarcity of rays typified by the region from 60 miles to 140 miles at 12,000 ft altitude and trapping of rays at 7000 ft. Moving the altitude knob so as to simulate positioning the transmitter just above the negative dn/dh layer yields Fig. 15. Here, rays striking layers of lowered refractive index at relatively steep angles are not affected, while rays striking at smaller angles are bent downward, thus causing a scarcity of rays in the region corresponding to a range of 100 to 160 miles at an altitude of 12,000 ft.

All of the pictures shown thus far were taken on the small basic ray tracer whose function generator will provide a good approximation of a refractive index profile at only one point on the propagation path. Numerous refractive index soundings taken at intervals along propagation paths show that for many localities we are justified in doing this unless a new mass of weather is moving in (Fig. 16). However, in some coastal areas where our propagation path starts over a land mass and moves out over the sea, we find that the layers are tilted and wavy (Fig. 17). Dr. Paul Taylor, WADC, has shown by analytical means that these layers can give rise to anomalous ground to air propagation.

To deal with these more complicated problems and to make the ray computer more versatile, a separate function generator has been built with several new features. Five generators capable of generating bell shaped curves are available. These circuits furnish a more convenient way of inserting a layer in the gradient of the index of refraction than by means of the step function generators. Any number (from 0 to 5) of the bell shaped curves can be incorporated in the gradient curve and each is adjustable for height, width, and intensity of the gradient. These curves are formed by a circuit which is cut-off except between prescribed limits of h set by the height and width controls. Between these limits the circuit generates an output which has a bell shape because of the characteristics of the transistors. This output is fed to the other portion of the gradient curve and forms part of it. The bell shape curve is probably a more realistic representation of the gradient of the index of the atmosphere in the neighbourhood of a layer than a sharp step function break. It corresponds to a layer with a smoother transition at the edges. The effect, however, is not great enough to make any difference in the ray patterns.

The function generator contains the usual step function generators to fill out the remainder of the gradient function. These are essentially the same as the circuits discussed previously.

A feature of the enlarged function generator is the incorporation of means of varying the index of refraction in range as well as in altitude. This allows a new and expanded body of problems to be set in the computer.

The range dependence is achieved by means of 10 segment generators, each consisting of 11 step generators. Each generator feeds a capacitor thereby producing a time function made up of 11 connected line segments. The range at which each step turns on is adjustable by a "Range Control". The output of

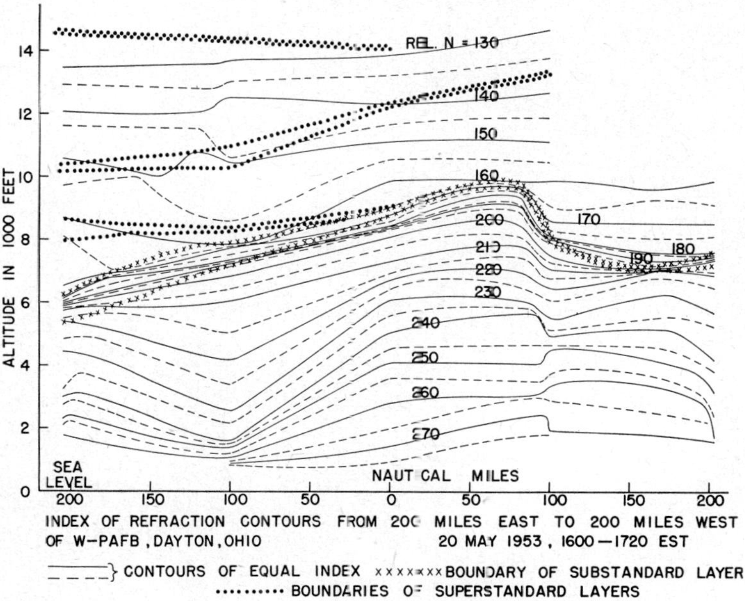

FIG. 16.

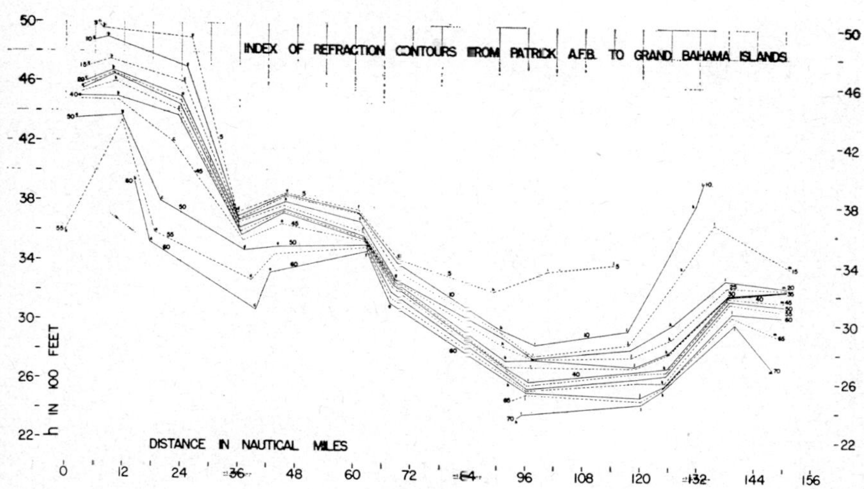

FIG. 17.

each segment is adjusted by a "Slope Control". This slope can be positive, zero, or negative so the functions generated have a high degree of arbitrariness. Each of these segmented functions may be used to control the characteristics of the bell shaped generators. Five of the voltages are used to affect the height of the 5 bell generators and the remaining 5 are used to control the intensity of the bells. Thus layers can be set into the computer which vary in height and intensity with range. Wavy layers, sloping layers and layers which terminate, are examples of conditions which may be studied. Slides showing the ray pattern with wavy layers will be shown.

The function generator also has provisions to generate the index curve as a function of height as well as the gradient. This is meant to be used in optical path length computations. In order to do this some changes must be made in the computing part of the machine. This has not been done to date so this feature of the function generator has not been used.

This increased versatility of the kind of gradients which can be handled is accomplished at the expense of a greatly increased complexity of function generating circuits. The function generator vastly overshadows the ray computation part of the equipment in size, bulk, and complexity. Continued evaluation will give answers to the question of how much function generating circuitry is needed.

The last three figures show some of the possibilities of the new function generators.

Figure 18. The ray trace represents the ray pattern from a transmitter located below a wavy super-standard layer at 7500 ft. The layer varies periodically about this altitude 500 ft above and below. The wave length of the variation is 7·5 miles. Such wavy layers have been observed and are analogous in many ways to ocean swell. For study on the ray tracer such a layer stays closer to real conditions than does a straight tilted layer.

Below the layer, the uniformly spaced rays indicate normal signal distribution. Abnormal signal is always found on the far side of the layer from the transmitter. In the region above the layer, is seen a periodic alteration of dense and sparse rays corresponding to strong and weak signals. The alternations are seen to correspond to the wave length of the layer waves. It may be seen that by refraction in the layer some rays are bent so severely that they cross other rays at a considerable angle. In regions where this occurs we have not only alternations of strong and weak signal due to alternate spreading and crowding of the rays but also we have interference (constructive or destructive) superposed on the pattern. To an observer who flew through the field the received signal would be so irregular that it would be very difficult to disentangle.

None of these effects would be produced by a level layer in this position. Only where the layer is tilted so as to meet a ray at a very low angle can the ray be much affected by the layer.

Figure 19. Here we have the transmitter located above the layer. For such relative positioning the layer would form a hole at about the region of the hole shown even though it were level. Here the waviness sends some rays into what would otherwise be a dead space, so that there is alternations of signal intensity within the hole.

Figure 20. Here we have a situation similar to that shown in Fig. 18, transmitter below the layer, closer than before but still not in the layer, in fact 3000 ft below it. There is a general pattern similar to that shown before but the ray

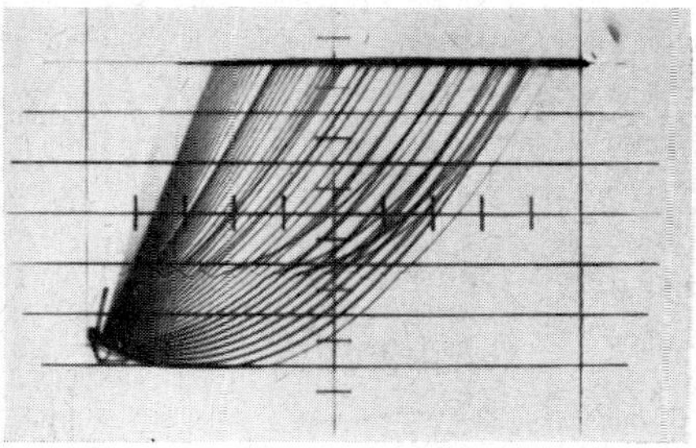

FIG. 18.

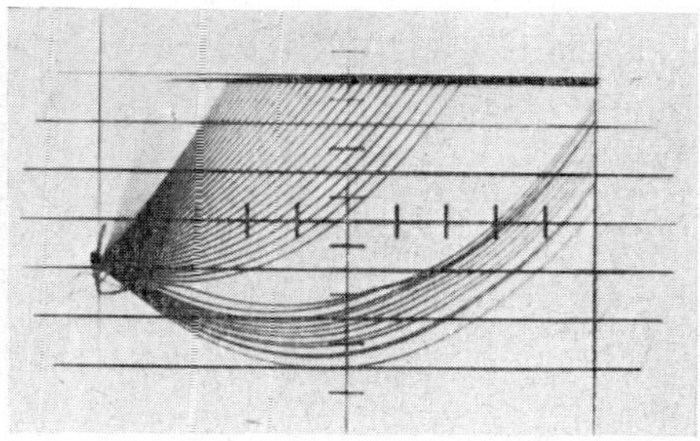

FIG. 19.

tracer has been fortunate enough to capture two most unusual rays. These are turned back and return to the level of the transmitter far beyond the normal horizon. Further they are returned in such position that they might again be returned to ground and this action might continue as far as the layer continued. This mechanism is seen to produce signal beyond the horizon. To accomplish it a continuous wavy layer is not needed only tilted lenticular patches.

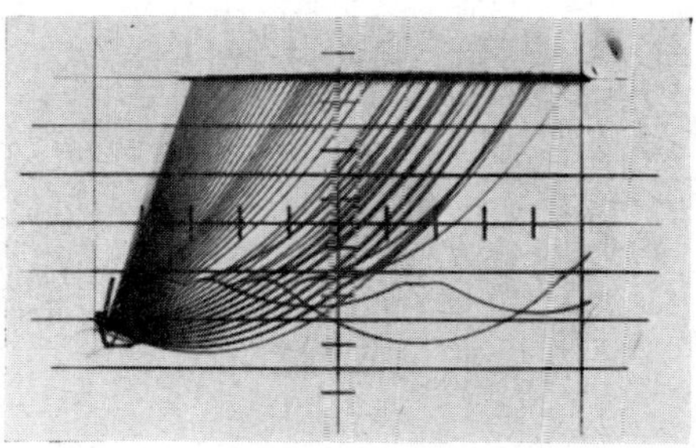

FIG. 20.

The ray tracer that I have described suffers from the same type of difficulties experienced in laboratory first models of many types of equipment. We have experienced the cluttering of the cathode ray screen previously described and have recently been troubled with equipment instability. The instruments stability has become worse with passing time. Most of this can be attributed to the fact that early types of transistors were used in its design. These transistors suffer characteristic changes under conditions of high humidity. We have experienced a very humid summer at Dayton which has aggravated this effect The Sperry Gyroscope Company has found that certain circuit changes, and the use of modern transistors greatly improve the stability of the instrument. Certainly present techniques would permit the design of an entirely satisfactory equipment of this type.

Small electronic ray tracers of this type are very useful for providing an immediate "picture" of propagation conditions where current refractive index information is available. The instrument provides information as to the existence of ducting, radio holes and anti-holes, and is accurate enough to provide a rough indication of angle and range errors.

The instrument is not accurate enough to provide great precision in range and angle error correction, and it does not seem likely that present cathode ray screens will provide such accuracy.

Invariant Imbedding and Wave Propagation in Stochastic Media

RICHARD BELLMAN and ROBERT KALABA

The RAND Corporation, Santa Monica, Calif.

In this paper we wish to describe an application of the theory of invariant imbedding to the propagation of electromagnetic waves through an inhomogeneous medium, especially where the inhomogeneity is stochastic in origin. Problems of this sort arise in a variety of fields including radio wave propagation and acoustics.

I. INTRODUCTION

There have been a large number of papers devoted to this topic, most using the routine perturbation approach, and a few the concept of a stratified medium; cf. Redheffer [22], Luneberg [19], Bremmer [13], Tatarskii [24], Gronwall [16], *et al.** Topics of current interest in radio wave propagation are discussed in [12, 14, 25, 27].

Our aim is first of all to illustrate the applicability of functional equations and principles of invariance to the study of various types of wave propagation. The basic ideas were sketched in our note [7], and fuller mathematical details will be given in papers shortly to appear. Secondly, we wish to point out the importance of functional equations in stochastic variables, prior to the appearance of any expected values. These relations can be used, as discussed below, to furnish numerical solutions along Monte Carlo lines. Finally, we wish to emphasize that the nonlinear aspects of the recurrence relations that are derived make it important not to use expected values, but rather to examine the actual probability distributions of the random variables that appear.

II. THE PHYSICAL PROCESS

We wish to consider the problem of determining the characteristics of waves reflected from a randomly inhomogeneous medium and the properties of waves transmitted through such a medium. In particular, we limit ourselves to the case in which a plane scalar wave is normally incident on a slab bounded by parallel planes, the wave number or index of refraction of which is a random function of distance from an interface. We assume that the media in which the slab is contained have constant wave numbers and, as usual, that time variations are simple harmonic. Ultimately we are concerned with random solutions of the reduced wave equation in one dimension

$$w''(x) + k^2(x)w(x) = 0 \tag{1}$$

*In his book on Radiative Transfer, Chandrasekhar refers to some earlier uses of invariance principles by Stokes and Rayleigh.

where the local wave number $k(x)$, is a random function of x. The index of refraction $n(x)$ is expressible as

$$n(x) = \frac{k(x)}{k_0} \tag{2}$$

Since it is the method of attack which we wish to emphasize, before taking up the stochastic case, we solve the reflection and transmission problems for the case in which $k(x)$ is deterministic, making use of the notion of invariant imbedding and a wave-localization principle which we first gave in [7]. Then we take up the problem of reflections from, and transmissions through, a medium which has random stratifications, each stratum being bounded by planes perpendicular to the direction of propagation of the incident wave and each having a random wave number. We make no assumption that these random wave numbers are independent, and no assumptions are made concerning "smallness" of departures of the wave numbers from their average values.

Some generalizations and indications of future work are provided in the discussion of Section VI.

III. Deterministic Wave Propagation

To illustrate some of the ideas which we shall employ in the treatment of the stochastic problem, we first discuss the problem of the reflection of a plane wave normally incident on an inhomogeneous slab, a problem of some difficulty and importance in itself [19]. Consider a plane wave, $\exp i(k_0 x - \omega t)$, arriving from the homogeneous space, $x < 0$, and approaching the inhomogeneous slab, $0 < x < b$. The space $x > b$ is assumed to be homogeneous.

Fig. 1. A plane wave incident on an inhomogeneous slab.

The medium $x < 0$ is characterized by the wave number k_0, and the medium $x > b$ by the wave number k_1. The inhomogeneous slab $0 < x < b$, is characterized by the wave number $k(x)$, which is assumed to be sectionally smooth. Our objective is the determination of the reflected wave, $Ae^{-ik_0 x}$, where here, and in what follows, we shall suppress the time factor $e^{-i\omega t}$.

We first imbed this problem within a class of problems in which the incident wave and the inhomogeneous slab are suitably altered. More precisely, we consider the problem of determining the wave reflected in the half-space $x \leqslant z$, where $0 \leqslant z \leqslant b$, as a result of having the wave $\exp [ik(z - 0)(x - z)]$ incident on the inhomogeneous slab $z \leqslant x \leqslant b$. As usual, $k(z - 0)$ denotes the limit of $k(x)$ as x approaches z from below and $k(z + 0)$ is the limit as x approaches z

from above. As before, we take the space $b < x$ to be homogeneous with the wave number k_1.

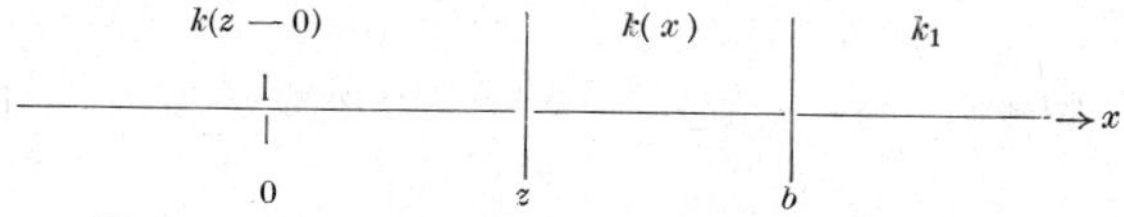

FIG. 2. The geometry of the class of slab problems.

For the modified problem the space $x < z$ has the constant wave number $k(z - 0)$, and we denote the reflected wave by $u(z) \exp [- ik(z - 0)(x - z)]$, where the explicit dependence of the coefficient of reflection $u(z)$, on the position of the left face of the slab z, is indicated.

Observe that when $z = 0$ we are confronted with the original problem, and when $z = b$ we are confronted with the simple problem of determining the reflection coefficient for waves incident on the plane interface of two homogeneous media. In the latter case we have

$$u(b) = \frac{k(b - 0) - k_1}{k(b - 0) + k_1} \tag{3}$$

which follows from the usual assumptions that are made concerning the continuity at the interface of the wave function and its first derivative with respect to x.

We now wish to show that a knowledge of the reflection coefficient for the inhomogeneous slab's extending from $z + \Delta$ to b enables one to find the reflection coefficient for the slab $[z, b]$. This is accomplished through use of the localization principle given in [7], valid under mild restrictions on the function $k(x)$, which in turn was suggested by Bremmer's paper [13]. Let us now indicate this principle.

Return for a moment to the original problem with $z = 0$ and consider a plane wave $\exp (ik_0 x)$ arriving from the homogeneous space $x < 0$ and entering the inhomogeneous slab. At the boundary the incident wave is split into a reflected wave and a refracted wave. There is an immediate reflected wave and an immediate refracted wave obtained by supposing that the inhomogeneous medium is actually homogeneous with wave number $k(+ 0)$. If the inhomogeneous medium is now taken to be the limit of a sequence of interfaces with this type of reflection and refraction occurring at each interface, we can obtain the total reflected wave and the disturbance within the inhomogeneous slab by adding up the effects of the reflected and refracted waves obtained in this way and then passing to the limit. This principle is valid regardless of the position of the left face of the slab z, $0 \leqslant z \leqslant b$.

To apply this principle, we consider the situation illustrated in Fig. 3, in which the wave numbers of the strata are as shown.

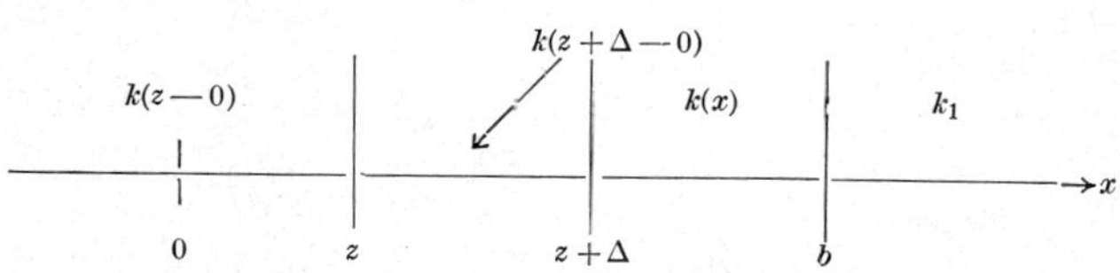

FIG. 3. The geometry for a class of modified problems.

To within terms of orders zero and one, the reflection coefficient for the slab $[z, b]$ is given by the sum of three terms. The first is the immediately reflected wave. The second is the wave that arises from transmission of the incident wave through the interface $x = z$, reflection at the interface $x = z + \Delta$ and transmission through the interface $x = z$. The third component arises from transmission of the incident wave through the interface at $x = z$, reflection at $x=z+\Delta$, reflection at $x = z$, reflection at $x = z + \Delta$, and finally transmission through the interface at $x = z$.

Mathematically, we are led to the equation

$$u(z) = \frac{k(z-0) - k(z+\Delta-0)}{k(z-0) + k(z+\Delta-0)} \tag{4}$$

$$+ \frac{2k(z-0)}{k(z-0) + k(z+\Delta-0)} u(z+\Delta)e^{2ik(z+\Delta-0)\Delta} \frac{2k(z+\Delta-0)}{k(z-0) + k(z+\Delta-0)}$$

$$+ \frac{2k(z-0)}{k(z-0) + k(z+\Delta-0)} e^{ik(z+\Delta-0)\Delta}u(z+\Delta)e^{ik(z+\Delta-0)\Delta}$$

$$\cdot \frac{k(z+\Delta-0) - k(z-0)}{k(z-0) + k(z+\Delta-0)} e^{ik(z+\Delta-0)\Delta}u(z+\Delta)e^{ik(z+\Delta-0)\Delta}$$

$$\cdot \frac{2k(z+\Delta-0)}{k(z+\Delta-0) + (z-0)} + o(\Delta).$$

A passage to the limit in which Δ tends to zero then shows that $u(z)$ satisfies the Riccati equation

$$u'(z) = \frac{k'}{2k} - 2iku - \frac{k'}{2k}u^2 \tag{5}$$

where the primes denote differentiation with respect to z.

Integration of this differential equation, which is easily accomplished numerically on a high speed digital computer or by hand, using the end condition of Equation (3), on the interval $(0, b)$ then yields the desired reflection coefficient, $u(0)$.

Similar Riccati equations for the impedance and reflection coefficient of non-uniform transmission lines have been given by Pierce [21], and by Walker and Wax [26]; see also the discussion given by Schelkunoff [23], and Osterberg [20]. The idea of concentrating attention on the reflection coefficient as a function of the thickness of the slab goes back to Ambarzumian [1], who applied it in certain radiative transfer problems involving diffuse reflection from a foggy medium. The notion was considerably extended and developed by Chandrasekhar [15] and Bellman and Kalaba [4]. In addition it forms the basis for the treatment by Bellman, Kalaba, and Wing [8–11] of various problems of neutron multiplication in fissionable material, including the determination of critical mass.

The point here is that it is easier to deal with the nonlinear Riccati equation and the end condition, insofar as numerical computation is concerned, than it is to solve the linear two-point boundary value problem for the wave within the slab. Furthermore, in many cases, it is the reflected wave rather than the wave within the slab which is of primary physical importance. Once the reflection coefficient $u(z)$ has been determined, though, for $0 \leqslant z \leqslant b$, the determination of the wave within the slab is also reduced to an initial value problem.

Similar considerations enable us to determine the transmission coefficient $v(z)$. For $z = b$ the problem reduces to that of determining the transmission through the interface between two homogeneous media having wave numbers $k(b - 0)$ and k_1. We have the usual end condition

$$v(b) = \frac{2k(b-0)}{k(b-0) + k_1} \tag{6}$$

To calculate the transmission coefficient for the case in which the inhomogeneous slab extends from z to b, from a knowledge of the transmission coefficient for the case in which the slab extends from $z + \Delta$ to b, we write the following equation:

$$v(z) = \frac{2k(z-0)}{k(z-0) + k(z+\Delta-0)} e^{ik(z+\Delta-0)\Delta} \left\{ v(z+\Delta) + u(z+\Delta)e^{ik(z+\Delta-0)\Delta} \right.$$
$$\left. \cdot \frac{k(z+\Delta-0) - k(z-0)}{k(z+\Delta-0) + k(z-0)} e^{ik(z+\Delta-0)\Delta} v(z+\Delta) \right\} + o(\Delta). \tag{7}$$

It states that excluding terms of order higher than the first in Δ the total transmitted wave consists of one component which is transmitted through the interface $x = z$, is transmitted through the interface $x = z + \Delta$, and is transmitted through the interface $x = b$, and a second component which arises from transmission through $x = z$, reflection at $x = z + \Delta$, reflection at $x = z$ and transmissions at $x = z + \Delta$ and $x = b$. A passage to the limit in which Δ tends toward zero then yields the differential equation

$$v' = \left(\frac{k'}{2k} - u \frac{k'}{2k} - ik \right) v \tag{8}$$

Assuming that $u(z)$ has been calculated on the interval $0 \leqslant z \leqslant b$, we see that the Equations (6) and (8) permit the determination of $v(z)$ on the same interval; in particular the transmission coefficient for the inhomogeneous slab, $v(0)$, can be calculated.

We wish to point out the strong similarity between Equations (5) and (8) above which describe reflection and transmission of waves and the equations in (6.2) of our previous paper [8], which describe reflection and transmission of neutrons in a rod of fissionable material. Abstractly the derivations of the equations for the wave and the particle cases are identical. We plan in the future to investigate the implications of this wave-particle analogy in more detail.

IV. Stochastic Wave Propagation

Let us now turn our attention to the case in which a wave is normally incident on a stratified slab, each stratum of which is characterized by a wave number which is a random variable. Our objectives will be the characterization of the stochastic reflected and refracted waves.

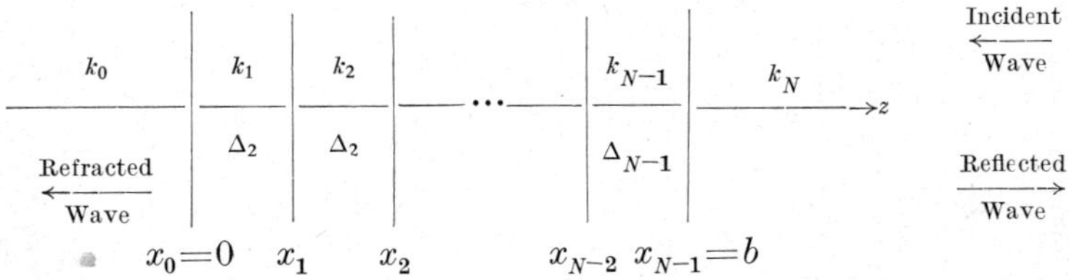

FIG. 4. A stochastic stratified slab of thickness b.

Referring to the above figure, we denote the wave number of the medium $z < 0$ by k_0, and that of the stratum between x_m and x_{m+1} by the random variable k_{m+1}, $m = 0, 1, \ldots, N-2$. For $z > x_{N-1} = b$, the medium has wave number k_N. Thus the slab $0 < z < x_{N-1}$ is divided into $N-1$ strata. We put $\Delta_m = x_m - x_{m-1}$. The incident wave, which arrives at the interface $z = x_{N-1}$ from $+\infty$, is specified by the function $e^{ik_N(z-x_{N-1})}$, where now the time factor $e^{i\omega t}$ is suppressed. The reflected wave is denoted by $R_N e^{-ik_N(z-x_{N-1})}$, $z > x_{N-1}$, and the refracted wave is denoted by $T_N e^{ik_0 z}$, $z < 0$.

Our attack on the problem of the determinationof the random variables R_N and T_N will be along the following lines. We first imbed the problem for a fixed value of N within the class of problems for which $N = 1, 2, \ldots$ Next, using the localization principle, we derive a relationship between the random variables R_N and R_{N-1}. Lastly, using analytic, Monte Carlo, or perturbation techniques we determine successively the distribution functions for the reflection coefficient, R_N, and the transmission coefficient, T_N. In this paper we do not assume that the random wave numbers k_m are independent, and no assumptions of smallness of variation of the random numbers k_m about their means need be made.

The simple case in which $N = 1$ is easily treated. The incoming wave is $e^{ik_1 z}$, $z > 0$, the reflected wave is $R_1 e^{-ik_1 z}$, $z > 0$, and the refracted wave is $T_1 e^{ik_0 z}$, $z < 0$. In terms of the wave numbers k_0 and k_1, the reflection and transmission coefficients are given by

$$R_1 = \frac{k_1 - k_0}{k_1 + k_0}, \tag{9}$$

$$T_1 = \frac{2k_1}{k_1 + k_0}. \tag{10}$$

These relations follow from the assumed continuity of the wave function and its first derivative at the interface $z = 0$.

We now turn to the case illustrated in Fig. 4. The total reflected wave is composed of an immediately reflected wave, a wave which arises from transmission through the interface at $z = x_{N-1}$, reflection at the face $z = x_{N-2}$, and retransmission through the interface at $z = x_{N-1}$, and so on. Expressed mathematically, this statement becomes

$$R_N = \frac{k_N - k_{N-1}}{k_N + k_{N-1}} + \frac{2k_N}{k_N + k_{N-1}} e^{ik_{N-1}(-\Delta_{N-1})} R_{N-1} e^{-ik_{N-1}\Delta_{N-1}} \cdot \frac{2k_{N-1}}{k_{N-1} + k_N} + \cdots \tag{11}$$

where the dots denote the contributions from terms involving more than one reflection from the medium $z < x_{N-2}$. To evaluate these terms we note that each component incident on the interface $z = x_{N-1}$ from the left gives rise to a reflected wave which is then partially reflected at the interface $z = x_{N-2}$ and is then partially transmitted through the interface $z = x_{N-1}$. These components are the terms of a geometric series with common ratio

$$\frac{k_{N-1} - k_N}{k_{N-1} + k_N} R_{N-1} e^{-2ik_{N-1}\Delta_{N-1}}.$$

Equation (11) becomes

$$R_N = \frac{k_N - k_{N-1}}{k_N + k_{N-1}} + \left(\frac{4k_N k_{N-1}}{(k_N + k_{N-1})^2} e^{-2ik_{N-1}\Delta_{N-1}} R_{N-1} \right) \cdot \left(\frac{1}{1 - \frac{k_{N-1} - k_N}{k_{N-1} + k_N} e^{-2ik_{N-1}\Delta_{N-1}} R_{N-1}} \right) \tag{12}$$

It is convenient to introduce the local reflection and transmission coefficients for waves incident from the right on the interface $x = x_{N-1}$:

$$r_N = \frac{k_N - k_{N-1}}{k_N + k_{N-1}} \tag{13}$$

$$t_N = \frac{2k_N}{k_N + k_{N-1}} \tag{14}$$

and the corresponding coefficients for waves incident from the left,

$$r'_N = \frac{k_{N-1} - k_N}{k_N + k_{N-1}} \tag{15}$$

$$t'_N = \frac{2k_{N-1}}{k_N + k_{N-1}} \tag{16}$$

These coefficients obviously satisfy Stokes' relations,

$$r_N + r'_N = 0 \tag{17}$$

$$t_N t'_N = 1 - r_N^2 \tag{18}$$

$$t_N + t'_N = 2 \tag{19}$$

so that Equation (12) becomes

$$\begin{aligned} R_N &= r_N + t_N t'_N e^{-2ik_{N-1}\Delta_{N-1}} R_{N-1} \frac{1}{1 - r'_N e^{-2ik_{N-1}\Delta_{N-1}} R_{N-1}} \\ &= \frac{r_N + e^{-2ik_{N-1}\Delta_{N-1}}(t_N t_N - r_N r'_N) R_{N-1}}{1 - r'_N e^{-2ik_{N-1}\Delta_{N-1}} R_{N-1}} \end{aligned} \tag{20}$$

If, finally, we make use of Stokes' relations and introduce

$$\delta_{N-1} = e^{-2ik_{N-1}\Delta_{N-1}} \tag{21}$$

we find

$$R_N = \frac{r_N + \delta_{N-1}R_{N-1}}{1 + r_N R_{N-1}}, \; N = 2, 3, \ldots \tag{22}$$

Equation (9) yields

$$R_1 = r_1 \tag{23}$$

Equations (22) and (23) constitute the desired stochastic functional equations for the reflection coefficients R_N.

Similar considerations for the transmission coefficients yield the relations

$$T_N = \frac{t_N e^{-ik_{N-1}\Delta_{N-1}}}{1 + r_N \delta_{N-1} R_{N-1}} T_{N-1}, \; N = 2, 3, \ldots \tag{24}$$

$$T_1 = t_1 \tag{25}$$

V. Applications of the Preceding Results

We now consider the uses to which we may put the stochastic functional equations derived in the last section. The basic problem that confronts us is that of determining the distribution of R_N and T_N, given the distribution of the random variables r_i and δ_i, $i = 1, 2, \ldots, N - 1$. Here, in speaking of the determination, we are primarily interested in the *numerical* determination.

We shall discuss three possible uses of the preceding results.

1. *Monte Carlo.* The formula seems ideally suited for a direct determination of the distribution for R_N and T_N using a Monte Carlo technique in conjunction with a digital computer which generates the random sequences $\{r_i\}$ and $\{\delta_i\}$. This will be easy to carry out even when the distributions of r_i and δ_i depend upon those of r_{i-1} and δ_{i-1}. Thousands of runs can be carried out in a matter of minutes.

2. *Recurrence Relations for Distribution Functions.* Since R_N is complex, we take as a fundamental distribution associated with this stochastic variable, the function

$$p_N(z, r) = \text{the probability that } R_N \text{ be in a circle about } z \text{ with radius } r \tag{26}$$

Since the transformation in Section (IV. 22) maps the interiors of circles into the interiors of circles, it is clear that we can obtain a relation for $p_N(z, r)$ in terms of the corresponding function for $N - 1$ [17].

Since, however, $p_N(z, r)$ depends upon the three real variables x, y and r, where $z = x + iy$, we see that this relation is not ideally suited for direct numerical computation.

Having obtained the formula, we can now introduce perturbation techniques. Assuming that r_i and δ_i may be written in the form

$$\begin{aligned} r_i &= r + \epsilon\rho_i, \\ \delta_i &= d + \epsilon\Delta_i, \end{aligned} \tag{27}$$

where ϵ is a small parameter and ρ_i and Δ_i are random variables with known distributions, we can write

$$p_N(z, r) = p_N^{(0)} + \epsilon p_N^{(1)} + \epsilon^2 p_N^{(2)} + \ldots \tag{28}$$

and use the recurrence relations to obtain relations for $p_N^{(0)}, p_N^{(1)}, \ldots$

3. *Stochastic Iteration.* The formula in Equation (22) yields a relation of the type

$$R_N = \frac{u_N + v_N R_0}{w_N + z_N R_0}, \tag{29}$$

where u_N, v_N, w_N, z_N are random variables determined by the sequences $\{r_i\}$ and $\{\delta_i\}$.

As a matter of fact, we know that Equation (22) is equivalent to the matrix relation

$$\begin{pmatrix} u_N & v_N \\ w_N & z_N \end{pmatrix} = \begin{pmatrix} r_N & \delta_{N-1} \\ 1 & r_N \end{pmatrix} \begin{pmatrix} u_{N-1} & v_{N-1} \\ w_{N-1} & z_{N-1} \end{pmatrix} \tag{30}$$

These relations can now be used to determine the individual and joint moments of the u_N, v_N, w_N, z_N, in terms of the moments of the r_i and δ_i, cf. [2].

Results of this type are useful in connection with the inverse problem of determining the distribution of the r_i and δ_i, given the observed stochastic values of the R_N.

Each of these methods will have its analog in the multi-dimensional case. From the standpoint of computational ease and effectiveness, we feel that the Monte Carlo method is the most promising.

VI. Discussion

It is clear that this paper is but a first step toward a more comprehensive theory of wave propagation in stochastic media based on the principle of invariant imbedding, a radical departure from current practice. In future papers, we plan to discuss applications to the case of oblique incidence, other geometries (see [5, 10]), and electromagnetic wave propagation, *per se*. In addition we propose to exploit the wave-particle similarity mentioned in Section III.

Finally, let us mention other techniques for treating nonlinear equations with stochastic elements, given in Bellman [3], and Kalaba [18], and which are relevant in wave propagation studies.

References

1. Ambarzumian, V. A. (1943). *C.R. Acad. Sci. U.R.S.S.* **38,** 229–232.
2. Bellman, R. (1954). *Duke math. J.* **21,** 491–500.
3. Bellman, R. (1958). *Proc. Amer. math. Soc.* **9,** 326–327.
4. Bellman, R. & Kalaba, R. (1956). *Proc. nat. Acad. Sci., Wash.* **42,** 629–632.
5. Bellman, R. & Kalaba, R. (1957). *Proc. nat. Acad. Sci., Wash.* **43,** 514–517.

6. Bellman, R. & Kalaba, R. (1957). *Proc. nat. Acad. Sci., Wash.* **43,** 930–933.
7. Bellman, R. & Kalaba, R. (1958). *Proc. nat. Acad. Sci., Wash.* **44,** 317–319.
8. Bellman, R., Kalaba, R. & Wing, G. M. (1957). *Proc. nat. Acad. Sci., Wash.* **43,** 517–520.
9. Bellman, R., Kalaba, R. & Wing, G. M. (1958). *J. Math. Mech.* **7,** 149–162.
10. Bellman, R., Kalaba, R. & Wing, G. M. (1958). "Invariant imbedding and generalized transport theory—a basic stochastic functional equation." The RAND Corporation, Paper P–1390.
11. Bellman, R. Kalaba, R. & Wing, G. M. (I959). *J. Math. Mech.* **8,** 575–584.
12. Booker, H. G. & Gordon, W. E. (1950). *Proc. Inst. Radio Engrs, N.Y.* **38,** 401–412.
13. Bremmer, H., "The Theory of Electromagnetic Waves, a Symposium," pp. 105–115. Interscience Publishers, Inc., New York (1951).
14. Carroll, T. J. & Ring, R. M. (1955). *Proc. Inst. Radio Engrs, N.Y.* **43,** 1384–1390.
15. Chandrasekhar, S. "Radiative Transfer." Oxford (1950).
16. Gronwall, T. H. (1926). *Phys. Rev.* **27,** 277–285.
17. Hurwitz, A. & Courant, R. "Allgemeine Funktionentheorie und Elliptische Funktionen, Geometrische Funktionentheorie" (in one volume), distributed by Interscience Publishers, Inc., New York (1929). See especially pp. 347–354.
18. Kalaba, R. (1959). *J. Math Mech.* **8,** 519–574.
19. Luneberg, R. K. (1947). *Res. Rep.* No. 172–173. New York University, Washington Square College.
20. Osterberg, H. (1958). *J. opt. Soc. Amer.* **48,** 513–521.
21. Pierce, J. (1943). *Bell Syst. tech. J.* **22,** 263–265.
22. Redheffer, R. (1954). *J. Rat. Mech. Anal.* **3,** 271–279.
23. Schelkunoff, S. A. (1951). "The Theory of Electromagnetic Waves, A Symposium", pp. 117–128. Interscience Publishers, Inc., New York (1951).
24. Tatarskii, V. I. (1956). *C. R. Acad. Sci. U.R.S.S.* **107,** 245–248.
25. Villars, F. & Weisskopf, V. F. (1955). *Proc. Inst. Radio Engrs, N.Y.* **43,** 1232–1239.
26. Walker, L. R. & Wax, N. (1946). *J. appl. Phys.* **17,** 1043–1045.
27. Wheelon, A. D. & Muchmore, R. B. (1955). *Proc. Inst. Radio Engrs, N.Y.* **43,** 1450–1458.

The Propagation over an Inhomogeneous Earth considered as a Two-dimensional Scattering Problem

H. Bremmer

Philips Research Laboratories, N. V. Philips' Gloeilampenfabrieken, Eindhoven

This article deals with a two-dimensional integral equation for the distribution of field-strength across a flat inhomogeneous earth. The equation in question is based on an approximative boundary condition, which accounts for the local distribution of the electrical constants. An infinite series representing the solution is interpreted in terms of scattering effects. The result applying to a special situation (two homogeneous regions separated by a straight boundary) is compared with that derived from the conventional one-dimensional equation which constitutes a saddlepoint approximation of the above two-dimensional equation.

I. Introduction, Leontovich's Boundary Condition

The propagation over a smooth inhomogeneous earth has been discussed by many authors with the aid of a proper boundary condition at the earth's surface. The field inside the earth can then be left out of consideration. The condition in question, often termed "Leontovich's boundary condition," has the following form in the case of a flat earth:

$$\frac{\partial \Pi}{\partial z} = \gamma(x, y)\Pi \tag{1}$$

In the example of a vertical dipole Π represents the amplitude of a Hertzian vector which has a vertical direction throughout the space ($z > 0$) above the earth's surface ($z = 0$). Further, the function $\gamma(x, y)$ is defined by:

$$\gamma(x, y) = -ik_0 \frac{\{n^2(x, y) - 1\}^{\frac{1}{2}}}{n^2(x, y)}$$

in which $n^2(x, y) = \epsilon(x, y) + i4\pi\sigma(x, y)/\omega$ fixes the local value of the refractive index of the earth in terms of its relative dielectric constant ϵ and its conductivity σ (in e.s.u.).

More general situations (such as a corrugated surface) can also be formulated with the aid of a surface admittance $A(x, y)$ varying from point to point; the parameters γ and A are connected by the relation:

$$A(x, y) = -\frac{i}{k_0}\gamma(x, y).$$

The most general problem of propagation over a flat inhomogeneous earth deals with the solution of Maxwell's equations in the half space $z > 0$, under the assumption of the boundary condition Equation (1), for an arbitrarily given distribution of the parameter γ.

II. The Basic Integral Equations

An application of Green's integral theorem to the half space $z > 0$ leads to an integral relation in which occur the values at $z = 0$ of both Π and $\partial\Pi/\partial z$. By substituting Equation (1) for the derivative we next obtain a two-dimensional integral equation for the function $\Pi(x, y, 0)$, that is for the distribution $\Pi(x, y)$ across the earth's surface of the amplitude of the Hertzian vector.

For simplicity's sake we restrict ourselves to a short vertical dipole of moment $\mathscr{K}$ which is placed at a point T just above the earth. Its undisturbed primary field, the free-space field, can be derived from the primary distribution

$$\Pi_{pr}(P, T) = \frac{\mathscr{K}}{c} \frac{e^{ik_0 TP}}{TP}, \tag{2}$$

TP being the distance from the transmitter to the point of observation P. For this transmitter the integral equation in question for the distribution

$$\Pi(x, y) = \Pi(P, T)$$

reads as follows:

$$\Pi(P, T) = 2\frac{\mathscr{K}}{c} \frac{e^{ik_0 TP}}{TP} - \frac{1}{2\pi}\int\int \gamma(Q)\Pi(Q, T)\frac{e^{ik_0 QP}}{QP}\, dO_Q \tag{3}$$

the integration extends over the complete surface ($z = 0$) of the earth, dO_Q being a surface element near the integration point Q which has a distance QP to the transmitter.

The conventional treatment [1] of the propagation problem starts from another one-dimensional integral equation which is obtained from Equation (3) by applying a saddlepoint approximation. The latter is connected with the exponential function $\exp\{ik_0(TQ + QP)\}$; it constitutes a factor of the integrand in Equation (3), the factor $\exp(ik_0\, TQ)$ being the rapidly varying constituent in $\Pi(Q, T)$. The new one-dimensional integral equation reads as follows:

$$W(P, T) = 1 - \left(\frac{iPT}{2\pi k_0}\right)^{\frac{1}{2}} \int_T^P \frac{\gamma(Q) W(Q, T)}{(TQ \,.\, QP)^{\frac{1}{2}}} ds_Q \tag{4}$$

when expressed in terms of the function $W(P, T) = 2\Pi(P, T)/\Pi_{pr}(P, T)$ which represents the attenuation relative to propagation over a perfectly conducting earth. The integration in Equation (4) extends along the section TP of the straight line connecting the transmitter at T with the point of observation P; ds_Q is a line element of this section.

A physical consequence of the simple form of the one-dimensional integral Equation (4) (which is of the Volterra type) is the complete independence of the field at some point P of the γ distribution *off* the path TP, and even of its distribution along the part of the path that is beyond P. This shows that Equation (4) may only be applied under restricted conditions; for instance, well pronounced inhomogeneities near the path, with boundaries not reaching the latter, should not occur. Therefore, a general consideration of the original two-dimensional integral Equation (3), which only assumes the validity of the Leontovich boundary condition Equation (1), is most important in order to cover all possible circumstances. Some consequences of the latter equation will be discussed in the present paper after having derived a more convenient equivalent form of it in the next section. So far, only one paper by Wait [4] concerns a two-dimensional treatment which can be compared with that dealt with in this article.

III. An Alternative Form of the Basic Two-dimensional Integral Equation

Both the two-dimensional integral Equation (3) and its one-dimensional approximation Equation (4) connect the distribution $\Pi(PT)$ to be determined with the primary distribution given by Equation (2). On the other hand, Feinberg's [3] analysis starts from an one-dimensional integral equation that links the unknown $\Pi(P, T)$ distribution with the special distribution $\Pi_{\gamma_0}(P, T)$ corresponding to a homogeneous earth characterized by a constant value γ_0 of γ. A similar integral equation connecting $\Pi(P, T)$ and $\Pi_{\gamma_0}(P, T)$ exists in our two-dimensional treatment; it reads:

$$\Pi(P, T) = \Pi_{\gamma_0}(P, T) - \frac{c}{4\pi\mathscr{K}} \iint \{\gamma(Q) - \gamma_0\} \Pi(Q, T) \Pi_{\gamma_0}(Q, P) dO_Q \qquad (5)$$

in which $\Pi_{\gamma_0}(Q, P)$ represents the solution for a homogeneous earth with $\gamma = \gamma_0$ when the transmitter [described by Equation (2)] is placed at P instead of T.

The equivalency of the two integral Equations (3) and (5) can be proved, as indicated hereafter, with the aid of two-dimensional operational calculus. In this calculus the "image" $f(p, q)$ of some original function $h(x, y)$ of two variables is defined by

$$f(p, q) = pq \int_{-\infty}^{\infty} dx \int_{-\infty}^{\infty} dy\, e^{-px-qy} h(x, y) \qquad (6)$$

This relation is abbreviated symbolically by

$$h(x, y) \doteqdot f(p, q).$$

The operational derivation of Equation (5) from Equation (3) is facilitated by expressing these integral equations in terms of convolution products of two variables. The latter are of the type:

$$h_1(x, y) * h_2(x, y) \equiv \int_{-\infty}^{\infty} d\xi \int_{-\infty}^{\infty} d\eta\, h_1(\xi, \eta) h_2(x - \xi, y - \eta).$$

We take the origin of coordinates at T. Taking into account the symmetry relation $\Pi_{\gamma_0}(Q, P) = \Pi_{\gamma_0}(x_P - x_Q, y_P - y_Q)$, we find that the integral Equations (3) and (5) for $\Pi(P, T) = \Pi(x, y)$ and $\Pi_{\gamma_0}(P, T) = \Pi_{\gamma_0}(x, y)$ can be put in the following forms:

$$\Pi(x, y) = \frac{2\mathscr{K}}{c} \frac{e^{ik_0\sqrt{(x^2+y^2)}}}{\sqrt{(x^2 + y^2)}} - \frac{1}{2\pi} \{\gamma(x, y)\Pi(x, y)\} * \frac{e^{ik_0\sqrt{(x^2+y^2)}}}{\sqrt{(x^2 + y^2)}}, \qquad (7)$$

$$\Pi(x, y) = \Pi_{\gamma_0}(x, y) - \frac{c}{4\pi\mathscr{K}} \Big[\{\gamma(x, y) - \gamma_0\}\Pi(x, y)\Big] * \Pi_{\gamma_0}(x, y) \qquad (8)$$

We further define the "images":

$$\Pi(x, y) \doteqdot F(p, q)\,;\ \gamma(x, y)\Pi(x, y) \doteqdot f_\gamma(p, q)\,;\ \Pi_{\gamma_0}(x, y) \doteqdot F_{\gamma_0}(p, q).$$

We next apply the general operational rule

$$h_1(x, y) \ast h_2(x, y) \doteqdot \frac{f_1(p, q)f_2(p, q)}{pq},$$

(f_1 and f_2 being the "images" of h_1 and h_2), and also the special operational relation:

$$\frac{e^{ik_0\sqrt{(x^2+y^2)}}}{\sqrt{(x^2+y^2)}} \doteqdot 2\pi i \frac{pq}{\sqrt{(p^2+q^2+k_0^2)}} \tag{9}$$

This relation holds for $Tm\, k_0 > |\, Re \sqrt{(p^2+q^2)}\,|$, the sign of the square root being defined by $|\sqrt{(p^2+q^2+k_0^2)}+k_0| > |\sqrt{(p^2+q^2)}|$.

We thus obtain the following operational "images" of the complete Equations (7) and (8):

$$\mathrm{F}(p, q) = 4\pi i \frac{\mathscr{K}}{c} \frac{pq}{\sqrt{(p^2+q^2+k_0^2)}} - \frac{if_\gamma(p, q)}{\sqrt{(p^2+q^2+k_0^2)}} \tag{10}$$

$$F(p, q) = F_{\gamma_0}(p, q) - \frac{c}{\mathscr{K}} \frac{\{f_\gamma(p, q) - \gamma_0 F(p, q)\}F_{\gamma_0}(p, q)}{4\pi pq} \tag{11}$$

These latter relations prove to be identical provided that

$$F_{\gamma_0}(p, q) = 4\pi i \frac{\mathscr{K}}{c} \frac{pq}{\sqrt{(p^2+q^2+k_0^2)}+i\gamma_0} \tag{12}$$

The equivalency of Equations (3) and (5) can thus be concluded from that of their images Equations (10) and (11) if we verify the correctness of the image Equation (12). However, this latter image is obtained by substituting $\gamma = \gamma_0$ in the image Equation (10) of the original Equation (3); this substitution involves the other relation $f\gamma = \gamma_0 F_{\gamma_0}$ and then leads to the value Equation (12) for F_{γ_0}. This completes the derivation of the new integral Equation (5) from the original Equation (3).

IV. A Formal Solution of the Integral Equation (5)

We may represent the distribution of $\gamma(x, y)$ by:

$$\gamma_0(x, y) = \gamma_0 + \delta\gamma(x, y),$$

in which γ_0 should correspond to the electrical constants of an average soil. The basic integral Equation (5) thus becomes:

$$\Pi(P, T) = \Pi_{\gamma_0}(P, T) - \frac{c}{4\pi\mathscr{K}} \iint \delta\gamma(Q)\Pi(Q, T)\Pi_{\gamma_0}(Q, P)dO_Q \tag{13}$$

This equation can be solved formally by the expansion

$$\Pi(P, T) = \sum_{n=0}^{\infty} \Pi_n(P, T) \tag{14}$$

if we take $\Pi_0(P, T) = \Pi_{\gamma_0}(P, T)$, whereas the further terms should be derived in succession with the aid of the recurrence relation:

$$\Pi_n(P, T) = -\frac{c}{4\pi\mathscr{K}} \int\int \delta\gamma(Q)\Pi_{n-1}(Q, T)\Pi_{\gamma_0}(Q, P)dO_Q \,. \, (n \geqslant 1) \tag{15}$$

The individual term Π_n can also be represented as a $2n$-fold integral; it results after n successive applications of Equation (15). The integrand of this multidimensional integral contains a factor consisting of the product of the $\delta\gamma$ values at n points $Q_1, Q_2, \ldots, Q_n$. It shows that Π_n will be of the order of $(\delta\gamma)^n$ with respect to $\delta\gamma$. Therefore, Π_n can be interpreted as the contribution produced by n successive scatterings in places at which $\delta\gamma \neq 0$. This physical interpretation is clearest when $\delta\gamma$ does differ from zero only in small isolated regions. However, the interpretation in terms of multiple scatterings can still be maintained if adjacent scattering regions cover, e.g., a complete half plane with a constant value γ_1 of γ which differs from the other value γ_0 in the remaining part of the earth's surface. An illustrative example dealing with such a simple distribution is discussed in the next section.

V. Two Homogeneous γ- Regions Separated by a Straight Boundary, with the Special Case of one Perfectly Conducting Medium

This case can be characterized as follows, if the boundary is given by $x = x_0$:

$$\gamma(x, y) = \begin{cases} \gamma_0 \text{ for } x < x_0. \\ \gamma_1 \text{ for } x > x_0 \end{cases}$$

In view of Equation (15) the contribution of first-order scatterings is represented by

$$\Pi_1(P, T) = -\frac{c}{4\pi\mathscr{K}}(\gamma_1 - \gamma_0) \int_{x_0}^{\infty} dx_Q \int_{-\infty}^{\infty} dy_Q \Pi_{\gamma_0}(Q, T)\Pi_{\gamma_0}(Q, P) \tag{16}$$

In order to arrive at some simple numerical result we further assume a transmitter at infinity, in a direction perpendicular to the boundary, and a perfectly conducting earth to the left of this boundary. We therefore can take $x_T \to -\infty$, $y_T = 0, \gamma_0 = 0$. Moreover, the primary field $\Pi_0(P, T)$ here equals twice the free-space field, so as to have

$$\Pi_{\gamma_0}(Q, T) = \frac{2\mathscr{K}}{c}\frac{e^{ik_0 QT}}{QT}; \tag{17}$$

this is in accordance with Equations (2), (9), and (12) for $\gamma_0 = 0$. However, the last expression can also be approximated by

$$\Pi_{\gamma_0}(Q, T) = 2\frac{\mathscr{K}}{c}\frac{e^{ik_0(x_Q - x_T)}}{|x_T|},$$

since we assumed the transmitter at a very distant point to the left. On the other hand, the original formula Equation (17) has to be applied when deriving $\Pi_{\gamma_0}(Q, P)$ in Equation (16) since Q and P are in general both at finite distances there. The corresponding substitutions into Equation (16) yield:

$$\Pi_1(P, T) = -\frac{\gamma_1 \mathscr{K}}{\pi c \mid x_T \mid} e^{ik_0 x_T} \int_{x_0}^{\infty} dx_Q e^{ik_0 x_Q} \int_{-\infty}^{\infty} dy_Q \frac{e^{ik_0 QP}}{QP}$$

$$= -\frac{\gamma_1 \mathscr{K}}{\pi c \mid x_T \mid} e^{-ik_0 x_T} \int_{x_0}^{\infty} dx\, e^{ik_0 x} \int_{-\infty}^{\infty} dy \frac{e^{ik_0 \sqrt{((x_p - x)^2 + (y_p - y)^2)}}}{\sqrt{((x_p - x)^2 + (y_p - y)^2)}}$$

$$= \frac{\gamma_1 \mathscr{K}}{ic} \frac{e^{ik_0(x_0 - x_T)}}{\mid x_T \mid} \int_0^{\infty} d\xi\, e^{ik_0 \xi} H_0^{(1)}(k_0 \mid \xi + x_0 - x_p \mid) \tag{18}$$

The integral containing a Hankel function can be evaluated with the aid of the identity

$$\int_0^{\infty} d\xi\, e^{ik_0 \xi} H_0^{(1)}(k_0 \mid \xi - a \mid) = a\{H_0^{(1)}(k_0 \mid a \mid) \pm i H_1^{(1)}(k_0 \mid a \mid)\} \text{ for } a \gtrless 0 \tag{19}$$

This relation can be verified, e.g., by showing that both sides satisfy the differential equation

$$\frac{dT}{da} - ik_0 T = H_0^{(1)}(k_0 \mid a \mid),$$

and, moreoever, vanish at $a = \pm \infty$ (the imaginary part of k_0 has then to be assumed as positive, though it may be very small).

The first two terms of the expansion Equation (14) are now obtained by adding Equation (17) (for Q replaced by P) and Equation (18), the integral of the latter being evaluated with the aid of Equation (19). We obtain:

$$\Pi(P, T) = \frac{2\mathscr{K}}{c} \frac{e^{ik_0(x_P - x_T)}}{\mid x_T \mid}$$

$$\left[1 + \frac{\gamma_1}{2}(x_P - x_0) e^{-ik_0(x_P - x_0)} \{iH_0^{(1)}(k_0 \mid x_P - x_0 \mid) \pm H_1^{(1)}(k_0 \mid x_P - x_0 \mid)\} \ldots \right],$$

$$\text{for } x_P \gtrless x_0 \tag{20}$$

in which the terms not written out represent the contributions due to higher-order scatterings. The common effect of the latter will be smallest for small values of γ_1, $\Pi_n(P, T)$ being of the order of γ_1^n.

VI. Discussion of the Solution, eq. (20) for the Propagation over a Perfectly Conducting Half Plane with an Adjacent Region having a Constant Conductivity

We can substitute the asymptotic expressions for the Hankel functions in Equation (20); it shows the oscillatory behaviour of $\Pi(P, T)$ (as a function of x_P)

in the region near the boundary $x = x_0$ but only at the side of the transmitter. This oscillatory character has also been noticed by Wait [4] who derived an expression similar to Equation (20) for a more general case ($\gamma_1 \neq 0$).

It is interesting to compare Equation (20) with the corresponding result derived from the one-dimensional integral Equation (4). This result reads as follows [as can be verified by substituting $\mu_1 = 0$ and $\mu_2 = -\gamma_1/(2ik_0)^{\frac{1}{2}}$ in the limit for $d_1 \to \infty$ of Equation (26) in reference [5]]

$$\Pi(P, T) = \frac{2\mathscr{K}}{c} \frac{e^{ik_0(x_P - x_T)}}{|x_T|} \left\{1 - \left(\frac{i}{2\pi k_0}\right)^{\frac{1}{2}} \gamma_1 \int_0^{x_P - x_0} y\left(\frac{\gamma_1^2}{2ik_0}\xi\right) \frac{d\xi}{\sqrt{(\xi)}}\right\}; \ (x_P > x)_0 \tag{21}$$

$y(\rho)$ here represents the well-known Sommerfeld function for the attenuation (relative to that over a perfectly conducting earth) in the case of propagation over a homogeneous earth. Apart from substituting the asymptotic expressions for the Hankel functions in Equation (20), we can also derive the first two terms of the Taylor expansion of Equation (21) with respect to $(x_P - x_0)^{\frac{1}{2}}$. [See reference [5], formula below Equation (42)]. It turns out that both procedures lead to one and the same approximation; the latter holds in the region near the boundary at $x_P = x_0$ but beyond it ($x_P > x_0$). The approximation in question reads:

$$\Pi(P, T) = 2\frac{\mathscr{K}}{c} \frac{e^{ik_0(x_P - x_T)}}{|x_T|} \left\{1 - \left(\frac{2i}{k_0 \pi}\right)^{\frac{1}{2}} \gamma_1 (x_P - x_0)^{\frac{1}{2}} \ldots\right\} \tag{22}$$

When deriving these two terms from the expression Equation (20) (resulting from the two-dimensional treatment) we realize the restriction $k_0(x_P - x_0) \gg 1$; it is a consequence of the limited applicability of the asymtotic representations of the Hankel functions. On the other hand, the validity of Equation (22) as the beginning of the Taylor expansion of Equation (21) (corresponding to the one-dimensional treatment) requires a small value of the next term of this expansion; this proves to lead to the other restriction $x_P - x_0 \ll k_0/\gamma_1^2$. Hence both the more rigorous two-dimensional treatment and the simpler one-dimensional derivation lead to the common approximative result (22) in the overlapping domain characterized by

$$\frac{1}{k_0} \ll x_P - x_0 \ll \frac{k_0}{\gamma_1^2} \tag{23}$$

In this domain the terms in Equation (21) that are proportional to $\gamma_1^2, \gamma_1^3 \ldots$ prove to be negligible. These powers of γ_1 also determine the orders of magnitude of the terms $\Pi_2(P), \Pi_3(P) \ldots$ in Equation (15) [compare Section (IV) for $\delta\gamma = \gamma_1$]. This suggests that all higher-order scatterings produce very small contributions in the domain of Equation (23). The one-dimensional treatment also leads to a reasonable approximation there. The domain in question extends in the γ_1 medium from a few wavelengths beyond the boundary with the perfectly conducting medium, up to a distance of the order of k_0/γ_1^2 away from this boundary. Hence this particular domain becomes widest for small values of γ_1.

We observe that the oscillations in front of the boundary $x_P = x_0$ could not be derived from the one-dimensional treatment. As a matter of fact the latter involves an undisturbed field having the properties of the first medium up to its boundary with the second medium. We expect similar properties in situations more complicated than that discussed in the previous section. In particular fieldstrength oscillations might in general occur near boundaries separating almost homogeneous sections of the earth's surface.

References

1. A detailed list of references is given in Z. Godzinski, *Inst. Elect. Engrs*, Monograph 299R, 1958.
2. Bremmer, H. (1954). *Physica 's Grav.* **20**, 441.
3. Feinberg, E. (1945). *J. Phys., Moscow* **9**, 1.
4. Wait, J. R. (1956). *Rep. nat. Bur. Stand.* No. 5014.
5. Wait, J. R. (1956). *Rep. nat. Bur. Stand.* No. 5014.

Low Frequency Propagation in the Ionosphere

BERNARD FRIEDMAN
University of California, Berkeley, Calif.

The propagation of low frequency radio waves between a conducting earth and a constant impedance layer is studied. For a highly conducting ionosphere, the attenuation is a minimum at a frequency of 260 H^{-1} kc where H is the height of the atmosphere in kilometers.

I. INTRODUCTION

The work of Budden [2] has shown that, for ionospheric propagation of radio waves with frequencies below 100 kc, the effect of the earth's magnetic field can be neglected. Using this fact, Budden [3], Al'pert [1] and Wait [5] have studied the extremely long-range propagation of low-frequency radio waves and have shown that the long-range propagation can be explained in terms of the modes that are produced when a radio wave travels between the earth and the ionosphere.

In this paper, we shall consider a simpler model of the ionosphere than has been used previously and we shall discuss the modes that exist in this model and how they affect long-range propagation. The advantages of this simpler model are two-fold:

(1) A mathematically rigorous treatment of the field produced by an electric dipole can be obtained.

(2) The modal equation and the dependence of the modes on the characteristics of the ionosphere can be studied analytically without extensive numerical computation.

The model we use is essentially that of a "leaky" spherical wave guide. We make the following assumptions:

(a) The earth is a perfect sphere surrounded by a sharply bounded ionosphere at a constant height H above the surface of the earth.

(*b*) The ionosphere can be replaced by a surface layer of constant impedance Z at the height H above the earth's surface. (Note this implies no electromagnetic field above the constant impedance layer.) We shall see later that very long range propagation is due to the lowest order modes and we can show that for such modes it makes little difference whether the electromagnetic field is restricted to the region below a height H (as we do in this paper) or whether the electromagnetic field is considered to exist also in the ionosphere above the height H.

(*c*) The magnetic field of the earth can be neglected. This assumption is justified by Budden's work [2].

(*d*) The earth is a perfect conductor (the finite conductivity of the earth can be taken into account but the analysis then becomes more complicated).

In this paper, the field produced by a vertical electric dipole at a height h above the earth is obtained. After some approximations, it is found that the field at a distance d from the source can be expressed as a product of three factors. The first factor is the field amplitude produced by a vertical electric dipole above a flat conducting earth at a distance d from the dipole. The second factor is a modification produced by the sphericity of the earth and by the fact that the radiation is channeled between the earth and the ionosphere. The third factor is a sum over the modes which propagate in the region between the earth and the atmosphere. These results are similar to those obtained by Al'pert [1] and Wait [5].

A careful study of the modal equation leads to the following conclusions:

(1) At a sufficiently large distance from the transmitter and for a highly conducting ionosphere, most of the radio energy is carried by either the zeroth or the first mode. It is shown that the attenuation in the first mode is always more than the attenuation in the zeroth mode. To explain the observed facts about the propagation of atmospherics, we assume that an atmospheric is represented not by a vertical electric dipole but by a vertical electric quadrupole. Because of this assumption, the amplitude of the zeroth mode is very small compared to the amplitude of the first mode; consequently, above the cut-off frequency (equal to 150 H^{-1} kc if H is measured in km) the first mode is responsible for the propagation; below the cut-off frequency, the zeroth mode takes over. This seems to agree with the results of Liebermann [4].

(2) The negative modes considered by Budden [3] are probably incorrect for two reasons. First, they are obtained by shifting the contour for a complex integral. This shifting is legitimate in only one half of the complex plane and not in the other half-plane; consequently, only the positive modes and not the negative modes can be obtained in this manner. Second, in solving the modal equation for the negative modes, if Budden had kept the physically necessary condition that the imaginary part of S_n be negative (thus ensuring finite fields at infinite distance), the attenuation of the negative modes would have been the same as the attenuation for the corresponding positive mode.

(3) Again, for a highly conducting ionosphere, it is found in Section III that the attenuation in nepers per unit distance has a minimum at the frequency 260 H^{-1} kc, where H is measured in kilometers. For $H = 100$ km, this would indicate that the attenuation is a minimum at 2·6 kc.

(4) If the ionosphere is highly conducting it behaves like a short circuit to the E.M. field (E, the electrical field, is approximately zero). If the ionosphere is only moderately conducting, then for the low-numbered modes it behaves like an open circuit (the magnetic field is approximately zero). However, even for moderate conductivity, the ionosphere behaves like a short circuit for the higher numbered modes. This peculiar behavior of the ionosphere explains the fact noticed by Wait [5] that calculating the modes by perturbing those of a perfectly conducting wave guide gave unsatisfactory results.

II. Mathematical Formulation

Consider the field produced by a vertical electric dipole located at a height h above the surface of the earth. We shall assume that the earth is a perfect conductor and that the ionosphere can be replaced by a layer of constant impedance Z at a height $H(> h)$ above the surface of the earth. We use spherical coordinates ρ, θ, ϕ with the origin at the center of the earth and the polar axis through the axis of the dipole. If we assume that all the field quantities have a time factor $e^{-i\omega t}$, Maxwell's equations for the region between $\rho = a$ and $\rho = a + H$ are as follows:

$$\begin{aligned} \text{curl } E &= i\omega\mu_0 H \\ \text{curl } H &= -\, i\omega\epsilon_0 E + J. \end{aligned} \tag{1}$$

Here ϵ_0, μ_0 are the permittivity and permeability of free space, respectively, and

$$J = M \frac{\delta(\rho - a - h)\delta(\theta)}{2\pi\rho^2 \sin\theta} i_\rho \tag{2}$$

where M is the electric moment of the dipole and i_ρ is a unit vector in the ρ-direction.

Because the only non-zero component of J is the ρ-component and because of the symmetry around the polar axis, we shall express the electromagnetic field quantities in terms of a Hertz vector whose only non-zero component is the ρ-component. First, we write Maxwell's equations in spherical coordinates as follows:

$$\frac{1}{\rho \sin\theta} \frac{\partial}{\partial\theta} (H_\phi \sin\theta) = -\, i\omega\epsilon_0 E_\rho + J_\rho \tag{3}$$

$$\frac{1}{\rho} \frac{\partial}{\partial\rho} (\rho H_\phi) = i\omega\epsilon_0 E_\theta \tag{4}$$

$$\frac{\partial}{\partial\rho} (\rho E_\theta) - \frac{\partial E_\rho}{\partial\theta} = i\omega\mu_0 \rho H_\phi \tag{5}$$

Next, we define a Hertzian potential U by the equation

$$H_\phi = \frac{1}{\rho} \frac{\partial U}{\partial\theta} \tag{6}$$

From Equation (4) we find

$$E_\theta = \frac{Z_0}{ik\rho} \frac{\partial^2 U}{\partial\theta\partial\rho} \tag{7}$$

where k, the free space wave number of the radiation, and Z_0, the impedance of free space, are defined by the following relations:

$$k = \omega(\epsilon_0\mu_0)^{\frac{1}{2}}, \quad Z_0 = (\mu_0/\epsilon_0)^{\frac{1}{2}} = 120\pi \text{ ohms}.$$

Using Equations (6) and (7) in (5) and integrating with respect to θ, we find that

$$E_\rho = \frac{Z_0}{ik} \left[\frac{\partial^2 U}{\partial\rho^2} + k^2 U \right] \tag{8}$$

Substituting Equations (6) and (8) into (3), we obtain the following partial differential equation for U:

$$\frac{\partial^2 U}{\partial \rho^2} + \frac{1}{\rho^2 \sin\theta}\frac{\partial}{\partial\theta}\left[\sin\theta\frac{\partial U}{\partial\theta}\right] + k^2 U = J_\rho \tag{9}$$

where

$$J_\rho = M\frac{\delta(\rho - a - h)\delta(\theta)}{2\pi\rho^2 \sin\theta} \tag{10}$$

The assumption that the earth is a perfect conductor implies that $E_\theta = 0$ for $\rho = a$, or what is equivalent, that

$$\frac{\partial U}{\partial \rho} = 0 \tag{11}$$

for $\rho = a$. Our assumption that the ionosphere can be replaced by a constant impedance sheet means that

$$E_\theta / H_\phi = Z,$$

where Z is a complex constant whose value depends on the properties of the ionosphere. Expressing this relation in terms of U, we get that

$$\frac{\partial U}{\partial \rho} = \frac{ikZ}{Z_0} U \tag{12}$$

for $\rho = a + H$.

Our mathematical formulation is now complete. We must solve Equation (9) subject to boundary conditions expressed by Equations (11) and (12). The solution will be obtained by the method of separation of variables, but first we shall introduce some approximations.

III. Solution of the Equation

Put

$$\rho = a + zH, \quad h = z_0 H, \quad U = V(\sin\theta)^{-\frac{1}{2}} \tag{13}$$

into Equation (9). After some simplifications it becomes

$$\frac{\partial^2 V}{\partial z^2} + \left[\frac{H}{a + zH}\right]^2 \left[\frac{\partial^2 V}{\partial \theta^2} + \frac{1}{4}(1 + \operatorname{cosec}^2\theta)V\right] + (kH)^2 V \tag{14}$$

$$= \frac{\delta(z - z_0)\delta(\theta)}{2\pi(\sin\theta)^{\frac{1}{2}}}\frac{M}{H}\left[\frac{H}{a + zH}\right]^2,$$

and the boundary conditions in Equations (11) and (12) become

$$\frac{\partial V}{\partial z} = 0 \tag{15}$$

for $z = 0$ and

$$\frac{\partial V}{\partial z} = ikHBV \tag{16}$$

for $z = 1$ where

$$B = Z/Z_0, \tag{17}$$

the inverse of the index of refraction of the ionosphere. Since a, the radius of the earth, is very much larger than H, the height of the ionosphere, we may approximate $a + zH$ by a and therefore Equation (14) becomes

$$\frac{\partial^2 V}{\partial z^2} + \left[\frac{H}{a}\right]^2 \left[\frac{\partial^2 V}{\partial \theta^2} + \frac{1}{4}(1 + \operatorname{cosec}^2 \theta) V\right] + (kH)^2 V = \frac{MH}{a^2} \frac{\delta(z - z_0)\delta(\theta)}{2\pi (\sin \theta)^{\frac{1}{2}}}. \tag{18}$$

We now complete the solution by the method of separation of variables.

Consider the following eigenvalue problem: To find non-zero solutions $V_n(z)$ of the equation

$$\frac{d^2 V_n}{dz^2} + \lambda_n^2 V = 0 \tag{19}$$

such that

$$\frac{dV_n}{dz} = 0 \qquad \text{for } z = 0 \tag{20}$$

$$\frac{dV_n}{dz} = ikHBV_n \quad \text{for } z = 1.$$

The solution of Equation (19) satisfying the first equation of (20) is

$$V_n = \alpha_n \cos \lambda_n z \tag{21}$$

where α_n is some normalization constant. To satisfy the second equation of (20), we must have

$$-\lambda_n \tan \lambda_n = ikHB \tag{22}$$

The roots of this equation are the eigenvalues of Equation (19), and the functions defined in Equation (21) are the eigenfunctions of Equation (19). It is easy to show that the eigenfunctions are mutually orthogonal. For normalization, we put

$$1 = \int_0^1 V_n^2 dz = \alpha_n^2 \int_0^1 \cos^2 \lambda_n z dz = \alpha_n^2 \left[\frac{1}{2} + \frac{\sin 2\lambda_n}{4\lambda_n}\right];$$

and therefore

$$\alpha_n = \left[\frac{1}{2} + \frac{\sin 2\lambda_n}{4\lambda_n}\right]^{-\frac{1}{2}}. \tag{23}$$

Because the eigenfunctions of (19) are complete, we may write the solution $V(z, \theta)$ of Equation (18) as follows:

$$V = \sum c_n(\theta) V_n(z_0) V_n(z) \tag{24}$$

where

$$c_n(\theta) V_n(z_0) = \int_0^1 V(z, \theta) V_n(z) dz \tag{25}$$

Multiply Equation (18) by $V_n(z)dz$ and integrate between 0 and 1. With the help of integration by parts and the use of Equations (15), (16), and (20), it becomes

$$\frac{d^2c_n}{d\theta^2} + \left\{[(kH)^2 - \lambda_n^2]\left[\frac{a}{H}\right]^2 + \frac{1}{4} + \frac{1}{4}\operatorname{cosec}^2\theta\right\}c_n \tag{26}$$

$$= \frac{M}{2\pi H}\frac{\delta(\theta)}{(\sin\theta)^{\frac{1}{2}}}.$$

In this equation we use the approximation $\sin\theta \approx \theta - \dfrac{\theta^3}{6}$ and neglect the 1/4 and $\theta^3/6$ compared to the term multiplied by $(a/H)^2$. If we put

$$\lambda_n = kHC_n = kH\cos\theta_n, \quad S_n^2 = 1 - C_n^2 \tag{27}$$

then Equation (26) becomes

$$\frac{d^2c_n}{d\theta^2} + \left[(kaS_n)^2 + \frac{1}{4}\theta^{-2}\right]c_n = \frac{M}{2\pi H}\frac{\delta(\theta)}{\theta^{\frac{1}{2}}} \tag{28}$$

The solutions of Equation (28) are the Bessel functions of order zero and argument $kaS_n\theta$, multiplied by $\theta^{\frac{1}{2}}$. Because of the fact that the field must be outgoing from the line $\theta = 0$, the solution must be a Hankel function. Taking into account the presence of the δ-function in Equation (28), we find that its solution is

$$c_n(\theta) = \frac{M\theta^{\frac{1}{2}}}{4iH}H_0^{(1)}(kaS_n\theta).$$

Using this in Equation (24) we get

$$V = \frac{M\theta^{\frac{1}{2}}}{4iH}\Sigma V_n(z_0)V_n(z)H_0^{(1)}(kaS_n\theta).$$

Since $ka\theta \gg 1$, we may use the asymptotic formula for the Hankel function and thus reduce the formula for V to this:

$$V = \frac{Me^{-3\pi i/4}\theta^{\frac{1}{2}}}{H(8\pi ka\theta)^{\frac{1}{2}}}\Sigma V_n(z_0)V_n(z)e^{ikaS_n\theta}S_n^{-\frac{1}{2}} \tag{29}$$

Finally, from Equation (13) we get

$$U = \frac{M}{H}\left[\frac{\theta}{\sin\theta}\right]^{\frac{1}{2}}\frac{e^{-3\pi i/4}}{(8\pi ka\theta)^{\frac{1}{2}}}\Sigma V_n(z_0)V_n(z)e^{ikaS_n\theta}S_n^{-\frac{1}{2}} \tag{30}$$

Using Equations (6), (7), (8), and (21), we get the following formulas for the field components:

$$H_\phi = \frac{Mk}{H}\left[\frac{\theta}{\sin\theta}\right]^{\frac{1}{2}}\frac{e^{-5\pi i/4}}{(8\pi ka\theta)^{\frac{1}{2}}}\Sigma\,\alpha^2\cos(kHC_nz_0)\cos(kHC_nz)S_n^{\frac{1}{2}}e^{ikaS_n\theta}$$

$$E_\theta = \frac{-MkZ_0}{H}\left[\frac{\theta}{\sin\theta}\right]^{\frac{1}{2}}\frac{e^{-3\pi i/4}}{(8\pi ka\theta)^{\frac{1}{2}}}\Sigma\alpha_n^2S_n^{\frac{1}{2}}C_n\cos(kHC_nz_0)\sin(kHC_nz)e^{ikaS_n\theta} \tag{31}$$

$$E_\rho = \frac{MkZ_0}{H}\left[\frac{\theta}{\sin\theta}\right]^{\frac{1}{2}}\frac{e^{\pi i/4}}{(8\pi ka\theta)^{\frac{1}{2}}}\Sigma\alpha_n^2S_n^{\frac{3}{2}}\cos(kHC_nz_0)\cos(kHC_nz)e^{ikaS_n\theta}$$

We can simplify these formulas somewhat by introducing E_0, the amplitude of the field produced by a vertical electric dipole on a flat perfectly conducting earth at a distance d. We have

$$E_0 = \frac{Z_0}{\lambda}\frac{M}{2d} \tag{32a}$$

Note that $d = a\theta$. Introduce also a symbol F where

$$F = \left[\frac{\theta}{\sin\theta}\right]^{\frac{1}{2}} \frac{(d/\lambda)^{\frac{1}{2}}}{(H/\lambda)} \tag{32b}$$

Then Equation (31) becomes

$$\begin{aligned} E_\rho &= E_0 F e^{\pi i/4} \Sigma \alpha_n^2 S_n^{\frac{3}{2}} \cos\left[\frac{2\pi H}{\lambda} C_n z_0\right] \cos\left[\frac{2\pi H}{\lambda} C_n z\right] e^{i2\pi S_n d/\lambda} \\ E_\theta &= -\, E_0 F e^{-3\pi i/4} \Sigma \alpha_n^2 S_n^{\frac{1}{2}} C_n \cos\left[\frac{2\pi H}{\lambda} C_n z_0\right] \sin\left[\frac{2\pi H}{\lambda} C_n z\right] e^{i2\pi S_n d/\lambda} \\ H_\phi &= \frac{E_0 F}{Z_0} e^{-5\pi i/4} \Sigma \alpha_n^2 S_n^{\frac{1}{2}} \cos\left[\frac{2\pi H}{\lambda} C_n z_0\right] \cos\left[\frac{2\pi H}{\lambda} C_n z\right] e^{i2\pi S_n d/\lambda}. \end{aligned} \tag{33}$$

IV. The Attenuation of the Modes

From Equation (33), it is seen that the nth mode in the expression for the field has a factor $\exp\,|\,2\pi i S_n d/\lambda|$, where

$$S_n = (1 - C_n^2)^{1/2} \tag{34}$$

and where, by Equations (22) and (27), C_n is the root of the following equation:

$$-\, C_n \tan(kHC_n) = iB.$$

If the tangent is replaced by its expression in terms of exponential functions, this equation becomes

$$e^{2ikHC_n} = \frac{C_n + B}{C_n - B} \tag{35}$$

The attenuation of the field in nepers per unit length is clearly given by

$$\beta_n = 2\pi \mathrm{Im}(S_n/\lambda) \tag{36}$$

The discussion of the attenuation will depend on a study of the roots of Equation (35). We consider two limiting cases. First, we have the case where B is small compared to C_n. This is the case if the ionosphere is highly conducting and behaves like a short circuit. Second, we have the case where B is large compared to C_n. This is the case if the ionosphere is moderately conducting and we are considering only the lower numbered modes. In this case, the ionosphere behaves like an open circuit.

Before discussing these cases, we shall consider how B depends on the conductivity. The complex dialectric constant for a medium with conductivity σ is given by

$$\frac{\epsilon'}{\epsilon_0} = 1 + \frac{i\sigma}{\epsilon_0 \omega} \tag{37}$$

Here ϵ_0 is the free space permittivity and ω is the circular frequency of the radio wave. We have

$$B = \frac{Z}{Z_0} = \frac{(\mu_0/\epsilon')^{\frac{1}{2}}}{(\mu_0/\epsilon_0)^{\frac{1}{2}}} = (\epsilon_0/\epsilon')^{\frac{1}{2}}.$$

We put

$$\sigma/\epsilon_0 = \omega_r,$$

a characteristic frequency of the ionosphere, and in Equation (37) we neglect the one on the right-hand side. Then, approximately,

$$B = (\omega/\omega_r)^{\frac{1}{2}} e^{-\pi i/4} \tag{38}$$

Let us suppose now that B is small compared to C_n. We write Equation (35) as

$$e^{4\pi i H C_n/\lambda} = 1 + \frac{2B}{C_n - B} \approx 1 + \frac{2B}{C_n}.$$

Taking the natural logarithm of both sides, we get

$$4\pi i H C_n/\lambda \approx \frac{2B}{C_n} + 2n\pi i \; (n = 0, 1, 2, \ldots)$$

approximately. From this equation we get

$$C_n \approx \frac{n\lambda/2}{H} - \frac{iB}{C_n}\frac{\lambda}{2\pi H} \tag{39}$$

For $n = 0$, we have

$$C_0 \approx \left(\frac{B\lambda}{2\pi H}\right)^{\frac{1}{2}} e^{-\pi i/4} \tag{40}$$

For this approximation to be valid, we must have

$$|B| \ll |C_0| \tag{41}$$

We introduce a characteristic wavelength λ_r for the ionosphere where $\lambda_r = \frac{1}{2\pi}\omega_r/c$, and we put $D = \frac{H}{\lambda_r}\left(\frac{\omega}{\omega_r}\right)^{\frac{3}{2}}$. Using this and Equation (38) in Equation (41), we find that

$$(2\pi D)^{\frac{1}{2}} \ll 1 \tag{42}$$

is the condition necessary for Equation (40) to hold.

For $n \neq 0$, we replace C_n on the right-hand side of Equation (39) by the first term on the right-hand side. In this way, we get

$$C_n \approx \frac{n\lambda/2}{H} - \frac{iB}{\pi n} \tag{43}$$

This approximation is valid if

$$|B| \ll \frac{n\lambda/2}{H},$$

that is, if

$$D \ll n/2 \tag{44}$$

The attenuation in the zeroth mode is given by

$$\beta_0 = 2\pi \operatorname{Im} S_0/\lambda = 2\pi \operatorname{Im} \lambda^{-1}\left[1 + \frac{iB\lambda}{2\pi H}\right]^{\frac{1}{2}} = \frac{1}{2H}\left(\frac{\omega}{2\omega_r}\right)^{\frac{1}{2}} \tag{45}$$

The attenuation in the nth mode is given by

$$\beta_n = 2\pi \operatorname{Im} S_n/\lambda = 2\pi \operatorname{Im} \lambda^{-1}\left[1 - \left(\frac{n\lambda/2}{H}\right)^2 + \frac{iB\lambda}{\pi H}\right]^{\frac{1}{2}}.$$

If we assume $n\lambda/2 < H$ so that the nth mode is a propagating mode, and put

$$1 - \left(\frac{n\lambda/2}{H}\right)^2 = \gamma_n^2,$$

then

$$\beta_n = 2\pi \operatorname{Im} \lambda^{-1}\left| \gamma_n + \frac{iB\lambda}{2\pi H\gamma_n} \right| = \frac{1}{H\gamma_n}\left(\frac{\omega}{2\omega_r}\right)^{\frac{1}{2}} \tag{46}$$

Because γ_n is always less than one, we have

$$H\gamma_n < 2H \tag{47}$$

and therefore

$$\beta_1 > \beta_0 \tag{48}$$

The value of β_1 will be a minimum for that value of ω which makes

$$\frac{\omega^{\frac{1}{2}}}{\gamma_1}$$

a minimum. We find that the minimum of β_1 occurs when $H = \lambda(\frac{3}{4})^{\frac{1}{2}}$. When this result is expressed in terms of frequency ν, it becomes

$$\nu = \frac{c(\frac{3}{4})^{\frac{1}{2}}}{H} = \frac{260}{H}\ \text{kc}$$

if H is in kilometers.

The preceding results were obtained on the assumption that B was small compared to C_n. Let us suppose now that B is large compared to C_n. We write Equation (35) as

$$e^{4\pi iHC_n/\lambda} = -1 + \frac{2C_n}{C_n - B} \approx -1 - \frac{2C_n}{B} \tag{49}$$

Again taking the logarithms of both sides, we get

$$4\pi iHC_n/\lambda \approx \frac{2C_n}{B} + (2n+1)\pi i$$

or

$$C_n \approx \frac{(2n+1)}{4H}\left[1 - \frac{\lambda}{2\pi i\, HB}\right]^{-1} \tag{50}$$

This approximation to C_n is valid if

$$| C_n | \ll | B |$$

that is, if

$$\left| \frac{(2n+1)\lambda}{4HB} \right| \ll 1 \tag{51}$$

that is, if

$$\frac{2n+1}{4} \ll D \tag{52}$$

It is clear that this condition is surely violated for large values of n. This shows that the number of such "open-circuit" modes is limited. Suppose that Equation (52) is satisfied for $n = 0$. The attenuation in this case will be

$$2\pi \mathrm{Im}\, S_0/\lambda = 2\pi\lambda^{-1}\left[1 - \left(\frac{\lambda}{4H}\right)^2 + \frac{i}{16\pi B}\left(\frac{\lambda}{H}\right)\right]^{\frac{1}{2}} \tag{53}$$

$$\approx \frac{1}{16H\gamma_{\frac{1}{2}}} \left(\frac{\lambda}{H}\right)^2 \left(\frac{2\omega_n}{\omega}\right)^{\frac{1}{2}}$$

where

$$\gamma_{\frac{1}{2}}^2 = 1 - \left(\frac{\lambda}{4H}\right)^2. \tag{54}$$

Acknowledgment

This work was done under the sponsorship of the Boeing Airplane Company, Seattle, Washington, U.S.A., and I wish to thank them for permission to publish.

References

1. Al'pert, Ia. L. "Low Frequency Electromagnetic Wave Propagation over the Surface of the Earth." Moscow (1955).
2. Budden, K. G. (1951). *Phil. Mag.* **42,** 1–19.
3. Budden, K. G. (1951). *Phil Mag.* **42,** 833; (1952) **43,** 1179.
4. Liebermann, L. N. (1956). *J. appl. Phys.* **27,** 1477.
5. Wait, J. R. (1957). *Proc. V.L.F. Symposium,* Paper No. 6, Boulder, Colorado. (1957). *Proc. Inst. Radio Engrs, N.Y.* **45,** 760 and 768.

Inexistence d'une Onde de Surface sur une Terre Plane et Diélectrique. Comparaison aux Cas des Cylindres Diélectrique et Infiniment Conducteur *

C. JAUQUET

Laboratoire de Recherches Physiques, Ateliers de Constructions Electriques de Charleroi, Charleroi

On montre que les cylindres diélectrique parfait et infiniment conducteur ont sur le champ créé par un cornet coaxial, un effet nettement different de celui qu'a un plan diélectrique et homogène sur le champ créé par un dipôle de Hertz. Si on donne le nom de guide d'ondes de surface aux deux cylindres, on ne peut plus employer la même expression pour désigner le plan diélectrique. Ces conclusions ne sont pas modifiées fondamentalement par la présence de pertes dans le diélectrique ou le métal.

I. INTRODUCTION

Certains corps de forme et de constitution déterminées ont la propriété, lorsqu'ils sont éclairés par une antenne bien choisie, de concentrer au voisinage de leur surface le champ électromagnétique créé par la source et de provoquer la propagation de l'onde dans une direction préférentielle de la surface.

Le but de la communication est de donner le résultat de calculs qui fournissent la loi de variation du champ le long d'une génératrice d'un cylindre circulaire, infiniment long et diélectrique parfait et d'un cylindre infiniment conducteur, dans le cas où ils sont éclairés par un cornet électromagnétique coaxial. D'autre part, on rappellera la loi de variation du champ en fonction de la distance radiale sur une terre plane et diélectrique, avec ou sans perte, éclairée par un dipôle de Hertz.

On verra que les cylindres diélectrique et infiniment conducteur ont sur le champ un effet nettement différent de celui de la terre plane et diélectrique. Si l'on considère ces deux cylindres comme des guides d'ondes de surface, comme nous verrons qu'il est logique de le faire, on ne pourra pas admettre que la terre plane et diélectrique guide une onde de surface; tout au moins, ces expressions ne peuvent pas avoir la même signification dans les deux cas.

II. CAS DES CYLINDRES DIELECTRIQUE ET INFINIMENT CONDUCTEUR

Considérons (Fig. 1) le cylindre circulaire s'étendant de $z = 0$ à $z = \infty$ et éclairé par un cornet électromagnétique coaxial. Ce dernier est un câble coaxial dont le conducteur intérieur se trouve dans le prolongement du cylindre et le conducteur extérieur s'évase en forme de cornet. Celui-ci débouche dans un plan infiniment conducteur. Un théorème d'équivalence permet de remplacer le

* Cette étude a été partiellement faite au Laboratoire de Radio-électricité de l'Université de Bruxelles et a fait l'objet d'une dissertation de doctorat.

cornet excité selon son mode principal par une nappe de courant magnétique coïncidant avec son embouchure, le cylindre s'étendant alors de $z = -\infty$ à

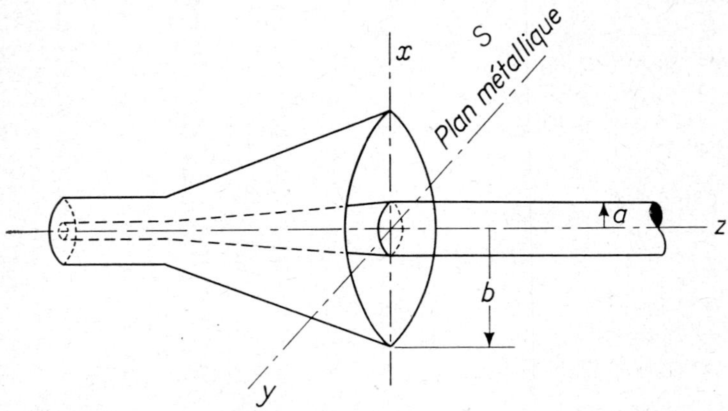

FIG. 1.

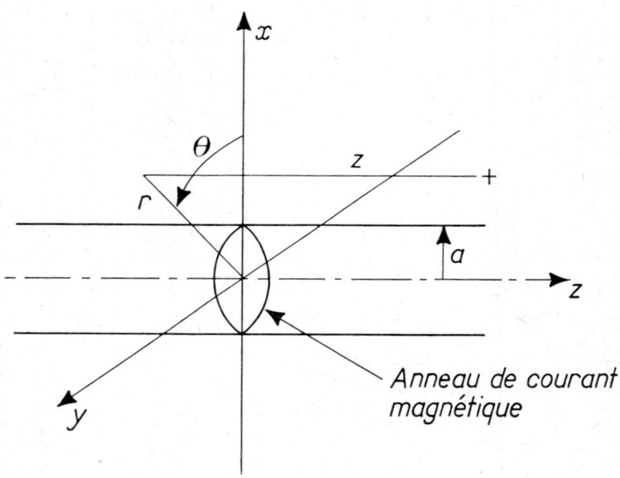

FIG. 2.

$z = +\infty$. La seule composante non nulle de la densité superficielle du courant magnétique de la nappe est circulaire. C'est pourquoi, pour les calculs, nous idéalisons la source qu'est le cornet coaxial par un anneau de courant magnétique infiniment mince et de même rayon que le cylindre (Fig. 2).

A. Cylindre diélectrique

Nous nous proposons de calculer le champ créé par un tel anneau dans l'hypothèse d'un cylindre diélectrique (milieu 1, $r < a$). Pour ce faire, on écrit les équations de Maxwell où les seconds membres dépendent de la nature de la source, c'est-à-dire traduisent l'existence de l'anneau de courant magnétique. On résoud ces équations, compte tenu des conditions aux limites, par la méthode des transformées de Fourier. On obtient dans l'air (milieu 2, $r > a$) pour la composante circulaire du champ magnétique H_θ (la seule non nulle dans un système cylindrique de coordonnées) l'expression suivante dans le cas où le cylindre est un diélectrique parfait [1],

$$H_\theta = -\frac{k_0}{2\pi} i\omega\epsilon_0 a \int_C \frac{H_1^{(1)}(\lambda_2 r)e^{ihz}e^{-i\omega t}dh}{\lambda_2 a H_0^{(1)}(\lambda_2 a) - (\lambda_1 a/\epsilon_r)(J_0(\lambda_1 a)/J_1(\lambda_1 a))\, H_1^{(1)}(\lambda_2 a)}. \quad (1)$$

pour $r \geqslant a$.

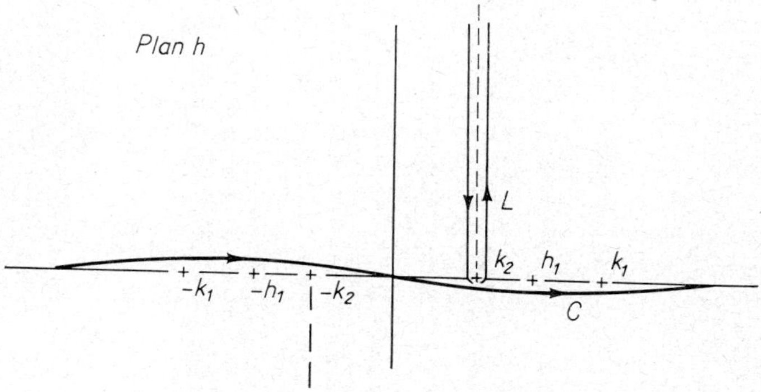

FIG. 3.

Dans cette expression, les symboles ont la signification suivante:

C = contour d'intégration représenté sur la Fig. 3.
k_0 = amplitude du courant magnétique source (en V).
ω = pulsation du courant magnétique supposé sinusoïdal.
ϵ_0, μ_0 = permittivité et perméabilité de l'air.
ϵ_1, μ_1 = permittivité et perméabilité du diélectrique ($\epsilon_r = \epsilon_1/\epsilon_0$).
a = rayon du cylindre diélectrique.
$H_j^{(1)}$ = fonction de Hankel de première espèce d'ordre j.
J_j = fonction de Bessel d'ordre j.
k_1 = constante de propagation du diélectrique.
k_2 = constante de propagation du vide.
$\lambda_2 = (k_2^2 - h^2)^{\frac{1}{2}}$, $\lambda_1 = (k_1^2 - h^2)^{\frac{1}{2}}$.

Les points singuliers de l'intégrand dans (1) sont les points de branchement $\pm k_2$ et les pôles. Lorsque la longueur d'onde dans le vide λ_0 se trouve dans les limites ci-après:

$$\frac{2\pi a(\epsilon_r - 1)^{\frac{1}{2}}}{2{\cdot}405} > \lambda_0 > \frac{2\pi a(\epsilon_r - 1)^{\frac{1}{2}}}{5{\cdot}52}, \quad (2)$$

l'intégrand a une paire de pôles $\pm\, h_1$ situés sur le plan h comme indiqué sur la Fig. 3. Dans ce cas, l'application du théorème de Cauchy donne:

$$H_\theta = \int_C \ldots = \int_{\mathscr{L}} \ldots + 2\pi i \text{ (Résidu au pôle } h_1\text{), pour } z > 0, \tag{3}$$

$\mathscr{L}$ étant un lacet entourant la ligne de coupure émanant du point de branchement k_2. La première contribution à H_θ est appelée "onde de coupure," la seconde "onde de résidu."

Si l'on calcule la valeur asymptotique de l'onde de coupure (z grand) à la surface du cylindre ($r = a$), on trouve une loi de variation en fonction de z de la forme $1/z^2$. La loi de variation de l'onde de résidu est de la forme e^{ih_1z}, celle-ci prédomine donc rapidement devant la première. Or, en l'absence du cylindre, la loi de variation du champ magnétique dans la direction axiale est de la forme $1/z^2$. Il en résulte que le cylindre diélectrique excité par un anneau de courant magnétique provoque dans cette direction une loi de décroissance de l'onde beaucoup plus lente que celle obtenue en son absence. Pour cette raison, on dit qu'il constitue un guide d'ondes de surface. Il faut remarquer que dans ce cas, le caractère d'onde guidée est conféré à l'onde totale par la présence du terme de résidu.

B. Cylindre infiniment conducteur

Supposons maintenant que le cylindre est infiniment conducteur. Dans ce cas, on trouve [2]:

$$H_\theta = \frac{-k_0}{2\pi}\, i\omega\epsilon_0 a \int_C \frac{H_1^{(1)}(\lambda_2 r)e^{ihz}dh}{\lambda_2 a H_0^{(1)}(\lambda_2 a)}. \tag{4}$$

Cette fois, l'onde de résidu est nulle. L'onde totale égale à l'onde de coupure, vaut pour $r = a$ et z grand:

$$H_\theta = -\frac{k_0}{a\left(\dfrac{\mu_0}{\epsilon_0}\right)^{\frac{1}{2}}}\, \frac{e^{ik_2z}}{\ln\left(\dfrac{z}{a}\cdot\dfrac{1}{k_2a}\right)} \tag{5}$$

Le champ magnétique décroit donc en fonction de z dans la direction axiale beaucoup plus lentement en la présence du cylindre infiniment conducteur qu'en son absence; ce dernier se comporte donc aussi comme un guide d'ondes de surface. Un calcul plus complet montre que la plus grande partie de l'énergie rayonnée par la source est concentrée dans un voisinage très immédiat du cylindre.

Il faut noter ici que bien que l'onde de résidu est nulle, l'onde totale se comporte comme une onde de surface. On remarquera encore que la présence de pertes dans les cylindres diélectrique ou infiniment conducteur ne modifie pas fondamentalement les phénomènes.

III. Cas de la Terre Plane, Homogene et Dielectrique

Soit un dipôle de Hertz situé dans l'air (demi-espace 2, $z > 0$) à la surface d'une terre plane et diélectrique occupant le demi-espace $z < 0$ (milieu 1). Le champ créé dans l'air par ce dipôle dérive d'un potentiel de Hertz π purement axial. La résolution des équations de Maxwell par la méthode des transformées de Bessel-Hankel fournit pour $z > 0$ l'expression suivante:

$$\pi = A \int_\Gamma H_0^{(1)}(\lambda r)\lambda d\lambda e^{-\mu_2 z} \frac{\epsilon_r'}{\epsilon_r'\mu_2 + \mu_1}, \; z > 0. \tag{6}$$

Dans cette expression, on utilise les nouveaux symboles suivants:

A = constante proportionnelle au moment du dipôle.

$\epsilon_r' = \frac{\epsilon_1}{\epsilon_0}\left(1 + i\frac{\sigma_1}{\omega\epsilon_1}\right) = k_1^2/k_2^2.$

$\mu_2 = (\lambda^2 - k_2^2)^{\frac{1}{2}}, \; \mu_1 = (\lambda^2 - k_1^2)^{\frac{1}{2}}.$

Γ = contour d'intégration représenté sur la Fig. 4, faite dans l'hypothèse où la terre a des pertes ($\sigma_1 \neq 0$).

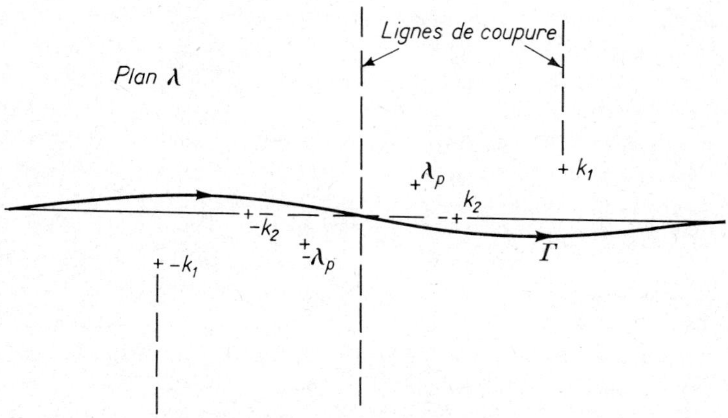

Fig. 4.

Les points de branchement de l'intégrand dans (6) sont $\pm k_2$ et $\pm k_1$. Avec les lignes de coupure représentées sur la Fig. 4, les points $\pm \lambda_p$, racines de $(\epsilon_r'\mu_2 + \mu_1)$, sont des pôles de l'intégrand. Lorsqu'on trace la ligne de coupure émanant de k_2 parallèlement à l'axe imaginaire (Fig. 5), le point λ_p n'est plus un pôle sur ce nouveau feuillet de Riemann. Cet exemple montre que la décomposition de l'onde totale en onde de résidu et onde de coupure est purement arbitraire; c'est la raison pour laquelle on ne peut pas attribuer une signification physique particulière au terme de résidu.

Utilisant le théorème de Cauchy, on remplace le contour d'intégration Γ par des lacets entourant les lignes de coupure représentées sur la Fig. 5; il n'y a alors pas de terme de résidu. On trouve à la surface de la terre diélectrique avec perte, pour r grand, une loi de variation de π de la forme e^{ik_2r}/r^2. Par contre, dans le vide, le champ créé par de dipôle de Hertz varie dans son plan équatorial en

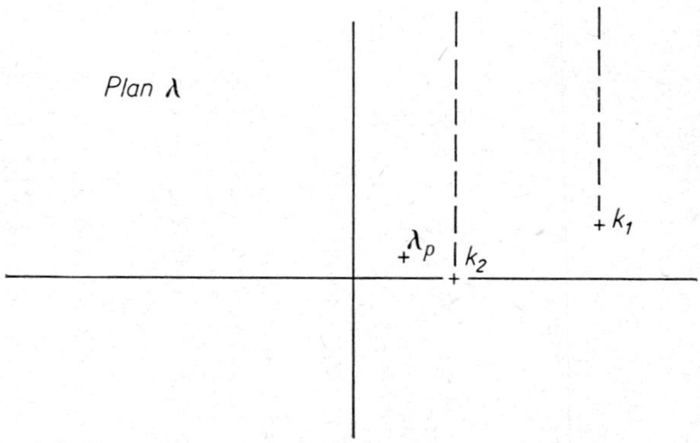

Fig. 5.

fonction de la distance radiale r selon la loi e^{ik_2r}/r. Comme même à faible distance du dipôle, la décroissance du champ à la surface de la terre est plus rapide que celle de la fonction $1/r$, il en résulte que la terre ne guide pas une onde de surface, tout au moins si l'on donne à cette expression le même sens que dans le cas des cylindres.

IV. Conclusion

Un guide d'ondes de surface a été défini comme un corps qui, éclairé par une source donnée, provoque le long d'une direction privilégiée de sa surface une loi de décroissance du champ plus lente que celle obtenue en son absence. Avec cette définition, les cylindres diélectrique et infiniment conducteur sont des guides d'ondes de surface alors que le plan diélectrique n'en est pas un. Ces conclusions restent valables dans le cas où il y a des pertes. Il faut remarquer que la définition adoptée a un sens physique puisqu'elle revient à dire qu'un guide d'ondes de surface placé entre l'antenne et un récepteur provoque une meilleure réception que celle obtenue en son absence.

Enfin, contrairement aux conclusions obtenues par la méthode qui consiste à étudier les modes de propagation possibles—et qui revient en fait à négliger la contribution des ondes de coupure—il est montré qu'un cylindre métallique se comporte comme un guide d'ondes de surface, même dans l'hypothèse où il est infiniment conducteur.

References

1. Jauquet, C. (1957). *Ann. Télécomm.* **12,** no. 6, 217–233.
2. Jauquet, C. (1956). *Bull. Acad. Sci. Belg.* pp. 1178–1183.

Statistical Characteristics of Forward Scattered Radio Echoes from Meteor Trails

T. J. Keary and H. J. Wirth

U.S. Navy Electronics Laboratory, San Diego, Calif.

I. Introduction

For many years there has been considerable interest in increasing the knowledge of meteor astronomy and meteor physics by measuring the characteristics of echoes received when radio waves are scattered from meteor trails.

Kaiser [11] has summarized theoretical treatments of backscattering of radio waves from the ionized trails of meteors. Eshleman and Manning [4] have extended this theory to the forward scattering of radio waves so as to pertain to a situation in which the receiver may be located several hundred kilometers from the transmitter.

This paper examines the statistical characteristics of decay type echoes observed in the forward scattering of radio waves from the ionized trails of meteors and in particular examines the diurnal variation in these statistics.

II. Equipment

Since the summer of 1955 the authors have intermittently observed and recorded at San Diego, California, bursts of signal received from a 43·5 Mc/s transmitter at Stanford, California. The transmitter and receiver were 690 km apart; the transmission path was oriented NW to SE. During the tests the transmitter at Stanford was operated unmodulated with a continuous radiated power of about 700 W broadly directed toward San Diego by a horizontally polarized three-element Yagi antenna having an azimuth beam width at half power of 65°. For reception of signals at San Diego a similar three-element Yagi antenna was used with its beam directed toward Stanford. The receiving equipment consisted of a cascode preamplifier, Hammarlund SP600 receiver with intermediate frequency bandwidth of 3 kc/s, deflection amplifier and Brush recorder. The receiver was operated without automatic gain control. The deflection on the records was proportional to the electric field strength incident on the antennas and was linear for antenna output voltages of 0·2 μV to above 10 μV. The receiving equipment threshold level of sensitivity was at — 154 db re 1 W, limited by cosmic noise background. The overall time response of the receiving and recording equipment was such that with square waves impressed in the input to the receiver the recorder pen attained 90 % of its ultimate amplitude in less than 20 msec.

The Yagi antennas at both ends of path were suitably elevated above the foreground terrain so that ideally their main lobes in the elevation plane intersect in a common volume located approximately 85–110 km above the surface of the earth, a height at which the ionized trails of meteors are known to be produced. Figure 1 is a vertical section through the transmitter-receiver axis showing

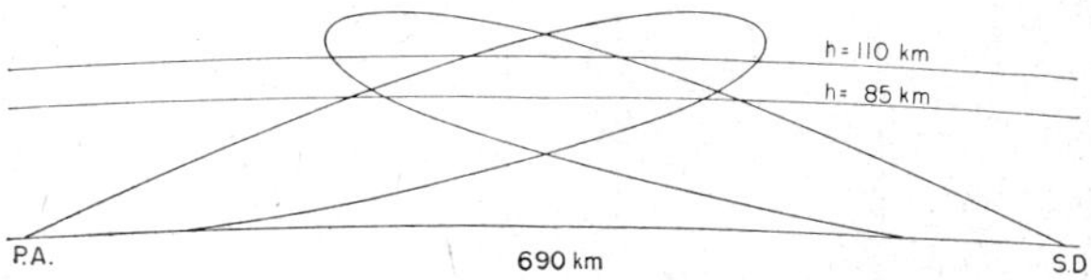

FIG. 1. Vertical section through transmitter-receiver axis.

ideally the intersection of the transmitter and receiver beams. Actually the foreground terrain was somewhat irregular so that this ideal condition was not completely attained.

III. Interpretation of Received Signal Bursts

Although normally no signals are received in San Diego, occasionally a burst of signal does erupt above the noise level. Such a signal burst is interpreted to mean that a meteor has entered the upper atmosphere and produced an ionized trail with suitable orientation and adequate linear electron density to scatter the transmitted radiation to the receiver. Figure 2 is a portion of a Brush chart, taken at a speed of 5 mm/s, showing typical meteor echoes.

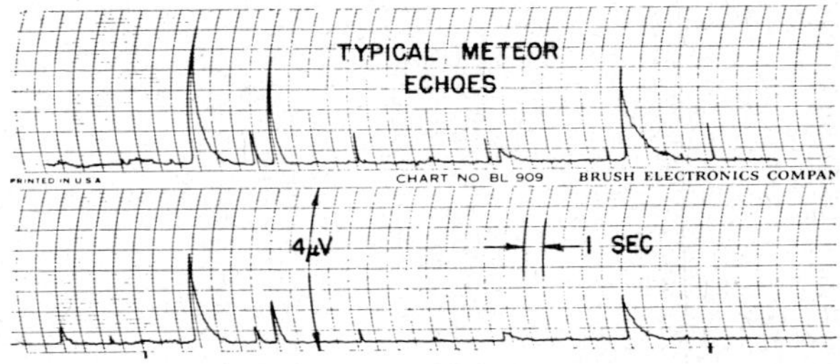

FIG. 2. Chart showing typical bursts of signal.

Each observed signal burst may be classified according to whether or not it shows a precipitous rise from the noise level. Those signal bursts which show a sharp rise (specular bursts) may be further classified according to whether the signal amplitude immediately decays from its initial peak value (decay-type burst associated with an under dense trail) or remains high for some time after reaching its initial value (persistent-type burst associated with an overdense trail).

In addition to the above, the signal amplitude may decay smoothly (no wind distortion present) or may fluctuate (wind distortion present). Figure 3 illustrates these several types of signal bursts.

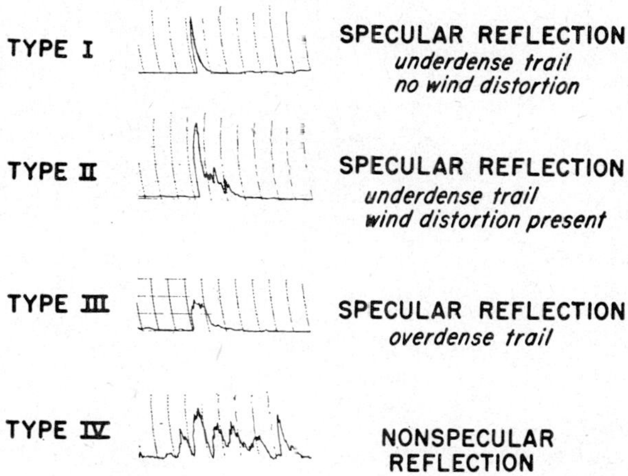

FIG. 3. Classification of meteor echoes.

IV. ANALYSIS OF RECORDS

It was desired to obtain from the records of the received signal bursts data from which the following statistics might be obtained:

(1) The statistical frequency of the intervals between successive specular signal bursts.
(2) The distribution of the peak amplitudes for several periods during the day.
(3) The statistical frequency of the intrinsic durations of the specular decay type signal bursts (i.e., durations at 1/e times the peak amplitude) for several periods during the day.
(4) The statistical frequency of durations of decay type bursts measured at fixed received signal thresholds.

The Brush records were projected on a Telecomputor Corp. Telereader, and the associated Telecordex was used to measure the following:

(1) The time interval from the start of the preceding specular signal burst to the specular signal burst being measured.
(2) The amplitude of the signal burst.
(3) The duration of the signal burst at 1/e times its peak amplitude.
(4) The duration of the signal burst at several fixed received signal thresholds such as $\frac{1}{2}$, 1, and 2 uV.

These measured quantities were entered on IBM (International Business Machine) cards using an IBM 514 reproducing punch, one card being used for the data of each signal burst. Lists in order of time intervals preceding the signal bursts, amplitudes, and durations were then prepared from the cards using an IBM sorter and tabulator. From such lists it was a straightforward procedure to plot the graphs discussed in the subsequent sections.

V. Time Intervals between Starts of Successive Signal Bursts

The frequency distribution of time intervals preceding the signal bursts have been studied by Briggs [2], Wylie and Castillo [14], Bowden and Davies [1], and Keary and Wirth [12]. The results reported in the latter paper are summarized here.

If one assumes the occurrence of meteors in the upper atmosphere to be completely random then the expected number of signal bursts in the time interval t to $t + \Delta t$ after the start of a signal burst is

$$n(t)\Delta t = \frac{N}{T} e^{-t/T} \Delta t$$

where T is the average interval between starts of successive signal bursts and N is the total number of signal bursts in the sample.

The intervals preceding 778 signal bursts observed during the perior 04.00–06.00 hr on 3 August 1956 were used to test the validity of the preceding equation. As explained in detail in the above reference the total number of signal bursts observed in the 2-hr interval was corrected from 778 to 825 to account for the masking effect of large bursts on small bursts occurring immediately thereafter. Figure 4 compares the observed with the expected frequency of time

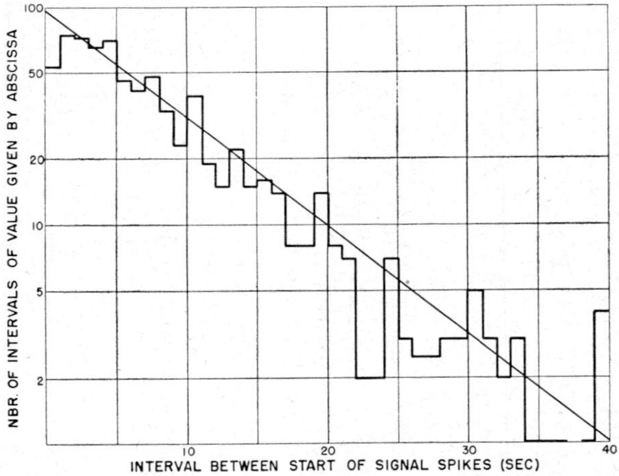

Fig. 4. Comparison of observed with expected statistical frequency of time intervals between starts of specular signal bursts for period 04.00—06.00 hr on 3 August 1956.

intervals between the starts of signal bursts for the data of the above period. The predicted frequency of time intervals is given by the straight-line plot. The experimentally observed numbers of intervals of value on the abscissa are plotted as a histogram. One observes general agreement between the experimentally observed counts and the number predicted on the basis of random occurrence of the bursts. The number of observed intervals under 1 sec is less than predicted since, when the interval between signal bursts is so small as to be comparable with the duration of the bursts, some signal bursts are missed. Thus the count of short intervals is less than the true count.

One may also test for the grouping of bursts in triplets by examining the number of pairs of successive time intervals substantially equal and comparing this number with that expected if the bursts occurred at random. If a second interval be counted equal to the first if it differs in length from the first by not more than $\pm\ \delta t$, then the expected number of pairs of substantially equal intervals is $N\delta t/T$ where T is the average time interval between successive signal bursts and N is the total number of time intervals in the period.

Table I compares the expected and observed number of pairs of equal intervals. One observes substantial agreement between the observed number of pairs of equal intervals and the number expected if the bursts occurred at random.

TABLE I

COMPARISON OF EXPECTED AND OBSERVED NUMBER OF PAIRS OF EQUAL INTERVALS

Hours	Date 1956	N	T sec	δt sec	Number of pairs Expected	Observed
04.00–06.00	3 August	778	9·25	0·05	4·2	4
				0·15	12·6	13
				0·25	21·0	27
05.00–07.00	6 November	446	15·31	0·05	1·5	2
				0·15	4·4	4
				0·25	7·3	8

VI. Peak Amplitude of Signal Bursts

The mechanism of radio reflections from the ionized trails of meteors in the range of radio wavelengths, including the one used in this investigation ($\lambda = 6{\cdot}9$ m), has been discussed by Eshleman [3]. One distinguishes two situations according to whether the linear electron density, q, of the ionized trail of the meteor is appreciably less than 10^{14} electrons/m (underdense trail) or appreciably more than 10^{14} electrons/m (overdense trail). Between the two extreme situations is an intermediate condition in which the ionized trail cannot be considered as either underdense or overdense.

For the underdense trail situation ($f < 10^{14}$ electron/m) the formula derived by Lovell and Clegg [13] applies. The radiation scattered from the meteor trail and impinging on the receiving antenna has a field strength proportional to the linear electron density in the trail. If the recorder amplitude, A, is proportional to the field strength at the antenna then $A \sim 1/q$.

For the overdense trail situation ($q > 10^{14}$ e/m) the formula of Greenhow [5] applies The radiation scattered from the meteor trail and impinging on the receiving antenna has a field strength proportional to the fourth root of the linear electron density in the trail. Again if the recorder amplitude, A, is proportional to the field strength at the antenna then $A \sim 1/q^{\frac{1}{4}}$.

If one now assumes that the number of meteor trails formed with a linear electron density exceeding a value q is proportional to $1/q$, i.e., $N_{q'>q} \sim 1/q$, it

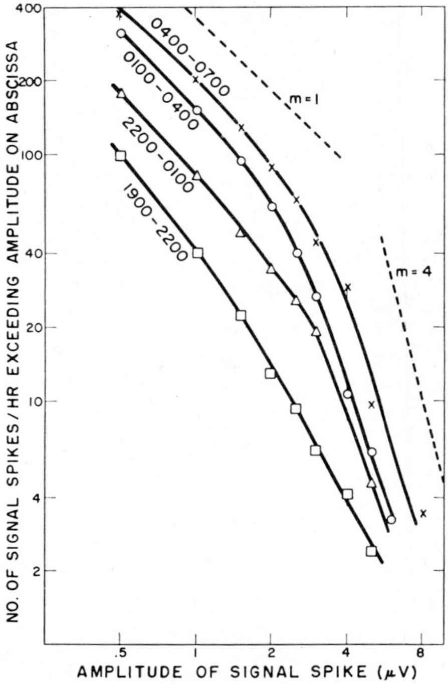

FIG. 5a. Distribution of peak amplitudes of specular signal bursts observed during night of 2–3 August 1956.

follows that for underdense trails the numbers of signal bursts having an amplitude greater than A is proportional to $1/A$, i.e., $N_{A'>A} \sim 1/A$, and for overdense trails the number of signal bursts having an amplitude greater than A is proportional to $1/A^4$, i.e., $N_{A'>A} \sim 1/A^4$.

The above arguments were developed for back-scattered radiation from the ionized trails of meteors. The arguments have been extended to forward-scattered radiation from the ionized trails of meteors by Hines and Forsyth [10] and Hines [8].

In order to be assured that the signal-burst peak amplitudes as recorded represented faithfully the peak field strengths of the burst radiation incident on the antenna, certain precautions were taken with the receiving equipment. First the recorder deflection was made accurately proportional to the input voltage to the receiver for input voltages from 0·2 μV to more than 10 μV. Next the response of the receiver and recorder to impulse signals were examined to make sure that the recorder pen would attain essentially peak amplitude within the time that the meteor takes to form the first Fresnel zone of its ionized trail.

For the most adverse combination of (1) a meteor trail orientation and location giving the shortest first Fresnel zone and (2) a meteor of highest geocentric

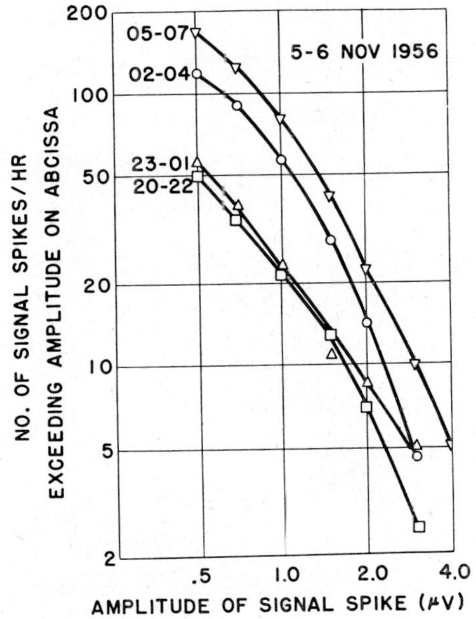

FIG. 5b. Distribution of peak amplitudes of specular decay-type signal bursts observed during night of 5–6 November 1956.

velocity consistent with a non hyperbolic orbit about the sun (72 km/s), the meteor would taken 20 msec to describe this first Fresnel zone. In this time the recorder pen would attain 90 % of the true peak amplitude. For other orientations and locations of the meteor trail or lower meteor velocities the time for the meteor to describe the first Fresnel zone would be longer and the recorder pen would still more closely attain the true peak amplitude.

Figure 5a shows the distribution of peak amplitudes of specular signal bursts observed on 2–3 August 1956. The distributions are shown for four periods of 3 hr each and extend from 19.00 on 2 August to 07.00 on 3 August. Night-time records were used for this analysis since they were essentially free from any man-made noise and the limitation in observing signal bursts was the cosmic noise background.

One notes incidentally that at any fixed signal level the rates of occurrence of meteors increase during the night. This increase corresponds to the fact that the rotation of the earth during the night brings the transmitter-receiver path around more nearly to the apex of the earth's way where the apparent rate of incidence of meteors is high.

On this log-log plot of the time rate of occurrence of signal bursts exceeding a given burst amplitude versus amplitude, one observes that for the evening periods the plot shows little change in slope between the large-burst amplitude and the small-burst amplitude ends of the plot. As one progresses toward morning the slope at the large-burst amplitude end of the plot steepens and the slope at the small-burst amplitude end of the plot flattens so that there is a pronounced difference in slopes and the curve assumes a bilinear appearance. The dotted lines on the figure indicate the negative slopes $m = 1$ and $m = 4$ for comparison with the observed plots.

Figure 5b shows the distribution of peak amplitudes of specular decay type signal spikes observed during the night of 5–6 November. Data for this plot included only those signal spikes which rose precipitously from the noise level and started to decay immediately after reaching the initial peak amplitude—i.e., decay type signal spikes. The distributions are shown for four periods of 2 hr each and extend from 20.00 on 5 November to 07.00 on 6 November. Although these curves are less steep at lower signal amplitudes, than at higher signal amplitudes they do not have the bilinear appearance of the curves of Fig. 5a.

VII. Intrinsic Durations of Decay Type Signal Bursts

The decay of signal bursts associated with the back-scattered radiation from underdense trails of meteors was examined by Herlofson [7]. Due to diffusion in the ionized trail the echo amplitude decays exponentially, falling to $1/e$ times the peak amplitude in a time $\tau = \lambda^2/16\pi^2 D$ where λ is wavelength of radiation (meters) and D is diffusion coefficient (m^2/s). Eshleman and Manning [4] have

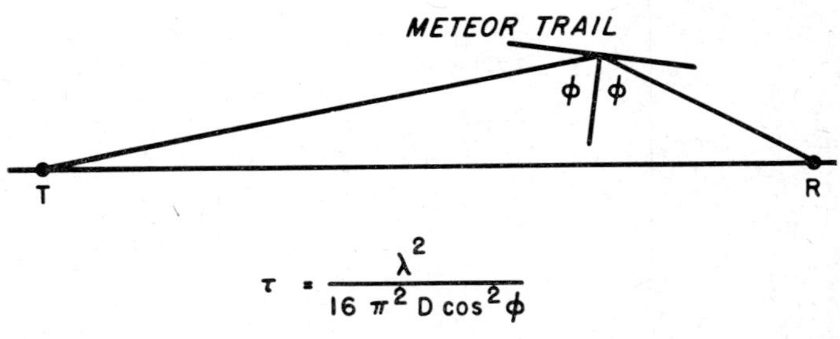

Fig. 6. Intrinsic duration of decay-type meteor echoes observed in forward scatter.

extended the theory to forward-scattered echoes from meteor trails and find that the echo decays to $1/e$ times its peak amplitude in a time $\tau = \lambda^2 \sec^2 \phi/16\pi^2 D$ where 2ϕ is the forward-scattering angle—i.e., the angle subtended at the specular point of reflection on the meteor trail by the transmitter-receiver path as illustrated in Fig. 6. The intrinsic durations, τ, of forward-scattered echoes from meteor trails at a fixed wavelength thus depend upon two factors:

(1) The height at which the specular reflection from the meteor trail takes place, since the diffusion coefficient D is a function of height.
(2) The lateral location in the plane at height h at which the point of specular reflection occurs, since the forward-scattering angle 2ϕ is a function of lateral location.

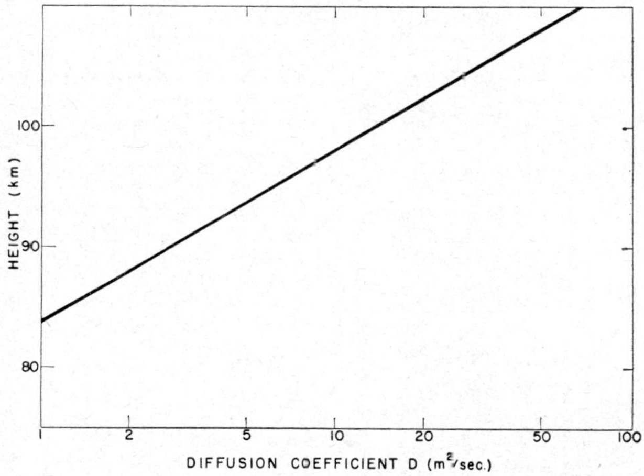

FIG. 7. Variation of diffusion coefficient with height (Greenhow & Neufeld).

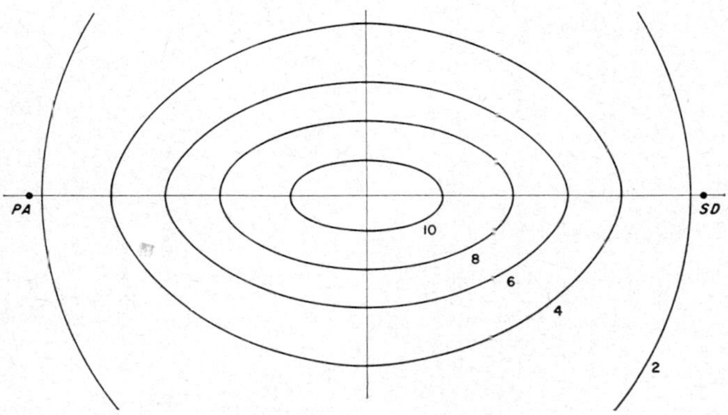

FIG. 8. Contours of calculated values of $\sec^2\phi$ in plane at height of 100 km.

Greenhow and Neufeld [6] by simultaneously measuring the height and duration of individual back-scattered decay-type echoes have determined the diffusion coefficient as a function of height. Figure 7 shows a plot of their values of diffusion coefficient versus height.

Figure 8 shows the calculated contours of $\sec^2\phi$ in a plane at a height of 100 km or the 690-km path between Stanford and San Diego. One observes a large

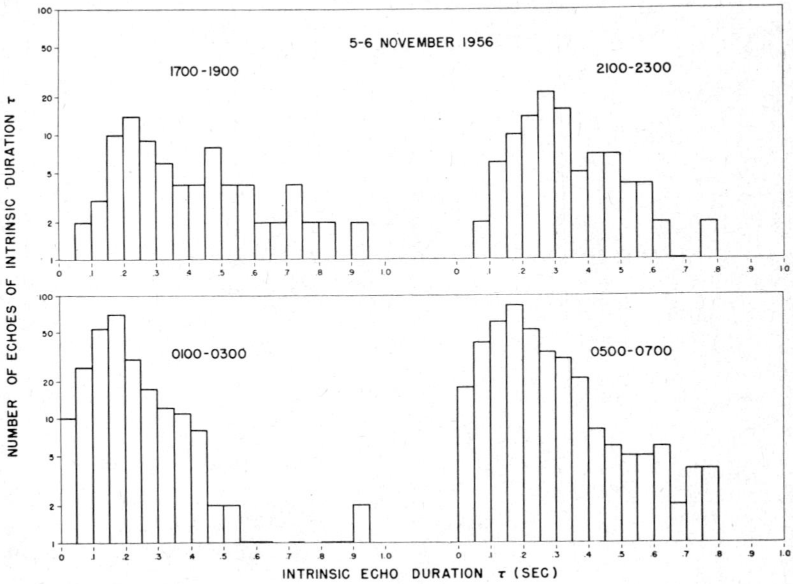

FIG. 9. Observed intrinsic duration of signal spikes for periods on 5–6 November 1956.

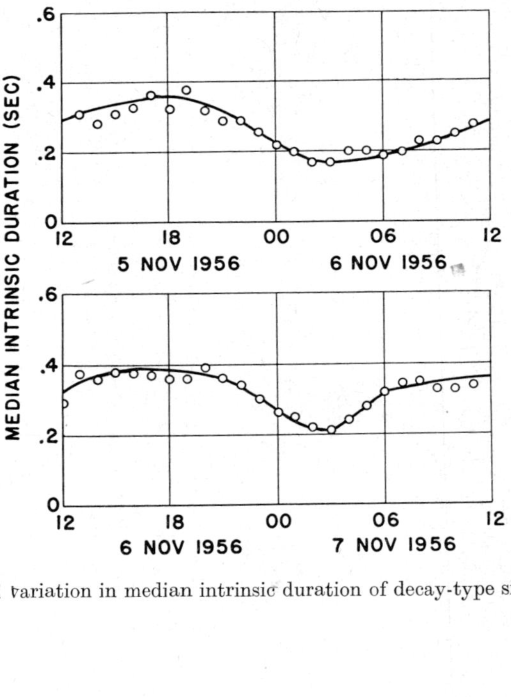

FIG. 10. Diurnal variation in median intrinsic duration of decay-type signal spikes.

variation in $\sec^2\phi$ depending on lateral position in the h plane. Eshleman and Manning [4] and Hines [9] have independently shown that, in making measurements of the forward-scattered radiation from meteor trails, those trails formed either side of the great circle path are more observable than those on the great circle path. Thus the values of $\sec^2\phi$ most appropriate to the signal bursts observed in this experiment lie in the range 4 to 6.

Figure 9 presents histograms showing the statistical frequencies of durations experimentally observed for four periods of 2 hr each during the night of 5–6 November 1956. One observes that in the evening for the period 21.00–23.00 hr the histogram shows a most probable intrinsic duration of 0·28 sec. As the night progresses the histogram becomes more peaked with the most probable intrinsic duration moving to a lower value. By 01.00–03.00 hr this most probable duration is 0·16 sec.

If one assumes that on the average the same values of $\sec^2\phi$ are effective throughout the night then one attributes the decrease in most probable duration by a factor of 1·8 to an increase in the diffusion coefficient by a factor of 1·8. By reference to the Greenhow-Neufeld values of diffusion coefficient versus height (Fig. 7) one notes that this increase in diffusion coefficient would imply that the most probable height for production of the ionized trails increased by 4 km during the period from 22.00 to 02.00 hr.

Now the rotation of the earth during the night brings the transmitter-receiver path more nearly to the apex of the earth's way where the geocentric velocities of the meteors are higher. These higher velocity meteors produce their ionization higher in the atmosphere where diffusion is more rapid. Consequently as the night progresses the intrinsic duration of the signal burst becomes less.

This decrease in the intrinsic duration of meteor trails during the night may also be examined on a plot of the median intrinsic duration versus the time of day. In Fig. 10 is plotted for each hour of the day the median intrinsic duration of a 2-hr period centred about that hour. One observes that the median intrinsic duration decreases by a factor of 2 in the period from 18.00 hr to 03.00 hr.

VIII. Duration of Signal Bursts at a Fixed Receiver Threshold

Figure 11 shows the distribution of the durations of signal bursts on a log-linear plot. One observes that the plot can be approximated by a straight line, implying that the distribution of durations of signals at a fixed level is essentially exponential.

IX. Conclusions

The statistical frequency of time intervals between the start of successive signal bursts is consistent with that expected on the hypothesis that the signal bursts are produced by scattered radiation from the ionized trails of meteors which enter the upper atmosphere at random. The observed number of triplets of equally spaced bursts substantially agrees with the number expected on the hypothesis that the bursts occur at random.

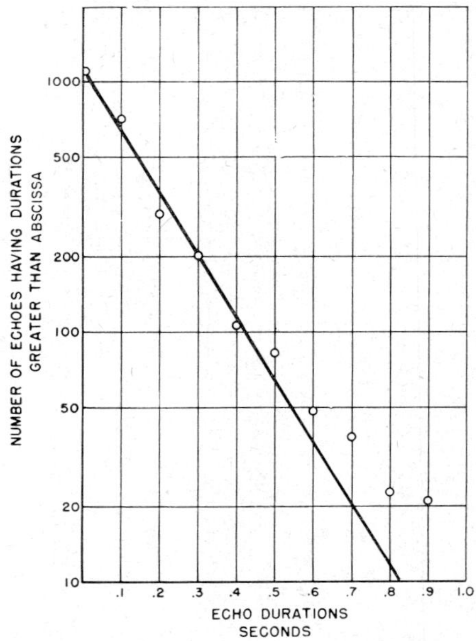

FIG. 11. Statistical distribution of durations of signal bursts at a fixed receiver threshold.

A consideration of the radiation scattered from underdense and from overdense meteor trails leads one to conclude that a log-log plot of the time rate of occurrence of specular signal bursts exceeding a given burst amplitude versus the burst amplitude should, for underdense trails, be a line of slope of—1 and for overdense trails a line of slope of — 4. The observed distributions of signal burst amplitudes are in qualitative agreement with the above consideration in that the curves have less steep slopes at small burst amplitudes than at large burst amplitudes. On some occasions the difference in slopes between the small burst amplitude and large burst amplitude ends of the plot become more accentuated as the night progresses from evening to morning hours, the slopes approaching more closely the values of — 1 and — 4 anticipated from theoretical considerations.

Observed histograms of the statistical frequencies of the intrinsic durations show as night progresses from evening to morning hours the most probable duration (and also the median duration) becomes shorter. This observation is consistent with the fact that as the night progresses the rotation of the earth brings the transmitter-receiver path around toward the apex of the earth where the geocentric meteor velocities become higher. The higher velocity meteors ionize higher in the atmosphere and the ionized meteor trails diffuse more rapidly.

The observed distribution of durations of signal bursts at a fixed receiver threshold reveals that the number of bursts exceeding a given duration increases exponentially as the given duration is made less.

References

1. Bowden, K. R. R. & Davies, J. G. (1957). *J. atmos. terr. Phys.* **11**, 62–66.
2. Briggs, B. H. (1956). *J. atmos. terr. Phys.* **8**, 171.
3. Eshleman, V. R. (1953). *Trans. Inst Radio Engrs, N.Y. AP*–**1**, 37–42.
4. Eshleman, V. R. & Manning, L. A. (1954). *Proc. Inst. Radio Engrs, N.Y.* **42**, 530–536.
5. Greenhow, J. S. (1952). *Proc. phys. Soc. Lond.* **65**, 169–181.
6. Greenhow, J. S. & Neufeld, E. L. (1955). *J. atmos. terr. Phys.* **6**, 133–140.
7. Herlofson, N. (1948). *Rep. Progr. Phys.* **11**, 444–454.
8. Hines, C. O. (1958). *Canad. J. Phys.* **36**, 539–554.
9. Hines, C. O. (1955). *Canad. J. Phys.* **33**, 493–503.
10. Hines, C. O. & Forsyth, P. A. (1957). *Canad. J. Phys.* **35**, 1033–1041.
11. Kaiser, T. R. (1953). *Advanc. Phys.* **2**, 495–544.
12. Keary, T. J. & Wirth, H. J. (1958). *J. geophys. Res.* **63**, 67–75.
13. Lovell, A. C. B. & Clegg, J. A. (1948). *Proc. phys. Soc. Lond.* **60**, 491–498.
14. Wylie, L. R. & Castillo, H. T. (1956). *Ohio J. Sci.* **6**, 339–347.

The Measurement of Electron Densities in the Outer Ionosphere

JOHN M. KELSO

Space Technology Laboratories, Los Angeles 45, Calif.

In a Doppler-shift experiment to determine the low electron densities found well above the principal ionospheric layers, errors due to time variations in the electron density along the ray path below the vehicle may be an order of magnitude greater than the quantity to be measured. Following an analysis of this source of error, it is shown that the simultaneous measurement of the rate of Faraday rotation can be used to remove the major portion of the error in question.

I. DOPPLER ERROR CAUSED BY IONOSPHERIC VARIATIONS

The presence of ionization around a space vehicle results in a Doppler shift which is different from that which would be observed in the absence of ionization. The difference in Doppler shifts observed at two frequencies can be used to form the basis for a method for measuring the electron density in the neighbourhood of the vehicle. Experiments of this nature have been undertaken for a number of years at the Naval Research Laboratory by Seddon and Jackson [6].

Because of the imminent use of long-range space vehicles, it is of interest to investigate the limitations of the Doppler method in measuring very low electron densities high in the outer ionosphere, or in cislunar or interplanetary space.

Consider a signal of frequency $f_0 = \omega_0/2\pi$ transmitted along some ray path (in general, curved) between the ground at O and the vehicle at some point Q. The received signal, E, has the form

$$E = E_0 e^{i\omega_0(t-P/c)} \tag{1}$$

where E_0 is the amplitude (assumed to be constant), c is the velocity of light in free space (in meters per second), and P is the phase path given by

$$P = \int_{\text{path}} \mu ds \tag{2}$$

where μ is the refractive index, and where the integral is taken along the ray path.

The observed frequency, $f = \omega/2\pi$, is given by the time derivative of the exponent of Equation (1), i.e.,

$$f = f_0\left[1 - \frac{1}{c}\frac{dP}{dt}\right].$$

The Doppler shift, $\Delta f = f_0 - f$, is given by

$$\Delta f = \frac{f_0}{c}\frac{dP}{dt}. \tag{3}$$

In the Doppler experiment it is usually assumed that during a small time interval, dt, the only significant change in phase path is that due to the component, ds, of the vehicle motion along the ray path. This assumption is well justified when the electron density in the neighbourhood of the vehicle is large or when the total electron density between the vehicle and the ground is relatively small. Thus, the Doppler shift is essentially a property of the velocity and the local ionization about the vehicle for vehicle positions from the low-density lower layers up to heights somewhat above the maximum of the F_2 layer. For these regions, we can write

$$\Delta f = \frac{f_0 \bar{\mu}}{c} \frac{ds}{dt} = \frac{f_0 \bar{\mu} v_{11}}{c} \tag{4}$$

where $\bar{\mu}$ is the average refractive index over the interval ds, and $v_{11} = ds/dt$ (in meters per second) is the component of the velocity along the ray path. When the Doppler shift is suitably measured at two frequencies, the quantity v_{11}/c can be eliminated since it is independent of frequency, while the index of refraction and the Doppler shift are frequency dependent. Therefore, the measurement establishes the ratio of refractive indices at two frequencies, and the electron density can be extracted from this ratio.

We wish now to return to Equation (3) and to consider the derivative dP/dt more extensively. Time variations in the phase path, P, arise from three sources:

(1) The component of motion of the vehicle along the ray path, as discussed in the preceding paragraph.
(2) Time variations in electron density along the ray path.
(3) The component of vehicle motion across the initial ray path. This motion establishes a new ray path through a different part of the ionized region.

In the following analysis, the source (1) is the local effect discussed earlier, and we consider the sources (2) and (3) together as a partial derivative with respect to time, $\partial P/\partial t$. Then we can rewrite Equation (3) as

$$\Delta f = \frac{f_0 \bar{\mu} v_{11}}{c} + \frac{f_0}{c} \frac{\partial}{\partial t} \int_{\text{path}} \mu \, ds \tag{5}$$

For a frequency of 108 Mc/s and even for electron densities of 10^{12} electrons per cubic meter (a density corresponding to the F_2 layer maximum), the refractive index varies only between 1 and 0·99. Thus, μ can be approximated very well by the expression

$$\mu = 1 - \frac{40N}{f^2} \tag{6}$$

where N is measured in electrons per cubic meter, and where f is measured in cycles per second. Then Equation (5) can be written

$$\Delta f = \frac{f_0 v_{11}}{c} \left[1 - \frac{40\overline{N}}{f_0^2} \right] - \frac{40}{c f_0} \frac{\partial}{\partial t} \int_{\text{path}} N \, ds. \tag{7}$$

The first term is the quantity considered in the Doppler method, where we write $\overline{N}$ to denote the average over a small interval, ds. The second term is usually neglected, and therefore represents an error term. We can re-order the terms in Equation (7) to read

$$\Delta f = \frac{f_0 v_{11}}{c} - \frac{40 v_{11}}{f_0 c}\left[\overline{N} + \frac{1}{v_{11}}\frac{\partial}{\partial t}\int_{\text{path}} N\, ds\right].$$

If the presence of derivative term is neglected, the result of the measurement will be interpreted as an electron density N^*, where N^* is given by

$$N^* = \overline{N} + \frac{1}{v_{11}}\frac{\partial}{\partial t}\int_{\text{path}} N\, ds \tag{8}$$

we therefore write the error, $\Delta N = N^* - \overline{N}$, as

$$\Delta N = \frac{1}{v_{11}}\frac{\partial}{\partial t}\int_{\text{path}} N\, ds. \tag{9}$$

The error, ΔN, is a rather complicated function of the motion of the vehicle and the earth, and of the time variations of the ionosphere. It is sufficient for present purposes to assume a horizontally stratified ionosphere. Then, for frequencies of the order of 100 Mc/s or more, we may write

$$\int_{\text{path}} N\, ds = \sec\chi \int N\, dh \tag{10}$$

where χ is the angle between the ray path and the vertical, and dh is an element of vertical displacement. The integral on the right-hand side is taken along a vertical line from the ground to the height of the vehicle.

Then,

$$\frac{\partial}{\partial t}\int_{\text{path}} N\, ds = \sec\chi\frac{\partial}{\partial t}\int N\, dh + \left(\sec\chi \tan\chi\frac{d\chi}{dt}\right)\int N dh. \tag{11}$$

As indicated above, $\partial P/\partial t$ is composed of two terms, the first due to the time variations of the ionosphere along a fixed path, and the second term due to movement of the path through a fixed ionosphere. As a first example, we consider a vehicle moving vertically above the observation point. Then $\sec\chi = 1$ and $\tan\chi = 0$, so that,

$$\frac{\partial}{\partial t}\int N\, ds = \frac{\partial}{\partial t}\int N\, dh.$$

The total electron content in a vertical column has been computed by Evans [3], from measurements of the Faraday rotation of radar signals reflected from the moon. For morning or afternoon hours, when the ionosphere is varying rapidly, Evans obtains average time variations in a one-square meter vertical column having values from $2{\cdot}5 \times 10^{16}$ electrons/sq. m column per hour to

5×10^{16} electrons/sq. m column per hour. Using the larger of these values, and assuming the vehicle to be directly above the observation point, we may write

$$\frac{\partial}{\partial t}\int N\,dh = 1{\cdot}4 \times 10^{13} \text{ electrons m}^{-2} \text{ sec}^{-1}.$$

Assuming the vehicle to have escape velocity, we have $v_{11} = 1{\cdot}1 \times 10^4$ m sec^{-1}, so that

$$\Delta N = \frac{1{\cdot}4 \times 10^{13}}{1{\cdot}1 \times 10^4} = 1{\cdot}3 \times 10^9 \text{ electrons, } m^{-3}$$

$$= 1300 \text{ electrons cm}^{-3}.$$

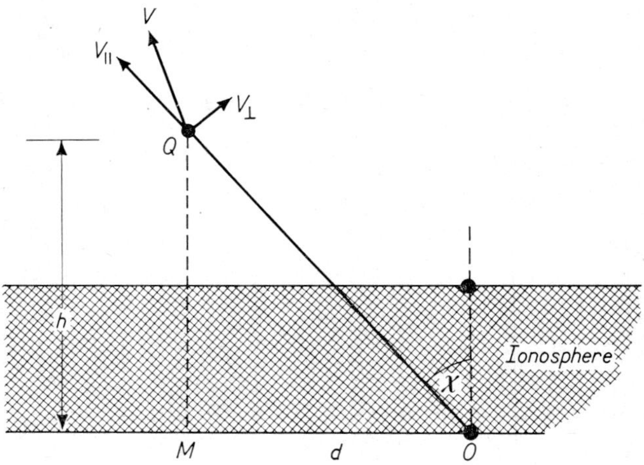

FIG. 1.

Storey [8], in his work on whistlers, estimates the electron density at two earth's radii to be 400 electrons cm^{-3}, while Van de Hulst [9], from considerations of the zodiacal light, obtains a value of about 600 electrons cm^{-3} at points just inside the earth's orbit. Thus, the error term calculated above is greater than the estimated value of electron density at heights of one earth's radius or more.

A more general case is shown in Fig. 1 where we consider the vehicle to be moving in an arbitrary direction in a vertical plane through the vehicle at Q and the observation point at O. The velocity, v, is resolved into components v_{11} and $v_{\perp}$ along and perpendicular to the direction OQ. From Equations (9) and (11), we obtain

$$\Delta N = \sec\chi\left[\frac{1}{v_{11}}\frac{\partial}{\partial t}\int N\,dh + \frac{v_{\perp}}{v_{11}}\frac{\tan\chi}{r}\int N\,dh\right] \tag{12}$$

where $r = \overline{OQ}$.

The presence of v_{11} in the denominator of both terms, and $v_{\perp}$ in the numerator of the second term shows that it is desirable to make the observations from a position such that $v_{\perp}$ is minimized and v_{11} is maximized.

In Fig. 2, we apply Equation (12) to the case of a vertically moving vehicle and plot ΔN versus h for parametric values of $d = \overline{OM}$. We choose $v = 1{\cdot}1 \times 10^4$ m sec^{-1}, $\frac{\partial}{\partial t}\int N\,dh = 1{\cdot}4 \times 10^{13}$ electrons m^{-2} sec^{-1}, and use $\int N\,dh = 1{\cdot}5 \times 10^{17}$ electrons m^{-2} as typical values found by Evans [3]. We assume that the sense of both terms is the same, so that the terms add instead of tending to cancel. We do not plot the curves for $h < 500$ km, since otherwise it would be necessary to correct $\int N\,dh$ and its derivative for the part of the ionosphere not included in the integrals, which are assumed to cover the range from $0 \leqslant h \leqslant 500$ km.

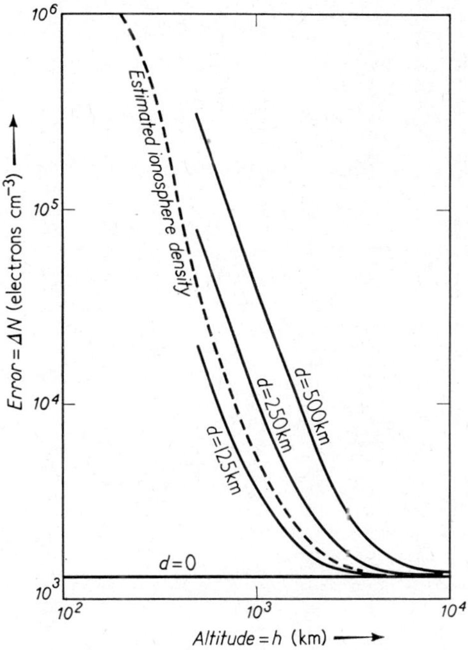

FIG. 2.

Figure 2 shows that large errors ΔN are obtained, especially for large values of d, i.e., for large values of χ at a given h. The dotted curve on Fig. 2 gives a rough estimate of the electron density of the ionosphere over the region.

The results shown in Fig. 2 show that the Doppler method of determining electron densities can lead to errors at least an order of magnitude greater than the quantity to be measured when the major part of the ionosphere lies below the vehicle.

The situation illustrated in Fig. 2 represents an extreme case, especially for $h = 500$ km. It serves, however, as rather dramatic evidence of the advantage of choosing the observation point in such a way as to minimize the velocity normal to the ray path. These curves also show that the absolute error becomes relatively small for large heights above the earth. Although the case in Fig. 2 has neglected the vehicle's angular velocity about the earth, it is not as inaccurate for long-range vehicles as might at first be supposed. First, we note that the term of $\partial/\partial t \int N\, ds$, which causes the principal part of the error for h very large, is the result of the earth's rotation, i.e., the apparent motion of the sun. For a long-range space vehicle, the angular velocity about the earth's center decreases with increasing radial distance from the earth's center. With increasing radial distance, the apparent vehicle motion approaches more and more closely to that of a fixed celestial body. Hence, in the limit, the error contribution due to the earth's rotation and the vehicle motion across the ray path will approach the magnitude of the term given by $\partial/\partial t \int N\, ds$. The actual value of the error will depend on the zenith angles of the vehicle and of the sun.

The preceding special cases are sufficient to establish the need for the correction of Doppler measurements in the outer ionosphere. Additional information on Doppler errors may be obtained from the analyses given by Jackson [5] and by Al'pert [1].

II. Doppler Faraday Experiment

The preceding paragraphs have shown that a pure Doppler experiment to measure electron densities is subject to an error ΔN proportional to the time rate of change of total electron content along the path, and inversely proportional to the component v_{11} of vehicle velocity along the ray path. The angle of rotation of the plane of polarization, due to the Faraday effect, has been shown by Hatanaka [4] to be equal to

$$\phi = 2{\cdot}97 \times 10^{-2} f^{-2} \int NH \cos \theta_p \, ds \tag{13}$$

where ϕ is the angle of rotation of the plane of polarization (radians), θ_p is the angle between the direction of propagation and the earth's magnetic field (degrees), and H is the earth's magnetic field (ampere turns per meter).

Then, taking the total derivative with respect to time,

$$\frac{d\phi}{dt} = 2{\cdot}97 \times 10^{-2} f^{-2} \left[(H \cos \theta_p N v_{11})_{\text{vehicle}} + \frac{\partial}{\partial t} \int NH \cos \theta_p \, ds \right] \tag{14}$$

where the term in parentheses is evaluated at the vehicle. If the principal contributions to the integral occur over a restricted height range, we can move $H \cos \theta_p$ outside the integral sign. When the vehicle is well above the ionosphere layers, the magnetic field at the vehicle will be much less than that in the ionosphere. If, also, $\Delta N \geqslant N$, then Equation (14) becomes, approximately,

$$\frac{d\phi}{dt} = 2{\cdot}97 \times 10^{-2} f^{-2} H \cos \theta_p \frac{\partial}{\partial t} \int N \, ds \tag{15}$$

where $H \cos \theta_p$ is evaluated in the ionosphere.

Thus, a measurement of the time rate of Faraday rotation, $d\phi/dt$, together with a knowledge of v_{11} from the trajectory, can be used to estimate the error ΔN under the conditions noted above.

At a frequency of 108 Mc/s, the value $\partial/\partial t \int N\, dh = 1{\cdot}4 \times 10^{13}$ electrons m^{-2} sec^{-1} corresponds to a rotation of the plane of polarization through about 260° per hour or, roughly, 4° per min. The Faraday rotation rates corresponding to the errors shown on Fig. 2 range up to 12°/s, or 2 r.p.m. for the worst case. (We assume that $H \cos \theta_p = 35$ amp turns/m).

Several methods are available for the measurement of the angle made by the plane of polarization. The simplest of these uses the fading rate as observed by a linear polarized antenna. This is inadequate for the present purpose, since we require to know the sense of the rotation to distinguish between increases or decreases in ionization along the path. The use of a second antenna oriented at 45° to the first in the wave front is sufficient to determine the sense of rotation in the fading method.

A second method is an adaptation of a polarimeter such as that described by Benner and Nearhoof [2] for use at a frequency of 150 kc. In this device, the outputs of two orthogonal plane polarized antennas are received on two matched receivers, and are displayed orthogonally on an oscilloscope.

A third method would be to measure the phase difference between the outputs of two circularly-polarized antennas of opposite senses of rotation. This quantity is directly proportional to the angle of rotation of the plane of polarization.

Although the accuracy of the measurement of the polarization angle will depend heavily upon the exact instrumentation used, it appears reasonable to suppose that the instantaneous value of this angle can be determined to within an error of about 5°, when a good signal-to-noise ratio is available. As an example, Seddon [7] used the Faraday rotation directly to measure electron densities in the lower ionosphere, and obtained agreement to within 1 to 2 % with simultaneous Doppler measurements.

Thus, it is reasonable to suppose that measurements of the Faraday rotation rate can be made with errors of about 5 to 10 % under good conditions of signal-to-noise ratio. Hence, if the observation point is selected so that the error ΔN is minimized, the residual error in the Doppler measurement can be reduced to perhaps 10 to 25 % of the quantity to be measured. For example, for heights above 10^4 km in Fig. 2, the residual error in electron density would be less than 60 to 130 electrons cm^{-3}, an amount less than the expected value of $N > 400$ electrons cm^{-3}.

Estimates of the credibility of the corrected results can be made on the basis of the continuity of the observed Faraday rotations, and on a comparison of the variations in the corrected value of N with the predicted errors in the measurement of the Faraday rotation rate.

A limitation to the proposed type of experiments will be found in ionospheric disturbances. It does not appear probable that very useful measurements of the type described can be made when the ionosphere is severely disturbed (as indicated, for example, by the presence of very severe spread F echoes, and similar manifestations).

III. Conclusion

It has been shown that time variations in electron density along the ray path can lead to substantial errors in a Doppler measurement of the electron density in the regions above the major part of the ionosphere. It has also been shown that this error can be reduced by a factor of 10 to 20 by using corrections obtained from the observation of the rate of Faraday rotation of the received signals when the signal-to-noise ratio is reasonably large. By proper selection of the geometry between the vehicle trajectory and the ray path, the resulting measurement can be made to yield errors considerably smaller than the electron density to be measured.

References

1. Al'pert, Ia. L. (1958). *Advanc phys. Sci., Moscow* **64,** 3.
2. Benner, A. H. & Nearhoof, H. J. (1950). *Rev. sci. Instrum.* **21**, 830.
3. Evans, J. V. (1957). *J. atmos. terr. Phys.* **11,** 259.
4. Hatanaka, T. (1955). *Sci. Rep.* No. 5, Contract No. AF–19(604)–73, Air Force Cambridge Research Center.
5. Jackson, J. E. (1957). *Upper Atmosphere Res. Rep.* No. XXIX, NRL Report 4960.
6. Jackson, J. E. & Seddon, J. C. (1958). *J. geophys. Res.* **63**, 197.
7. Seddon, J. C. (1958). *J. geophys. Res.* **63,** 209.
8. Storey, L. R. O. (1953). *Phil. Trans.* **246,** 113.
9. Van de Hulst, H. C. (1956). *Vistas in Astronomy*, **2,** 998. London and New York: Pergamon Press.

Ionospheric Scintillation of Cosmic Radio Noise

SHELDON GRUBER

Air Force Cambridge Research Center, Bedford, Mass.

Measurements of scintillation of signals from the radio source in Cygnus were made at 50 and 200 Mc. Both phase difference and amplitude fluctuations were recorded and processed. Autocorrelation functions and power density spectra were computed. It was observed that the usual turbulent scattering theories do not explain the processed fluctuations.

I. INTRODUCTION

It has been proposed by Booker [1] that turbulence in the ionosphere is the cause of radio star scintillation at meter wavelengths. This paper summarizes the results of an experimental program of recording the phase and amplitude scintillations of the radio star in the constellation of Cygnus and compares the records with those which might have been obtained if the signals had traversed a turbulent medium.

Studies of scattering [2, 10] of electromagnetic waves from a turbulent medium assume a continuous spectrum of turbulent energy in that the spatial autocorrelation function of the dielectric constant fluctuations is generally assumed to be exponential or Gaussian in form.

The permittivity of the ionosphere is represented as having an average value plus a time varying component whose average value is zero. One then defines a spatial autocorrelation function of the temporally and spatially varying component.

$$\epsilon(\vec{r}, t) = \epsilon_0 + \Delta\epsilon(\vec{r}, t) \quad \text{(the total permittivity)}$$

$$\overline{\epsilon(\vec{r}, t)} = \epsilon_0$$

and the spatial autocorrelation functions are given as:

$$\rho(\vec{\xi}) \triangleq \frac{1}{\text{Vol}} \int_{\text{Vol}} \Delta\epsilon(\vec{r}, t)\Delta\epsilon(\vec{r} + \vec{\xi}, t)d^3r$$

$$\lim \text{Vol} \rightarrow \infty$$

or

$$\rho(\vec{\xi}) \triangleq \frac{1}{2T} \int_{-T}^{T} \Delta\epsilon(\vec{r}, t)\Delta\epsilon(\vec{r} + \vec{\xi}, t)dt.$$

$$\lim \text{T} \rightarrow \infty$$

It is implied in the above that the result of the first integration is time independent and therefore the process is stationary. Here it is assumed measurements are made during an interval short enough so that the statistics do not significantly vary in the second definition of the autocorrelation function. The medium is assumed to be homogeneous.

In both the papers of Booker and Gordon [2] and Villars and Weisskopf [10] the spectrum of the spatial autocorrelation function of the dielectric fluctuations is determined. While Booker and Gordon assume a form for the autocorrelation functions, Villars and Weisskopf compute a spectrum by an analysis based upon the works of Komolgoroff and G. K. Batchelor.

Booker-Gordon Theory assumes an exponential form for the correlation function which yields a continuously decreasing spectrum. Villars and Weisskopf use an eddy decay scheme which inherently yields a continuously decreasing spectrum. Functions of this type will not lead to spectra of scattered fields having emphasis (or peaks) at non-zero frequencies.

In what follows, it will be shown that some aspects of the statistics of the phase and amplitude scintillations of the Cygnus source do not agree with those derived from present theories of scattering from a turbulent ionosphere.

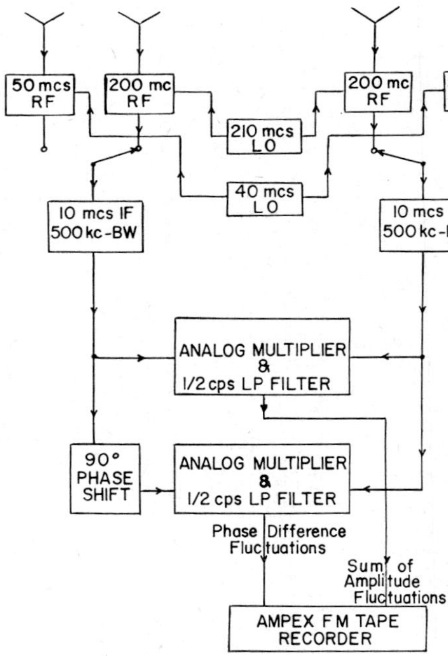

Fig. 1. Scintillation recorder.

II. Equipment Description

The measuring equipment (see Fig. 1) consisted of two receivers, one for each site. The sites were separated by either 100, 250 or 400 m on a north-south baseline. The radio frequency sections were either tuned to 50 Mc or 200 Mc and were switched between these frequencies simultaneously from a central processing point. The 50 Mc antennas were a section of a parabolic cylinder fed by a half wave dipole with a theoretical gain of about 8 db. The 200 Mc antennas were 28 ft parabolic reflectors fed by a dipole and two reflectors with a gain of approximately 25 db. A single local oscillator was employed to provide phase coherence. An intermediate frequency amplifier tuned to 10·5 Mc was used for both 50 Mc and 200 Mc. The output of the I.F. amplifiers was sent over coaxial cables to the central processing station where a multiplication of the two receiver I.F. amplifier outputs and short time (analog) integration of the product was performed. Two such "correlators" were employed. One correlator had a 90° phase shifting network in one of the I.F. amplifier channels. The outputs of the "correlators" were then recorded on a two channel Sanborn pen recorder, for a visual record and also on an Ampex F.M. magnetic tape recorder to provide a processible record for analysis.

III. Description of Signals Based on Ionospheric Model

The radio star signal can be considered as being composed of a set of statistically independent oscillators with random phase angle.

$$S'_A(t) = \tfrac{1}{2}R_e \sum_{n=1}^{M} S'_{An}(t) = \tfrac{1}{2}R_e \sum_{n=1}^{M} c_n\, e^{j(\omega_n t - \theta_n)}$$

where θ_n is a random variable uniformly distributed from 0 to 2π. This is a convenient way of representing Gaussian random noise and it has been discussed by Rice [8].

It was mentioned that scattering from ionospheric inhomogeneities can be represented as causing amplitude and phase modulation of the incoming signal, $S'_A(t)$, for small scattering,

$$S_A(t) = \tfrac{1}{2}R_e \sum_{n=1}^{M} [1 + \alpha_n(t)]S'_{An}(t)e^{-j\phi_n(t)}$$

after neglecting a phase factor that is immaterial for the overhead star case.

It should be noted that the subscript, n, has been attached to $\alpha(t)$ and $\phi(t)$ to indicate possible frequency dependence of these quantities. With an intermediate frequency amplifier bandwidth of 500 kc and a radio frequency of at least 50 Mc, the percentage bandwidth is quite small enough to preclude any important frequency dependence of the scattering for a given center frequency.

$$S_A(t) \approx \tfrac{1}{2}R_e \left\{[1 + \alpha(t)]e^{-j\phi(t)} \sum_{n=1}^{M} C_n\, e^{j(\omega_n t - \theta_n)}\right\}$$

$$\approx \tfrac{1}{2}R_e \left\{[1 + \alpha(t) - j\phi(t)] \sum_{n=1}^{M} C_n\, e^{j(\omega_n t - \theta_n)}\right\}$$

where it is now understood that the summation is such that only values of ω_n which fall within the band $2\pi A$ of center frequency ω_0 are considered and that

$A/f_0 \ll 1$. The total field received at the receiver for single scatter at a frequency, ω_n, may be written as

$$S_{An}(t) \approx E_n e^{j[\omega_n t - \theta_n - \overrightarrow{k_{n_0}} \cdot (\overrightarrow{R_0} + \overrightarrow{R_1})]} \left\{ 1 + \frac{k^2}{4\pi\epsilon_0 R_1} \int_{\text{Vol}} \Delta\epsilon\, (\overrightarrow{r_1}, t) \exp \{jg(\overrightarrow{r_1})\} d^3 r_1 \right\}$$

where $g(\overrightarrow{r_1})$ is a phase function of position and the integration is performed over the scatter volume. Now $\alpha(t)$ and $\phi(t)$ can be identified.

$$\alpha(t) = R_e \left\{ \frac{k^2}{4\pi\epsilon_0 R_1} \int_{\text{Vol}} \Delta\epsilon(\overrightarrow{r_1}, t) \exp \{jg(\overrightarrow{r_1})\} d^3 r_1 \right\}$$

$$\phi(t) = \text{Im} \left\{ \frac{k^2}{4\pi\epsilon_0 R_1} \int_{\text{Vol}} \Delta\epsilon(\overrightarrow{r_1}, t) \exp \{jg(\overrightarrow{r_1})\} d^3 r_1 \right\}$$

where $$k = \frac{2\pi}{\lambda_0}.$$

That the foregoing analysis pertains to small scattering must be emphasized. When strong fluctuations in dielectric "constant" exist, a multiple scattering approximation must be employed. The difficulties in the approach to this problem are great.

A. Fraunhofer Zone

When the scatter geometry satisfies the conditions for Fraunhofer scattering the scattered field is Gaussianly distributed with uniform phase distribution from 0 to 2π radians.

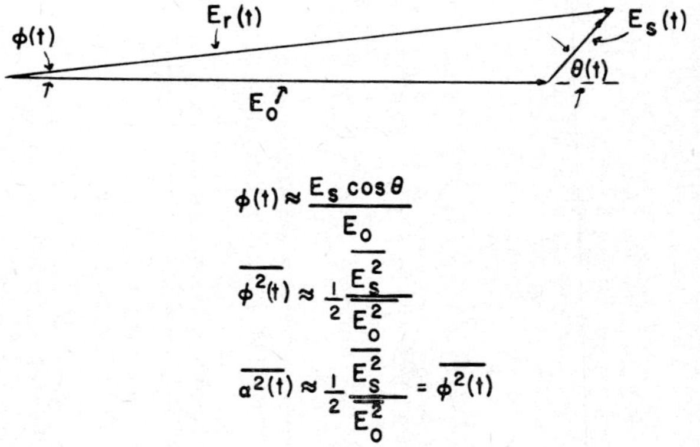

FIG. 2. Phasor diagram.

The scattered field, E_s, may be represented by two statistically independent Gaussian random vectors, one in phase with the direct field $E_0 e^{j\overrightarrow{k_0} \cdot (\overrightarrow{R_0} + \overrightarrow{R_1})}$, and the other at right angles to it. Each of these vectors has the same variance, $\frac{1}{2}\overline{E_s^2}$.

This is shown in Fig. 2. From this figure one can see that for small fluctuations the phase $\phi(t)$ is given by

$$\phi(t) \approx \frac{E_s \cos \theta,}{E_0}$$

therefore

$$\overline{\phi_s(t)} \approx \tfrac{1}{2}\frac{\overline{E_s^2}}{\overline{E_0^2}}.$$

Similarly the amplitude fluctuations have a mean square value given by

$$\overline{\alpha^2(t)} \approx \tfrac{1}{2}\frac{\overline{E_s^2}}{\overline{E_0^2}} = \overline{\phi^2(t)}.$$

B. *Fresnel Zone*

Scattering from a Fresnel region contributes mainly to phase fluctuations in the total received signal. This may be seen by the following argument:

In the Fresnel zone the scattering is predominantly from a region of linear dimensions in the order of l, the effective correlation length. The fact that coherence of dielectric fluctuation exists over a very large region restricts the first order scattered field to have its phase either $\pm\,\pi/2$ from the direct field.* The incident wave on the ground now appears as if it traversed a region of infinite stacked dielectric sheets of thickness l, and dielectric constant $\epsilon_0 \pm \Delta\epsilon(t)$.

The mean square phase fluctuations in the Fresnel zone are given approximately as

$$\overline{\phi^2(t)} = \overline{E_s^2}/\overline{E_0^2}.$$

IV. Autocorrelation Functions and Power Density Spectra of the Amplitude and Phase Fluctuation Detector Outputs

In general, for small fluctuations, one may analyze the "correlator" outputs as follows:

(1) The fluctuating part of the filtered analog multiplier output for the star overhead is $\alpha_1(t) + \alpha_2(t)$. The autocorrelation function of this is given by,

$$\begin{aligned} R_{A0}(\tau) &= \overline{[\alpha_1(t) + \alpha_2(t)][\alpha_1(t+\tau) + \alpha_2(t+\tau)]} \\ &= \overline{\alpha_1(t)\alpha_1(t+\tau)} + \overline{\alpha_2(t)\alpha_2(t+\tau)} + \overline{\alpha_1(t)\alpha_2(t+\tau)} + \overline{\alpha_2(t)\alpha_1(t+\tau)} \\ &\triangleq R_{A1}(\tau) + R_{A2}(\tau) + R_{A12}(\tau) + R_{A21}(\tau) \end{aligned}$$

here the numerical subscripts refer to the two receivers.

*Equivalently by not using the scattering approach, the field is either advanced or retarded in phase depending upon whether the dielectric constant is smaller or larger than the average dielectric constant.

Since the medium is regarded as stationary and homogeneous

$$R_{A1}(\tau) = R_{A2}(\tau) \triangleq R_A(\tau)$$

$$R_{A12}(\tau) = R_{A21}(-\tau)$$

so that

$$R_{A0}(\tau) = 2R_A(\tau) + R_{A12}(\tau) + R_{A12}(-\tau)$$

where $R_A(\tau)$ is the autocorrelation function of the amplitude fluctuation at one receiver.

The power density spectrum of the cosine detector output is given by

$$G_{A0}(\omega) = \int_{-\infty}^{\infty} R_{A0}(\tau)e^{-j\omega\tau}\, d\tau = 2\{G_A(\omega) + R_e G_{A12}(\omega)\}.$$

(2) The filtered output of the multiplier with the 90° phase shift under the same conditions, is $\phi_1(t) - \phi_2(t)$ and its autocorrelation function is given by

$$R_{P0}(\tau) = 2R_P(\tau) - R_{P12}(\tau) - R_{P12}(-\tau)$$

where $R_P(\tau)$ is the autocorrelation function of the phase fluctuations at one receiver and $R_{P12}(\tau)$ is the cross correlation function.

The power density spectrum of the phase difference detector output is given by

$$G_{P0}(\omega) = 2\{G_P(\omega) - R_e G_{P12}(\omega)\}\cdot$$

It now remains to find the functional form of these auto-correlation functions and power density spectra for the different models of ionospheric permittivity fluctuations.

If one assumes that the fluctuations in amplitude and phase are produced by a stable system of electrons moving in the ionosphere at a steady rate as the region causing radio star scintillations appears to be, a simple and useful tool in the analysis of the data results.

The derivation proceeds as follows: A drift direction with respect to the preferred turbulence axis is chosen. After an elapse of time τ, the displacement of the point of observation is noted and this displacement is substitute[illegible] $(\vec{\xi})$, the correlation coefficient for the scattered fields at spaced receivers. This yields $R_P(\tau)$. $R_{P12}(\tau)$ may be obtained in a similar manner by taking into account the receiver displacement.

From the size of effective correlation length as derived by Hewish, it appears that the Earth's surface lies in the Fresnel zone of the scatter region. It can be shown that the autocorrelation function of the scattered field in the Fresnel zone is the same as the spatial auto-correlation function of the permittivity fluctuations $\beta(\vec{\xi}) \triangleq \rho(\vec{\xi})$.

For this case, the analysis will be made for anisotropic turbulence. Fig. 3 shows the situation of the receivers and an ellipse represents a contour† of

†This is true for a Gaussian model of the fluctuating dielectric correlation function. The contours are diamond shaped for an exponential model,

$$e^{-|r_x/l_x| - |r_y/l_y| - |r_z/l_z|}.$$

constant correlation coefficient on the ground for a non-drifting pattern with one receiver at the center. The major axis of the ellipse is chosen as the y axis and the minor as the x axis of the figure. The drift velocity, v, is parallel to the ground and an angle ϕ clockwise from the major axis of the ellipse. The line between the receivers is of length D, and at an angle θ clockwise from the y axis.

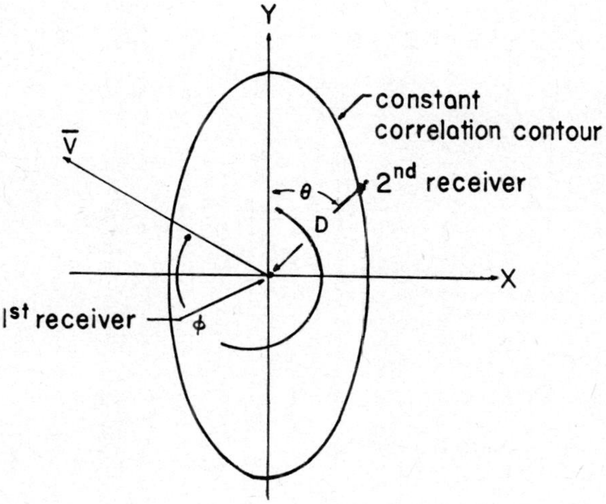

FIG. 3. Two-receiver drift geometry.

The phase difference detector autocorrelation function will be investigated. After a time, τ, the pattern drifts a distance in the y direction,

$$y_0 = v\tau \cos \phi$$

and in the x direction

$$x_0 = -v\tau \sin \phi$$

then since

$$R_p(\tau) \triangleq \overline{E_{s1}E_{s2}^*}/\overline{E_0^2} = \rho(x_0 = -v\tau \sin \phi, y_0 = v\tau \cos \phi).$$

The function $R_{p12}(\tau)$ may similarly be determined. The x distance between receiver 2 and the drifted system center after a time τ is

$$x_{12} = D \sin \theta - v\tau \sin \phi$$

and the y distance is

$$y_{12} = D \cos \theta - v\tau \cos \phi$$

hence

$$R_{p12}(\tau) = \rho(x_{12} = D \sin \theta - v\tau \sin \phi, y_{12} = D \cos \theta - v\tau \cos \phi).$$

One situation of interest is when the receiver separation D is sufficiently small so that the function, ρ, does not change much with D. Under these conditions one may expand ρ in a Taylor series and retain only the first few terms. For the autocorrelation function of the "sine" detector output then

$$R_P(\tau) \approx -\left[D^2 \sin^2\theta \frac{\partial^2\rho}{\partial x^2} + 2D^2 \sin\theta\cos\theta\,\frac{\partial^2\rho}{\partial x\partial y} + D^2\cos^2\theta\,\frac{\partial^2\rho}{dy^2}\right]_{\substack{x = v\tau\sin\phi \\ y = v\tau\cos\phi.}}$$

Similarly one may find

$$R_A(\tau) \approx \left[4\rho(x, y) + R_P(\tau)\right]_{\substack{x = v\tau\sin\phi \\ y = v\tau\cos\phi}}$$

where both of the above apply when $\dfrac{D\sin\theta}{l_x} < 1$ and $\dfrac{D\cos\theta}{l_y} < 1$.

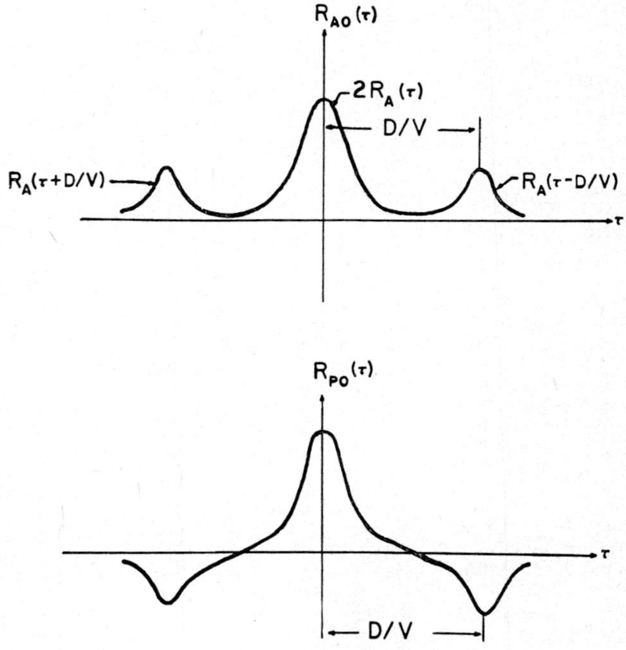

FIG. 4. Detector autocorrelation function (D $\gg l$).

If one now assumes that the correlation length, l_y, is very large as compared to l_x so that the fluctuations are due mainly to changes in dielectric constant in the x direction.

$$R_P(\tau) \approx -\,D_x^2\,\frac{\partial^2\rho}{\partial x^2}$$

$$R_A(\tau) \approx 4\rho(x)$$

and the power density spectrum of the phase fluctuations can be written in terms of the power density spectrum for amplitude as

$$G_P(\omega) \sim \omega^2 G_A(\omega).$$

A very important assumption to this discussion is that the amplitude and phase fluctuations are caused by the same irregularities in the medium.

One may use the exponential and Gaussian models to obtain specific results to compare to the experimental records.

The Guassian model for $\rho(r)$, $e^{-\left[\left(\frac{r_x}{l_x}\right)^2+\left(\frac{r_y}{l_y}\right)^2+\left(\frac{r_z}{l_z}\right)^2\right]}$ yields

$$R_{P0}(\tau) \approx 2\exp\left\{-\left(\frac{v\tau\sin\phi}{l_x}\right)^2-\left(\frac{v\tau\cos\phi}{l_y}\right)^2\right\}$$
$$-\exp\left\{-\left(\frac{D\sin\theta-v\tau\sin\phi}{l_x}\right)^2-\left(\frac{D\cos\theta-v\tau\cos\phi}{l_y}\right)^2\right\}$$
$$-\exp\left\{-\left(\frac{D\sin\theta+v\tau\sin\phi}{l_x}\right)^2-\left(\frac{D\cos\theta+v\tau\cos\phi}{l_y}\right)^2\right\}.$$

For the Fraunhofer region analysis the turbulence is considered to be isotropic. Since both the amplitude and phase fluctions have identical statistics

$$R_A(\tau) = R_P(\tau) = \rho(v\tau)$$

and

$$R_{A12}(\tau) = R_{P12}(\tau) = \rho(D - v\tau).$$

Hence

$$R_{A0}(\tau) = 2\rho(v\tau) + \rho(D - v\tau) + \rho(D + v\tau)$$

and

$$R_{P0}(\tau) = 2\rho(v\tau) - \rho(D - v\tau) - \rho(D + v\tau).$$

Figure 4 shows $R_{A0}(\tau)$ and $R_{P0}(\tau)$. A similar plot is valid in the Fresnel zone.

V. Experimental Results

The outputs of both the "phase" and "amplitude" fluctuation detectors were recorded on an Ampex FM tape recorder at a 0·3 in/s tape speed. These recorded signals were then sampled three times per sec and quantized to 64 levels. The digital information was then punched on paper tape in binary form. Autocorrelation functions were computed from the quantized data on Massachusetts Institute of Technology's Whirlwind I digital computer.* Fourier transforms of the autocorrelation function were performed on this computer using a weighting function to minimize the effects of finite function length [9].

What is most obvious upon inspection of the processed data is that the simple models used in the mathematical analysis do not generally apply. The following characteristics of the autocorrelation functions and power density can be noted:

*The computation time was generously contributed by Lincoln Laboratory. The autocorrelation program was written by W. C. Mason of that Laboratory.

1. The fluctuations frequently contain a random part and also a non-random component. The non-random character is evidenced by the oscillatory nature of the autocorrelation function tails (this component is much too large to be attributed to statistical errors generated by using a finite number of samples) and peaks in the power density spectra.

2. Differences between 50 Mc and 200 Mc records are apparent in the random part of the autocorrelation functions.

3. The "phase detector" fluctuations autocorrelation functions are very nearly the same as those of the "amplitude detector" fluctuations on most occasions. A faster drop off of the random part of the "phase" autocorrelation function is nearly always apparent.

4. The general character of the autocorrelation functions does not vary with different receiver spacings.

The first thing placed in evidence by the experimental autocorrelation functions is that the amplitude and phase fluctuation records have autocorrelation functions with very nearly the same characteristics. The initial drop off can be approximated by very nearly the same functional form and the tails likewise have very much the same appearance.

This fact also appears to refute an argument by Booker [1] (based upon a relationship between the amplitude at a given height in the ionosphere as derived by Hewish [5]) that, assuming the same scale of turbulence throughout the ionosphere, the phase fluctuations maximize at about 100 km, while the amplitude fluctuations occur mainly at a level of 150 km.

It should be noted that a difference in the random component does appear between the phase fluctuation autocorrelation function. It is conceivable that this might be due to a different level of occurrence.

Earlier it was noted that the autocorrelation functions indicate that the fluctuations have both a random and non-random component. Present day theories of radio scattering from a turbulent medium do not lead to a solution where the received wave could be periodic or contain periodicities. The theoretical spectrum of the scattered radio wave is continuous. It therefore seems likely that the regularities in the observed fluctuations are not caused by a medium that is composed of many scatter centers as supposed.

A possible mechanism is that of spatially sinusoidal traveling wave of electron density. This traveling wave would most always have to be composed of more than one frequency in order to explain the power density spectra observed. This model has the virtue of explaining the radio frequency independent nature of these fluctuations.

The existence of traveling waves of electron density has been shown by a number of workers in the field of ionospheric physics [4, 6, 7]. The observations of Bramley indicate a wavelength of 25 to 450 km and a velocity, predominantly towards the west, of from 25 to 350 m/s. These long waves in themselves cannot explain the observed periodicities in the radio star scintillations but they do indicate that shorter waves could exist with the same range of measured velocity of propagation.

It is quite possible that a true sinusoidal wave structure does not exist. What might very well exist, however, is a structure of rather equally spaced blobs. (This explains the large line width.)

The analysis of such a model is quite straight forward if one employs the theorem derived by Booker, Ratcliffe and Shinn [3] about the relationship between the autocorrelation function of the electric field immediately below a phase changing screen and that of the electric field at any distance from this screen. This theorem says that they are identical functions providing there are no periodicities in the screen smaller than a wavelength. Since, the region causing the scintillation is not considered to be very deep (the thickness is very much smaller than the first Fresnel zone distance) only phase and not amplitude changes appear at the lower edge so that it fits the description of phase changing screen. The screen is assumed to vary only in the x direction (horizontally) and only two frequency components of the electron density wave are considered. The field just below the screen at $z = 0$ using their notation is

$$E(x, o, t) = E_0 \sin [\omega_0 t - B_1 \sin (\alpha_1 x + \theta_1) - B_2 \sin (\alpha_2 x + \theta_2)]$$

$$= E_0 \sum^{\infty}_{m,\, n=-\infty}\sum J_m(B_1) J_n(B_2) \sin [\omega_0 t + (m\alpha_1 + n\alpha_2)x + m\theta_1 + n\theta_2]$$

$$E(x + \xi, o, t)$$

$$= E_0 \sum^{\infty}_{k,\, l=-\infty}\sum J_k(B_1) J_l(B_2) \sin [\omega_0 t + (k\alpha_1 + n\alpha_2)(x + \xi) + k\theta_1 + l\theta_2]$$

$$\rho(\xi) = \overline{E(x + \xi)E(x)} = \tfrac{1}{2}E_0^2 \sum^{\infty}_{m,\, n=-\infty}\sum J^2_m(B_1) J^2_n(B_2) \cos (m\alpha_1 + n\alpha_2)\xi \alpha_1 \neq \alpha_2.$$

If the velocity in the x direction is v, then

$$R(\tau) = \rho(v\tau) = \tfrac{1}{2}E_0^2 \sum_{m,\, n} J^2_m(B_1) J^2_n(B_2) \cos (m\alpha_1 + n\alpha_2)v\tau.$$

The power density spectrum is

$$G(\omega) = \tfrac{1}{2}E_0^2 \sum_{m,\, n}\sum J^2_m(B_1) J^2_n(B_2)\, \delta[\omega - (m\alpha_1 + n\alpha_2)v].$$

If the arguments of the Bessel functions B_1 and B_2 are in the order of unity then the number of the important components of the spectrum is about the observed number. (If $B_1 = 1$ and $B_2 = 0{\cdot}5$ then about six component frequencies are important. This is a fairly typical case when discrete frequencies appear to any extent.)

VI. Examples of Processed Data

The computer was programmed to produce an oscilloscope plot of both the autocorrelation functions and the power density spectra. The following photographs are some typical examples. The plots in the left column are the normalized autocorrelation functions with the marks on the horizontal axis indicating eight thirds of a second intervals. The power density spectra are in the right column, each mark on the horizontal axis here being $\frac{1}{50}$ c/s.

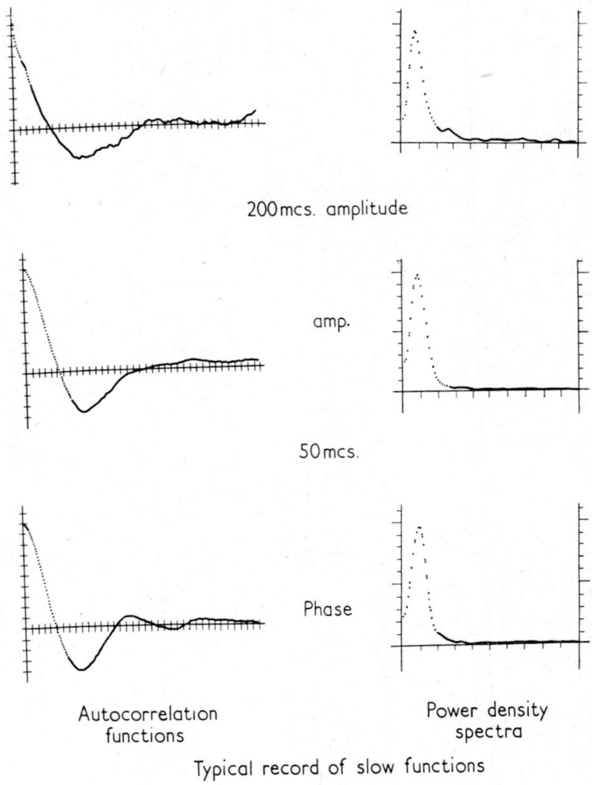

Fig. 5.

The first photograph entitled "Typical Record of Slow Fluctuations" is self explanatory. It should be noted that the amplitude and phase records are very nearly identical on 50 Mc (and on 200 Mc although the phase record for 200 Mc is not shown) and that the autocorrelation function and power density spectrum for the amplitude fluctuations on 200 Mc are very similar to the 50 Mc record. Note that the noise on 200 Mc is higher and that the drop off near the origin of the autocorrelation function for this frequency is different being more like a time constant than the apparent zero slope functions for 50 Mc. This is most likely the autocorrelation function of the system noise after being passed through the R–C integrators. The low frequencies in the data have been attenuated below about $\frac{1}{50}$ c/s. It is conceivable therefore that this filtering could modify the amplitude records to appear similar to the phase records if the condition $D/l \gg 1$ were satisfied.

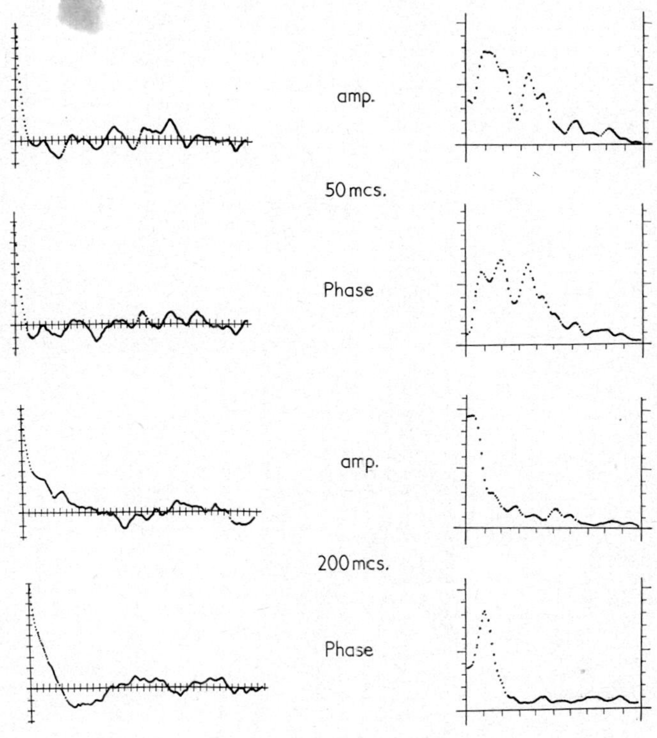

20 August 1957

FIG. 6.

The records of the 14th August (Fig. 5) illustrate the point made above that sometimes the phase and amplitude records are different in character. The emphasis is towards the lower frequencies in the amplitude whereas the phase records emphasize the frequencies from $\frac{2}{50}$ c/s to about $\frac{4}{50}$ c/s. Note should be taken of the appearance of the sharp spikes in the spectra and that different frequencies receive emphasis in the amplitude and phase records. There again similarity in the phase records and also amplitude records at both frequencies is apparent.

The record of 20 August (Fig. 6) indicates a different situation. On this night

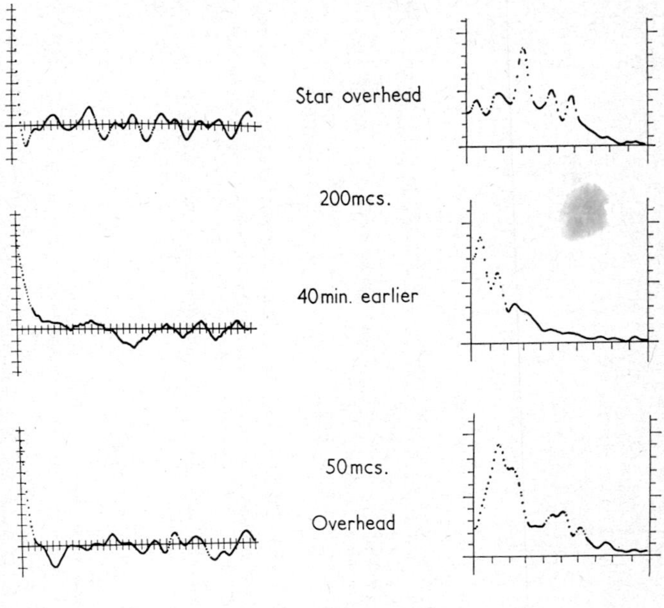

Amplitude records 3 September 1957

FIG. 7.

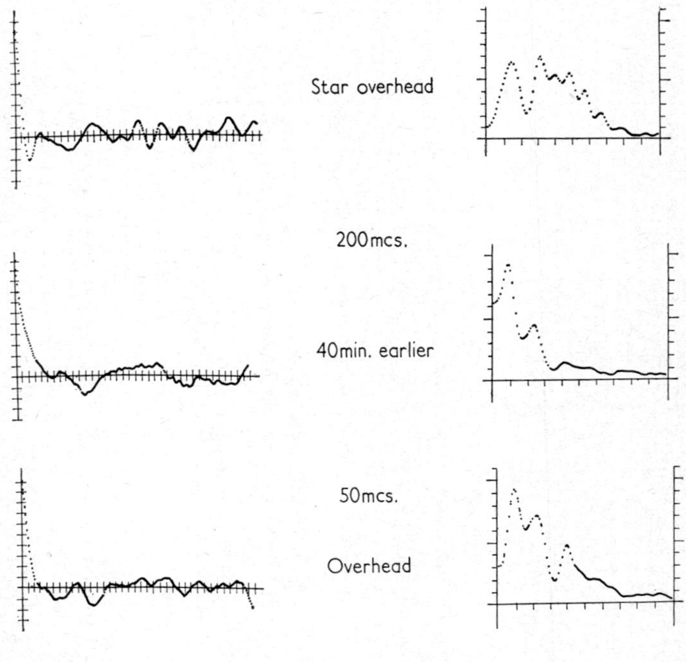

Phase records 3 September 1957

FIG. 8.

the fluctuations on 50 Mc are evidently dissimilar. The initial fall off of the autocorrelation function on 200 Mc resembles that of the 50 Mc. However, the emphasis is upon the low frequencies on the 200 Mc fluctuations. Again amplitude and phase records are quite similar. The 200 Mc records appear to be close

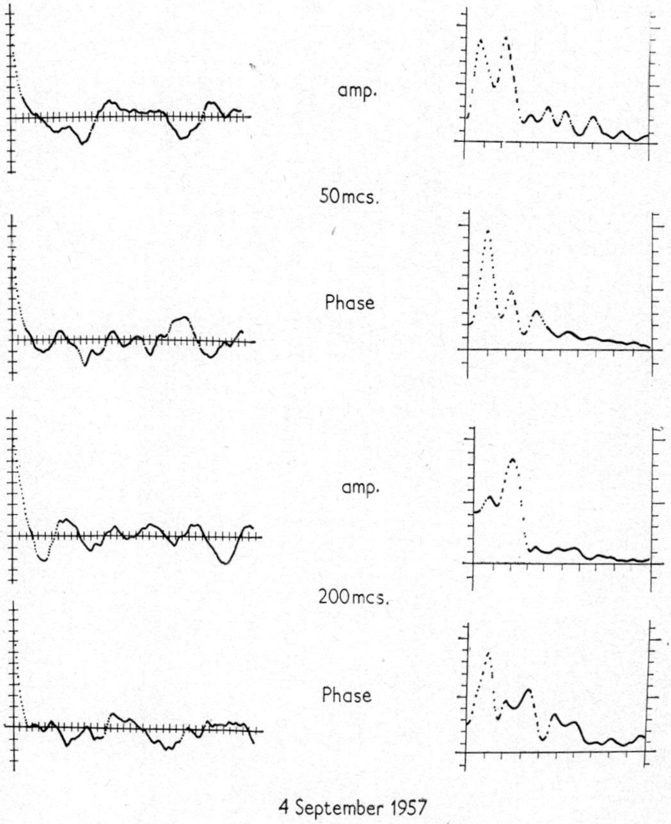

FIG. 9.

to theory in that they do not have significant peaks in the spectra and the phase autocorrelation function crosses the abscissa as would happen if $D/l \gg 1$.

Figures 7 and 8 present a record of the amplitude and phase fluctuations for 3 September 1957. Here again one can note the similarity of amplitude and phase fluctuations at least as far as the random component. The spectra, with the peaks imagined removed, have the same frequency dependence. A record of the fluctuations on 200 Mc (Fig. 9) 40 min prior to the upper culmination is presented to indicate another typical condition, that of the non-stationarity of the fluctuations. One notes the decided difference between this early record and the overhead record on 200 Mc as well as the difference between the 200 Mc and 50 Mc overhead records.

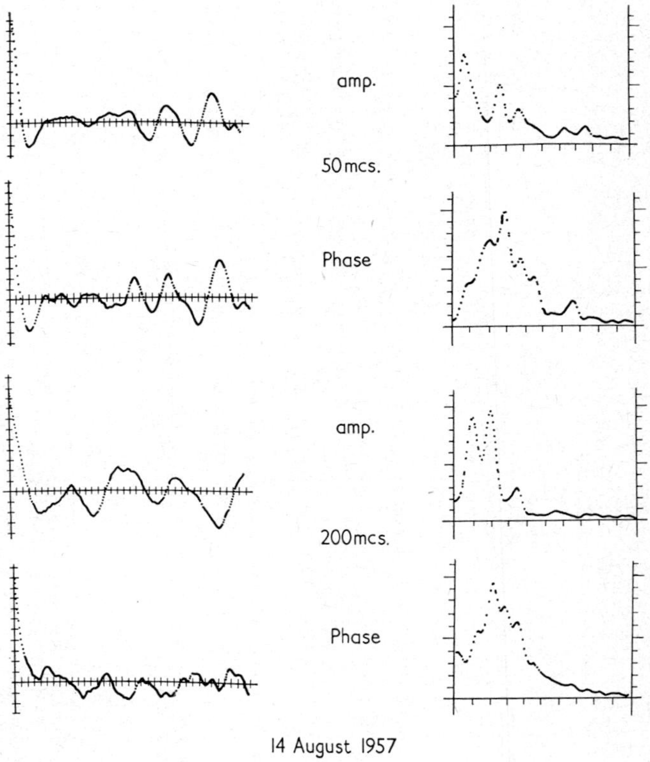

FIG. 10.

Figure 10 is indicative of the situation where the phase records on the two frequencies are similar and the amplitude records are also somewhat similar in shape. One can note the faster drop off in the phase records.

ACKNOWLEDGMENTS

The author wishes to acknowledge the advice and encouragement of Prof. J. B. Wiesner; the help of Lincoln Laboratory Division 3 during the experimental part; Arthur E. Berg whose unselfish assistance during the experimental work greatly aided in the completion of this work; the staff of the Research Laboratory of Electronics at M.I.T. for their support and patience; and the Air Force Cambridge Research Center for the help and time consumed in preparing this paper.

References

1. Booker, H. G. (1956). *J. geophys. Res.* **61**, 673.
2. Booker, H. G. & Gordon, W. E. (1950). *Proc. Inst. Radio Engrs, N.Y.* **38**, 401.
3. Booker, H. G., Ratcliffe, J. A. & Shinn, D. H. (1950). *Phil. Trans. A* **242**, 579.
4. Bramley, E. N. (1953). *Proc. roy. Soc. A* **220**, 39.
5. Hewish, A. (1952). *Proc. roy. Soc. A* **214**, 494.
6. Munro, G. H. (1950). *Proc. roy. Soc. A* **202**, 208.
7. Pierce, J. A. & Mimno, H. R. (1940). *Phys. Rev.* **57**, 95.
8. Rice, S. O. "Selected Papers on Noise and Stochastic Processes." Ed. Wax. Dover (1954).
9. Ross, D. T. *Servomechanisms Lab. Rep.* no. 7138–R–5. Mass. Inst Technol.
10. Villars, F. & Weisskopf, V. F. (1954). *Phys. Rev.* **94**, 232.

Experimental Studies of Meteor Echoes at 200 Megacycles

J. L. HERITAGE, S. WEISBROD and W. J. FAY

Smyth Research Associates, San Diego, Calif.

Forward scatter studies were made at 200 Mc using a high power pulsed transmitter and highly directive transmitting and receiving antennas. At distances from the transmitter of 1300 and 1800 km, six receiving sites were occupied, three on either side of the main beam great circle path. Diurnal burst rate curves given for each site show the dependence of burst rate on "off path" geometry.

At 940 km, a count was made of Doppler shifted signals. The diurnal variation of the "Doppler" count is compared with the "specular" count at the same receiving site.

At sites close to the region where specular reflection is possible from ionization strung out along lines of the Earth's magnetic field, many rapid fading signals were received. These signals, named H_E scatter, were marked by frequent enhancements lasting from a few seconds to many minutes.

I. INTRODUCTION

The literature contains many reports on experimental studies of the characteristics of meteor bursts both at forward and back scatter. These studies, however, were usually carried out in the 20 to 100 Mc region, with antennas of only moderate gain-directivity. Transmitters of moderate or low power were usually employed.

In 1957 an opportunity arose to carry out a fairly extensive forward scatter experiment at 200 Mc using high power and highly directive antennas. Since the transmitter beams were fixed, the receiving units were made mobile, and carried antennas as directive as possible considering the mobile function. This arrangement made possible considerable departure from conventional great circle transmission geometry while insuring that the region of the intersection of the transmitter and receiver beams provided a sensible majority of received signals.

II. EXPERIMENTAL PROCEDURE

A. Apparatus

The experimental setup consisted of a high power pulse transmitter located in Southern Texas. The transmission path was roughly from south-east to north-west. Two transmitting beams labeled "one" and "three," each with $2\frac{1}{2}°$ elevation angle and offset from each other by 5° azimuth, were switched alternately in 10-minute intervals, doubling the number of paths under study. We regret that we are not at liberty to describe transmitter parameters in more detail.

On the receiving end of the circuit we had two mobile units, each consisting of a $1\frac{1}{2}$ ton truck and a pickup, shown in Fig. 1, carried the antenna tower and the dissembled antenna. The truck in Fig. 2 housed the electronic equipment and also towed a portable 5 kW a.c. power unit.

Figure 3 shows the mobile unit on location. The receiving antenna consists of a 4 × 4 array of 10 element Yagi antennas, spaced approximately one wavelength. The array is mounted on a telescopic tower which can be raised to any

FIG. 1. Disassembled antenna and pickup truck.

FIG. 2. Truck with receiving equipment.

height from 12 to 38 ft. The array can be oriented in any direction by remote control from the truck. It takes two men an average of 4 hr to get the mobile unit in operation after arriving on a site. The dismantling time is somewhat less.

FIG. 3. Receiving station in operation.

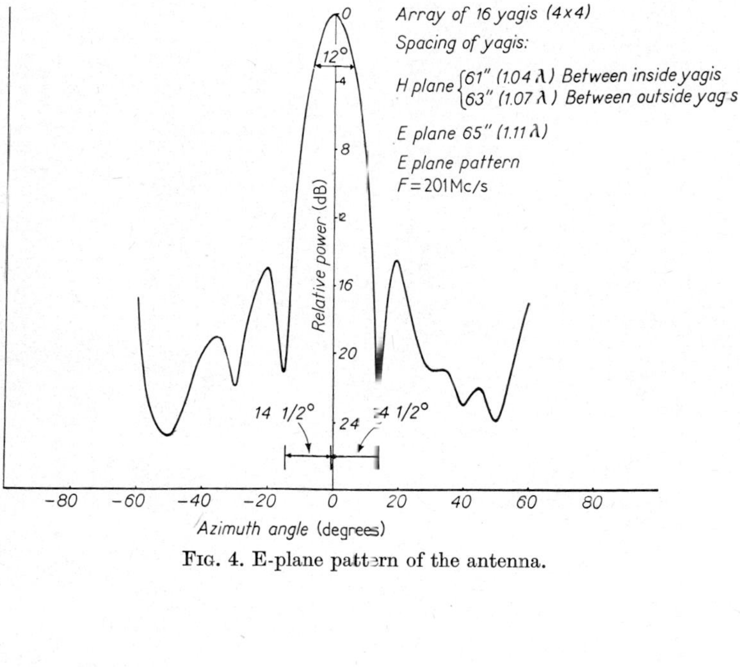

FIG. 4. E-plane pattern of the antenna.

The array has a 12° beam width between half power points in the E plane and 13° in the H plane. The E-plane pattern is shown in Fig. 4. The gain of the array is $21\frac{1}{2}$ db over an isotropic source.

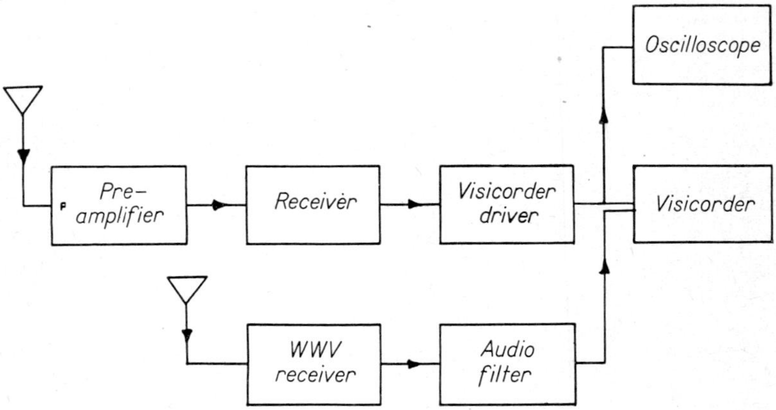

FIG. 5. Block diagram of the receiving equipment.

FIG. 6. Electronic equipment inside the truck.

The block diagram of the receiving equipment is shown in Fig. 5. The signal from the antenna is fed through a low-noise preamplifier into a crystal controlled double superheterodyne receiver. A 3·1 kc mechanical filter centered on 455 kc sec I.F. controls the overall bandwidth. The recorder is the Highland "Visicorder," which writes out the received signal pulse by pulse with no integration. The record is written with a mirror-reflected beam of ultraviolet light on sensitized paper which moves at about one-half inch per second. The paper is developed by exposure to fluorescent light. The WWV timing information is recorded along the side of the test signal. The base line of the WWV trace can be manually keyed to enter additional information as needed.

Figure 6 shows some of the electronic equipment inside the truck. On the left there is the communication equipment used for coordinating the work between two mobile units. In the center, from top to bottom, are two 200 Mc receivers, a WWV receiver, Visicorder amplifiers, and a power supply. On the right we have a monitoring scope, antenna direction control panel, and a signal generator used for calibrating.

B. *Choice of Sites*

The receiving sites for transverse cuts to the great circle path were selected in a manner illustrated by Fig. 7. At 1300 and 1800 km from the transmitter, two sites, designated by S1 and N1, are spotted on the great circle path of beams 1 and 3. The letters S and N refer to the southern and the northern series of sites.

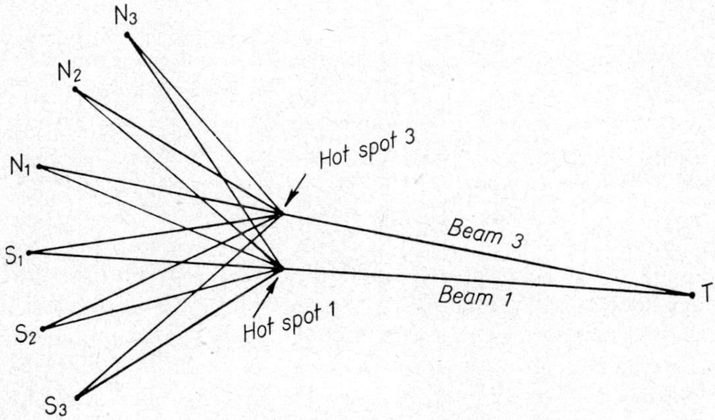

FIG. 7. Selection of sites.

The "hot spot" regions are defined by the intersection of the beams with the hypothetical "h plane" 100 km above and parallel to the Earth. The site S2 was located by specifying congruity between the triangles formed by S1-Hot Spot 3-T and S2-Hot Spot 1-T. When the receiving site was moved from S1 to S2, the geometry of one of the two available transmission paths remained the same. The remaining sites were located in an analogous manner. The corresponding S and N sites were occupied simultaneously by the two mobile units. The naming of the various sites was completed by appending approximate distance from the transmitter, i.e., S1–1300, N2–1800, etc.

At any one site, observations were made over a period of 24 hr. The first series of observations was made in November 1957, when the 1300 km sequence of sites was occupied. In this particular sequence, due to an unfortunate circumstance, the beam 3 data turned out to be of questionable value and was subsequently deleted from further analysis. The second sequence of measurements was made in March 1958, where sites S1–850, N1–850, S1–940, and N1–940 were investigated. The last sequence, the 1800 km series, was made in May 1958. In addition to the data gathered at above sites, some interesting data were also obtained at Ocotillo, California, where the experimental equipment and procedure were first checked out in October 1957.

III. Experimental Results

A. Typical Signals

First, consider some of the typical bursts received at Ocotillo and the 1300 km sites. Figure 8 shows samples of record obtained at N1–1300 site looking at beam 1. The direction of advancing time is shown by the arrows. The 1-sec WWV time ticks are visible most of the time. The upper trace of Fig. 8(a) shows an echo with a slow rise and a fast fall, followed by three smaller ones with a fast rise and slow decay. The lower trace is a classical example of an overdense meteor trail with wind distortion. The echo lasted about seven seconds. In Fig. 8(b) the echo rises in about one-fourth the decay time and has a smooth envelope.

Figure 8(c) illustrates a frequently observed steeple-like shape of total duration of about $\frac{1}{2}$ sec. This particular one is followed by a 2-sec burst with a stepped leading edge. The approximate intensity of these echoes may be estimated from the fact that the strong echoes were on the order of 50 db above noise and that the vertical scale is almost linear in db.

The shapes of echoes were found to be extremely varied. There were a few which followed the classic shape of the Lovell-Clegg formula [5] of a fast rise and an exponential decay. There were also some which approximated the bell-like shape of the Eshleman high frequency formula [2]. Most of the echoes, however, had shapes which fell between the two extremes, and lasted less than one second. There were, however, a few very long ones which lasted on the order of a minute or more. For statistical analysis of the short bursts, all echoes which exceeded in duration 10 sec were arbitrarily excluded since it was felt that the data sample was not sufficiently large to allow the inclusion of these relatively rare events.

B. Short Burst Statistics

For statistical analysis of meteor echoes, all identifiable signals, exceeding 125 dbm and lasting less than 10 sec, were counted. The burst rate found at the six 1300 km sites is shown in Fig. 9. Consideration of system sensitivity, size of the h plane area illuminated by the transmitting and receiving antenna beams, and the probability of specular alignment factor [3], lead to a predicted average

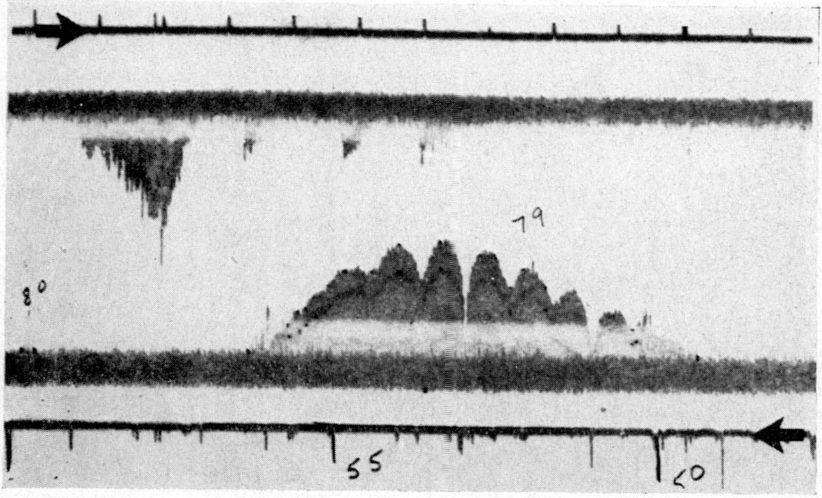

(a) N1-1300 0845 above-slow rise, fast fall below-overdense, interference fading.

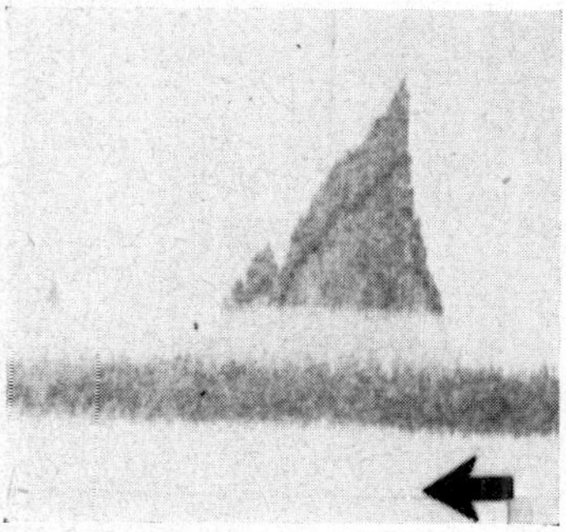

(b) N1-1300 0621.

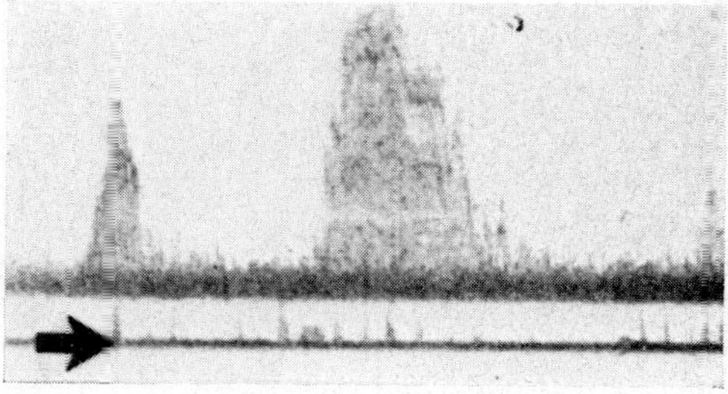

(c) Typical burst followed by burst with step-like leading edge.

FIG. 8. Samples of short meteor echoes at 1300 km.

burst rate of 16 per min. The observed rate varied from 4 to 31 bursts per min and thus appeared to be in good quantitative agreement with the predicted average values. The site S1–1300, located on the path of beam 1, exhibited a relatively low rate. The rate increased rapidly with the departure from the great circle path and then gradually began to diminish beyond the S2 and N2 sites.

In the curves shown here, the diurnal effects predictable from consideration of the motion of the Earth through space, in sweeping up more meteor particles on the morning side, and of the resulting concentration of meteor radiants toward

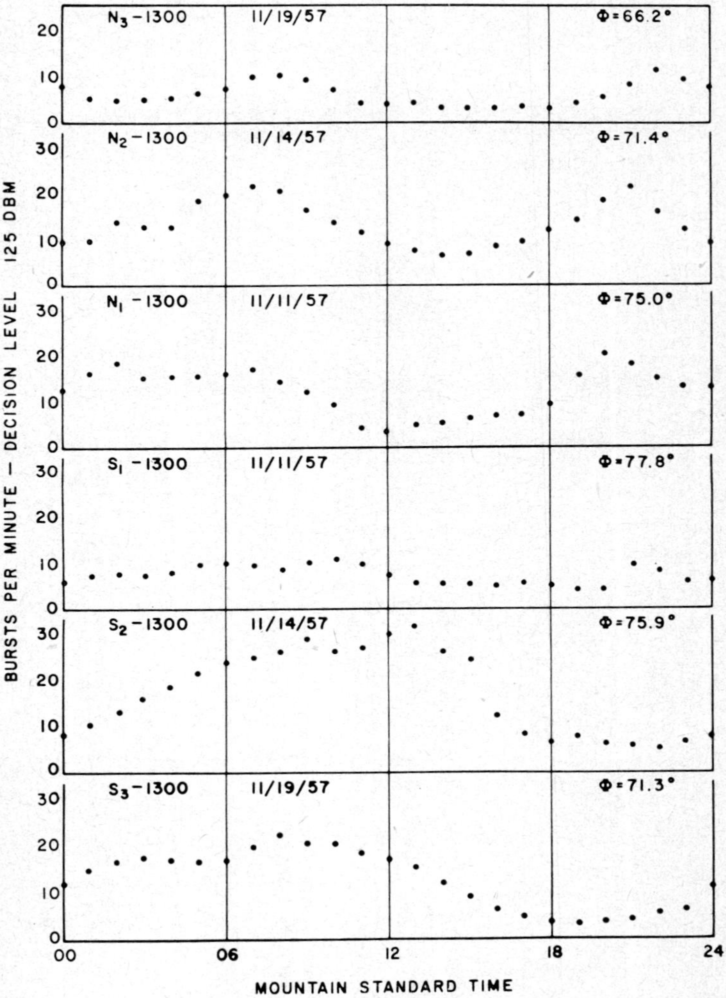

FIG. 9. Echo rates at 1300 km.

the apex of the Earth's way, can be seen. (The motion of the Earth imparts to the meteors a velocity relative to itself. Meteors already having radiants close to the apex of the Earth's way are thus closer aligned toward the apex and that radiant tends to dominate.) The N1, N2 and N3 sites, all lying north and east of the main beam great circle path, show an afternoon and evening peak as expected from the radiant consideration. The S2 and S3 sites, lying south and west of the great circle, combine the predicted daytime and morning peaks due

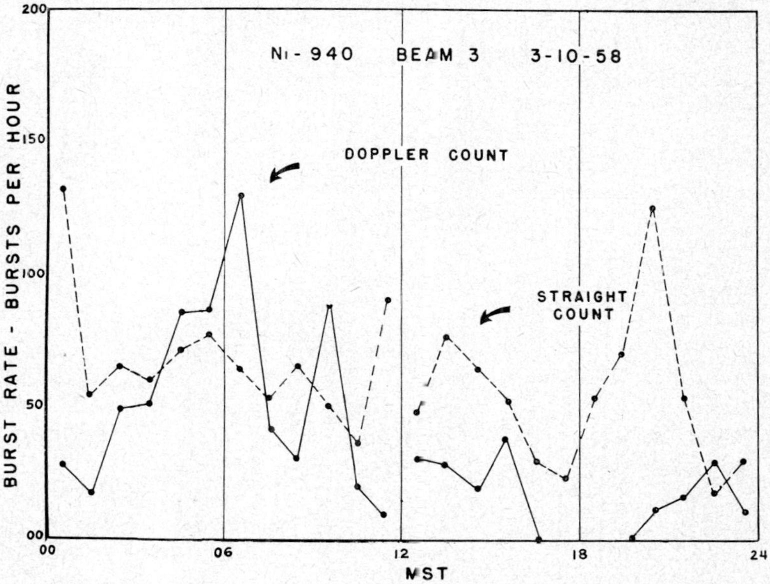

FIG. 10. Specular and Doppler echo rates on the great circle path, 940 km.

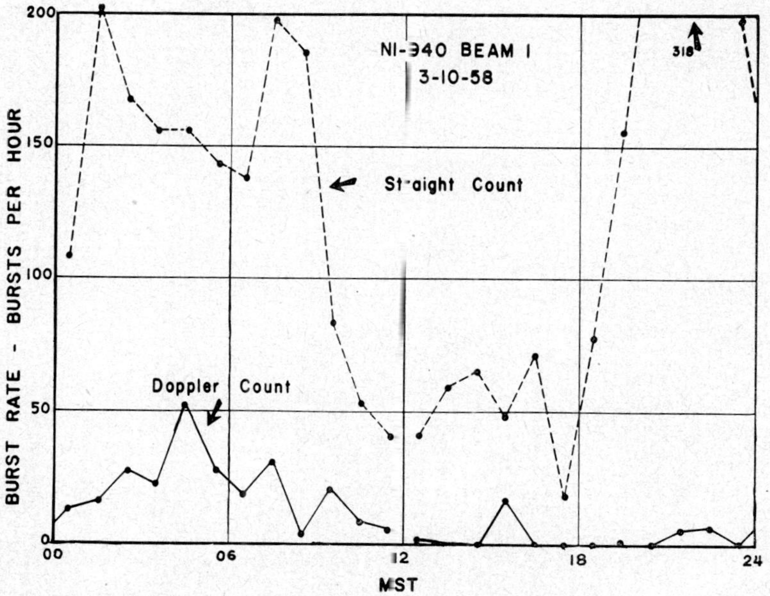

FIG. 11. Specular and Doppler echo rates slightly off the great circle path.

to radiant distribution to give a broad region of high burst rate through the daytime and early morning. It should be remembered that each curve is only one

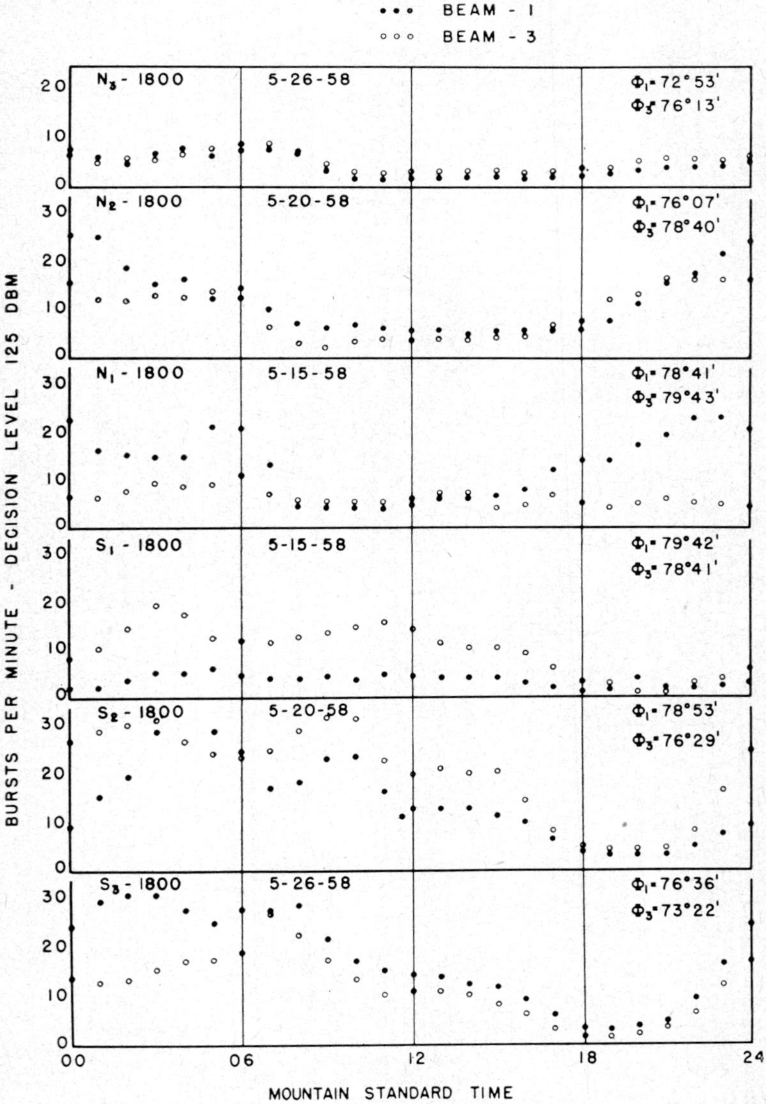

FIG. 12. Echo rates at 1800 km.

day's data and that the N1 and S1 data are simultaneous, as is the N2 and S2, etc.

At 940 km, measurements were made on the great circle path of beam 3.

The diurnal burst rate for the on circle path can be seen in Fig. 10, and for the off circle path in Fig. 11. It can be seen from Fig. 10 that the burst rate is considerably lower at 940 km than at 1300 km. This difference is what one could expect from the reduction in the H plane area common to the transmitter and receiver antenna beams. From Fig. 11 it can be seen that the "off-path" specular count considerably exceeds the "on-path" count.

The last sequence of measurements was made at the distance of 1800 km. Just as in the case of the 1300 km series, three sites were occupied on each side of the great circle paths. The overall results were quite similar to those at 1300 km. Figure 12 shows the diurnal variation of meteor burst rate as a function of

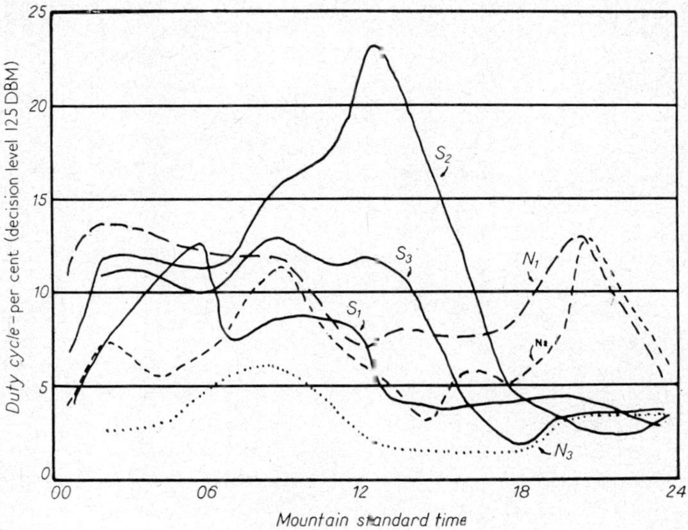

FIG. 13. Duty cycle at 1300 km.

time for all six sites and the two beams. Just as in the case of the 1300 km series, there is a marked and very rapid increase in the echo rate as one departs from the great circle path. This effect appears to be maximum at the S2 and the N2 sites, and gradually decreases for greater off-the-path departures. As expected, the overall diurnal shapes of the various curves agree quite well with the 1300 km series.

One of the statistical quantities of interest for practical applications is the fraction of time that signal is detectable. Figure 13 shows a duty cycle plot as obtained from the 1300 km observations excluding bursts longer than 10 sec. The duty cycle curves are superimposed on a single chart to facilitate comparison between the various paths throughout the day. The general trend, as expected, follows the straight rate picture: the S sites dominate throughout the day, while the N sites take over in the evening.

Eshleman [2] developed an all wavelength expression for the "intrinsic duration" τ for specular echoes, which is defined as the total energy contained in the pulse divided by the peak received power. The value of τ was actually worked out only for the two cases where the wavelength was long and short compared with the so-called "transition wavelength." Since the results for the two extreme cases were identical, Eshleman felt that it was reasonable to conclude that the expression for τ is valid for all wavelengths.

The "transition wavelength" and the "intrinsic duration" are functions of path geometry, meteor velocity, and physical properties of the medium. In Table I the computed values of τ are compared with the experimental 1300 km

TABLE I

Duration of Meteor Echoes

		Transition Wavelength in in Meters		Calculated τ Eshleman Specular all λ		Measured τ Median	
Site	ϕ	$\beta = 0°$	$\beta = 90°$	$\beta = 0°$	$\beta = 90°$	0600	1800
S1–1300	77·84°	1·45	0·496	0·106	0·053	0·190	0·245
N3–1300	66·19°	2·38	1·03	0·059	0·023	0·115	0·200

results for the geometry of S1 and N3 sites. The values of the transition wavelength are also tabulated and it is seen that the present case falls well within the theoretical transition region. The median meteor velocity was assumed to be 40 km/s and the diffusion coefficient was taken as $4m^2/s$. In Table I, 2ϕ is the angle between the transmitter and receiver antenna beams; $\beta = 0°$ and $\beta = 90°$ refer to meteor trails in, and perpendicular to, the plane of propagation. For comparison with the calculated durations, the duration of bursts for morning and evening samples of data at the indicated sites were read and the median values recorded. The method used was to read the duration 10 db down from the maximum. The segregation into morning and evening groups was to observe the effect of the generally higher morning velocities. The morning and evening values of τ at the great circle site S1 are approximately twice as large as the $\beta = 0°$ calculation. This is a very good agreement considering that the classical low frequency formulae give values almost ten times smaller than the Eshleman formula. It can be shown that for very high frequencies the level at which the duration should be read is $(1 + \pi^2)^{-1}$ or nearly 10 db down [1]. For the low frequency exponential shape the level is exp (— 1). Since most of the echoes that we observed lie between the two extremes, a 6 or 7 db level might have been more reasonable and would have resulted in a better agreement with the calculated values. The significant point, however, is that as we moved from S1 to N3 site, the angle ϕ changed by almost 12°, and both the calculated duration from the Eshleman all wavelength formula and the observed duration dropped by about a factor of two. The classic low frequency formula which involves $\sec^2\phi$ calls for a decrease in duration of a factor of 4 for this reduction in ϕ. The observed afternoon values of τ are longer than the morning values, as might be expected from the lower values of median meteor velocity.

C. Doppler Observations

From theoretical considerations [2], we expected to observe only specularly oriented trails, since the ratio of echoing power of the non-specular to specular echo should have been several orders of magnitude down for this geometry and wavelength. Consequently, since we anticipated only very infrequent head echoes from the largest meteor particles, we initially instrumented to record only the specular echoes. In the course of analysis of the 1300 km data, we began to wonder if some of the difficult-to-understand echo shapes that were observed could be caused by a sufficient Doppler shift to cause the signal frequencies to move to the edge of the pass band and be shaped by the steep skirt of the mechanical filter. We therefore instrumented one receiver for a rough count of non-specular echoes. The overall result with the improvised circuit was that the signal had to be at least 15 db stronger than the minimum detectable specular signal in order to operate the limiter and register its correct Doppler shift. The minimum Doppler shift measurable with this instrumentation was $\pm$ 5 kc.

Figure 10 shows a comparison between the specular or "straight" and the Doppler channels at the N1–940 site using beam 3. At this on-the-path site the Doppler count was surprisingly high and even exceeded the straight count during the morning hours. Figure 11 shows the count at N1–940 site using beam 1. The specular count on this slightly-off-the-great-circle-path always dominated the Doppler count.

Using beam 3 at the N1–940 site, which was the most active geometry for the Doppler signals, only 6 % of the total specular signals has a Doppler shift associated with them. Presumably in these cases the meteoric particle is close to the specular point when the Doppler signal appears. Also, only 14 % of the Doppler signals observed were associated with a specular signal. This means that 86 % of the Doppler signals observed were presumably due to particles traveling along radiants well off the specular point. Of the 295 Doppler bursts recorded at the N1–940 site, only 8 were down Dopplers, presumably due to particles which have passed the specular point. None of the 8 down Dopplers had a specular signal associated with them.

Some examples of the Doppler signals are shown in Fig. 14. The upper trace is the specular channel and the lower one is the Doppler channel. The downward deflection in the Doppler channel indicates an "up" Doppler. In samples A, B, and C the specular signal is preceded by the "up" Doppler as might be expected. In sample D there is a "down" Doppler with no apparent specular echo associated with it.

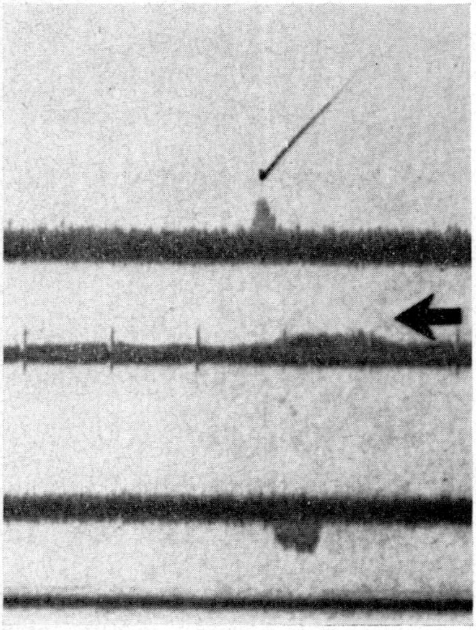

(a) N1 940 2151 up Doppler 14 kc max.

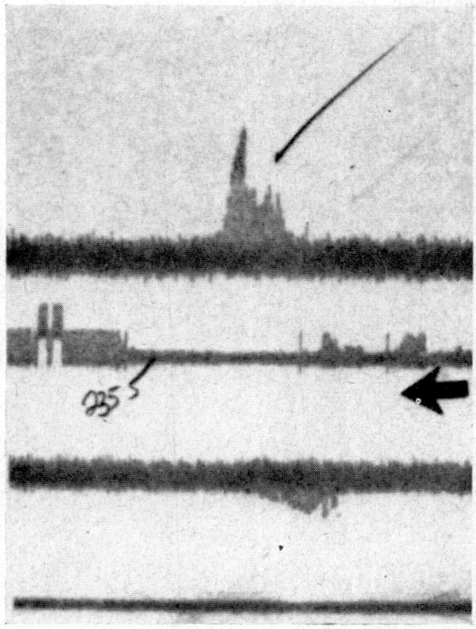

(b) N1 940 2355 up Doppler 13 kc max.

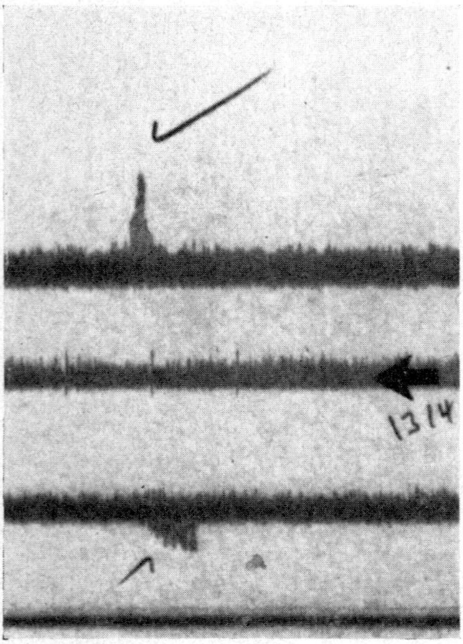

(c) N1 940 1314 up Doppler 10 kc max.

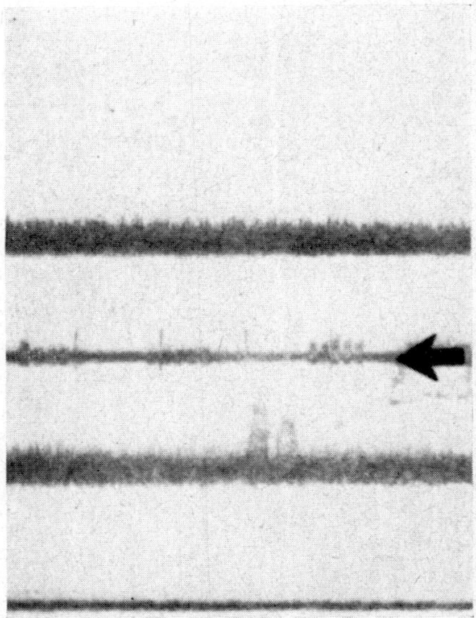

(d) N1 940 2255 down Doppler 24 kc max.

FIG. 14. Samples of Doppler echoes.

D. Multiple Bursts

Another interesting feature which was observed was the fairly frequent appearance of multiple bursts apparently evenly spaced, as illustrated in Fig. 17. A statistical analysis has been made to determine whether the number of these evenly spaced bursts exceeds the expected value, assuming a random occurrence of meteor echoes [4]. The result was that the number of observed triples agreed closely with the expected value.

E. H_E Scatter

Two different types of long lasting signals were observed. The first is illustrated in Fig. 15(b) and is presumably a classic overdense meteor trail: smooth

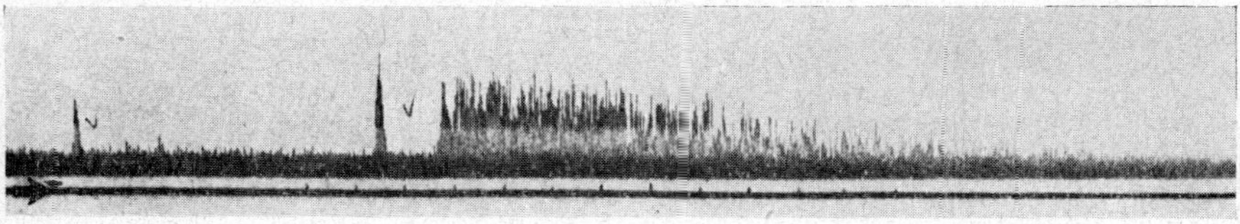

(a) Ocotillo 1115 Long signal-ragged envelope after what looks like meteor burst.

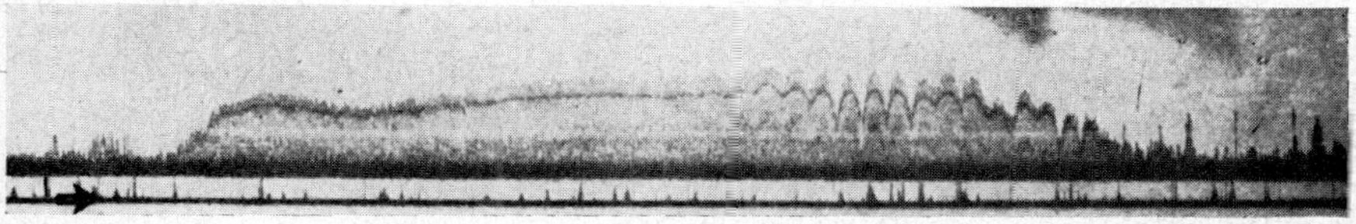

(b) Ocotillo 1155 Long signal-smooth envelope throughout.

FIG. 15. Samples of long echoes.

envelope at first, breaking into a well defined but smooth interference pattern. Wind shears, imparting different velocity to different parts of the trail, supposedly cause the interference phenomenon. A similar type trail is illustrated in Fig. 16(b) and is shown to belong to the non-specular type by its simultaneous transmission of signal to both N2 and S2 sites. The other type of long lasting signal was observed at the S3–1300 km site and at Ocotillo, California. Ocotillo is located at 1800 km from the transmitter on a line from hot spot 1 through site S3–1300. These signals are illustrated in Figs. 15(a) and 16(a). Distinguishing features are the rapid fading throughout the life of the echo and duration considerably greater than underdense meteor echoes. The signal in Fig. 16(a) lasted 28 min. The rate of occurrence of these signals appeared to maximize during the night and early morning hours, and were much more frequent at Ocotillo than at the S3–1300 site. We have recently gathered considerable evidence which suggests that these echoes are caused by reflection from ionized trails oriented along the Earth's

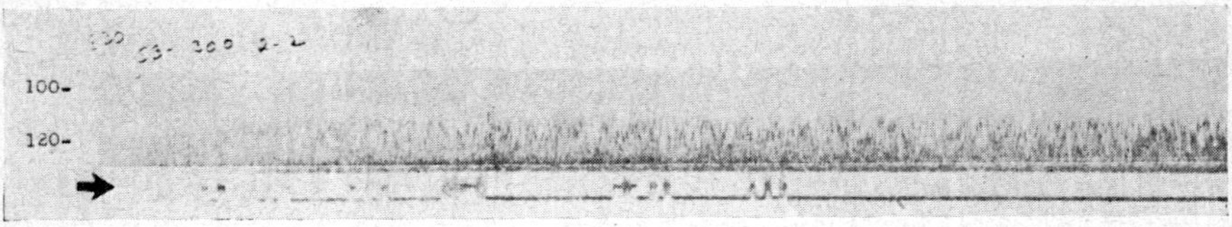

(a) S3-1300-1830 . . . Very long lasting signal that maintains this character and amplitude for 28 minutes. It fades gradually the last eight minutes. This signal did not appear at the N3 site.

(b) N2 & S2-1300-0740 . . . Long signal appearing almost simultaneously at N2 and S2. The signal lasted 1·0 minutes at N2 and 1·5 minutes at S2.

FIG. 16. Samples of long echoes.

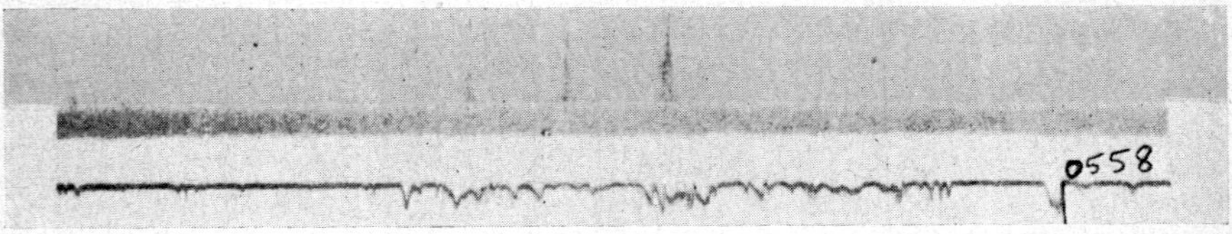

(a) N3-1300 0558 triple.

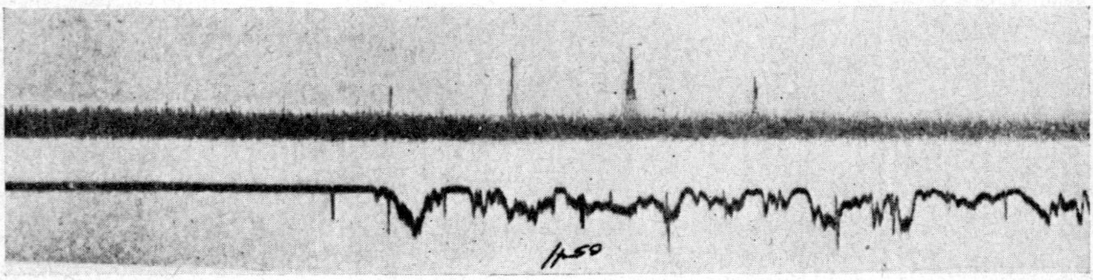

(b) N3-1300 0541 quadruple.

FIG. 17. Evenly spaced multiple meteor bursts.

magnetic field. These signals, which we have named H_E scatter in reference to their relation to the magnetic field and the area of origin, showed considerable aspect sensitivity. At locations where such specularly reflected signals could be observed, the H_E scatter was almost continuously visible throughout the night and well into the daytime hours. The H_E scatter is characterized by so-called enhancements where the signal intensity is abruptly increased and then slowly decays over periods ranging from several seconds to many minutes. The H_E scatter differs in character from the usual auroral echoes and from the E_a scatter observed and described by Gallacher at Stanford University [1]. A paper describing experimental work in connection with the H_E scatter is now being prepared by the authors [7].

III. Conclusions

To summarize the important conclusions drawn from the experimental data:

(a) Burst rates increased with increasing distance from the hot spots.

(b) Lower burst rates and duty cycles were found *on* the great circle path than *off* the path on either side. The increase is very rapid with the departure from the path, reaching a maximum near the N2 and S2 sites and then slowly decreasing.

(c) The diurnal behavior of burst rate was quite different at the S. and N. sites, but was consistent with the idea that the dominant effects result from the Earth's rotation and orbital motion.

(d) Median duration of echoes compared favorably with Eshleman's theory of meteor echoes at VHF.

(e) Off-specular Doppler shifted meteor bursts exceeded the specular burst rate on the great circle path during morning hours.

(f) Long lasting signals, named H_E scatter, were observed at the southernmost sites. These rapidly fading echoes are associated with reflections from ionized columns aligned along the Earth's magnetic field.

References

1. Gallagher, P. B. (1956). *Stanf. Univ. Tech. Rep.* No. 107. Radio Propagation Laboratory.
2. Eshleman, Von R. (1956). *Stanf. Univ. Tech. Rep.* No. 5. Radio Propagation Laboratory.
3. Eshleman, Von R. & Manning, L. A. (1954). *Proc. Inst. Radio Engrs, N.Y.* **42,** 533.
4. Keary, T. J. & Wirth, H. J. (1958). *J. geophys. Res.* **63,** 67.
5. Lovell, Bernard & Clegg, J. A. (1948). *Proc. phys. Soc. Lond. A* **60,** 491.
6. Villard, O. G. Jr., Eshleman, V. R., Manning, L. A. & Peterson, A. M. (1955). *Proc. Inst. Radio Engrs, N.Y.* **43,** 1473.
7. Heritage, J. L., Weisbrod, S. & Fay, W. J. (1959). *J. geophys. Res.* (in press).

Radiation from Idealized Shock Excitation Currents in a Straight Conductor Rising from a Perfect Earth at an Arbitrary Angle

E. A. Lewis

Electronics Research Directorate, Air Force Cambridge Research Center, Bedford, Mass.

If a long straight antenna wire inclined at an arbitrary zenith angle above a perfectly conducting ground has a constant voltage suddenly applied to its lower end, the resulting transient currents produce radiation. Neglecting certain dispersive effects in the antenna, these radiation fields were calculated from retarded potential theory. It was found that the positive and negative peaks of the radiation waveforms are in general unequal, both in duration and in magnitude. For example, with the antenna tilted at a zenith angle of 60° and the antenna wave velocity 99 per cent or more of the freespace value, the computed ratios of the front-to-back amplitude and duration are about 13 : 1.

I. Introduction

Newman [1] has described a method of producing lightning-like radiation transients by shock-exciting a very long wire held aloft by balloons or by a helicopter. The excitation consists in charging a bank of condensers in parallel and discharging them in series (thus effectively obtaining voltages up to 1 or 2 million V) into the wire antenna.

The current in the wire is initially determined by its surge impedance Z_a, which in the following treatment is assumed to be independent of height above the ground.* The current then proceeds along the wire at a velocity arbitrarily close to the velocity of electromagnetic waves in the medium, here assumed to be freespace. The propagation of this transient wave along the wire sets up radiation fields.

The case treated in this paper is that of a long straight antenna wire inclined at an arbitrary zenith angle α above a perfectly conducting ground. The disturbance in the wire causing radiation has a velocity component toward the observer on the ground in the direction of antenna inclination, thus producing a nonisotropic directional pattern. Owing to differences in travel time of the radiation from the different elements of the current waves bouncing back and forth in the antenna, the successive positive and negative peaks of the radiation waveforms are, in general, not equal in duration (see part IV).

This distortion of time is analogous to that which would be caused by ordinary Doppler effect if a short radiating dipole were mechanically oscillated back and forth perpendicularly to its axis with the same period and phase as the electric oscillations.

*In the cases considered by Newman *et al.*, Z_a is on the order of 800 ohms.

II. Transient Waves in the Antenna

In Fig. 1 the antenna wire of length L is represented making an angle α with the zenith. At time $t = 0$, the positive side of a constant source of d.c. voltage V is connected to the end of the antenna, the negative side being connected to the (perfectly conducting) ground. The following idealizations are made:

1. The propagation characteristics along the wire are assumed to be independent of height above the ground, although actually, for example, the characteristic impedance Z_a varies slightly with height.

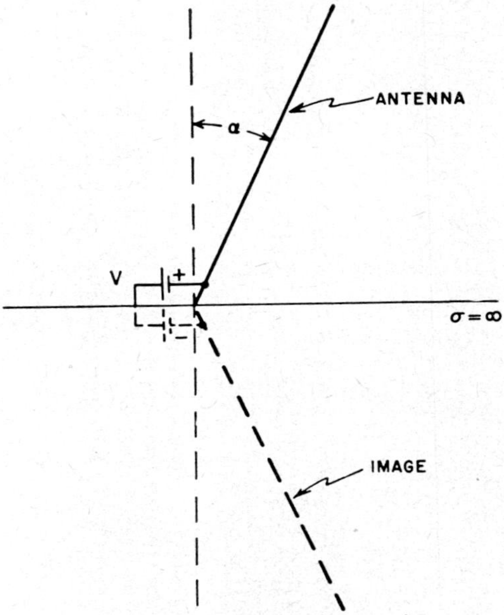

Fig. 1. Antenna and image.

2. It is assumed that propagation along the wire is free of dispersive and other losses so that, for example, a steep wavefront will propagate with no change in shape or amplitude, with a velocity v very nearly the velocity c of electromagnetic waves in the medium (freespace).
3. Radiation reaction is neglected—a condition already implicitly contained in (2).

At time $T = 0$, a voltage wave of amplitude V and current of amplitude V/Z_a starts up the antenna with velocity v. At time $t = L/v$, this wave reaches the upper end of the wire where the "open circuit" boundary condition of zero current flow guarantees a reflected wave of equal current in the opposite direction. In this wave the potential is also V, leaving a total potential of $2V$ in its wake.

On arrival back at the source at time $t = 2L/v$, the condition that the potential be V demands an upgoing wave of potential $-V$, and so on. The various waves are given in Table I, in which potentials are shown positive when the wire is positive, and currents due to positive potentials are shown as positive together with the direction of flow, up or down.

TABLE I

WAVES IN ANTENNA

	Potential		Current		Charge density	
	Wave	Total	Wave	Total	Wave	Total
$t = 0$ wave starts up	$+V$	$+V$	$+V/Z_a$ (up)	$+V/Z_a$ (up)	$+V/(Z_a v)$	$+V/(Z_a v)$
$t = L/v$ wave starts down	$+V$	$+2V$	$+V/Z_a$ (down)	0	$+V/(Z_a v)$	$+2V/(Z_a v)$
$t = 2L/v$ wave starts up	$-V$	$+V$	$-V/Z_a$ (up)	$-V/Z_a$ (up)	$-V/(Z_a v)$	$V/(Z_a v)$
$t = 3L/v$ wave starts down	$-V$	0	$-V/Z_a$ (down)	0	$-V/(Z_a v)$	0
$t = 4L/v$ wave starts up			same as no. 1			

By symmetry, the electromagnetic boundary conditions at the surface of the perfectly conducting plane are satisfied by assuming an image antenna having equal but opposite excitation, as indicated in Fig. 1. The corresponding waves are given in Table II.

TABLE II

WAVES IN ANTENNA IMAGE

	Potential		Current		Charge density	
	Wave	Total	Wave	Total	Wave	Total
$t = 0$ wave starts down	$-V$	$-V$	$-V/Z_a$ (down)	$-V/Z_a$ (down)	$-V/(Z_a v)$	$-V/(Z_a v)$
$t = L/v$ wave starts up	$-V$	$-2V$	$-V/Z_a$ (up)	0	$-V/(Z_a v)$	$-2V/(Z_a v)$
$t = 2L/v$ wave starts down	$+V$	$-V$	$+V/Z_a$ (down)	$+V/Z_a$ (down)	$+V/(Z_a v)$	$-V/(Z_a v)$
$t = 3L/v$ wave starts up	$+V$	0	$+V/Z_a$ (up)	0	$+V/(Z_a v)$	0
$t = 4L/v$ wave starts down			same as no. 1			

III. General Solution for the Radiation Fields

In Fig. 2, the antenna and its image are in the plane YOZ, and unit vectors in these directions are denoted by $\bar{a}_+$ and $\bar{a}_-$ respectively. The observation point P has coordinates (x, y, z), or, in the corresponding spherical coordinate system (R, θ, ϕ).

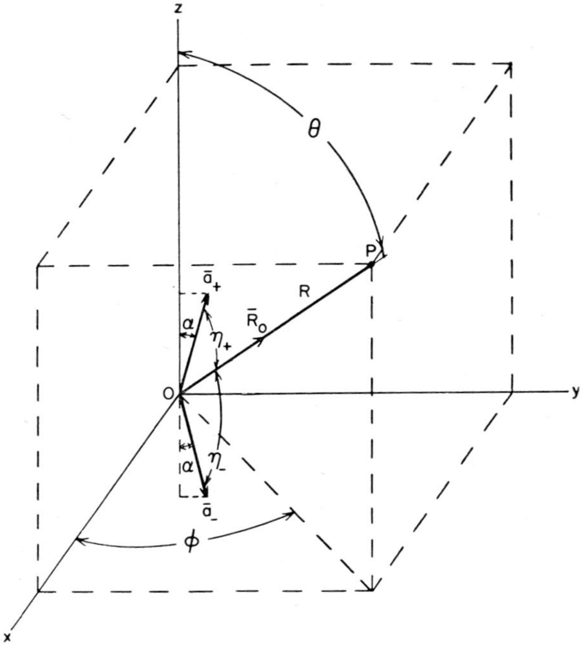

Fig. 2. Coordinate system.

If we denote unit vectors in the coordinate directions by $\bar{x}_0$, $\bar{y}_0$, $\bar{z}_0$ in the rectangular system, and by $\bar{R}_0$, $\bar{\theta}_0$, $\bar{\phi}_0$, in the spherical system, then by geometry,

$$\left.\begin{aligned} \bar{R}_0 &= \bar{x}_0 \sin\theta\cos\phi + \bar{y}_0 \sin\theta\sin\phi + \bar{z}_0\cos\theta \\ \bar{\theta}_0 &= \bar{x}_0 \cos\theta\cos\phi + \bar{y}_0\cos\theta\sin\phi - \bar{z}_0\sin\theta \\ \bar{\phi}_0 &= -\,\bar{x}_0\sin\phi + \bar{y}_0\cos\phi \end{aligned}\right\} \tag{1}$$

and

$$\bar{a}_\pm = \bar{y}_0 \sin\alpha \pm \bar{z}_0\cos\alpha. \tag{2}$$

With the angle between the observer and the antenna denoted by η_+, and that of the image by η_-,

$$\cos\eta_\pm = \bar{a}_+ \cdot \bar{R}_0 = \sin\theta\sin\phi\sin\alpha \pm \cos\theta\cos\alpha. \tag{3}$$

Further, for $0 < \theta < 180°$,

$$\begin{aligned} \bar{a}_\pm &\equiv \bar{R}_0(\bar{R}_0 \cdot \bar{a}_\pm) + \theta_0(\theta_0 \cdot \bar{a}_\pm) + \phi_0(\phi_0 \cdot \bar{a}_\pm) \\ &= \bar{R}_0\cos\eta_\pm + \bar{\theta}_0(\sin\alpha\cos\theta\sin\phi \mp \sin\theta\cos\alpha) + \phi_0(\sin\alpha\cos\phi). \end{aligned} \tag{4}$$

In the retarded potential solution of Maxwell's equations,[2] the electromagnetic fields given by

$$\bar{E} = -\nabla\Phi - \mu\frac{\partial\bar{A}}{\partial t} \tag{5}$$

$$\bar{H} = \nabla\times\bar{A} \tag{6}$$

where

$$\Phi = \frac{1}{4\pi\epsilon_0}\int_{-L}^{L}\rho^*\frac{dl}{r} \tag{7}$$

$$\bar{A} = \frac{1}{4\pi}\int_{-L}^{L}\bar{J}^*\frac{dl}{r} \tag{8}$$

where l is the running coordinate along the length of the antenna, increasing upward; r is the distance to the observer from the element dl; ρ^* and $\bar{J}^*$ are respectively the charge density and current in the element at the "retarded time"

$$\tau \equiv t - \frac{r}{c}. \tag{9}$$

This allows for the propagation travel time from the element of integration to the observation point.

For very distant observation points, that is, $R \gg 1$,

$$r = R - l\cos\eta_+. \tag{10}$$

The procedure will be to calculate (1) the fields due to the antenna alone, (2) the fields of the image, and (3) their vector sums.

A. Fields from the First Antenna Wave

When the first current wave in the antenna has progressed to the point l_1 and the wavelet from l_1 has just reached the observer, the time will be

$$t = \frac{l_1}{v} + \frac{R - l_1\cos\eta_+}{c} \tag{11}$$

whence

$$l_1 = v\frac{t - R/c}{1 - \beta\cos\eta_+}, \quad (l_1 \leqslant L). \tag{12}$$

For those combinations of t, R, and η_+ for which l_1 as found from Equation (12) exceeds L, the pertinent fields will be those due to one of the succeeding waves in the antenna.

So far as the first-wave field at P at time t is concerned, only values of $l < l_1$ can contribute to the integrals of Equations (7) and (8) since for values of $l > l_1$ both $\bar{J}^*$ and ρ^* are zero. For $l \leqslant l_1$, the (constant) values of $\bar{J}^*$ and ρ^* found from Table I are such that

$$\Phi = \frac{V}{4\pi\epsilon_0 Z_a v}\int_0^{l_1}\frac{dl}{r} - \frac{V(t - R/c)}{4\pi\epsilon_0 Z_a R}. \tag{13}$$

$$\bar{A} = \frac{V\bar{a}_+}{4\pi Z_a}\int_0^{l_1}\frac{dl}{r}. \tag{14}$$

The second term in Equation (13) accounts for the negative charge that builds up on the generator as current is supplied to the antenna.

Next,

$$\nabla\Phi = \frac{V}{4\pi\epsilon_0 Z_a v}\left[\int_0^{l_1}\nabla\left(\frac{1}{r}\right)dl + \frac{\nabla l_1}{r(l_1)}\right] - \frac{V}{4\pi\epsilon_0 Z_0}\nabla\left[\frac{t-R/c}{R}\right]$$

$$= \frac{V}{4\pi\epsilon_0 R Z_a v}\nabla l_1 + \frac{V\bar{R}_0}{4\pi\epsilon_0 R Z_a c} + \text{terms in } \frac{1}{R^2} \quad \ldots{}^*$$

In spherical coordinates,

$$\nabla l_1 = \bar{R}_0\frac{\partial l_1}{\partial R} + \frac{\bar{\theta}_0}{R}\frac{\partial l_1}{\partial\theta} + \frac{\bar{\phi}_0}{R\sin\theta}\frac{\partial l_1}{\partial\phi}.$$

By the use of Equations (12) and (13),

$$\nabla l_1 = -\frac{(v/c)\bar{R}_0}{1-\beta\cos\eta_+} + \beta\frac{\bar{\theta}_0}{R}v(t-R/c)\frac{\cos\theta\sin\phi\sin\alpha - \sin\theta\cos\alpha}{(1-\beta\cos\eta_+)^2}$$
$$+\beta\frac{\bar{\phi}_0}{R}v(t-R/c)\frac{\cos\phi\sin\alpha}{(1-\beta\cos\eta_+)^2}.$$

Since for the times of interest, $v(t - R/c)$ is of the order L, the coefficients of $\bar{\theta}_0$ and $\bar{\phi}_0$ are of order $1/R$, and thus contribute terms of order $1/R^2$ in the expression for $\Delta\Phi$.

Thus,

$$\nabla\Phi = \frac{V\bar{R}_0}{4\pi\epsilon_0 R Z_a c}\cdot\frac{1}{1-\beta\cos\eta_+} + \frac{V\bar{R}_0}{4\pi\epsilon_0 R Z_a c} + \text{terms in } \frac{1}{R^2} \quad \ldots . \tag{15}$$

From Equation (14)

$$\mu\frac{\partial\bar{A}}{\partial t} = \frac{\mu V\bar{a}_+}{4\pi Z_a}\frac{1}{r(l_1)}\frac{\partial l_1}{\partial t}$$

$$= \frac{V\beta}{4\pi\epsilon_0 R Z_a c}\cdot\frac{\bar{a}_+}{1-\beta\cos\eta_+} + \text{terms in } 1/R^2 \quad \ldots . \tag{16}$$

in which use is made of the relation

$$\mu = \frac{1}{\epsilon c^2}. \tag{16a}$$

Using Equation (5), it follows that R, θ, and ϕ components of the electric vector of the radiation field (terms in $1/R$) are:

*This result, and also Equation (16), can be obtained more rigorously by evaluating the integral in Equations (13) and (14) and then performing the differentiations.

$$\left.\begin{aligned} E_R &= 0, \\ E_\theta &= -K\beta\frac{\bar{a}_+ \cdot \bar{\theta}_0}{1-\beta\cos\eta_-} \\ E_\phi &= -K\beta\frac{\bar{a}_+ \cdot \bar{\phi}_0}{1-\beta\cos\eta_-} \end{aligned}\right\} \tag{17}$$

where
$$K \equiv \frac{V}{4\pi\epsilon R Z_a c}. \tag{18}$$

In spherical coordinates, by well-known formulas for the curl of a vector, the components of the magnetic $\overline{H} = \nabla \times \overline{A}$ are

$$\left.\begin{aligned} H_R &= \frac{1}{R\sin\theta}\left[\frac{\partial(A_\phi\sin\theta)}{\partial\theta} - \frac{\partial(A_\theta)}{\partial\phi}\right] \\ H_\theta &= -\frac{1}{R\sin\theta}\left[\frac{\partial(A_\phi R)}{\partial R}\cdot\sin\theta - \frac{\partial A_\theta}{\partial\phi}\right] \\ H_\phi &= \frac{1}{R}\left[\frac{\partial(A_\theta R)}{\partial R} - \frac{\partial A_R}{\partial\theta}\right] \end{aligned}\right\}, \tag{19}$$

where A_r, A_θ, and A_ϕ are the R, θ, and ϕ components of the vector $\overline{A}$. Since it may be seen by inspection of Equations (14) and (12) that the components of $\overline{A}$ and their derivatives must contain terms in $1/R$ or its higher powers, so far as the radiation, or $1/R$, fields are concerned, Equation (19) reduces to

$$\left.\begin{aligned} H_R &= 0 \\ H_\theta &= -\frac{\partial A_\phi}{\partial R} \\ H_\phi &= \frac{\partial A_\theta}{\partial R} \end{aligned}\right\}. \tag{20}$$

By Equation (14),

$$\left.\begin{aligned} A_\phi &= \frac{V}{4\pi Z_a}(\bar{a}_+ \cdot \bar{\phi}_0)\int_0^{l_1}\frac{dl}{r} \\ A_\theta &= \frac{V}{4\pi Z_a}(\bar{a}_+ \cdot \bar{\theta}_0)\int_0^{l_1}\frac{dl}{r} \end{aligned}\right\}. \tag{21}$$

Correct in terms of $1/R$, the essential variable in Equations (21) is l, which by Equation (12) contains t and R in such a way that

$$\frac{\partial}{\partial R} = -\frac{1}{c}\frac{\partial}{\partial t} \tag{21a}$$

and

$$H_\theta = \frac{1}{c}\frac{\partial A_\phi}{\partial t},$$

$$H_\phi = -\frac{1}{c}\frac{\partial A_\theta}{\partial t}.$$

By Equations (5) and (15), Equation (21) becomes:

$$H_R = 0,$$
$$H_\theta = -\frac{E_\phi}{Z_0} \tag{22}$$
$$H_\phi = \frac{E}{Z_0},$$

where Z_0 is the so-called impedance of freespace:

$$Z_0 \equiv \mu_0 c = \sqrt{\frac{\mu_0}{\epsilon_0}}. \tag{22a}$$

It is clear that Equation (22) holds whenever $\bar{A}$ is of the form of Equation (14). On reaching the upper end of the antenna, the original current wave no longer radiates; the radiation then comes from the reflected wave.* At point P the radiation field due to the original wave is zero up to time $t = R/c$, whereupon it jumps to the full values given in Equations (17) and (22), and remains constant until time corresponding to $l_1 = L$ in Equation (11), namely,

$$t_{12} = \frac{L}{v} + \frac{R - L\cos\eta_+}{c}$$
$$= \frac{R}{c} + \frac{L}{v}(1 - \beta\cos\eta_+). \tag{23}$$

At this time, the onset of the fields from the return current wave in the wire occurs.

B. Fields from the Second Antenna Wave

If l_2 is the instantaneous position of the downward current wavefront, a time $L/v - (L - l_2)/c$ has elapsed since time zero. After a propagation time delay $(R - l_2 \cos\eta_+)/c$, the radiation arrives at the observer at time

$$t = \frac{L}{v} + \frac{L - l_2}{v} + \frac{R - l_2\cos\eta_+}{c} \tag{24}$$

whence

$$l_2 = -v\frac{t - R/c - 2L/v}{1 + \beta\cos\eta_+}. \tag{25}$$

*The integrals in Equations (7) and (8) may then be written as the sum of two integrals: $\int_0^L$ for the first wave, plus $\int_{l_2}^L$ for the reflected wave. The first integral gives rise to no terms in $1/R$.

Instead of Equations (13) and (14), we then have

$$\left.\begin{aligned} \Phi &= \frac{V}{4\pi\epsilon_0 Z_a v}\int_{l_2}^{L}\frac{dl}{r} - \frac{V(t - R/c)}{4\pi\epsilon_0 Z_a R} \\ A &= \frac{V(-\bar{a}_+)}{4\pi Z_a}\int_{l_2}^{L}\frac{dl}{r} \end{aligned}\right\} . \tag{25a}$$

In a manner similar to the previous case,

$$\left.\begin{aligned} \nabla\Phi &= -\frac{V}{4\pi\epsilon_0 R Z_a v}\nabla l_2 + \frac{V\bar{R}_0}{4\pi\epsilon_0 R Z_a c} \\ \mu\frac{\partial\bar{A}}{\partial t} &= -\frac{V\beta}{4\pi\epsilon_0 R Z_a c}\frac{\bar{a}_+}{1 + \beta\cos\eta_+} \end{aligned}\right\} , \tag{26}$$

where, by Equation (24)

$$\nabla l_2 = \frac{\beta\bar{R}_0}{1 + \beta\cos\eta_+} .$$

It follows that

$$\left.\begin{aligned} E_R &= 0 \\ E_\theta &= K\beta\frac{\bar{a}_+ . \bar{\theta}_0}{1 + \beta\cos\eta_+} \\ E_\phi &= K\beta\frac{\bar{a}_+ . \bar{\phi}_0}{1 + \beta\cos\eta_+} \end{aligned}\right\} . \tag{27}$$

The magnetic fields are given by Equation (22).

C. *Fields from the Third Antenna Wave*

At point P, Equation (27) no longer describes the field after the time corresponding to $l_2 = 0$ in Equation (24), namely, $t = 2L/v + R/c$ when fields from the third current wave in the antenna arrive. If l_3 is the instantaneous position of this third current wave, the corresponding radiation field arrives at P at time

$$t = \frac{2L}{v} + \frac{l_3}{v} + \frac{R - l_3\cos\eta_+}{c} \tag{27a}$$

whence

$$l_3 = \frac{v(t - 2L/v - R/c)}{1 - \beta\cos\eta_+} . \tag{28}$$

Referring to Table I we find, in analogy to Equations (13) and (14), that

$$\left.\begin{aligned} \Phi &= -\frac{V}{4\pi\epsilon_0 Z_a v}\int_0^{l_3}\frac{dl}{r} + \frac{V(t - R/c)}{4\pi\epsilon_0 Z_a R} \\ \bar{A} &= -\frac{V\bar{a}_+}{4\pi Z_a}\int_0^{l_3}\frac{dl}{r} \end{aligned}\right\} . \tag{29}$$

The electric fields, which can be obtained by inspection from Equation (17) are:

$$\left.\begin{aligned} E_R &= 0 \\ E_\theta &= K\beta \frac{\bar{a}_+ \,.\, \bar{\theta}_0}{1 - \beta \cos \eta_+} \\ E_\phi &= K\beta \frac{\bar{a}_+ \,.\, \bar{\phi}_0}{1 - \beta \cos \eta_+} \end{aligned}\right\} . \tag{30}$$

D. Fields from the Fourth Antenna Wave

At point P, Equation (30) no longer describes the field after the time corresponding to $l_3 = L$ in Equation (27a), namely,

$$t = R/c + 2L/v + (L/v)(1 - \beta \cos \eta_+), \tag{31}$$

when fields from the fourth wave begin to arrive. If l_4 is the instantaneous position of this wave, the corresponding radiation arrives at P at time

$$t = \frac{3L}{v} + \frac{L - l_4}{v} + \frac{R - 1_4 \cos \eta_+}{c} \tag{32}$$

whence

$$l_4 = \frac{-v(t - R/c - 4L/v)}{1 + \beta \cos \eta_+}. \tag{33}$$

Referring to Table I in analogy to Equations (13) and (14), we find

$$\left.\begin{aligned} \Phi &= -\frac{V}{4\pi\epsilon_0 Z_a v}\int_{l_4}^{L} \frac{dl}{r} + \frac{V(t - R/c)}{4\pi\epsilon_0 Z_a R} \\ \bar{A} &= \frac{V\bar{a}_+}{4\pi Z_a}\int_{l_4}^{L} \frac{dl}{r} \end{aligned}\right\} . \tag{34}$$

Comparing with Equations (25) and (27) for the second wave, we find

$$\left.\begin{aligned} E_R &= 0 \\ E_\theta &= -K\beta \frac{a_+ \,.\, \bar{\theta}_0}{1 + \beta \cos \eta_+} \\ E_\phi &= -K\beta \frac{a_+ \,.\, \bar{\phi}_0}{1 + \beta \cos \eta_+} \end{aligned}\right\} . \tag{35}$$

The magnetic fields can be obtained by using Equation (22). At point P, Equation (35) no longer describes the field after a time corresponding to $l_4 = 0$ in Equation (32), namely,

$$t = R/c + 4L/v. \tag{36}$$

At this time, fields from the fifth current wave arrive, these fields being the same as for the first wave, and so on.

The fields from the various waves in the antenna are shown in Table III.

TABLE III

FIELDS FROM ANTENNA

Wave	Time interval	E_R	E_θ	E_ϕ
First	R/c to $R/c + (L/v)(1 - \beta \cos \eta_+)$	0	$-K\beta \dfrac{\bar{a}_+ \cdot \bar{\theta}_0}{1 - \beta \cos \eta_+}$	$-K\beta \dfrac{\bar{a}_+ \cdot \bar{\phi}_0}{1 - \beta \cos \eta_+}$
Second	$R/c + (L/v)(1 - \beta \cos \eta_+)$ to $R/c + 2L/v$	0	$K\beta \dfrac{\bar{a}_+ \cdot \bar{\theta}_0}{1 + \beta \cos \eta_+}$	$K\beta \dfrac{\bar{a}_+ \cdot \bar{\phi}_0}{1 + \beta \cos \eta_+}$
Third	$R/c + 2L/v$ to $R/c + 2L/v + (L/v)(1 - \beta \cos \eta_+)$	0	$K\beta \dfrac{\bar{a}_+ \cdot \bar{\theta}_0}{1 - \beta \cos \eta_+}$	$K\beta \dfrac{\bar{a}_+ \cdot \bar{\phi}_0}{1 - \beta \cos \eta_+}$
Fourth	$R/c + 2L/v + (L/v)(1 - \beta \cos \eta_+)$ to $R/c + 4L/v$	0	$-K\beta \dfrac{\bar{a}_+ \cdot \bar{\theta}_0}{1 + \beta \cos \eta_+}$	$-K\beta \dfrac{\bar{a}_+ \cdot \bar{\phi}_0}{1 + \beta \cos \eta_+}$

E. Combined Fields of Antenna and Image

Referring to Table II and retracing the steps used in finding the antenna fields gives us the image fields by inspection (see Table IV).

TABLE IV

FIELDS FROM IMAGE

Wave	Time interval	E_R	E_θ	E_ϕ
First	R/c to $R/c + (L/v)(1 - \beta \cos \eta_-)$	0	$K\beta \dfrac{\bar{a}_- \cdot \bar{\theta}_0}{1 - \beta \cos \eta_-}$	$K\beta \dfrac{\bar{a}_- \cdot \bar{\phi}_0}{1 - \beta \cos \eta_-}$
Second	$R/c + (L/v)(1 - \beta \cos \eta_-)$ to $R/c + 2L/v$	0	$-K\beta \dfrac{\bar{a}_- \cdot \bar{\theta}_0}{1 + \beta \cos \eta_-}$	$-K\beta \dfrac{\bar{a}_- \cdot \bar{\phi}_0}{1 + \beta \cos \eta_-}$
Third	$R/c + 2L/v$ to $R/c + 2L/v + (L/v)(1 - \beta \cos \eta_-)$	0	$-K\beta \dfrac{\bar{a}_- \cdot \bar{\theta}_0}{1 - \beta \cos \eta_-}$	$-K\beta \dfrac{\bar{a}_- \cdot \bar{\phi}_0}{1 - \beta \cos \eta_-}$
Fourth	$R/c + 2L/v + (L/v)(1 - \beta \cos \eta_-)$ to $R/c + 4L/v$	0	$K\beta \dfrac{\bar{a}_- \cdot \bar{\theta}_0}{1 + \beta \cos \eta_-}$	$K\beta \dfrac{\bar{a}_- \cdot \bar{\phi}_0}{1 + \beta \cos \eta_-}$

TABLE V

COMBINED FIELDS

Phase	Time Interval	E_R	$E_\theta/K\beta = H_\phi(Z_0/K\beta)$	$E_\theta/K\beta = - H_\theta(Z_0/K\beta)$
1	R/c to $R/c + (L/v)(1 - \beta \cos \eta_+)$	0	$-\frac{\bar{a}_+ \cdot \bar{\theta}_0}{1 - \beta \cos \eta_+} + \frac{\bar{a}_- \cdot \bar{\theta}_0}{1 - \beta \cos \eta_-}$	$-\frac{\bar{a}_+ \cdot \bar{\phi}_0}{1 - \beta \cos \eta_+} + \frac{\bar{a}_- \cdot \phi_0}{1 - \beta \cos \eta_-}$
1 +	$R/c + (L/v)(1 - \beta \cos \eta_+)$ to $R/c + (L/v)(1 - \beta \cos \eta_-)$	0	$\frac{\bar{a}_+ \cdot \bar{\theta}_0}{1 + \beta \cos \eta_+} + \frac{\bar{a}_- \cdot \bar{\theta}_0}{1 - \beta \cos \eta_-}$	$\frac{\bar{a}_+ \cdot \bar{\phi}_0}{1 + \beta \cos \eta_+} + \frac{\bar{a}_- \cdot \bar{\phi}_0}{1 - \beta \cos \eta_-}$
2	$R/c + (L/v)(1 - \beta \cos \eta_-)$ to $R/c + 2L/v$	0	$\frac{\bar{a}_+ \cdot \bar{\theta}_0}{1 + \beta \cos \eta_+} - \frac{\bar{a}_- \cdot \bar{\theta}_0}{1 + \beta \cos \eta_-}$	$\frac{\bar{a}_+ \cdot \bar{\phi}_0}{1 + \beta \cos \eta_+} - \frac{\bar{a}_- \cdot \bar{\phi}_0}{1 + \beta \cos \eta_-}$
3	$R/c + 2L/v$ to $R/c + 2L/v + (L/v)(1 - \beta \cos \eta_+)$	0	(Negative of phase 1)	(Negative of phase 1)
3 +	$R/c + 2L/v + (L/v)(1 - \beta \cos \eta_+)$ to $R/c + 2L/v + (L/v)(1 - \beta \cos \eta_-)$	0	(Negative of phase 1 +)	(Negative of phase 1 +)
4	$R/c + 2L/v + (L/v)(1 - \beta \cos \eta_-)$ to $R/c + 4L/v$	0	(Negative of phase 2)	(Negative of phase 2)

Table V shows the combined fields of the antenna and image, giving due regard to the time intervals or "phases" in which the various solutions are valid. In general, owing to the nonanalytic nature of the excitation current and the differences in travel time, six different functions are required to describe the complete cycle of fields observed at given point.

Together with the geometric relations of Equations (3) and (4), Table V constitutes the formal solution for the radiation fields. It may be seen that the electric and magnetic fields are purely transverse and at right angles to each other.

IV. Discussion of the Solution

A. Duration of the Phases

The length of the time intervals during which the observed fields are due to the various phases can be obtained from Table V. On the coordinate planes

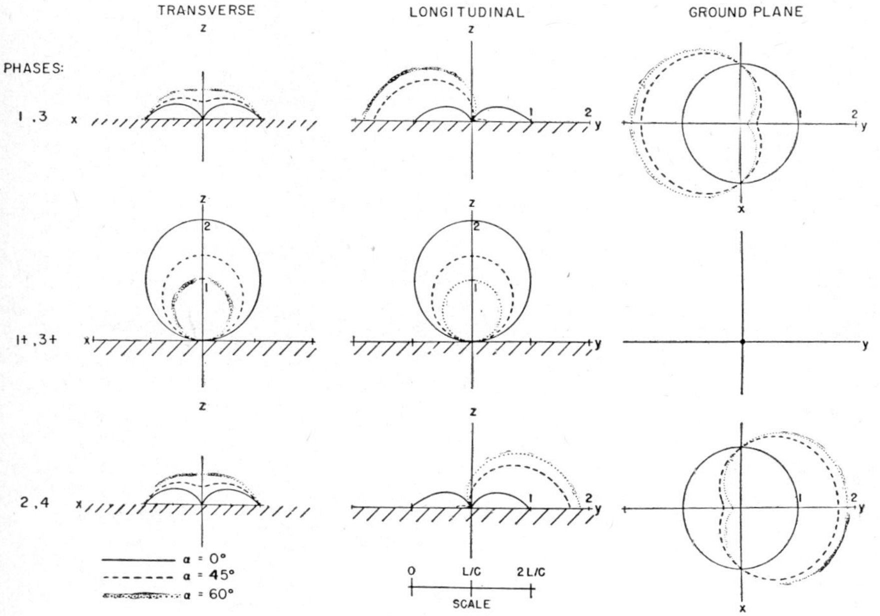

Fig. 3. Duration of the phases in the coordinate planes parametric in α.

XOZ, YOZ and XOY, the durations simplify (see Table VI). Figure 3 shows polar plots of these durations, in units of L/c, for an antenna with zenith angles

$\alpha = 0°$, 45°, and 60°. A larger-scale plot of the durations on the important plane $\theta = 90°$ is shown in Fig. 4.

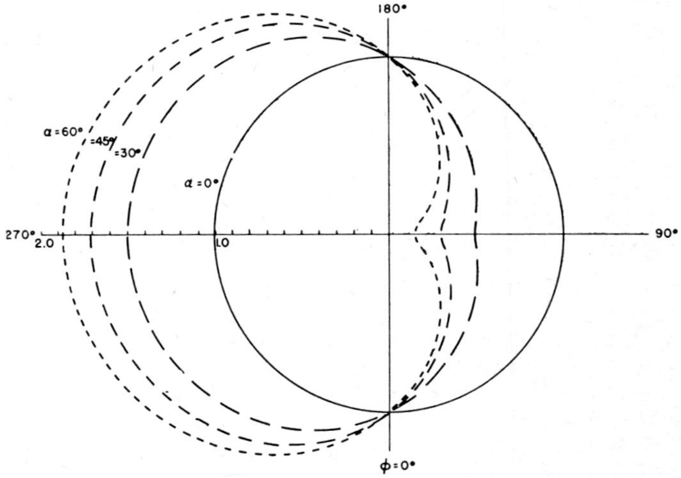

FIG. 4. Duration of phase 1 in plane $\theta = 90°$ ($\beta = 1$) (units of L/c).

TABLE VI

DURATION OF PHASES ON COORDINATE PLANES

Phase	$\phi = 0°, 180°$	$\phi = {+90° \atop -90°}$	$\theta = 90°$
1	$(L/v)(1 - \beta \cos\theta \cos\alpha)$	$(L/v)[1 - \beta \cos(\theta \mp \alpha)]$	$(L/v)(1 - \beta \sin\phi \sin\alpha)$
1 +	$(2L/c)(\cos\theta \cos\alpha)$	$(2L/c)(\cos\theta \cos\alpha)$	0
2	$(L/v)(1 - \beta \cos\theta \cos\alpha)$ (same as phase 1)	$(L/v)[1 - \beta \cos(\theta \pm \alpha)]$	$(L/v)(1 + \beta \sin\phi \sin\alpha)$
3	(same as phase 1)	(same as phase 1)	(same as phase 1)
3 +	(same as phase 1 +)	(same as phase 1 +)	(same as phase 1 +)
4	(same as phase 1)	(same as phase 2)	(same as phase 2)

B. The E_ϕ Fields

Using Equations (3) and (4), the alternate forms of E_ϕ shown in Table VII are obtained. For $\beta \to 0$, these fields all vanish. Also, for $\beta < 1$ the E_ϕ fields vanish

TABLE VII

ALTERNATE EXPRESSIONS FOR E_ϕ

Phase	E_ϕ $(0° < \theta < 180°)$
1	$-2K\beta \sin\alpha \cos\phi \cdot \dfrac{\beta \cos\alpha \cos\theta}{(1 - \beta \cos\eta_+)(1 - \beta \cos\eta_-)}$
1 +	$2K\beta \sin\alpha \cos\phi \cdot \dfrac{1 + \beta \cos\alpha \cos\theta}{(1 + \beta \cos\eta_+)(1 - \beta \cos\eta_-)}$
2	$-2K\beta \sin\alpha \cos\phi \cdot \dfrac{\beta \cos\alpha \cos\theta}{(1 + \beta \cos\eta_+)(1 + \beta \cos\eta_-)}$
3	(negative of phase 1)
3 +	(negative of phase 1 +)
4	(negative of phase 2)

TABLE VIII

E_ϕ ON THE TRANSVERSE PLANE $(\phi = 0°)$

Phase	Time Interval	E_ϕ $(0° < \theta < 180°)$
1	R/c to $R/c + (L/v)(1 - \beta \cos\theta \cos\alpha)$	$-\dfrac{K\beta^2 \sin 2\alpha \cos\theta}{1 - \beta^2 \cos^2\theta \cos^2\alpha}$
1 +	$R/c + (L/v)(1 - \beta \cos\theta \cos\alpha)$ to $R/c + (L/v)(1 + \beta \cos\theta \cos\alpha)$	$\dfrac{2K\beta \sin\alpha}{1 + \beta \cos\alpha \cos\theta}$
2	$R/c + (L/v)(1 + \beta \cos\theta \cos\alpha)$ to $R/c + 2L/v$	(same as phase 1)
3	$R/c + 2L/v$ to $R/c + 2L/v + (L/v)(1 - \beta \cos\theta \cos\alpha)$	(negative of phase 1)
3 +	$R/c + 2L/v + (L/v)(1 - \beta \cos\theta \cos\alpha)$ to $R/c + 2L/v + (L/v)(1 + \beta \cos\theta \cos\alpha)$	(negative of phase 1 +)
4	$R/c + 2L/v + (L/v)(1 + \beta \cos\theta \cos\alpha)$ to $R/c + 4L/v$	(negative of phase 1)

everywhere when $\alpha \rightarrow 0$,* and may therefore be regarded as being due directly to the antenna tilt. By Table VII, the E_ϕ fields vanish on the ground plane $\theta = 90°$ no matter what the tilt, as might be expected in view of the boundary condition. (Phases 1+ and 3+ actually do not exist in the case $\theta = 90°$, as may be noted from Table VI.)

As might be expected from symmetry, the E_ϕ fields also vanish on the $\pm$ YOZ plane ($\phi = \pm 90°$), except that singularities may occur for $\theta = \alpha$.

Table VIII gives the E_ϕ fields on the transverse plane XOZ ($\phi = 0$).

On the plane $-$ XOZ ($\phi = 180°$), E_ϕ has the same magnitude but opposite sign. The magnitudes of $|E_\phi|/K$ in the transverse plane are shown in Fig. 5

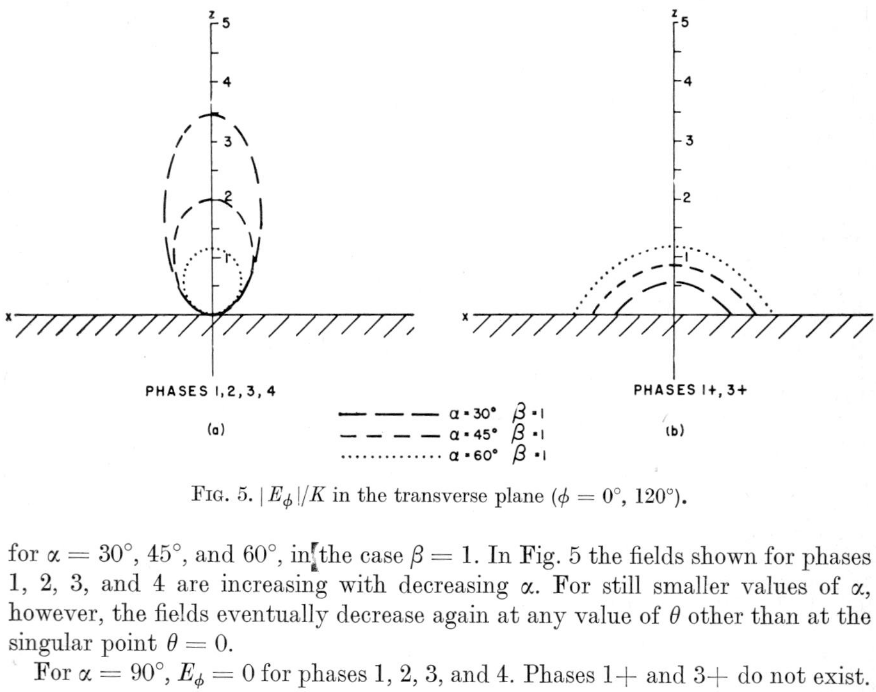

FIG. 5. $|E_\phi|/K$ in the transverse plane ($\phi = 0°$, 120°).

for $\alpha = 30°$, 45°, and 60°, in the case $\beta = 1$. In Fig. 5 the fields shown for phases 1, 2, 3, and 4 are increasing with decreasing α. For still smaller values of α, however, the fields eventually decrease again at any value of θ other than at the singular point $\theta = 0$.

For $\alpha = 90°$, $E_\phi = 0$ for phases 1, 2, 3, and 4. Phases 1+ and 3+ do not exist.

C. The E_θ Fields

1. *Special Case*: $\beta \rightarrow 0$

By Table V, the E_θ fields during phases 1+ and 3+ have the magnitude

$$|E_\theta| = 2K\beta \sin\alpha \cos\theta \sin\phi, \quad \beta \rightarrow 0. \tag{37}$$

The duration $(2L/c) \cos\theta \cos\alpha$ of these phases becomes negligible compared with the duration $(L/\beta c)$ of phases 1, 2, 3, and 4 as $\beta \rightarrow 0$.

*The statement is true even for $\beta = 1$ except that a singular point develops at $\theta = 0$.

The magnitude of the E_θ fields during phases 1, 2, 3, and 4 is

$$|E_\theta| = 2K\beta \cos\alpha \cos\theta, \quad \beta \to 0 \tag{38}$$

or by Equations (18) and (16a),

$$|E_\theta| = (\mu/4\pi)(\sin\theta/R)(dM'_i/dt) \tag{39}$$

where

$$M'_i = (2V/Z_a)vt \cos\alpha \tag{40}$$

is the vertical component of the total current "moment" of the antenna and its image. Equation (39) is equivalent to the conventional expressions [3] for the radiation E fields for a differential dipole in freespace.

2. *E_θ on the Coordinate Planes*

In the general case ($\beta < 1$) the nature of the θ fields can be understood by examining them in the coordinate planes, for which the expressions of Table V reduce to those of Tables IX, X, and XI.

TABLE IX

E_θ ON THE TRANSVERSE PLANE ($\phi = 0°, 180°$)

Phase	Time Interval	E_θ $(0° < \theta < 180°)$
1	R/c to $R/c + (L/v)(1 - \beta\cos\theta\cos\alpha)$	$\dfrac{2K\beta\sin\theta\cos\alpha}{[1 - \beta\cos\theta\cos\alpha][1 + \beta\cos\theta\cos\alpha]}$
1 +	to $R/c + (L/v)(1 + \beta\cos\theta\cos\alpha)$	0
2	to $R/c + 2L/v$	(negative of phase 1)
3	to $R/c + 2L/v + (L/v)(1 - \beta\cos\theta\cos\alpha)$	(negative of phase 1)
3 +	to $R/c + 2L/v + (L/v)(1 + \beta\cos\theta\cos\alpha)$	0
4	to $R/c + 4L/v$	(same as phase 1)

TABLE X

E_θ ON THE LONGITUDINAL PLANE ($\phi = \pm 90°$)

Phase	Time Interval	E_θ ($0° < \theta < 180°$)
1	R/c to $R/c + (L/v)\,[1 - \beta \cos(\theta \mp \alpha)]$	$2K\beta \cos\alpha \dfrac{\sin\theta \mp \beta \sin\alpha}{[1 - \beta\cos(\theta \mp \alpha)][1 + \beta\cos(\theta \pm \alpha)]}$
1 +	$R/c + (L/v)\,[1 - \beta \cos(\theta \mp \alpha)]$ to $R/c + (L/v)\,[1 + \beta \cos(\iota \mp \alpha)]$	$\pm 2K\beta \sin\alpha \dfrac{\cos\theta + \beta\cos\alpha}{[1+\beta\cos(\theta - \alpha)][1 + \beta\cos(\theta + \alpha)]}$
2	$R/c + (L/v)\,[1 + \beta \cos(\iota \mp \alpha)]$ to $R/c + 2L/v$	$-2K\beta \cos\alpha \dfrac{\sin\theta \pm \beta\sin\alpha}{[1 + \beta\cos(\theta \mp \alpha)][1 - \beta\cos(\theta \pm \alpha)]}$
	$R/c + 2L/v$ to $R/c + 2L/v + (L/v)\,[1 - \beta\cos(\theta \pm \alpha)]$	(negative of phase 1)
+	$R/c + 2L/v + (L/v)\,[1 - \beta\cos(\theta \pm \alpha)]$ to $R/c + 2L/v + (L/v)\,[1 + \beta\cos(\theta \pm \alpha)]$	(negative of phase 1 +)
4	$R/c + 2L/v + (L/v)\,[1 + \beta\cos(\theta \pm \alpha)]$ to $R/c + 4L/v$	(negative of phase 2)

(a) On the transverse plane ($\phi = 0°$, $180°$),

$$|\,E_\theta\,| = \frac{2K\beta_1 \sin\theta}{1 - \beta_1^2 \cos\theta} \tag{41}$$

where

$$\beta_1 \equiv \beta\cos\alpha. \tag{42}$$

For $\beta_1 \geqq 1/\sqrt{2}$, $|\,E_\theta\,|$ has the maximum value

$$|\,E_\theta\,|_{\max} = \frac{K}{\sqrt{(1 - \beta_1^2)}} \tag{43}$$

when

$$\sin_\theta = \frac{\sqrt{(1 - \beta_1^2)}}{\beta_1} \tag{44}$$

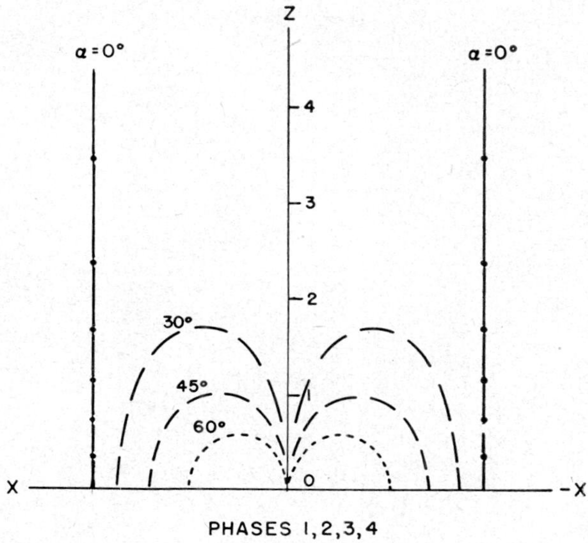

FIG. 6. $|E\phi|/K$ on the transverse plane ($\beta \approx 1$).

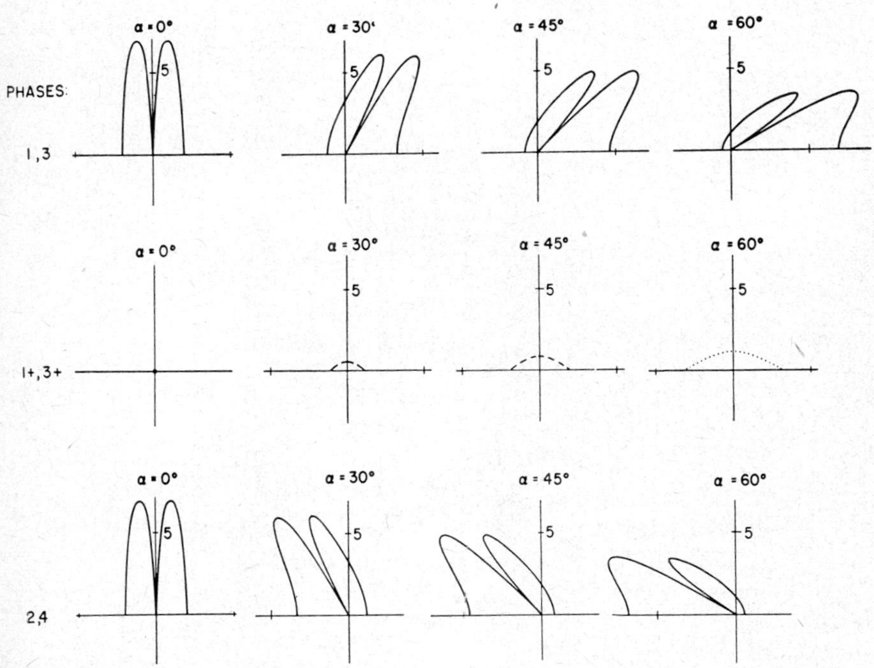

FIG. 7. Polar patterns of $|E\phi|/K$ in the longitudinal plane $\phi = \pm 90°$ ($\beta = 0{\cdot}99000$).

At $\theta = 0$, $E_\theta = 0$ for all $\beta < 1$. The behavior of $| E_\theta |/K$ on the plane $\phi = 0$ in the case $\beta \approx 1$ is illustrated in Fig. 6. For $\alpha = 0$ and $\beta \approx 1$, E_θ is critically sensitive to the value of β, particularly for the smaller values of θ.

(*b*) On the longitudinal plane ($\phi = \pm 90°$) containing the antenna itself, E_θ for phases 1, 2, 3, and 4 is critically sensitive to the value of β, for $\beta \approx 1$, and for values of $\theta \approx \alpha$. In Fig. 7, which shows typical polar patterns for $| E_\theta |/K$, the value $\beta = 0{\cdot}99$ was chosen.

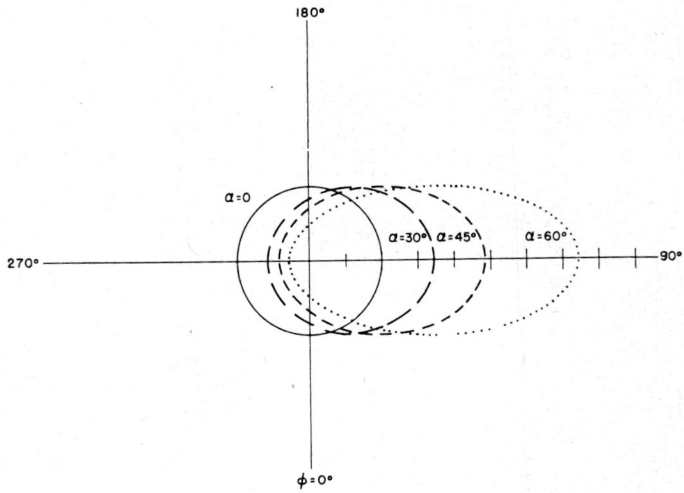

FIG. 8. $| E\phi |/K$ in the Plane $\theta = 90°$, phases 1 and 3 ($\beta \approx 1$).

(*c*) The horizontal plane ($\theta = 90°$) is a particularly interesting one from the standpoint of a possible experiment. By Table XI the phases 1+ and 3+ are nonexistent, and the expressions for phases 1, 2, 3, and 4 are simple. Selected field patterns for $\beta \approx 1$ are shown in Fig. 8 for phases 1 and 3. The patterns for phases 2 and 4 are identical except for a 180° rotation about the $\phi = 0°$, 180° axis.

From Table XI it follows that the front-to-back ratio for $\phi = +90°$ to $\phi = -90°$ is

$$\frac{1 + \beta \sin \alpha}{1 - \beta \sin \alpha}. \tag{45}$$

Some typical ratios are given in Table XII.

It will be seen that when the current wave in the antenna has a velocity within 1 % of the velocity c of freespace waves, the front-to-back ratio in the $\alpha = 60°$ case is at least 13 : 1. Even for a current wave 10 % slower than c, the front-to-back ratio is larger than 7 : 1.

TABLE XI

E_θ ON THE HORIZONTAL PLANE ($\theta = 90°$)

Phase	Time Interval	E_θ
1	R/c to $R/c + (L/v)(1 - \beta \sin \phi \sin \alpha)$	$\dfrac{2K\beta \cos \alpha}{1 - \beta \sin \phi \sin \alpha}$
1 +	(phase 1 + non-existent)	
2	$R/c + (L/v)(1 - \beta \sin \phi \sin \alpha)$ to $R/c + 2L/v$	$-\dfrac{2K\beta \cos \alpha}{1 + \beta \sin \phi \sin \alpha}$
3	$R/c + 2L/v$ to $R/c + 2L/v + (L/v)(1 - \beta \sin \phi \sin \alpha)$	(negative of phase 1)
3 +	(phase 3 + non-existent)	
4	$R/c + 2L/v + (L/v)(1 - \beta \sin \phi \sin \alpha)$ to $R/c + 4L/v$	(negative of phase 2)

TABLE XII

FRONT-TO-BACK RATIOS

Tilt Angle	Front-to-Back Ratios		
$\alpha(°)$	$\beta = 1$	$\beta = 0·99$	$\beta = 0·9$
0	1·00	1·00	1·00
30	3.00	2·96	2·64
45	5·83	5·68	4·51
60	13·9	13·0	7·26

By using the equations in Table VI for the durations of the phases, and equations in Table XI for the magnitudes, waveforms for points on the ground plane ($\theta = 90°$) can be readily constructed. Some sample waveforms for the case $\beta \approx 1$ are shown in Fig. 9. The effect of the various tilt angles is readily seen,

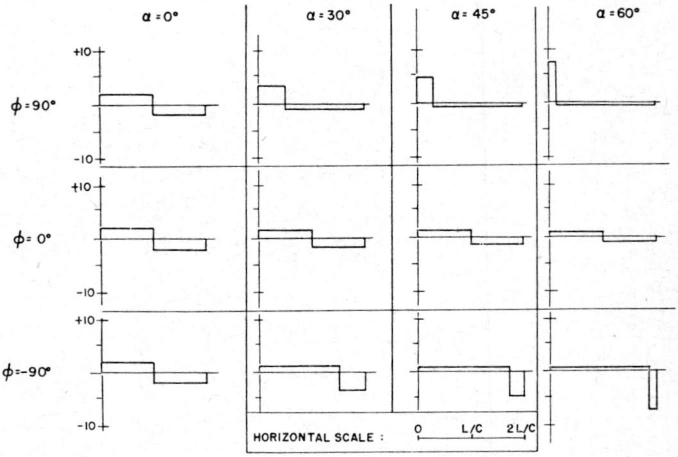

FIG. 9. Sample E_θ/K waveforms in the plane $\phi = 30°$ ($\beta \approx 1$).

the distortion in the $\alpha = 60°$ case being very marked. From Tables VI and VII it will be noted that on the $\theta = 90°$ plane the product of field strength in a phase times the duration of the phase has the constant value $2K(L/c) \cos \alpha$.

By Equation (18) and Table XI numerical values for E_θ on the $\theta = 90°$ plane for the case $\beta \approx 1$ and $\alpha = 0$ can be obtained from the relation

$$|E_\theta| = 60 \frac{I}{R} \tag{46}$$

in which E_θ will be in millivolts/meter if $I = V/Z_a$ is in amperes, and R is in kilometers. For example, at a distance of 1000 km, a current of 1000 amp would give a field of 60 mV/m.

Acknowledgments

Dr. E. L. Hill, Dr. M. M. Newman, and Dr. H. Poeverlein provided stimulating discussions and suggestions during the course of this work. Professor Hill was especially helpful in pointing out an oversight in the original draft.

References

1. Newman, M. M. (1957). *VLF Symposium* at Nat. Bur. Stand., Boulder, Colorada.
2. Stratton, J. A. "Electromagnetic Theory" (1st edn.), p. 248. McGraw-Hill, New York (1941).
3. Ramo, S. & Whinnery, S. "Fields and Waves in Modern Radio," p. 431, Equation (4). Wiley (1944).

Pulse Phase-change Signaling in the Presence of Ionospheric Multipath Distortion

S. G. Lutz, F. A. Losee and A. W. Ladd

Communications Division, Hughes Aircraft Company, Los Angeles, Calif.

Cine recordings of millisecond rectangular pulses show multipath stratching and sectional interference fading which agree with the concept of gradually changing path lengths. The $d\phi/dt$ for individual paths is negligible compared to the information phase-changes and all apparent phase discontinuities result from multipath interference. Phase-change information is recovered by a running phase comparison of successive pulses. Test results show superiority over frequency-change keying, even at kilobit rates.

I. Introduction

One can communicate digital information by changing either the amplitude, frequency, or phase of a wave. Phase keying permits transmission of maximum power, as with frequency shift keying, without requiring excess bandwidth. Use of phase keying for long-distance communication has been retarded, however, by the phase instability of ionospheric propagation, which characteristically involves multiple paths of different and gradually changing lengths.

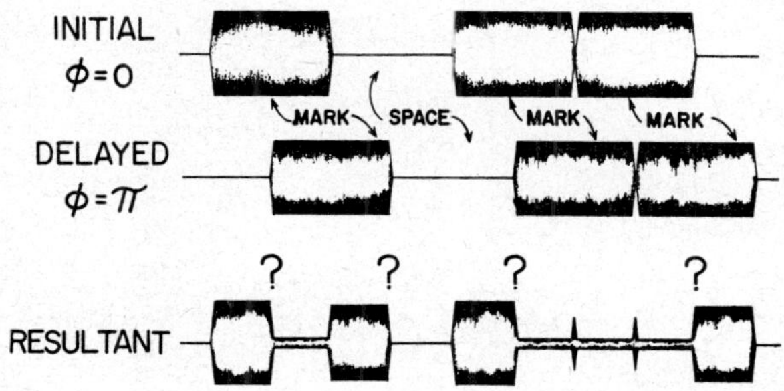

Fig. 1.

Interference fading is the most familiar manifestation of ionospheric multipath distortion, and is the predominant manifestation at slow keying speeds. When one attempts to increase keying speeds or data rates much above 100 bits/sec, for instance, the multipath delay differences come into prominence and tend to "smear" the information bits, thereby confusing the mark/space decisions. To illustrate this, Fig. 1 shows reception of a high-speed single-symbol (on-off) teletype signal over two paths differing in delay by half the duration of a baud (bit); the two components have nearly equal amplitude but opposite phase.

There would be corresponding confusion with frequency or phase keying. Recognition of this smearing seems to have established the practice of reducing the keying speed and thereby lengthening the bauds to combat multipath smearing, relying on diversity reception to control the fading. Acceptance of this practice has retarded the development of high-speed digital communication by directing it primarily toward frequency-division multiplexing of slow-speed channels.

Long bauds are not essential for ionospheric communication. It has long been known [1] that bauds can be as short as desired if they are separated by at least the maximum probable multipath delay difference so they can stretch into this separating gap instead of smearing over subsequent bauds. Exploitation of this technique requires greater knowledge of multipath distortion than has appeared in the literature, especially so if spaced phase-keyed pulses are to be employed. Recognition of this fact motivated the research effort reported in this paper and has resulted in the development of a successful system for pulse-phase digital communication.

II. Short-long Pulse-sounding Technique

Oblique incidence ionospheric sounding techniques [2, 5] provide the necessary information on maximum probable delay differences if the recorded virtual

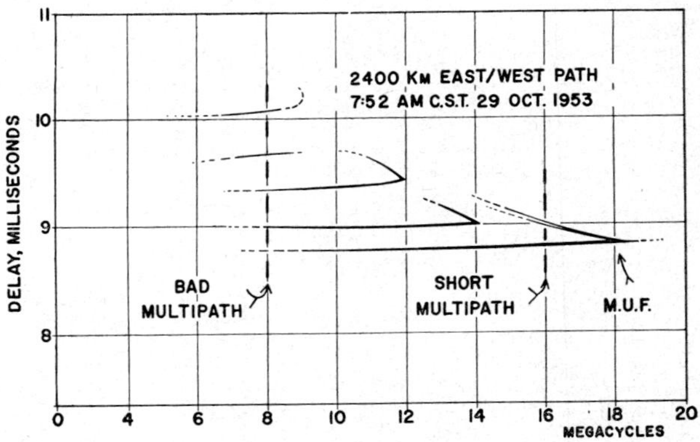

Fig. 2.

heights are converted back to delays. Figure 2 shows a delay-frequency plot corresponding to Sulzer's Fig. 4. One sees that delay differences are greatest and most complex at low frequencies, but that they vanish as the maximum useable frequency is approached. If operation could be maintained just below the maximum useable frequency the multipath protective gap could be only a millisecond or less. Bearing in mind that this recording is only illustrative, and that there can be more multiple reflection modes on longer paths, it is seen that gaps of several milliseconds are required if frequency selection is limited.

Oblique incidence records do not provide all the information needed. They tell the approximate delay separation of the multipath components as a function of frequency at the time the record was made, but they reveal nothing about the relative phases of these components or how they change with time. For communication, one is interested in longer pulses and in how they are distorted by overlapping of the multipath components. This additional information has been obtained through photographically recording long-distance receptions of alternate short and long pulses, as will be described.

Figure 3a shows the wave train transmitted. Each short pulse is followed by a millisecond rectangular pulse, whose leading and trailing edges are as steep as

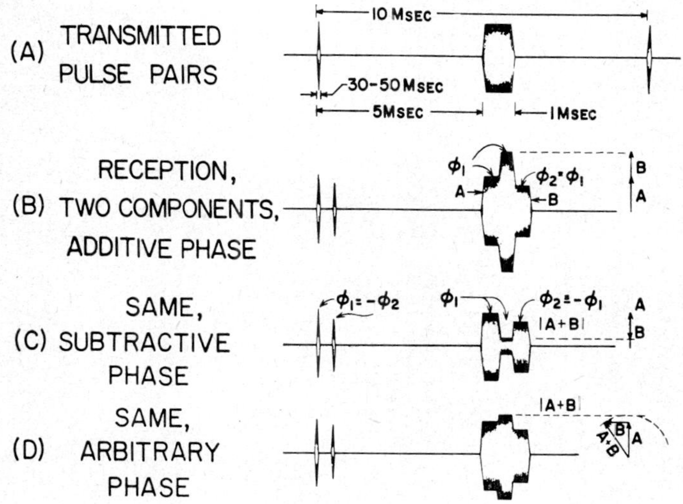

FIG. 3.

those of the short pulse. The purpose of the short pulse is to resolve the multipath components and to establish their relative amplitudes as in oblique incidence sounding, therefore, these pulses should be as short as practicable. Limited transmitter bandwidth and power determined a minimum width of approximately 20 μsec, which had to be increased for tests over the longer circuits to increase the energy per pulse. The long pulses permit these multipath components to overlap in time and, therefore, to beat and fade in accordance with their changing phase. Figure 3b represents the received wave with two components in phase and displaced 0·5 millisec. The long pulse has been stretched an amount equal to this delay difference. Its amplitude at the beginning and end corresponds to the short-pulse amplitudes, whereas the sum of these amplitudes is attained over the central half millisec during the period that the long-pulse components overlap and are in phase. In Fig. 3c the short pulses appear exactly the same as the beginning and end of the long pulse, but the amplitude of the center portion is now the amplitude difference of the two components, showing that the short pulses have reached phase opposition. Figure 3d represents an intermediate condition with the components in phase quadrature, as shown in the associated phasor diagram.

Although Fig. 3 has been over-simplified to achieve clarity, one recognizes the interference fading tends to be less serious with millisecond pulse signaling than conventional long-pulse telegraphy because of the relatively greater pulse stretching and reduced overlap. Consequently, space diversity reception becomes unnecessary.

III. Test Records and Analyses

Cine recordings were made of these short-long pulse receptions and the time-fading cycles were plotted from film frame measurements. Figure 4 shows such a

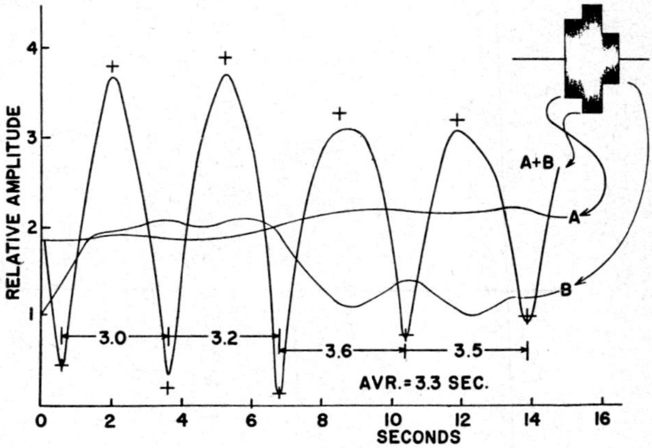

Fig. 4.

plot for a typical sample of two-component reception. The fading cycle is essentially sinusoidal. One notes that the first arriving component (A) is more stable in amplitude and presumably, therefore, more stable in phase than the second (B) component. These components probably are the low and high rays of one-hop F_2 propagation, and greater interference between the unresolved O and X magnetoionic components of the high ray may account for its instability. The $+$ marks denote the sum and difference of A and B and they coincide well with the minima of $|A + B|$. Amplifier nonlinearity may explain the $+$'s lying slightly above the maxima. The fading period is $3{\cdot}3 \pm 0{\cdot}3$ sec and corresponds to a Doppler shift difference of $1/3$ c/s. Examination of oblique incidence recordings shows that this shift can reach $\pm$ 2 c/s or more at sunrise and sunset.

Figure 5 shows three-component reception from Dayton with relatively well-defined overlapping of components in the long pulse. One notes that the amplitudes of first and last short-pulses are less than the amplitudes of corresponding sections (beginning and end) of the long pulse because the transmitter bandwidth did not permit full buildup of such short pulses. Figure 6 shows a similar case which is idealized slightly to clarify its interpretation via a phasor diagram. The initial component is represented by the phasor (rotating vector) A, the second by B, and the last by C. Their amplitudes are proportional to the heights of the short pulses, a, b, and c. The phase of A is taken as reference. A $\frac{1}{2}$ millisec following the A component, the B component is superimposed which results in $|A + B|$ of the amplitude shown. The phasor diagram shows that $\angle(A + B) = 78°$* with

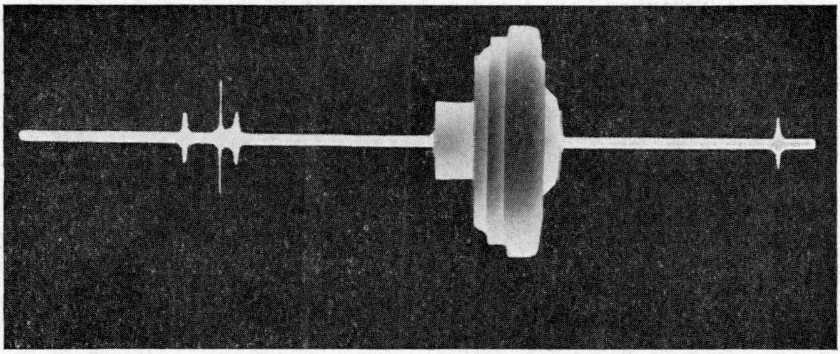

FIG. 5.

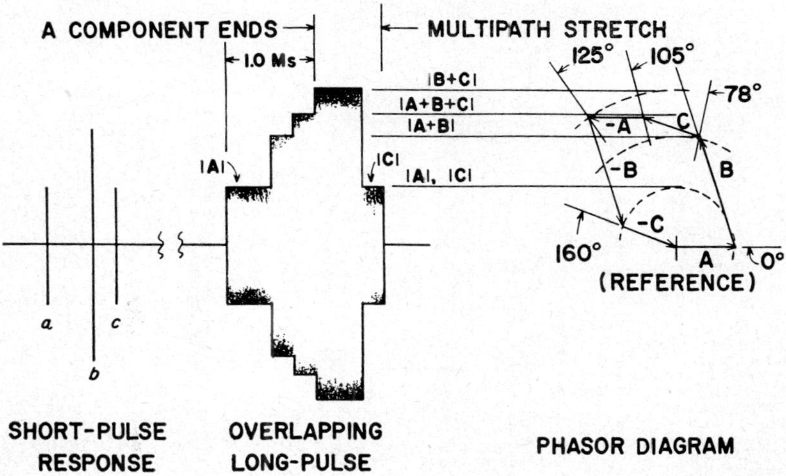

FIG. 6.

respect to the reference, $\angle A = 0$. This pulse section, therefore, would exhibit two-component fading which would be quasi-periodic as previously shown. When the C component is superimposed a $\frac{1}{4}$ millisec later, C adds to $A + B$ at such an angle that $|B + A + C|$ has the value shown for this third section, from which $\angle(A + B + C) = 125°$. This section exhibits three-component fading, which is less periodic. In fact, since these components gradually change amplitude, this three-component fading appears more random than it generally is.

*Actually, it can be $-78°$. The sign is ambiguous unless it is determined by decrease or increase of the delay difference.

Finally, it is also observed that the relative phases of the B and C components themselves are obtainable from the phasor diagram, for the case shown,

$$B = \pm 109° \quad \text{and} \quad C = \pm 160°.$$

Figure 7 shows an expanded-sweep oscillogram of the 20-μsec pulses during typical, clean, three-component propagation.

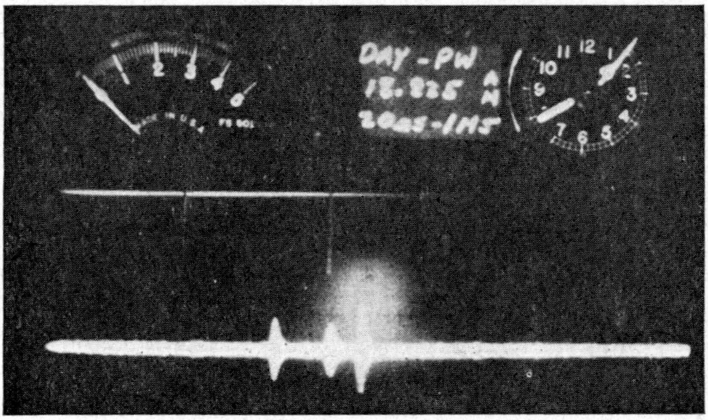

FIG. 7. Three-component short pulses, expended.

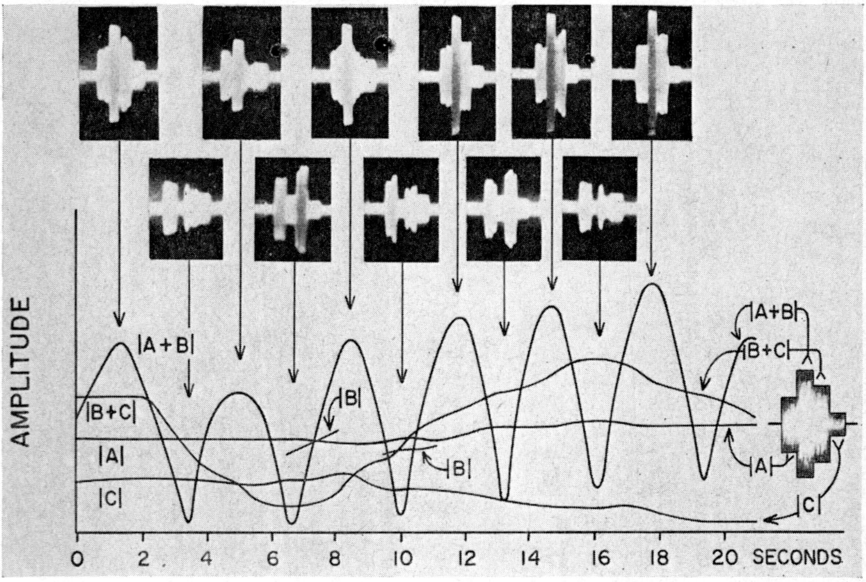

FIG. 8.

Figure 8 shows a sequence of millisecond pulse waveforms and plots of their sectional amplitudes for an example of three-component propagation, which is similar to the above, but interesting in that the delays differed by so nearly a $\frac{1}{2}$ millisec that the A component ended as C started. Therefore, there was no

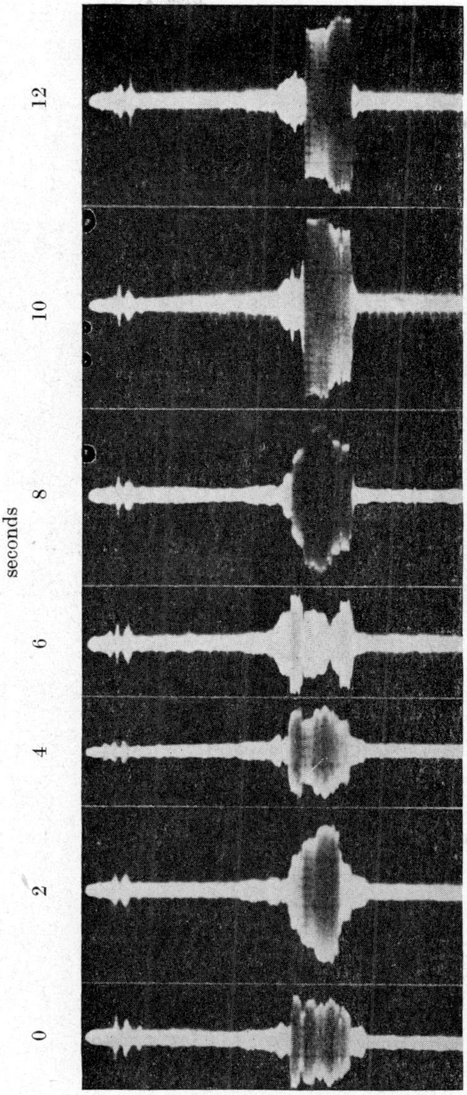

FIG. 9. Complex multipath at 18·825 Mc. Dayton-Los Angeles.

$|A + B + C|$ portion, as in Figs. 5 and 6, and only two brief instants when $|B|$ alone could be measured with considerable uncertainty. This 21-sec recording was made at 18·825 Mc, Dayton to Los Angeles at 13.39, 18 July 1957. One notes that the A component remained essentially constant while C gradually died out.

Throughout the record $| A + B |$ shows characteristic two-component fading, with an average period of 3·2 sec. This again corresponds to a Doppler shift of about $\frac{1}{3}$ c/s between A and B. One infers from the maxima and minima of $| A + B |$, together with $| B + C |$ and $| C |$, that $| B |$ approximated $| A |$ at first but was less constant, increasing considerably toward the end of the recording. From $| B + C |$ one sees that these two components were additive initially and again after about 16 sec, but were in phase opposition at 6 to 7 sec with a broad minimum. Thus, the frequency difference between these components must have averaged less than 0·1 c/s.

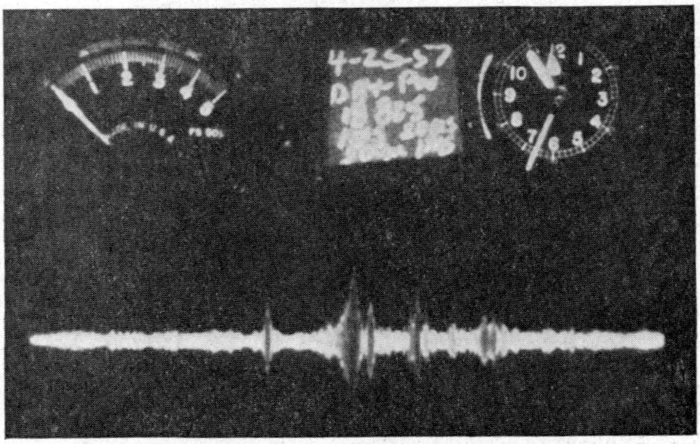

FIG. 10. Expanded short pulse recording, complex multipath.

Figure 9 shows an example of multipath propagation too complex, smeared, and noisy for satisfactory analysis by this technique. This record also was made at 18·825 Mc, Dayton to Los Angeles at 11.53, 25 April 1957. The seven long-pulse waveforms shown were approximately 2 sec apart. Figure 10 is a recording of the 20-μsec pulse made a few seconds later. The preceding long-pulse waveforms are at least consistent with this short-pulse structure. On watching expanded sweeps of these short pulses, one frequently observed fine-structure phenomena not otherwise evident or explainable, such as scatter-broadening and presence of both O and X magnetoionic components.

Sharply defined multipath components are not always observed. Smearing is common, especially with multihop propagation over long paths. Scattering from the earth reflections probably accounts for much of this smearing because both land and ocean surfaces are diffuse reflectors. Consequently, energy reaches the receiver via reflection from the rough earth in some area surrounding the point of geometric reflection. As a result, there is no single path length and delay time but a distribution of delay about a mean delay time. This causes even the shortest pulses to be broadened when received. Turbulence of the ionosphere may account for similar smearing observed even with single-hop propagation.

A short film has been assembled of clippings from recordings of Dayton transmissions received at Los Angeles (1660 nautical miles) and at Honolulu (3200 nautical miles). These show the character of pulses received with increasingly complex multipath structure.

A single-trace oscilloscope was used for all recordings. This permitted an expanded-sweep presentation of either the short-pulse or long-pulse waveforms, but only a compressed presentation of both simultaneously. For further work, a dual-trace oscilloscope display of the type shown in Fig. 11 is recommended. If a dual-beam tube is used, one can correct for any incomplete build-up of the short

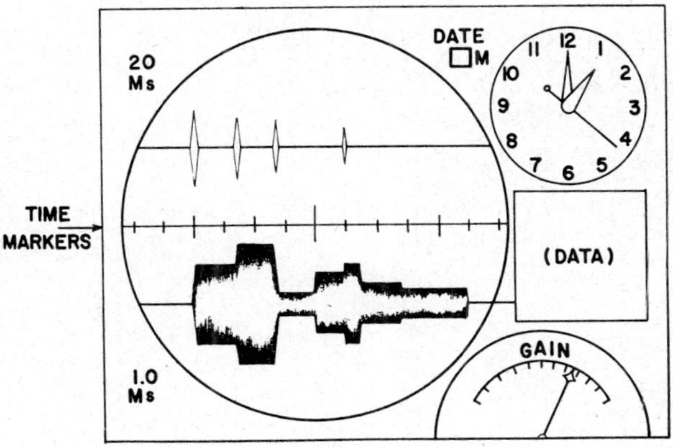

FIG. 11.

pulses by amplification of the short-pulse trace until the initial and final component amplitudes of both traces agree. One can then use the amplitudes of the intermediate short pulses for analysis. It also is helpful to be able to request change of the long-pulse width to suit observed delay differences. For the example of Fig. 8, a longer pulse would have given $|A + B + C|$, typifying the fading of a CW signal, whereas a shorter pulse would have shown $|B|$.

IV. Multipath Propagation Conclusions

Only a small fraction of the footage recorded during these tests has been analyzed to determine rates of change of phase difference or delay difference; partly because interfering signals, noise, and smearing made analysis too difficult. Chiefly, one learned what to look for in the films and thus obtained adequate qualitative information which supported the following conclusions:

(1) Multipath components may have well-defined delays or may be smeared in delay by as much as 0·1 millisec, diffuse reflection being the presumed cause of this smearing.

(2) Individual components fade gradually and nonperiodically without the abrupt deep fades characteristic of their resultant in a CW signal and without

apparent correlation between components. The amplitude of the initial component is generally the most stable.

(3) The beginning and end of the received long pulse correspond to the initial and final multipath components, and show only the above single-component gradual fading.

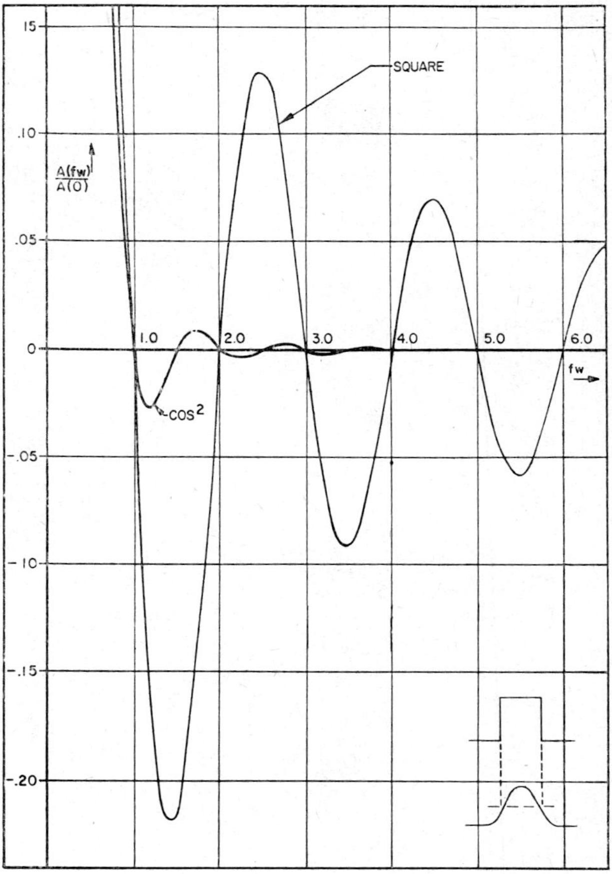

FIG. 12.

(4) Sections of the long pulse in which only two components over-lap show simple interference fading in agreement with the phasor sum of the two components. This fading is not random and deviates from being cyclic only to the extent that the component amplitudes vary. The depth of this fading depends on the difference between these component amplitudes and becomes complete only when both components are of equal amplitude and opposite phase.

(5) When three or more components overlap, their resultant still agrees with their phasor sum, which changes in a less apparent cyclic manner. This addition is modified of course by the gradually changing amplitudes of the individual components, so that the resultant amplitude variation in these sections of the long pulse may seem random.

(6) The periodicity of two-component fading confirms that delays of these components never change abruptly, and probably not more rapidly than would agree with results of oblique incidence soundings. Consequently, the rate of change of phase in single components is correspondingly gradual, with no abrupt phase changes.

(7) As a consequence, the phase of the resultant of two or more components cannot change abruptly except as the resultant amplitude goes through zero.

(8) With relatively unsmeared components (of definite delay times), there can be correspondingly abrupt changes of phase from section to section of the received rectangular long pulse, as successive components start and stop. There is correspondingly little correlation between the phases of these sections.

(9) For pulse repetition rates which are high compared to the fading rates of any of these sections (i.e., repetition rates of 100 pulses/sec or above), corresponding sections of successive pulses differ in phase only slightly, except as a section fades through zero between successive pulses. Even in this latter case, the amplitudes of this section, immediately before and after the complete fade, are low compared to other sections of the pulse.

This final conclusion was the one needed to assure feasibility of the pulse phase-change signaling system to be described. First, however, it is recognized that the rectangular pulses thus far considered have broad frequency spectra, undesirable for practical communication because of interference to adjacent channels. Furthermore, narrow-band reception destroys their rectangularity and obscures the sectional multipath effects described. To conserve bandwidth and control adjacent channel interference, pulse envelopes approximating $\cos^2$ shape are preferred, as evident from the enlarged comparison of these spectra in Fig. 12. This pulse shaping also obscures the sectional multipath effects described. Nonetheless, conclusion (9) above still applies, in that the multipath-produced phase changes between corresponding parts of successive pulses is small regardless of whether the nonrectangularity results from pulse shaping or from narrow-band reception or both.

V. Development of Phase-change System

We turn next to the utilization of this propagation information in a practical phase-keyed communication system. In comparison with other methods, digital phase modulation combines the bandwidth advantages of on-off amplitude modulation with the dual-symbol detection advantages of frequency shift keyed FM, leading to least probability of error at any given signal carrier to noise ratio. Figure 13 shows Losee's [3] comparison of amplitude, frequency, and phase modulation in the presence of Rayleigh-distribution fading, derived from

Montgomery's work [4]. Though the preceding study of multipath pulse distortion leads one to question whether pulse fading follows a Rayleigh distribution, it is at least a safely pessimistic assumption. Less severe fading leads to much better error rates, especially at s/n ratios of 10 db or more, and phase modulation retains its superiority to the other systems regardless of fading.

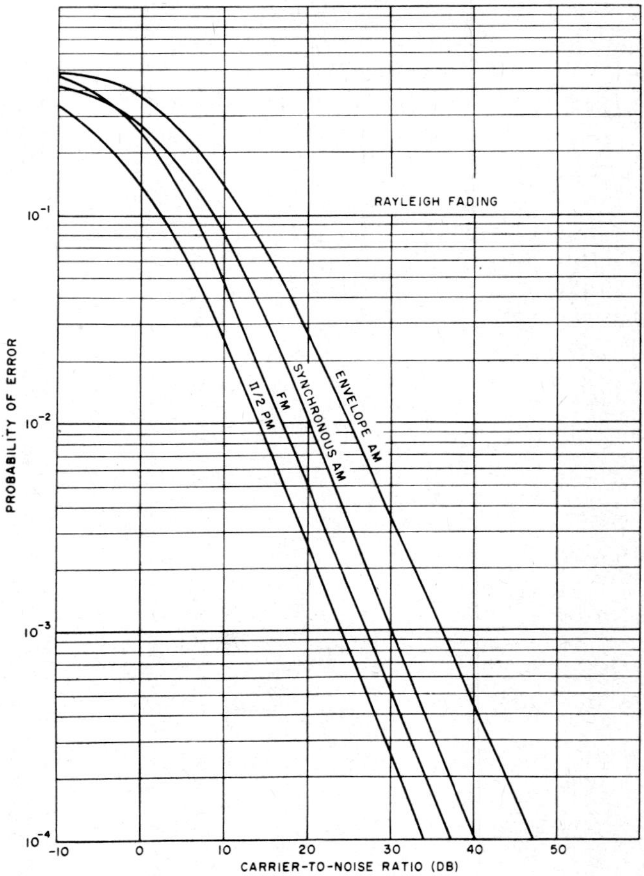

FIG. 13.

Unfortunately, Montgomery's and most other comparisons have assumed detection against an accurate phase reference available at the receiver. Even neglecting the impracticality of attaining the necessary oscillator stabilities at frequencies in the order of 10 Mc, it has been shown that ionospheric instability and multipath distortion rule out use of a fixed phase reference. However, since the ionospheric phase changes of each propagation component are very slow in comparison with the information phase changes keyed into the signal, a remedy lies in the use of a "running" phase reference derived from the received signal, with each pulse serving as the reference for the next one. With such a system, information is conveyed by phase *change* rather than by phase *state*.

A major problem in developing a satisfactory spaced-pulse, phase-change detection system lies in obtaining the pulse-to-pulse phase comparison while

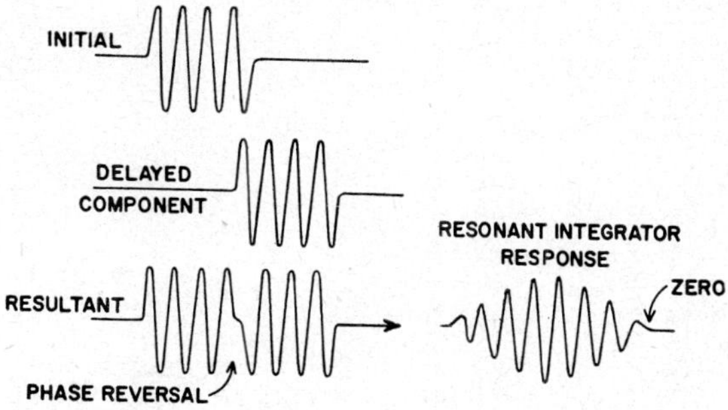

FIG. 14.

ignoring the multipath-produced phase changes from beginning to end of the received pulses. Perhaps one thinks of "remembering" the average phase of the preceding pulse with a resonant integrator of infinite-Q, as with an earlier long-baud system, but the multipath phase changes during an otherwise strong and useful pulse can cause it to integrate to zero, as shown in a simplified manner by Fig. 14.

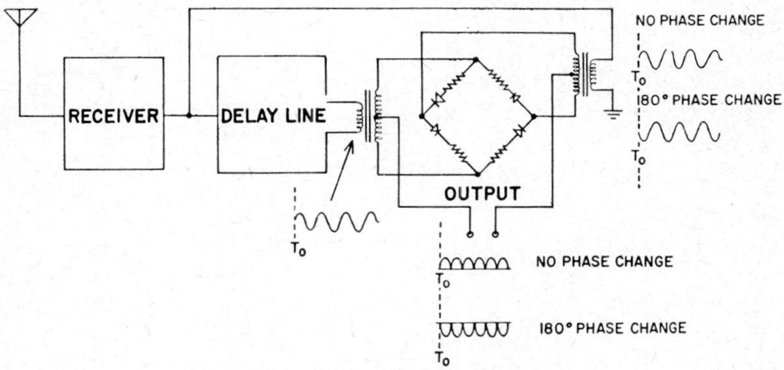

FIG. 15.

One needs to perform a continuing phase comparison or correlation between each section of a pulse and the corresponding section of the preceding reference pulse. This has been accomplished by accurately delaying each pulse a full period then comparing it in a phase detector with each succeeding pulse, as indicated in Fig. 15. Instant by instant, this detector output follows the cosine of the phase difference between its two inputs. Therefore, it is positive when the pulses are in phase, negative when they are of opposite phase, and zero only for phase-quadrature. This detection technique is not impaired by use of shaped pulses or by narrow-band filtering, both of which smear the multipath structure unrecognizably but similarly for successive pulses.

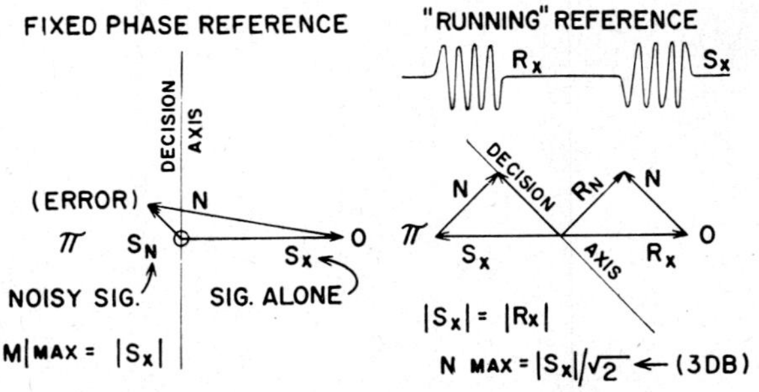

FIG. 16.

Delaying one pulse into accurate phase coincidence with the succeeding one presented obvious difficulties which retarded this project until stable broad-band magnetostrictive delay lines became available. Even with a perfect line of constant delay, any frequency error produces a corresponding phase error. For example, with a 5 millisec delay, a 50 c/s frequency error produces a 90° phase error, irrespective of the conversion frequency employed. One is limited in reducing this frequency sensitivity through reducing the delay because of the pulse spacing required for multipath protection. Though ionospheric Doppler shifts are negligible, 50-cycle Doppler is experienced at 25 Mc in Mach-2 aircraft. Manual correction via vernier adjustment of the conversion frequency is simple, and automatic frequency control can be provided by addition of a quadrature phase detector as disclosed by Losee[3].

Phase-change modulation with pulse-to-pulse comparison is slightly penalized in comparison with an ideal phase-reference system such as was assumed in Fig. 11, in that both pulses contribute noise, which is generally uncorrelated. For the phase-reversal case thus far considered, one perhaps recognizes that this doubling of noise energy should entail a 3 db s/n penalty. However, it may be helpful to confirm and clarify this condition by phasor diagrams, Fig. 16, admitting the limitations of this means of representing effects of noise. For an accurate noise-free reference, the left-hand diagram shows that a decision error will be made only if the noise, N, exceeds and opposes the noise-free signal, S_x, enough to throw the resultant noisy signal S_n, across the decision axis. Assuming no error in the phase reference, so that S_x is perpendicular to the decision axis, as shown, errors can occur only if $| N^x_{\max} | \geqslant | S_x |$.

For the running phase-reference case shown at the right, the decision axis is established by the noisy reference, R_n. This axis may be displaced $\pm$ 45° from the noise-free reference, R_x, in the limiting case shown. Remembering that the noise-free signal, S_x, must be exactly in or out of phase with R_x, one sees that noise of equal magnitude, $|N|$, can place the noisy signal, S_n, on the decision axis. Thus, from the 45° right-angle relationship shown, $|N_{\max}|$ can be 3 db less than $|R_x|$ or $|S_x|$. This 3 db certainly represents a reasonable loss from theoretical perfection in obtaining a workable multipath-resistant system with the other advantages of phase modulation.

One is tempted to explore transmission of more than one information bit per pulse by resorting to smaller phase-change increments, perhaps more so than with digital frequency-shift or amplitude modulation with which the limitations of multibit coding are more evident. One recognizes that a 4-phase system can be regarded as a phase-quadrature multiplex of two 2-phase systems, requiring double power (3 db) but no increase in bandwidth. Either with a fixed or running phase reference, the two quadrature channels should be independent and, therefore, their error rates at given s/n ratios should not deteriorate. A four-phase system operating at the same power and bandwidth as a 2-phase system, but at twice the bit-rate, therefore should be only 3 db less resistant to noise. In practice, the 4-phase running-reference system is apt to suffer more than this 3-db penalty because of its tighter frequency tolerance, greater difficulty of automatic frequency control, tighter linearity requirements, and similar problems which will require further development effort.

Attempting to achieve 3 bits per pulse would require eight 45° phase states, 4 bits per pulse would require 16 of 22·5° each, etc. Analyses of these and higher bit cases, which assumed fixed phase reference, no fading, and Gaussian noise, confirmed that there would be an increasing s/n penalty which would lead to prohibitive error rates except over very *clean* low-noise circuits.

VI. Test Results

The phase-change systems have been tested in comparison with an equivalent pulse frequency shift system (also termed FM). The latter designates each binary digit either by a pulse on a *mark* (1) channel or by a pulse on a companion *space* (0) channel.

Pulse phase-change modulation first was tested as a single-channel 2-phase experimental system from Thule, Greenland to Los Angeles in the fall of 1957. During these tests it consistently yielded lower error rates than an equivalent pulse frequency shift system. These encouraging results prompted the development of more versatile equipment for use in subsequent tests.

This test system employs four channels each 500 cycles wide and separated 400 cycles. The transmitted pulses are approximately bell-shape, or "rounded triangles," 1 millisec wide at half voltage, and occurring every 4 millisec. The same equipment provided either frequency shift or phase-change modulation, with the latter being readable either as 2-phase or 4-phase. For phase-change modulation, all four channels were pulsed simultaneously to deliberately create maximum interchannel interference and to test correlation of simultaneous errors at the closely spaced frequencies. The same channels were pulsed in mark/space pairs simultaneously for the frequency shift tests. The transmitter employed a linear amplifier which was operated at about 500 W, or about 100 W per phase keyed channel.* The same or only slightly more power per channel was used for frequency shift tests. The frequency shift information rate was half that of the 2-phase system and a quarter that of the 4-phase system, all in the same total

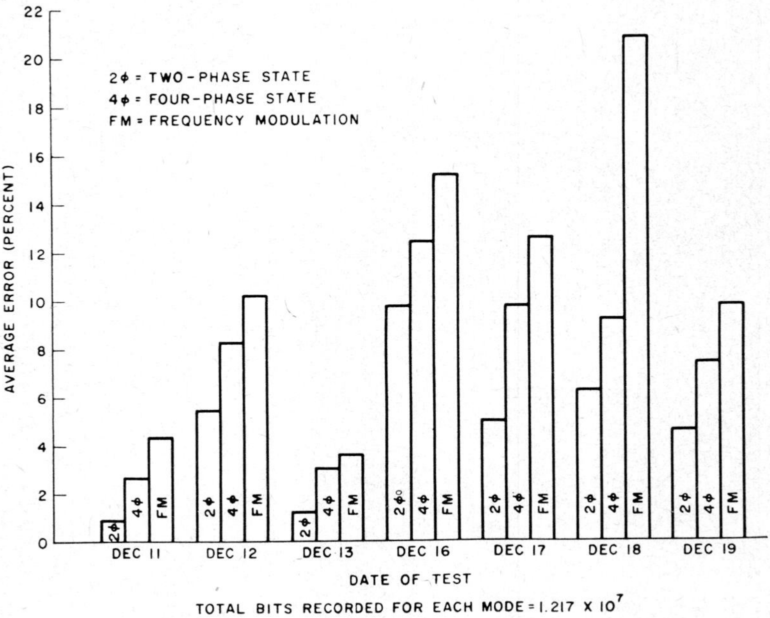

FIG. 17.

3·5 kc bandwidth. A single rhombic antenna of approximately 12 db gain was used for reception (no diversity).

This system was first tested from Dayton, Ohio, to Los Angeles, a distance of about 3000 miles (3300 km). A 7-day series of tests was conducted to compare the three types of modulations at high error rates. This was accomplished by operating through the dawn, on frequencies with bad noise and multipath, by reducing transmitter power and even by operating off the back of a directive antenna. The day-by-day results, Fig. 17, consistently showed lowest error rates with 2-phase, highest with frequency shift, 4-phase being intermediate.

The better relative performance of 4-phase in these tests than in subsequent ones may have resulted partly from the better linearity at reduced power.

Next, an 8-hour test of the phase systems was made at normal power and with a

*Theoretically, four frequencies of equal peak voltage, E, will beat to peaks of $4E$, or peak powers proportional to $(4E)^2$. Thus, for maximum linearity, the power per channel should have been limited to 1/16th, the peak linear power capability of the transmitter. In practice, this can lead to weak-signal errors worse than the nonlinearity errors. Consequently, the power per channel was adjusted for minimum two-phase error rate. This condition probably raised the 4-phase error rate.

directive transmitting antenna (steerable Alford) of about 12-db gain. Total bits and errors were counted for each 40-sec period simultaneously for 2-phase and 4-phase. Figure 18 shows that no 4-phase scores had errors exceeding 5 %, while 80 % showed errors of 0·5 % or less. Taken alone, these results would have seemed good, but such is not the case when they are compared with the much better 2-phase results. Here, no scores showed errors exceeding 0·5 %, all but 10 %

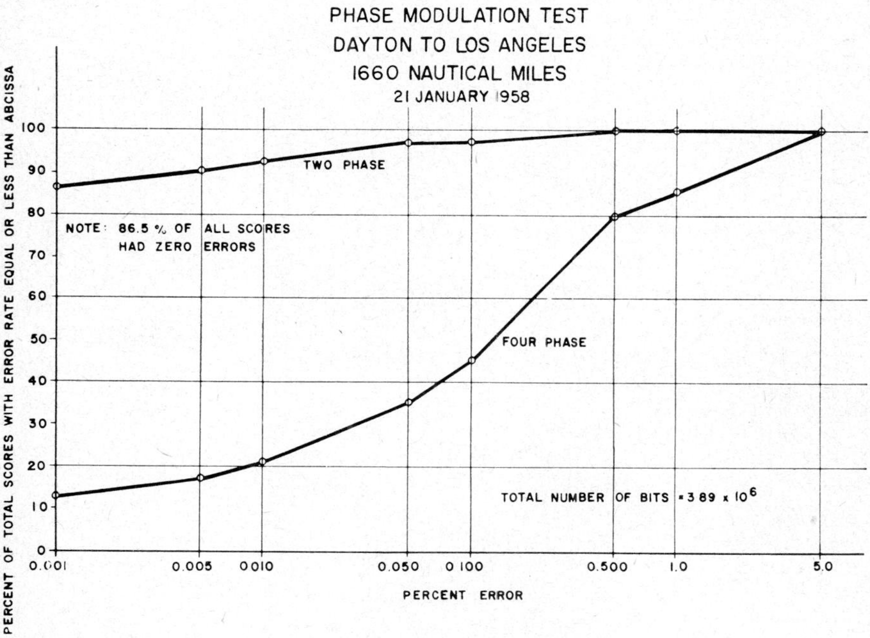

FIG. 18.

showed errors of 0·005 % or less, and 86·5 % had zero errors. Most of these few errors occurred during the dawn transition, that is, after sunrise at Dayton but while it was still dark in Los Angeles.

Following these tests the transmitting equipment was installed in a test aircraft which was flown to Puerto Rico for tests from there and beyond. The location of these ground tests was 2970 nautical miles (5500 km) from the Los Angeles receiving site, and most test flights were made in a direction to increase this distance to a maximum of nearly 4000 nautical miles, or more than twice the distance between Dayton and Los Angeles.

Figure 19 shows the percentage distribution of bit error rates during the 14 periods of airborne operation, these periods being about 2 hours apart throughout the 24 hours in order to include any diurnal effects. Each period included 7 min. of the phase modulation and 3 min. of frequency shift. One notes that all of the phase scores had error rates less than 5 %, with 2 % or less errors 85 % of the periods and 1 % or less errors 57 % of the periods. For the 22 periods of ground operation, no scores exceeded 5 % error, although the error rate was 1 % or less for only 50 % of the periods. Considering the extended range, poor antenna, low power, and the difficulty of keeping experimental equipment in proper adjustment under the conditions prevailing, these results are considered successful.

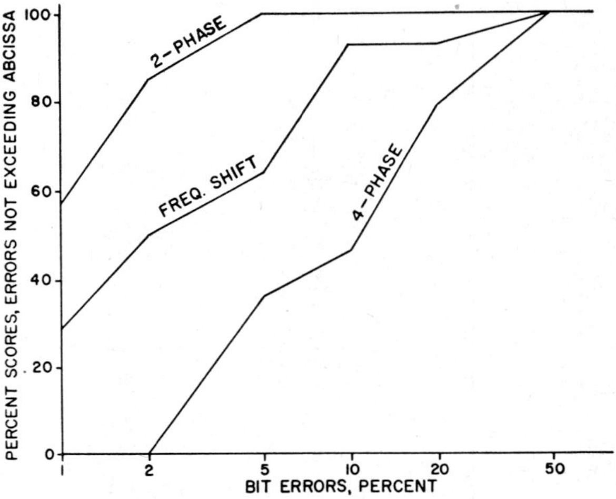

Fig. 19.

All tests have shown that the 2-phase system consistently outperforms the pulse frequency shift system, despite its operating in half the bandwidth for a given bit rate. This superiority over pulse frequency shift results largely from the correlation detection, and is achieved in exchange for greater sensitivity to frequency drift. The less efficient detection in the pulse frequency shift system can tolerate drifts of $\pm$ 100 cycles, so that automatic frequency control is not required.

Though results with the 4-phase system thus far have only been comparable with pulse frequency shift in bit-accuracy, one should consider its four-times greater bit-rate-per-unit bandwidth and the good probability of its further improvement.

Acknowledgment

The authors wish to express their indebtedness to their many coworkers who contributed to this program, and their special thanks to the Communication and Navigation Laboratory, Wright Air Development Center, which sponsored this development and conducted the test transmissions.

References

1. Crosby, U.S. Patent, No. 2,401,618, filed 30 April 1941.
2. Letter by P. G. Sulzer & E. E. Ferguson, *Proc. Inst. Radio Engrs, N.Y.* September 1952.
3. Losee, F. A *Proc. Inst. Radio Engrs, Aeronautical Electronics Conf., 1959.* Dayton, Ohio.
4. Montgomery, G. F. (1954). *Proc. Inst. Radio Engrs, N.Y.* **42**, 447–454.
5. Sulzer, P. G. (1955). *J. geophys. Res.* 411–420.

Low and Medium Frequency Radio Propagation

KENNETH A. NORTON

National Bureau of Standards, Boulder, Colo.

During recent years, extensive progress has been made in understanding the nature of ionospheric propagation in the frequency range from 30 to 1000 kc. In particular, Wait and Conda [30] have developed suitable methods for determining the effect of the curvature of the earth on the illumination of the ionosphere by antennas radiating at angles near and below grazing incidence, Belrose [6] has developed empirical formulas for ionospheric absorption in the range 70 to 250 kc, Watt [33] has analyzed data which makes possible the extension of these formulas to still lower frequencies, and Bean [1] has developed a radio standard tropospheric atmosphere which makes possible good estimates of the bending of the radio waves in propagation to and from the ionosphere. Making use of these results, together with earlier studies made by the author of the absorption of radio waves in the band 540 to 1600 kc and some new results on focusing by the curved surface of a rough ionosphere, predictions are made of the propagation loss expected in ionospheric propagation between short vertical electric dipoles for waves propagated by *m* reflections at the ionosphere. The predictions appear to be in general agreement with the available experimental data, although the paper clearly points up the desirability of more definitive studies of the influence of absorption and polarization, particularly as regards the larger distances which involve propagation by more than one ionospheric reflection. The methods in this paper indicate that the propagation loss increases only very slowly as the angle of departure approaches and goes well below grazing incidence with the earth; this provides an explanation for recent experimental data obtained with a pulse system as reported by Doherty [9] which indicates very large intensities for one-hop sky waves on 100 kc both day and night at ranges up to 1800 miles.

I. TRANSMISSION LOSS IN RADIO SYSTEMS STUDIES

The advantages of the concept of transmission loss in the study of radio systems have been discussed and illustrated in earlier papers by the author [16, 17, 20]. This concept will be used in the present analysis, and it will be convenient to define at the outset the terminology to be used since some additional terms and concepts have been found to be useful in the lower frequency ranges which are the concern of the present paper.

The system loss of a radio circuit consisting of a transmitting antenna, receiving antenna, and the intervening propagation medium is defined as the ratio, p_t/p_a, where p_t is the radio frequency power input to the terminals of the transmitting antenna and p_a is the resultant radio frequency signal power available at the terminals of the receiving antenna. Both p_t and p_a are expressed in watts. The system loss is usually expressed in decibels:

$$L_s = 10 \log_{10} (p_t/p_a) = P_t - P_a \tag{1*}$$

*Throughout this paper, capital letters are used to denote the ratios expressed in decibels, of the corresponding quantities designated with lower-case, type; e.g., $P_t = 10 \log_{10} p_t$ is the input power to the transmitting antenna expressed in decibels above 1 W.

Note that the system loss, as above defined, excludes any transmitting or receiving antenna transmission line losses since it is considered that such losses are readily measurable. On the other hand, the system loss includes all of the losses in the transmitting and receiving antenna circuits, including not only the transmission loss due to radiation from the transmitting antenna and re-radiation from the receiving antenna, but also any ground losses, dielectric losses, antenna loading coil losses, terminating resistor losses in rhombic antennas, etc. The inclusion of all of the antenna circuit losses in the definition of system loss provides a quantity which can always be accurately measured and which is directly applicable to the solution of radio system design problems.

For many applications, however, particularly in the study of radio wave propagation, it is convenient to have a definition of a system loss which excludes all of the antenna circuit losses except those associated with the antenna radiation resistances; thus, the transmission loss of a radio circuit consisting of a transmitting antenna, receiving antenna, and the intervening propagation medium is defined as the dimensionless ratio, p_r/p'_a, where p_r is the radio frequency power radiated from the transmitting antenna, and p'_a is the resultant radio frequency signal power which would be available from the receiving antenna if there were no circuit losses other than those associated with its radiation resistance. The transmission loss is usually expressed in decibels:

$$L \equiv 10 \log_{10} (p_r/p'_a) = L_s - L_{tc} - L_{rc} \qquad (2)$$

where L_{tc} and L_{rc} are the losses, expressed in decibels, in the transmitting and receiving antenna circuits, respectively, excluding the losses associated with the antenna radiation resistances: $L_{t,rc} = 10 \log_{10} (r'/r)$ where r' is the actual resistive component of the antenna circuit and r is the radiation resistance.

In order to separate the effects of the transmitting and receiving antenna gains and circuit losses from the effects of the propagation, it is convenient to define the basic transmission loss, L_b (sometimes called path loss), as the transmission loss expected between fictitious loss-free isotropic transmitting and receiving antennas at the same locations as the actual transmitting and receiving antennas. This serves also to define the path antenna gain, G_p:

$$G_p \equiv L_b - L \qquad (3)$$

$$L_b = L + G_p = L_s + G_p - L_{tc} - L_{rc}. \qquad (4)$$

In some cases it may be quite difficult to measure the antenna circuit losses; thus it is convenient to define the path antenna power gain, G_{pp}, as:

$$G_{pp} \equiv L_b - L_s = G_p - L_{tc} - L_{rc}. \qquad (5)$$

We see that the path antenna power gain is the increase in the system loss when loss-less isotropic antennas are used at the same locations as the actual antennas; note that G_{pp} will be negative when the antenna circuit losses exceed the path antenna gain.

In some idealized situations the path antenna power gain, G_{pp}, is simply the sum $(G_{tp} + G_{rp})$ of the free space power gains G_{tp} and G_{rp} of the transmitting and receiving antennas relative to loss-less isotropic antennas. However, in most practical situations G_{pp} is less than $G_{tp} + G_{rp}$ because of the complex nature of the received field. The path antenna power gain may be measured by determining the increase in the system loss when both the transmitting and receiving antennas are replaced *simultaneously* by simple standard antennas such as short electric or magnetic dipoles, and then adding calculated path antenna power gain corresponding to the use of the standard antennas. In the case of ionospheric propagation, the path antenna power gain is sometimes substantially smaller than the sum of the free space power gains G_{tp} and G_{rp}; in such cases the path antenna power gain cannot be determined accurately as the sum of the effective power gains of the transmitting and receiving antennas (as determined by replacing first one antenna by a standard antenna and then the other antenna by a standard antenna) since such effective power gains depend upon the gain of the antenna used at the other terminal.

As an example of the simplicity of transmission loss calculations in some cases, we may consider the transmission loss between two isotropic antennas in free space. At a distance d, very much greater than the wavelength λ, the field intensity, expressed in watts per square meter, is simply $p_r/4\pi d^2$ since the power is radiated uniformly in all directions. Since the effective absorbing area of the receiving antenna is $\lambda^2/4\pi$, the available power at the terminals of the loss-free isotropic receiving antenna is given by:

$$p'_a = (\lambda^2/4\pi) \,.\, (p_r/4\pi d^2) \tag{6}$$

consequently, the basic transmission loss in free space may be expressed:

$$L_{bf} = 10 \log_{10} (4\pi d/\lambda)^2 \qquad \text{(Free space; } d \gg \lambda\text{)}. \tag{7}$$

Since the free space gain of a short electric dipole is $g_t = g_r = 1{\cdot}5$, the path antenna gain for two optimally oriented short electric dipoles in free space is:

$$G_p = G_t + G_r = 3{\cdot}52 \text{ db}. \tag{8}$$

Consequently, the transmission loss between two optimally oriented short electric loss-less dipoles in free space is:

$$L = 10 \log_{10} (4\pi d/\lambda)^2 - 3{\cdot}52. \tag{9}$$

II. The Effect on the Transmission Loss of the Antenna Impedance

In order to illustrate the influence of changes in the impedances of the antennas, we will consider the transmission loss between two short vertical electric loss-less dipoles at heights h_t and h_r, respectively, above a plane perfectly conducting surface and separated by a distance large with respect to the wavelength. In free space the field strength e, expressed in volts per meter at a distance d, expressed in meters, in the equatorial plane of a short electric dipole radiating p_r watts is given by:

$$\frac{e^2}{z} = \frac{1{\cdot}5 p_r}{4\pi d^2} \tag{10}$$

where $z = 4\pi c \cdot 10^{-7}$ = impedance of free space expressed in ohms, and $c = 2{\cdot}997925 \cdot 10^{8}$ m/s = velocity of light in free space. Note that the factor (1·5) is just the free space gain of the transmitting dipole antenna. The radiation resistance of a short vertical electric dipole of length l and at a height h_t above a perfectly conducting plane is given by:

$$r = \frac{2\pi z l^2}{3\lambda^2}[1 + \Delta_t] = r_f[1 + \Delta_t] \tag{11}$$

$$\Delta_t = \frac{3}{(2kh_t)^2}\left[\frac{\sin(2kh_t)}{2kh_t} - \cos(2kh_t)\right]. \tag{12}$$

In the above $k = 2\pi/\lambda = 2\pi f/c$, i.e., λ is the wavelength in free space. Note that Δ_t approaches zero at large heights above the surface and r approaches its free space value, r_f. On the other hand, $\Delta_t = 1$ for $h_t = 0$, and the radiation resistance is then just twice its free space value. The field intensity for a short vertical electric dipole over the perfectly conducting surface may be expressed:

$$\frac{e^2}{z} \simeq \frac{1{\cdot}5\, p_r [2\cos^2\psi \cos(k\, h_t \sin\psi)]^2}{4\pi d^2[1 + \Delta_t]}. \tag{13}$$

In the above expression $\tan\psi = h_r/d$ and the distance, d, along the surface must be large with respect to both λ and h_t; in this case the distance between the antennas is approximately $d/\cos\psi$. Equations (11), (12) and (13) were derived by Schelkunoff in Chapters VI and IX of Reference [23]. Since $\Delta_t = 1$ for $h_t = 0$, the field intensity is 3 db greater when $\psi = 0$ for a short vertical electric dipole on the surface of a perfectly conducting plane than for a short dipole in free space. In more familiar units, when $h_t = h_r = 0$, Equation (13) may be expressed

$$e(\mu\text{V/m}) = 2{\cdot}998962 \cdot 10^5 \sqrt{(p_r\,(\text{kW}))}/d(\text{km}). \tag{14}$$

The effective absorbing area of a short vertical electric dipole receiving antenna at a height h_r above the perfectly conducting plane may be expressed:

$$a_e = \frac{1{\cdot}5\lambda^2 \cos^2\psi}{4\pi[1 + \Delta_r]} \tag{15}$$

where Δ_r is defined by Equation (12) with h_t replaced by h_r. Since $p_a = e^2 a_e/z$, we find by combining Equations (13) and (15) that the transmission loss between two short vertical electric loss-less dipoles at heights h_t and h_r above a plane perfectly conducting surface may be expressed:

$$L \simeq 10 \log_{10}\left\{\frac{[4\pi d/\lambda \cos\psi]^2\,[1 + \Delta_t]\,[1 + \Delta_r]}{(1{\cdot}5)^2\,[2\cos^2\psi \cos(kh_t \sin\psi)]^2}\right\} \tag{16}$$

or
$$L = L_{bf} - G_p - 6{\cdot}02 + A + L_{tg} + L_{rg} \equiv L_p + L_{tg} + L_{rg} \tag{17}$$

where $L_{t,rg} = 10 \log_{10}[1 + \Delta_{t,r}]$, $G_p \simeq 20 \log_{10}[(3/2)\cos^2\psi]$ and $A \simeq -20 \log_{10}[\cos(kh_t \sin\psi)]$. The factor $L_{t,rg}$ is shown graphically as a function of (h/λ) in Fig. 1. It is of interest to note that the transmission loss between vertical electric dipoles on a plane perfectly conducting surface, $h_t = h_r = 0$, is the same as if the dipoles were separated by the same distance in free space, although the field intensity is 3 db greater for the same power radiated. Equation (17) illustrates the definition of L_p, the propagation loss, i.e., the transmission loss expected if the antennas had the same impedances they would have had if they were in free space. More generally, L_{tg} and L_{rg} are defined as:

$$L_{t,rg} = 10 \log_{10} (r/r_f) \tag{18}$$

where r is the radiation resistance of the antenna in the presence of a perfectly conducting environment, and r_f is the radiation resistance it would have had if it were in free space.

The above is not the only logically consistent method of describing the gains and losses in the presence of a perfectly conducting ground. Thus Schelkunoff in the reference cited considered the perfectly reflecting surface to be an integral part of the transmitting and receiving antennas, and set $G_{ts} = 10 \log_{10} \{3/[1 + \Delta_t]\}$ and $G_{rs} = 10 \log_{10} \{3/[1 + \Delta_r]\}$. In other words, he referred his maximum gains to those expected for isotropic antennas with the earth removed; this leads to a path antenna gain, referred to that expected between isotropic antennas with the earth removed, given by $G_{ps} = G_p + 6{\cdot}02 - L_{tg} - L_{rg} - A$, and his transmission loss is then simply $L = L_{bf} - G_{ps}$. However, this method of approach is not recommended since (a) it is impracticable to remove the earth in order to measure G_{ps} and (b) it would lead to antenna gains 3 db larger than their free space values even when they are many wavelengths above a perfectly reflecting plane surface, and this is in inconsistent with the present usage of the concept of antenna gain in the higher frequency ranges.

Suppose now that we use small loop antennas of area S, with their axes normal to the plane of propagation, parallel to the perfectly conducting surface and at heights h_t and h_r, respectively. In this case [23]:

$$r'_m = \frac{8\pi^3 z S^2}{3\lambda^4} [1 + \Delta'_r] = r_{mf} [1 + \Delta'_r]. \tag{19}$$

The subscript m refers to the magnetic dipole, and the prime to the effect of the surface.

$$\Delta'_r = (3/2) \left[\left(1 - \frac{1}{(2kh_r)^2} \right) \frac{\sin (2kh_r)}{2kh_r} + \frac{\cos (2kh_r)}{(2kh_r)^2} \right] \tag{20}$$

$$a_e = \frac{\lambda^2 (3/2)}{4\pi [1 + \Delta'_r]} \tag{21}$$

$$G'_p = 20 \log_{10} [3/2] = 3{\cdot}52 \tag{22}$$

$$A' = -20 \log_{10} [\cos (kh_t \sin \psi)] \tag{23}$$

$$L'_{t,rg} = 10 \log_{10} [1 + \Delta'_{t,r}]. \tag{24}$$

The factor $L'_{t,rg}$ is also shown graphically on Fig. 1. Note that Δ'_r approaches zero at large heights and $\Delta'_r = 1$ for $h_r = 0$.

Consider next the transmission loss between two small loop antennas at heights h_t and h_r, respectively, above a perfectly conducting surface with their axes normal to this surface. In this case [23]:

$$r''_m = \frac{8\pi^3 z S^2}{3\lambda^4}[1-\Delta] = r_{mf}[1-\Delta] \tag{25}$$

$$G''_p = 20 \log_{10} [(3/2) \cos^2 \psi] \tag{26}$$

$$A'' = -20 \log_{10} [\sin (kh_a \sin \psi)] \tag{27}$$

$$L''_{t,rg} = 10 \log_{10} [1 - \Delta_{t,r}] \tag{28}$$

The factor $L''_{t,rg}$ is also shown as a function of (h/λ) on Fig. 1.

Finally, consider the transmission loss between two short horizontal electric dipoles of length l, normal to the plane of propagation and at heights h_t and h_r, respectively, above a perfectly conducting plane surface. In this case [23]:

$$r''' = \frac{2\pi z l^2}{3\lambda^2}[1-\Delta'] = r_f[1-\Delta'] \tag{29}$$

$$G'''_p = 20 \log_{10} (3/2) = 3{\cdot}52 \tag{30}$$

$$A''' = -20 \log_{10} [\sin (kh_a \sin \psi)] \tag{31}$$

$$L'''_{t,rg} = 10 \log_{10} [1 - \Delta'_{t,r}]. \tag{32}$$

The factor $L'''_{t,rg}$ is also shown graphically on Fig. 1. Note that L''' and L'' both approach $(-\infty)$ as h approaches zero, but the radiation resistance simultaneously approaches zero, and it would be difficult in practice to keep the radiated power constant as the antennas are brought nearer and nearer to the surface. When h_t and h_r are both much less than a wavelength, $A'' + L_{tg} + L''_{rg}$ for horizontal loops becomes independent of these heights and equal to $20 \log_{10} (kd/10)$; similarly $A''' + L'''_{tg} + L'''_{rg}$ for horizontal electric dipoles approaches $20 \log_{10} (kd/5)$ for h_t and h_r much less than a wavelength.

All of the above results refer to the case of a perfectly conducting plane surface and to distances $d \gg \lambda$. For a finitely conducting ground, the attenuation A in the above expressions will be substantially modified. For example, for vertical electric dipoles over a flat earth of finite conductivity and with $h_t = h_r = 0$:

$$A = -20 \log_{10} | 1 + i \sqrt{(\pi p)} \exp(-p) \operatorname{erfc}(-i\sqrt{p}) |. \tag{33}$$

Here p denotes Sommerfeld's numerical distance as defined in Reference 18 where a comprehensive discussion is given of the radiation fields of electric and magnetic dipoles over a finitely conducting plane earth. Furthermore, Δ and Δ' will be modified when the antennas are located over a finite ground [27, 28, 36, 37], but this difference will largely be cancelled if a large ground screen is used under the antennas. Although Equation (25) indicates that r_m approaches zero as the vertical magnetic dipoles approach the perfectly conducting surface, Wait [27] has shown that r_m actually becomes very large when such loops are brought near a finitely conducting ground.

When we note that L_{tg}, L_{rg}, L_{tc} and L_{rc} may vary substantially from one antenna installation to the next, depending upon polarization, ground conditions, whether or not a ground screen is used, and upon other environmental factors such as the presence of trees or over-head wires, it becomes clear why it is

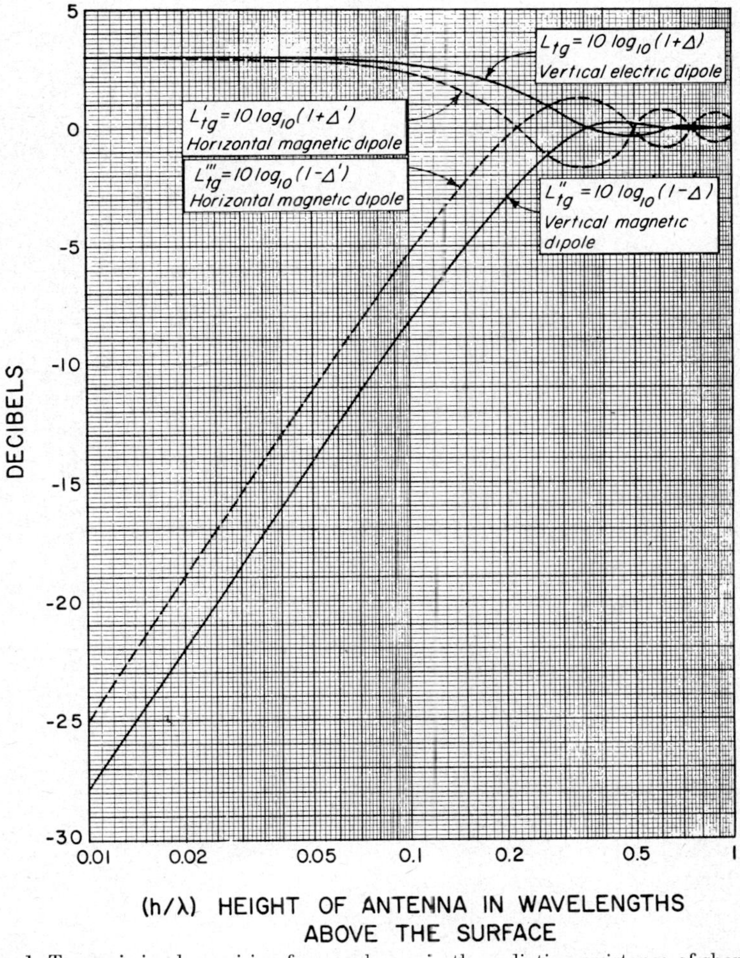

FIG. 1. Transmission loss arising from a change in the radiation resistance of short dipole antennas near a perfectly conditioned surface.

desirable to separate these components from the system loss and to have a propagation loss, L_p, independent of these antenna environmental conditions. Under many circumstances, however, particularly at the higher frequencies where the antennas are usually located several wavelengths away from the ground or other environmental disturbing conditions, L_{tg}, L_{rg}, L_{tc} and L_{rc} will be very small, and in those cases it is not necessary to distinguish between L_p and L or L_s.

It will often be useful to use the combined quantities: $L_t = L_{tc} + L_{tg}$ and $L_r = L_{rc} + L_{rg}$ where $L_{t,r} = 10 \log_{10}(r'/r_f)$ and r' is the actual antenna resistance, while r_f is the radiation resistance it would have if it were in free space. Thus $L_s = L_p + L_t + L_r$.

III. Relation between Propagation Loss, L_p, and Field Strength, E

The results of many previous studies of radio wave propagation in the low and medium frequency range have been expressed in terms of the field strength, E, expressed in decibels above one microvolt per meter for an unattenuated field, I, at a unit distance, i.e., to the radiation field expected at a unit distance on a perfectly conducting plane surface. It is convenient to express this in decibels above 1 V, e.g., 1 V/m at 1 m. We will relate such values to the propagation loss, L_p.

In free space the inverse distance field, I_f, may be obtained from Equation (10):

$$I_f = 10 \log_{10} (p_r c \, . \, 10^{-7} g_t) = P_r + G_t + 14{\cdot}77. \qquad (34)$$

Over a perfectly conducting ground plane the inverse distance field, I, may be obtained from Equation (13):

$$I = I_f + 6{\cdot}02 - L_{tg} = P_r + G_t + 20{\cdot}79 - L_{tg}. \qquad (35)$$

The relation between the available power, p'_a, from a loss-less receiving antenna and the field strength, e, may be expressed in decibels as follows:

$$E = 10 \log_{10} (4\pi z p'_a [1 + \Delta_r] \, . \, 10^{12}/\lambda^2 g_r) = P'_a + 20 \log_{10} f_{Mc} - G_r + L_{rg} + 107{\cdot}22. \qquad (36)$$

If we solve Equation (35) for P_r and Equation (36) for P'_a and combine the results, we obtain the following expression:

$$L = P_r - P'_a = I - E - G_p + 20 \log_{10} f_{Mc} + 86{\cdot}43 + L_{tg} + L_{rg}. \qquad (37)$$

In the literature E is often referred to an inverse distance field of 300 V, i.e., $3 \times 10^5 \, \mu$V/m at 1 km (i.e., by Equation (14) to 1 kw radiated from a short vertical electric dipole over a perfectly conducting plane surface, and in this case $I = 49{\cdot}54$ db so that:

$$L_p = L - L_{tg} - L_{rg} = 135{\cdot}97 - G_p + 20 \log_{10} f_{Mc} - E. \qquad (38)$$

In the remainder of this paper, we will be concerned only with the propagation loss, L_p, from which the system loss may be determined for practical systems by adding $L_t + L_r$.

IV. Very Low Frequency Ionospheric Propagation

At the very low frequencies below 30 kc, the ionosphere reflects the waves at relatively low heights, about 70 km in the daytime and of the order of 90 km at night. At these low heights the ionization gradients are sufficiently large so that the ionosphere behaves as a sharp boundary, and it is convenient to use wave guide theory for determining the phase and amplitude of the received waves; good discussions of this theory are presented in the June 1957 issue of the *Proc. Inst. Radio Engrs, N.Y.* The transmission loss as a function of distance exhibits minima and maxima which are caused by interference between the ground wave and ionospheric wave modes at distances less than about 1000 miles, and are caused by interference between the several ionospheric modes at the larger distances. At these long wavelengths the fading of the received waves is caused by a gradual shift from midday to midnight conditions and consequently has a very long period; thus at certain distances the received field may remain weak throughout the day or the night. The comparison between the calculated and observed locations and magnitudes of such anomalies provides a useful means for determining the effective constants of the ionosphere. At large distances, there is usually only one wave guide mode, and the curves then do not have the maxima and minima which characterize the shorter distances.

V. Low and Medium Frequency Ionospheric Propagation

As we increase the radio frequency above 30 kc, the ionosphere behaves much less as a sharp boundary and instead gradually refracts the waves back to the receiving point only after they have penetrated many kilometers into it, this penetration being greater the higher the radio frequency. The available evidence appears to indicate that the D and E regions of the ionosphere, which extend from 70 to 110 km, are turbulent, consisting of "blobs" of ionization which drift with the mean wind with velocities often in excess of 100 miles per hour. An interesting discussion of these irregular ionospheric motions is given in a recent article by Gautier [11]. The radio waves will travel along many different paths through this turbulent ionized medium, the received field being the resultant vector sum of the waves received after propagation along these different paths. At sufficiently high frequencies the received field for a single ionospheric mode may be considered to be the sum of a steady specularly-reflected component plus a random Rayleigh-distributed component, and the resultant received field will have an amplitude distribution as shown on Fig. 2; on this figure, $K \equiv 10 \log_{10} k^2$ and represents the ratio in decibels between the available power received from the random ionospheric waves and from the specularly-reflected component. Thus it becomes convenient, particularly for frequencies above 30 kc, to determine the propagation loss separately for each of the ionospheric modes and the ground wave mode of propagation. When K is very large, the distributions of the amplitude and phase of the received field for a single ionospheric mode approach those expected for a Rayleigh distribution; furthermore, even when the individual modes have large specular components, the resultant field from several modes received simultaneously also tend to approach a Rayleigh distribution. Figures 3 and 4 show the propagation loss expected at 100 kc between short vertical electric dipole antennas for the ground wave and several ionospheric wave modes of propagation over land and over the sea, and for day and night conditions. The method of calculation used in determining the results shown on Figs. 3 and 4 involves a combination of ray and wave theory. Figure 5 illustrates the geometry of our model and some of the assumptions made in the calculations. The waves are refracted in the troposphere down towards the earth and, as a consequence, the distance d_1, traveled for a given ray angle of elevation, ψ, before the waves arrive back at the earth is substantially larger than if there were no atmosphere.

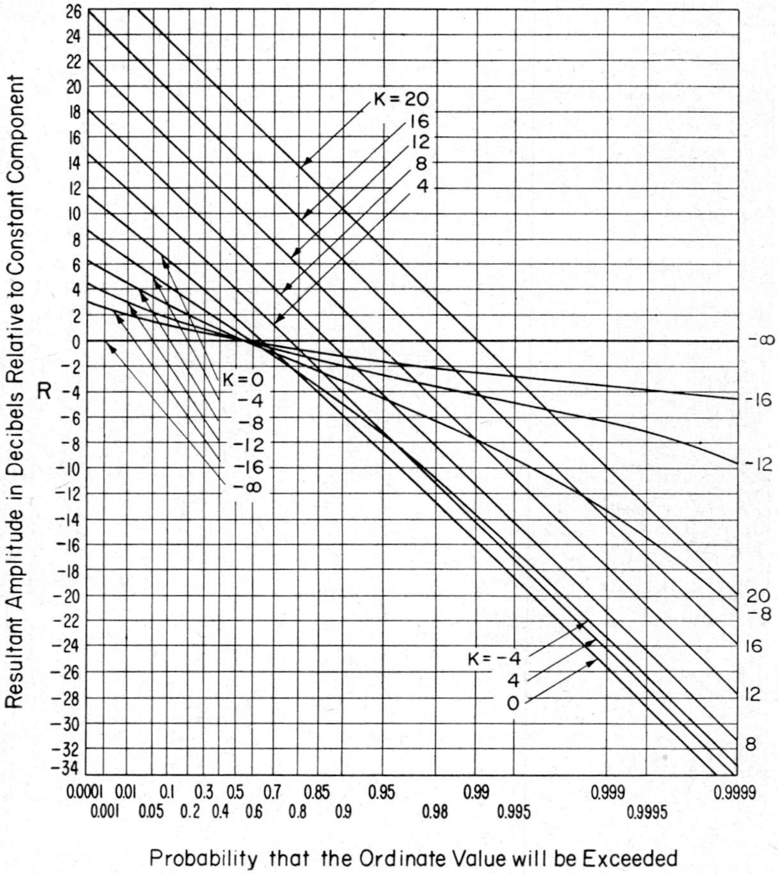

FIG. 2. Distribution of the Resultant Amplitude of a constant vector plus a Rayleigh distributed vector. Power in Random component is K decibels relative to power in constant component.

We have idealized our problem by assuming for all points along the path that the ionosphere has the same height h, and the same reflection coefficient, while the ground is assumed to have the same electrical constants even for propagation all the way to the antipode at a distance of about 12,500 miles. The actual ionosphere and ground reflection conditions over particular propagation paths are obviously much different from these idealized paths, but our present model seems better for expository purposes. The principle of stationary phase (essentially the same as Fermat's principle) leads to the conclusion that the received waves may be considered to travel along several discrete ray paths between the transmitter and the receiver. All of these paths are great circle paths, the shortest

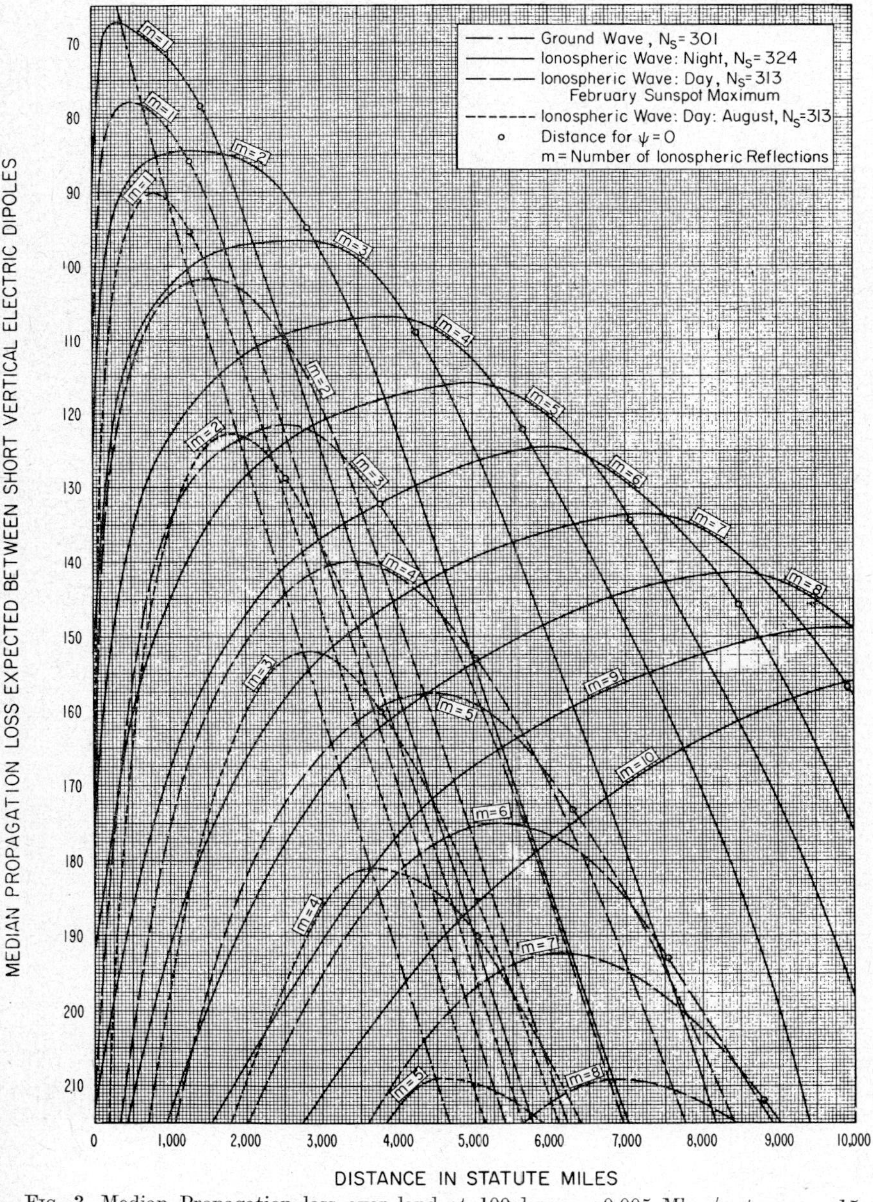

FIG. 3. Median Propagation loss over land at 100 kc. $\sigma = 0{\cdot}005$ Mhos/meter; $\epsilon = 15$; $h_t = h_r = 30$ feet. Day h = 70 km; Night h = 90 km.

corresponding to the ground wave mode of propagation. The other paths involve m reflections at the ionosphere, and the waves propagated along these other paths arrive at the receiving point at successively later times. By transmitting short

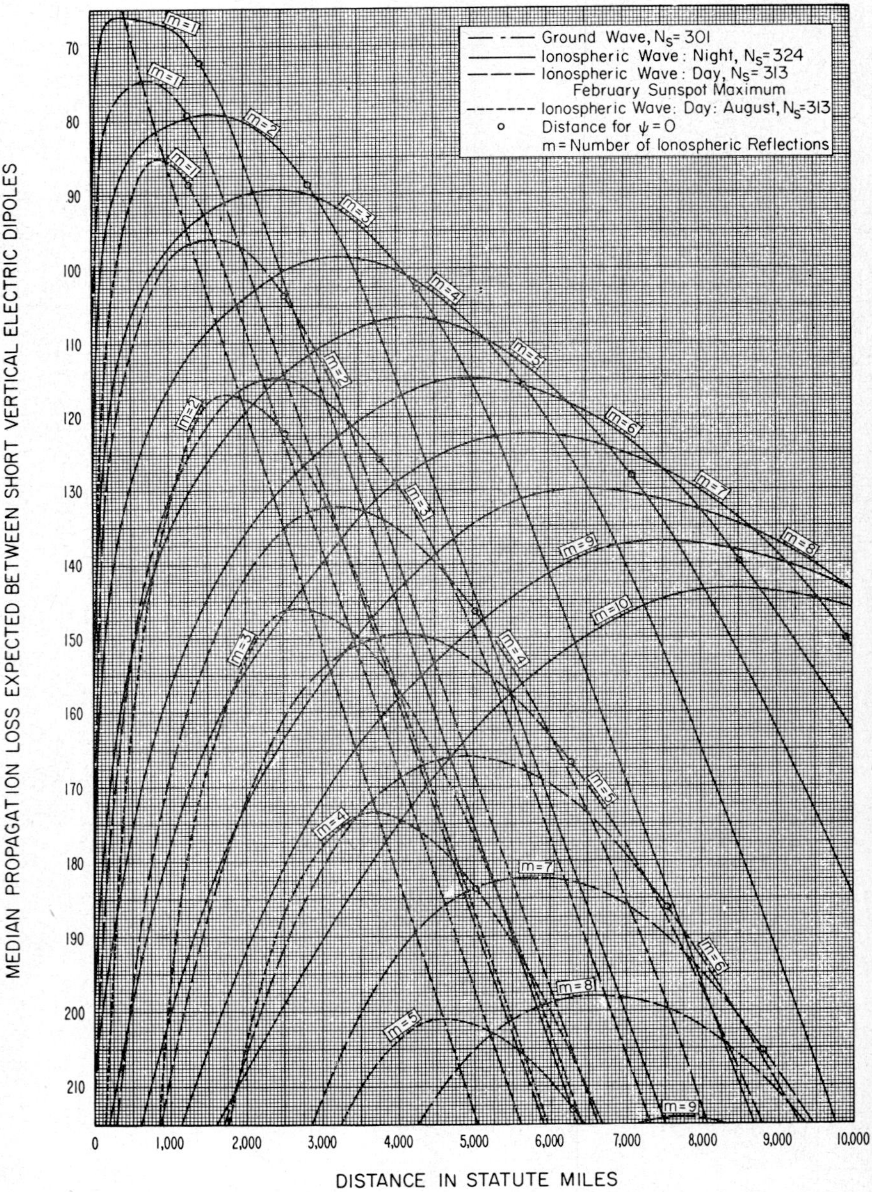

FIG. 4. Median Propagation loss over sea at 100 kc. $\sigma = 5$ mhos/meter; $\epsilon = 80$; $h_t = h_r = 30$ feet; Day h = 70 km; Night h = 90 km.

pulses, it is possible to observe these several modes independently at a distant receiving point, and in this way their physical reality has been verified. The term "mode of propagation" here, and in the remainder of this paper, refers to the waves propagated along one of these ray paths, and has a distinctly different meaning from the usage in the previous section where the modes of propagation

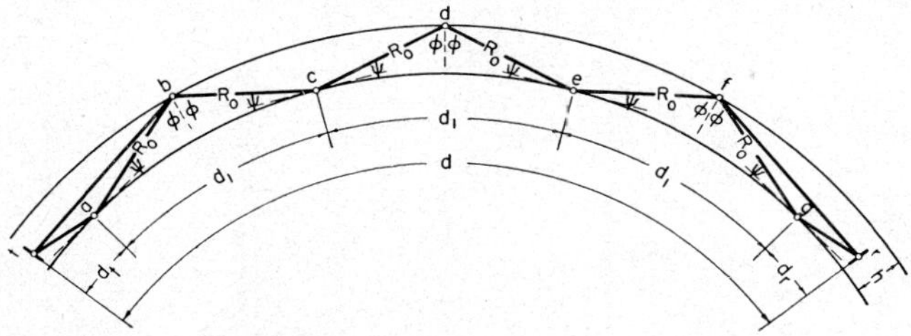

Case (a) ψ positive and m = 3.

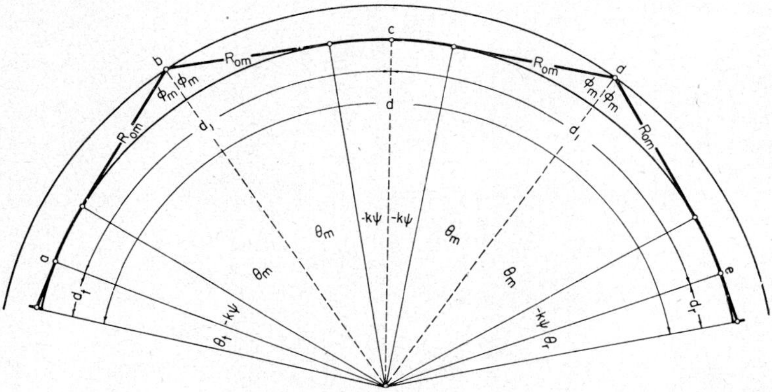

Case (b) ψ negative and m = 2; $k = \left(1 - \frac{d\tau}{d\psi}\right)\psi = 0$.

FIG. 5

were the wave guide modes which are simply the successive terms in a mathematical expression for the field. The use of short pulses to make possible the separate reception of each of these modes is a very useful device for radio navigation and, in this connection, the estimation of the time of arrival of the successive modes becomes of great practical importance. These time delays have been studied both theoretically [13, 32, 35] and experimentally [9, 12, 21, 25] and the reader is referred to the references for information of this kind; here we will be primarily interested only in their transmission losses.

Consider the phase of a radio wave for a given ray path, $\Omega(\psi) = 2\pi R/\lambda$, where λ is the wavelength and R is the total length of the ray path between the transmitter and the receiver. Now consider the variation of this phase for all possible adjacent paths between the transmitter and the receiver as we vary the angle of elevation, ψ, and azimuth, χ; it should be clear (a) that $\Omega(\psi)$ will be a minimum with respect to variations in χ for the set of rays lying in the great circle plane, i.e., for $\chi = 0$ and (b), of this set there will be $m + 1$ points of stationary phase, $\Omega'(\psi) = 0$, for the ray representing the ground wave and for the m rays reflected at the ionosphere in such a way that the angle of incidence, ϕ, at the ionosphere (for example, at b, d, and f on Fig. 5) is equal to the angle of reflection, and also (for $m > 1$) that the angle of incidence, $(90° - \psi)$, at the ground (for example, at

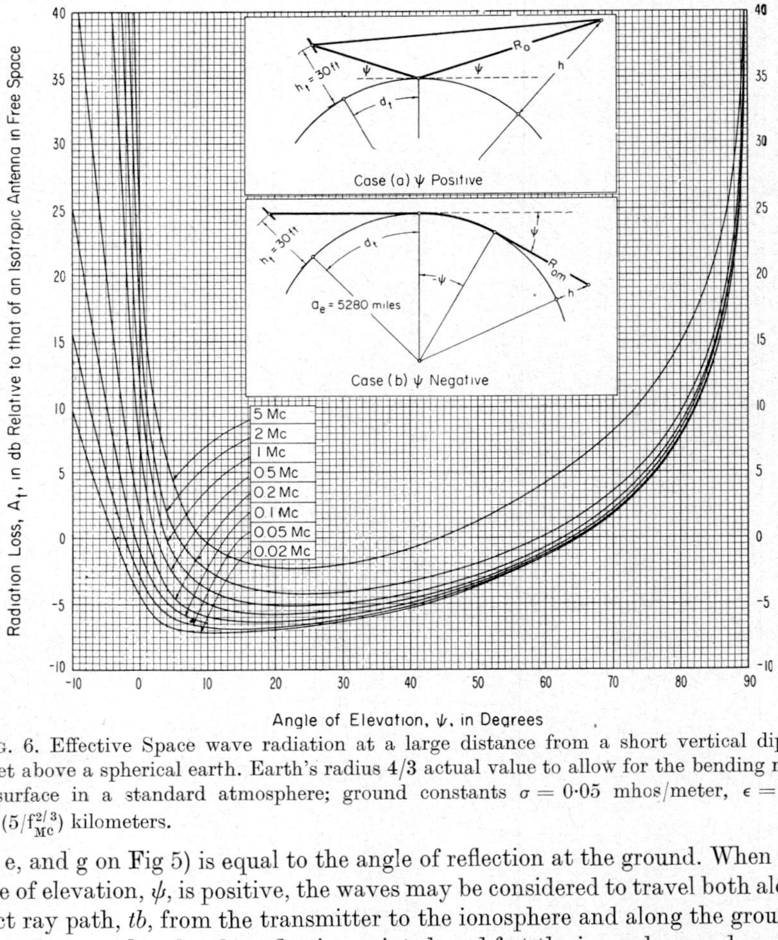

FIG. 6. Effective Space wave radiation at a large distance from a short vertical dipole 30 feet above a spherical earth. Earth's radius 4/3 actual value to allow for the bending near the surface in a standard atmosphere; ground constants $\sigma = 0{\cdot}05$ mhos/meter, $\epsilon = 15$; $h > (5/f_{\text{Mc}}^{2/3})$ kilometers.

a, c, e, and g on Fig 5) is equal to the angle of reflection at the ground. When the angle of elevation, ψ, is positive, the waves may be considered to travel both along direct ray path, *tb*, from the transmitter to the ionosphere and along the ground-reflected ray path, *tab* ; the reflection points b and f at the ionosphere and c and e at the ground will be very slightly different for the direct and ground-reflected ray paths, but this small difference is ignored in our calculations. Note that the angle of elevation, ψ, can be negative as is illustrated on Fig. 5, case (b).

The following formula may be used to calculate the median propagation loss of an ionospheric mode of propagation involving m reflections at the ionosphere and a ray path of length, R:

$$L_{pm} = L_{bf}(R) + A_t(\psi) + A_r(\psi) + (m - 1)A_g(\psi) - C_m(R, 0{\cdot}5) + P + mA(\phi, 0{\cdot}5). \quad (39)$$

Each of the terms in the above is expressed in decibels; $L_{bf}(R)$ denotes the basic free space transmission loss for the ray distance, R; Equation (7) may also be expressed in the following form which is convenient for numerical work:

$$L_{bf} = 36{\cdot}58 + 20 \log_{10} R + 20 \log f_{Mc} \tag{40}$$

where R is expressed in statute miles. We see by Fig. 5 that we may calculate R as follows:

$$R \simeq 2mR_0 + \frac{(d_t + d_r)}{\cos \psi} \qquad (\psi > 0) \tag{41}$$

$$R \simeq 2m\,(R_{om} - ka\psi) + \sqrt{(2kah_t)} + \sqrt{(2kah_r)} \qquad (\psi \leqslant 0) \tag{42}$$

$$d_t = \sqrt{(ka \tan \psi)^2 + 2kah_t)} - ka \tan \psi. \tag{43}$$

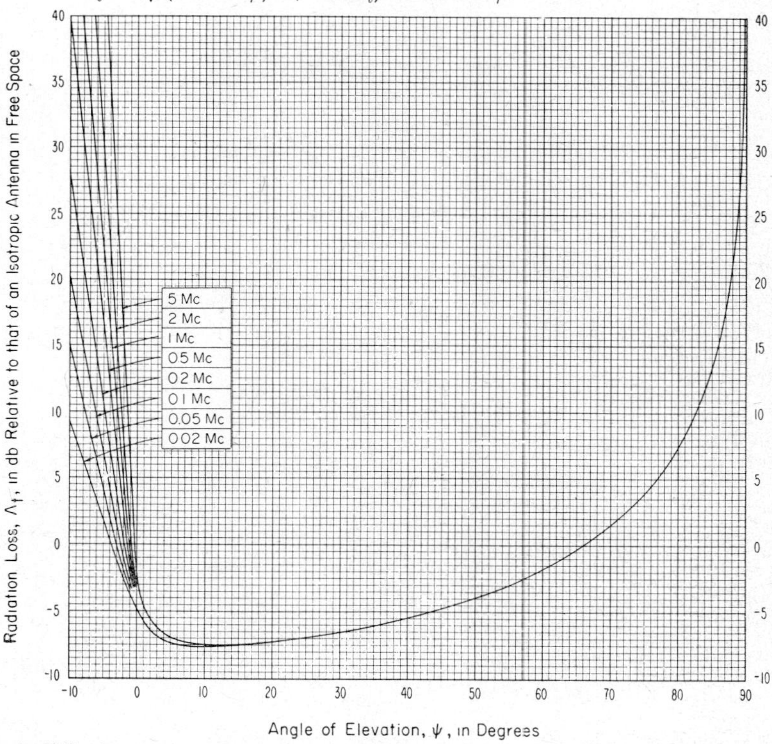

FIG. 7. Effective space wave radiation at a large distance from a short vertical dipole 30 feet above a spherical earth. Earth's radius 4/3 actual value to allow for the bending near the surface in standard atmosphere; ground constants $\sigma = 5$ mhos/meter, $\epsilon = 80$; h > $(5/f_{MC}^{2/3})$ kilometers.

The distance, d_r, is determined by a formula similar to Equation (43) with h_t replaced by h_r; in this equation ka is the effective earth's radius, and k has been chosen equal to 4/3 in our ionospheric examples. Methods for estimating k as a function of time and geographical location are given in a later section. It is convenient to choose several values of ψ at conveniently spaced intervals and then to calculate all of the remaining factors at these particular values of ψ.

The space wave radiation factors, $A_t(\psi)$ and $A_r(\psi)$ include the gains of the transmitting and receiving antennas, respectively, relative to that of an isotropic antenna in free space, and allow for the radiation patterns of the antennas and the loss arising from the proximity of the antennas to the curved earth. The magnitude of $A_t(\psi)$ can be determined from $A_t(\psi) = L_i(\psi) - L_{bf}(R_o + d_t)$, where $L_i(\psi)$ denotes the transmission loss expected for the ground wave mode propagated between the actual transmitting antenna and an isotropic receiving antenna placed at the first point of reflection in the ionosphere, while $L_{bf}(R_o + d_t)$ is the corresponding basic free space transmission loss at this distance. Figures 6 and 7 give typical values of $A_t(\psi)$ expected for short vertical electric dipoles 30 ft above the ground; in this case we may express $A_t(\psi)$ as follows:

$$A_t(\psi) = 20 \log_{10} | F | - 1{\cdot}761 - 20 \log_{10} \cos \psi - 20 \log_{10} f(q). \qquad (44)$$

In the above $| F |$ is a "cut-back" factor. When ψ is large and positive, $| F |$ is just $| 1 + R_v(\psi) |$ where R_v is the complex Fresnel reflection coefficient for plane vertically polarized waves incident on the ground at the grazing angle ψ; when ψ is small or negative, the curvature of the earth becomes important and the values of $| F |$ have then been determined by formulas recently developed by Wait [30]. The term 1·761 is just the gain of the short dipole; the term $20 \log_{10} \cos \psi$ allows for the cosine pattern of the dipole; and finally $f(q)$ is the height gain factor given by Equation (19) in Reference 19 which allows for the effect of the height, h_t, of the antenna above the surface. The "cut-back" factor $| F |$ was calculated for a spherical surface of radius, $a_e = ka$, with $k = 4/3$; this provides approximately for the effect of air refraction.

The factor $(m - 1)A_g(\psi)$ allows for loss on reflection at the ground, for example at c and e on Fig. 5 in Case (a) and at c in Case (b). The amount of this loss will depend on the polarization of the downcoming waves. Since the waves reflected from the ionosphere will have both vertically and horizontally polarized components, even when the incident waves are linearly polarized, it becomes necessary to know the relative amounts of energy associated with each polarization in the downcoming waves. This problem is not easy to solve precisely, and we have obtained an estimate for $A_g(\psi)$ by assuming, quite arbitrarily, that the energy in the downcoming waves is equally divided between the two polarizations. Thus, for angles $\psi > 2°$, we have:

$$A_g(\psi) \simeq - 10 \log_{10} [(| R_v^2 | + | R_h^2 |)/2] \qquad \psi > 2 \cdot \qquad (45)$$

where R_v and R_h are the complex Fresnel reflection coefficients for vertical and horizontal polarization, respectively. Since the value determined by Equation (45) represents only a few decibels, the use of the above approximate expression will not lead to serious errors.

It has been shown by Rice [22] and by Fock [10] that a $A_g(0) = 6{\cdot}021$ db when ψ equals zero (in the limit as R_{om}/λ is very large, i.e., for $h > (5/f_{Mc}^{2/3})$ km) regardless of the polarization or ground constants; their results can also be used to compute $A_g(\psi)$ for other values of ψ, but we have instead assumed that $A_g(\psi)$ can be calculated for $\psi \leqslant 0$ by the following approximate formula:

$$A_g(\psi) \simeq 6{\cdot}021 - 2A_t(0) + 2A_t(\psi) \qquad (\psi \leqslant 0). \qquad (46)$$

For values of ψ between 0 and 2°, it is easy to sketch in a smooth curve between the results given by Equations (45) and (46). See also References [29] and [31].

We turn next to a consideration of the convergence factor $C_m(R, p)$ which provides a measure of the focusing of the energy on reflection at the curved surface of the ionosphere. Figure 8 is a geometrical construction which demonstrates the

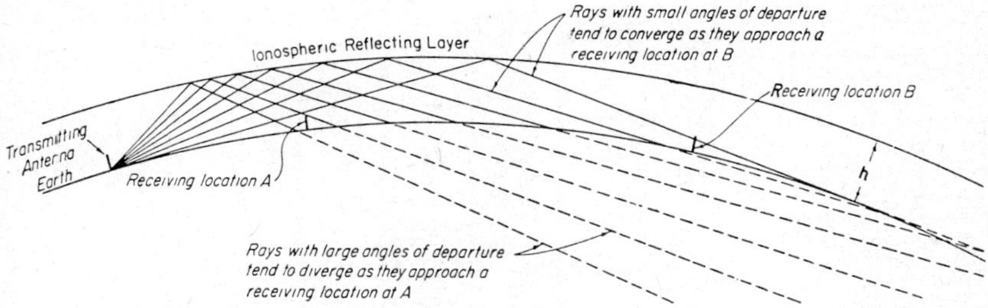

FIG. 8. Geometrical construction demonstrating the convergence of rays near grazing incidence.

nature of this focusing of rays in the vertical plane for ψ near zero. At the antipode of the transmitter, half way around the earth, the rays are also focused in the horizontal plane. A detailed discussion of this phenomenon is given in Appendix I. For rays leaving the earth's surface at grazing incidence ($\psi \leqslant 0$) and at the antipode ($m\theta = 90°$), it is necessary to use a wave treatment of the problem, the amount of the focusing then being a function of the frequency. At points substantially removed from these caustics, geometrical optics leads to the following formula for $C_m(R)$ which provides for a smooth ionosphere a measure in decibels of the expected increase in the received field due to this focusing.

$$C_m(R) = 10 \log_{10} c_m = 10 \log_{10} \left\{ \frac{R_0}{- a \sin \psi (d\theta/d\psi)} \right\}_v \cdot \left\{ \frac{2mR_0 \cos \psi}{a \sin 2m\theta} \right\}_h . \quad (47)$$

$$\frac{d\theta}{d\psi} = - \frac{\sin (\theta - \tau)}{\cos \psi \cos \phi} + \frac{d\tau}{d\psi}. \quad (48)$$

In the above τ denotes the total bending of a radio wave in passing through the troposphere, and methods of calculating τ are given in a later section. The two factors in Equation (47) correspond to the focusing in the vertical and horizontal planes, respectively. Equation (47) may be used except near the caustic ($\psi \leqslant 0$) and ($m\theta = 90°$). When $\psi \leqslant 0$ and $m\theta \neq 90°$, we may use:

$$C_m(R) = 20 \log_{10} [R_{om} - 92{\cdot}153\, \psi \text{ (degrees)}] + (10/3) \log_{10} f_{kc} - 10 \log_{10} [\sin 2m (\theta_m - 4/3\psi)] + (40/3) \log_{10} m - 60{\cdot}694. \quad (49)$$

When $\psi > 0$ and we are at a distance in wavelengths (ρ/λ) from the antipode, we may use:

$$C_m(\rho) = 20 \log_{10} (R_a/\pi a) + 10 \log_{10} - \left\{ \frac{\cos^2\psi}{\sin \psi\, (d\theta/d\psi)} \right\} - 10 \log_{10} m + 10 \log_{10} f_{kc} + 10 \log_{10} [J_0(2\pi \cos \psi\, \rho/\lambda)]^2 + 36{\cdot}172 \; (\psi \leqslant 0). \quad (50)$$

The above may be used for $m \geqslant 9$ at night since ψ is greater than zero for these modes with $h = 90$ km. For $m \leqslant 8$, $\psi < 0$ near the antipode at night, and we may then use the following formula:

$$C_m(\rho) = 20 \log_{10}(R_a/\pi a) - (20/3) \log_{10} m + (40/3) \log_{10} f_{kc} + 10 \log_{10}[J_0(2\pi\rho/\lambda)]^2 + 44{\cdot}422 \qquad (\psi \leqslant 0). \qquad (51)$$

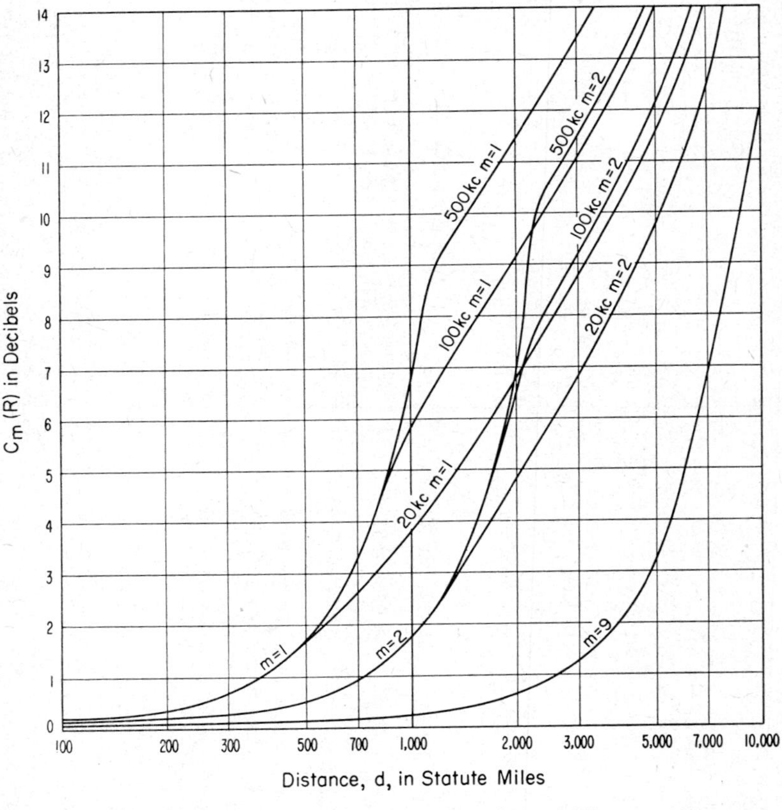

FIG. 9. Convergence factor in ionospheric propagation. h = 70 kilometers; N_s = 313 and h_s = 0 and k^2 (1) = 0.

In Equations (50) and (51), R_a denotes the ray distance to the antipode. If a horizontal magnetic dipole is used for reception of the field radiated from the vertical electric dipole, then the J_0 in Equations (50) and (51) is to be replaced by J_1; here J_0 and J_1 are Bessel functions. Very near $\psi = 0$, the values of $C_m(R)$ determined by Equation (47) will exceed those given by Equation (49), particularly at the lower frequencies, and in this region a smooth curve may be drawn between the values for $\psi \geqslant 0$ and the values given by Equation (49); similarly, at the antipode the values of $C_m(\rho)$ given by Equation (50) will exceed those given by Equation (51), and a smooth curve may be drawn between the values for $\psi \geqslant 0$ and those given by Equation (51). Figure 9 gives examples of $C_m(R)$ calculated in this way for $m = 1$, 2 and 9 for daytime propagation ($h = 70$ km), for a smooth ionosphere, and typical refraction conditions.

It is clear from Fig. 8 that this focusing will be fully realized in practice only to the extent that the ionosphere presents a smooth surface to the radio waves. A discussion is presented in Appendix I which indicates how allowance may be made

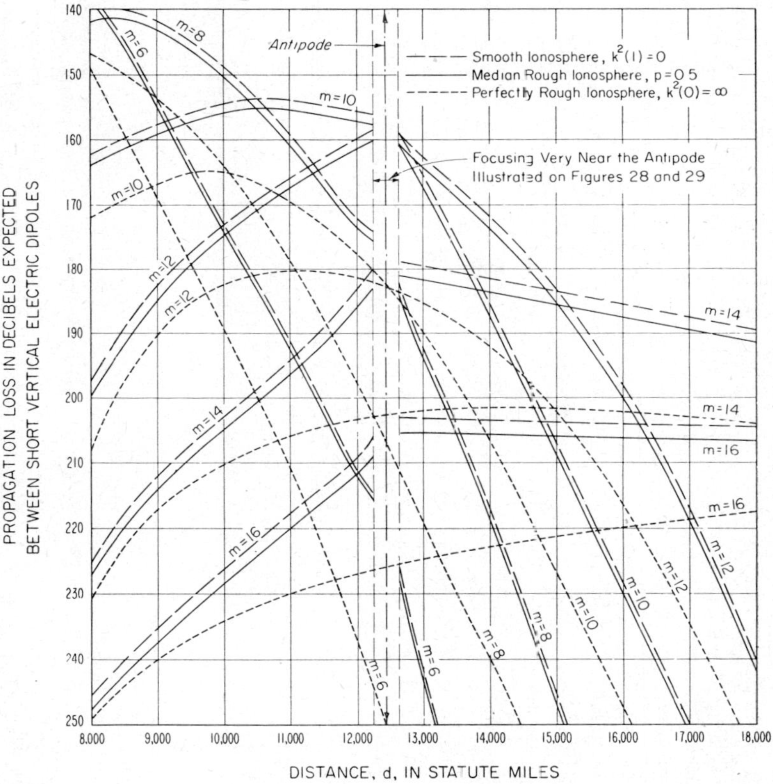

FIG. 10. Propagation loss expected near the antipode at night for 100 kc. Over land $= \sigma = 0{\cdot}005$ Mhos/meter, $\epsilon = 15$.

for ionospheric roughness. It is shown that an individual ionospheric mode of propagation consists of a steady specularly-reflected component plus a random Rayleigh distributed component. If we let k^2 denote the ratio of the power in the random component relative to that in the specularly-reflected component, then we suggest that we may estimate the convergence $C_m(R, p)$ exceeded 100 p % of the time in terms of the values of $k^2(1 - p)$ exceeded 100 $(1 - p)$% of the time as follows:

$$C_m(R, p) = 10 \log_{10} \left[\frac{c_m + k^2(1 - p)}{1 + k^2(1 - p)} \right]. \tag{52}$$

Here c_m (see Equation 47) denotes the ratio of the received power with and without focusing for a smooth ionosphere. As the probability varies from 0 to 1, $k^2 (1 - p)$ will vary from zero for a smooth ionosphere to ∞ for a perfectly rough ionosphere, and $C_m(R, p)$ will vary from the values given by Equations (47), (49), (50) and (51) for a smooth ionosphere to zero for a perfectly rough ionosphere. The propagation losses given in this report correspond to median values, i.e., $p = 0{\cdot}5$. The random variable k^2 depends upon the radio frequency, angle of incidence, ϕ, and time of day.

As an illustration of the effects of focusing near the antipode and of the influence of ionospheric roughness, Fig. 10 shows the propagation loss expected between vertical electric dipoles in the range from 8000 to 18,000 miles for 100 kc at night over land. Three values are shown for each of the modes $m = 6, 8, 10, 12, 14$, and 16 corresponding (a) to a smooth ionosphere ($p = 0$, $k^2(1) = 0$), (b) to an ionosphere of medium roughness ($p = 0{\cdot}5$), and (c) to a perfectly rough ionosphere ($p = 1$, $k^2(0) = \infty$).

In the immediate vicinity of the antipode, i.e., within a few wavelengths, the focusing for a smooth concentric ionosphere is very large. Thus Fig. 11 shows the value of $C_m(R_a, p)$ expected right at the antipode at night for a smooth, concentric ionosphere and for a rough, concentric ionosphere; the values expected for $h = 70$ km during the daytime would be only slightly different; Fig. 12 shows the rapid decrease of the focusing as we leave the antipode and, at distances greater than 100 λ, the envelope will be just 6 db above the values of propagation loss shown on Fig. 10, i.e., the values shown on Fig. 10 correspond to the single wave expected from a directive transmitting antenna with an infinite front-to-back ratio. We see on Fig. 12 that the field expected from the non-directive dipole oscillates with increasing distance from the antipode; this oscillation is caused by the interference between the waves arriving at the receiving point along the short and long great circle paths. Thus there will be concentric rings around the antipode at which the expected field will be equal to zero. The radii of these concentric rings are the same for a concentric ionosphere, regardless of the number, m, of ionospheric reflections, and are determined by the zeros of the Bessel functions; for the electric field, the first two such rings have radii equal to $0{\cdot}38\ \lambda$ and $0{\cdot}88\ \lambda$.

The actual ionosphere will never be concentric with the surface of the earth. In practice, as the sun rises and sets, or as the geomagnetic latitude of the reflection point is varied, the surface of the ionosphere will undoubtedly change in such a way that its radius of curvature and slope relative to a tangent plane on the earth will vary over appreciable ranges, and this will cause $C_m(R)$ to vary up and down relative to the values expected on the basis of the above analysis. However, except near the antipode, it seems plausible to assume that the median values of $C_m(R)$ may not be much influenced by such changes. The magnitude of the antipodal anomaly will be substantially reduced by these macroscopic perturbations of the spherical concentric shell model assumed for our calculations. Also, for the actual non-concentric ionosphere, the geographical location of the antipode may be expected to vary with time over a fairly large area, and this should result in a net increase in the fading range in this region. Furthermore,

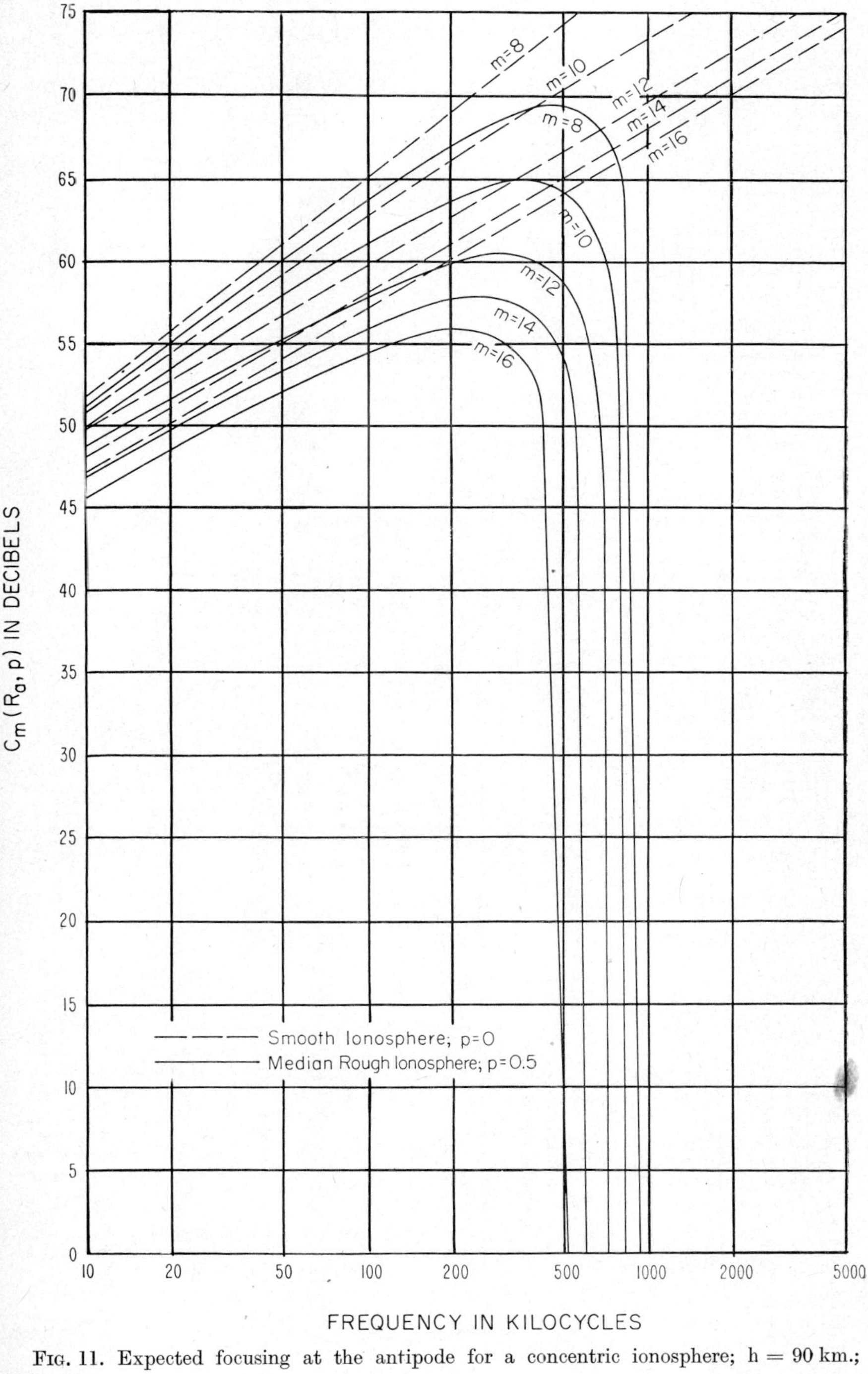

FIG. 11. Expected focusing at the antipode for a concentric ionosphere; h = 90 km.; $N_s = 324$.

the antipodal locations may be expected at any given time to be different for the different modes, and thus the zeros predicted by Equations (50) and (51) and shown on Fig. 12 are not likely to be observable unless some means is used to exploit the different times of arrival of the individual modes.

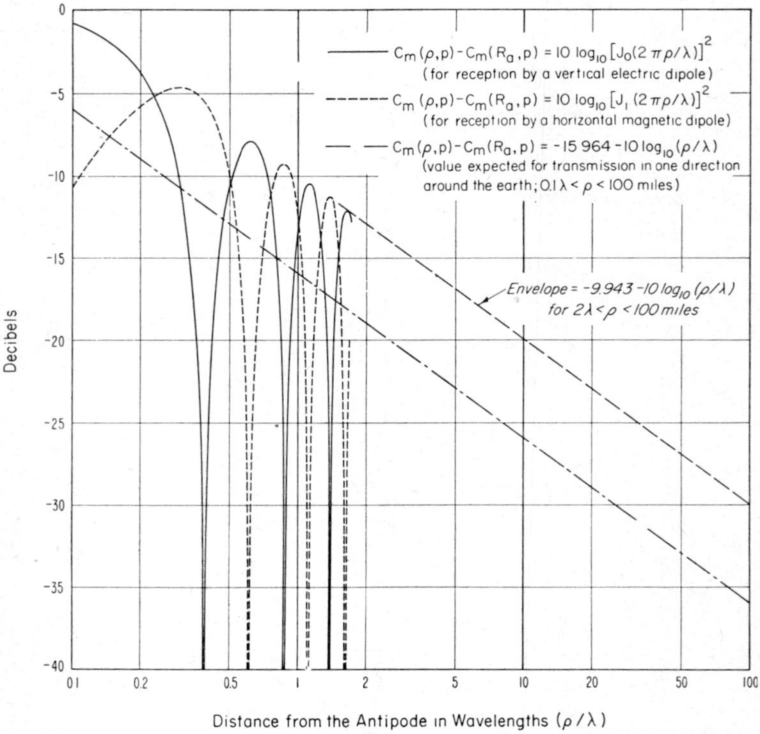

FIG. 12. The behaviour near the antipode of the convergence factors for the radiation from a vertical electric dipole for a concentric ionospheric shell around a spherical earth.

Consider next the loss, P, arising from the polarization characteristics of the downcoming ionospheric waves. An incident linearly-polarized wave will be reflected as two component waves, an ordinary and an extraordinary wave, each of which will be elliptically polarized. These two component waves, which will have roughly the same amplitudes except on frequencies near the gyrofrequency (near 1·5 Mc in the United States) will mutually interfere, and this causes the rapidly varying polarization characteristics of the observed downcoming waves. The polarization loss, P, arises from the fact that typical receiving antennas will respond to only one polarization. The amount of this loss will depend principally upon the transmission frequency, the penetration frequency for the layer involved, and the intensity and direction of the earth's magnetic field relative to the path; it can be calculated [8, 18, 34] with some accuracy when these parameters are known. For more than one reflection at the ionosphere, the polarization loss becomes a very complex function of the reflection coefficients for the parallel and perpendicular component of the incident fields and is difficult to separate from the absorption loss, $A(\phi, 1 - p)$. All of the low frequency examples of ionospheric

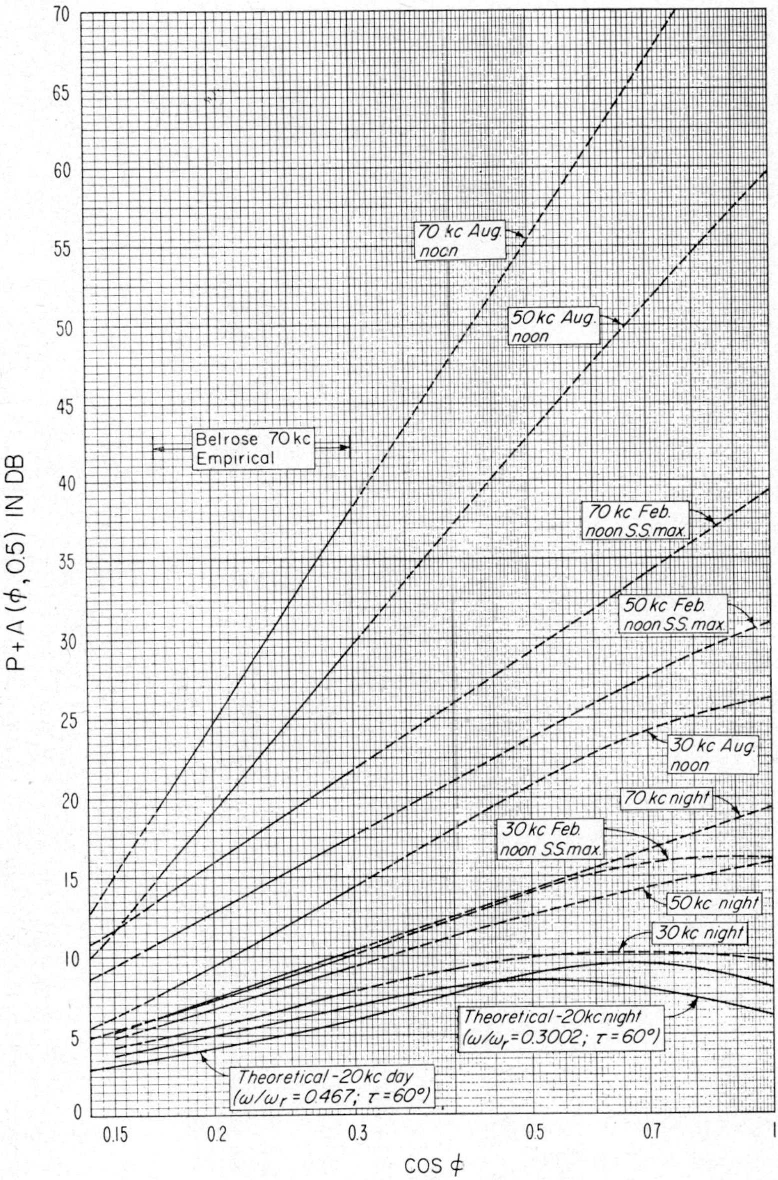

FIG. 13. Median loss for one reflection at the ionosphere.

wave propagation in this report have been obtained using the empirical estimates described below for $P + A(\phi, 1 - p)$ which thus includes the polarization loss P; consequently we have calculated the total reflection loss as $m\{P + A(\phi, 1 - p)\} - (m - 1)P$ for $m > 1$. The calculations at $f = 100$ kc have been made in this report by setting $P = 3{\cdot}01$ db, but this estimate is now believed to be substantially too large, except near vertical incidence. The calculations for all of the other frequencies from 20 kc to 1000 kc were calculated with $P = 0$ for all values of m, although this assumption probably leads to somewhat more propagation loss than would be expected for $m > 1$ since the appropriate value of P probably lies between 0 and 3 db, approaching this latter value near vertical incidence; however, we are usually more interested in the values near oblique incidence for our applications, and this latter assumption should yield more nearly correct results for the solution of these problems.

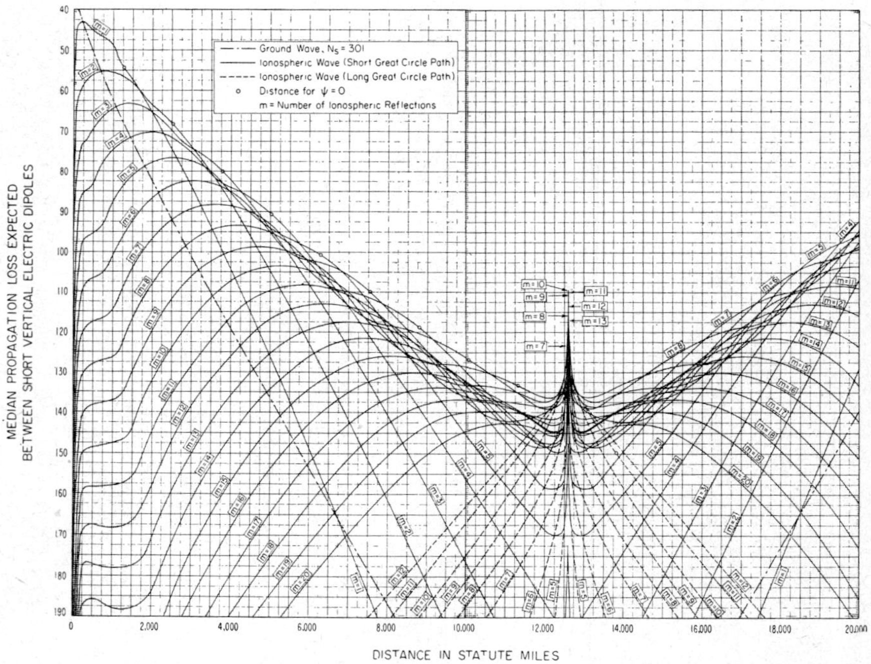

FIG. 14. Median propagation loss over land in the daytime at 20 kc; $\sigma = 0{\cdot}005$ Mhos/meter; $\epsilon = 15$; $h_t = h_r = 30$ feet; $h = 70$ km.

Finally we will consider the loss, $A(\phi, 1 - p)$, on reflection at the ionosphere, exceeded for $100(1 - p)$ % of the time; this depends on the angle of incidence, ϕ, at the ionosphere, the radio frequency, time of day, season of the year, phase of the sunspot cycle and the geomagnetic latitude of the reflection point. For the examples developed in this report, we have used some empirical evaluations of $P + A(\phi, 1 - p)$ made by Belrose [7] on transmission paths between stations in

England, the Scandinavian countries, and Germany. Using the results in his doctorial thesis, we may express the median values of $P + A(\phi, 0{\cdot}5)$ as follows:

$$P + A(\phi, 0{\cdot}5) = 17{\cdot}2 \log_{10}(f_{kc} \cos \phi) - 12{\cdot}4 \qquad (70 < f_{kc} < f_D) \qquad (53)$$

(Night)

$$P + A(\phi, 0{\cdot}5) = 30{\cdot}8 \log_{10}(f_{kc} \cos \phi) - 22{\cdot}6 \qquad (70 < f_{kc} < f_D) \qquad (54)$$

(February, Noon, Sunspot Minimum)

$$P + A(\phi, 0{\cdot}5) = 33{\cdot}6 \log_{10}(f_{kc} \cos \phi) - 22{\cdot}6 \qquad (70 < f_{kc} < f_D) \qquad (55)$$

(February, Noon, Sunspot Maximum)

$$P + A(\phi, 0{\cdot}5) = 77{\cdot}3 \log_{10}(f_{kc} \cos \phi) - 64{\cdot}0 \qquad (70 < f_{kc} < f_D) \qquad (56)$$

(August, Noon)

The above formulas were determined empirically from data extending only over

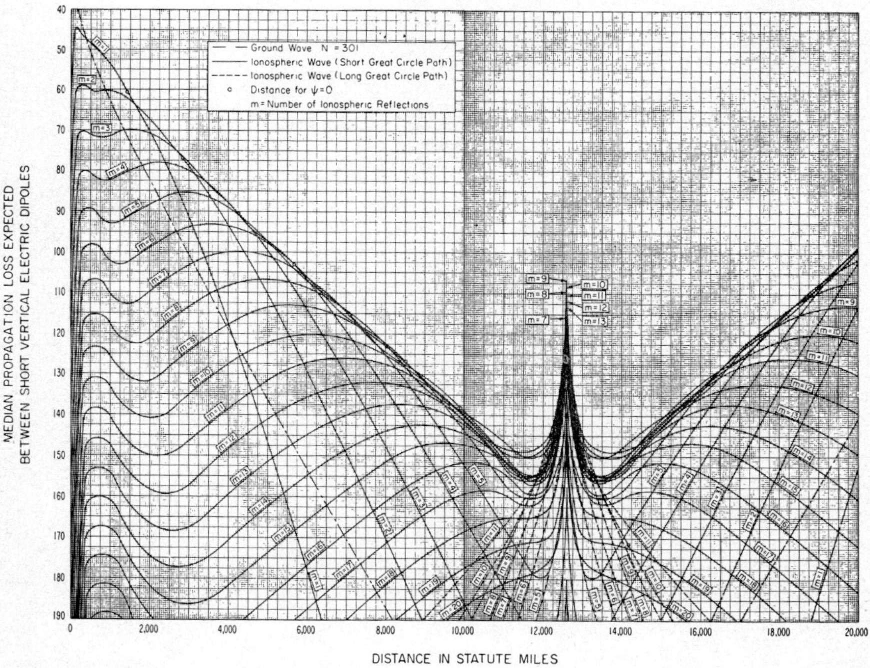

FIG. 15. Median propagation loss over land at night at 20 kc; $\sigma = 0{\cdot}005$ Mhos/meter; $\epsilon = 15$; $h_t = h_r = 30$ feet; h = 90 km.

the range of frequencies from 70–250 kc and the range of distances from 350 to 900 miles; a recent analysis of lower frequency data by Watt, Maxwell and Whelan [33] indicates that the absorption is greater at frequencies less than 70 kc than would be predicted by the above formulas. Consequently, as shown on Fig. 13, we have used the theoretical results of Wait and Murphy [34] at 20 kc and then interpolated linearly on a logarithmic frequency scale to obtain values for

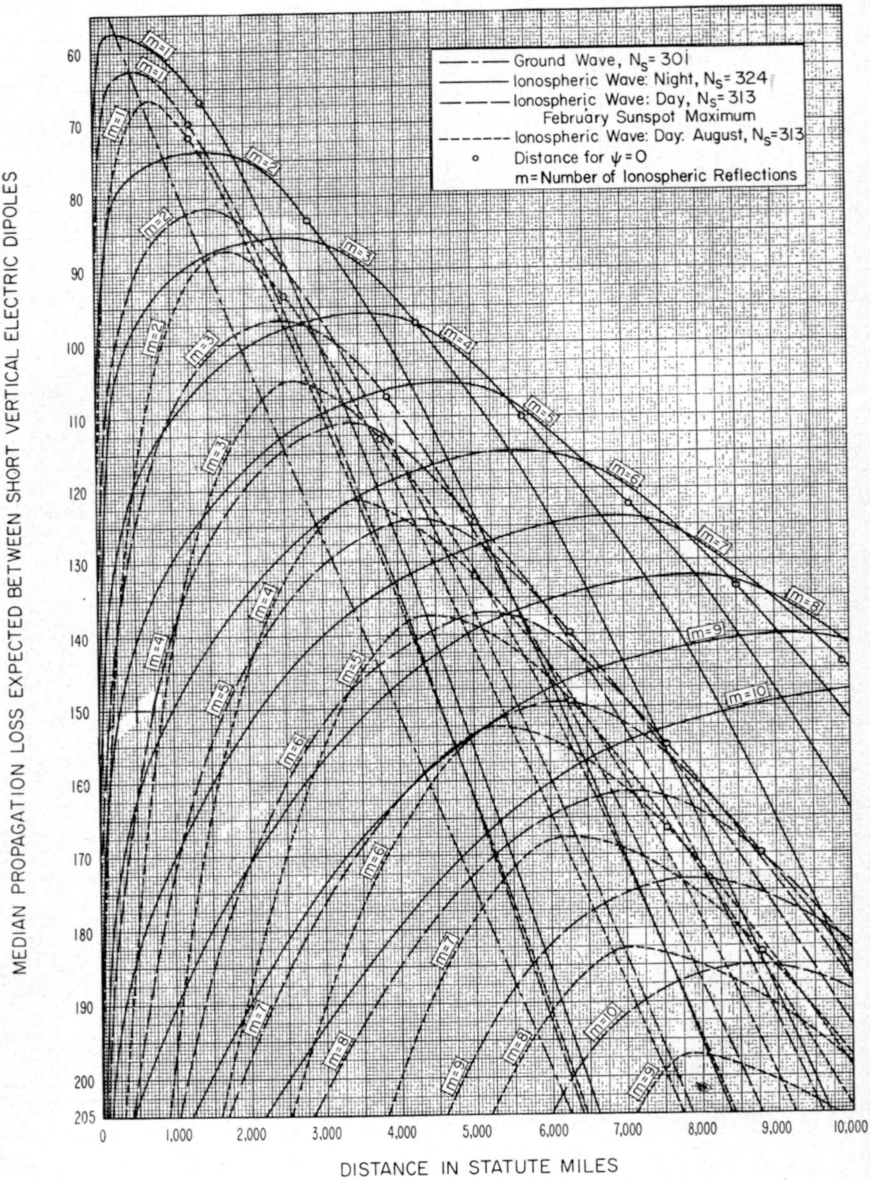

FIG. 16. Median propagation loss over land at 50 kc; $\sigma = 0{\cdot}005$ Mhos/meter; $\epsilon = 15$; $h_f = h_r = 30$ feet; Day h = 70 km; Night h = 90 km.

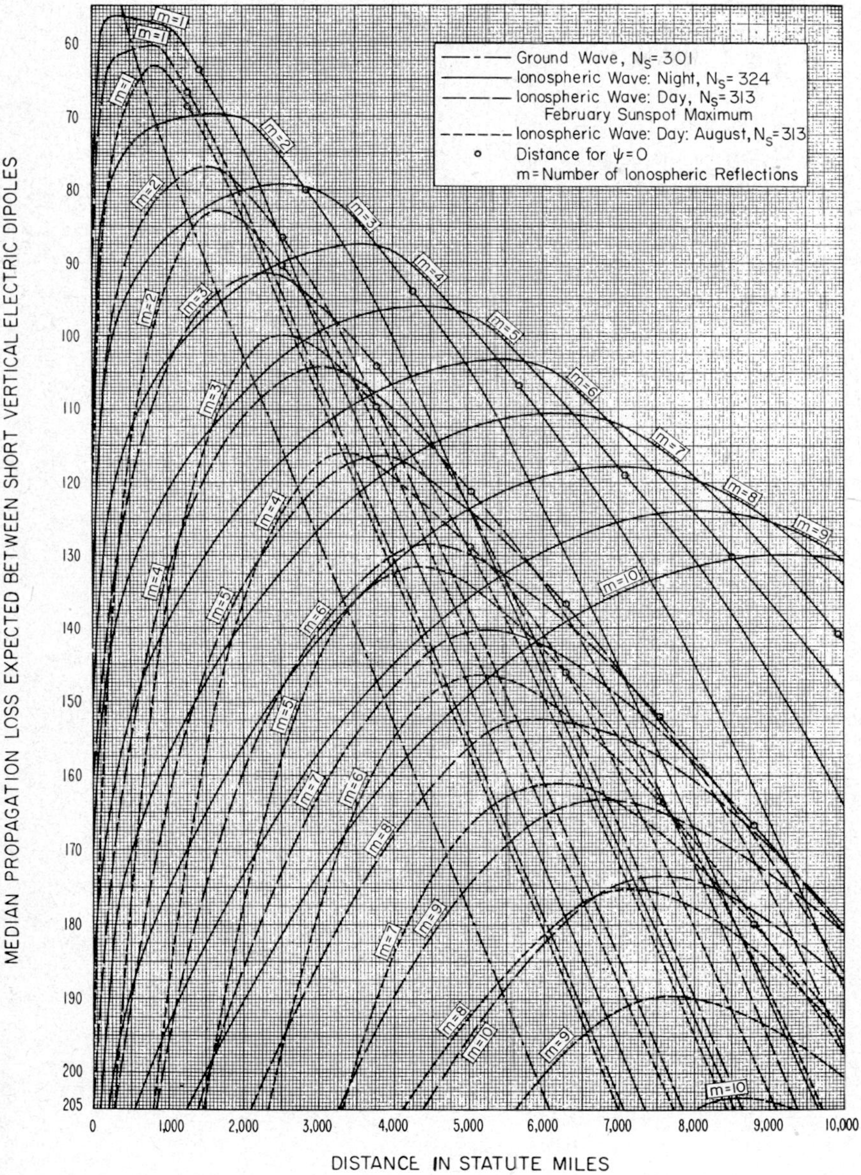

FIG. 17. Median propagation loss over sea at 50 kc: $\sigma = 5$ mhos/meter; $\epsilon = 80$; $h_f = h_r = 30$ feet. Day h = 70 km; Night h = 90 km.

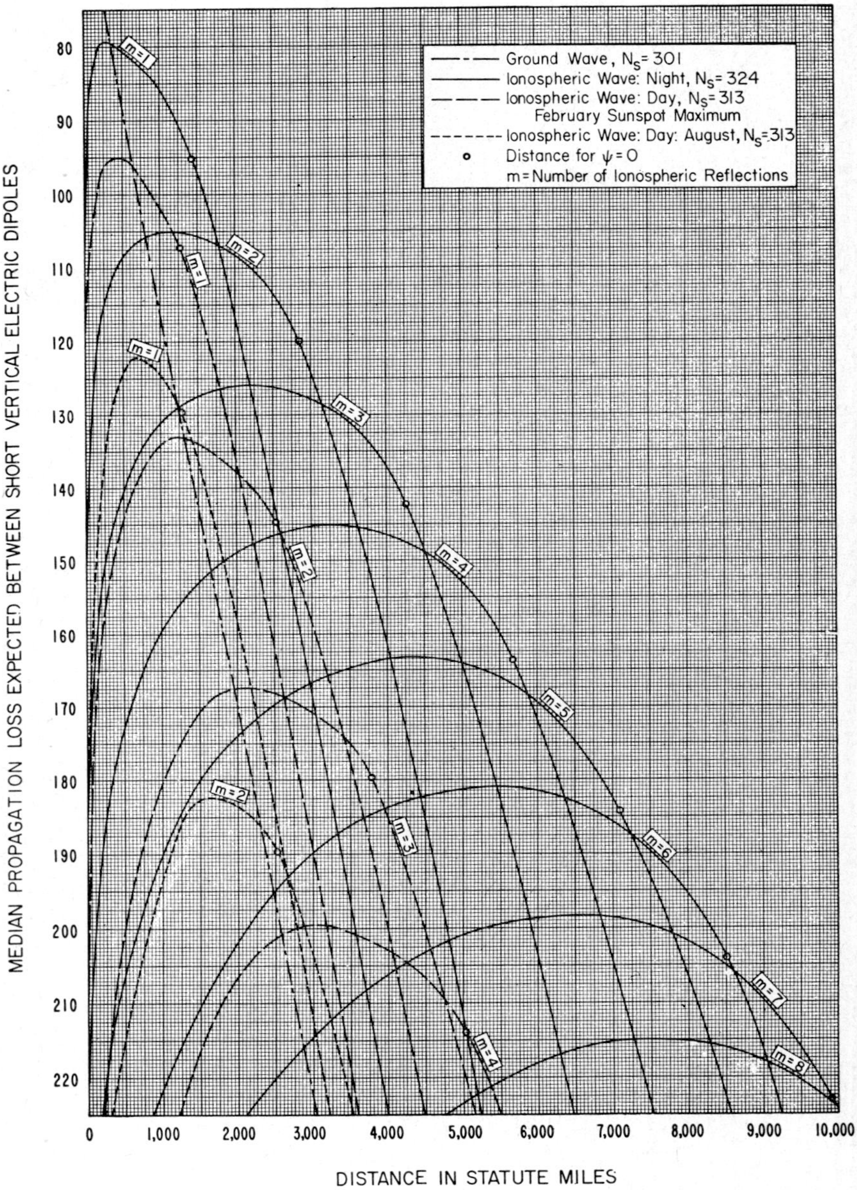

FIG. 18. Median propagation loss over land at 200 kc; $\sigma = 0{\cdot}005$ Mhos/meter; $\epsilon = 15$; $h_t = h_r = 30$ feet; Day h = 70 km; Night h = 90 km.

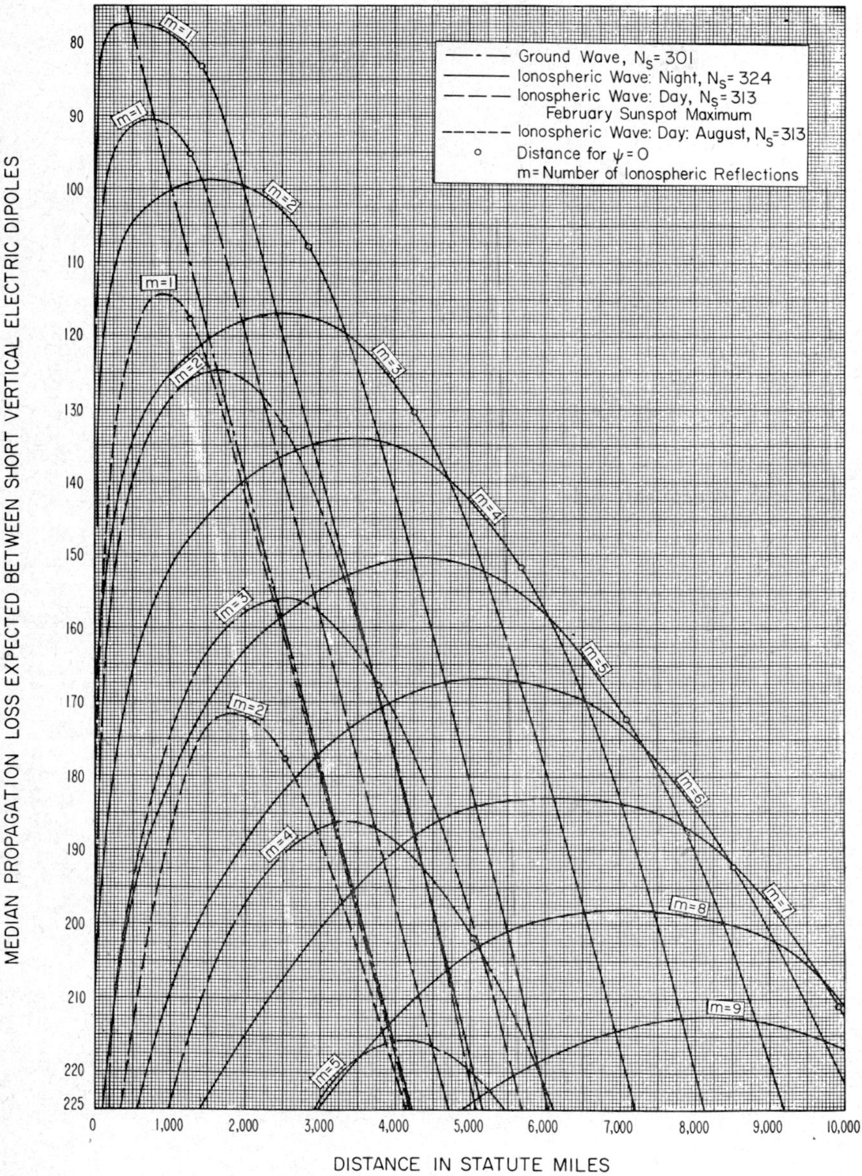

FIG. 19. Median propagation loss over sea at 200 kc; $\sigma = 5$ Mhos/meter; $\epsilon = 80$; $h_t = h_r = 30$ feet; Day h = 70 km; Night h = 90 km.

intermediate frequencies; the ionospheric parameters $L/H = 0{\cdot}1$ by day (i.e., $\omega/\omega_r = 0{\cdot}467$) and $L/H = 0{\cdot}05$ at night (i.e., $\omega/\omega_r = 0{\cdot}3002$) were used in these calculations at $f = 20$ kc and the index τ for the earth's magnetic field was set equal to 60°.

Figures 14 to 19 give the median propagation loss expected in accordance with the above methods of calculation at 20, 50, and 200 kc in over-land and over-sea propagation and for day and night conditions.

Figures 14 and 15 show the large decrease in the propagation loss near the antipode, and illustrate the fact that each mode of propagation has two important branches at points somewhat removed from the antipode, corresponding to propagation via the short and long great circle paths, respectively; actually there are still other branches for each mode corresponding to propagation more than once around the earth, but these are not shown. The decrease in propagation loss shown at the antipode is the value for an idealized concentric ionosphere and will, in practice, undoubtedly be somewhat smaller. Propagation loss curves for the separate modes are not given at 20 kc for sea water since they differ so little from those for over-land.

Throughout this section the formulas and graphs refer to the median values of propagation loss for individual modes of propagation. This form of presentation was used since it is more useful in application such as the design of navigation systems or of systems to avoid multipath distortion. To determine the expected median propagation loss for a continuous wave transmission, it is necessary to convert the propagation losses for the individual modes to power ratios, and then add these power ratios; at 500 kc, and possibly even as low as 50 kc, it is reasonable to asume that the several ionospheric modes will have random relative phases so that the median power of the resultant will be equal to the sum of the median powers of the individual modes. For example, if there were two modes with equal median propagation losses, the median propagation loss for the sum of these two modes would be 3 db less, and for three equal modes the sum would have 4·77 db less propagation loss than each individual mode.

The author has studied the behaviour of $P + A(\phi, 0{\cdot}5)$ at night in the United States for frequencies in the standard broadcast band from 500 to 1500 kc over a very wide range of distances, and has found the following semi-empirical formula:

$$P + A(\phi, 0{\cdot}5) = \frac{26 \cos \phi}{(f_{\mathrm{Mc}} \cos \phi)^{0{\cdot}4}} \cdot \qquad (57)$$

(Night)

Note that Equation (53) indicates an increasing loss with increasing frequency, presumably because of a deeper penetration of the D layer as the frequency is increased, while Equation (57) indicates that the loss decreases with increasing frequency. At the higher frequencies where Equation (57) was established, the waves penetrated the D layer and, as shown by Martyn [15] this behaviour of $P + A(\phi, 0{\cdot}5)$ with frequency and angle of incidence is to be expected. By virtue of the method used for its determination, Equation (57) includes the polarization loss P; since the extraordinary waves are much weaker than the ordinary waves

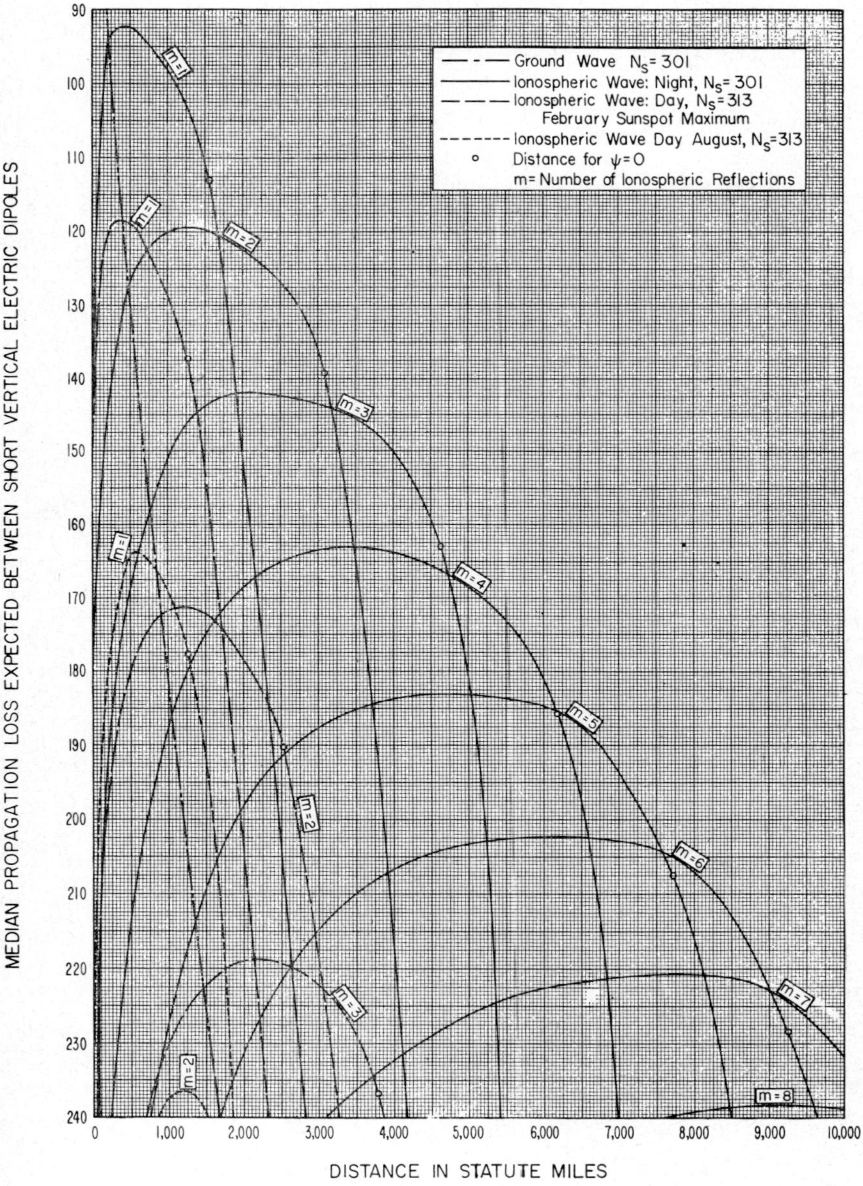

FIG. 20. Median propagation loss over land at 500 kc; $\sigma = 0{\cdot}005$ mhos/meter; $\epsilon = 15$; $h_t = h_r = 30$ feet; Day h = 70 km; Night h = 110 km.

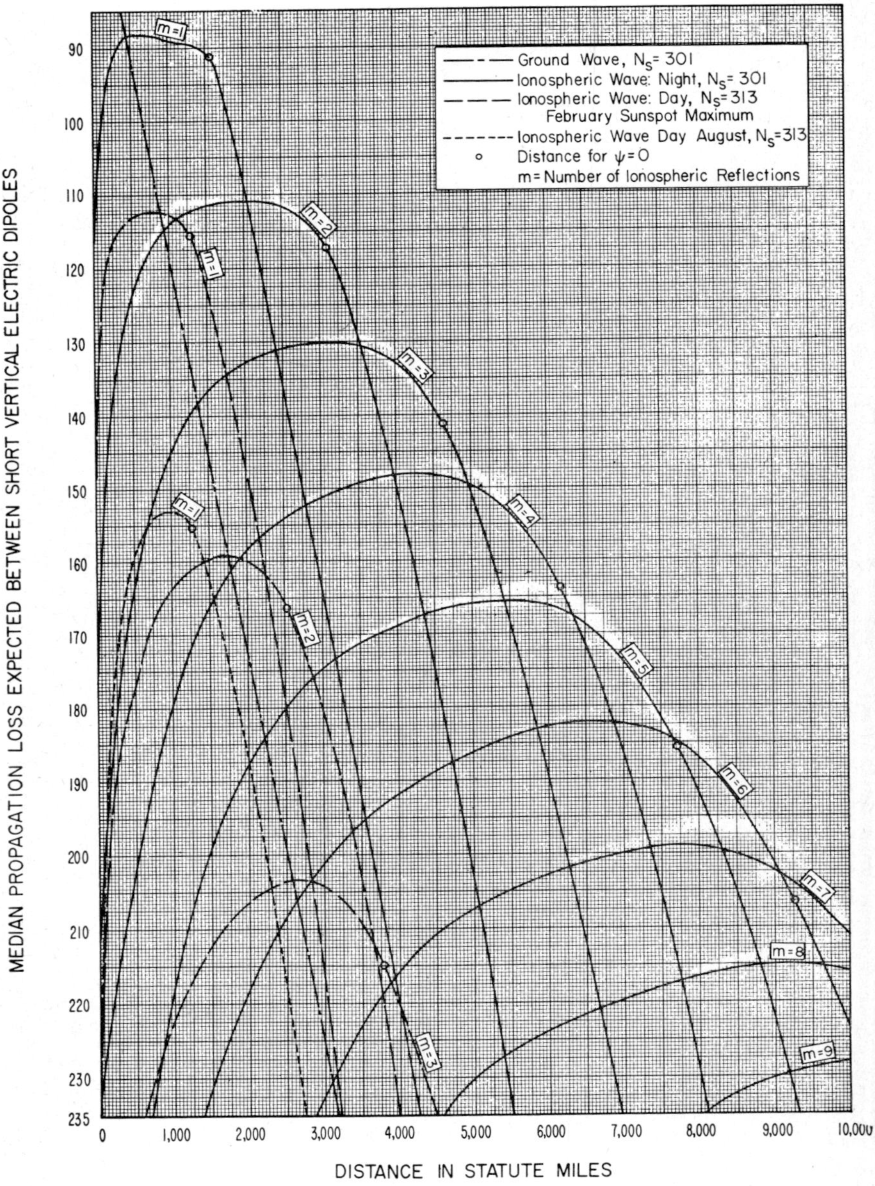

FIG. 21. Median propagation loss over sea at 500 kc. $\sigma = 5$ mhos/meter; $\epsilon = 80$; $h_f = h_r = 30$ feet; Day h = 70 km; Night h = 110 km.

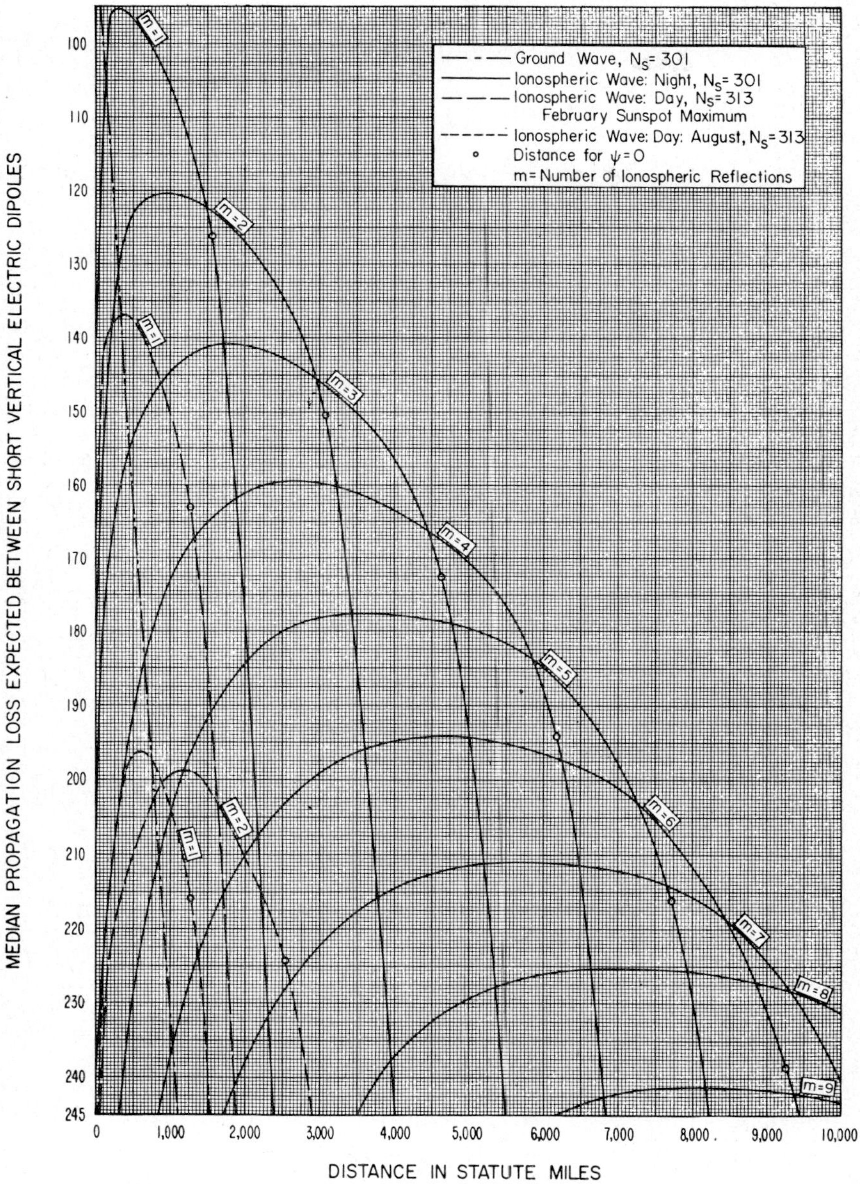

FIG. 22. Median propagation loss over land at 1000 kc. $\sigma = 0{\cdot}005$ mhos/meter; $\epsilon = 15$; $h_t = h_r = 30$ feet; Day h = 70 km; Night h = 110 km.

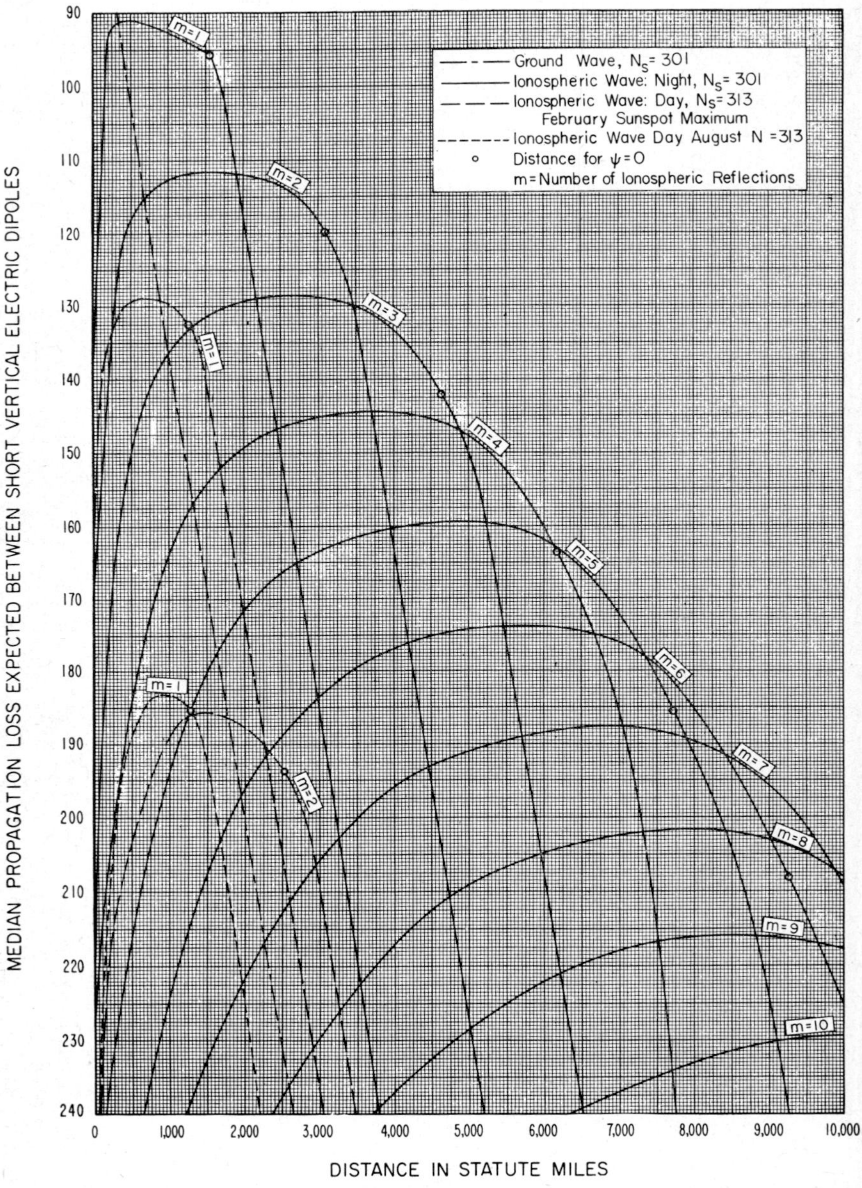

FIG. 23. Median propagation loss over sea at 1000 kc; $\sigma = 5$ mhos/meter; $\epsilon = 80$; $h_f = h_r = 30$ feet; Day h = 70 km; Night h = 110 km.

in this frequency range, there will be additional polarization loss at each reflection from the ionosphere. With the above discussion in mind, it seems appropriate to define the penetration frequency of the D layer at night as that frequency for which the absorption is the same for the two mechanisms, and this frequency may then be approximately determined as the frequency for which the absorption is the same as calculated by Equations (53) and (57):

$$P + A(\phi, 0{\cdot}5)_{(53)} = P + A(\phi, 0{\cdot}5)_{(57)}. \qquad (58)$$

(At the D layer penetration frequency, f_D, at night)

As determined in this way, the D layer penetration frequency at night varies from about 500 kc at vertical incidence to about 250 kc with $\cos\phi = 0{\cdot}164$, the minimum value expected for a 90 km layer height; the anomalous behaviour of this penetration frequency suggests that neither of our empirical absorption formulas are very dependable in this intermediate range of frequencies. Since nothing better is readily available, it was decided to calculate the propagation loss at night at frequencies greater than the above-defined D layer penetration frequency by arbitrarily using $m\ 26 \cos\phi/(f_{Mc} \cos\phi)^{0{\cdot}4}$ as the total loss on reflection; the reflection height was assumed to be 110 km at night, but the value of $\cos\phi$ to be used in the absorption equation was determined on the assumption that the absorption takes place at a height of 100 km.

Figure 20 to 23 give the median propagation loss expected between short vertical electric dipoles at 500 kc and at 1000 kc in over-land and over-sea propagation and for day and night conditions. The absorption at night was determined by Equation (57) as described above, but, in the daytime, Equations (55) and (56) were used since radio waves in this frequency range are then presumably reflected and absorbed by the D layer at an assumed height of 70 km.

At still higher frequencies during the daytime, the radio waves will penetrate the D layer and be reflected by the E layer at a height of about 110 km. The ionospheric absorption is so great during the daytime in the range of frequencies from, say 500 kc to 4 Mc, and the reflection phenomena so complex, with reflections taking place sometimes at the D layer, sometimes at the E layer and sometimes at the F layer, that useful, simple absorption formulas are not available; in the absence of better information, the calculations of propagation loss given in this report for this range of frequencies have been made by extrapolating Equations (54), (55) and (56) to these higher frequencies.

VI. The Bending of Radio Waves by the Troposphere

Since the density as well as the absolute humidity of the air decrease with the height, h, above sea level, the refractive index, n, also usually decreases with h, and this causes radio waves leaving an antenna at a given angle, ψ, to bend down towards the earth, the amount of this bending being larger, the smaller the value of ψ. This is illustrated on Fig. 24 which shows the total bending, τ, of a radio wave travelling entirely through the troposphere and subsequently being reflected at an ionospheric layer.

Since the refractive index, n, departs from unity by only a few parts in 10^{-4}, it is convenient to describe n in terms of the refractivity, N, which is defined:

$$N = (n - 1) \times 10^6. \tag{59}$$

If the value of N were known as a function of time at every point in the atmosphere between two radio antennas, it should be possible, in principle, to predict the instantaneous behaviour of the propagation loss between these antennas.

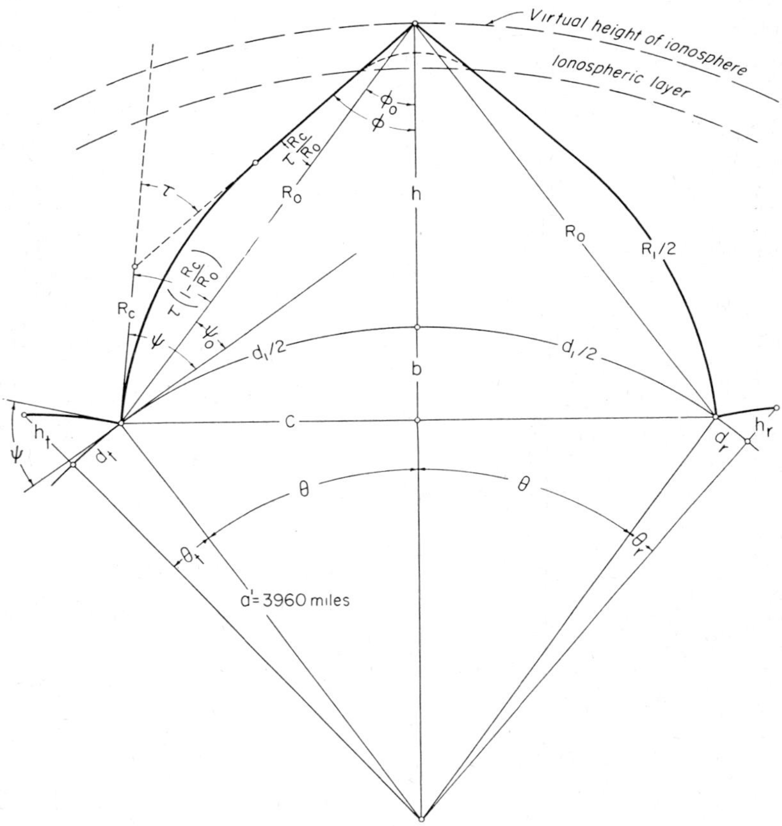

FIG. 24.

Actually, of course, this is not feasible because of the complexity of the solution of such an electromagnetic problem. Furthermore, even if the engineer could be provided with this instantaneous information, he would normally be forced to describe it in some statistical terms before he could use it effectively in the design or use of radio systems. Consequently, we are led to the description of N in statistical terms, with the hope that these statistical characteristics of N may be used for the prediction of the more important statistical parameters describing the propagation loss. At a given instant N will vary considerably with height above the surface, and, to a lesser extent, with distance along the path. If, however, we average the values of N over a period of an hour, then $\bar{N}$ will, on typical propagation paths, be more nearly constant along the path at a given height, but will normally decrease monotonically with increasing height above the surface. For the solution of most radio prediction problems, it is permissible and, indeed, desirable to average $\bar{N}$ over still longer periods of time in order to obtain mean conditions useful in radio systems design. Thus, for predicting the diurnal, seasonal, or geographical variations of the median propagation loss, we may further average the values of $\bar{N}$ as determined from day to day over a period of many years for a particular hour of the day, month of the year and geographical location. The resulting values $\bar{\bar{N}}$ will vary still less along the path and, on most paths, $\bar{\bar{N}}$, for a given time of day and season of the year, may be taken to be a function only of the height, $h - h_s$, above the earth's surface where h represents the height, expressed in kilometers, above sea level, while h_s represents the height of the surface above sea level. There will be some paths for which N will also vary appreciably in the horizontal plane; examples are paths with one terminal over land and the other over the sea, and these will undoubtedly require special treatment. Since the average values $\bar{\bar{N}}$ tend for most paths to be very nearly horizontally homogeneous, it should only be necessary to know the vertical profile of $\bar{\bar{N}}$ at one point along the path for a successful prediction of the median propagation loss at a particular time of day and season of the year. For very long paths on which $\bar{\bar{N}}$ does vary appreciably along the path, we may base our radio predictions on the two $\bar{\bar{N}}$ profiles at the intersection of the two radio horizons with the great circle path.

From the above discussion it appears to be desirable to study these $\bar{\bar{N}}$ profiles, and it will be convenient in the following analysis to omit the superscripts and simply let $N(h)$ denote these long-term average values. The most generally reliable single parameter for the description of the profile as it affects radio propagation is the difference, ΔN, in the values, N_1, at a height of 1 km above the surface and N_s the value at the surface:

$$\Delta N \equiv N(h_s + 1) - N(h_s) \equiv N_1 - N_s. \tag{60}$$

Note that ΔN is a negative quantity. Most of the diurnal, seasonal, and geographical variations in propagation between antennas at heights of less than 1 km above the surface may be predicted on the assumption that $N(h)$ decreases linearly with height above the surface up to a height of 1 km:

$$N(h) = N_s + \Delta N(h - h_s) \qquad [h_s \leqslant h \leqslant h_s + 1]. \tag{61}$$

For radio propagation predictions at the higher frequencies above, say 50 Mc, the above assumption of linearity for the initial decrease of $N(h)$ with height is not adequate for some times of the day or for some geographical locations; in some of these special cases there may be ducting with a resulting substantial increase [2] in the transmission loss for paths just short of the radio horizon, and a very large decrease [14] in the transmission loss on paths just beyond the radio horizon. Since these appreciably non-linear profiles occur in only a very small percentage of all cases [1], and are unimportant in the frequency range here under consideration, we will not consider them further in this paper.

The assumption of a linear profile makes possible the introduction of a great simplification in radio propagation predictions. Thus it has been shown [24] that the behaviour of radio waves in an atmosphere with a linear gradient is the same as that expected with no atmosphere for an earth with effective radius $a \equiv ka'$ where a' denotes the actual earth's radius, expressed in kilometers, and a is defined by:

$$\frac{1}{a} = \frac{1}{ka'} = \frac{1}{a'} + \frac{\Delta N \,.\, 10^{-6}}{(1 + N_s \,.\, 10^{-6})}. \tag{62}$$

Bean and Meaney [4] have given maps of the monthly median values of ΔN for the United States for several months of the year. The values of ΔN must be determined from radio-sonde observations and are, as a consequence, not as readily available at all hours of the day nor for as many geographical locations as the surface values, N_s. Fortunately, ΔN may be predicted [1] with quite good accuracy from N_s and, in the absence of observations of ΔN, the following empirical formula may be used to determine the predicted value $\Delta N'$:

$$\Delta N' = -\ 7{\cdot}32 \exp\{0{\cdot}005577\ N_s\}. \tag{63}$$

If now we combine Equations (61) and (63), we have the following expression for the initial behaviour of $N(h)$ in what will be referred to as the CRPL Standard Radio Refractivity Atmosphere as recently proposed and studied by Bean and Thayer [1, 5]:

$$N(h) = N_s - (h - h_s)\, 7{\cdot}32 \exp\{0{\cdot}005577\ N_s\} \qquad [h_s \leqslant h \leqslant h_s + 1]. \tag{64}$$

Note that the only parameters in Equation (64) are the surface refractivity, N_s, and the height h_s, of the surface above sea level.

Note that Equation (63) may also be used to predict N_s in terms of known values of ΔN:

$$N_s' = 412{\cdot}87 \log_{10}(-\,\Delta N) - 356{\cdot}93. \tag{65}$$

When values of ΔN and of N_s are both available, and when the actual value of N_s differs from the value predicted by Equation (65), it is very slightly better to use N_s' for predictions when $\psi < 3°$ rather than the actual value of N_s; in other words, ΔN is slightly better than N_s as a predictor of propagation conditions for small values of ψ. On the other hand, it is at present easier and usually also more accurate to predict N_s for some particular time of day, season of the year and geographical location, and then use Equation (63) for determining $\Delta N'$, than it is to use the available maps [4] directly for the prediction of ΔN. It should also be noted that, even when ΔN is available, N_s is a better predictor [5] of the bending at high elevation angles: $\psi > 3°$. It is expected that the recent study program [26] proposed by the International Radio Consultative Committee (CCIR) will tend to expedite the gathering of the data on ΔN required for the development of suitable prediction methods; however, it is unfortunate that emphasis was given in that proposal to the gathering of data at only two hours of the day, 0200 and 1400 U.T., since it is precisely the large diurnal variation of ΔN, occurring at many locations, which is most difficult at present to predict with adequate accuracy.

The effect on N_s of the height of the surface above sea level may be determined from the relation:

$$N_s = N_0 \exp(-c_s h_s) \tag{66}$$

where $c_s = 0{\cdot}1057/\text{km} = 0{\cdot}1701/\text{statute mile} = 0{\cdot}03222/\text{thousand ft}$. Bean and Horn [3] give maps which may be used to estimate the value of N_0 averaged throughout the day for the months of February and August at any geographical location in the world, together with a map of the annual range of N_s. A comprehensive climatological study of N_s for the United States is in preparation at CRPL; this study will make available charts useful for the prediction of N_0 (and thus of N_s by means of Equation (40) above) at 0200, 0800, 1400 and 2000 for February, May, August and November and, in addition, gives the detailed statistical characteristics of N_s at several representative weather stations in the United States.

Summarizing the above, we see that the average value of N up to a height of 1 km above the surface may be predicted preferably, when ψ is small, in terms of a measured mean gradient, ΔN, or, alternatively and with negligibly less accuracy, in terms of the mean value of the surface refractivity, N_s. Above 1 km, N decreases exponentially with height, and Bean and Thayer [5] give the following formulas for $N(h)$ in this range:

$$N(h) = N_1 \exp[-c_i(h - h_s - 1)] \qquad (h_s + 1 \leqslant h \leqslant 9 \text{ km}) \tag{67}$$

$$c_i = \frac{1}{8 - h_s} \log_e (N_1/105) \tag{68}$$

$$N(h) = 105 \exp[-0{\cdot}1424\,(h - 9)] \qquad h \geqslant 9 \text{ km}. \tag{69}$$

Since the constant c_i depends only on N_s and h_s, it appears that the mean atmosphere may be described in most cases in terms of these two parameters or, alternatively, in terms of h_s and N_s' when ΔN is known. The CRPL Model Radio Refractivity Atmospheres have arbitrary values of N_s, h_s, and ΔN; in the CRPL Standard Radio Refractivity Atmospheres N_s and ΔN are related by Equations (63) and (65), but h_s is arbitrary; and in the CRPL Reference Radio Refractivity Atmospheres N_s, ΔN, and h_s have the values given in Reference [1]. Note that h_s influences the description of the atmosphere [see Equation (68)] only in the range from 1 km above the surface to 9 km above sea level, and then only slightly for the range of values of h_s normally encountered in practice; consequently, we may, for practical purposes, consider that the atmosphere is well defined by the single parameter N_s. The success of this model in predicting the bending of radio waves has been examined by Bean and Thayer [5] and leaves little to be desired except in the small percentage of cases involving non-linear profiles.

The Bean and Thayer paper [5] gives the elevation angle error $\epsilon \equiv \psi - \psi_0$ as a function of electrical path length, R_e, for rays in the reference atmospheres as well as $\Delta R_e \equiv R_e - R_0$.

$$R_e = \int_0^{(R_1/2)} n dR = ct_e \tag{70}$$

where c is the velocity of light in a vacuum and t_e is the time of transit along the ray path.

VII. The Influence of Tropospheric Bending on Ionospheric Propagation

The bending of the radio waves by the troposphere has the effect of extending the range of ionospheric propagation, and although the effect is small, it is not negligible near oblique incidence. Tables I, II, III, IV, V and VI show the influence of N_s on R_1 d_1, and cos ϕ for ionosphere heights $h = 70$, 90, 110, 225, 350, and 475 km. The calculations were made by tracing rays in the CRPL reference atmospheres with $N_s = 0$, 301 and 400. Since the variations with N_s are comparatively small, it would appear to be reasonable for most applications to use only the single value $N_s = 301$ in ionospheric calculations, and tables VII, VIII, and IX give the convergence factor c_m for several angles, ψ, and for $m = 1$ to 20. The values of c_m are given since they involve the geometry in a rather complex way; however, since the ionosphere probably appears rough to the radio waves reflected at the higher layers, c_m is given only for $h = 70$, 90, and 110 km.

The results presented in this section were all obtained by ray tracing methods, and such methods are reliable only when the following two conditions are satisfied: (a) the index of refraction, n, must not change appreciably in a distance equal to a wavelength, and (b) the fractional change in the spacing between neighboring rays in a wavelength along the ray must be small compared with unity. Both of these conditions require that resort must be made to wave solutions of the problem at the lower frequencies. Condition (b) above is always violated at a caustic, and we have shown in Appendix I how such cases may be treated. It might be supposed, since n changes only from about 1·0003 to 1 for a 70 km change in h, that condition (a) would be well satisfied at frequencies even as low as 5 kc; however, it must be remembered that this small change in n actually causes appreciable bending when ψ is small, and we should, instead, require that $\Delta N/N < 0{\cdot}1(2\pi/\lambda) = 0{\cdot}002 f_{\text{kc}}$ if we are to expect ray tracing to apply. This more stringent requirement is met for $f > 65$ kc for the $N_s = 301$ atmosphere, and it appears that a wave solution will be required at lower frequencies for a precise treatment of the bending. Until an adequate wave solution becomes available, it would seem that the ray tracing solution here given should be used even for frequencies as low as 10 kc since the alternate assumption of $N_s = 0$ would undoubtedly yield an even poorer approximation to the actual bending.

VIII. The Total Tropospheric Bending

If we concern ourselves with the non-ionized portion of the atmosphere, we find that, above a height of about 70 km, it no longer bends the radio waves appreciably, and the designation τ has been given to the total bending of radio waves passing entirely through it. Tables X and XI give τ, the critical range, R_c, and $R_e - (R_1/2)$ as a function of ψ for the CRPL Reference Radio Refractivity Atmospheres. The results provide a convenient means for determining the true elevation angle ψ_0 and true range R_0 of a satellite at very high heights, say $h > 70$ km,* in terms of its observed elevation angle ψ, its observed radio range R_e and the surface value, N_s, of refractivity at the observing point:

$$\psi_0 = \psi - \tau[1 - (R_c/R_e)] \tag{71}$$

$$R_0 = (R_1/2) - R_c[0{\cdot}27 - 0{\cdot}5(R_c/R_e)] \sin^2 \tau. \tag{72}$$

The constants 0·27 and 0·5 were determined empirically by averaging the values of these coefficients found on seven profiles; the theory indicated that these constants should be $\frac{1}{3}$ and $\frac{1}{2}$, respectively, for a circular arc.

IX. Concluding Remarks

An analysis has been given of methods for estimating the propagation loss expected for radio waves received after m reflections at the ionosphere. It is shown that some of the components of this loss may be estimated precisely, whereas others have very doubtful magnitudes. The development of methods for improving the accuracy of estimation of reflection losses at the ionosphere depends upon having available accurate estimates of the other components of the transmission loss. The significant roles of diffraction around the curved earth and of refraction in the troposphere are pointed out. It is hoped that the methods of analysis presented in this paper may lead in the future to better empirical determinations of the ionospheric reflection loss than those presently available.

References

1. Bean, B. R. (1958). *Cong. Int. Propagation of Radio Waves*, Belgium.
2. Bean, B. R. (1954). *Proc. Inst. Radio Engrs, N.Y.* **42**, 848–853.
3. Bean, B. R. & Horn, J. D. (1959). "On the climatology of the surface values of radio refractivity of the earth's atmosphere." *J. Res. nat. Bur. Stand*, **63D**, Nov.
4. Bean, B. R. & Meaney, F. M. *Proc. Inst. Radio Engrs, N.Y.* **43**, 1419–1431.
5. Bean, B. R. & Thayer, G. (1959) "A model radio refractivity atmosphere." *Proc. Inst. Radio Engrs, N.Y.* **47**, May.
6. Belrose, J. S. Some investigations of the lowest ionosphere. A dissertation submitted for the degree of Doctor of Philosophy in the University of Cambridge in England, December 1956.
7. Belrose, J. S. Some investigations of the lowest ionosphere. A dissertation submitted for the degree of Doctor of Philosophy in the University of Cambridge in England, December 1956.

* These results may actually be used without appreciable error whenever $R_e > 2R_c$.

8. Bremmer, H. *Terrestrial Radio Waves,* chapter XI. Elsevier Publishing Co. (1949).
9. Doherty R. H. Pulsed sky wave phenomena observed at 100 kc. Unpublished report.
10. Fock, V. A. (1951). *Adv. phys. Sci., Moscow* **43,** 587–599. Also, Paper No. IX, Astia Document No. AD 117276, A series of translation of 13 papers by Fock, U.S. Department of Commerce, Office of Technical Services, Washington 25, D.C. (June 1957).
11. Gautier, T. N. (1955). *Sci. Amer.* **193,** 126–137.
12. Hefley, G., England, C., Kellar, W. & Dyer, R. Skywave delays at 100 kc and noise signal interference in the 100 kc navigation band. Unpublished report.
13. Johler, J. R., Kellar, W. J. & Walters, L. C. *Nat. Bur. Stand. Circ.* No. 573.
14. Katzin, M., Bauchman, R. W. & Binnian, W. *Proc. Inst. Radio Engrs, N.Y.* **35,** 891–905.
15. Martyn, D. F. (1935). *Proc. phys. Soc. Lond.* **47,** 323–339.
16. Norton, K. A. (1953). *Proc. Inst. Radio Engrs, N.Y.* **41,** 146–152.
17. Norton, K. A. (1952). *Trans. Inst. Radio Engrs, N.Y. AP*–**3,** 152–166.
18. Norton, K. A. The Polarization of downcoming ionospheric radio waves. Report prepared in connection with a National Bureau of Standards Project on Direction Finding sponsored by the National Defense Research Committee, 1942.
19. Norton, K. A. (1941). *Proc. Inst. Radio Engrs, N.Y.* **29,** 623–639.
20. Norton, K. A., Rice, P. L. & Vogler, L. E. (1955). *Proc. Inst. Radio Engrs, N.Y.* **43,** 1488–1526.
21. Pressey, B. G., Ashwell, G.E. & Fowler, C. S. (1953). *Proc. Instn elect. Engrs* **100,** 73–84.
22. Rice, S. O. (1954). *Bell Syst. tech. J.* **33,** 417–504.
23. Schelkunoff, S. A. "Electromagnetic Waves," D. Van Nostrand Co. (1943).
24. Schelling, J. C., Burrows, C. R. & Ferrell, E. B. (1933). *Proc. Inst. Radio Engrs, N.Y.* **21,** 427–463.
25. Schneider, A. B. (1955). *J. Brit. Instn Radio Engrs* **12,** no. 3, 181–194.
26. Study Programme No. 90. "Documents of the VIIIth Plenary Assembly, Warsaw, 1956," **1,** 538–539. Published by the International Telecommunication Union, Geneva, 1957.
27. Wait, J. R. (1953). *J. appl. Phys.* **24,** No. 5, 646–649.
28. Wait, J. R. (1956). *Trans. Inst. Radio Engrs, N.Y. AP*–**4,**
29. Wait, J. R. *Congr. Proc. of the Int. on the Propagation of Radio Waves.* October 1958.
30. Wait, J. R. & Conda, A. M. (1958). *Trans. Inst. Radio Engrs, N.Y. AP*–**5.**
31. Wait, J. R. & Conda, A. M. *Congr. Int. on the Propagation of Radio Waves* October 1958.
32. Wait, J. R. & Howe, H. H. (1956). *Nat. Bur. Stand. Circ.* No. 574.
33. Watt, A. D., Maxwell, E. L. & Whelan, E. H. (1959) Observations on some low frequency paths in arctic areas, *J. Res. nat. Bur. Stand.*, **63D,** July.
34. Wait, J. R. & Murphy, A. (1956). *Geofis. pur. Appl.* **35,** 61–72.
35. Wait, J. R. & Murphy, A. (1956). *Geofis. pur. Appl.* **35,** 61–72.
36. Wait, J. R. & Pope, W. A. (1955). *Wireless Engr* **32,** 131–138.
37. Wait, J. R. & Surtees, W. J. (1954). *J. appl. Phys.* **25,** 5, 553–555.

List of the More Important Symbols Used

$P_t = 10 \log_{10} p_t$, the radio frequency power input to the terminals of the transmitting antenna; p_t expressed in watts.

$P_a = 10 \log_{10} p_a$, the resultant radio frequency signal power available at the terminals of the receiving antenna; p_a expressed in watts.

$P_n = 10 \log_{10} p_r$, the radio frequency power radiated from the transmitting antenna; p_r expressed in watts.

$P'_a = 10 \log_{10} p'_a$, the available power from a loss-less receiving antenna.

L_{tc} and L_{rc}, losses, expressed in decibels, in the transmitting and receiving antenna circuits, respectively, excluding the losses associated with the antenna radiation resistances.

$L_{tc} = 10 \log_{10}(r'/r)$, with r' the resistive component of the transmitting antenna circuit and r the radiation resistance.

$L_s = P_t - P_a$, the system loss.

$L = P_r - P'_a$, the transmission loss.

$L_b = L + G_p$, the basic transmission loss.

$L_{bf} = 36{\cdot}58 + 20 \log_{10} f_{Mc} + 20 \log_{10} d_{mi}$, the basic transmission loss in free space.

$G_p = L_b - L$, the path antenna gain.

$G_{pp} = L_b - L_s = G_p - L_{tc} - L_{rc}$, the path antenna power gain.

$L_p = L - L_{tg} - L_{rg}, = L_s - L_e - L_n$, the propagation loss.

$G_t = 10 \log_{10} g_t$ and $G_r = 10 \log_{10} g_r$, the free space gains of the transmitting and receiving antennas, respectively.

G_{tp} and G_{rp}, the power gains of the transmitting and receiving antennas, respectively.

L_{tg} and L_{rg}, environmental loss factors.

$L_{tg} = 10 \log_{10}(r/r_f)$, with r the actual transmitting antenna radiation resistance and r_f the radiation resistance it would have if it were in free space.

L_t and L_r, antenna loss factors for the transmitting and receiving antennas, respectively.

$L_t = 10 \log_{10}(r'/r_f)$, with r the resistive component of the transmitting antenna circuit and r_f the radiation resistance the transmitting antenna would have if it were in free space.

L_{pm}, the median value of the propagation loss.

$C_m(R) = 10 \log_{10} c_m$, the convergence factor for a wave reflected m times at a smooth ionosphere over a path of length R.

P, the polarization loss.

$A(\phi, 1 - p)$, the absorption loss arising from a single reflection at the ionosphere exceeded for $100(1 - p)$ % of the time.

n, the refractive index of the troposphere.

$N = (n - 1) \,.\, 10^6$, the refractivity.

N_s, the refractivity at the surface of the earth.

N_0, the surface refractivity reduced to sea level.

ΔN, the difference in refractivity at the surface and at a height one kilometer above the surface.

τ, the total bending of a radio wave in passing entirely through the troposphere.

TABLE I $h = 70$ km

ψ		$N_s = 0$			$N_s = 301$			$N_s = 400$	
m.r.	$R_1/2$ miles	d_1 miles	$\cos \phi$	$R_1/2$ miles	d_1 miles	$\cos \phi$	$R_1/2$ miles	d_1 miles	$\cos \phi$
0·0	588558 53*	116852 54	0·147007	628625 53	124865 54	0·144985	658709 53	130881 54	0·144319
0·5	586581 53	116457 54	0·147007	625995 53	124339 54	0·144986	655233 53	130186 54	0·144320
1·0	584611 53	116063 54	0·147010	623384 53	123816 54	0·144989	651804 53	129500 54	0·144323
2·0	580691 53	115279 54	0·147020	618219 53	122783 54	0·144999	645083 53	128156 54	0·144333
3·0	576797 53	114500 54	0·147036	613131 53	121766 54	0·145016	638543 53	126848 54	0·144350
4·0	572930 53	113727 54	0·147060	608118 53	120763 54	0·145039	632178 53	125575 54	0·144374
5·0	569090 53	112959 54	0·147090	603179 53	119776 54	0·145070	625984 53	124336 54	0·144404
6·0	565277 53	112196 54	0·147126	598313 53	118802 54	0·145107	619952 53	123130 54	0·144441
7·0	561490 53	111439 54	0·147169	593518 53	117843 54	0·145151	614075 53	121954 54	0·144486
8·0	557729 53	110687 54	0·147219	588791 53	116898 54	0·145201	608345 53	120809 54	0·144536
9·0	553996 53	109940 54	0·147276	584132 53	115966 54	0·145259	602754 53	119690 54	0·144594
10·0	550288 53	109198 54	0·147339	579538 53	115047 54	0·145323	597294 53	118599 54	0·144658
12·0	542953 53	107731 54	0·147485	570537 53	113247 54	0·145471	586737 53	116487 54	0·144807
15·0	532148 53	105570 54	0·147753	557475 53	110635 54	0·145743	571708 53	113482 54	0·145081
20·0	514666 53	102073 54	0·148331	536771 53	106494 54	0·146329	548453 53	108831 54	0·145670
25·0	497832 53	987062 53	0·149071	517255 53	102591 54	0·147080	527018 53	104544 54	0·146424
30·0	481640 53	954666 53	0·149971	498799 53	988990 53	0·147992	507075 53	100554 54	0·147340
40·0	451128 53	893608 53	0·152235	464702 53	920770 53	0·150287	470841 53	933056 53	0·149646
50·0	423022 53	837342 53	0·155096	433914 53	859148 53	0·153187	438614 53	868558 53	0·152558
65·0	385085 53	761346 53	0·160433	393083 53	777370 53	0·158591	396365 53	783948 53	0·157985
80·0	351773 53	694550 53	0·166911	357770 53	706578 53	0·165145	360147 53	711348 53	0·164564
100·0	313660 53	618020 53	0·177094	317856 53	626450 53	0·175437	319468 53	629690 53	0·174892
150·0	242840 53	475288 53	0·208471	244755 53	479160 53	0·207082	245461 53	480588 53	0·206627
200·0	195780 53	379728 53	0·245413	196770 53	381746 53	0·244255	197128 53	382478 53	0·243876
300·0	139660 53	263992 53	0·327194	140003 53	264710 53	0·326370	140125 53	264968 53	0·326100
400·0	108492 53	197703 53	0·412287	108642 53	198029 53	0·411679	108695 53	198145 53	0·411481
600·0	761588 52	124352 53	0·577531	762006 52	124453 53	0·577183	762154 52	124489 53	0·577069
900·0	553385 52	680510 52	0·788639	553476 52	680800 52	0·788494	553508 52	680906 52	0·788447

*588558 53 is to be read $0{\cdot}588558 \times 10^3 = 588{\cdot}558$ miles

TABLE II $h = 90$ km

	$N_s = 0$			$N_s = 301$			$N_s = 400$		
ψ m.r.	$R_1/2$ miles	d_1 miles	cos ϕ	$R_1/2$ miles	d_1 miles	cos ϕ	$R_1/2$ miles	d_1 miles	cos ϕ
0·0	667884 53*	132327 54	0·166304	708924 53	140534 54	0·164530	739332 53	146615 54	0·163948
0·5	665906 53	131931 54	0·166304	706293 53	140007 54	0·164530	735857 53	145920 54	0·163948
1·0	663935 53	131537 54	0·166306	703681 53	139485 54	0·164533	732426 53	145234 54	0·163951
2·0	660010 53	130752 54	0·166315	698511 53	138451 54	0·164542	725700 53	143888 54	0·163960
3·0	656109 53	129972 54	0·166330	693415 53	137432 54	0·164556	719151 53	142579 54	0·163974
4·0	652231 53	129196 54	0·166350	688390 53	136427 54	0·164577	712775 53	141304 54	0·163995
5·0	648377 53	128425 54	0·166377	683437 53	135436 54	0·164604	706565 53	140062 54	0·164022
6·0	644546 53	127659 54	0·166409	678553 53	134459 54	0·164636	700515 53	138852 54	0·164054
7·0	640738 53	126898 54	0·166447	673736 53	133496 54	0·164675	694617 53	137672 54	0·164093
8.0	636954 53	126141 54	0·166491	668985 53	132546 54	0·164719	688862 53	136521 54	0·164137
9·0	633194 53	125389 54	0·166540	664298 53	131609 54	0·164769	683243 53	135398 54	0·164188
10·0	629456 53	124641 54	0·166596	659673 53	130684 54	0·164825	677752 53	134299 54	0·164244
12·0	622052 53	123160 54	0·166724	650600 53	128869 54	0·164955	667122 53	132174 54	0·164374
15·0	611121 53	120974 54	0·166960	637407 53	126231 54	0·165194	651960 53	129141 54	0·164614
20·0	593366 53	117423 54	0·167469	616421 53	122033 54	0·165708	628420 53	124433 54	0·165130
25·0	576188 53	113986 54	0·168120	596546 53	118058 54	0·166367	606622 53	120074 54	0·165792
30·0	559579 53	110663 54	0·168913	577659 53	114280 54	0·167169	586242 53	115997 54	0·166596
40·0	528039 53	104352 54	0·170915	542496 53	107245 54	0·169192	548930 53	108532 54	0·168627
50·0	498669 53	984726 53	0·173452	510401 53	100821 54	0·171757	515380 53	101818 54	0·171200
65·0	458477 53	904212 53	0·178211	467239 53	921768 53	0·176564	470775 53	928856 53	0·176024
80·0	422594 53	832262 53	0·184029	429274 53	845662 53	0·182438	431879 53	850888 53	0·181916
100·0	380770 53	748280 53	0·193256	385544 53	757870 53	0·191747	387348 53	761498 53	0·191253
150·0	300551 53	586604 53	0·222196	302828 53	591206 53	0·220901	303653 53	592876 53	0·220478
200·0	245253 53	474312 53	0·256972	246466 53	476786 53	0·255873	246897 53	477668 53	0·255514
300·0	177164 53	333886 53	0·335520	177599 53	334796 53	0·334721	177752 53	335116 53	0·334460
400·0	138410 53	251460 53	0·418461	138604 53	251880 53	0·417866	138671 53	252026 53	0·417671
600·0	976108 52	158890 53	0·581086	976655 52	159022 53	0·580741	976845 52	159068 53	0·580629

*667884 53 is to be read $0{\cdot}667884 \times 10^3 = 667{\cdot}884$ miles

TABLE III $h = 110$ km

ψ		$N_s = 0$			$N_s = 301$			$N_s = 400$	
m.r.	$R_1/2$ miles	d_1 miles	cos ϕ	$R_1/2$ miles	d_1 miles	cos ϕ	$R_1/2$ miles	d_1 miles	cos ϕ
0·0	738948 53*	146104 54	0·183431	780680 53	154449 54	0·181833	811318 53	160577 54	0·181311
0·5	736971 53	145709 54	0·183432	778049 53	153923 54	0·181834	807842 53	159881 54	0·181312
1·0	734999 53	145314 54	0·183434	775436 53	153400 54	0·181836	804411 53	159195 54	0·181314
2·0	731071 53	144529 54	0·183442	770263 53	152366 54	0·181844	797682 53	157849 54	0·181322
3·0	727164 53	143747 54	0·183455	765161 53	151345 54	0·181857	791127 53	156538 54	0·181335
4·0	723278 53	142970 54	0·183473	760128 53	150339 54	0·181876	784742 53	155261 54	0·181354
5·0	719413 53	142197 54	0·183497	755164 53	149346 54	0·181900	778522 53	154017 54	0·181378
6·0	715570 53	141429 54	0·183526	750267 53	148367 54	0·181929	772459 53	152805 54	0·181407
7·0	711748 53	140664 54	0·183560	745436 53	147401 54	0·181964	766546 53	151622 54	0·181442
8·0	707947 53	139904 54	0·183600	740667 53	146447 54	0·182004	760773 53	150468 54	0·181482
9·0	704167 53	139148 54	0·183644	735960 53	145506 54	0·182049	755135 53	149340 54	0·181527
10·0	700409 53	138396 54	0·183694	731313 53	144576 54	0·182099	749621 53	148238 54	0·181578
12·0	692955 53	136906 54	0·183810	722189 53	142751 54	0·182216	738940 53	146102 54	0·181695
15·0	681933 53	134701 54	0·184023	708903 53	140094 54	0·182431	723684 53	143050 54	0·181910
20·0	663984 53	131111 54	0·184482	687717 53	135857 54	0·182894	699940 53	138302 54	0·182375
25·0	646557 53	127625 54	0·185070	667587 53	131831 54	0·183487	677885 53	133891 54	0·182970
30·0	629649 53	124242 54	0·185786	648391 53	127991 54	0·184210	657194 53	129752 54	0·183695
40·0	597364 53	117781 54	0·187596	612462 53	120802 54	0·186037	619108 53	122132 54	0·185527
50·0	567071 53	111717 54	0·189897	579417 53	114188 54	0·188358	584600 53	115226 54	0·187855
65·0	525210 53	103331 54	0·194227	534541 53	105201 54	0·192725	538266 53	105947 54	0·192234
86·0	487390 53	957482 53	0·199546	494590 53	971920 53	0·198087	497366 53	977492 53	0·197611
100·0	442705 53	867754 53	0·208034	447930 53	878252 53	0·206641	449883 53	882178 53	0·206186
150·0	354913 53	690810 53	0·235005	357490 53	696020 53	0·233788	358414 53	697892 53	0·233391
200·0	292603 53	564280 53	0·267934	294010 53	567150 53	0·266886	294506 53	568162 53	0·266544
300·0	213764 53	401672 53	0·343570	214284 53	402760 53	0·342795	214465 53	403140 53	0·342542
400·0	167896 53	304110 53	0·424489	168130 53	304620 53	0·423906	168211 53	304796 53	0·423716
600·0	118933 53	193007 53	0·584586	119000 53	193169 53	0·584246	119023 53	193226 53	0·584135
900·0	867957 52	106078 53	0·791582	868105 52	106125 53	0·791440	868156 52	106142 53	0·791393

*738948 53 is to be read $0{\cdot}738948 \times 10^3 = 738{\cdot}948$ miles

TABLE IV $h = 225$ km

ψ m.r.	$N_s = 0$ $R_1/2$ miles	d_1 miles	$\cos\phi$	$N_s = 301$ $R_1/2$ miles	d_1 miles	$\cos\phi$	$N_s = 400$ $R_1/2$ miles	d_1 miles	$\cos\phi$
0·0	106179 54*	207428 54	0·258920	110549 54	216166 54	0·257826	113678 54	222422 54	0·257474
0·5	105981 54	207032 54	0·258920	110286 54	215640 54	0·257827	113330 54	221728 54	0·257475
1·0	105783 54	206638 54	0·258922	110024 54	215116 54	0·257828	112987 54	221040 54	0·257476
2·0	105390 54	205850 54	0·258927	109506 54	214080 54	0·257834	112313 54	219692 54	0·257481
3·0	104997 54	205066 54	0·258936	108994 54	213056 54	0·257843	111656 54	218378 54	0·257490
4·0	104607 54	204284 54	0·258949	108489 54	212046 54	0·257855	111015 54	217096 54	0·257503
5·0	104217 54	203504 54	0·258965	107989 54	211046 54	0·257872	110390 54	215846 54	0·257519
6·0	103829 54	202728 54	0·258985	107496 54	210060 54	0·257892	109780 54	214628 54	0·257539
7·0	103443 54	201956 54	0·259008	107009 54	209086 54	0·257915	109184 54	213436 54	0·257563
8·0	103058 54	201186 54	0·259035	106527 54	208122 54	0·257942	108602 54	212272 54	0·257590
9·0	102674 54	200420 54	0·259066	106051 54	207170 54	0·257973	108032 54	211132 54	0·257621
10·0	102292 54	199655 54	0·259100	105580 54	206228 54	0·258007	107475 54	210018 54	0·257655
12·0	101533 54	198136 54	0·259179	104653 54	204374 54	0·258087	106392 54	207852 54	0·257735
15·0	100405 54	195879 54	0·259325	103298 54	201664 54	0·258233	104840 54	204748 54	0·257882
20·0	985538 53	192177 54	0·259639	101122 54	197312 54	0·258549	102408 54	199885 54	0·258198
25·0	967396 53	188549 54	0·260043	990359 53	193140 54	0·258955	100129 54	195327 54	0·258605
30·0	949620 53	184993 54	0·260536	970279 53	189124 54	0·259450	979709 53	191011 54	0·259100
40·0	915160 53	178098 54	0·261785	932132 53	181493 54	0·260705	939392 53	182946 54	0·260358
50·0	882139 53	171488 54	0·263382	896308 53	174324 54	0·262310	902087 53	175481 54	0·261964
65·0	835250 53	162098 54	0·266413	846311 53	164313 54	0·265355	850602 53	165173 54	0·265014
80·0	791441 53	153316 54	0·270181	800264 53	155085 54	0·269140	803572 53	155748 54	0·268805
100·0	737602 53	142508 54	0·276293	744298 53	143853 54	0·275279	746732 53	144342 54	0·274953
150·0	623696 53	119559 54	0·296436	627364 53	120300 54	0·295502	628644 53	120559 54	0·295202
200·0	534648 53	101482 54	0·322278	536832 53	101927 54	0·321435	537580 53	102080 54	0·321163
300·0	409770 53	757298 53	0·385379	410680 53	759200 53	0·384709	410987 53	759844 53	0·384494
400·0	330065 53	587778 53	0·456640	330504 53	588739 53	0·456115	330651 53	589050 53	0·455946
600·0	239192 53	381496 53	0·603728	239326 53	381820 53	0·603409	239370 53	381930 53	0·603306
900·0	176605 53	212096 53	0·799691	176636 53	212194 53	0·799554	176646 53	212226 53	0·799510

*106179 54 is to be read $0{\cdot}106179 \times 10^4 = 1061{\cdot}79$ miles

TABLE V $h = 350$ km

ψ m.r.	$N_s = 0$ $R_1/2$ miles	$N_s = 0$ d_1 miles	$N_s = 0$ $\cos\phi$	$N_s = 301$ $R_1/2$ miles	$N_s = 301$ d_1 miles	$N_s = 301$ $\cos\phi$	$N_s = 400$ $R_1/2$ miles	$N_s = 400$ d_1 miles	$N_s = 400$ $\cos\phi$
0·0	133071 54*	256682 54	0·318449	137533 54	265602 54	0·317591	140690 54	271916 54	0·317318
0·5	132873 54	256286 54	0·318449	137269 54	265076 54	0·317591	140343 54	271222 54	0·317318
1·0	132676 54	255890 54	0·318450	137008 54	264552 54	0·317592	139999 54	270534 54	0·317319
2·0	132281 54	255102 54	0·318455	136489 54	263514 54	0·317596	139325 54	269186 54	0·317324
3·0	131888 54	254316 54	0·318462	135976 54	262490 54	0·317603	138667 54	267870 54	0·317331
4·0	131496 54	253532 54	0·318472	135470 54	261476 54	0·317613	138025 54	266586 54	0·317341
5·0	131106 54	252750 54	0·318484	134969 54	260474 54	0·317626	137399 54	265334 54	0·317353
6·0	130716 54	251972 54	0·318500	134474 54	259484 54	0·317642	136787 54	264110 54	0·317369
7·0	130328 54	251196 54	0·318518	133985 54	258506 54	0·317660	136189 54	262916 54	0·317387
8·0	129941 54	250420 54	0·318539	133500 54	257538 54	0·317681	135605 54	261746 54	0·317409
9·0	129555 54	249648 54	0·318563	133022 54	256580 54	0·317705	135033 54	260602 54	0·317433
10·0	129170 54	248880 54	0·318590	132548 54	255632 54	0·317732	134472 54	259482 54	0·317460
12·0	128404 54	247348 54	0·318652	131614 54	253766 54	0·317794	133383 54	257304 54	0·317522
15·0	127263 54	245066 54	0·318766	130247 54	251032 54	0·317909	131818 54	254174 54	0·317637
20·0	125386 54	241312 54	0·319013	128045 54	246628 54	0·318156	129360 54	249258 54	0·317884
25·0	123539 54	237618 54	0·319329	125925 54	242388 54	0·318474	127047 54	244634 54	0·318202
30·0	121721 54	233980 54	0·319716	123876 54	238290 54	0·318862	124848 54	240234 54	0·318590
40·0	118172 54	226880 54	0·320697	119958 54	230452 54	0·319846	120713 54	231962 54	0·319576
50·0	114740 54	220012 54	0·321954	116245 54	223022 54	0·321107	116851 54	224234 54	0·320838
65·0	109808 54	210134 54	0·324347	110999 54	212518 54	0·323508	111455 54	213434 54	0·323241
80·0	105130 54	200758 54	0·327335	106094 54	202690 54	0·326506	106451 54	203406 54	0·326242
100·0	992777 53	189010 54	0·332213	100024 54	190509 54	0·331399	100292 54	191048 54	0·331140
150·0	864457 53	163158 54	0·348535	868760 53	164027 54	0·347769	870245 53	164328 54	0·347525
200·0	758979 53	141748 54	0·369968	761667 53	142296 54	0·369259	762577 53	142481 54	0·369034
300·0	601625 53	109296 54	0·424129	602833 53	109548 54	0·423541	603235 53	109633 54	0·423355
400·0	494569 53	865258 53	0·487522	495183 53	866586 53	0·487047	495386 53	867028 53	0·486896
600·0	365833 53	572904 53	0·622816	366031 53	573382 53	0·622517	366096 53	573540 53	0·622422
900·0	273231 53	322086 53	0·807951	273277 53	322234 53	0·807821	273293 53	322284 53	0·807780

*133071 54 is to be read $0{\cdot}133071 \times 10^4 = 1330{\cdot}71$ miles

TABLE VI $h = 475$ km

		$N_s = 0$			$N_s = 301$			$N_s = 400$	
ψ m.r.	$R_1/2$ miles	d_1 miles	$\cos\phi$	$R_1/2$ miles	d_1 miles	$\cos\phi$	$R_1/2$ miles	d_1 miles	$\cos\phi$
0·0	155762 54*	296720 54	0·365944	160276 54	305746 54	0·365222	163450 54	312094 54	0·364996
0·5	155564 54	296324 54	0·365944	160012 54	305220 54	0·365223	163102 54	311398 54	0·364996
1.0	155366 54	295930 54	0·365945	159751 54	304696 54	0·365224	162759 54	310712 54	0·364997
2·0	154972 54	295140 54	0·365949	159232 54	303658 54	0·365227	162084 54	309362 54	0·365001
3·0	154578 54	294354 54	0·365955	158719 54	302632 54	0·365233	161426 54	308046 54	0·365007
4·0	154186 54	293568 54	0·365963	158211 54	301618 54	0·365241	160783 54	306760 54	0·365015
5·0	153794 54	292786 54	0·365974	157710 54	300614 54	0·365252	160156 54	305506 54	0·365026
6·0	153404 54	292004 54	0·365987	157214 54	299622 54	0·365265	159544 54	304282 54	0·365039
7·0	153014 54	291226 54	0·366002	156723 54	298642 54	0·365281	158945 54	303084 54	0·365054
8·0	152626 54	290448 54	0·366020	156238 54	297670 54	0·365298	158359 54	301912 54	0·365072
9·0	152238 54	289674 54	0·366040	155758 54	296710 54	0·365319	157785 54	300766 54	0·365092
10·0	151852 54	288902 54	0·366062	155282 54	295760 54	0·365341	157223 54	299642 54	0·365115
12·0	151082 54	287362 54	0·366114	154345 54	293886 54	0·365393	156130 54	297456 54	0·365167
15·0	149935 54	285068 54	0·366210	152971 54	291136 54	0·365489	154559 54	294312 54	0·365263
20·0	148043 54	281284 54	0·366417	150753 54	286702 54	0·365697	152085 54	289366 54	0·365471
25·0	146176 54	277550 54	0·366683	148614 54	282424 54	0·365963	149753 54	284702 54	0·365737
30·0	144334 54	273866 54	0·367007	146542 54	278280 54	0·366288	147530 54	280256 54	0·366063
40·0	140727 54	266648 54	0·367832	142564 54	270322 54	0·367115	143335 54	271864 54	0·366890
50·0	137219 54	259628 54	0·368888	138774 54	262738 54	0·368174	139396 54	263984 54	0·367950
65·0	132145 54	249466 54	0·370903	133385 54	251950 54	0·370194	133857 54	252896 54	0·369972
80·0	127291 54	239738 54	0·373425	128304 54	241766 54	0·372722	128676 54	242514 54	0·372502
100·0	121158 54	227426 54	0·377554	121951 54	229018 54	0·376862	122233 54	229586 54	0·376644
150·0	107432 54	199773 54	0·391480	107903 54	200724 54	0·390820	108064 54	201050 54	0·390613
200·0	958097 53	176182 54	0·409999	961124 53	176799 54	0·409380	962142 53	177007 54	0·409186
300·0	777739 53	138983 54	0·457769	779170 53	139282 54	0·457242	779643 53	139381 54	0·457077
400·0	649497 53	111704 54	0·515028	650251 53	111867 54	0·514593	650499 53	111921 54	0·514457
600·0	488805 53	751980 53	0·640345	489059 53	752594 53	0·640064	489141 53	752796 53	0·639976
900·0	368904 53	427016 53	0·815687	368966 53	427214 53	0·815562	368986 53	427280 53	0·815523

*155762 54 is to be read $0{\cdot}155762 \times 10^4 = 1557{\cdot}62$ miles

TABLE VII

$h = 70$ km: $N_s = 301$

ψ m.r.	c_1 c_{11}	c_2 c_{12}	c_3 c_{13}	c_4 c_{14}	c_5 c_{15}	c_6 c_{16}	c_7 c_{17}	c_8 c_{18}	c_9 c_{19}	c_{10} c_{20}
0·5	24508 53	25768 53	28079 53	31842 53	37845 53	47734 53	65429 53	10280 54	21931 54	40469 56
	27115 54	15500 54	12184 54	11151 54	11353 54	12721 54	15868 54	23072 54	46036 54	40467 56
1·0	12321 53	12949 53	14100 53	15971 53	18950 53	23841 53	32545 53	50745 53	10597 54	25172 55
	14217 54	79361 53	61971 53	56187 53	56864 53	63283 53	78210 53	11192 54	21425 54	25175 55
2·0	61999 52	65103 52	70784 52	79996 52	94600 52	11842 53	16038 53	24642 53	49488 53	45994 54
	78319 53	41486 53	31671 53	28424 53	28428 53	31223 53	37916 53	52705 53	94002 53	46033 54
3·0	41598 52	43645 52	47386 52	53435 52	62985 52	78468 52	10546 53	15982 53	30990 53	18851 54
	58110 53	28993 53	21679 53	19200 53	18980 53	20584 53	24588 53	33295 53	56024 53	18893 54
4·0	31400 52	32919 52	35690 52	40160 52	47189 52	58516 52	78079 52	11679 53	21939 53	10245 54
	49105 53	22859 53	16721 53	14610 53	14277 53	15296 53	17991 53	23791 53	38115 53	10289 54
5·0	25282 52	26484 52	28674 52	32197 52	37717 52	46561 52	61695 52	91158 52	16638 53	64461 53
	44946 53	19280 53	13778 53	11874 53	11472 53	12146 53	14079 53	18218 53	27986 53	64905 53
6·0	21202 52	22193 52	23996 52	26890 52	31406 52	38602 52	50808 52	74203 52	13192 53	44376 53
	43713 53	16993 53	11844 53	10064 53	96147 52	10063 53	11504 53	14592 53	21613 53	44825 53
7·0	18286 52	19127 52	20654 52	23098 52	26900 52	32925 52	43057 52	62194 52	10794 53	32470 53
	44916 53	15459 53	10489 53	87855 52	82981 52	85889 52	96902 52	12067 53	17307 53	32922 53
8·0	16098 52	16826 52	18146 52	20254 52	23522 52	28673 52	37264 52	53267 52	90420 52	24829 53
	48951 53	14410 53	94980 52	78375 52	73194 52	74936 52	83487 52	10220 53	14245 53	25283 53
9·0	14394 52	15034 52	16193 52	18040 52	20894 52	25369 52	32774 52	46385 52	77140 52	19630 53
	57516 53	13704 53	87515 52	71108 52	65658 52	66506 52	65332 52	88197 52	11981 53	20085 53
10·0	13078 52	13650 52	14685 52	16330 52	18863 52	22815 52	29306 52	41086 52	67039 52	15990 53
	76308 53	13309 53	82097 52	65645 52	59924 52	60064 52	77562 52	77562 52	10294 53	16448 53
12·0	11051 52	11519 52	12363 52	13699 52	15743 52	18903 52	24019 52	33084 52	52165 52	11205 53
	85170 54	13083 53	74412 52	57496 52	51281 52	53052 52	53592 52	61878 52	78854 52	11665 53
15·0	90180 51	93817 51	10035 52	11064 52	12624 52	15003 52	18780 52	25263 52	38131 52	72940 52
	39945 53	14258 53	68663 52	50032 52	43027 52	40986 52	42303 52	47096 52	57076 52	77588 52

20·0	68980 51	71552 51	76147 51	83313 51	94031 51	11006 52	13477 52	17533 52	24975 52	41920 52
	11045 53	25794 53	68677 52	43962 52	35316 52	31970 52	31467 52	33260 52	37757 52	46611 52
25·0	56154 51	58092 51	61538 51	66869 51	74743 51	86321 51	10373 52	13125 52	17873 52	27474 52
	54971 52	52439 53	83406 52	43181 52	31852 52	27294 52	25665 52	25924 52	27943 52	32248 52
30·0	47905 51	49439 51	52153 51	56319 51	62408 51	71224 51	84198 51	10406 52	13668 52	19712 52
	33882 52	98466 52	14694 53	47729 52	31251 52	25123 52	22540 52	21824 52	22507 52	24655 52
40·0	36953 51	37975 51	39769 51	42487 51	46386 51	51888 51	59699 51	71071 51	88395 51	11683 52
	16986 52	29675 52	93230 52	10525 53	37660 52	24916 52	19966 52	17735 52	16907 52	17062 52
50·0	30523 51	31255 51	32533 51	34448 51	37153 51	40892 51	46053 51	53282 51	63717 51	79532 51
	10544 52	15379 52	27046 52	88308 52	87711 52	32484 52	21492 52	17104 52	15027 52	14113 52
65·0	24503 51	24982 51	25812 51	27040 51	28743 51	31040 51	34111 51	38232 51	43849 51	51716 51
	63200 51	81051 51	11171 52	17452 52	36598 52	89830 54	41834 52	22530 52	16368 52	13531 52
80·0	20886 51	21223 51	21801 51	22649 51	23810 51	25346 51	27351 51	29961 51	33381 51	37928 51
	44118 51	52842 51	65777 51	86467 51	12392 52	20960 52	58219 52	95433 52	28647 52	17908 52
100·0	17965 51	18192 51	18579 51	19142 51	19901 51	20888 51	22146 51	23737 51	25748 51	28301 51
	31581 51	35865 51	45499 51	49536 51	61055 51	78972 51	11005 52	17565 52	39645 52	25566 53
150·0	14187 51	14292 51	14468 51	14721 51	15054 51	15477 51	15998 51	16631 51	17392 51	18303 51
	19392 51	20695 51	22261 51	24157 51	26471 51	29332 51	32925 51	37530 51	43588 51	51839 51
200·0	12648 51	12707 51	12806 51	12947 51	13131 51	13361 51	13640 51	13973 51	14364 51	14819 51
	15347 51	15957 51	16659 51	17469 51	18404 51	19486 51	20743 51	22212 51	23939 51	25987 51
300·0	11345 51	11370 51	11413 51	11472 51	11550 51	11646 51	11761 51	11895 51	12050 51	12227 51
	12426 51	12650 51	12899 51	13177 51	13485 51	13825 51	14201 51	14616 51	15074 51	15580 51
400·0	10889 51	10902 51	10925 51	10957 51	10998 51	11049 51	11109 51	11180 51	11260 51	11351 51
	11453 51	11566 51	11690 51	11827 51	11976 51	12138 51	12315 51	12506 51	12712 51	12935 51
600·0	10524 51	10529 51	10538 51	10550 51	10566 51	10585 51	10608 51	10634 51	10664 51	10698 51
	10735 51	10776 51	10821 51	10870 51	10922 51	10979 51	11040 51	11105 51	11175 51	11249 51

TABLE VIII

$h = 90$ km: $N_s = 301$

ψ m.r.	c_1 / c_{11}	c_2 / c_{12}	c_3 / c_{13}	c_4 / c_{14}	c_5 / c_{15}	c_6 / c_{16}	c_7 / c_{17}	c_8 / c_{18}	c_9 / c_{19}	c_{10} / c_{20}
0·5	27821 53	29656 53	33114 53	39007 53	49111 53	67810 53	10901 54	25004 54	21543 55	25104 54
	15589 54	12964 54	12607 54	13873 54	17396 54	26283 54	60726 54	21561 55	43519 54	27185 54
1·0	13989 53	14904 53	16628 53	19560 53	24574 53	33811 53	53990 53	12133 54	15308 55	12991 54
	79357 53	65489 53	63303 53	69214 53	86035 53	12798 54	28176 54	15314 55	23058 54	13993 54
2·0	70409 52	74943 52	83467 52	97920 52	12250 53	16740 53	26383 53	57092 53	44467 55	69480 53
	40981 53	33293 53	31797 53	34332 53	41971 53	60661 53	12299 54	44468 55	13029 54	74086 53
3·0	47253 52	50249 52	55870 52	65373 52	81455 52	11057 53	17212 53	35989 53	79118 54	49778 53
	28262 53	22593 53	21321 53	22742 53	27367 53	38532 53	73001 53	79132 54	99951 53	52594 53
4·0	35677 52	37904 52	42076 52	49106 52	60945 52	82208 53	12646 53	25627 53	26380 54	40332 53
	21962 53	17267 53	16102 53	16972 53	20121 53	27658 53	49488 53	26402 54	88301 53	42262 53
5·0	28731 52	30497 52	33799 52	39348 52	48648 52	65218 52	99204 52	19534 53	13637 54	35066 53
	18232 53	14091 53	12984 53	13528 53	15811 53	21260 53	36221 53	13664 54	85917 53	36469 53
6·0	24099 52	25558 52	28281 52	32844 52	40455 52	53913 52	81134 52	15559 53	84261 53	31978 53
	15792 53	11990 53	10916 53	11245 53	12966 53	17082 53	27889 53	84572 53	91271 53	33035 53
7·0	20788 52	22028 52	24338 52	28197 52	34605 52	45854 52	68304 52	12783 53	57555 53	30240 53
	14093 53	10503 53	94486 52	96262 52	10955 53	14161 53	22270 53	57891 54	10765 54	31053 53
8·0	18302 52	19378 52	21378 52	24710 52	30220 52	39821 52	58740 52	10747 53	41959 53	29480 53
	12861 53	94014 52	83553 52	84200 52	94638 52	12017 53	18279 53	42312 53	15032 54	30113 53
9·0	16367 52	17314 52	19074 52	21997 52	26809 52	35138 52	51349 52	91987 52	32027 53	29548 53
	11945 53	85564 52	75117 52	74890 52	83167 52	10386 53	15333 53	32394 53	32638 54	30043 53
10·0	14872 52	15720 52	17293 52	19899 52	24171 52	31516 52	45646 52	80175 52	25393 53	30560 53
	11298 53	79217 52	68687 52	67757 52	74375 52	91441 52	13138 53	25772 53	68898 54	30947 53
12·0	12568 52	13264 52	14551 52	16672 52	20122 52	25978 52	36997 52	62701 52	17116 53	35781 53
	10422 53	69744 52	58953 52	56951 52	61141 52	73029 52	10008 53	17512 53	76209 53	36009 53
15·0	10255 52	10799 52	11800 52	13438 52	16073 52	20466 52	28487 52	46102 52	10733 53	70031 53
	99128 52	61076 52	49582 52	46435 52	48321 52	55541 52	71819 52	11146 53	27139 53	70105 53

20·0	78422 51	82298 51	89385 51	10085 52	11898 52	14841 52	19987 52	30397 52	59469 52	39613 53
	10655 53	54059 52	40648 52	36126 52	35823 52	38979 52	46760 52	63741 52	10880 53	39685 53
25·0	63799 51	66742 51	72091 51	80660 51	93996 51	11515 52	15080 52	21836 52	38098 52	11728 53
	15021 53	53272 52	36372 52	30564 52	28942 52	30031 52	33981 52	42488 52	61360 52	11881 53
30·0	54364 51	56709 51	60945 51	67670 51	77993 51	94038 51	12022 52	16726 52	26883 52	61131 52
	52036 53	58791 52	35115 52	27685 52	25027 52	24866 52	26800 52	31427 52	40965 52	63188 52
40·0	41819 51	43400 51	46230 51	50650 51	57275 51	67220 51	82631 51	10811 52	15549 52	26631 52
	75966 52	12442 53	39315 52	25888 52	21083 52	19270 52	19130 52	20404 52	23415 52	29346 52
50·0	34408 51	35554 51	37587 51	40720 51	45324 51	52043 51	62039 51	77595 51	10378 52	15443 52
	28491 52	12029 53	67360 52	29579 52	20744 52	17273 52	15891 52	15730 52	16593 52	18657 52
65·0	27437 51	28198 51	29532 51	31555 51	34459 51	38556 51	44377 51	52857 51	65787 51	87048 51
	12679 52	22240 52	71250 52	75885 52	27701 52	18406 52	14780 52	13147 52	12548 52	12678 52
80·0	23207 51	23746 51	24685 51	26090 51	28068 51	30790 51	34526 51	39720 51	47138 51	58216 51
	75959 51	10783 52	17885 52	45466 52	11999 53	28972 52	17812 52	13706 52	11784 52	10879 52
100·0	19737 51	20104 51	20737 51	21672 51	22964 51	24697 51	26997 51	30057 51	34179 51	39862 51
	47983 51	60223 51	80247 51	11782 52	21035 52	75507 52	57236 52	22639 52	15030 52	11855 52
150·0	15215 51	15387 51	15678 51	16099 51	16664 51	17393 51	18315 51	19468 51	20905 51	22699 51
	24957 51	27832 51	31555 51	36493 51	43258 51	52951 51	67769 51	92804 51	14312 52	29160 52
200·0	13312 51	13410 51	13574 51	13808 51	14118 51	14510 51	14993 51	15580 51	16285 51	17129 51
	18136 51	19340 51	20786 51	22533 51	24662 51	27289 51	30580 51	34784 51	40292 51	47756 51
300·0	11693 51	11735 51	11805 51	11905 51	12034 51	12195 51	12390 51	12619 51	12887 51	13196 51
	13549 51	13952 51	14410 51	14928 51	15516 51	16181 51	16936 51	17794 51	18772 51	19891 51
400·0	11101 51	11124 51	11161 51	11214 51	11283 51	11368 51	11469 51	11587 51	11724 51	11879 51
	12053 51	12249 51	12466 51	12708 51	12974 51	13268 51	13592 51	13947 51	14337 51	14766 51
600·0	10636 51	10645 51	10659 51	10608 51	10705 51	10737 51	10775 51	10819 51	10869 51	10925 51
	10987 51	11057 51	11132 51	11215 51	11305 51	11402 51	11507 51	11619 51	11740 51	11869 51

TABLE IX

$h = 110$ km: $N_s = 301$

ψ m.r.	c_1 c_{11}	c_2 c_{12}	c_3 c_{13}	c_4 c_{14}	c_5 c_{15}	c_6 c_{16}	c_7 c_{17}	c_8 c_{18}	c_9 c_{19}	c_{10} c_{20}
0·5	30838 53	33323 53	38141 53	46752 53	62737 53	96854 53	20023 54	29060 55	30137 54	17234 54
	14188 54	14040 54	16117 54	21938 54	40050 54	29075 55	62344 54	32159 54	24893 54	23448 54
1·0	15507 53	16748 53	19151 53	23437 53	31367 53	48186 53	98345 53	10969 55	15605 54	87648 53
	71612 53	70439 53	80312 53	10819 54	19294 54	10979 55	33493 54	16580 54	12642 54	11783 54
2·0	78066 52	84223 52	96119 52	11725 53	15612 53	23755 53	47311 53	36860 54	83549 53	45168 53
	36339 53	35320 53	39741 53	52501 53	89706 53	36935 54	19560 54	88050 53	65055 53	59349 53
3·0	52401 52	56475 52	64330 52	78238 52	10364 53	15626 53	30420 53	18578 54	59918 53	31075 53
	24610 53	23639 53	26259 53	34057 53	56021 53	18645 54	15666 54	62679 53	44768 53	39967 53
4·0	39570 52	42603 52	48439 52	58735 52	77430 52	11571 53	22050 53	11237 54	48592 53	24082 53
	18766 53	17815 53	19545 53	24914 53	39606 53	11299 54	14694 54	50490 53	34763 53	30359 53
5·0	31871 52	34279 52	38905 52	47036 52	61714 52	91449 52	17081 53	75528 53	42282 53	19931 53
	15276 53	14333 53	15537 53	19482 53	30030 53	76118 53	15648 54	43667 53	28882 53	24660 53
6·0	26736 52	28728 52	32547 52	39238 52	51245 52	75318 52	13806 53	54386 53	38586 53	17205 53
	12963 53	12021 53	12879 53	15900 53	23829 53	23829 53	19406 54	39632 53	25074 53	20916 53
7·0	23065 52	24761 52	28004 52	33666 52	43772 52	63831 52	11494 53	41118 53	36512 53	15295 53
	11322 53	10378 53	10992 53	13371 53	19529 53	41677 53	32179 54	37319 53	22464 53	18291 53
8·0	20310 52	21782 52	24593 52	29486 52	38170 52	55241 52	97830 52	32234 53	35612 53	13899 53
	10101 53	91522 52	95860 52	11496 53	16398 53	32783 53	36774 55	36241 53	20617 53	16368 53
9·0	18163 52	19462 52	21938 52	26232 52	33815 52	48581 52	84684 52	25989 53	35707 53	12848 53
	91620 52	82041 52	84995 52	10056 53	14032 53	26529 53	29972 54	36200 53	19293 53	14915 53
10·0	16505 52	17670 52	19885 52	23715 52	30445 52	43431 52	74580 52	21506 53	36936 53	12089 53
	84506 52	74785 52	76649 52	89519 52	12238 53	22042 53	13037 54	37323 53	18422 53	13845 53
12·0	13950 52	14908 52	16724 52	19846 52	25276 52	35576 52	59417 52	15435 53	43231 53	11014 53
	73782 52	63753 52	63986 52	72918 52	96063 52	15963 53	54098 53	43459 53	17441 53	12311 53
15·0	11382 52	12134 52	13551 52	15966 52	20109 52	27782 52	44716 52	10291 53	84073 53	10223 53
	63652 52	53001 52	51594 52	56837 52	71373 52	10812 53	25195 53	84148 53	17657 53	11080 53

20·0	87030 51	92415 51	10249 52	11944 52	14791 52	19881 52	30384 52	60550 52	48748 53	10272 53
	54506 52	42439 52	39304 52	41152 52	48386 52	65661 52	11413 53	48820 53	23896 53	10716 53
25·0	70774 51	74886 51	82524 51	95222 51	11614 52	15242 52	22323 52	40185 52	14348 53	12419 53
	51181 52	36841 52	32438 52	32422 52	36107 52	45269 52	67286 52	14502 53	10530 54	12640 53
30·0	60263 51	63554 51	69628 51	79615 51	95796 51	12309 52	17394 52	29009 52	74654 52	21211 53
	52390 52	34184 52	28523 52	27274 52	28975 52	34154 52	45855 52	76724 52	25609 53	21298 53
40·0	46274 51	48514 51	52600 51	59192 51	69561 51	86274 51	11514 52	17232 52	32410 52	15265 53
	72027 52	33627 52	24620 52	21523 52	21043 52	22555 52	26544 52	35132 52	56385 52	15329 53
50·0	37978 51	39614 51	42568 51	47256 51	54451 51	65628 51	83858 51	11656 52	18701 52	42298 52
	39389 53	41385 52	24618 52	19375 52	17495 52	17364 52	18693 52	21890 52	28474 52	43748 52
65·0	30148 51	31244 51	33199 51	36240 51	40768 51	47502 51	57797 51	74473 51	10441 52	16973 52
	39988 52	20580 53	33539 52	20298 52	15899 52	14147 52	13719 52	14281 52	15923 52	19187 52
80·0	25365 51	26148 51	27532 51	29649 51	32728 51	37156 51	43611 51	53362 51	69049 51	97129 51
	15850 52	37875 52	16389 53	29742 52	17987 52	13948 52	12216 52	11593 52	11723 52	12566 52
100·0	21401 51	21938 51	22877 51	24289 51	26292 51	29079 51	32957 51	38451 51	46505 51	58993 51
	80177 51	12224 52	23932 52	17907 53	38395 52	19038 52	13611 52	11275 52	10184 52	97879 51
150·0	16202 51	16455 51	16890 51	17527 51	18396 51	19545 51	21038 51	22974 51	25498 51	28830 51
	33328 51	39593 51	48731 51	62995 51	87789 51	14007 52	31443 52	23197 53	27399 52	15528 52
200·0	13963 51	14107 51	14353 51	14707 51	15180 51	15788 51	16552 51	17501 51	18673 51	20122 51
	21922 51	24178 51	27044 51	30754 51	35677 51	42436 51	52159 51	67130 51	92748 51	14557 52
300·0	12042 51	12105 51	12210 51	12360 51	12556 51	12801 51	13100 51	13458 51	13880 51	14374 51
	14951 51	15620 51	16398 51	17303 51	18358 51	19593 51	21047 51	22772 51	24836 51	27334 51
400·0	11320 51	11354 51	11410 51	11489 51	11593 51	11721 51	11874 51	12055 51	12265 51	12506 51
	12780 51	13090 51	13439 51	13831 51	14271 51	14764 51	15315 51	15934 51	16627 51	17407 51
600·0	10756 51	10769 51	10790 51	10820 51	10859 51	10907 51	10963 51	11029 51	11105 51	11190 51
	11285 51	11391 51	11507 51	11635 51	11774 51	11926 51	12090 51	12268 51	12460 51	12667 51

TABLE X

THE TOTAL BENDING, τ, CRITICAL Range, R_c, and $R_e - (R_1/2)$ FOR THE C.R.P.L. REFERENCE RADIO REFRACTIVITY ATMOSPHERES

	$N_s = 200{\cdot}0$ $h_s = 10{,}000'$			$N_s = 250{\cdot}0$ $h_s = 5{,}000'$			$N_s = 301{\cdot}0$ $h_s = 1{,}000'$		
ψ	τ	R_c	$R_e - (R_1/2)$	τ	R_c	$R_e - (R_1/2)$	τ	R_c	$R_e - (R_1/2)$
m.r.	m.r.	km	meters	m.r.	km	meters	m.r.	km	meters
0·0	7·1845	2000·400	62·2	9·4504	198·752	78·1	12·1522	196·588	95·40
0·5	7·1022	199·007	61·5	9·3356	197·274	77·1	11·9871	195·049	94·1
1·0	7·0211	197·632	60·8	9·2226	195·817	76·2	11·8252	193·537	92·9
2·0	6·8630	194·937	59·4	9·0025	192·966	74·3	11·5109	190·592	90·5
3·0	6·7100	192·309	58·1	8·7899	190·195	72·5	11·2090	187·744	88·3
4·0	6·5622	189·743	56·7	8·5848	187·498	70·8	10·9194	184·987	86·1
5·0	6·4193	187·235	55·4	8·3869	184·870	69·1	10·6416	182·313	83·9
6·0	6·2813	184·782	54·3	8·1962	182·308	67·6	10·3754	179·717	82·0
7·0	6·1480	182·381	53·1	8·0124	179·807	66·1	10·1202	177·190	80·0
8·0	6·0194	180·027	51·9	7·8353	177·363	64·7	9·8757	174·730	78·1
9·0	5·8951	177·719	50·8	7·6645	174·973	63·2	9·6413	172·331	76·3
10·0	5·7751	175·454	49·8	7·4999	172·635	61·9	9·4165	169·988	74·6
12·0	5·5470	171·046	47·8	7·1882	168·104	59·3	8·9940	165·461	71·4
15·0	5·2326	164·720	45·0	6·7602	161·641	55·7	8·4207	159·027	67·0
20·0	4·7721	154·868	40·97	6·1379	151·664	50·58	7·5998	149·131	60·57
25·0	4·3771	145·812	37·44	5·6086	142·576	46·11	6·9125	140·141	55·19
30·0	4·0352	137·488	34·54	5·1539	134·283	42·31	6·3292	131·954	50·52
40·0	3·4745	122·821	29·49	4·4155	119·781	36·33	5·3943	117·660	43·17
50·0	3·0364	110·443	25·80	3·8448	107·623	31·50	4·6808	105·694	37·39
65·0	2·5379	95·332	21·46	3·2016	92·847	26·22	3·8847	91·165	31·17
80·0	2·1689	83·422	18·30	2·7295	81·232	22·29	3·3051	79·749	26·50
100·0	1·8076	71·142	15·30	2·2702	69·270	18·55	2·7442	67·997	21·96
150·0	1·2606	51·401	10·68	1·5794	50·052	13·02	1·9053	49·122	15·39
200·0	0·9586	39·975	8·19	1·2000	38·928	9·97	1·4464	38·201	11·80
300·0	0·6379	27·600	5·59	0·7978	26·879	6·80	0·9611	26·375	8·03
400·0	0·4693	21·156	4·26	0·5868	20·604	5·19	0·7068	20·217	6·12
600·0	0·2912	14·699	2·94	0·3641	14·316	3·59	0·4384	14·047	4·25
900·0	0·1584	10·630	2·13	0·1980	10·354	2·59	0·2384	10·158	3·06

TABLE XI

THE TOTAL BENDING, τ, CRITICAL RANGE, R_c, and $R_e - (R_1/2)$ FOR THE C.R.P.L. REFERENCE RADIO REFRACTIVITY ATMOSPHERES

	$N_s = 350{\cdot}0$ $h_s = 0'$			$N_s = 400{\cdot}0$	$h_s = 0'$		$N_s = 450{\cdot}0$ $h_s = 0'$		
ψ	τ	R_c	$R_e - (R_1/2)$	τ	R_c	$R_e - (R_1/2)$	τ	R_c	$R_e - (R_1/2)$
m.r.	m.r.	km	meters	m.r.	km	meters	m.r.	km	meters
0·0	15·6236	192·552	111·10	20·4292	189·878	128·70	27·5747	191·865	152·6
0·5	15·3818	190·840	109·40	20·0507	187·842	126·60	26·9134	189·128	149·3
1·0	15·1452	189·165	107·80	19·6822	185·861	124·30	26·2746	186·497	146·1
2·0	14·6878	185·921	104·80	18·9747	182·063	120·20	25·0635	181·540	140·2
3·0	14·2512	182·809	101·80	18·3060	178·467	116·40	23·9391	176·959	134·6
4·0	13·8349	179·818	99·00	17·6748	175·055	112·70	22·8975	172·716	129·4
5·0	13·4383	176·939	96·3	17·0798	171·811	109·3	21·9340	168·775	124·7
6·0	13·0605	174·163	93·8	16·5190	168·717	106·0	21·0429	165·099	120·3
7·0	12·7008	171·479	91·4	15·9905	165·759	103·0	20·2186	161·655	116·2
8·0	12·3583	168·881	89·1	15·4923	162·923	100·1	19·4552	158·414	112·4
9·0	12·0320	166·361	86·9	15·0224	160·199	97·2	18·7472	155·350	108·8
10·0	11·7210	163·914	84·8	14·5787	157·574	94·6	18·0892	152·443	105·5
12·0	11·1413	159·218	80·7	13·7626	152·593	89·8	16·9049	147·026	99·3
15·0	10·3651	152·608	75·4	12·6912	145·688	83·3	15·3978	139·701	91·4
20·0	9·2736	142·573	67·9	11·2232	135·398	74·3	13·4115	129·072	80·7
25·0	8·3768	133·571	61·49	10·0480	126·325	66·93	11·8776	119·907	72·17
30·0	7·6271	125·452	56·08	9·0849	118·249	60·80	10·6525	111·868	65.12
40·0	6·4460	111·434	47·70	7·6000	104·501	51·28	8·8124	98·383	54·48
50·0	5·5604	99·827	41·23	6·5103	93·278	44·12	7·4944	87·524	46·62
65·0	4·5872	85·861	34·05	5·3338	79·931	36·33	6·0979	74·749	38·22
80·0	3·8877	74·969	29·02	4·5005	69·621	30·82	5·1232	64·964	32·30
100·0	3·2172	63·818	24·05	3·7104	59·138	25·51	4·2089	55·077	26·79
150·0	2·2247	46·007	16·76	2·5542	42·519	17·71	2·8853	39·506	18·50
200·0	1·6860	35·747	12·83	1·9321	32·998	13·6	2·1789	30·628	14·13
300·0	1·1188	24·662	8·73	1·2803	22·745	9·21	1·4419	21·093	9·60
400·0	0·8224	18·899	6·65	0·9405	17·423	7·03	1·0588	16·153	7·33
600·0	0·5099	13·128	4·62	0·5830	12·100	4·86	0·6560	11·216	5·0
900·0	0·2772	9·494	3·33	0·3169	8·749	3·51	0·3566	8·108	3·66

Appendix

THE ATTENUATION OF RADIO WAVES PROPAGATED BETWEEN A PERFECTLY REFLECTING SPHERICAL IONOSPHERIC LAYER AND A SPHERICAL EARTH

The attenuation with which we are here concerning ourselves is that due to the spreading of the energy over larger and larger areas as it progresses further and further from the transmitting antenna.

For the sake of clarity in presentation, several simpler problems will be solved first in order to illustrate the principles involved. Consider first the attenuation of waves emanating from an isotropic radiator in free space as in Fig. I.1. The total

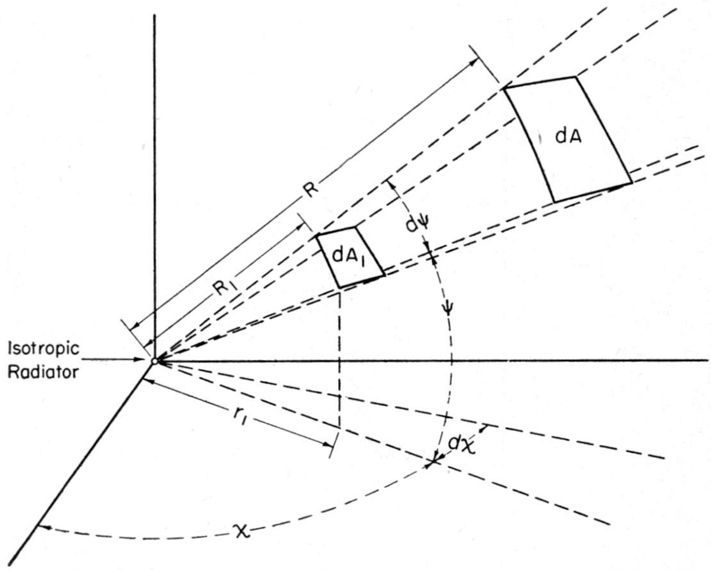

FIG. 1.1.

energy passing through the differential elements af area, dA_1 and, dA, normal to the radius vector and at the unit of distance R_1 and at R, respectively, will be equal

$$dA_1 = R_1 d\psi r_1 d\chi = R_1^2 \cos \psi d\psi d\chi \tag{1}$$

$$dA = R^2 \cos \psi d\psi d\chi . \tag{2}$$

Now, if we let p_1 and p_2 represent the energy density per unit area at the distances R_1 and R_2, we obtain:

$$p_1 dA_1 = p_2 dA \tag{3}$$

$$(p_2/p_1) = (dA_1/dA) = R_1^2/R^2 . \tag{4}$$

Thus we see that the field intensity (i.e., the energy density) is inversely proportional in free space to the square of the distance from the source. In terms of transmission loss, this result may be expressed:

$$L_{bf}(R) = L_{bf}(R_1) + 20 \log_{10}(R/R_1). \tag{5}$$

Consider next (see Fig. I.2) the attenuation of waves reflected from a plane perfectly conducting ionosphere at a height, h, above a plane earth and with no atmospheric refraction; in this case we have:

$$dA = - r d\chi \sin \psi dr = R^2 \cos \psi d\psi d\chi . \tag{6}$$

In this case again we find that the attenuation of waves reflected from a plane ionosphere is the same as in (I.5).

Finally consider (see Fig. I.3) the attenuation of waves reflected from a

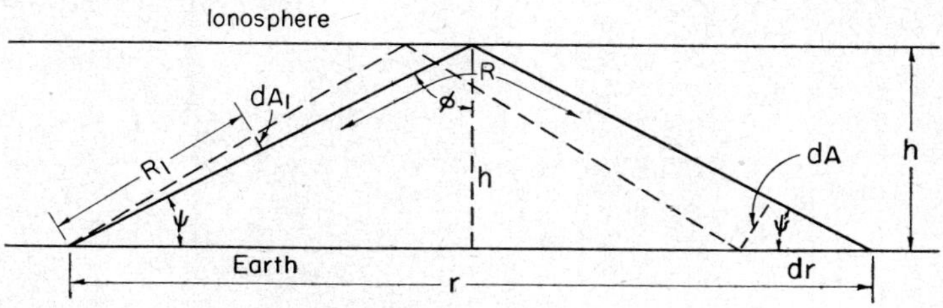

FIG. 1.2.

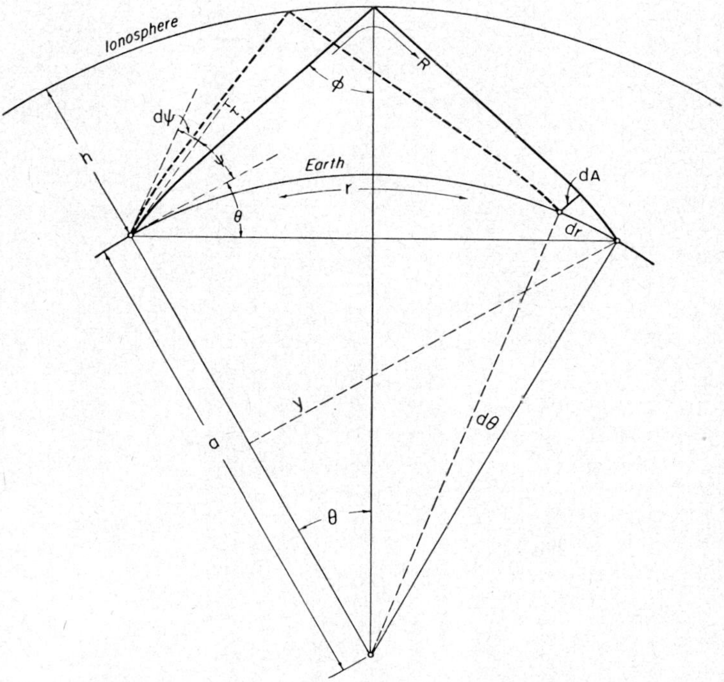

FIG. 1.3.

perfectly conducting spherical ionosphere and a perfectly conducting spherical earth with the effects of atmospheric refraction included:

$$dA = -\, y d\chi \sin \psi dr \tag{7}$$

$$dr = 2ad\theta \tag{8}$$

$$y = a \sin 2\theta \tag{9}$$

$$dA = -\, 2a^2 \sin 2\theta \sin \psi d\theta d\chi. \tag{10}$$

By Snell's law:

$$\frac{\cos(\psi+\theta-\tau)}{\cos\psi}=C=\frac{\{1+N_0\,.\,10^{-6}\exp(-c_s h_s)\}(a+h_s)}{\{1+N_0\,.\,10^{-6}\exp(-c_s h)\}(a+h)} \tag{11}$$

$$C=\cos(\theta-\tau)-\tan\psi\sin(\theta-\tau). \tag{12}$$

Since C is a constant, independent of θ, τ and ψ, we find:

$$dC=\frac{\partial C}{\partial\theta}d\theta+\frac{\partial C}{\partial\tau}d\tau+\frac{\partial C}{\partial\psi}d\psi=0. \tag{13}$$

Thus

$$(d\theta/d\psi)=-\left\{\frac{\partial C}{\partial\psi}\Big/\frac{\partial C}{\partial\theta}\right\}-\left\{\frac{\partial C}{\partial\tau}\Big/\frac{\partial C}{\partial\theta}\right\}\frac{d\tau}{d\psi} \tag{14}$$

$$(d\theta/d\psi)=-\frac{\sin(\theta-\tau)}{\cos\psi\cos\phi}+\frac{d\tau}{d\psi}. \tag{15}$$

In the particular case when $\psi=0$:

$$(d\theta/d\psi)_{\psi=0}=-1+(d\tau/d\psi)_{\psi=0}=-k \tag{16*}$$

$$L(R)=L_{bf}(R_1)+10\log_{10}\left(\frac{dA}{dA_1}\right)=L_{bf}(R)-C_1(R). \tag{17}$$

Thus, if we substitute Equations (10) and (1) in the above and solve for $C_1(R)$, we obtain:

$$C_1(R)=10\log_{10}\left(\frac{R^2dA_1}{R_1^2dA}\right)=10\log_{10}\left\{\frac{-R^2\cot\psi}{2a^2\sin 2\theta(d\theta/d\psi)}\right\}. \tag{18}$$

The function $C_1(R)$ is a convergence factor, expressed in decibels, which measures how much stronger the field intensity at the receiving point is for one reflection at a spherical ionosphere than it would be if it were plane.

The generalization of this expression to m reflections at the ionosphere may be obtained by noting that the only changes required in the above analysis are:

$$dr=2mad\theta \tag{19}$$

$$y=a\sin 2m\theta. \tag{20}$$

Thus the convergence factor for m ionospheric reflections may be written:

$$C_m(R)=10\log_{10}\left\{\frac{R}{-2ma\sin\psi(d\theta/d\psi)}\right\}_v\cdot\left\{\frac{R\cos\psi}{a\sin 2m\theta}\right\}_h. \tag{21}$$

The above expression has been divided into two factors with subscripts v and h so that the convergence of the rays in the vertical and horizontal planes, respectively, can be considered separately.

If we introduce the approximate expression $R\simeq 2ma\sin\theta/\sin\phi_0$ into Equation (21) above, we obtain:

$$C_m(R)\simeq 10\log_{10}\left\{\frac{\sin\theta}{-\sin\phi_0\sin\psi(d\theta/d\psi)}\right\}_v\cdot\left\{\frac{2m\sin\theta\cos\psi}{\sin\phi_0\sin 2m\theta}\right\}_h. \tag{21a}$$

* Note that k here refers to the ratio between the effective and actual radii of the earth.

The above expression indicates that the convergence in the vertical plane at points far removed from a caustic is independent of the number of ionospheric reflections for a given angle, ψ.

Note that the convergence in the vertical plane becomes infinite when ψ approaches zero; this infinite convergence is demonstrated on Fig. 8. Similarly, the convergence in the horizontal plane becomes infinite at the antipode of the transmitter where $2m\theta = \pi$. Actually, of course, the received energy is finite at these points, and we may use Airy's integral to evaluate the convergence in the vertical plane when $\psi \leqslant 0$. It can be shown [3] by the solution of the two dimensional wave equation that:

$$C_v(R) = 10 \log_{10} \left\{ \frac{-\Omega}{\Omega''(\psi)} \right\}_v \tag{22}$$

where $\Omega = 2\pi R/\lambda$ is the phase of the waves at the receiving point. If we multiply the numerator and denominator of the first term in Equation (21) by $2\pi/\lambda$ and compare the results with Equation (22), we find:

$$\Omega''(\psi) = 4\pi m(a/\lambda) \sin \psi (d\theta_0 d\psi).$$

Now we see that the above second derivative of the phase is equal to zero when the convergence factor becomes infinite; this is the definition of a caustic and the convergence at this caustic may be evaluated by means of the third derivative at this point:

$$\Omega'''(\psi) = -4\pi mk(a/\lambda) \qquad (\psi \leqslant 0).$$

Thus, at the caustic in the vertical plane, $C_v(R)$ may be expressed [3]:

$$C_v(R) = 10 \log_{10} \left\{ \frac{\Omega 2\pi \{A_i(0)\}^2}{[\frac{1}{2}\Omega''']^{2/3}} \right\}. \tag{23}$$

In the above $A_i(0)$ is the Airy integral [2] with argument zero:

$$2\pi\{A_i(0)\}^2 = 0{\cdot}79196357.$$

From the above results we obtain the following expression for $C_m(R)$ at the caustic in the vertical plane:

$$C_m(R) = 10 \log_{10} \left\{ \frac{2\pi(R/\lambda)\, 0{\cdot}792}{[2\pi mk(a/\lambda)]^{2/3}} \right\}_v \cdot \left\{ \frac{P}{a \sin 2m(\theta_m - k\psi)} \right\}_h (\psi \leqslant 0). \tag{24}$$

The convergence at the antipode of the transmitter is of a somewhat different nature. Note, in particular, that there is no point of stationary phase with respect to variations in the azimuth angle χ, since the waves appear to be arriving from all directions at this particular point. The following treatment of this problem is due to J. R. Wait [10].

Using a cylindrical coordinate system centred at the antipode (i.e., ρ, χ, z), we may obtain the following axially symmetric solution of the wave equation, for a time factor exp $(i\omega t)$:

$$E_z = A \exp[-ik(R_a + z \sin \psi)]J_0(k\rho \cos \psi) \tag{25}$$

$$H_\chi = B \exp[-ik(R_a + z \sin \psi)J_1(k\rho \cos \psi) \tag{25a}$$

where $k = 2\pi/\lambda$, R_a is the distance along the ray path to the antipode, A and B are constants, and J_0 and J_1 denote Bessel functions. For the ground wave and for those ionospheric modes for which m is sufficiently small so that ψ is negative at the antipode, ψ should be set equal to zero in Equations (25) and (25a). Note that, in addition to the oscillations with time, the magnitudes of E_z and H_χ oscillate with the distance ρ from the antipode, E_z having its maximum value at the antipode while H_χ is equal to zero at the antipode. Note also that the variation with ρ is the same, independent of the azimuth angle, χ; this would be expected since we have assumed that our source radiates uniformly in all directions. When $k\rho \cos\psi \gg 1$, we may replace the Bessel functions by the first terms in their asymptotic expansions and obtain:

$$E_z = A \exp(-ikz \sin\psi) \left\{\frac{\exp\{i[k(\rho\cos\psi - R_a) - \pi/4]\} + \exp\{-i[k(\rho\cos + R_a) - \pi/4]\}}{\sqrt{(2\pi k\rho\cos\psi)}}\right\} \tag{26}$$

$$H_\phi = -iB \exp(-ikz \sin\psi) \left\{\frac{\exp\{i[k(\rho\cos\psi - R_a) - \pi/4]\} - \exp\{-i[k(\rho\cos\psi + R_a) - \pi/4]\}}{\sqrt{(2\pi k\rho\cos\psi)}}\right\}. \tag{26a}$$

The two exponential terms in the above may be identified with waves arriving from opposite directions at a receiving point at a distance ρ from the antipode along great circle paths of lengths $R_a - \rho\cos\psi$ and $R_a + \rho\cos\psi$, respectively. It is the interference between these two waves which causes the oscillations in the magnitude of the field near the antipode.

To complete our solution we need only evaluate the constants A and B. Rather than doing this directly, we note by Equation (21) that the geometrical theory indicates that the focusing in the horizontal plane not too near the antipode is given by:

$$c_h = \frac{R\cos\psi}{a\sin 2m\theta} \quad (2m\theta < \pi) \tag{27}$$

and if we multiply Equation (27) by the square of the ratio of $|E_z|$ as given by Equation (25) and by the first term in Equation (26), we obtain the following expression for c_h which must be used instead of Equation (27) at points very near the antipode:

$$c_h = \frac{\mathrm{R}\cos\psi}{a\sin 2m\theta} \cdot [J_0(k\rho\cos\psi)]^2\, 2\pi k\rho\cos\psi. \tag{28}$$

When we note that $\rho = a\sin(\pi - 2m\theta)$, the above reduces to:

$$c_h = 2\pi kR\cos^2\psi\, [J_0(k\rho\cos\psi)]^2. \tag{29}$$

For the ground wave $R = \pi a$ and $\psi = 0$; thus, at the antipode $c_h = 2\pi^2 ka$ and $C_h = 10\log_{10} c_h = 34{\cdot}210 + 10\log_{10} f_{kc}$, and this clearly represents an extremely large focusing effect for the ground wave at and near this point. The focusing in the horizontal plane for the sky wave modes is only slightly different,* but we

*Note that $\cos\psi > 0{\cdot}995$ for $m \leqslant 16$, and $\cos\psi = 1$ for $m \leqslant 8$ when $h = 90$ km.

must add to this the focusing in the vertical plane to obtain the total focusing for these modes. Equation (29) is for a vertical electric dipole receiving antenna; if a horizontal magnetic dipole were used for reception, then the J_0 should be replaced by J_1.

All of the above discussion applies to the case when the effective reflecting surface of the ionosphere is smooth and concentric with the surface of the earth. In practice, as the sun rises and sets, or as the geomagnetic latitude of the reflection point is varied, the surface of the ionosphere will undoubtedly change in such a way that its radius of curvature and slope relative to a tangent plane on the earth will vary over appreciable ranges, and this will cause $C_m(R)$ to vary up and down relative to the values expected on the basis of the above analysis. However, except near the antipode, it seems plausible to assume that the median values of $C_m(R)$ may not be much influenced by such changes. The magnitude of the antipodal anomaly will be substantially reduced by these macroscopic perturbations of the spherical concentric shell model assumed for our calculations. Note also that there will be concentric rings around the antipode at which the expected field will be equal to zero. The radii of these concentric rings are the same, regardless of the number, m, of ionospheric reflections, and are determined by the zeros of the Bessel functions; for the electric field, the first two such rings have radii equal to $0{\cdot}38\lambda$ and $0{\cdot}88\lambda$. Note, however, that the geographical location of the centers of the antipodal anomalies may be expected to be somewhat different for the different modes for the actual non-concentric ionosphere, and thus these zeros are not likely to be observable unless some means is used to exploit their different times of arrival. The shifts in the geographical locations of the anomalies caused by these macroscopic changes in the ionosphere would be expected to result in a net increase in the fading range in the neighborhood of the antipode.

In addition to these systematic macroscopic changes in the ionosphere, the reflecting surface of the ionosphere will be locally rough, and we will see in the following analysis how this local roughness may be expected to reduce the median values of convergence as computed above for a smooth concentric ionosphere.

We have seen above that the convergence depends, at a given receiving point, upon the smooth variation of the phase of the received waves with changes in elevation angle and azimuth. If we let σ_Ω denote the standard deviation of the phase of the waves received via m hops, then we may use Rayleigh's criterion of ionospheric roughness to calculate σ_Ω in terms of the variance, σ_h^2, of the local effective reflection heights, h, of the ionosphere at the points of stationary phase.

$$\sigma_\Omega = \frac{720^\circ\, \sigma_h \cos\phi \sqrt{m}}{\lambda}. \tag{30}$$

Note that the variance, σ_Ω^2, of phase consists of components arising from (a) a drift of a fixed pattern of ionospheric irregularities relative to a reference great-circle-smooth-concentric-ionosphere path, (b) changes in the shape of these irregularities with time, and (c) changes in the location of the reflection points in the horizontal plane.

Brennan and Phillips [1] find that variations with time of the intensity and phase of a one-hop transmission at 543 kc over a 380 mile path at night indicate rather conclusively that they may be described adequately most of the time by assuming that the received waves consist of a steady component with constant phase and approximately constant amplitude plus a random Rayleigh distributed component of relative intensity k^2 and random relative phase; the amplitude

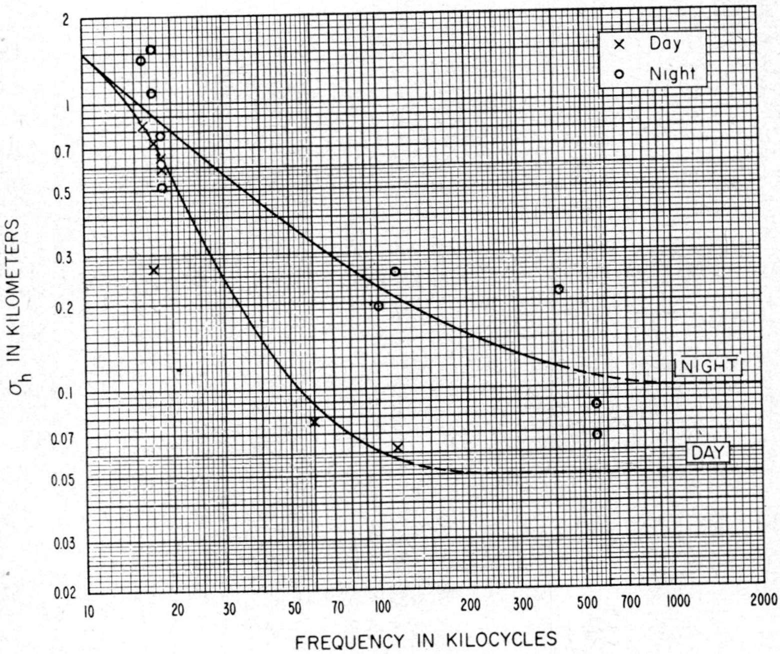

FIG. 1.4. Median effective ionospheric roughness parameter, σ_h obtained from observations of σ_Ω. σ_Ω (degrees) $\equiv 2{\cdot}4 f_{kc} \cos \phi \; \sigma_h$ (km).

distribution expected in this case is given on Fig. 2 with $K \equiv 10 \log_{10} k^2$. The fixed component may be identified with a specular reflection expected from the reference smooth surface while the random component arises from the surface roughness. It now becomes clear how ionospheric roughness may be expected to affect the convergence; the specular component will be increased $C_m(R)$ db whereas the random component, since its phase is random, will not be increased at all. If we write $C_m(R) \equiv 10 \log_{10} c_m$, then we find the following expression for the convergence factor $C_m(R, p)$ exceeded $100\, p$ % of the time in terms of the values of $k^2(1 - p)$ exceeded $100(1 - p)$ % of the time:

$$C_m(R, p) = 10 \log_{10} \left\{ \frac{c_m + k^2(1 - p)}{1 + k^2(1 - p)} \right\}. \tag{31}$$

TABLE I.1

ESTIMATES OF THE VARIANCE OF PHASE ON IONOSPHERIC PATHS

No.	Freq. kc	Distance Statute Miles	Source Reference Number	Time	Month	$\sigma_{obs.}$ Degrees	S λ	$\sqrt{(2(1-\rho^2))}$	σ_Ω	Mode m	$\cos\phi$	σ_h km_1
1	16·0	3230	Pierce [7]	Day	February	8·06			8·06	3	0·146	0·829
2	16·0	3230	Pierce [7]	Night	February	16·1			16·1	3	0·170	1·424
3	17·2	640	Redgment [8]	Day	September	1·9	10·0	1·00	1·9	1	0·174	0·265
4	17·2	640	Redgment [8]	Night	September	9·3	10·0	1·00	9·3	1	0·206	1·093
5	17·2	640	Redgment [8]	Day	January	5·2	10·0	1·00	5·2	1	0·174	0·726
6	17·2	640	Redgment [8]	Night	January	13·3	10·0	1·00	13·3	1	0·206	1·563
7	18·4	3488	Redgment [8]	Day	June July	7·1	10·0	1·00	7·1	3	0·144	0·643
8	18·4	3488	Redgment [8]	Night	June July	10·0	10·0	1·00	10·0	3	0·169	0·775
9	18·4	3488	Redgment [8]	Day	December	6·5	10·0	1·00	6·5	3	0·144	0·588
10	18·4	3488	Redgment [8]	Night	December	6·5	10·0	1·00	6·5	3	0·169	0·503
11	60·0	3230	Pierce [7]	Day	December	2·8			2·8	3	0·146	0·077
12	100·0	2350	Doherty [4]	Night	April	10·8	>40·0	1·41	7·64	1	0·165	0·193
13	115·0	780	Florman [5]	Night	March	13·3	11·0	1·00	13·3	1	0·189	0·255
14	115·0	780	Florman [5]	Day	March	2·7	11·0	1·00	2·7	1	0·158	0·062
15	418·0	1250	Florman [5]	Night	March	51·0	41·0	1·41	36·1	1	0·167	0·216
16	543·0	380	Brennan [1]	Night	January to April	34·0	3·39	0·83	41·0	1	0·371	0·085
17	556·0	454	Redgment [8]	Night	April May	19·0	1·85	0·63	30·2	1	0·334	0·068

Note that $k^2(1 - p)$ approaches zero as p approaches zero, and thus $C_m(R, p)$ approaches $C_m(R)$, the value expected for a smooth ionosphere, as p approaches zero. On the other hand, $k^2(1 - p)$ approaches ∞ for a perfectly rough ionosphere, and in this case there will be no convergence and $C_m(R, p)$ approaches zero.

The values of k^2 to be used in Equation (31) may, in principle, be determined from observations of the variations in either the amplitude or the phase of the waves corresponding to a single mode of propagation. However, it is ordinarily better in practice to use the variations in phase as an index to k^2 since the amplitudes of the received waves also vary with ionospheric absorption, and it is sometimes difficult to separate out these absorption variations from the amplitude variations arising from surface roughness alone. In Table I.1 are tabulated some experimental measurements of σ_Ω. In some cases the required variance was observed directly, but in other cases it had to be estimated from phase difference measurements made on paths with one common terminal, but with their other terminals separated by S wavelengths. In this latter case the observed phase difference was related to σ_Ω^2 as follows:

$$\sigma^2_{\text{obs.}} = 2(1 - \rho^2)\sigma^2_\Omega. \tag{32}$$

In the above ρ is the correlation between the phase variations along the two independent paths. The correlation will vary from $\rho = 1$ for $S = 0$ to $\rho \simeq 0$ for $S > 40\lambda$. Although direct measurements of ρ do not appear to be available in the literature, it seems reasonable to assume that ρ will be of the same order of magnitude as the correlation between the amplitude variations on paths separated a distance S at one end. Measurements of the latter correlation were reported in reference 3, and these data constituted the basis for the estimates in Table I.1.

Using Equation (30) estimates of σ_h can also be made, and these are also given in Table I.1 and shown on Fig. 1.4. Although the data are quite scattered, the curved lines labelled day and night, respectively, represent the estimates used in this report for calculating the median values of σ_Ω using Equation (30). It should be noted that σ_Ω is itself a random variable which changes over wide ranges from hour to hour and from day to day. For example, an analysis of the data in reference [3] shows that the observed phase differences, $\sigma_{\text{obs.}}$, ranged from 5·3° to more than 180°, 10 % of the values exceeded 118°, 50 % exceeded 34° and 90 % exceeded 14°. We will see below that the maximum value of $\sigma_{\text{obs.}}$ to be expected in practice is $103{\cdot}9 \times \sqrt{2} = 147°$, which corresponds to $k^2 = \infty$ and only 5 % of their observed values exceeded this value.

The relation between σ_Ω and k^2 has been obtained on the assumption that the data fit the Rice [9] distribution of a constant vector plus a Rayleigh distributed vector. By integrating over the joint probability distribution given by Rice for all of the variables except Ω, we obtain the following expression for the probability density function for Ω with k^2 as a parameter:

$$2\pi p(\Omega) = \{1 + z\sqrt{\pi} \exp (z^2)[1 + \text{erf} (z)]\} \exp (- 1/k^2) \tag{33}$$

where $z \equiv \cos \Omega/k$. Figure I.5 shows $p(\Omega)$ is symmetrically distributed about zero for all values of k^2. When k^2 is very small, sin Ω is distributed approximately normally about zero with variance $\sigma^2_{\sin\,\Omega} = k^2/2$. When k^2 approaches infinity,

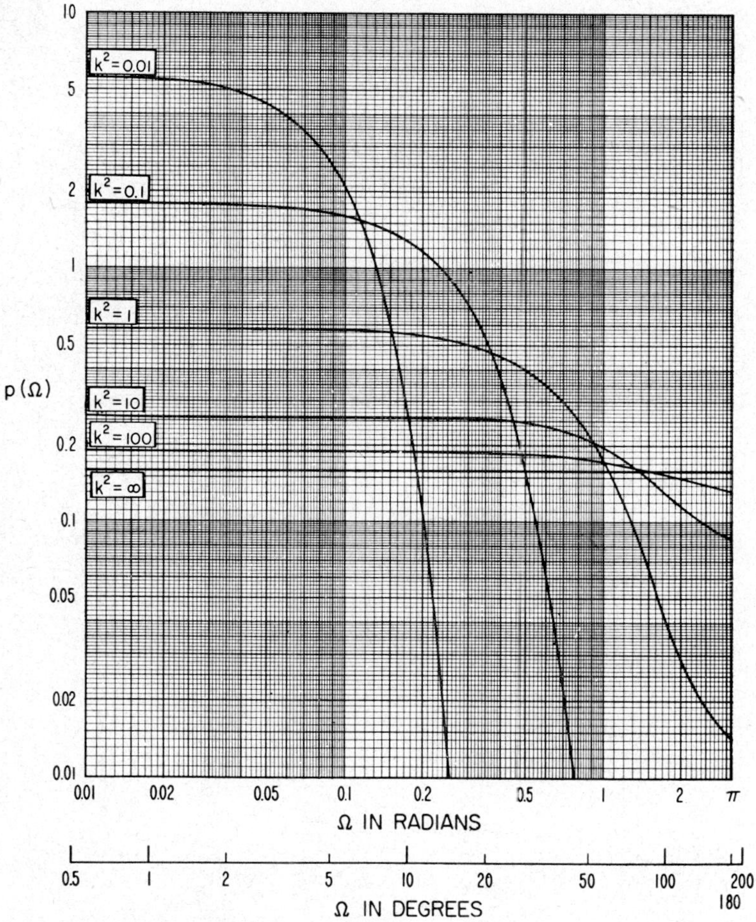

Fig. 1.5.

Ω is uniformly distributed between $-180°$ and $+ 180°$ and σ_Ω approaches 103·923°. Figure I.6 gives the cumulative distribution defined by:

$$P[\Omega(t) > \Omega(P)] = 1 - \int_{-\pi}^{\Omega(P)} p(\Omega) d\Omega. \tag{34}$$

The mean absolute value $|\overline{\Omega}|$ and variance σ^2_Ω are also of interest:

$$|\overline{\Omega}| = 2 \int_0^\pi \Omega p(\Omega) d\Omega \tag{35}$$

$$\sigma^2_\Omega = 2 \int_0^\pi \Omega^2 p(\Omega) d\Omega. \tag{36}$$

For many applications the cumulative distribution of the absolute value of $|\Omega(t)|$ is of greater interest:

$$P'[\,|\Omega(t)| > \Omega(P')] = 2P[\Omega(t) > \Omega(P)]. \tag{37}$$

The distribution of P' is given in a recent paper [6] and several of its percentage points, together with $|\overline{\Omega}|$ and σ_Ω are shown on Fig. I.7 as a function of k^2. By

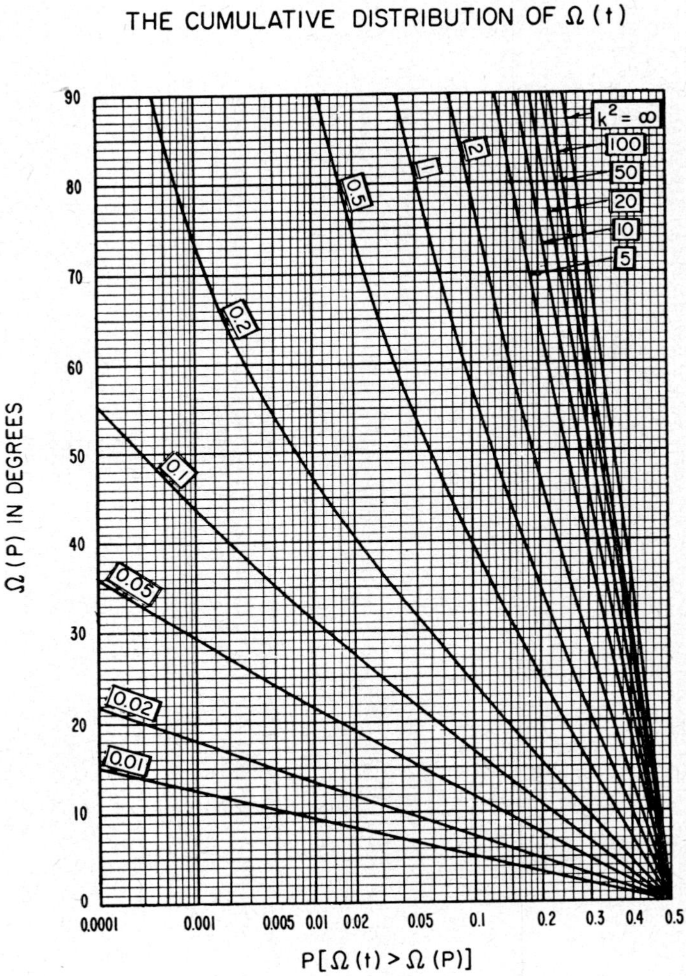

FIG. 1.6.

using the median values of σ_Ω determined from Fig. I.4, we may use the results shown on Fig. I.7 to determine the median values of $k^2(0{\cdot}5)$ required for the evaluation of $C_m(R, 0{\cdot}5)$.

Table I.2 below gives the values of $|\Omega|$ and of σ_Ω for several values o fk^2.

TABLE I.2

k^2	$\lvert\bar{\Omega}\rvert$ radians	σ_Ω radians	$\lvert\bar{\Omega}\rvert$ degrees	σ_Ω degrees
0·01	0·056514	0·070890	3·2380	4·0617
0·02	0·080059	0·10051	4·5871	5·7590
0·05	0·12726	0·16023	7·2915	9·1803
0·1	0·18172	0·23013	10·412	13·185
0·2	0·26330	0·34032	15·086	19·499
0·5	0·44605	0·60664	25·557	34·758
1	0·64346	0·87134	36·868	49·924
2	0·85196	1·1175	48·813	64·031
5	1·0876	1·3661	62·313	78·270
10	1·2217	1·4972	69·999	85·785
20	1·3212	1·5907	75·701	91·142
50	1·4119	1·6735	80·896	95·884
100	1·4582	1·7149	83·547	98·259
200	1·4911	1·7441	85·431	99·929
500	1·5203	1·7698	87·107	101·40
1000	1·5351	1·7827	87·953	102·14
∞	$\pi/2$	$\pi/\sqrt{3}$	90·000	103·92

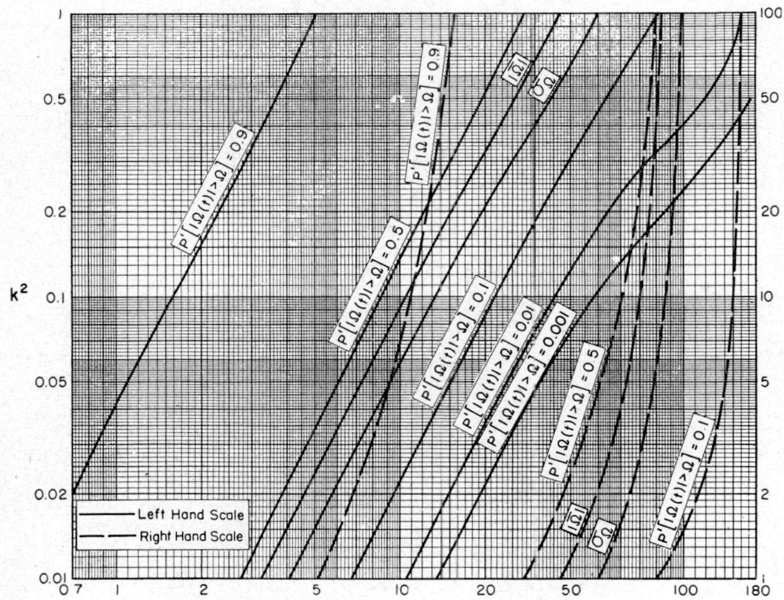

FIG. 1.7. Statistics of the distribution of the phase of a constant vector plus a Rayleigh distributed vector.

An analysis was made in reference [1] of both the amplitude variations on single paths as well as the phase differences between paths separated by 3·39 wavelengths at one end. Using their observed median standard deviation of $\sigma_{\text{obs.}} = 34°$, the median value $\sigma_{\Omega} = 41°$ given in Table I.1 was estimated; this value corresponds by Fig. I.7 to $k^2(0{\cdot}5) = 0{\cdot}68$. The analysis of their amplitude variations gives directly the estimate $k^2(0{\cdot}5) = 2/\bar{a}^2 = 0{\cdot}854$, and this latter estimate is somewhat larger, as might have been expected, since the amplitude variations are biased by changes in absorption.

References

1. Brennan, D. G. & Lindeman Phillips, M. *Tech. Rep.* No. 93, Mass. Inst. Tech. Lincoln Lab., September 1957. Note that the a in this report, which is used as a parameter describing the Rice distribution of a constant plus a Rayleigh distributed vector, is equal to $\sqrt{2}/k$ where k^2 denotes the relative intensity of the random component as used in this appendix.
2. Cambridge *British Association Mathematical Tables,* Part Volume B. University Press (1946).
3. Doherty, L. H. *Cornell Univ. Rep.* EE 138.
4. Doherty, R. H. (1957). *Rep. nat. Bur. Stand.* no. 6CB103.
5. Florman, E. F. Boulder Lab. Nat. Bur. Stand. Private communication.
6. Norton, K. A., Shultz, E. L. & Yarbrough, H. (1952). *J. appl. Phys.* **23**, 137–141. Note that the k in this reference is a power ratio rather than a voltage ratio and that the formulas and graphs give the distribution of $|\phi|$.
7. Pierce, John A. (1957). *Proc. Inst. Radio Engrs, N.Y.* **45**, 794–803.
8. Redgment, P. G. & Watson, D. W. *Monograph,* no. 836. Admiralty Signal and Radar Est., Lythe Hill House, Haslemere, Surrey, England. October 1948.
9. Rice, S. O. (1948). *Bell Syst. tech. J.* **27**, 109–157. The joint probability distribution is given by equation (4.6).
10. Wait, J. R. Boulder Lab. Nat. Bur. Stand. Private communication.

A Generalized Rayleigh Distribution and its Application to Tropospheric Propagation

PETR BECKMANN

Institute of Radio Engineering and Electronics, Czechoslovak Academy of Sciences, Prague

The paper derives the probability distribution of the amplitude of the vector sum of n random unit vectors with given (not necessarily uniform) phase distribution, which includes the Rayleigh distribution as a special case, and compares the result with experimentally measured distributions of the rapid fading in trans-horizon tropospheric propagation.

It is usually stated in papers on tropospheric propagation that the rapid fluctuations of the signal received far beyond the horizon are Rayleigh-distributed. This statement, however, is not always exact. Thus for example, when there is strong layer reflection present, the signal amplitude may show for relatively long time intervals almost no fluctuation at all; in this case it ceases to be a random variable and tends to a constant value, i.e. it is Dirac-distributed. On other days the cumulative distribution function lies between that of the Rayleigh and the Dirac distribution, whilst at some times the rapid fluctuations are purely Rayleigh-distributed. Figure 1 shows distribution curves measured on a 250 km path at 91 Mc/s in 4 min intervals. It is one of very many examples, where the slope of the measured distribution curves lies between that of the Rayleigh and that of the Dirac distribution; the same results were obtained using other paths and other time intervals of sampling.

If isotropic turbulence were the only mechanism responsible for trans-horizon propagation of VHF, the signal should be purely Rayleigh-distributed. The measured distribution can, however, be very well explained by considering the mechanism of reflection or scattering from rough layer boundaries or turbulent layers in the troposphere. According to this theory, which is described in detail in [1], the individual scattered wavelets arrive at the receiver with random phases; however, the phase distribution may or may not be uniform, depending on the surface roughness h compared to the wavelength λ and the grazing angle γ of the incident radiation. If $(h \sin \gamma/\lambda)$ is sufficiently large, the phases will be uniformly distributed between 0 and 2π and the resulting signal will be Rayleigh-distributed; as $(h \sin \gamma)/\lambda$ decreases, the variance of the resulting signal amplitude will decrease until with $(h \sin \gamma)/\lambda = 0$, incoherent scattering turns into specular reflection and the received signal amplitude turns from a random variable to a certain variable.

In determining the probability distribution of the resultant signal made up of elementary waves with random phases, we are thus faced with the following problem: to determine the probability distribution of the vector sum of n independent unit vectors with given (not necessarily uniform) phase distributions. This problem was solved in a special case of phase distribution, and only approximately, in [1]. The main point of this paper is an outline of the general and exact solution of the above problem.

Let the amplitude of the resulting vector be r, so that

$$r^2 = \left(\sum_{i=1}^{n} \cos \phi_i \right)^2 + \left(\sum_{i=1}^{n} \sin \phi_i \right)^2 = x^2 + y^2. \tag{1}$$

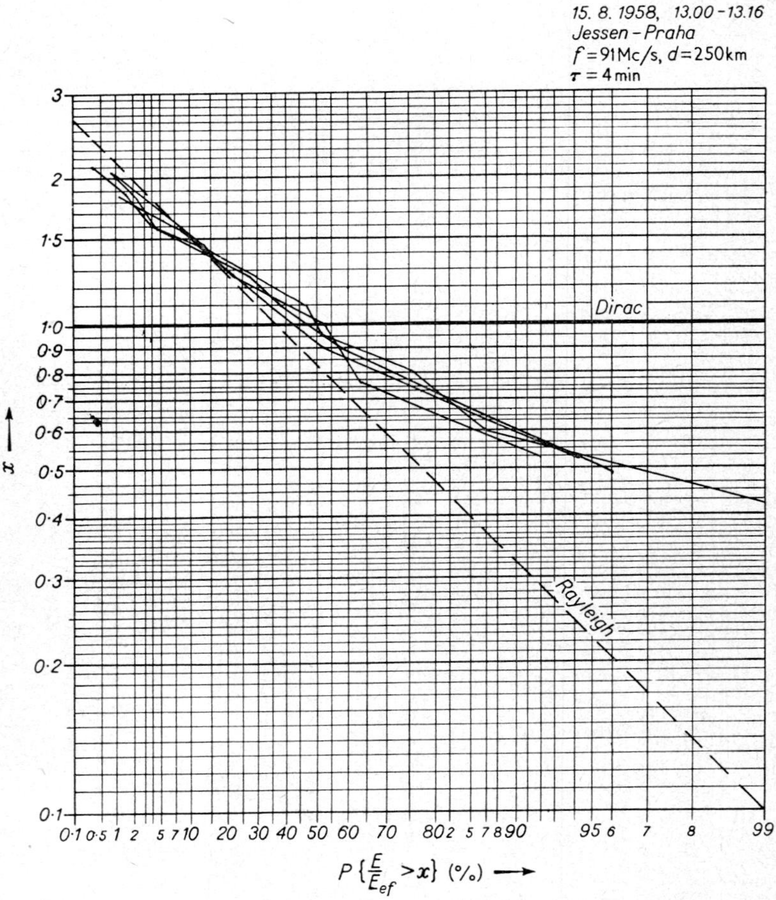

FIG. 1.

Let the probability densities of ϕ be $w_i(\phi)$; we shall assume that they are symmetrical about zero. Then it may easily be shown that the mean values of x and y are

$$\left. \begin{aligned} M\{x\} &= \sum_{i=1}^{n} \int_{-\infty}^{\infty} w_i(\phi) \cos \phi d\phi = \alpha \\ M\{y\} &= 0 \end{aligned} \right\} \tag{2}$$

and the variances of x and y are

$$D\{x\} = \sum_{i=1}^{n} \left(\int_{-\infty}^{\infty} w_i(\phi) \cos^2 \phi d\phi - \alpha_i^2 \right) = s_1 \quad (3)$$

$$D\{y\} = \sum_{i=1}^{n} \int_{-\infty}^{\infty} w_i(\phi) \sin^2 \phi d\phi = s_2.$$

Table I shows the values of α, s_1, and s_2 when all the $w_i(\phi)$ are mutually equal,

TABLE I

$w(\phi)$	α	s_1	s_1
Normal; mean O stand. dev. σ	$ne^{-\sigma^2/2}$	$\frac{n}{2}(1 - e^{-\sigma^2})^2$	$\frac{n}{2}(1 - e^{-2\sigma^2})$
Uniform from $-$ a to $+$ a	n sinc a	$\frac{n}{2}(1 + \text{sinc } 2a - 2 \text{ sinc}^2 a)$	$\frac{n}{2}(1 - \text{sinc } 2a)$
Simpson from $-$ 2a to $+$ 2a	n sinc^2 a	$\frac{n}{2}(1 + \text{sinc}^2 2a - 2 \text{ sinc}^4 a)$	$\frac{n}{2}(1 - \text{sinc}^2 a)$

for three cases, namely a normal distribution with mean value 0 and variance σ, a uniform distribution from $- a$ to $+ a$, and a Simpson distribution from $- 2a$ to $+ 2a$. The table uses the symbol sinc $a = (\sin a)/a$.

Now if n is sufficiently large, and the $w_i(\phi)$ obey the Central Limit Theorem (as they practically always will), then both x and y will be normally distributed with parameters given by Equations (2) and (3); transforming the resultant two-dimensional normal distribution to polar co-ordinates (r, θ), we therefore find the distribution of the amplitude of the resulting vector sum as

$$p(r) = \frac{r}{2\pi\sqrt{(s_1 s_2)}} \int_0^{2\pi} e^{-\frac{(r\cos\theta - \alpha)^2}{2s_1} - \frac{r^2 \sin^2\theta}{2s_2}} d\theta.$$

This integral has been treated in [2] and leads to a series of modified Bessel functions; normalizing for convenience as follows:

$$\rho = \frac{r}{\sqrt{(2s_2)}}; \; K = \sqrt{\left(\frac{s_2}{s_1}\right)}; \; v = \frac{\alpha}{\sqrt{(2s_1)}} \quad (4)$$

we find from (3)

$$p(\rho) = 2K\rho e^{-v^2 - \frac{K^2+1}{2}\rho^2} \sum_{m=0}^{\infty} (-1)^m \epsilon_m I_m \left(\frac{K^2 - 1}{2} \rho^2 \right) I_{2m}(2Kv\rho) \quad (5)$$

where I_m is the modified Bessel function of the first kind and order m, and

$$\epsilon_m = \begin{cases} 1 \text{ for } m = 0 \\ 2 \text{ for } m \neq 0 \end{cases}$$

This is the general and exact solution of our problem. It may be shown that Equation (5) includes several already known distributions. Thus if $w(\phi)$ is uniform from $-\pi$ to $+\pi$, we find $s_1 = s_2 = n/2$, $\alpha = 0$, making

$$p(\rho) = 2\rho e^{-\rho^2} \text{ or } p(r) = 2n^{-1} r e^{-r^2/n} \tag{6}$$

which is the Rayleigh distribution; for the distribution of a constant vector $\vec{C}$ plus a Rayleigh distributed vector we find $s_1 = s_2 = n/2$, $\alpha = |C|$, making

$$p(\rho) = 2\rho e^{-v^2-\rho^2} I_0(2v\rho); \; p(r) = \frac{2}{n} r e^{-\frac{C^2+r^2}{n}} I_0\left(\frac{2Cr}{n}\right) \tag{7}$$

in agreement with Rice as quoted in [5]. Finally, for $v = 0$ we obtain an asymmetrical χ^2 distribution

$$p(\rho) = 2K\rho e^{-\frac{K^2+1}{2}\rho^2} I_0\left(\frac{K^2-1}{2}\rho^2\right) \tag{8}$$

derived in another connection by Hoyt [4].

Further details, such as the mean and RMS values, curves, tables and the phase distribution of the resulting vector sum will be found in a paper to be published in the Czechoslovak journal *Acta Technica ČSAV* No. 4, 1959.

Since in the general case Equation (5) there is no such simple relationship between n and the mean power of the received signal as in the case of the Rayleigh distribution, experimentally measured distribution curves cannot be checked directly against Equation (5), because the number of component waves is unknown. This difficulty can be overcome by subtracting a reference signal with constant amplitude and phase from the received signal and checking the result against Equation (8). Since accurate phase measurements at VHF are rather difficult, it may take some time before this is carried out. However, in the meantime the following conclusions may be drawn:

1. Experimental measurements show that the VHF signal received far beyond the horizon is often, but by no means always, Rayleigh-distributed. This contradicts the theory of isotropic turbulence as the only propagation mechanism.
2. The distribution given by Equation (5) is at least qualitatively in agreement with experimental data inasfar as it explains the transition from the Rayleigh to the Dirac distribution as $w(\phi)$ changes from a uniform distribution to a distribution with privileged phase, which in turn corresponds to the transition from incoherent scattering from a very rough layer or from isotropic turbulence to specular reflection from a smooth layer or from strongly anisotropic turbulence. Accurate quantitative comparison is dependent on somewhat difficult phase measurements.
3. Phase measurements that have been carried out, even if only on a relatively short path [3], show that the scattered component is very often nearly in quadrature with the directly propagated (constant) component, again showing the presence of a privileged phase instead of the uniform phase distribution of a Rayleigh vector.

Finally I should like to say that in my opinion it is most important to study not only the mean values of the received signal, but its entire statistical distribution both with respect to time and space, since it is in the microstructure of the received signal that the various theories of the propagation mechanism differ most amongst each other.

References

1. Beckmann, P. (1957). *Acta tech. ČSAV* **2,** 311–355.
2. Beckmann, P. & Schmelovsky, K. H. Über ein bei Schwunderscheinungsuntersuchungen vorkommendes Integral. (1958). *Z. Hochschule Elektrotech. Ilmenau.* **4,** 167–171.
3. Herbstreit, J. W. & Thompson, M. C. (1955). *Proc. Inst. Radio Engrs, N.Y.* **43,** 1391–1401.
4. Hoyt, R. S. (1947). *Bell Syst. tech. J.* **26,** 318–359.
5. Norton, K. A., Vogler, L. E., Mansfield, W. V. & Short, P. J. (1955). *Proc. Inst. Radio Engrs, N.Y.* **43,** 1354–1361.
6. Beckmann, P. (1959). *Acta tech. ČSAV* **4,** 323–335.

Correlation of F2 Layer Electron Density and Solar Activity in the Years 1938-1944

F. Mariani

Istituto di Fisica dell'Università, Roma
Istituto Nazionale di Geofisica, Roma

The noon values of electron density in the F2 layer, during the decreasing stage of the solar cycle ended on 1944, exhibit as a whole a good double correlation with the Wolf number R and the area of hydrogen filaments A_F. The dependence on R is practically independent of the latitude; the dependence on A_F, instead, seems to indicate some latitudinal (likely asymmetric) variation in the two hemispheres. Similar results have also been obtained for the decreasing stage of the solar cycle ended on 1954.

I. Introduction

An extensive study of the correlations between ionospheric maximum electron densities in the F2 layer at noon, in about thirty Observatories, and solar activity during the decreasing stage of the last solar cycle (May 1947 to March 1954) has shown some very interesting characteristics, principally dependent upon the geomagnetic latitude [4, 5, 6].

In particular, it has shown that the double correlation of the electron density with the Wolf number R and the area A_F of hydrogen filaments seems to replace advantageously the usual simple linear correlation of N with R. At least in the studied time interval, the effect of R on N is practically independent on the latitude; instead, the dependence of N upon A_F exhibits a remarkable latitudinal variation, greater in the northern hemisphere and whose importance, with respect to the variation with R, increases with the northern geomagnetic latitude.

In order to test the above conclusion, we study in this paper the analogous correlations of N with some solar activity parameters for the decreasing stage of the solar cycle ended in March 1944. We have had at our disposal the monthly ionospheric data of Washington, Huancayo and Watheroo; concerning the Sun we have used the data of Zürich (for R), Arcetri (for areas of hydrogen filaments and flocculi, and calcium flocculi) and Meudon (for filaments activity).

II. The Results of the Correlations

The method of analysis of the monthly series of data is the periodogram analysis, we have used also in the precedent papers.

The symbols used are the following:

$N \; = (\text{foF2})^2 =$ maximum electron density of the F2 layer at noon;
$R \; =$ Wolf number;
$A_F \; =$ areas of hydrogen filaments at Arcetri;
$A_{\phi_H} \; =$ areas of hydrogen flocculi at Arcetri;
$A_{\phi_{Ca}} =$ areas of calcium flocculi at Arcetri;
$A_F^M \; =$ filament activity at Meudon;

$\overline{N}, \overline{R}, \overline{A}_F, \overline{A}_{\phi_H}, \overline{A}_{\phi_{Ca}}, \overline{A}_F^M =$ the secular variation of the above parameters; $N_{12} =$ seasonal variation of N.

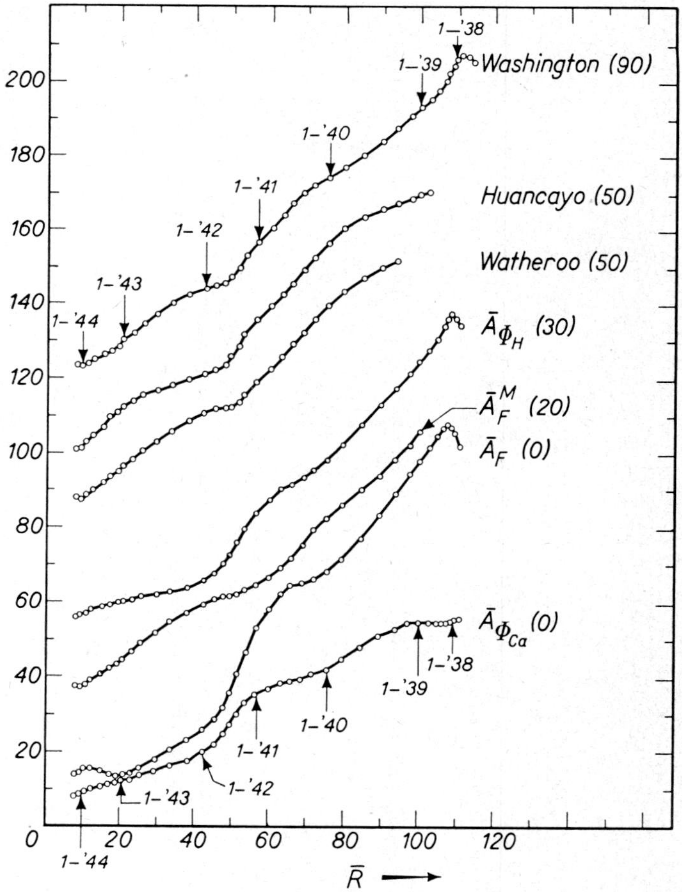

FIG. 1. The points are the calculated values for the odd months; the arrows indicate the date (for example, 1–'39 indicates January 1939). The areas $\overline{A}_F$ $\overline{A}_{\phi_{Ca}}$ are in 10^{-4} of the solar disc; the areas $\overline{A}_{\phi_H}$ in $2 \cdot 10^{-4}$ of the solar disc; the filament activities $\overline{A}_F^M$ are in conventional units. The numbers within parentheses indicate the ordinate corresponding to the zero ordinate for every graph.

In Fig. 1 we show the graphs of $\overline{N}$ for the three ionospheric observatories and those of $\overline{A}_F$, $\overline{A}_{\phi_H}$, $\overline{A}_{\phi_{Ca}}$, $\overline{A}_F^M$ as functions of $\overline{R}$.

As in the years 1947–54, some parallel variation of $\overline{N}(\overline{R})$ and $\overline{A}_F(\overline{R})$ is apparent in the years of higher solar activity; in the years of lower solar activity, instead, the correlation is not good.

A greater correlation of $\overline{N}$ with $\overline{A}_F^M$ is instead apparent during all the years 1939–44. The different behaviour of $\overline{A}_F$ and $\overline{A}_F^M$ cannot be considered contradictory; in effect, the values of Meudon, are desumed by maps of the cromospheric activity, as observed in various cooperating observatories, while the data of Arcetri give the results for one observatory; moreover, the definitions of $\overline{A}_F$ and $\overline{A}_F^M$ are rather different.

Concerning the behaviour of $\overline{A}_{\phi_H}$ and $\overline{A}_{\phi_{Ca}}$, a remarkable similarity of $\overline{A}_{\phi_H}$ and $\overline{A}_F$ is apparent; however, this was not the case for the years 1947–54.

Figure 1 shows two interesting features.

1. The presence of a "discontinuity" of the slope above all in the graphs of $\overline{N}$, $\overline{A}_F$, $\overline{A}_{\phi_H}$ in the summer months of 1940, which precedes of about eleven years an analogous discontinuity we have noted in the spring months of 1951, in the corresponding graphs for the years 1947–54.
2. A maximum of $\overline{N}$ at Washington, during maximum solar activity, preceding and likely corresponding to an analogous maximum of $\overline{A}_F$.

In order to compare the results of present analysis with those relating to the years 1947–54 we have considered the simple and double correlations of $\overline{N}$ with $\overline{R}$ and $\overline{A}_F$, and of $N - N_{12}$ with R and A_F, according to the regression laws:

$$\begin{cases} \overline{N} = \overline{N}_0(1 + \alpha\overline{R}) \\ \overline{N} = \overline{N}_0(1 + \delta\overline{A}_F) \\ \overline{N} = \overline{N}_0(1 + \beta\overline{R} + \gamma\overline{A}_F) \end{cases} \tag{1}$$

or

$$\begin{cases} N - N_{12} = (N - N_{12})_0\,[1 + \alpha' R] \\ N - N_{12} = (N - N_{12})_0\,[1 + \delta' A_F] \\ N - N_{12} = (N - N_{12})_0\,[1 + \beta' R + \gamma' A_F] \end{cases} \tag{2}$$

The subtraction of the seasonal variation of N in Equations (2) is due to the fact that, in our opinion, this variation, although *a priori* may be correlated with solar activity, however, it is strongly influenced by other causes (i.e., the variation of the zenithal angles of the Sun, the height distribution of the temperature, etc.).

The series of $\overline{N}$ and $N - N_{12}$ which we have used in the calculation of the correlations begin in the months of October 1937, November 1938 and April 1939

TABLE I

Observatory	$\overline{N}_0$	α	$\overline{N}_0$	δ	$\overline{N}_0$	β	γ
Washington	22·2	0·0371 ± 0·0004	28·2	0·0283 ± 0·0006	24·0	0·0224 ± 0·0017	0·0117 ± 0·0016
	20·6	*0·0310 ± 0·0006*	*24·9*	*0·0489 ± 0·0016*	*21·2*	*0·0199 ± 0·0015*	*0·0205 ± 0·0029*
Huancayo	43·1	0·0181 ± 0·0004	48·3	0·0159 ± 0·0004	44·5	0·0116 ± 0·0013	0·0060 ± 0·0013
	40·8	*0·0174 ± 0·0002*	*46·5*	*0·0287 ± 0·0012*	*41·1*	*0·0153 ± 0·0008*	*0·0040 ± 0·0014*
Watheroo	29·1	0·0263 ± 0·0004	33·7	0·0228 ± 0·0007	29·6	0·0212 ± 0·0017	0·0049 ± 0·0018
	27·3	*0·0238 ± 0·0004*	*32·7*	*0·0374 ± 0·0016*	*27·5*	*0·0215 ± 0·0014*	*0·0043 ± 0·0027*

TABLE II

Observatory	$(N - N_{12})_0$	α'	$(N - N_{12})_0$	δ'	$(N - N_{12})_0$	β'	γ'
Washington	29·7	0·0237 ± 0·0020	35·0	0·0189 ± 0·0014	27·3	0·0135 ± 0·0027	0·0148 ± 0·0025
	24·1	*0·0245 ± 0·0014*	*28·6*	*0·0395 ± 0·0022*	*22·4*	*0·0143 ± 0·0024*	*0·0271 ± 0·0045*
Huancayo	46·9	0·0150 ± 0·0014	55·8	0·0107 ± 0·0013	45·1	0·0115 ± 0·0018	0·0056 ± 0·0017
	43·7	*0·0154 ± 0·0008*	*51·2*	*0·0239 ± 0·0016*	*42·4*	*0·0108 ± 0·0015*	*0·0114 ± 0·0029*
Watheroo	32·1	0·0210 ± 0·0020	38·6	0·0165 ± 0·0021	30·3	0·0168 ± 0·0030	0·0081 ± 0·0032
	30·8	*0·0197 ± 0·0010*	*38·9*	*0·0284 ± 0·0020*	*29·8*	*0·0149 ± 0·0019*	*0·0125 ± 0·0036*

respectively, for the Observatories of Washington, Huancayo and Watheroo. This fact introduces some "inhomogeneity" in the numerical results.

The coefficient of the regression Equations (1) are shown in Table I, where, for comparison, we indicate in italics the corresponding values obtained for each Observatory in the years 1947–54. The coefficients α do not exhibit remarkable difference in the two different period; on the contrary *all* the values δ are reduced of about 40 % with respect to 1947–54.

The values of β, we have obtained, are equal, within the statistical errors, to the values for the years 1947–54; concerning instead the coefficients γ, we have obtained a reduced value at Washington and equal values, within the statistical errors, at Huancayo and Watheroo.

A significantly different effect of $\overline{A_F}$ in northern and southern hemispheres is apparent.

Similar conclusions one can draw from the consideration of the regression Equations (2), whose coefficients are shown in Table II.

The values of $\overline{N}_0$ and $(N - N_{12})_0$ do not differ from those relative to the years 1947–54.

In order to compare the relative importance of the two variables R and A_F, as they contribute to the variation of the dependent variable, we show in Table III

TABLE III

	Regressions (1)			Regressions (2)		
Observatory	β_s	γ_s	γ_s/β_s	β'_s	γ'_s	γ'_s/β'_s
Washington	0·0271	0·0145	0·535	0·0155	0·0185	1·194
	0·0308	*0·0164*	*0·532*	*0·0215*	*0·0214*	*0·999*
Huancayo	0·0147	0·0076	0·517	0·0134	0·0069	0·515
	0·0215	*0·0029*	*0·135*	*0·0143*	*0·0079*	*0·552*
Watheroo	0·0275	0·0062	0·227	0·0202	0·0091	0·451
	0·0328	*0·0034*	*0·104*	*0·0221*	*0·0097*	*0·439*

the standard partial regression coefficients β_s, γ_s and β'_s, γ'_s for the years 1938–44 and 1947–54, for the three observatories. The ratio of corresponding standard regression coefficients is significantly higher at Washington than at Huancayo and Watheroo, for both regression Equations (1) and (2), except than Equation (1) at Huancayo in 1938–44; this exception is likely an effect of the "inhomogeneity" of the data correlated.

Concerning the correlations of $N - N_{12} - \overline{N}$ with $R - \overline{R}$ and $A_F - \overline{A}_F$, we have obtained only results of poor statistical meaning. This fact is not surprising because the deviation of $N - N_{12}$ from the mean values $\overline{N}$ is largely determined also by other causes than the deviation of R and A_F from their mean values $\overline{R}$ and $\overline{A}_F$. On the other hand, our usual solar parameters have to be considered a qualitative "index," rather than a rigorous measure of the corresponding activities.

III. Conclusions

At the present stage, our results seem to indicate also for the years 1938–44 the different "sensibility" of the two hemispheres to the activity of the hydrogen filaments, which we have found for the years 1947–54.

Obviously, their validity is *a priori* confined only to the periods we have studied and any extrapolation to different periods has to be considered with caution, for their typical statistical character.

However an interesting fact is that the protuberances (whose proiections on solar disc are the filaments) have been indicated [1, 2] as the source of a slow solar corpuscular emission. Although some objections to this view have been recently raised [3, 7], we cannot disregard the possibility that some ionizing corpuscular radiation enters the upper atmosphere, where it is absorbed: no surprise that such a radiation, and hence its ionizing effects, may exhibit some latitudinal variation.

A complete worldwide study of the correlations for all the Observatories which have performed ionospheric measurements after 1937 is now in progress. We hope that a close examination of correlation of ionospheric density with *all* solar parameters, during different solar cycles (or different phases of a same solar cycle) will allow a discrimination of the effects having a physical (besides the statistical) meaning.

Acknowledgments

I am indebted to the Central Radio Propagation Laboratory of the National Bureau of Standard and the Carnegie Institution of Washington, which have furnished many of the ionospheric data used. I also am greatly indebted to Professor L. D'Azambuja who very kindly has sent me his "Cartes Sinoptyques de la chromosphere solaire".

At last, I wish to acknowledge the helpful suggestions which I have received from Professor G. Righini, Director of the Astrophysical Observatory of Arcetri.

References

1. Kiepenheur, K. O. (1947). *Astrophys. J.* **105,** 408.
2. Kiepenheur, K. O. (1952). *J. geophys. Res.* **57,** 113.
3. Leighton, H. I. & Billings, D. E. (1955). *J. atmos. terr. Phys.* **7,** 349.
4. Mariani, F. (1959). *Nuovo Cim.* **12,** 218.
5. Mariani, F. (1957) *J. atmos. terr. Phys.* **10,** 240.
6. Mariani, F. (1957). *Ann. Geofis.* **10, 71.**
7. Roberts, W. O. & Trotter, D. E. (1955). *J. atmos. terr. Phys.* **6,** 282.

Beyond-the-Horizon Propagation Characteristics at 3000 MHz

G. CARLSON

Research Institute of National Defence, Stockholm

I. INTRODUCTION

This paper will only give a brief summary of the experimental results from long distance measurements at 3000 MHz. The investigation covers a period of more than one year and the results may be referred to as typical for the middle of Sweden. Some characteristics as propagation loss, fading and pulse distortion will be discussed.

II. MEASURING FACILITIES

The transmitter, operating at 3000 MHz, has a peak pulse power of some 100 kW. The pulse frequency is crystal-controlled and amounts to 500 Hz with a duty cycle of 0·0005. The transmitting antenna has a free space gain of 37 db. The same type of antenna is used at the receiving end for fixed path measurements over a distance of 270 km. The bandwidth of this receiver is 6 MHz and the noise factor 8 db. A transportable equipment with a bandwidth of 2 MHz is used for measurements at various distances. Different antennas (29 and 37 db) have been used in this case. Only vertical polarization has been used. All the measurements are made over a land-path.

The recitified output of the receivers is fed to an Esterline Angus recorder and a Brush pen motor to facilitate the analysis of the median propagation loss and the rapid fading. The video signal is examined from pictures of oscilloscope traces taken as well by a single shot camera as by a high speed camera. Furthermore it is made possible to study the pulse-time-modulation over the fixed test path due to changes in the time of arrival of the pulses by comparing the video phase of the received signal to a crystal-controlled time base at the receiving end.

III. RESULTS

In the analysis of the propagation loss, the propagation factor F is introduced, giving the ratio between the measured field strength and the calculated free space field strength.

The free space power level is given by

$$P_{\mathrm{FS}} = P_{\mathrm{T}} \frac{G_{\mathrm{T}} A_{\mathrm{R}}}{4\pi d^2}$$

where

P_{T} = transmitted power; G_{T} = directivity of transmitting antenna; A_{R} = absorption cross section of receiving antenna; d = path distance.

The propagation factor F is then defined by

$$P_R = F^2 . P_{FS}$$

where P_R is the actually received power.

A. Medians of the Propagation Factor

The propagation factor has been determined from half hour medians of the received signal. The observations over the fixed path have been carried out

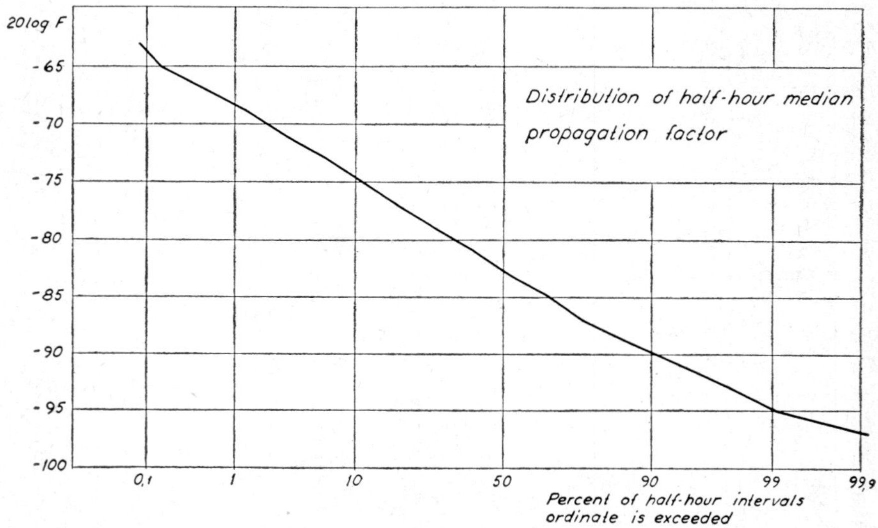

FIG. 1. Distribution of the half-hour medians of the propagation factor December 1956–November 1957.

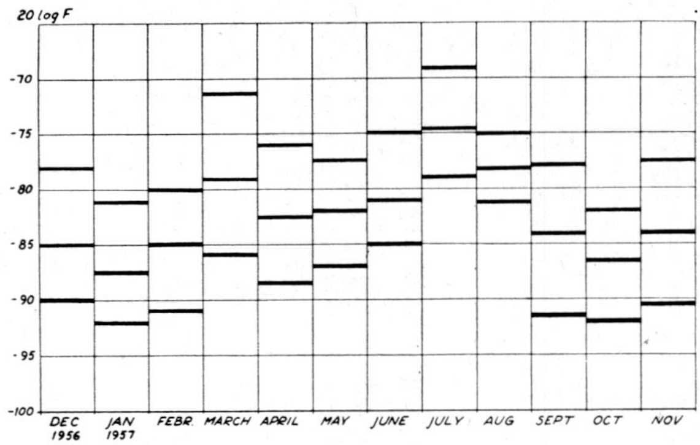

FIG. 2. Seasonal variation of the propagation factor (monthly 10, 50 and 90 % values).

continuously for more than one year, while the measurements at different distances from the transmitter have been made for short periods only, however, covering the whole year.

The median propagation loss over the fixed test path amounts to 83 db relative to the free space loss as can be seen from Fig. 1, where 20 log F is plotted in a

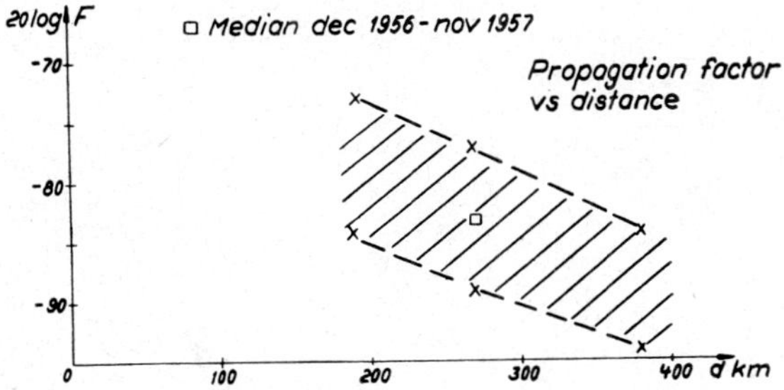

FIG. 3. Propagation factor versus distance.

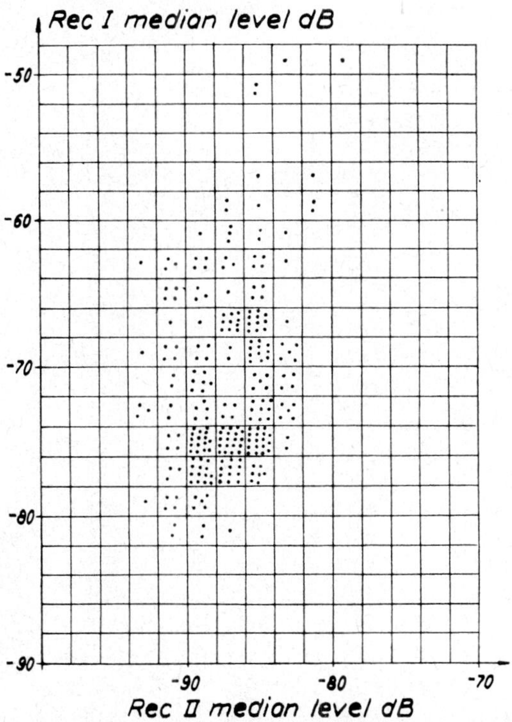

FIG. 4. Simultaneous hourly medians at receiving sites I and II (10 days August 58)
Rec. I: 270 km from transmitter. Rec. II: 380 km from transmitter.

normal distribution paper. The signal is found to be log-normal distributed with a standard deviation of 6 db. The seasonal variation of the propagation factor can be seen from Fig. 2, where the 10, 50 and 90 % levels are plotted for the period December 1956 to December 1957. The standard deviation within the different months varies between 3 and 6 db, the low value of which refers to the summer period. Sometimes the propagation loss is affected by abnormal meteorological conditions for short time during the summer period. This will rise the upper end of the distribution curve, thus giving a relatively great ratio between the 10 and 50 % values in spite of the small standard deviation.

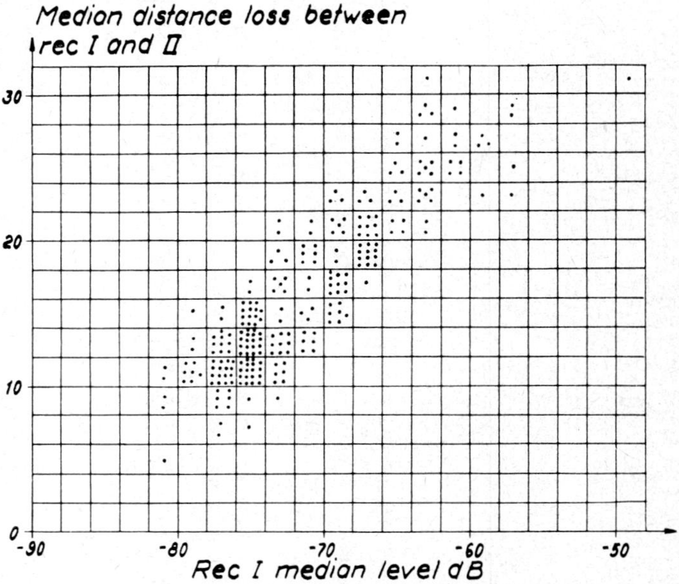

FIG. 5. Hourly median distance loss between receiving sites I and II versus median level at rec. I (10 days August 58).

Rec. I: 270 km from transmitter.
Rec. II: 380 km from transmitter.

The propagation factors for different path lengths as measured by the aid of the transportable equipment for short periods have been related to the propagation factor for the fixed path. Local variations at the receiving sites such as obstructions have made it difficult to estimate the distance dependence, however. Furthermore the signal-to-noise ratio sometimes has been too small to facilitate reliable measurements of the propagation loss for great path lengths, particularly when using the small antenna type. Eliminating the estimated influence of the obstructions the propagation factor has been plotted versus distance for three different places in Fig. 3. As can be seen from the following section, it is difficult to compare the propagation loss over various distances as measured only for short periods in order to give the distance dependence. That is why a definite curve has not been plotted in Fig. 3.

Hourly medians observed simultaneously at different distances do not seem to be well correlated, as can be seen from Fig. 4, where the signal levels at distances of 270 and 380 km are compared for a period of 10 days. The corresponding loss difference of the two paths is given in Fig. 5 as a function of hourly median levels for the fixed path. A decrease in the propagation loss over the shorter distance corresponds in most cases to an increase in the loss difference, as can be seen from Fig. 5, which shows the negligible correlation more clearly than Fig. 4. By taking the medians for periods of a few days, however, good correlation will be obtained.

B. Fading Characteristics

The rapid fading has been studied only for short periods covering, however, the whole year.

Statistical analysis of short time intervals of the signal shows a Rayleigh distribution. This distribution changes for longer intervals into a normal distribution (Fig. 6).

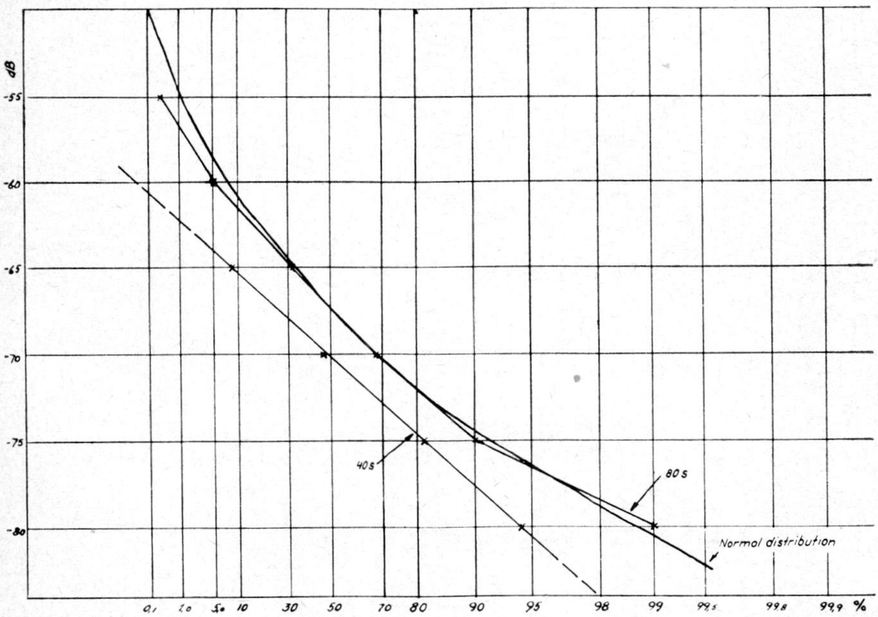

FIG. 6. Distribution of instantaneous signal-level (40 and 80 sec intervals respectively).

The nature of the fast fading has mostly shown to be independent of the prevailing average propagation characteristics. The fading range, as defined between 10 and 90 % levels, has been approximately 11 db discarding abnormal circumstances (Fig. 7).

The fading range has shown to be almost independent of the path length in the interval 200 to 400 km. The fading frequency (the number of times per second the median is passed in one direction) is approximately proportional to the path length (or distance angle) and amounts normally to a few cycles per second at a distance of 270 km from the transmitter.

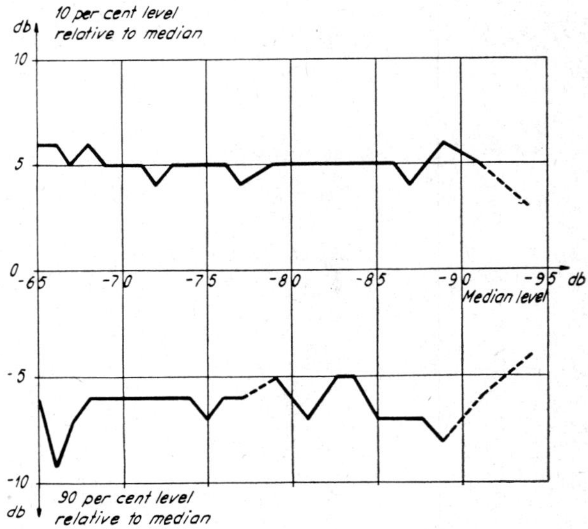

FIG. 7. Medians of 10 and 90 % values of instantaneous signal versus median level (500 samples from 270 km test path.)

C. Pulse Distortion

The pulse distortion has been studied over the 270 km test path. The shape of the transmitted pulse is shown in Fig. 8(a), pulse length being approximately

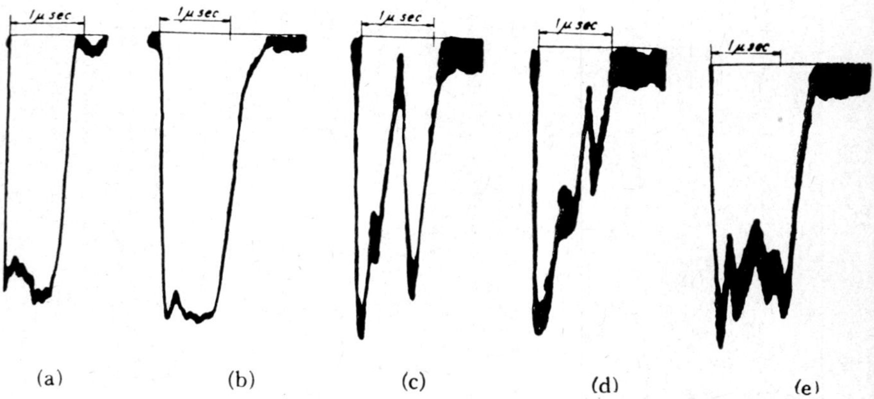

FIG. 8. Various types of pulse distortion.
(a) Transmitted pulse.
(b)—(e) Received pulses.

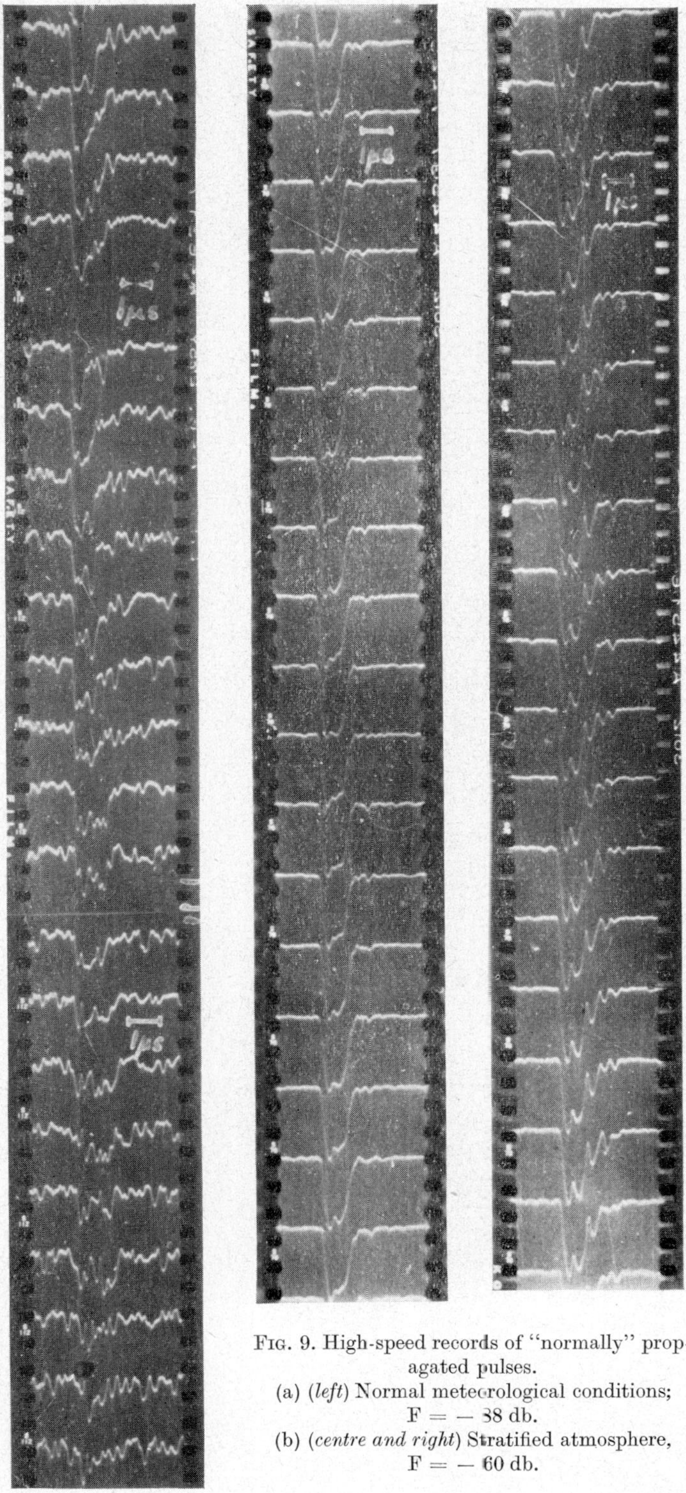

FIG. 9. High-speed records of "normally" propagated pulses.
(a) (*left*) Normal meteorological conditions; F = − 38 db.
(b) (*centre and right*) Stratified atmosphere, F = − 60 db.

$0{\cdot}8\,\mu s$. The received pulses have been photographed with $\frac{1}{25}$ sec exposure time, thus giving the average shape of 20 pulses. Normally a pulse broadening of 0·1 to

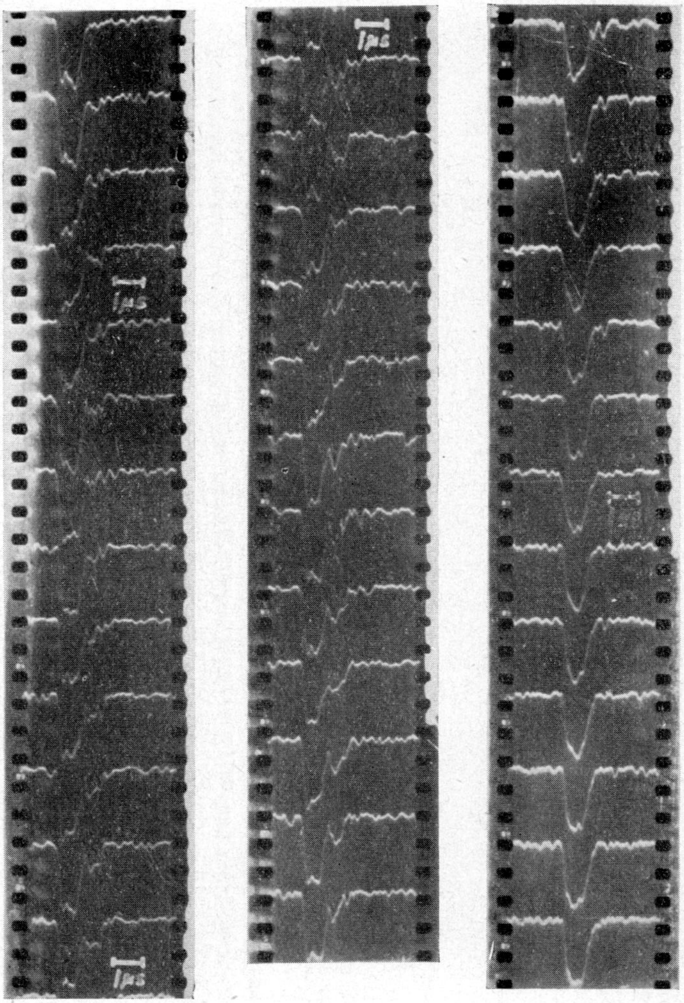

Fig. 10. High-speed records of "normally" propagated and aircraft reflected pulses. (a) (*left and centre*) Weak aircraft reflections from the common volume at 4300 m height. $F = -82$ db. (b) (*right*) Dominating aircraft reflection from the common volume at 1300 m height. $F = -82$ db.

$0{\cdot}2\,\mu s$ is obtained 8(b). As shown in Fig. 8(c) and 8(d) the pulses are often severely distorted due to selective fading. This type of distortion is very frequent when observed on the 6 MHz wide receiver, while the transportable equipment cannot be used for this purpose because of its smaller bandwidth (2 MHz). In a few cases heavily broadened pulses have been obtained as in Fig. 8(e).

Furthermore the pulse distortion has been examined by the aid of a high speed camera. Samples of these observations are given in Figs. 9 and 10. Fig. 9(a) shows the distortion obtained in a case with a propagation factor of 88 db and at normal

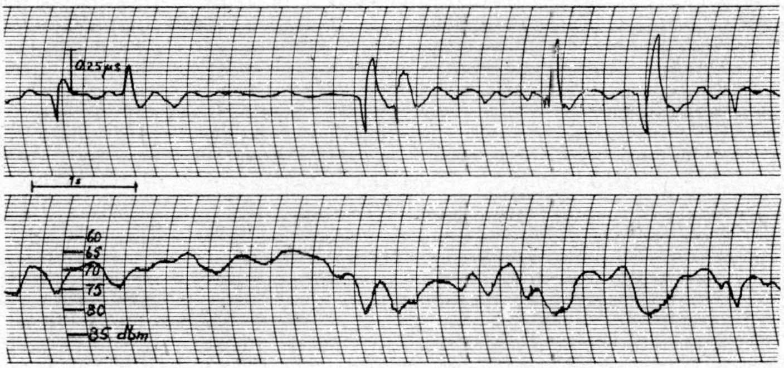

(a) Normally propagated signal.

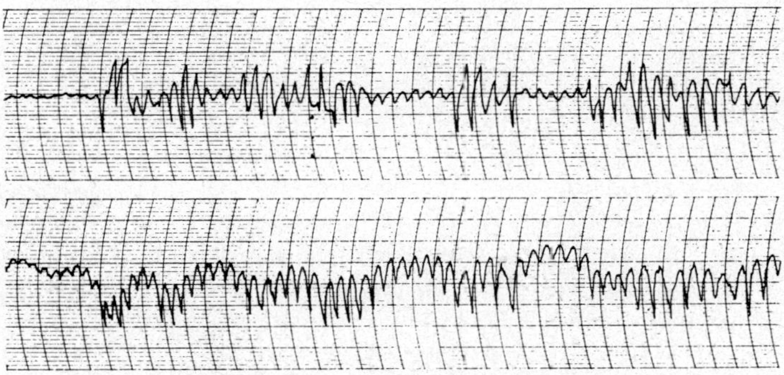

(b) Aircraft reflections. Free space level: + 3 dbm.

FIG. 11. Signal and video phase fluctuations on a 270 km path.

meteorological conditions. The pulse shape changes quite rapidly as can be seen from the film strip. A rather undistorted pulse shape is shown in Fig. 9(b). In this case the atmosphere was strongly stratified and the signal level was some 30 db above median level.

Aircraft reflections have also been observed. Various forms of distortion are obtained due to the position of the aircraft related to the "normal" propagation path. Figure 10 shows aircraft reflections, when the propagation factor was approximately 82 db (about median value for the whole period) and the aircraft was at a height of 4300 (a) and 1300 m (b) respectively. The reflected signal can be seen as a delayed pulse in Fig. 10(a). Calculations show that the normally propagated wave has taken the shortest way between transmitter and receiver.

Figure 11 shows two samples of video phase variation and signal records. The phase fluctuation amounts normally to about 0·2 μs and has a period of the same order as the fading of the received signal (a). Aircraft reflections have shown to give very strong and rapid fluctuations of both the phase and the signal level (b).

D. Signal Dependence on Meteorological Factors

Ground values of temperature and dew point have been observed at three places along the test path. These observations have been compared to the propagation loss on the fixed path of 270 km. Some correlation is evident as can be seen from

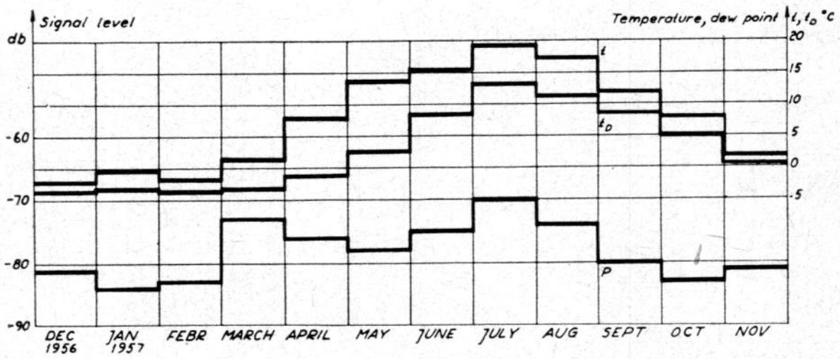

FIG. 12. Monthly median of signal level, temperature and dew point.

Fig. 12. The correlation factors are rather small, however, and amounts for day-observations to 0·7 as a maximum. Corresponding values for night-observations are only of the order 0·3 to 0·5.

From earlier oversea measurements [1] correlation factors of 0·5 to 0·6 are reported. It was stated, however, that those figures were quite uncertain in view of the short period analyzed at that time. After this more complete test program it can be said that the correlation between meteorological surface values and propagation loss is not very good. In any case not so good that it is possible to predict the propagation loss within narrow limits from the knowledge of meteorological surface values. It can be mentioned that the observed median loss is approximately 10 db larger than the predicted value according to the formula given by the National Bureau of Standards, U.S.A. [2].

E. Various Observations

The results from the long time test program have shown not to be able to be explained by existing theories and formulas. In order to look for various factors affecting the propagation special measurements have been carried out for a short period during August-September 1958. Of course, these observations have not been completely evaluated up to now, but some indications may be mentioned.

As nobody exactly knows, what is happening to the radio waves in the troposphere the idea was to measure as many factors as possible probably affecting the propagation loss.

The N-profile was observed up to a height of 500 m by the aid of a meteorological sonde mounted in a captive balloon. Furthermore the vertical antenna diagram of the transmitter was studied up to 1500 m to determine the height of the radio-horizon at a place 50 km far from the transmitter. These measurements have made it possible to take into account the ray bending or to evaluate the effective earth radius. An airborne refractometer of the Birnbaum type [3] has also been used measuring turbulence and N-profile in the common volume up to 5000 m.

As said above, time has been too short to evaluate all the data obtained during this program, why only the trend can be pointed out.

The equivalent earth radius evaluated from as well the vertical soundings of the N-profile as the antenna diagram seems to correlate very badly with the propagation loss. This fact is rather contradictory to what might be expected from the scatter theories, which would give a propagation loss depending on the angular distance (determined by geometry and ray bending). Quite large fluctuations of the signal median have been observed, while the height of the radio horizon 50 km far from the transmitter has remained fairly stable.

Maybe, these results seem to be a little confusing when thinking of scattered signals. Probably, long distance signals may not be explained only by scatter phenomena but also by reflections from different layers. This has been suspected from earlier observations, too [1]. Of course, it is too early to say that scattering is no reality, as these measurements are carried out only during summertime when the meteorological conditions may favour the stratification of the atmosphere. For that reason similar measurements have to be carried out during winter when the probability of important stratification is fairly low.

IV. Conclusions

The results presented in this paper are published to contribute to the experimental studies made all over the world. Today everybody knows that the "scattered" signals are helpful means to broadband communication over fairly long distances. Many questions are unanswered, however. So must the question of bandwidth capability be of great interest, as one can expect to have transmitters with very high output in the future, why it will be possible to transmit very broad bands from a signal-to-noise ratio point of view. The pulse studies may be a tool to determine the bandwidth capability experimentally.

Furthermore a method to perdict the propagation loss within narrow limits have to be developed when planning communication links in this field. It cannot be practical to permit deviations of the order of 5 to 10 db, which must be usual today to take into account local variations.

V. Acknowledgments

The author wishes to thank the Director General, Research Institute of National Defence, for permission to publish the paper. The assistance of Mr. L. Schörling, who, to a great deal, has been responsible for developing the measuring equipment, is acknowledged with gratitude.

The helpful assistance of Messrs K.-R. Larsson and L. G. Angantyr during the investigation and the analysis of the records is also appreciated.

References

1. *Trans. Inst. Radio Engrs, N.Y.* (1958). AP–**6,** 173–175.
2. *Trans. Inst. Radio Engrs, N.Y.* (1955). CS–**4,** 39–49.
3. *Rep. nat. Bur. Stand.* No. 5057.

Propagation Tests at 250, 500, 1000, 2000 Mc/s on a 189 Km Path

F. Carassa and P. Quarta

Laboratorio Centrale Radio della Fabbrica Italiana Magneti Marelli, Milano

Results are given of propagation tests carried out in 1954, 1955 and 1956 on a over land, line-of-sight, path, 189 Km long, simultaneously with the four frequencies of 250, 500, 1000 and 2000 mc/s: statistical analysis show that the propagation is more irregular in summer than in winter, and that fadings increase in depth and in number as the frequency is increased.

I. Introduction

Since 1950 a series of propagation tests at various frequencies has been carried out by the Laboratorio Centrale Radio of the F.I. Magneti Marelli on the line-of-sight path Monte Cimone-Milan. Figure 1 shows the locations and the profile of the path together with an enlarged portion of the profile near Milan. Characteristics of the path are the considerable length and the remarkable altitude of one of the terminals.

The path may be considered as one of the typical line-of-sight long paths which it is possible to find in Italy due to the presence of high mountains throughout the country. It may be noted, for example, that this path was used as the first hop of the first Italian radio link, put into service between Milan and Rome in 1939.

In previous papers [1, 2] the results were given of the tests at 250, 500, 1000 mc/s, carried out from 1950 to 1953; the present paper gives the results obtained on the same path and with the same frequencies in 1954, 1955 and 1956; it is to be noted that since June 1955 the tests were extended to include also the frequency of 2000 mc/s. Figure 2 shows the test periods for each wavelength.

Especially designed portable equipment has been used, its main characteristics being the following:

Quartz-controlled continuous wave transmitters with power output of: 30 W at 250 mc/s, 5 W at 500 mc/s, 10 W at 1000 mc/s, 3 W at 2000 mc/s.

3 m diameter parabolic transmitting antennas, one for 250 mc/s and 500 mc/s and another for 1000 and 2000 mc/s.

One 5 m diameter parabolic receiving antenna for both 250 and 500 mc/s.

One 3 m diameter receiving antenna for both 1000 and 2000 mc/s.

Receivers having quartz-controlled local oscillators, 12·5 mc/s intermediate frequency, 0·5 mc/s bandwidth, logarythmic response of the output d.c. current versus the input radio frequency voltage. This response allows the recordings to be calibrated directly in decibel.

Paper recorders printing a point every 10 sec were used until 1956, when a continuous trace recorder with one second time constant was introduced for the 1000 and 2000 mc/s waves. All the recorders used were of the dual trace type, so that only one recorder was used for the 250 and 500 mc/s, and another for the

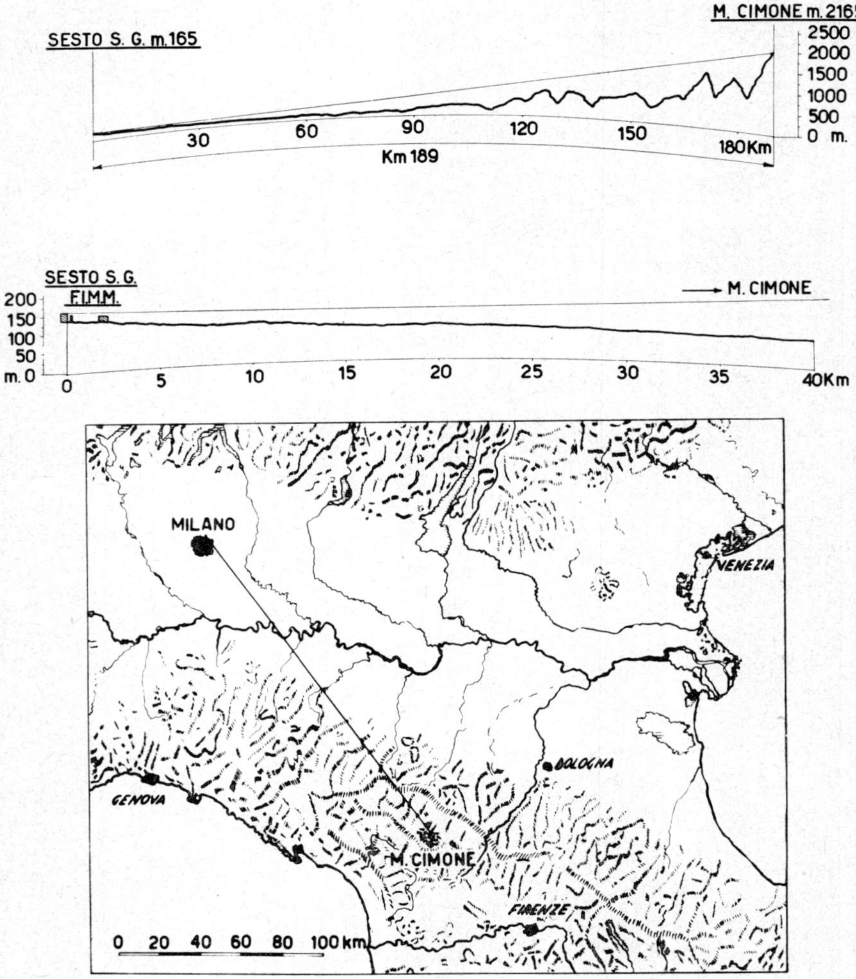
M. CIMONE m. 2165
SESTO S. G. m. 165
2500
2000
1500
1000
500
0 m.
30
60
90
120
150
180 Km
Km 189
SESTO S. G.
F.I.M.M.
M. CIMONE
200
150
100
50
m. 0
0
5
10
15
20
25
30
35
40 Km
MILANO
VENEZIA
BOLOGNA
GENOVA
M. CIMONE
FIRENZE
0 20 40 60 80 100 km

FIG. 1.

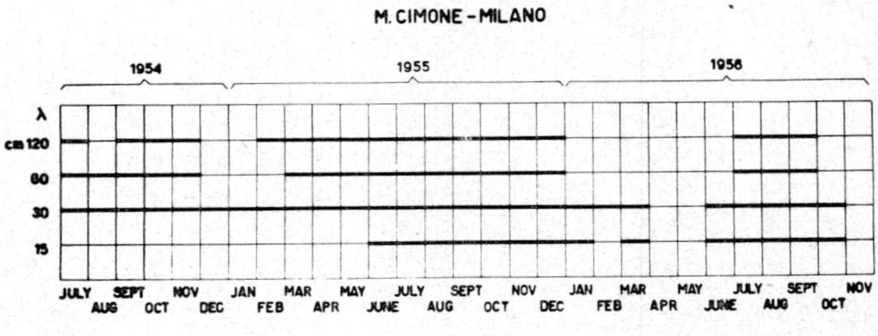
M. CIMONE - MILANO
1954
1955
1956
λ
cm 120
60
30
15
JULY
AUG
SEPT
OCT
NOV
DEC
JAN
FEB
MAR
APR
MAY
JUNE
JULY
AUG
SEPT
OCT
NOV
DEC
JAN
FEB
MAR
APR
MAY
JUNE
JULY
AUG
SEPT
OCT
NOV

FIG. 2.

1000 and 2000 mc/s waves. The time scales is therefore exactly the same for 250 and 500 mc/s as well as for 1000 and 2000 mc/s; slight uncertainties may be present between the time scales of the two pairs of recordings.

II. General Remarks on the Test Results

As in the previous tests the results are affected by an error of the order of 1 ÷ 2 db as far as the field variations are concerned; however a higher error (± 5 db) is to be taken into account in the measurement of the absolute path attenuation.

Therefore it seemed better to assume as reference attenuation value the monthly

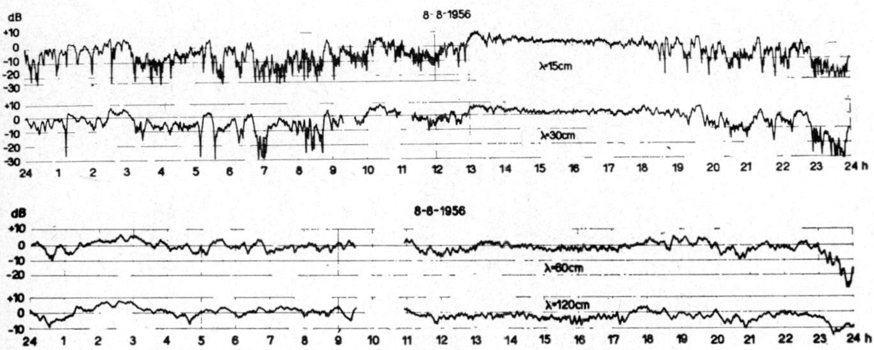

Fig. 3.

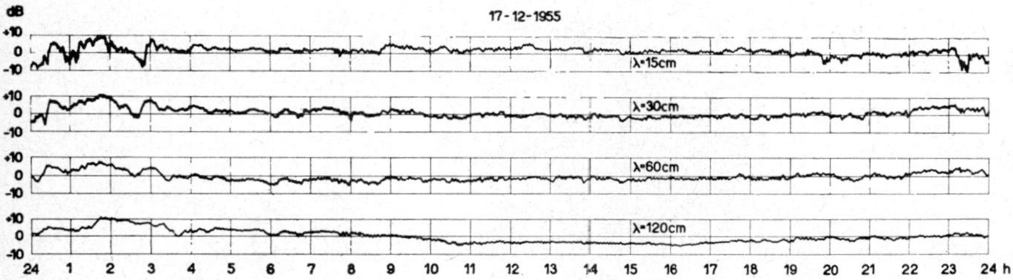

Fig. 4.

median value instead of the free space value, although the latter one is more suitable for practical applications.

Nevertheless it is to be noted that the free space values do not differ from the monthly median values more than the errors mentioned above. Figure 3 gives an example of recordings, which may be taken as typical for a summer day; in this example clearly appear the differences among the four frequencies and particularly how deeper and more frequent the fadings are as the frequency is increased. It may be noted that fadings are nearly absent during the central

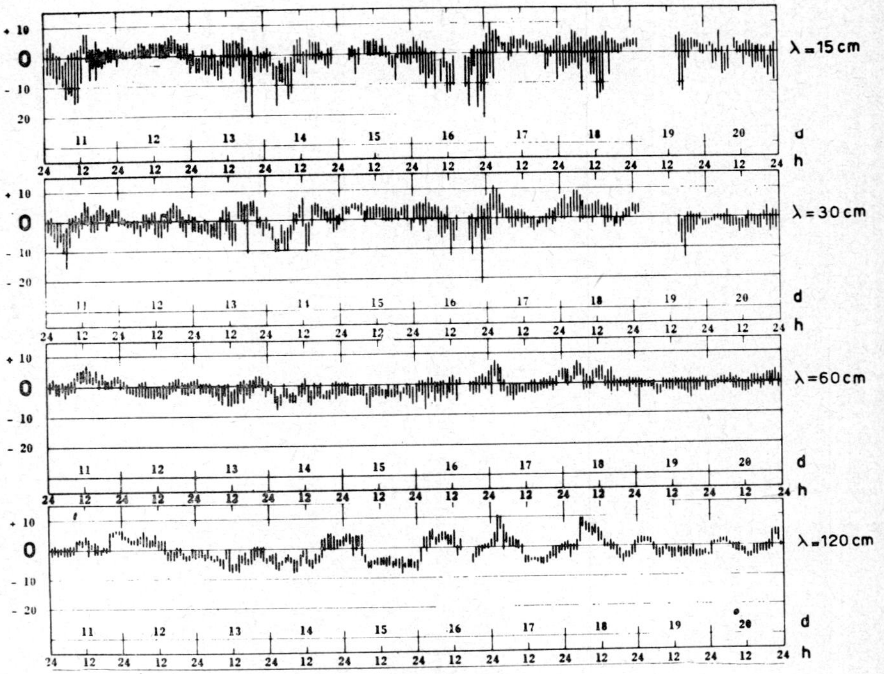

AUGUST 1956

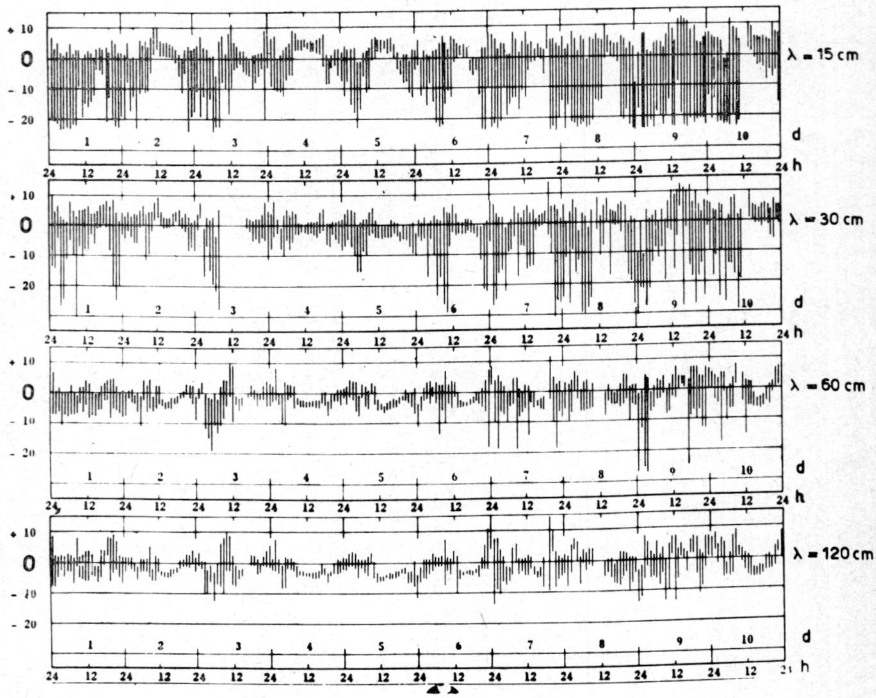

Fig. 5.

hours of the summer day, as generally noted in summer months on overland paths. In the evening, night and early morning both slow and fast fadings appear; the fast fadings seem to be due to interference by multipath propagation and, as far as these fadings are concerned, no correlations seem to exist among the various frequencies.

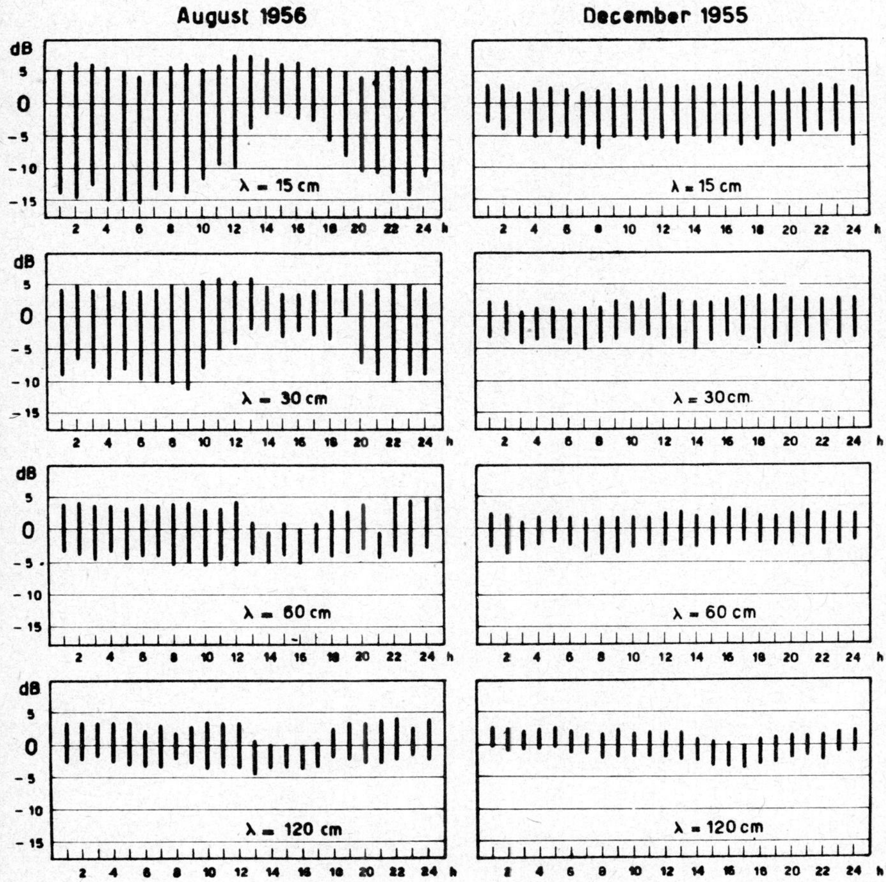

FIG. 6.

It is to be noted that the fast fadings are often observed at the bottom of a slow fading probably when a primary field becomes comparable with secondary multipath contributions.

It may be noted also that the slow field variations appear often nearly simultaneously for the various frequencies. Long periods of small field variations were observed in this season during bad weather conditions.

Fig. 4 shows an example of recording in a winter day; the fadings are smaller than in summer and it may be noted that the slow field variations are often similar in shape for the various frequencies.

III. Data Analysis and Statistical Results

From the original recordings were first deduced the maximum and minimum values in each hour; these were plotted in graphs of which Fig. 5 is an example, relative to 10 winter days and 10 summer days. In this compressed time scale it is possible to find in a larger number of days some of the propagation characteristics noted before.

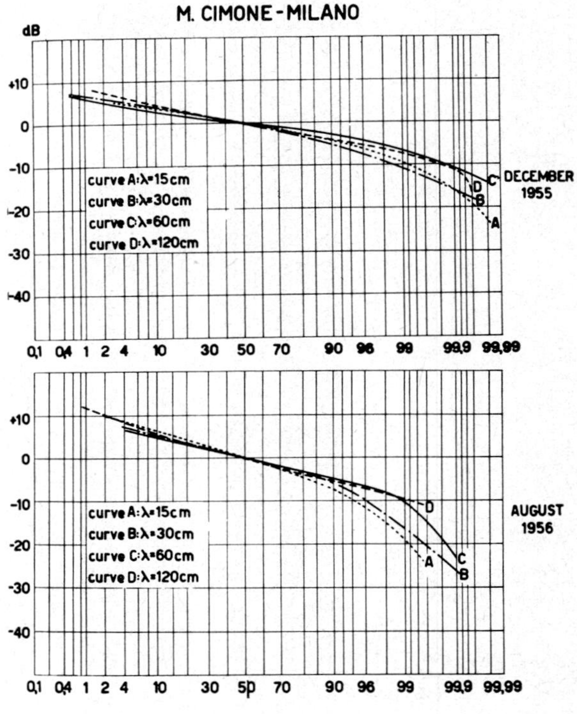

Fig. 7.

Figure 6 gives for 2 months, the monthly average values of the maximum and minimum values, hour by hour; in other words, these graphs are representative of an average summer day and of an average winter day.

Statistical graphs were also deduced for each months and Fig. 7 shows two examples of these graphs for a winter month and for a summer month.

These graphs give the attenuation below the median value not exceeded for the percentages of the time indicated in the abscissae. The graphs are plotted in a chart in which the gaussian distribution curves are straight lines.

The better propagation of the lower frequencies is evident; also evident are the better propagation condition during the winter. The unusual increase of the attenuation at the higher percentages of time for the 250 mc/s wave in December

1955 is due to a long fading, with a duration of nearly 20 min on 3 December at 20.30; this is to be considered as a very exceptional event.

Some typical points of the distribution curves are given for each month of the test period on Fig. 8; the points correspond to 90, 98 and 99·5 % of the time.

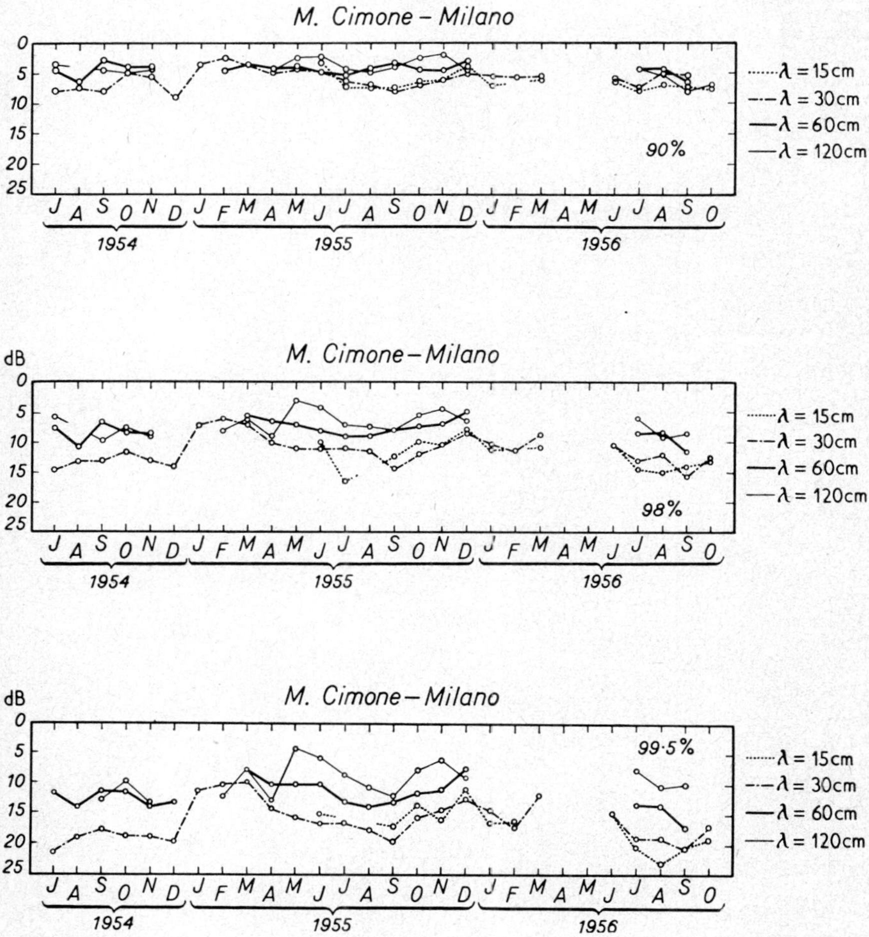

FIG. 8.

The differences among the various frequencies during the various periods of the year are evident. It may be noted again that the higher the frequency the larger are generally the attenuations exceeded for a given percentage of the time; actually, at least during this test period, the differences among the 1000 and the 2000 mc/s waves are rather slight: in fact the 2000 mc/s has undergone fadings more frequent but of shorter duration than the 1000 mc/s signal.

The differences in the number of fadings among the various frequencies may be seen in Fig. 9, which shows for a summer month the number of times the received signal dropped below the levels indicated in the ordinatae. It clearly appears how the number of fadings increases with the frequency.

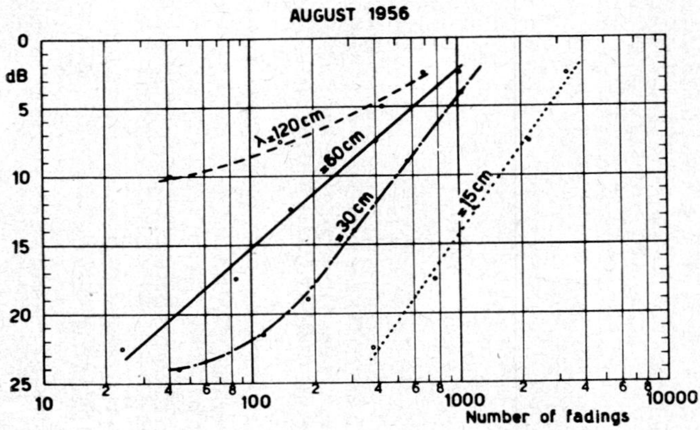

Fig. 9.

IV. Conclusion

The remarks made on the propagation along the path may be summarized as follows:

(a) For all frequencies, the propagation is more irregular during the summer months; however, a period of steady propagation is noted in the central hours of the summer days. Long periods of small field variations are observed in summer during bad weather conditions.

(b) The slow field variations show often a similar behavior for all the frequencies; the fast fadings, on the other hand, seem to be uncorrelated. These fast fadings increase in depth and specially in number as the frequency is increased.

(*c*) The attenuations not exceeded for a given high percentage of time increase as the frequency is increased.

References

1. Carassa, F. (1956). *Alta Frequenza* **25**, 378.
2. Vecchiacchi, F. (1956). *Alta Frequenza* **25**, 100.

Precipitation Static on Modern Aircraft

P. W. Couch

Communication and Navigation Laboratory, Wright Air Development Center, Wright-Patterson Air Force Base, Dayton, Ohio

Flight research has established the existence of three noise-generating mechanisms contributing to precipitation static, which frequently disrupts airborne radio reception at frequencies below 30 mc s. Above 250 knots, noise level increases at rates as high as the sixth power of speed, and requires careful antenna design in conjunction with high-performance dischargers to obtain acceptable all-weather performance of HF and LF radio systems. Noise-generating mechanisms, relation of noise level to charging rates and speed, and a sample analysis of effects on performance of an LF receiver are discussed.

Safe and rapid movement of the worldwide air traffic of today depends on reliable reception in flight of radio signals in many different frequency ranges. Expansion of air traffic and growth in number of faster aircraft in service increases the need for fast and accurate radio contact and for knowledge of noise and propagation factors which affect airborne reception. The subject of this discussion is a noise problem unique to aircraft in flight, a problem capable of exerting strong influence on the performance of aircraft radio systems intended to operate over distances greater than line-of-sight.

Strong and continuous radio noise which destroys reception at all frequencies below 30 Mc/s is sometimes encountered in flight through precipitation; in rare instances, the noise affects reception in the 100 to 150 Mc/s frequency range. It is called "precipitation static," and is one of the factors which has led to development of radio aids operating at frequencies above 100 Mc/s. Nevertheless, reception beyond line-of-sight distances is still necessary, particularly for intercontinental air traffic, and frequencies below 30 Mc/s are still in daily use; precipitation static is still an important consideration, especially for long-range aircraft.

Corona-resistant antennas and noise-free dischargers were developed by the United States military services during World War II [1, 3, 5]. These devices are effective on propeller-driven aircraft and are now in use throughout the world. However, the static problems grow worse at higher speeds, and faster aircraft require greater protection.

Precipitation static typically occurs in flight through hard-particle precipitation, such as snow or ice crystals in atmospheric temperatures below freezing, where noise may persist without ceasing for long intervals, sometimes more than an hour. Electrical activity of clouds in such regions is often too small to exert appreciable effects, and this type of situation will be the one discussed here. Charged clouds in regions of thunderstorm activity cause very erratic bursts of strong noise, even in rain or clear air. These effects produce a substantial portion of the total static encountered, but are extremely difficult to analyze and predict because they do not have single-valued relationships with quantities measured on the aircraft, and because they vary widely within short time intervals.

Precipitation particles which collide with frontal aircraft surfaces leave one sign of charge on the surface and carry away charge of opposite sign. This is the

FIG. 1. Particle and impact charging and exhaust discharging in jet aircraft.

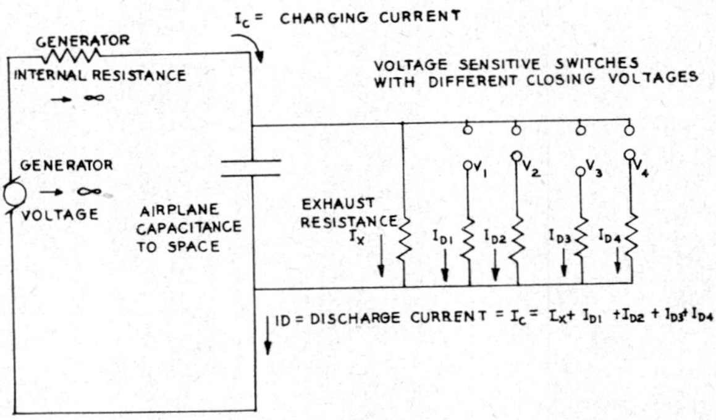

FIG. 2. Simplified approximation of aircraft charging.

fundamental phenomenon of precipitation static; it underlies all three noise-producing processes now known to occur in flight. Charge separation occurs at impact; particle charge prior to impact is many times smaller than the impact charge [3, 5, 7]. Millions of particles are intercepted each second and establish a unidirectional charge flow, a charging current, to the airframe. In most instances, but not always, negative charge flows to the aircraft and positive charge is carried away on the particle or its fragments. Magnitude of charging current is proportional to mean charge per impact and number of impacts per unit time, and is used as a measure of precipitation static qualities of aircraft and of weather for several reasons: the separation effect is partly dependent on properties of weather; it is a necessary condition for precipitation static; and no practical means of preventing the charge separation is known.

As current flows to the aircraft, airframe voltage rises until discharge currents from the airframe back to the atmosphere are in equilibrium with the charging current. Discharge takes place principally by two paths, the engine exhaust and airframe corona. The exhaust functions as a linear resistance; the current through it is proportional to airframe voltage. Figure 1 depicts the relation between impact charging and exhaust current. When airframe voltage reaches or exceeds a critical, or threshold, value, corona is initiated from airframe extremities having small radii, such as trailing edges of wingtip and tail, and bare antenna wires. The simplified equivalent circuit in Fig. 2 illustrates current and voltage relations and shows how more discharge paths come into play as voltage rises, although the real situation is more complicated.

Exhaust currents do not produce radio noise, but corona is an intense noise source, and its control or prevention is one of the major problems in designing noise-free radio systems. The corona discharge mechanism produces strong, short-duration pulses which induce noise into antennas having appreciable coupling to the discharge region. If the discharge takes place directly on an antenna conductor, very intense noise is produced in the associated receiver. Corona is responsible for most of the noise encountered on older type aircraft [1, 3, 5], and makes substantial contributions to noise on jet aircraft.

Noise generated by other mechanisms has become important on modern aircraft because they employ antennas mounted underneath exterior surfaces of dielectric material, including the windshield-canopy enclosure on military aircraft. Impact charges accumulate on the insulating surface until flashover to the airframe takes place. This process is known as streamering and produces strong pulses at fairly low repetition rates, pulses which couple noise into nearby antennas. Charge on the outer surface of the dielectric may also cause short bursts of corona from antenna conductors under the dielectric. Noise levels which reached 20,000 μV for 8 kc/s bandwidth at 300 kc/s were measured on a high-impedance antenna installed under a Plexiglas nose canopy on a four-engine jet aircraft, a B-45 [2].

The sudden change in particle voltage accompanying impact charging can also induce noise in antennas exposed to the electric field of the particle at the point of impact. This condition sometimes exists on antennas under or near dielectric surfaces exposed to particle impact, and the resultant noise is known as "particle impact noise." Both streamering and particle impact were studied in flight tests performed in a six-engine jet, a B-47 [7]. Recent exhaustive analysis of the flight data led to the conclusion that particle impact noise levels were less than streamering noise levels by at least 54 db in the test configuration [4].

The mean number of noise pulses per second is different for the different noise mechanisms, and is lowest for streamering, but is, in most instances, high enough to produce complete overlapping of the individual disturbances in a receiver of

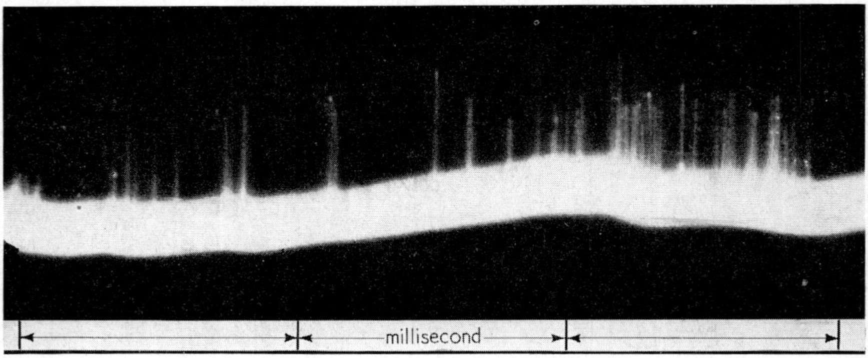

FIG. 3. Corona Pulses on aircraft antenna. Average corona current—one microampere.

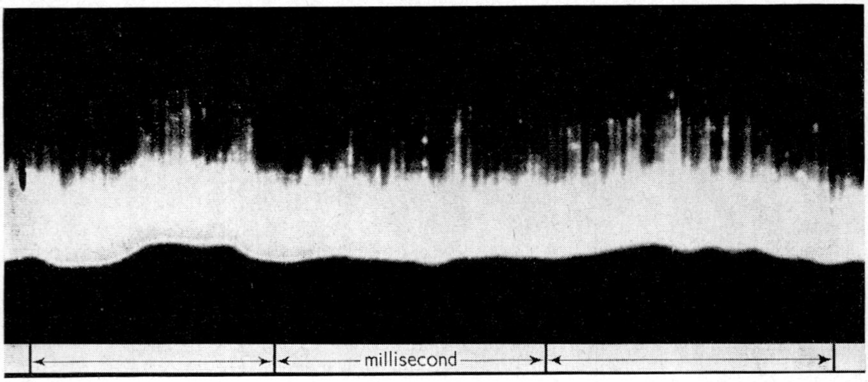

FIG. 4. Corona pulses on aircraft antenna. Average corona current—15 microamperes.

normal bandwidth. Thus, the relation between bandwidth and noise level is the conventional one for random noise, for receiver bandwidths up to about 30 kc/s; at bandwidths much greater than this, some types of precipitation static begin to depart from this relation. Almost all aircraft receivers employed at frequencies below 30 Mc/s have bandwidths of 8 kc/s or less, and the familiar relation of noise power proportional to bandwidth will hold, which can also be expressed as the r.m.s. value of noise current or voltage proportional to square root of bandwidth.

Figure 3 shows corona pulses on a bare copper antenna wire 1 mm in diameter, when the average corona current was 1 or 2 μA, barely above threshold. The antenna was terminated in a 1500 ohm resistance shunted by 20 $\mu\mu$F capacitance, which reduced pulse reflection to levels not discernible on the 5 Mc/s bandwidth oscilloscope connected to the antenna. The photograph was taken in flight through snow at an altitude of 3·1 km (10,000 ft) and true air velocity of 90 m/s (175 knots). Irregular time spacing of the pulses is clearly evident, and several pulses can be seen within a single 1 ms time interval. Figure 4 is a photograph taken a few

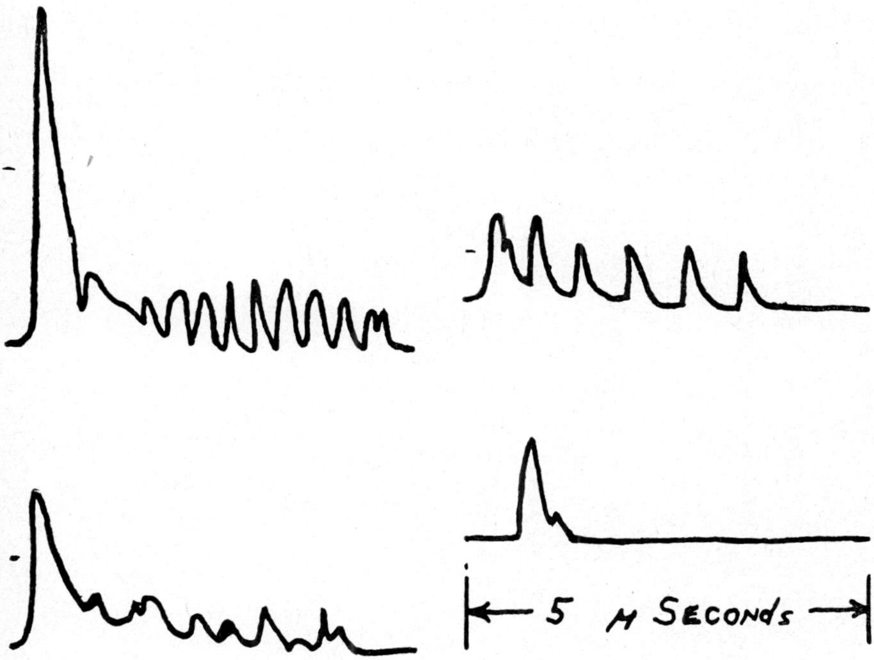

FIG. 5. Corona pulses on aircraft antenna.

seconds later, when the average corona current increased to 15 μA. A sharp increase in number of pulses is easily discernible. Peak pulse amplitudes in these photographs are in the range of 1 to 2 V. From data such as these, and from studies of pulse count versus average corona current made in the Laboratory, it is apparent that corona can easily generate pulses at rates up to 100,000 per sec.

Figure 5 is a tracing made from another scope photograph in the same flight run, and shows the waveform of a single corona pulse as viewed on a 5 μs sweep. It is apparent the pulse duration is much less than a microsecond, and that the 5 Mc/s bandwidth was probably insufficient to resolve the actual waveform. Laboratory studies show corona pulse rise times as short as 6 mμsec, with an exponential decay falling to 50 % of peak amplitude in 30 mμsec, when atmospheric pressure is 760 mm of mercury. Both rise and decay time increased in inverse proportion to pressure, reaching a rise time of 20 mμsec and decay time of

100 mμsec at a pressure of 200 mm, which is the pressure equivalent of an altitude of 9·9 km, or 32,000 ft [6].

Results of Laboratory measurements of the spectrum of corona noise are shown in Fig. 6. Data was taken as shown in the sketch, using a small corona generator

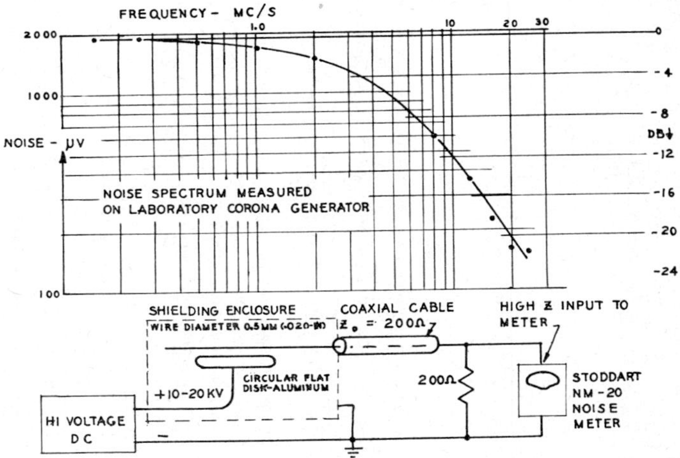

FIG. 6. Setup for corona spectrum measurement-noise meter bandwidth 8kc/s.

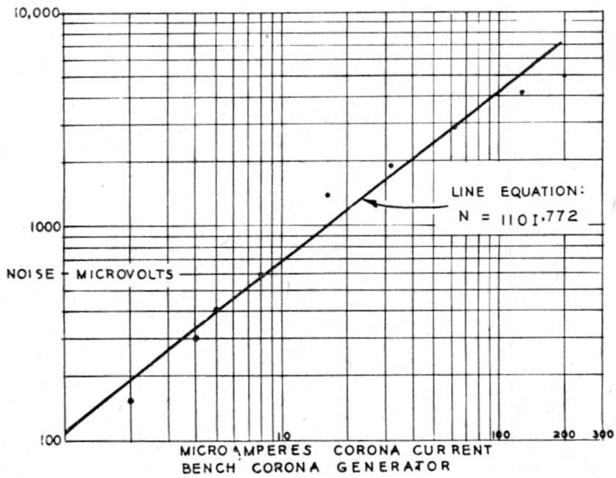

FIG. 7. Noise amplitude at 250 kc/s vs average corona current. Noise meter bandwidth = 8 kc/s.

designed to duplicate amplitude and number of corona pulses observed on aircraft antenna wire. Shape of the spectrum curve is in good agreement with the wave-shape just discussed.

Variation of noise voltage with average corona current at a frequency of 250 kc/s, as measured in the same setup, is shown in Fig. 7. At 200 μA average corona current, the noise level is 5000 μV across the 200 ohm resistor, which is equivalent to a noise current of 25 μA within the 8 kc/s bandwidth of the noise meter. Low frequency aircraft receivers are frequently used with high impedance antennas terminated in high impedances, where 1 μA of noise current will produce several thousand microvolts of noise. It can be readily appreciated that corona on an antenna of this type is indeed a prolific noise source.

At low frequencies, noise induced on typical long-wire antennas by corona on wingtips and tail is less than antenna corona noise by 40 to 60 db. At certain frequencies above 3 Mc/s, airframe resonances can increase coupling between

FIG. 8. (1) Induced charge pick-up. (2) Precipitation charging patch. (3) Insulated noise probe. (4) Electrostatic field detector head.

antenna and discharge region and increase the noise level from physically remote discharges. Wingtip corona on a B-47 has been observed to cause false indications of a radio compass operating in the 200-400 kc/s frequency range. The compass antennas were installed in the canopy [2].

Information obtained from the B-47 studies of streamering and impact noise yielded the conclusion that the spectrum of impact noise is such that the microvolt level decreases inversely with frequency, and the spectrum of streamering noise is nearly flat up to a frequency of 160 kc/s, rolling off slowly to about 3 db down at 480 kc/s [4].

The intensity of precipitation static as a function of frequency on a particular aircraft antenna is very strongly influenced by such factors as the variation of antenna impedance and the variation of coupling between antenna and noise source as a function of frequency. Aircraft size, shape, and antenna location all have important bearing on these factors, and hence influence the noise levels likely to be encountered at particular frequencies.

To determine the relation of aircraft speed to noise, tests were made to determine the variation of charging rate with speed, and the variation of noise with charging

FIG. 9. (1) Airfoil charging patch. (2) Conventional frontal surface patch. (3) External electric field detector head. (4) Plexiglass test area (graphite coated).

FIG 10. (1) Canopy current collector strips. (2) Windshield current collector strips. (3) Windshield current pick-up probe.

rate. Previous work had substantiated increase of charging rate in proportion to the cube of the speed, for the range below 250 knots (128 m/s) [3, 5].

A single-engine jet fighter, an F-94 equipped with afterburner, was instrumented and flown to obtain measurements of charging and discharging rates, and noise levels. The test aircraft is shown in various aspects in the next three pictures.

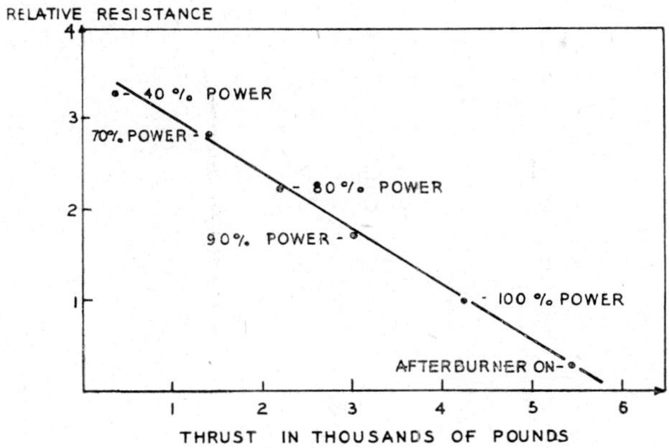

FIG. 11. Relative exhaust resistance vs thrust in F-94A aircraft.

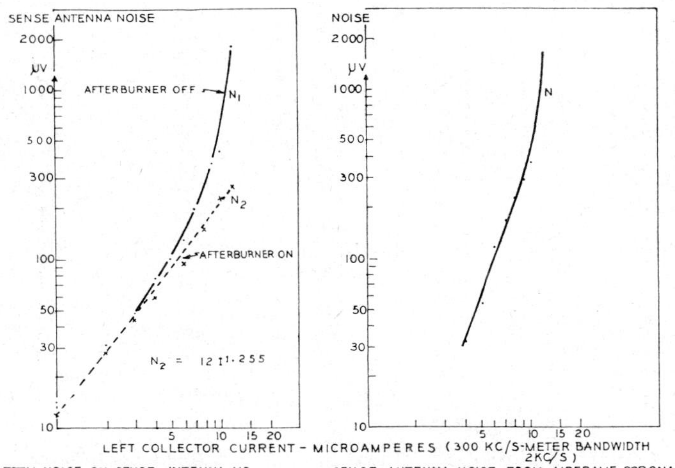

TOTAL NOISE ON SENSE ANTENNA VS CHARGING RATE F-94 AIRPLANE

SENSE ANTENNA NOISE FROM AIRFRAME CORONA CALCULATED FROM $N^2 = N_1^2 - N_2^2$

FIG. 12.

Charging rate to the left half of the plexiglas nose section, to an isolated segment of aluminum within the metal strip dividing the two halves of the nose, and to an isolated aluminum airfoil section mounted on the top of the fuselage forward of the windshield were measured. Noise on the canopy antenna was measured at

300 k/cs with a receiver of 2 kc/s bandwidth. Airframe electric field, proportional to airframe voltage, was measured by means of a rotating-vane field meter mounted on the left side of the fuselage forward of the wing.

Measurements of the discharging characteristics of the engine were made at different power settings, including operation with afterburner on. Values of

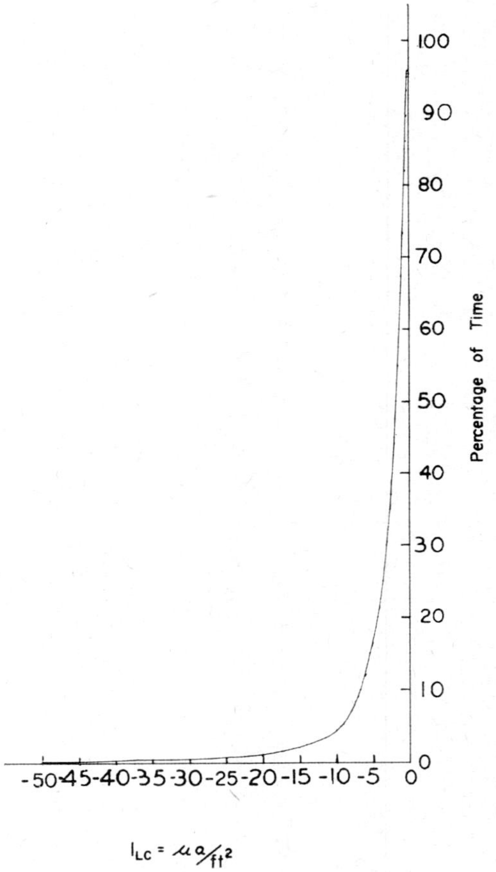

FIG. 13.

exhaust resistance, relative to resistance at 100 % power setting, are shown in Fig. 11. A linear relation between thrust and resistance is evident, but the point of interest to this discussion is that operation of the afterburner afforded a means of reducing airframe voltage by a factor of 4 or more for constant charging current, and thus made possible some investigation of the effect of airframe voltage on total noise level measured on the sense antenna.

Charging current to the left half of the nose radome, the half coated with conducting graphite paint, was used to evaluate noise as a function of charging rate. Total aircraft current was about 20 times the measured current, for the conditions prevailing during these flights. Individual noise level readings from a series of flights were averaged to obtain the data points shown in Fig. 12. For collector currents up to 4 μA, noise levels are substantially the same with afterburner on or off. Noise level with afterburner off is noticeably higher for collector

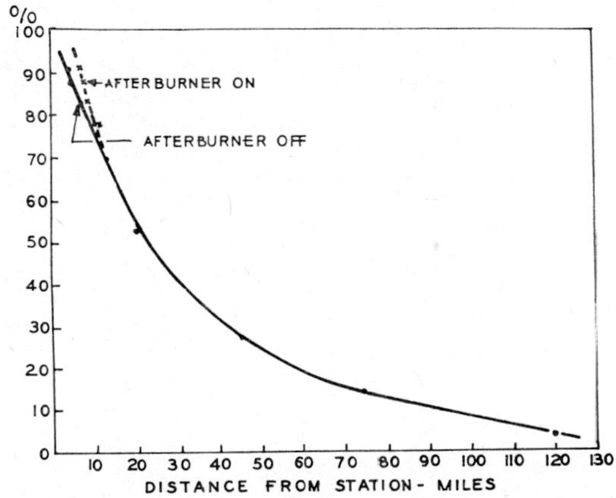

FIG. 14. Radio compass installation in F-94—% reliability vs distance —400 W. range station —242.

currents greater than 4. Apparently the change in airframe voltage produced by operating the afterburner was sufficient to extinguish airframe corona previously established at the higher collector current values with afterburner off. Figure 12 shows airframe corona noise as a function of collector currents, calculated from the other two curves. It is quite probable that sporadic corona from the sense antenna underneath the canopy accounts for some of the noise increase at collector currents in the range of 4 to 8 μA, because other data indicates that the threshold of corona from the airframe extremities was reached only at higher currents. Nevertheless, it is apparent that reduction of the airframe voltage produced a substantial decrease in noise at the higher collector currents. In the most favorable case of afterburner on, noise level increases somewhat faster than the collector current, and the noise is by no means negligible.

A curve of the percentage distribution of charging rates was obtained from analysis of data from 17 flights in ice crystals at altitudes in excess of 6·8 km (22,000 ft). The resulting curve is shown in Fig. 13. The ordinate of each point is the percentage of total time that charging was present that the charging current to the collector exceeded the corresponding current value on the abscissa. For example, a collector current of 5 μA was exceeded 17·5 % of the total time in weather which produced charging. Similar curves for other aircraft and other types of weather, such as frontal snow, have been obtained, and show that this particular curve corresponds to rather mild conditions, that is, low charging rates. All such curves, however, follow the trend of less frequent occurrence of higher charging rates.

By use of the relation of charging current to noise previously shown, in conjunction with the percentage distribution curve, an analysis of the effect of ice-crystal weather on operation of the radio compass in the F-94 was made. Figure 14 shows the percentage time the compass could be expected to operate, if in ice-crystal weather, as a function of distance from a 400 W range station.

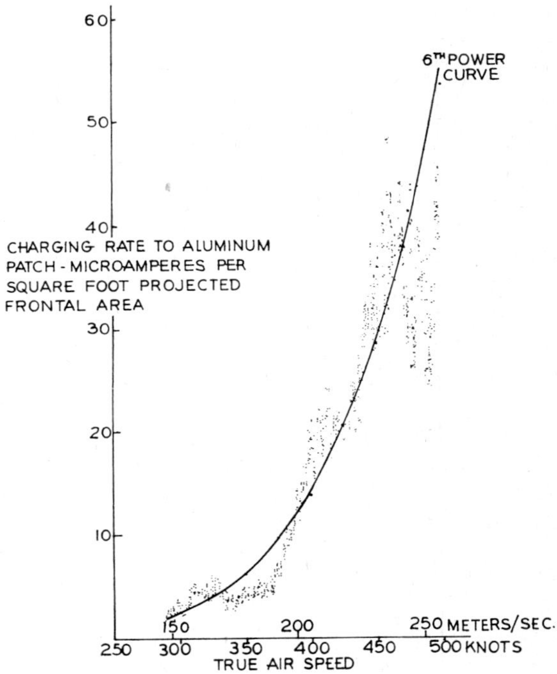

FIG. 15.

Effect of afterburner operation on this function is also plotted, and it is plain that it has virtually no effect. This serves to emphasize the point that, in the range of frequencies below 500 kc/s in particular, little advantage can be gained in system operation by use of effective dischargers, unless effective steps to eliminate streamering and impact noise are also taken.

The variation of charging rate to the aluminum airfoil collector as a function of speed, obtained in flight through frontal snow at an altitude of 6·1 km (22,000 ft) is shown in Fig. 15, with individual data points plotted to convey some notion of the amount of scatter. The best fit to the plotted points is a sixth power curve, as shown. A sixth power curve had been obtained previously on the B-45, but was considered an isolated datum, without much validity [2]. The F-94 data is considered valid for frontal snow, because it is comprised of a great many individual points which form a meaningful pattern unlikely to be a chance result.

A strong dependence of charging rate on speed was also evident in the fact that slightly faster aircraft were often much more noisy, and charging rate measurements were consistently higher on aircraft with faster cruising speeds.

From this we can conclude that precipitation static noise levels on a given aircraft will tend to increase as the sixth power of the speed, since noise levels increase as the first power of the charging rate. It is immediately obvious that all forms of precipitation static will tend to be more intense on the newer types of jet aircraft because their cruising speeds will be in excess of 500 knots, charging rates will be much higher, and the levels of streamering and particle impact noise will be increased.

Streamering and particle impact noise can be successfully controlled by proper antenna design. If dielectric housings are located where the surface is not bombarded by precipitation particles, streamering and impact noise are greatly reduced. Where visibility is not a problem, as it is in the windshield-canopy enclosure, resistive coatings can be applied to the external surfaces of dielectric antenna housings to eliminate streamering.

The prevention of corona at greatly increased charging rates requires enormously effective dischargers, although this problem is slightly alleviated by designing the aircraft exterior for high corona threshold, insofar as possible. The engine with afterburner on the F-94 was a very effective discharger by previous standards, but it could not keep the airframe below corona threshold during the speed run, where charging rate to the 4·3 sq. m of projected frontal area reached values of almost 2·5 mA, more than 500 μA/sq. m.

Efforts to develop more effective dischargers are continuing, but it is not yet certain that devices of adequate capability will be achieved.

In summary, adequate control of precipitation static on modern aircraft will require careful attention to details of antenna design and placement and the development of adequate noise-free dischargers.

References

1. Ayres, R. C. Aircraft Precipitation Static Investigation. Contract W33–106–sc–70; TWA, Inc. 1944.
2. Bartlet, C. A. (1951). *WADC Tech. Rep.* No. 52–34, Part 1.
3. Langmuir, Irving. Final Report on Investigation of Fundamental Phenomena of Precipitation Static. Contract W–33–106–sc–65; General Electric Company. (1945).
4. Nanevicz, J. E. (1957). *Stanf. Res. Inst. Rep.* 62; Contract AF19(604)–1296.
5. Stimmel, Rogers, Waterfall & Gunn (1946). *Inst. Radio Engrs, N.Y.*
6. Tanner, R. L. (1953). *Stanf. Res. Inst. Rep.* No. 37; Contract AF19–(604)–266.
7. Tanner, R. L. & Nanevicz, J. E. Radio Noise Generated on Aircraft Surfaces. Contract AF33(616)–2761; Stanford Research Institute, (1956).

Turbulence in the Lower E-Region from Meteor Echo Observations

J. S. Greenhow and E. L. Neufeld

University of Manchester, Jodrell Bank Experimental Station

This paper describes measurements of turbulence at heights of 80–100 km using radio echoes from meteor trails. The wind velocity is measured simultaneously at two points of variable separation along a meteor trail. Wind shears, and the fall off in correlation between the wind velocity at two points in space as their separation is increased, can be measured. The median value of the turbulent wind shear is 10 m sec^{-1} km^{-1}, although values exceeding 100 m sec^{-1} km^{-1} are observed occasionally. The large eddies are found to be anisotropic with vertical and horizontal scales of 7 km and 150 km. The r.m.s. wind velocity associated with the large eddies is 25 m sec^{-1}, and their time constant 6×10^3 sec. Values deduced for the scale, time constant and r.m.s. velocity of the smallest eddies are $\eta \sim 20$ m, $t_2 \sim 20$ sec and $V_2 \sim 1{\cdot}5$ m sec^{-1}. Energy is supplied to the turbulence at a rate of $\epsilon \sim 10^3$ ergs. gm^{-1} sec^{-1}.

I. Introduction

The radio echo observation of meteor trails provides a powerful method for the investigation of winds in the region 80 to 100 km above the earth. Measurements can be made at all times of the day, and with apparatus of reasonable sensitivity the mean wind velocity can be determined within a few minutes. This method has been used to investigate the diurnal and seasonal variations of upper atmosphere winds [3, 4, 5, 7] and regular diurnal, semi-diurnal and prevailing wind components have been measured. This paper is concerned with the irregular winds at heights of about 90 km, and measurements of turbulence using radio echoes from meteor trails are described.

II. Technique

A. Single-station Observations

A detailed description of the coherent pulse technique for the measurement of drifts of meteor trails has been given previously [4]. The transmitter operates at a frequency of 36 Mc/s with a peak power of 100 kW, pulse length 30 μs and P.R.F. 150. Echo characteristics are displayed on four cathode ray tubes (Fig. 1a). Tube A is a range amplitude display, with 50 and 200 km range markers, while tube B shows the variation of echo amplitude with time on a time base of 0·3 sec duration. The undetected receiver output is also mixed with a continuous wave reference signal of the same frequency, so that changes in phase of the echo due to drifts of the meteor trail may be determined. A change in range of $\lambda/2$ (4·1 m) produces a change in phase of 2π and gives one complete beat in echo amplitude. This phase measuring channel is displayed on tube C, and enables radial components of wind velocity to be determined by counting the number of echo pulses in a complete cycle or fraction of a cycle. In order to determine the sense of the drift (towards or away from the observer) a second phase measuring channel in which the reference signal differs in phase by $\pi/2$ from the first is displayed on tube D.

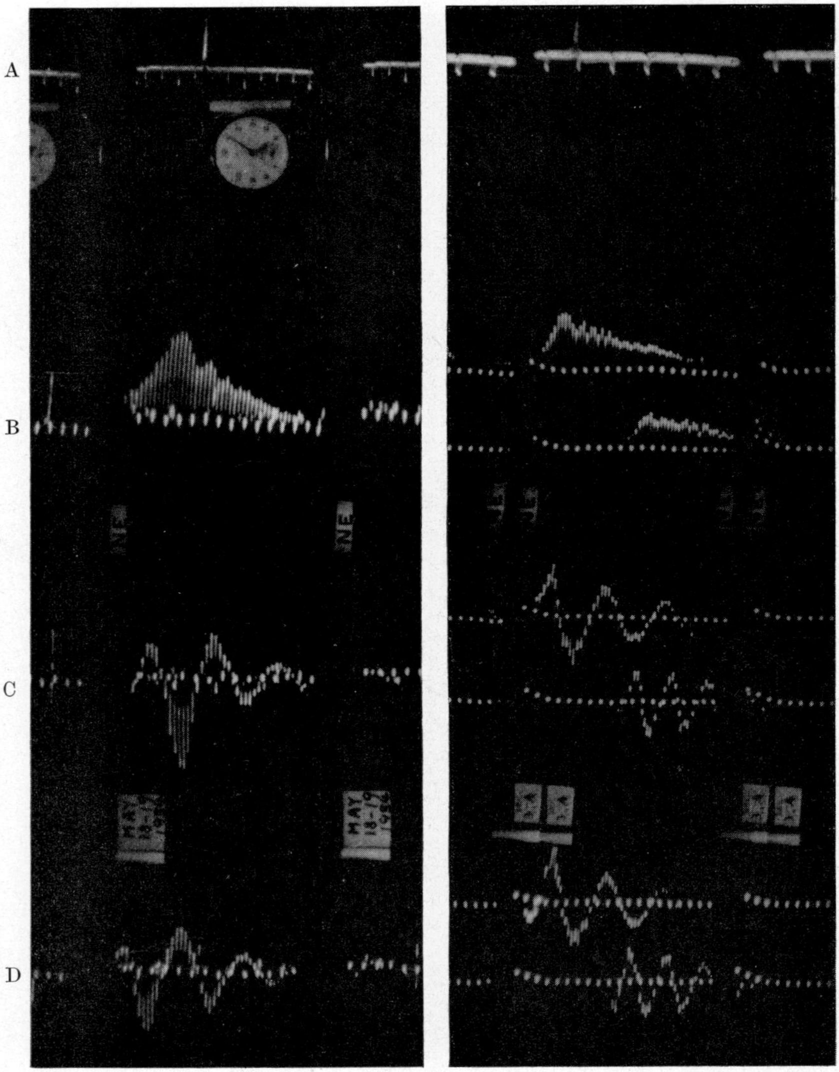

FIG. 1. Film records.

(*a*) Single-station observations, showing measurements of meteor echo ranges, amplitude and phase.

(*b*) Spaced station methods. Echoes are obtained from two points along the same meteor trail.

The mean wind velocity over any period of time is obtained by averaging all the individual meteor drifts which are recorded. Only the radial components of the wind are measured in this way, and in order to determine the resultant wind vector a beamed aerial is directed for alternate periods of 10 min in two directions at right angles. An example of the results obtained is given in Fig. 2, for a 24-hour run during 22–23 September, 1955. North-South and East-West components of the wind are plotted, and on this day the mean wind is seen to consist primarily of a semi-diurnal periodic component with an amplitude of 27 m sec^{-1}, and a prevailing wind towards East of 17 m sec^{-1}.

Each hourly wind measurement in Fig. 2 is the mean of up to 200 meteor drifts,

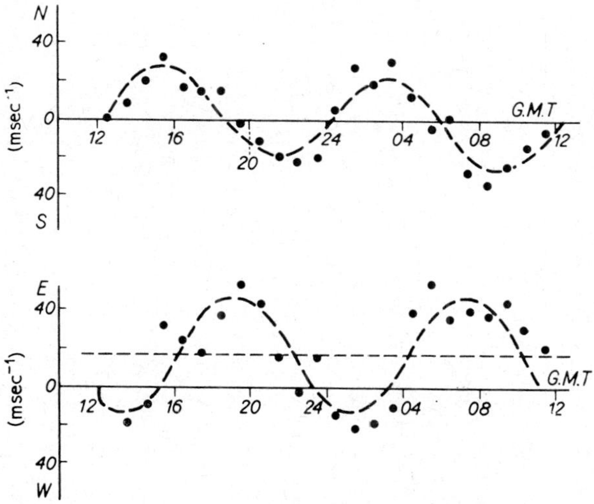

FIG. 2.

and although successive points lie on a fairly smooth curve there is a very large spread in the individual measurements due to turbulence. Root mean square wind speeds as high as 40 m sec^{-1} are found even when the mean wind speed is zero. In order to investigate these turbulent winds, therefore, a spaced station technique has been evolved to measure the wind at the same time at two points of variable separation.

B. Two-station Observations

Because of the initial linearity of a meteor trail echoes are received by specular reflection. When the transmitter and receiver are situated at the same place O (Fig. 3), the echo is obtained from the first Fresnel zone at X on the meteor trail. If, now, a second receiving station is set up at C, the echo will be received from a point Y further along the trail. As the meteor has a finite velocity V, there will be a small time delay T between the appearance of the echo at O and C. V can be determined from the diffraction effects at the beginning of an echo [2], and if T is measured, the separation XY of the two reflecting points along the trail can be determined. In the experiment a coherent radio link was used to transmit the signal from C to O, for phase measuring and recording purposes. Beam splits were applied to cathode ray tubes B, C and D and in Fig. 1(b) the echo characteristics at O are displayed on the upper traces, and at C on the lower traces.

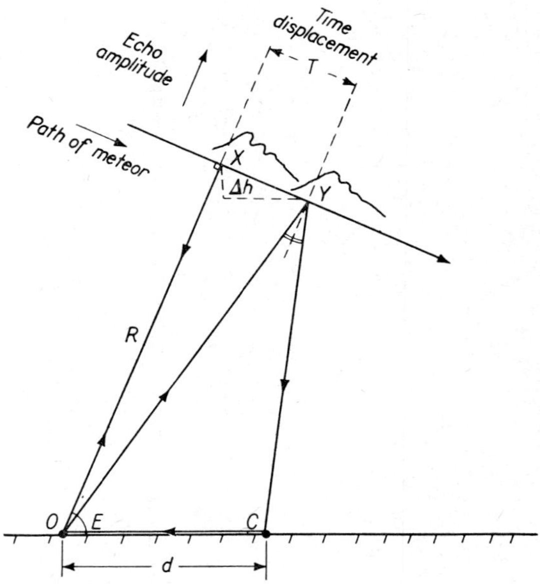

FIG. 3.

Thus, the wind velocity can be measured at the same time at two points along the same meteor trail [8]. By varying the distance between the receiving stations, or by observing meteor trails at different orientations to the plane of the paper, this separation may be varied. Two station separations "d" of 20 km and 3·6 km were used, giving a maximum value of XY = 10 km for the spatial separation of the reflecting points, and 5 km for their maximum height difference Δh.

A beamed transmitting aerial consisting of an array of four 6-element Yagis restricted echoes to within $\pm$ 10° in azimuth of the line OC. Δh can thus be determined approximately for trails with any orientation, from the measured value of XY and the elevation E, using the relation

$$\Delta h = \frac{2(VT)^2}{d} \cot E. \tag{1}$$

Elevation is determined from the direct range measurement, and height deduced from the echo duration [6].

III. Results

A. Wind Shear

The difference in wind velocity over a given difference in height enables the wind shear to be determined directly. The observations have been divided into a number of groups with different values of Δh, and the mean wind shear over various height separations investigated. Figure 4(a) is a histogram showing the

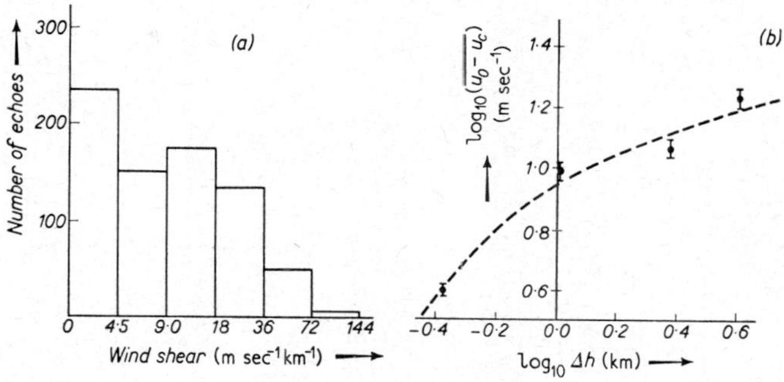

Fig. 4.

distribution of wind gradients measured over a height separation of approximately 0·4 km (these gradients may be positive or negative). The median value of the turbulent wind shear is 10 m sec^{-1} km^{-1}, although values as high as 140 m sec^{-1} km^{-1} are observed. Seven per cent. of the measurements exceed 36 m sec^{-1} km^{-1}.

The variation of the median value of the difference in wind velocity between two points, as a function of their height separation, is shown in Fig. 4(b). For the case of isotropic turbulence Kolomogroff's spectrum law would be expected to hold, in which case $\overline{u_0 - u_c} \propto \Delta h^{\frac{1}{3}}$ [11]. This does not appear to be the case for turbulence at heights of about 90 km.

B. Scale of Turbulence

1. Correlation between Velocities at two points of Variable Separation

The scale of the large eddies is determined by investigating the fall off in correlation between the turbulent wind velocities at two points, as their separation is increased. In addition to the turbulent winds, regular periodic and prevailing components are also present. In order to allow for these, the mean horizontal wind $\bar{u}$ during any hour is obtained by averaging all the measured velocities [4]. The deviations of the individual measurements u_0 and u_c at O and C, from the radial components $\bar{u}_r$ of the steady wind, are then obtained for each echo pair in that hour. This gives for the turbulent wind components δu_0 and δu_c at O and C.

$$\delta u_0 = \bar{u}_r - u_0, \quad \delta u_c = \bar{u}_r - u_c$$

The lateral correlation coefficient g is then given by

$$g = \frac{\sum \delta u_0 \delta u_c}{(\sum \delta u_0^2 \sum \delta u_c^2)^{\frac{1}{2}}}. \tag{2}$$

Approximately 900 echo pairs have been obtained in the spaced station experiment, and for each pair values of δu_0 and du_c have been determined, together with the spatial separation and height separation of the reflecting points.

The correlation coefficient g is plotted as a function of spatial separation VT in Fig. 5(a). g(VT) decreases from unity at zero separation to 0·2 at a separation of 6 km. For values of VT $>$ 6 km, instead of decreasing through zero to small negative values as expected for the case of isotropic turbulence, g(VT) remains practically constant at 0·2. The significance of this behavior will be considered in terms of anisotropy of the large scale turbulence.

2. Variation of Correlation Coefficient with Height Separation

The maximum value of VT occurs for a meteor travelling horizontally along the line OC (Fig. 3). For such a meteor, however, the height difference between the

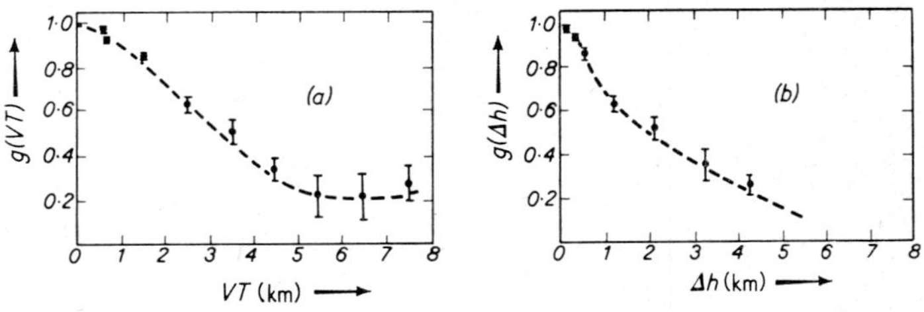

FIG. 5.

reflecting points falls to zero. Thus for a meteor trail lying in the plane of the paper, this height difference increases from zero for a vertical trail to a maximum of 5 km when $E = 45°$, decreasing once more to zero for a horizontal trail. The spatial separation XY = VT, on the other hand, increases steadily from zero to 10 km. The difference in height between the two reflecting points for a spatial separation of 9 km may therefore be no higher than that for a spatial separation of only 5 km.

In Fig. 5(b) the variation of correlation coefficient with height, irrespective of the horizontal separation of the reflecting points, is shown. The unusual behavior of the correlation curve is removed, and extrapolation of this new curve to zero shows that the vertical extent of the large eddies is of the order of 7 km. Thus the apparent levelling off in g(VT) can be explained if the turbulence is anisotropic, with the fall off in correlation mostly in the vertical direction rather than horizontally.

Because of the slow fall off in g horizontally, it is difficult to make an accurate estimate of the horizontal dimensions of the large eddies. However, if echo pairs of the same height separation but different horizontal separations are compared, an estimate of the rate of fall in correlation horizontally may be made. This shows that the horizontal scale of the turbulence is of the order 100 to 200 km, an order of magnitude greater than the vertical depth.

C. *Time Constant of the Large Eddies*

If turbulence of the scale discussed in the previous section exists in the 80–100 km region, then it should be possible to detect the large eddies as irregularities in the wind measured at a single station. Thus if a beamed aerial system is used, and echoes are limited to a narrow range of heights, then the wind measurements over a short interval of time will refer to a small volume of a single large eddy. Fluctuations of the wind velocity with time will be observed, and the time constant of the fluctuations given by the interval over which the auto-correlation function falls to zero, determines the time constant of the large scale turbulence.

During the course of routine single station wind observations, meteors are observed in a given direction during alternate periods of 10 min [4]. In Fig. 6(a),

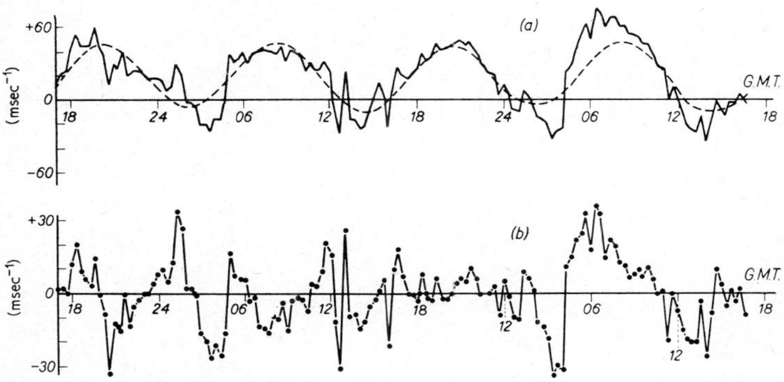

FIG. 6.

for example, the variation of wind velocity resolved along a fixed azimuth (315°) for a 2 day period is shown. The measurements refer to heights between approximately 90 and 95 km. Regular tidal and prevailing components of the wind are present, and these can be removed by harmonic analysis in the manner described in previous papers. The broken line in Fig. 6(a) is the sum of the first three terms of the Fourier series, representing the prevailing and the 24 and 12 hour periodic wind components. The deviations of the individual points from this smooth curve, representing the turbulent component of the wind, are shown in Fig. 6(b).

The auto-correlation function $g(\tau)$ which correlates the turbulent velocity u_s at time s, with the velocity $u_{s+\tau}$ at time $s - \tau$ is given by

$$g(\tau) = \frac{\sum u_s u_{s+\tau}}{[\sum u_s^2 \sum u_{s+\tau}^2]^{\frac{1}{2}}}. \tag{3}$$

The variation of $g(\tau)$ with τ for the turbulent wind component illustrated in Fig. 6(b), is shown in Fig. 7, where the auto-correlation function is seen to fall to zero in about 2 hr. It is clear from the slow fall off in $g(\tau)$, that very little smoothing has occurred through the use of a finite time interval of 20 min, and on this occasion 2 hr may be taken as the time constant t_1 of the large scale turbulence.

The r.m.s. turbulent velocity determined from the velocity fluctuations in Fig. 6(b) is 18 m sec^{-1}; that is of the same order as the mean r.m.s. turbulent velocity determined from the spaced station measurements (Section III E).

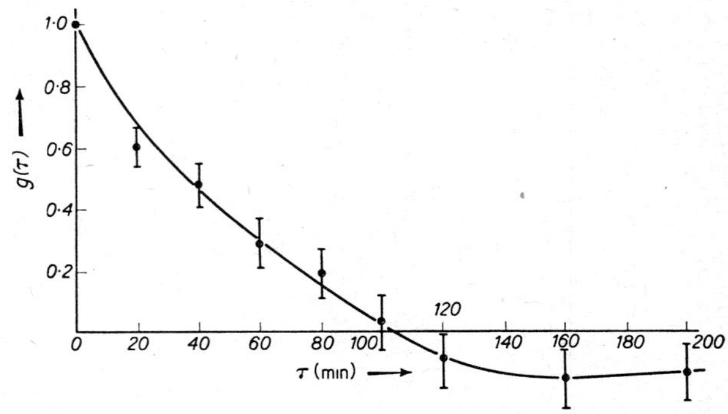

FIG. 7.

This also shows that very little smoothing of the large scale turbulent structure has occurred because of the finite volume of space over which the measurements were made.

The mean value of t_1, determined from a number of such auto-correlation curves, is found to be approximately 6×10^3 sec.

D. *Observations of Large Scale Irregularities in the Wind*

Direct evidence for the existence of irregularities in the wind with a horizontal scale greater than 100 km, a vertical extent of the order of 10 km, and a time constant of over an hour, is given by observations of the variation of wind velocity with height [7]. These measurements were designed to investigate the height variations in amplitude and phase of the uniform wind components, and North-South and East-West components of the wind were measured at heights of 85, 90, 95 and 100 km. As an example, the measurements for a 24 hour period between 12.00 U.T. 12 January to 12.00 U.T. 13 January, 1954 are shown in Fig. 8. Once again the wind shows a large 12 hr periodic component, but the variations are by no means sinusoidal and large irregularities occur. One such irregularity is observed at about 06.00 U.T., and is especially marked between 05.00 and 08.00 U.T. at 90 km. in an Easterly direction. At this height its amplitude is such as distort completely the regular 12 hr variation. At greater and lesser heights the disturbance is much less noticeable. The same irregularity is also present in the measurements of the North-South component of wind velocity, even though the observations refer to a region of sky several hundred kilometres away from the East-West measurements.

The horizontal scale of this irregularity may be estimated from its observed duration of approximately 4 hr, and a wind velocity of 30 m sec^{-1}. These figures give a length dimension of approximately 400 km, while the height measurements show that the vertical extent is only of the order of 10 km. Large flat irregularities of this type are clearly responsible for the variations of time and spatial correlation coefficients discussed previously, although in these height

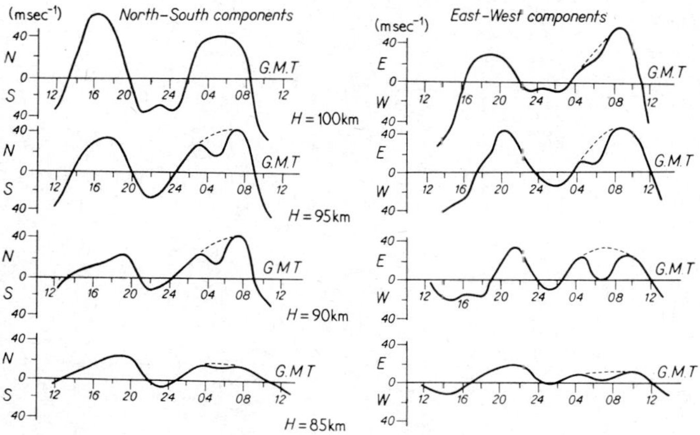

FIG. 8.

measurements only the eddies with time constants greater than about 2 hr can be resolved.

E. The Root Mean Square Turbulent Wind Velocity

The r.m.s. turbulent wind velocity V_{rms} may be determined by measuring the deviations of the individual meteor drift velocities from the mean drift velocity.

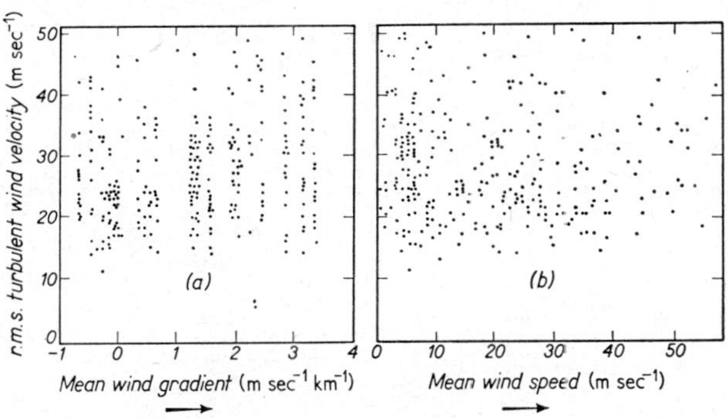

FIG. 9.

The results have been grouped in hourly intervals for which $V_{\text{rms}} = \left(\frac{\sum \delta u_c^2}{n}\right)^{\frac{1}{2}}$, where n is the number of individual meteor drifts and $\delta u_c = \bar{u}_r - u_0$ (using the nomenclature of Section III B).

Of particular interest is the dependence of the turbulent velocities upon the gradient of the mean wind and the mean wind velocity.

The variation of V_{rms} with wind gradient is shown in Fig. 9(a) for negative values of the gradient of 1 m sec^{-1} km^{-1}, up to positive values of 3·5 m sec^{-1} km^{-1} (which are about the highest observed). V_{rms} varies between approximately 15 and 45 m sec^{-1}, with a median value of 25 m sec^{-1}. A surprising result is that the turbulent velocity does not vary significantly with wind gradient. Even when the wind gradient is zero r.m.s. velocities up to 45 m sec^{-1} are observed, with a median value similar to that when the highest gradients are present. Further, Fig. 9(b) shows that when the mean wind velocity is also zero, the turbulent velocities are not reduced. Thus, over the whole range of mean wind speeds from 0 to 60 m sec^{-1}, V_{rms} does not vary significantly. There is no significant difference between the turbulent velocities or eddy sizes observed during day and night time.

F. The Small Eddies

It is very difficult to observe the smallest eddies at heights of 80–100 km, as their effects are usually overshadowed by larger eddies. An estimate of their size may be obtained from the shape of the correlation curves at small separations. Thus the dissipative length parameter λ, obtained by fitting a parabola to the early part of the correlation curve, is found to be 2·4 km. λ relates the scale of the large eddies L to that of the small eddies η by the expression [10]

$$\eta^2 = \frac{\lambda^3}{10\sqrt{15}\,L}. \tag{4}$$

For $L = 150$ km, this gives $\eta = 49$ m.

An estimate for a lower value of t_2 can be obtained from the delays in appearance of discrete radio echo components from bright meteor trails [9]. An initially linear meteor trail becomes completely rough at radio wavelengths in about the time constant of the small eddies. As the trail distorts echoes from parts of the trail removed from the specular region can therefore be delayed in appearance by up to a time t_2. (Although large scale turbulence would normally cause the echoes to appear earlier.) Delays of 30 sec have been observed, showing that $t_2 > 30$ sec.

IV. Discussion

A. The Large Scale Turbulence

In any turbulent system there exists a whole spectrum of eddy sizes from the largest of scale L, lifetime t_1, and r.m.s. velocity V_1, to the smallest of scale η, time constant t_2, and r.m.s. velocity V_2. Equations of the form $L = V_1 t_1$ and $\eta = V_2 t_2$ relate the length, velocity and time parameters of eddies of any size. The large eddies contain most of the energy of the turbulence, and a measure of their size is given by the distance over which the wind velocities lose correlation. Direct measurements of the correlation coefficient show that their vertical scale is approximately 7 km, while the horizontal size may be as high as 100 to 200 km. The time constant of the large eddies, determined from the auto-correlation function concerned with the time variations of wind velocity at a fixed point, is found to be 6×10^3 sec. The r.m.s.turbulent velocity is 25 m sec^{-1}. Combining these values of velocity and time, we obtain 150 km for the length parameter of the large eddies. This is in agreement with the horizontal scale suggested by the correlation curves, rather than the more apparent 7 km. vertical dimension, which is primarily responsible for the loss of correlation.

Thus an equation of the form $L = V_1 t_1$ holds for $L = 150$ km, $V_1 = 25$ m sec^{-1} and $t_1 = 6 \times 10^3$ sec, and these values have been taken as appropriate to the large eddies.

Energy is supplied to the turbulence at a rate $\epsilon = V_1^2/t = V_1^3/L$ erg. gm^{-1} sec^{-1} (ϵ is the turbulence power), and substitution of the above values of V_1 and t gives $\epsilon \sim 10^3$ erg gm^{-1} sec^{-1}.

B. Small Scale Turbulence

The energy supplied to the turbulence is finally dissipated in the form of heat by viscous forces. From theories of homogeneous turbulence the length and time scales of the smallest eddies, which are responsible for most of the dissipation, should depend only upon the rate of supply of energy and the kinematic viscosity ν [1]. Thus

$$\eta = (\nu^3/\epsilon)^{\frac{1}{4}} \tag{5}$$

$$t_2 = (\nu/\epsilon)^{\frac{1}{2}} \tag{6}$$

η and t_2 can therefore be determined if ϵ and ν are known. The kinematic viscosity at a height of 90 km can be calculated from kinetic theory using measured air densities and temperatures. Alternatively ν is equal to the molecular diffusion coefficient D, and measurements of the decay of meteor echoes show that $D \sim 4 \times 10^4$ cm^2 sec^{-1} at a height of 90 km [6]. However, the collision cross sections for meteor ions are somewhat greater than for neutral molecules, and this value of D should be increased by a factor of about 5 to give the appropriate value of ν . $\epsilon \sim 10^3$ erg gm^{-1} sec^{-1} from paragraph 4(a)

Thus Equations (5) and (6) give $\eta = 17$ m and $t_2 = 14$ sec, when $V_2 = 1{\cdot}2$ m sec^{-1}. These values are in good agreement with the estimates of η and t_2 given in paragraph 3(f), the differences of factors of 2 to 3 can easily be explained by errors in ϵ or ν; or in the use of Equation (4).

Geometric means of $\eta \sim 30$ m and $t_2 \sim 20$ sec have therefore been taken for the best values of the scale and time constant of the smallest eddies, giving $V_2 \sim 1{\cdot}5$ m sec^{-1}.

C. Richardson's Criterion

Richardson's number, R_i, is often used as a criterion to decide whether turbulence should increase or decrease. Turbulence increases if $R_i < 1$, and is damped if $R_i > 1$, where

$$R_i = \frac{g/T(dT/dh + \Gamma)}{(dV/dh)^2} \tag{7}$$

g is the acceleration due to gravity, dT/dh the temperature gradient, T the adiabatic lapse rate, dV/dh the wind gradient and T the temperature. At a height of 90 km $\Gamma = 10°$. km^{-1}, $dT/dh = 1{\cdot}3°$. km^{-1}, $g = 950$ cm sec^{-1} and $T = 210°$. The wind gradient varies between 3·5 and 0 m sec^{-1} km^{-1}, and we are therefore concerned with values of R_1 between 45 and ∞. Thus the turbulence should be rapidly damped. However, it has been shown that even when $dV/dh = 0 (R_i = \infty)$, intense turbulence with an r.m.s. velocity of 25 m sec^{-1} is present. Also, the intensity of the turbulence does not vary with wind gradient over the range 0 to 3·5 m sec^{-1} km^{-1}. It is therefore difficult to account for the existence of turbulence at heights of 90 km, and Richardson's criterion clearly cannot be applied.

However, the turbulence appears to be of a most unusual form. The large eddies are very anisotropic and the turbulent winds are confined to the horizontal plane. Thus, the correlation in wind velocity is maintained for the order of 100 km in a horizontal direction, while correlation is lost in only 7 km vertically. The vertical components of turbulence are therefore an order of magnitude less than the horizontal velocities, and this vertical damping would be expected from the inherently stable conditions present at the heights of the order of 90 km.

References

1. Batchelor, G. K. "Homogeneous Turbulence." Cambridge Univ. Press (1953).
2. Davies J. G. & Ellyett, C. D. (1949). *Phil. Mag.* **40,** 614.
3. Elford, W. G. & Robertson, D. S. (1953). *J. atmos. terr. Phys.* **4,** 271.
4. Greenhow, J. S. (1954). *Phil. Mag.* **45,** 471.
5. Greenhow, J. S. & Neufeld, E. L. (1955a). *Phil. Mag.* **46,** 549.
6. Greenhow, J. S. & Neufeld, E. L. (1955b). *J. atmos. terr. Phys.* **6,** 133.
7. Greenhow, J. S. & Neufeld, E. L. (1956). *Phil. Mag.* **1,** 1157.
8. Greenhow, J. S. & Neufeld, E. L. (1959a). *Proc. phys. Soc. Lond.* **74,** 1.
9. Greenhow, J. S. & Neufeld, E. L. (1959b). *J. atmos. terr. Phys.* In press.
10. Townsend, A. A. "The Structure of Turbulent Shear Flow." Cambridge Univ. Press (1956).
11. Weizsäcker, Von (1949). *Proceedings of the Symposium on the Motions of Gaseous Masses of Cosmical Dimensions.* Paris.

Ionosphärische Reflexionskoeffizienten im Langwellenbereich

E. A. LAUTER

Observatorium für Ionosphärenforschung, Kühlungsborn

Ionospheric reflection data of a ten years period on 245 kc/s are analysed. Night-time values of reflection-coefficient have a slight maxima in summer. At constant solar heights there are maxima in April and October. The noon values have a clear asymmetry in shape between spring and autumn. High summer absorption is present until September. Noon, evening and night-time values of reflection-coefficient decrease with increasing solar activity. Evening and night-time reflection coefficients have a log-normal distribution under undisturbed ionospheric conditions. During periods with disturbances there is a remarkable deviation from the normal distribution mainly caused by the after-effect which occurs 3–5 days after the primary storm in long wave propagation. The depression of reflection-coefficients during this effect is strongest near sunset and is coming in most frequently and more intensively after storms in summer time. The summer anomaly of high absorption and the aftereffect of magnetic storms are two unsolved problems in low frequency propagation.

I. JAHRESZEITLICHE GÄNGE DER MITTELWERTE

Am Observatorium Kühlungsborn sind seit 1948 regelmäßige Registrierungen des Verhältnisses Raumwelle/Bodenwelle auf 245 kHz (Rundfunksender Kalundborg) im 180 km Senderabstand vorgenommen worden. Da die Bodenwellenausbreitungsbedingungen als konstant angenommen werden können (Ausbreitung über die Ostsee) wurden aus diesen Messungen die mittleren nächtlichen Reflexionskoeffizienten (oder genauer die Conversions-Koeffizienten) bestimmt. Seit 1952 liegen solche Messungen auch für den gesamten Tagesgang vor, die von Sprenger [3] 1958 analysiert worden sind. Die Ergebnisse über die mittleren jährlichen Variationen des Reflexionskoeffizienten (RK) seien an der Abb. 1 erläutert:

1. Die Nachtwerte des RK zeigen nur einen schwachen jahreszeitlichen Gang mit einem Maximum im Sommer (— 16 db bezogen auf RK = 1) und einem winterlichen Minimum (— 20 db). Je nach dem Grad der erdmagnetischen Aktivität sind auch um die Äquinoktien Minima angedeutet.

2. Die bei konstanter Sonnenhöhe (Meßintervall $h = 5°–10°$) gemessenen RK zeigen einen doppelten Jahresgang mit Maxima im April (— 24 db) und Oktober (— 27 db). Die sommerlichen und winterlichen Minima liegen bei — 32 db. Noch nicht klar erkannte Strukturänderungen in der sommerlichen tiefen Ionosphäre scheinen die Sommeranomalie hoher Dämpfung zu verursachen.

3. Der Jahresgang der Mittagswerte ist mit — 35 db im Winter und — 75 db im Sommer am stärksten ausgeprägt, doch ist der jährliche Gang nicht mit einem einfachen cos $^{n}\chi$ –Gesetz in Übereinstimmung zu bringen. Der Abfall vom Frühjahr zum Sommer ist nicht so steil wie der Anstieg vom Oktober zum November. Die sommerliche Anomalie zu hoher Dämpfung reicht bis in den September hinein.

4. Der Vergleich zwischen dem mittleren Verlauf der RK 1952–55 und 1956–57 läßt deutlich den Einfluß der Sonnenaktivität in einer Erniedrigung der RK erkennen. Die Nachtwerte sind im Mittel um 2, 5 db abgesunken. Wie die Korrelation zwischen den Monatsmittelwerten des RK und des erdmagnetischen Aktivitätsgrades A_K zeigt, sind dafür die Korpuskularstörungen verantwortlich (vgl. Tabelle 1).

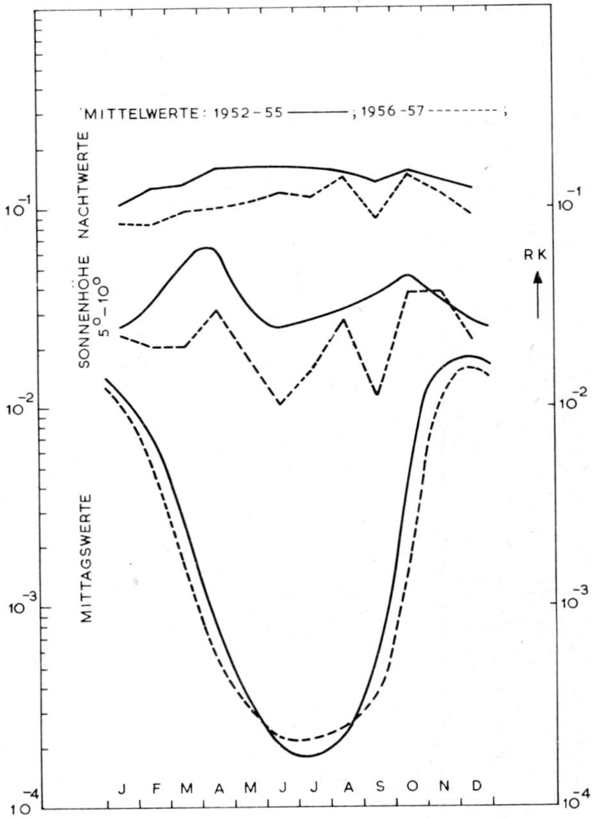

ABB. 1. Jahresgang der Reflexionskoeffizienten auf 245 kHz.

Bei den RK-Werten, die bei 5°–10° Sonnenhöhe ermittelt wurden, liegt die Abweichung mit 4 db noch höher. Hieran sind sowohl die Korpuskulareinbrüche als auch die mit der Sonnenaktivität steigende Tagesabsorption beteiligt. Die Mittagswerte zeigen ebenfalls bei niedrigen und mittleren Sonnenhöhen fast über das ganze Jahr eine verstärkte Dämpfung um etwa 3 db. Nur bei höherem Sonnenstand vom Mai-August ist mit einer Zunahme von 1 db ein Überwiegen der verbesserten Tagesreflektion über die verstärkte Absorption im Sonnenfleckenmaximum erkennbar.

TABELLE I

STATISTISCHE PARAMETER DER VERTEILUNGEN DER ABENDLICHEN UND NÄCHTLICHEN REFLEXIONSKOEFFIZIENTEN AUF 245 kHz. (Der Störungsgrad ist aus den Verhältnissen der 10 % und 90 %—Werte zum Medianwert (50 %) gebildet, so daß eine Normalverteilung dem Wert 1·00 entsprechen würde.)

		Medianwert des RK in 10^{-2}		Korrelation RK/A_K	Parameter der Häufigkeitsverteilung					
					Nachtwerte			bei 5°–10° Sonnenhöhe		
		nachts	5°–10° Sonnenhöhe	(Monats-mittel) r	10 %/50 %	50 %/90 %	Störungsgrad	10 %/50 %	50 %/90 %	Störungsgrad
Sommer	1948–57	15·0	2·1	— 0·83	0·62	0·76	1·22	0·22	0·46	2·19
(Mai-Aug.)	1953–55	15·9	2·9		0·75	0·76	1·01	0·53	0·59	1·11
Winter (Nov.-Febr.)	1948–57	9·6	2·2	— 0·66	0·60	0·65	1·08	0·39	0·40	1·01
Äquinoktien (März, April,	1948–57	12·1	3·25	— 0·75	0·46	0·68	1·48	0·20	0·40	2·00
Sept.-Okt.)	1953–55	13·9	4·70		0·60	0·70	1·17	0·33	0·42	1·29
Erdmagn. Stürme	5 Tg. vorher	13·3	3·1	—	0·70	0·72	1·03	0·44	0·44	1·00
(A_K 40)	5 Tg. nachher	9·0	1·2		0·34	0·64	1·88	0·15	0·32	2·14

II. Häufigkeitsverteilungen der abendlichen und nächtlichen RK auf 245 kHz

Um das allgemeine Verhalten der RK noch weiter analysieren zu können, sind in den Abb. 2 und 3 die Verteilungen der abendlichen und nächtlichen RK

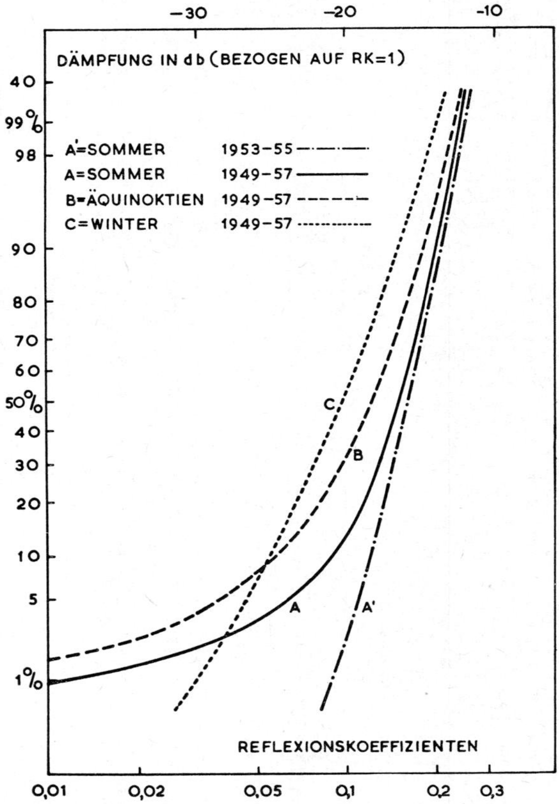

Abb. 2. Nächtliche Reflexionskoeffizienten 245 kHz.

in verschiedenen Jahreszeiten in der bekannten Summenhäufigkeitsdarstellung im logarithmischen Wahrscheinlichkeitsnetz gezeichnet. Allgemein ergibt sich eine Abweichung von der log.-Normalverteilung in dem Sinne, daß niedrige Werte des RK häufiger auftraten als es der Normalverteilung entsprechen würde. Eine Ausnahme machen die Winterwerte und auch die Sommerwerte RK für 1953–55, die der Normalverteilung völlig entsprechen. Da man bei ionosphärisch ungestörten Zeitperioden fast immer gut angenäherte log.-Normalverteilung erhält, liegt es nahe, die Abweichungen ausschließlich den immer eine Depression der RK verursachenden Ionosphärenstürmen zuzuschreiben. Jedoch zeigt sich, daß die Tage mit erdmagnetisch nachweisbaren Korpuskulareinbrüchen allein nicht solche starken Abweichungen von der Normalverteilung im Gesamtkollektiv verursachen können. Hier ist noch eine andere Erscheinung beteiligt, die anscheinend nur im Lang- und Längstwellenbereich ausgeprägt und nachweisbar ist.

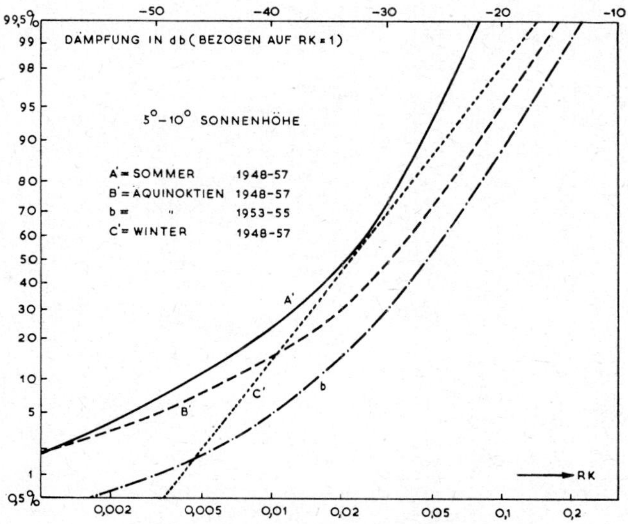

ABB. 3. Reflexionskoeffizienten 245 kHz.

III. DER NACHWIRKUNGSEFFEKT BEI STARKEN IONOSPHÄRENSTÜRMEN

Bei der Untersuchung der zwanzig stärksten erdmagnetischen Stürme seit 1948 nach der Methode der Überlagerung gleicher Epochen ergibt sich für die abendlichen und nächtlichen RK die in Abb. 4 gezeigte mittlere prozentuale Abweichung

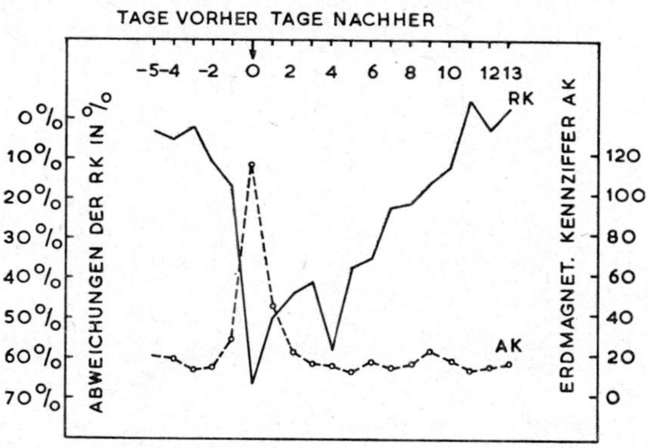

ABB. 4. Medianwerte der Abweichungen der abendlichen Reflexionskoeffizienten auf 245 kHz bei den 20 stärksten erdmagnet. Störungen (1948–1958).

des RK. Die Reflexionskoeffizienten fallen am Störungstag auf etwa ein Drittel des Normalwertes ab und erholen sich erst im Verlaufe der nächsten sieben Tage langsam. Besonders typisch ist ein erneuter markanter Rückgang des RK, der am 4. Tag nach der Störung einsetzt, und den wir als sogenannte Nachwirkung des Sturmes bezeichnen. Die Nachwirkung kann bis zu 10 Tagen dauern, doch beschränkt sie sich häufig auf ein bis zwei Tage im Intervall des 3.–5. Tages nach der Störung. Es handelt sich dabei nicht um ein langsames Abklingen der Primärstörung, sondern um eine neue Störungsform, die sich nur in einer verstärkten Absorption äußert und nicht von dem beim Primäreinbruch bekannten starken Fading begleitet ist. Diese Erscheinung wurde bereits von Lauter [2] beschrieben und auch von Straker [4] in den Phasenbeobachtungen auf 16 kHz gefunden.

IV. Häufigkeitsverteilungen bei Ionosphräen Stürmen

Die Erscheinung des Nachwirkungseffektes hat den wesentlichsten Anteil an den Abweichungen der RK von der log.–Normalverteilung. In Abb. 5 sind die

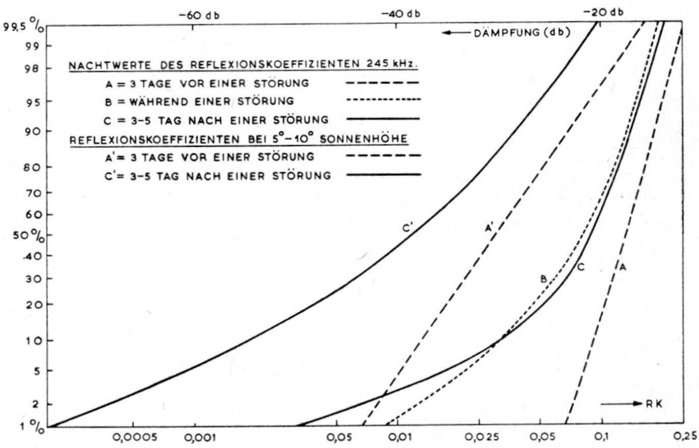

Abb. 5.

resultierenden Verteilungen bei 80 erdmagnetischen Stürmen ($A_K > 40$) dargestellt. Im ungestörten Intervall, 5 Tage vor Sturmbeginn, erhält man wieder die Normalverteilung (A u A'). Während des Sturmtages (B) und in der Zeit vom 3.–5. Tag nach dem Sturm (C u C') treten jetzt sehr starke Abweichungen von der Normalverteilung auf. Die Häufigkeit kleiner Werte des RK ist im Nachwirkungseffekt noch größer als am Sturmtage selbst. Besonders stark wird die Abweichung von der Normalverteilung im Nachwirkungseffekt bei den abendlichen 5°–10° Werten. Damit liegt es nahe, die Abweichungen von der Normalverteilung als ionosphärischen Störungsgrad des betreffenden Zeitintervalls zu interpretieren. Ermittelt man aus den gegebenen Verteilungen die RK bei 90 %, 50 % und 10 %, so wird für eine log-Normalverteilung der Ausdruck

$$\frac{RK_{50\%}}{RK_{10\%}} \cdot \frac{RK_{50\%}}{RK_{90\%}} = 1.$$

Die Abweichungen von 1 beschreiben also die durch Ionosphärenstörungen verursachten Verteilungen. Bildet man dieses Störungsmaß für die mittleren Verteilungen der abendlichen und nächtlichen RK der einzelnen Monate für die Jahre 1948 bis 57, so erhält man den Jahresgang in Abb. 6. Danach sind die

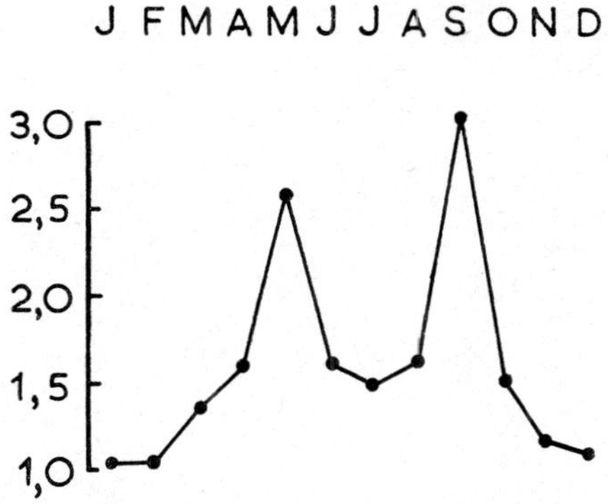

ABB. 6. Jahresgang des Störungsgrades (Abweichungen von der log. Normalverteilung).

Einflüsse von Ionosphärenstörungen im Winter am kleinsten und erreichen im Mai und September ein Maximum. Da die Abweichungen, wie ausgeführt, im wesentlichen durch den Nachwirkungseffekt zustande kommen, kann man bei Berücksichtigung des Jahresganges der erdmagnetischen Aktivität auch sagen, daß der Nachwirkungseffekt im Sommerhalbjahr am stärksten zur Auswirkung kommt. Vergleicht man die in Tabelle 1 zusammengestellten Daten über die RK auf 245 kHz, so bemerkt man, daß die RK bei 5°–10° Sonnenhöhe immer einen stärkeren Störungsgrad haben als die Nachtwerte, und daß besonders hier die Sommerwerte einen verstärkten Einfluß der Störungen zeigen.

V. Der Tagesgang des RK beim Nachwirkungseffekt

Der stärkere Störungsgrad der RK bei 5°–10° Sonnenhöhe wird verständlicher, wenn man die Beispiele der mittleren Tagesgänge in Nachwirkungsperioden (in Abb. 7) betrachtet. Während die Tageswerte sich vom normalen Ablauf (gestrichelte Kurven) kaum unterscheiden, tritt der Effekt etwa $2\frac{1}{2}$ Stunden vor Sonnenuntergang am Boden in Erscheinung.

Die hohe Tagesdämpfung bleibt bis über den Sonnenuntergang hinaus bestehen, so daß etwa eine halbe Stunde nach Sonnenuntergang die Abweichung vom Normalwert am größten ist. Erst $\frac{1}{2}$ bis $1\frac{1}{2}$ Stunden nach Sonnenuntergang erfolgt der Anstieg zum nächtlichen Wert, der aber meist auch unter dem Normalwert bleibt. Dieser typische Verlauf erklärt zumindest das statistische Erscheinungsbild der stärkeren Beeinflussung der RK-Werte bei niedrigen Sonnenhöhen gegenüber den Nachtwerten. Es sei noch erwähnt, daß sich beim Sonnenaufgang der gleiche Effekt im Verlauf des RK in der Nachwirkung einstellt.

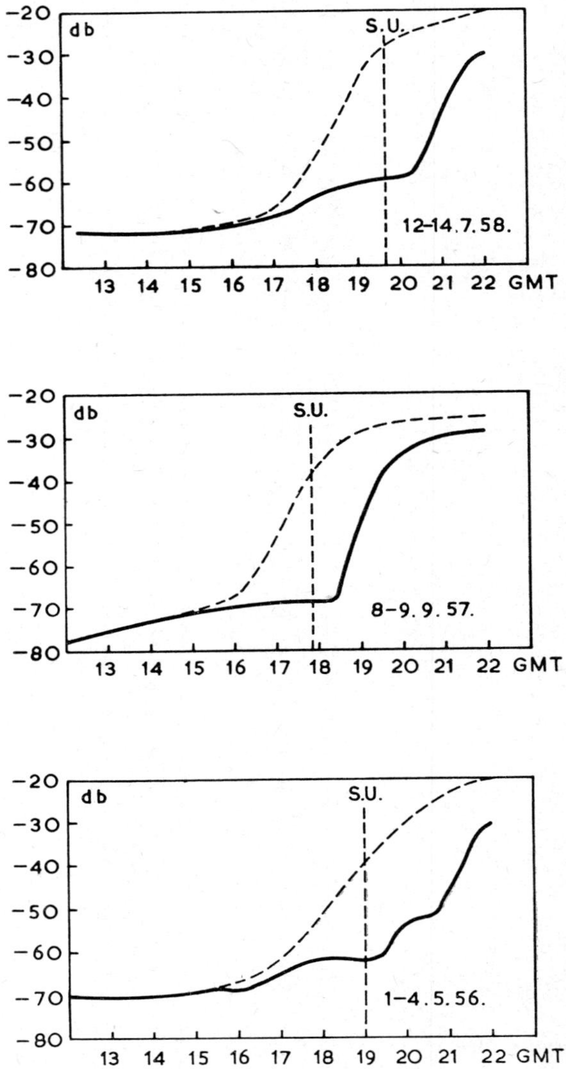

Abb. 7. Beispiele für den Nachwirkungseffect im Tagesgang des Reflexionskoeffizienten auf 245 kHz.

VI. Schlussbemerkungen

Die ionosphärischen Ursachen dieses Nachwirkungseffektes magnetischer Stürme sind noch hypothetisch. Man hat den Eindruck, daß sich während der Störung in der Polarlichtzone in der tiefen Ionosphäre leichtionisierbare Bestandteile des hochatmosphärischen Gases bilden, die langsam nach mittleren Breiten driften. Diese Bestandteile werden offenbar noch im sichtbaren Bereich der solaren Strahlung ionisiert. Der zusätzliche Ionisationsbeitrag ist offenbar nur gering und wird nur in den Abend- und Morgenstunden durch die Lang- und Längstwellenausbreitung erfaßt. Belrose [1] hat vermutet, daß die verhältnismäßig langlebigen Sauerstoffionen für die notwendige Elektronenerzeugung in Betracht zu ziehen seien. Aber noch liegen nicht genug Anhaltspunkte für die örtlichen und zeitlichen Eintrittszeiten dieses Effektes fest, um eine Transporthypothese zu festigen. Das im Rahmen des IGJ in Mitteleuropa eingerichtete Langwellenausbreitungsnetz scheint allerdings in einer ersten Übersicht zu bestätigen, daß diese Störungsform in Richtung N-S wandert, und daß die Nachwirkungseffekte mit abnehmender Breite immer schwächer und seltener werden.

Zusammenfassend haben wir also in den Langwellen-Messungen zwei bisher ungeklärte Effekte zu verzeichnen. Einmal die Anomalie der hohen sommerlichen Absorption und den Nachwirkungseffekt starker magnetischer Stürme. Mit einfachen statischen Modellen sind beide Effekte nicht zu erklären. Vielleicht können die neuerdings vorliegenden Driftmessungen im Langwellenbereich in Zukunft einiges zur Klärung der Vorgänge in der tiefen Ionosphäre beitragen.

References

1. Belrose, I. (1957). Dissertation. Cambridge.
2. Lauter, E. A. (1950). *Z. Met.* **4**, 234–35.
3. Sprenger, K. (1958). Dissertation. Leipzig.
4. Straker, T. W. (1955). *Proc. Instn elect. Engrs, N.Y.* Part C, 1–12.

Round-the-World Echoes

G. A. ISTED

Marconi's Wireless Telegraph Company, Ltd., Research Division, Great Baddow, Essex

Much is already known about the phenomenon of round-the-world echoes but, so far, no completely satisfactory explanation has been given for its unique behavior. It is known for example that even after four or five circuits round the world the signal suffers very little distortion and that the attenuation between successive echoes is only of the order of a few decibels. The observed facts indicate that the mode of propagation is unlikely to depend upon the conventional multihop reflection between the earth and the ionosphere.

Observations, made during an investigation into ionospheric scatter propagation at a frequency of 37 Mc/s, have shown that round-the-world echoes were present along a south-north route even when both the transmitting and receiving terminals were well inside the daylight hemisphere. Under the conditions of the experiment the signal seemed to have been propagated through areas where, in the normally accepted sense, the F_2 region could not have been the supporting mechanism.

Round-the-world echoes have also been observed from the signals radiated at 20 Mc/s by Satellite 1957 Alpha. At the time of the observation square-wave signals of 100 msec duration were being radiated with an interval between them of approximately 240 msec. Analysis of the records revealed the existence of two distinctly different beat notes, one from the square-wave signals themselves the other occurring during the interval between them. The maximum time-delay between the two components was of the order of 140 msec, which fairly identifies the component observed in the intervals between the radiated signals as a round-the-world echo. The difference in the beat note of the two components can be explained in terms of Doppler frequency shifts and divergent propagation paths.

The suggestion is put forward that transmission round the world is made possible by launching a radio wave against the ionosphere at an extremely small angle of incidence to produce a succession of reflections which are confined to the ionosphere itself. Such a shallow launching angle would be easily achieved by transmission from satellites flying just below the ionosphere, but a tilted layer would be necessary initially to achieve the required launching angle from the ground.

I. INTRODUCTION

Soon after the systematic exploitation of short radio waves for world-wide communication it was observed that, from time to time, undesirable echoes followed the main signal received from a distant transmitting station; the time-delays of these echoes are quoted by Eckersley [2] to have been of the order of 1/20–1/50 sec and 1/7 sec. In the light of our present knowledge it seems likely that the shorter time-delays could have been caused by F_2 back-scatter in the manner described by Crow *et al* [1]; on the other hand similar short time-delays would be observed if signals transmitted from the antipodes could be propagated simultaneously round each hemisphere. The use of directive aerials would, under such conditions, eliminate the deleterious effects of the signal propagated along the unwanted route, and it is interesting to note that under similar conditions the

Beam Service from the United Kingdom to Australia was maintained for several years on a single frequency by the simple expedient of switching the beam arrays through an angle of 180°; the need for separate day and night operating frequencies was thus avoided.

Eckersley [2] in 1927 pointed out that the echoes having time-delays of 1/7 sec on the main signal had undoubtedly been propagated once round the world; this type of echo, he stated, had been observed by J. G. Robb in the United Kingdom during the interception of transmissions from Poldhu, England, in 1925.

Transmission completely round the world, by the multi-hop mode of propagation, would seem to be perfectly feasible under suitable conditions of the ionosphere; there appears to be, however, another mode of transmission capable of propagating a radio signal round the world. The object of this paper is to review what is already known about round-the-world echoes or circulating signals as they will now be called, to resuscitate interest in the phenomenon and to draw certain conclusions from observations made of circulating signals received from the U.S.S.R. satellite (1957 Alpha) and from VHF Ionospheric Scatter transmissions.

II. Review of Past Investigations

Much is already known about the phenomenon of circulating signals but, so far, no completely satisfactory theory has been put forward which would account for many of the observed facts. The bibliography on the subject is very short and some of the references to the phenomenon are of a cursory nature and add little to our knowledge. Probably the most intensive investigation was carried out during World War II by von Schmidt and Hess, based on the very early work of Quaeck and Moegel in 1926; this investigation, carried out in Germany and Denmark, was described in detail by Hess [4] in 1948. A later investigation on a more modest scale was carried out in Holland by de Vooght [12] and described by him in 1950.

The propagation of short radio waves to great distances is sufficiently understood to state that true circulating signals possess certain unique characteristics; these appear to be very different to those which would be expected by the orthodox mode of multi-hop propagation.

A. Unique Characteristics of Circulating Signals

Among the many known unusual characteristics of circulating signals, their ability to pass several times round the world without suffering serious distortion, and without being subjected to serious attenuation, are probably the most outstanding; nevertheless other characteristics are of great interest and it is worthwhile reviewing the situation in more detail.

1. Distortion

Hess [4] and von Schmidt [11] have both pointed out that the circulating signals are much less mutilated than a direct signal which has travelled a much shorter distance by a series of reflections or hops between the ionosphere and the ground; circulating signals which have passed three and four times round the world have failed to show appreciable distortion as shown by Hess [4], Humby *et al.* [3] and Isted [5]; this is in sharp contrast to the destructive phase interference and dispersive effects suffered by signals propagated by the multi-hop mechanism. Hess, in his paper, shows convincing records which demonstrate the difference which exists between the two modes of propagation.

2. Attenuation

Hess first drew attention to the low attenuation suffered by successive circulating signals round the world. This low attenuation has been confirmed more recently by Humby *et al.* and Isted who have made measurements of circulating signal amplitudes; it would appear from their results that, although the first circulating signal is heavily attenuated, each successive circulating signal thereafter is attenuated only by some 5–10 db.

3. Circulating Time

Hess investigated at great length the time taken for signals to circulate round the world and he claims to have measured it to very great accuracies; he found that the average time taken in one complete circuit was 0·137788 sec. De Vooght states that his measurements indicate time delays between 137 and 139 msec. Quaeck and Moegel, von Schmidt and Isted all mention a figure of 138 msec. but Luscombe [8] finds a value of 140 msec. It is clear that most investigations have shown that the circulating time is very close to 138 msec. The influence of the radio frequency employed, time of day, season, ionospheric conditions and geographical position upon the circulating time is, by general agreement, small or negligible.

4. Activity

Quaeck and Moegel state that the propagation path followed by circulating signals is confined exclusively to the twilight belt round the world, but Marconi [9] and other investigators have found that this statement is not strictly true because circulating signals have been observed at places well displaced from twilight. The general opinion has been formed, however, that the greatest activity is likely to occur when the great circle path joining the transmitter and receiver lies within or near the twilight zone. These observations have led to the belief that the F_2 region of the ionosphere is implicated in the propagation mechanism because the highest mean density of ionization would be expected around the twilight belt. The opinion has also been expressed that circulating signal activity will be most likely to be observed for longer periods in the higher latitudes because of the longer twilight period there.

5. Direction of Arrival

Both Hess and de Vooght have carried out tests with directional aerials and they have both found that, generally speaking circulating signals can be observed from both great-circle directions but on frequent occasions the signals arrive from only one direction of the great circle route. Luscombe also carried out measurements on the direction of arrival of circulating signals along a route where the great-circle joining the transmitter and receiver was displaced 12° from True North; he found that the direction of arrival was close to the great-circle path. As regards the vertical angle of arrival von Schmidt states that his measurements indicate angles of the order of 20°; this is confirmed by Katowski, Shuettloeffel & Vogt [7] who have measured angles between 15° and 25°.

B. Theories of Circulating Signals

So far as the author knows only two theories have been put forward to account for the circulating signal. The first is probably the most obvious—that of multi-hop reflections between the ionosphere and the ground; this, as de Vooght rightly points out, involves on the average some 16 hops at a frequency of 20 Mc/s; he shows at the same time that the period taken by a signal to encircle the world by this mode of propagation would vary throughout the seasons from 138·5 to 145·5 msec.

In the second theory von Schmidt has suggested that propagation of certain radio waves in the ionosphere is analogous to the propagation of seismic waves in the ground. It is known from experiments in explosion seismology that, under certain conditions, shock-waves are propagated along a boundary separating two different media beneath the surface of the ground; it has been shown for example that if the velocity of wave propagation in the medium above the boundary is slower than that in the medium below it, rays with angles of incidence greater than the cut-off angle will be propagated along the boundary and will be continuously radiated into the upper-medium. von Schmidt then views the mechanism upside down and, making the assumption that the velocity of radio wave propagation is greater in a higher altitude medium than in one below it, shows that rays with an angle of incidence smaller than the cut-off angle enter the upper medium and are lost, those rays with angles of incidence greater than the cut-off angle will, however, be propagated along the boundary and will be radiated continuously downwards towards the ground. von Schmidt, considering the theory in relation to circulating signals, claims that it provides an accurate assessment of the F_2 skip distance and the period of circulation of a radio wave round the world.

Hess concludes that, in order to account for the highly constant time interval which he has measured between successive circulating signals, the mean height of the boundary required in von Schmidt's theory is between 203–204 km above the surface of the earth.

III. Circulating Signals at a Frequency of 37 Mc/s

For the purpose of investigating the propagation characteristics of ionospheric scatter signals a test circuit was set up between Gibraltar and the United Kingdom [6]. The conditions of the test were as follows:

Transmitter located at Gibraltar: 36°8′N, 5°19′W. Receiver located at East Hanningfield, Essex: 51°41′N, 0°34′E. Distance of Gibraltar to East Hanningfield: 1110 miles. Transmitter power (cw): 30–40 kW.

Transmitting aerial: 4 × 4 curtain array with parasitic reflectors; plane-wave gain of 16 db relative to a half-wave dipole. Horizontal polarization.

Receiving aerial: 1 × 8 stack of half-wave dipoles with parasitic reflectors; plain-wave gain of 12 db relative to a half-wave dipole.

Orientation of transmitting array: 12° E of N.

Operating frequencies: 37·3 Mc/s, 48 Mc/s and 70 Mc/s.

During the tests at the frequency of 37·3 Mc/s, conducted mainly during 1956, a troublesome source of self-interference was found to be due to the effects of F_2 back-scattered signals having time-delays between 30 and 80 msec on the main

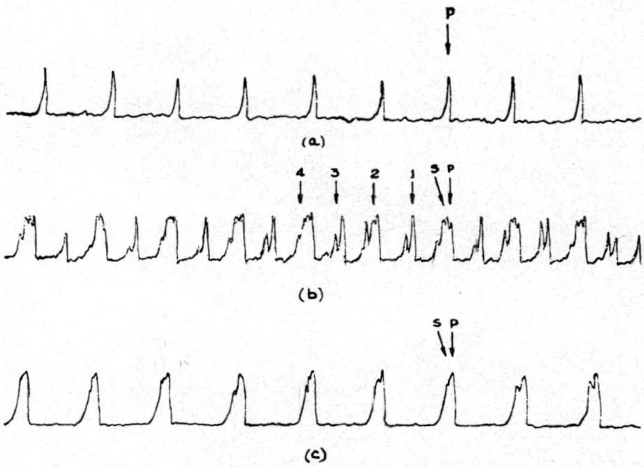

FIG. 1. High-speed amplitude recording of circulating signals.
(a) Forward scatter signal alone. (b) Forward scatter signal together with circulating and F_2 back-scatter signals. (c) Forward scatter signal together with F_2 back-scatter signal.

Pulse duration: 10 msec.
Repetition rate: 4/s.
Frequency: 37 Mc/s.
Path: Gibraltar—East Hanningfield, Essex.

signal [1]; at times, however, an additional serious form of self-interference was observed to be present. When the transmitted pulse width was reduced to 10 msec and the repetition rate decreased to 4 per sec, this latter form of interference was found to be due to signals which, having passed over the receiving point, continued onwards round the world and arrived again at the receiver after an interval of 138 msec. At about the time of the autumnal equinox, 1956, these circulating signals were observed for about 50 % of the days between 09.00 and 11.00 hours G.M.T. On many occasions the propagation conditions were such that the circulating signal was observed to have been repeated four or five times, each successive delay being 138 msec.

Figure 1 is a reproduction of a high-speed record, made on 19 November 1956, between 09.15 and 12.07 hours G.M.T., on which the effects of both F_2 back-scatter and circulating signals are clearly shown. Increases in amplitude are denoted on the record by vertical deflections of the pen from the base-line. A range of about 30 db in amplitude can be accommodated between the base-line and the saturation point of the apparatus. Time advances from right to left. Strip (a) in the Figure is typical of the signal pulses received by the ionospheric forward-scatter mode of propagation in the absence of other spurious components. The sequence of events under conditions of F_2 back-scatter and circulating signals, resulting from the transmission of a single pulse, "P," will be described with the aid of Strip (b).

The pulse propagated by the forward scatter mechanism is marked by an arrow at "P"; a further clean pulse displaced by 138 msec from "P" is marked by an arrow at "1," this pulse represents the first circuit round the world resulting from the original transmitted pulse "P"; successive circulating signals are indicated by arrows marked 2, 3 and 4, all of which are separated successively by a time interval of 138 msec. There are indications that a fifth circulating signal was sometimes present; this can be seen as a small amplitude pulse falling in position on the left-hand side of the circulating signal marked "3" in the figure. The signal resulting from F_2 back-scatter is indicated by the arrow marked "S"; this follows the initial transmitted pulse after a time interval of 10–15 msec. Strip (c) of the Figure was recorded about 2 hours after strip (b) and shows clearly that the circulating signals have disappeared leaving only the relatively short time-delayed signal due to F_2 back-scatter.

It should be pointed out that the complicated pattern of the signal as shown in the Figure is due mainly to the fact that the transmitted pulse was repeated at 250 msec intervals and that the circulating signal was repeated at intervals of 138 msec. The second circulating signal would therefore be received 276 msec after the initial transmitted pulse, i.e., about 26 msec after the succeeding transmitted pulse.

In a previous paper [5] a more detailed reproduction of the record was given which demonstrated that the onset of the circulating signal was coincident with the commencement of F_2 back-scatter. This present record serves to illustrate that F_2 back-scatter can be present after the circulating signal activity has ceased; furthermore it can be seen from this record that the small amount of distortion and the low attenuation suffered by succeeding circulating signals, described by other investigators, is readily observable.

IV. Circulating Signals from Satellite 1957 ALPHA

What are considered to be circulating signals resulting from transmissions, radiated at a radio-frequency of 20 Mc/s, from the U.S.S.R. satellite 1957 Alpha were recorded at the Marconi Research Laboratories, Great Baddow, England (51°45′N, 0°30′E) at approximately 00.30–00.33 hours G.M.T. on 8 October 1957.

A standard communication receiver connected to a half-wave dipole aerial installed $3/4\lambda$ above the ground formed the basis of the receiving apparatus. The incorporated beat frequency oscillator of the receiver was adjusted to give a convenient but arbitrary audible note at the output of the receiver; a permanent record of the signal was made by means of a magnetic tape recorder connected to the output of the receiver. During the period to which these observations refer, the satellite was radiating square-wave signals of 100 msec duration at a repetition rate of approximately 3 per sec.

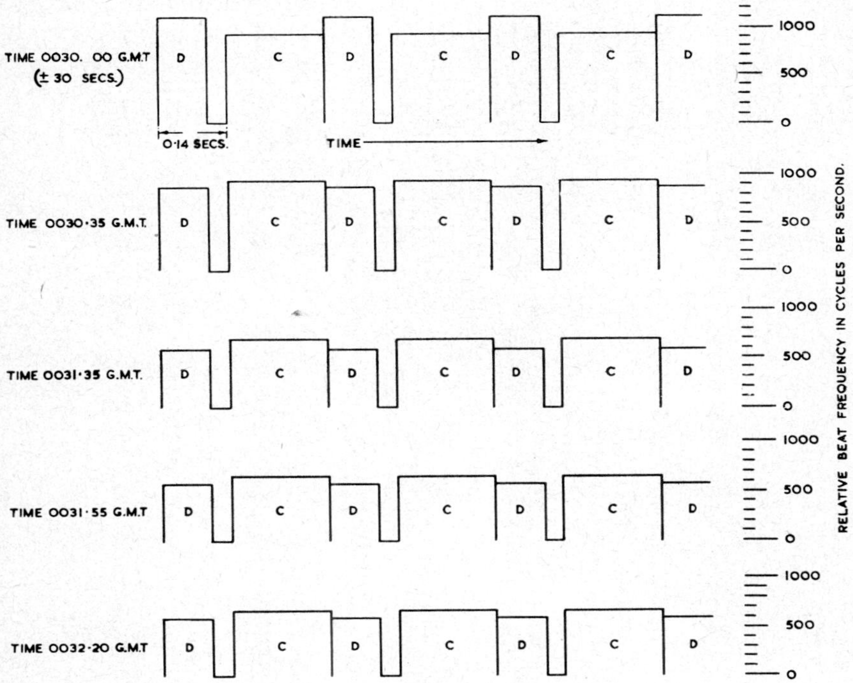

FIG. 2. Illustration of relative beat frequencies of direct and circulating signals received from U.S.S.R. Satellite

8 October 00.30–00.33 G.M.T.

The orbit of the satellite on this particular occasion tracked up from South America in the dark hemisphere, and passed between Iceland and Scotland at approximately 00.30 hours G.M.T., continued over Norway and entered the twilight zone at approximately 00.43 hours G.M.T. over Russia. The experimental timing arrangements at this very early stage of the satellite investigation were rather elementary, consequently times cannot be quoted to greater accuracies than $\pm$ 30 sec.

Subsequent analysis of the record obtained on this particular occasion disclosed the fact that two distinct signals, each with a different beat frequency, were present; one beat frequency corresponded to the direct signal radiated from the satellite, the other occurred in the intervals between the radiated signals. Figure 2 is an illustration of the signal made by playing back the original magnetic tape recording into a chart-recorder which gave an amplitude deflection proportional to frequency. In the portion of the record for 00.30 hours G.M.T. the direct signals are shown at "D" and it will be seen that their beat frequency was 1100 c/s; the signals received in the intervals between the direct signals and marked "C" in the Figure gave a beat frequency of 920 c/s. The interesting point to note is that the time-interval between the commencements of the "D" and "C" signals was approximately 140 msec; this time-interval strongly suggests that the "C" signal was the original transmitted signal which had circulated once round the

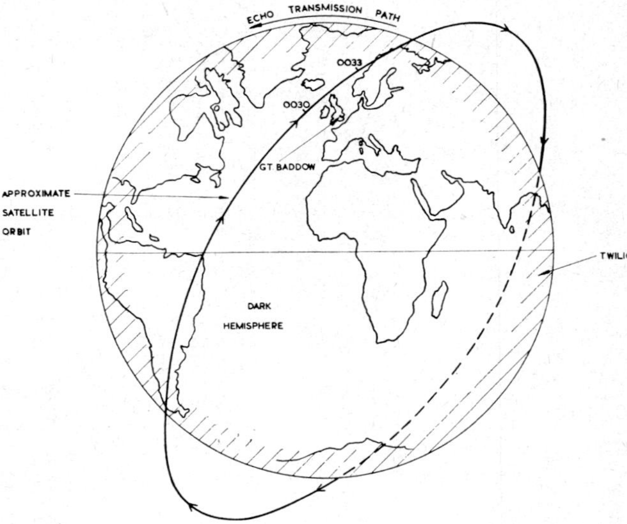

FIG. 3. Diagram illustrating relation between orbit of satellite and twilight zone 8 October 00.30–00.33 G.M.T.

world. The reason why the circulating signal was about twice the length of the original direct signal is not immediately apparent but the original record suggests that a second circulating signal followed the first one. The illustration shows that by 00.30·35 hours G.M.T. the beat frequency of the direct signal had decreased to 870 c/s but the beat frequency of the circulating signal had remained practically unchanged. By 00.31·35 hours G.M.T. the beat frequency of the direct and circulating signals had decreased to 580 c/s and 640 c/s respectively. At 00.31·55 and 00.32·20 hours G.M.T. the respective beat frequencies remained unchanged.

The decrease in the beat frequency of the direct signal was due to the Doppler frequency shift associated with the satellite's nearest approach to the observer; this effect is well-known and need not be enlarged upon here. The different rate of change in the beat frequencies of the direct and circulating signals suggests that different transmission paths were followed by each of the signals. There is good reason to believe that the transmission path of circulating signals is often associated with the twilight zone and it might therefore be expected that, if the satellite crossed this fixed transmission path, a Doppler frequency shift similar to that observed would be superimposed upon the circulating signal. Reference to Fig. 3 will show that the satellite was in fact near to the twilight zone at the time of the observation and that it crossed it obliquely. However, if the transmission path was in fact the twilight zone, it might have been expected that there would have been two components of circulating signals arriving from opposite directions. The evidence suggests that in this particular case the transmission path was unidirectional and, because of the fact that the Doppler beat frequently of the circulating signal decreased, the direction of transmission must have been approximately as indicated by the arrow in Fig. 3.

V. Discussion of Recent Obseevations

An examination of the new data concerning circulating signals brings to light one or two important points which, up to the present, have not been commented upon in the relevant literature.

The first points now to be commented upon concern the pulse transmissions from Gibraltar.

The receiving site, being less than 2,000 km from the transmitter was within the "skip-distance" by normal F_2 propagation; therefore there was no possibility of receiving the short-path direct signal propagated by the F_2 mechanism although, as previously described, the effects of F_2 back-scatter were readily observable.

It has long been suspected that the circulating signal mode of propagation is intimately associated with the F_2 region of ionization and the fact that they have been observed at frequencies as high as 37 Mc/s during a period of record high sunspot activity may therefore cause no undue surprise.

In describing the conditions under which the tests were carried out it was stated that signals, in the form of pulses, were radiated from Gibraltar by a directional aerial system oriented 12° East of North; Luscombe, who contemporaneously observed circulating signals resulting from the Gibraltar transmissions, confirmed that their azimuthal angle of arrival did not deviate from the great-circle path. It would therefore appear that the circulating signals had passed approximately through latitudes 80° North and South of the equator. An inspection of the published Maximum Usable Frequency (MUF) prediction charts for the appropriate time and period shows that the F_2 normal mode of propagation could not have been supported at latitudes higher than 50° North and South of the equator. The MUF in the region of Latitude 80° North was no greater than 22–24 Mc/s, while at Latitude 80° South it was no greater than 16–20 Mc/s. Thus the fact emerges that, in spite of the restricted MUF for normal F_2 propagation, circulating signals were nevertheless propagated at 37 Mc/s through regions near to the Poles. It might of course be argued that, in the case of a multi-hop mode of propagation, it is unnecessary to rely upon a high MUF over the Poles provided that a ground reflection occurs within the Polar Regions. If this condition was indeed being fulfilled evidence of a signal propagated by the multi-hop mechanism should have been observed; applying the criterion that a signal so propagated would exhibit considerable distortion due to destructive phase interference and dispersive effects one must conclude that, because this distortion was not observed, reflections between ground and ionosphere were not involved.

The record shown in Fig. 1 indicates that the average attenuation between successive circulating signals is approximately 5 db; this low attenuation is in sharp contrast to the great attenuation involved between the initial radiated signal and the first circulating signal. The inference to be drawn from the observed attenuation characteristics is that a transmission mechanism encircles the world in which the attenuation is very low but a wave which succeeds in energizing it is first subjected to a very high attenuation.

From the evidence provided by the Gibraltar-U.K. experiments the influence of F_2 ionization upon circulating signals cannot be determined with any degree of certainty. The most that can be said is that whenever circulating signals were recorded the ionization density of the F_2 region was sufficient to support the normal mode of F_2 propagation at 37 Mc/s in the region of the transmitter; this is demonstrated by the simultaneous presence of F_2 back-scattered components of the transmitted signal. The converse, however, does not hold because the F_2 back-scattered component could be recorded for many hours of the day while the circulating signal would be completely absent.

The final point of comment here concerns the twilight zone. Much of the published literature associates circulating signals with the twilight zone but, during the Gibraltar-United Kingdom experiments, it was observed that circulating signals were received 4 to 5 hours after local sunrise; this represents a considerable displacement of the twilight belt from the great-circle path followed by the circulating signal. It can therefore be concluded that the great-circle path followed by the 37 Mc/s circulating signal included large extremes of solar illumination and it is therefore demonstrated that the association of circulating signals with the twilight belt is not very rigid.

In considering the observations on the signal radiated from the satellite, an examination of the F_2 MUF prediction charts shows that at the time of the observations the satellite was flying through an area in which the MUF was only of the order of 10–14 Mc/s and therefore the signal radiated at 20 Mc/s was unlikely to have been propagated to great distances by the normal F_2 multi-hop mode of transmission, nevertheless the signals observed in the intervals between the direct signals radiated from the satellite had undoubtedly been transmitted round the world. In most other areas of the world, however, the MUF values at that time would have supported propagation at frequencies greater than 20 Mc/s.

The most interesting aspect of the round-the-world transmission of signals radiated from the satellite lies in the fact that, in spite of the very low power radiated, the conditions necessary for circulating signal propagation were achieved; bearing in mind that a very high attenuation has normally been observed between a signal radiated from the ground and its first circulating component, it would seem that in this particular case the circulating signal observed from the satellite could not have been subjected to that initial attenuation; it is concluded therefore that the satellite was at that time actually flying within or near the medium responsible for the circulating signal mode of propagation. From the published data [10] it would appear that the height of the satellite at the time of the observation was 240 $\pm$ 13 km above the earth; it was therefore within the lower part of the F_2 ionized region.

From the foregoing it seems probable that the F_2 region is in some way the supporting mechanism for circulating signals; on the other hand there is no evidence which suggests that the normally accepted mode of multi-hop transmission between the ionosphere and the ground is in any way involved; indeed, the observed low attenuation suggests that the signals are not propagated to any great extent through the absorbing D region.

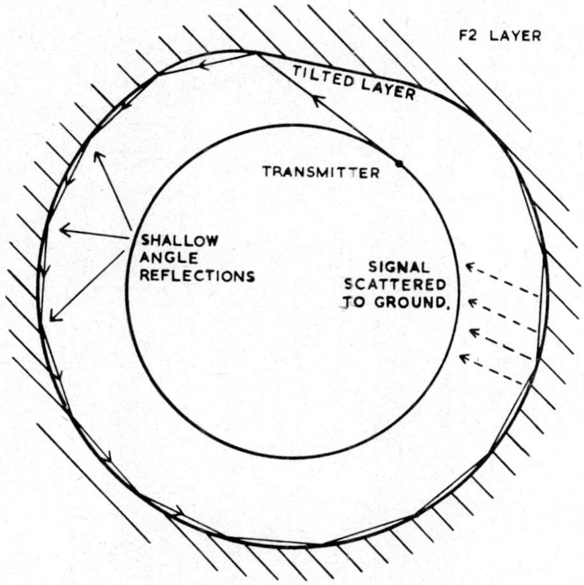

FIG. 4. Suggested mechanism of circulating signals by a tilted layer and shallow ionospheric reflections.

Many of the unique characteristics of circulating signals can be explained if it is assumed that a radio wave can be launched against the ionosphere at an extremely shallow angle of incidence, such as might have happened in the case of the satellite when it was at a height of 240 km, a succession of glancing reflections confined to the ionosphere might then follow even when the density of ionization was very low. To achieve the same necessary glancing angle at the ionosphere by a radio transmission from the ground a tilted ionized layer would have to be invoked (see Fig. 4). A tilted layer is a well-known feature of the F_2 region and its occurrence at twilight might be expected; a tilted layer, furthermore, would provide a ready explanation for the unidirectional characteristics so often observed in circulating signal propagation.

The signal which is observed on the ground would be due to a small amount of energy scattered downwards each time the circulating signal passed overhead.

It is possible, therefore, that a series of reflections confined to the ionosphere itself could explain all the known characteristics of circulating signals with the possible exception of the highly constant time interval of circulation measured by Hess; this, it would seem, should be dependent upon frequency, time of day and season.

VI. Conclusion

It has been shown that circulating signals are not restricted to frequencies below the normal F_2 MUF as at present generally understood but that they are capable of being propagated through areas of low ionization densities.

It is possible that the observation of signals from the satellite may provide an important clue to the understanding of the mechanism responsible for circulating signals and the suggestion that this mechanism is associated with reflections at glancing angles of incidence which are confined solely to the ionosphere may be worthy of further consideration.

It is possible that some of the other unusual propagation phenomena observed on the signals radiated by satellites [10, 13] may be explained in terms of circulating signals.

Although, up to the present, circulating signals have not caused serious disruption to orthodox modes of communication the phenomenon assumes considerable importance in connection with the operation of ionospheric scatter communication links.

Acknowledgments

The author wishes to thank the Engineer-in-Chief, Marconi's Wireless Telegraph Company Limited for permission to publish the paper. The author gratefully acknowledges the assistance given by many of his colleagues in making the observations discussed in the paper; he is particularly grateful to Dr. D. H. Shinn for very helpful discussions and suggestions. The ionospheric scatter investigation mentioned in the paper was sponsored by the Admiralty.

References

1. Crow, D. A., Kitchen, F. A., Isted, G. A. & Millington, G. (1956). *Nature, Lond.* **178,** 1280.
2. Eckersley, T. L. (1927). *J. Instn elect. Engrs* **65,** 600.
3. Humby, A. M., Minnis, C. M. & Hitchcock, R. J. (1955). *Proc. Instn elect. Engrs* **102**B, 513.
4. Hess, H. A. (1948). *Proc. Inst. Radio Engrs, N.Y.* **36,** 981.
5. Isted, G. A. (1958). *Proc. Inst. elect. Engrs* **105**B, Supp. no. 8, p. 27.
6. Kitchen, F. A. & Millington, G. (1958). *Proc. Instn. elect. Engrs* **105**B, Supp. no. 8, p. 2.
7. Kotowski, Schuettloeffel & Vogt(1940). *Telefunken Lab. Rep.*
8. Luscombe, G. W. (1957). *Nature, Lond.* **180,** 138.
9. Marconi, G. (1931). *Marconi Rev.* no. 32, p. 1.
10. Royal Aircraft Establishment (Staff), Farnborough (1957). *Nature, Lond.* **180,** 937.
11. von Schmidt, O. (1936). *Z. tech. Phys.* **17,** 433.
12. de Vooght, A. H. (1950). *Onde élect.* **30,** 443.
13. Wells, H. W. (1958). *Proc. Inst. Radio Engrs, N.Y.* **46,** 610.

A Calculation-Method of Ionospheric Propagation Conditions for Very High and Antipode Distance

E. Harnischmacher

Ionosphären Institut Breisach im Fernmeldetechnischen Zentralamt der Deutschen Bundespost, Breisach (Rhein)

A calculation method is given for a fixed elevation angle. Mixed propagation modes over E and F layers are considered. Some examples show the determination of the MUF and the LUF. Very high distance and antipode-distances are calculated for different azimuth angles. An intercomparison between prediction and experience for the antipodal distance Dakar-Noumea and also for "round the world echoes" is made. The agreement is highly satisfying.

I. Philosophy of the Method

Whilst the CRPL prediction method for high distances only considered the conditions of the ionosphere at two control points, there has been established a method by Rawer and his collaborators at Zentralstelle für Funkberatung and later at SPIM (Service de Prévision Ionosphérique Militaire) taking account of all possible propagation modes from 1E to nE and from 1F to mF. Theissen later on introduced mixed propagation modes nE + mF into the prediction for East-West distance at sunrise and sunset.

The application of this method on very large distances, however, would need too much calculation work; because there should be calculated additionally such propagation modes as nF + mE + 1F. We tried to generalize the analytic method in order to get a usable prediction after some simplification and with a reasonable lot of calculation work.

1. Considering now wave propagation at low elevation (small angle versus the horizon) we see that the MUF increases with decreasing *elevation angle*. As to the LUF we must make some distinction between absorption and blanketing. The absorption per hop increases with decreasing elevation angle, the total attenuation for one mode, however, increases with increasing number of hops; i.e., as smaller the elevation angle as weaker the effective attenuation when different modes are taken into account. Blanketing always increases with decreasing elevation angle. But in almost every case of blanketing, propagation continues by means of a "shunt mode" over the reflecting E- or Es-layer. In such cases the F-hop is effectively replaced by an E-hop. This replacement implies a slight correction of the effective elevation angle. It may also happen that the F-hop must be replaced by a 2E-hop or a E + F-hop. As there are two further penetration points at the D-layer, the total absorption increases in these cases, though usually this increase is less important than the blanketing effect. Therefore the effective LUF is somewhat lower than the value which is calculated simply with blanketing of the F-hop.

For our practical calculations we suppose that blanketing can be avoided by shunt modes via the E-layer. Then for large distances, modes corresponding to a small elevation angle are preferable with respect to the MUF as well as to the LUF. The sequence of usable elevation angles is rather dense for large distances. Therefore, in a small range of elevation angles a certain number of modes is present. It seemed therefore adequate to consider the propagation of a small bundle of rays starting with a fixed elevation angle, seeing what happens at the successive penetration and reflexion points. We introduce into our calculations only a few standard elevation angles, viz. 2° for angles between 1° and 3°, and 5° for angles between 4° and 6°. Higher elevation angles are rarely useful for very large distances.

2. An elevation angle of 2° implies F-hops of about 3500 km and E-hops of about 2200 km, according to the respective layer-height. The MUF for very large distances, we may call if MUF eff 2° is now defined as the highest of the following three MUF values: The MUF F2 3500, the MUF E 2200 and the MUF Es 2200; the numerical values can be obtained from Ionization maps. For our MUF prediction the MUF eff 2° is always used.

3. The LUF is determined according to Rawer's method, first for the following limiting cases:

(1) (Lowest) F-absorption supposing the distance to be covered only by F-hops of about 3500 km.
(2) (Lowest) E-absorption supposing only E-hops of about 2200 km.
(3) Blanketing in the case when the distance is covered only by F-hops of about 3500 km.

For all following calculations we use the standard elevation angle of 2°. Some simple rules are stated how the LUF can be determined or interpolated from the three values mentioned above.

II. Fundamental Prediction Data

1. The set of MUF eff 2° maps.

According to Rawer's method a continuous representation of MUF all over the world can be obtained for the F2-layer as well as the Es-layer separately. Actually three capital lines are used, which nearly follow a meridian and for which a sufficient number of observations is available (Europe-Africa, Far East-Australia, America). An ionization map dependent on latitude and time LMT is established for each capital line. In the intermediate regions between two capital lines values are interpolated at equal LMT following a certain number of fixed interpolation lines. It is a simple interpolation task to transform the LMT-map into a TU-map; this can be done by automatic means which are already used for the LMT interpolation. In the equatorial zone, the sequence of interpolation lines follows the magnetic isoclines; both Es as well as F2-layer being under geomagnetic control identical interpolation can be used without any scruple.

Prior to 1955 we used a (corrected) zonal system, interpolation was only necessary for the intermediate zones. As an example Fig. 1 shows a F2–3500–MUF map for May 1953, East-zone. (As the examples of observations given below are some years old we reproduce a corresponding prediction map of our old zonal system.) Figure 2 shows the E/Es–2200–MUF map for the same zone and same month. The significant difference between both maps is the strong auroral-E ionization in the North and South during the night on one side and the high MUF

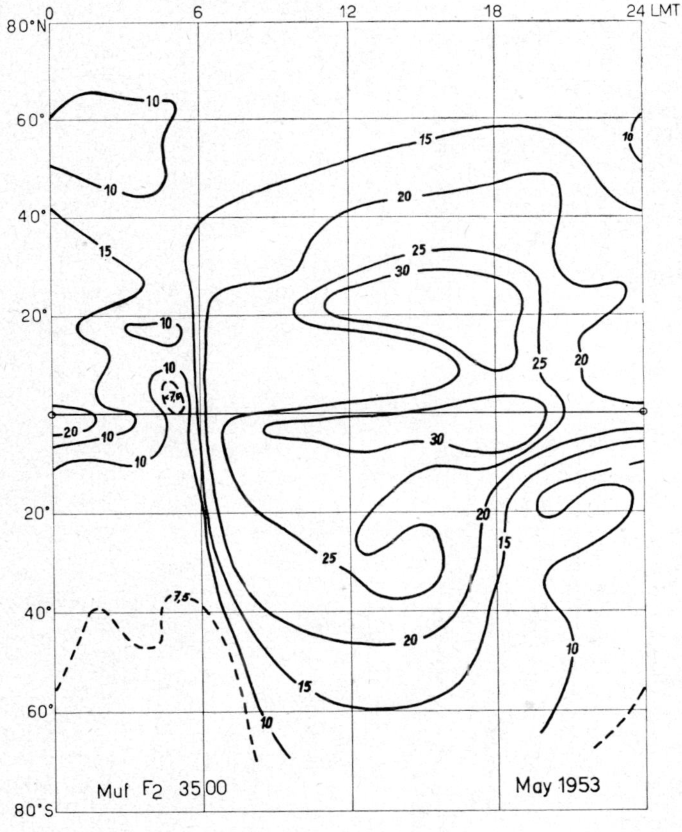

Fig. 1.

values near the magnetic equator at daytime on the other. We know now that the Es-ionization along the magnetic equator is essentially higher than it had been thought in 1953. At that time only extremely few ionospheric stations existed in the equatorial belt. Figure 2a shows the E/Es map as it would have been established the equatorial effect being adequately accounted for. Besides high ionization maxima are of no interest for prediction for very large distances; as they are too much limited in space they can not become efficacious for the MUF value of a long path. Of course these Es–2200–MUF cannot be established without some ambiguity. The characteristic fEs, unfortunately is power dependent;

during a long period its interpretation was different for different stations. Considerable improvement has recently been obtained by the introduction of foEs for the International Geophysical Year but even now some inhomogeneity of basic data may still exist. With these limitations, having critically tested the results of some doubtful stations it was possible to get a reasonable information

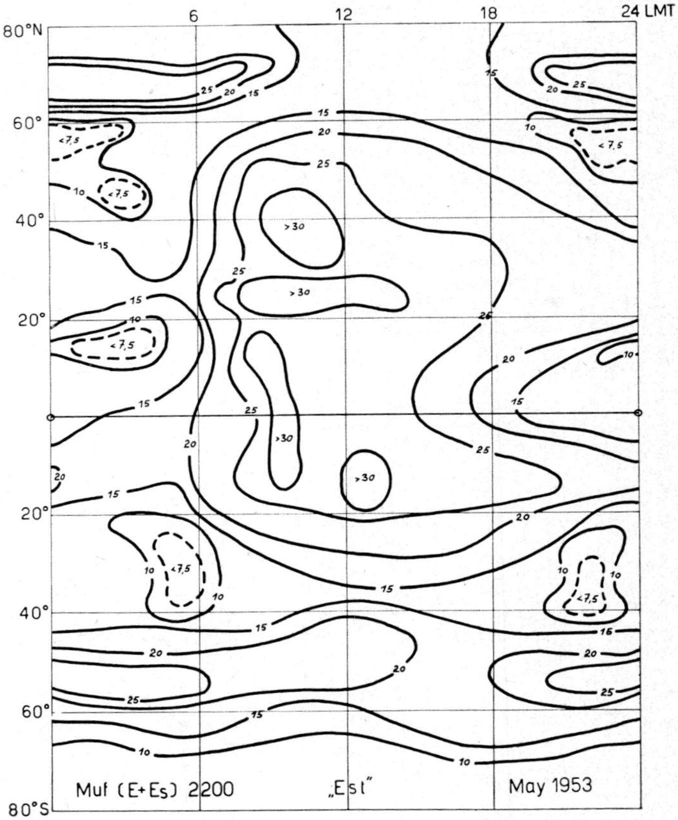

FIG. 2.

on the worldwide distribution of Es-ionization. Figure 3 shows the superposition of the maps of Fig. 1 and 2; the larger of both values has always been plotted in Fig. 3. This is called "MUF eff 2°." This composition corresponds to our general idea according to which the course of a ray starting with a given angle of elevation is considered all-over a great-circle path. It is interesting to see that a number of ionization minima occurring in one or the other of the basic maps have been sensibly smoothed out or have even disappeared in the composite map.

2. The *MUF for a certain distance* was formerly determined by graphical methods. One great-circle (in Ocagne's form of presentation) has been introduced

into Fig. 3. It is for the distance Saigon-Buenos-Aires. The position of the end-points A and B determines the great circle. The first reflection points on the F- or E-region are indicated by crosses near *A* and *B*, they are called A_W, A_E, B_W and B_E. These are the well known "control points." But, instead of inserting the lowest MUF-value found at these points, like does C.R.P.L., we look for the

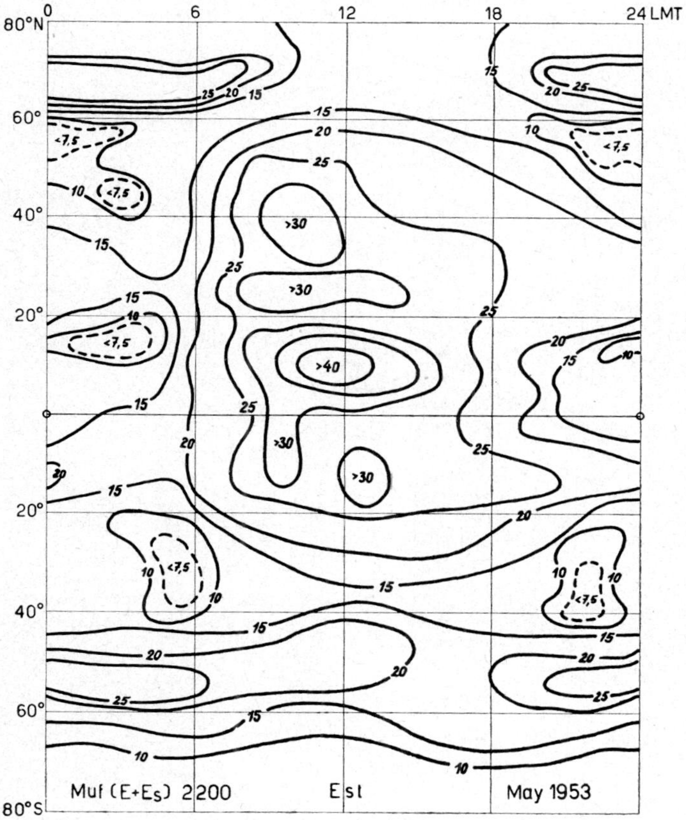

FIG. 2*a*.

lowest value *between* the control-points, i.e., we determine the minimum of MUF eff for the great circle paths $A_E - B_W$ and $B_E - A_W$ respectively. The upper limit of the usable frequency range was formerly determined by this way.* With modern automatic calculation methods, this determination can now be made by calculation.

*In Fig. 3 dotted and dashed lines along the great circle indicate, respectively the validity of the eastern and western ionozation map in our old zonal presentation. The minimum of the MUF was determined on each zonal map for the validity range only, with interpolation in the intermediate zone.

The MUF-value obtained with our rules is somewhat pessimistic as the points of reflection are not usually identical with the minimum, and as the geometry of ray refraction is even such that the points of lowest ionization will never become reflection points. Taking account of these facts we have to introduce a correcting factor which is about 1·2.

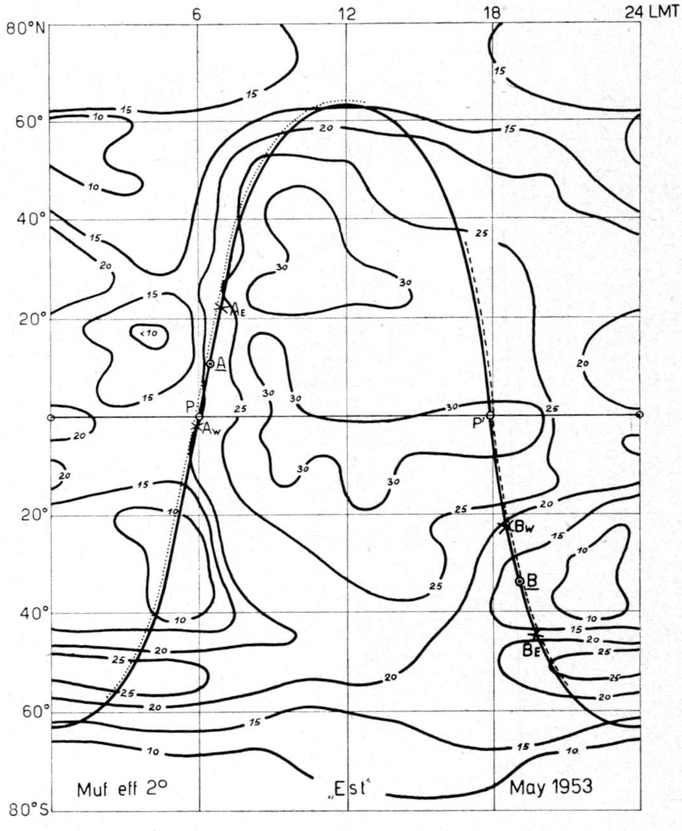

FIG. 3.

Besides it was found to be very helpful to use for all calculations the local mean time of P; this being the point where the great circle crosses the equator in north-eastern direction. Starting from P and following the great circle towards the East we first reach A then B and return to P. (This is our standard nomenclature.) It is most important for predictions on very large distance that both are taken into account, the larger as well as the smaller great circle path.

3. Figure 4 shows once more the Ocagne diagram now with the information we need for the *determination of LUF*.

Following Rawer we have in daytime when the LUF depends essentially on absorption the following expression for the LUF for a given path:

$$f_{\min} = C_A \sqrt{\sum \cos \tfrac{3}{4} \chi} - f_L \tag{1}$$

(χ solar zenith angle, f_L longitudinal component of gyrofrequency). Blanketing is given in the most simple (thin layer-) formula by:

$$f_A = C_0 \cdot \cos \chi^{0,3} \tag{2}$$

C_A, C_0 and f_L being constants for the respective prediction and depending on distance, elevation angle, season, solar activity, transmitter output, antenna gain and receiver sensitivity.

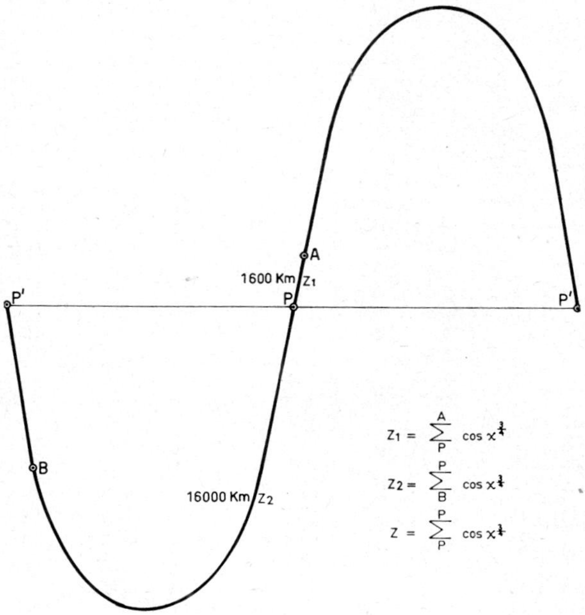

FIG. 4.

We also need:

$$(1)\ \sum \cos \tfrac{3}{4} \chi$$

$$(2)\ \mathrm{Max}\ (\cos \chi^{0,3}).$$

For a certain number of great circles we have tabulated these values as a function of hour and season and for different endpoints, the starting point being always P, time being given as LMT of the point P. For every elevation angle and reflection height another one-hop-distance is obtained. Strictly speaking one set of values should be prepared for each of these combinations. Practically it is sufficient to use only one set of tables for a medium one-hop-distance; the transformation for other parameters can be made by correction factors. These of course are different for E- and F-transmission. Our set of tables corresponds to an one-hop-distance of 2500 km.

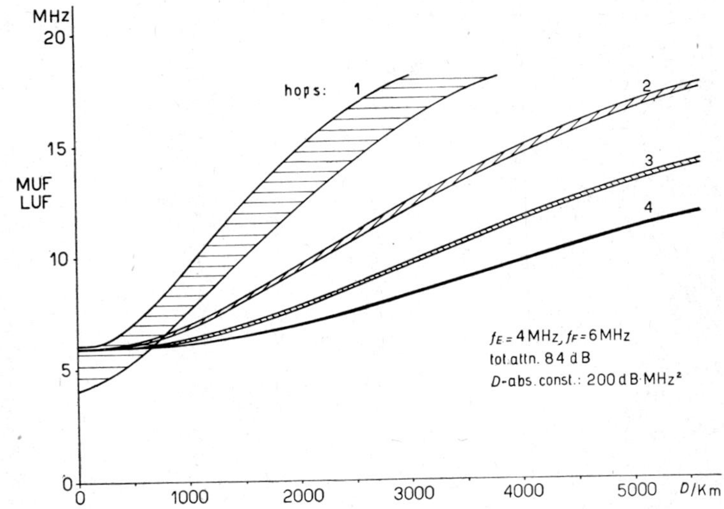

January
04 LMT

Great circles (left: Z_1) — Great circles (right: Z_2)

	0°	10°	20°	30°	40°	50°	60°	70°	80°	Distance from P	0°	10°	20°	30°	40°	50°	60°	70°	80°
	Z_1									P	Z_2								
	0	0	0	0	0	0	0	0	0	2500	10,21	9,64	9,00	8,30	7,34	6,59	6,02	5,92	6,21
	24	18	05	0	0	0	0	0	0	5000	9,97	9,46	885	8,30	7,34	6,59	6,02	5,92	6,21
	1,28	1,10	83	56	25	0	0	0	0	7500	8,93	8,54	8,17	7,47	7,09	6,59	6,02	5,92	6,21
	2,78	2,48	2,05	1,62	1,07	55	14	0	0	10000	7,43	7,16	6,95	6,68	6,27	6,04	5,88	5,92	6,21
	4,54	4,13	3,57	2,97	2,15	1,37	67	16	0	12500	5,67	5,51	5,43	5,33	5,19	5,22	5,35	5,76	6,21
	6,39	5,85	5,21	4,45	3,42	2,43	1,51	75	30	15000	3,82	3,79	3,79	3,85	3,92	4,16	4,51	5,17	5,91
	8,07	7,47	6,74	5,88	4,73	3,63	2,53	1,64	1,05	17500	2,14	2,17	2,26	2,42	2,61	2,96	3,49	4,28	5,16
	9,42	8,79	8,02	7,10	5,90	4,77	3,61	2,67	2,03	20000	79	85	98	1,20	1,44	1,82	2,41	3,25	4,18
	10,21	9,59	8,87	8,00	6,84	5,75	4,61	3,73	3,18	22500	0	05	13	30	50	84	1,41	2,19	3,03
	10,21	9,64	9,00	8,30	7,34	6,42	5,48	4,75	4,38	25000	0	0	0	0	0	17	54	1,18	1,83
	10,21	9,64	9,00	8,30	7,34	6,59	6,00	5,55	5,46	27500	0	0	0	0	0	0	02	37	75
	10,21	9,64	9.00	8,30	7,34	6,59	6,02	5,92	6,21	30000	0	0	0	0	0	0	0	0	0
	10,21	9,64	9,00	8,30	7,34	6,59	6,02	5,92	6,21	32500	0	0	0	0	0	0	0	0	0
	10,21	9,64	9.00	8,30	7,34	6,59	6,02	5,92	6,21	35000	0	0	0	0	0	0	0	0	0
	10,21	9.64	9,00	8,30	7.34	6,59	6,02	5,92	6,21	37500	0	0	0	0	0	0	0	0	0
Z	10,21	9,64	9,00	8,30	7,34	6,59	6,02	5,92	6,21	40000	0	0	0	0	0	0	0	0	0

FIG. 5.

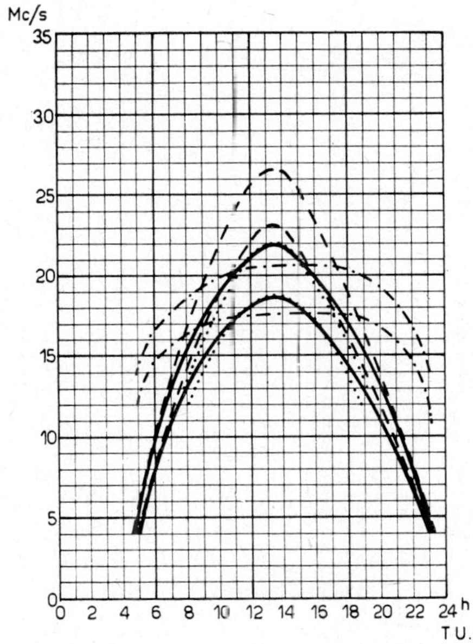

Absorption F .... Occultation —·—·—·—·
,, E — — — — Lut eff. ————

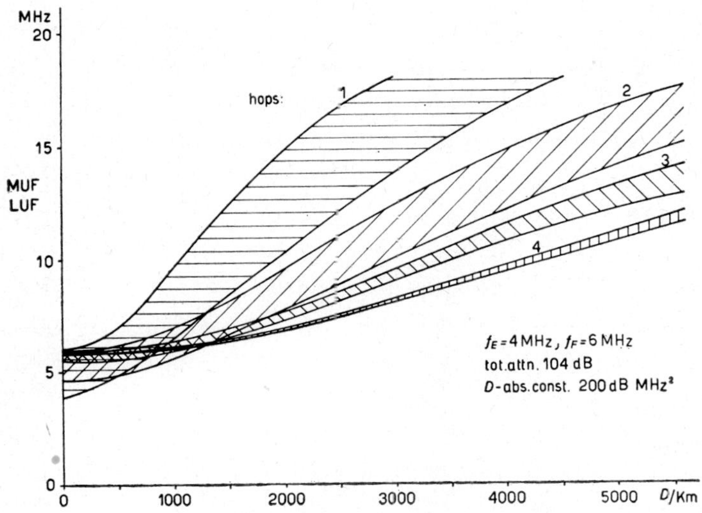

FIG. 6.

4. We may give further *details of the procedure*: From the Ocagne diagram we obtain the distance PA and BP; we then read from the table:

$$\sum_{P}^{A} \cos \tfrac{3}{4} \chi = Z_1$$

$$\sum_{B}^{P} \cos \tfrac{3}{4} \chi = Z_2$$

and

$$\sum_{P}^{P} \cos \tfrac{3}{4} \chi = Z.$$

Figure 5 as an example represents one page of the tables. Thanks to a simple nomogram representing Equation (1) the absorption for the distance BPA can be calculated with the above mentioned factors and the sum $Z_1 + Z_2$, the absorption for the distance AB with the difference $Z - (Z_1 + Z_2)$.

The blanketing frequency is derived in a similar way. The values of Max ($\cos \chi^{0,3}$) for the distance AB and BPA are also read from the tables, the transformation corresponding to Equation (2) is also done by a nomogram.

Figure 6 shows the lower limits for a calculated prediction. We always give two upper and lower limits viz. corresponding to a probability of 90 % or 30 %. We thus take account of the statistical fluctuation of basic data from one day to the other.

On Fig. 6 we see (for a given elevation angle of 2°):

(1) LUF for multiple F-hops, 3500 km (dotted curves).

(2) LUF for multiple E-hops, 2200 km (dashed curves).

(3) The blanketing limit for multiple F-hops (dot-dashed curves).

With respect to intermediate modes the following rules are used:

(a) If the F-LUF exceeds the blanketing limit the first determines the effective LUF.

(b) If the blanketing limit exceeds the F-LUF the effective LUF must approach more and more the E-LUF.

With these rules the effective LUF is easily determined (see, e.g., full lines on Fig. 6).

III. Experiences, Results, Comparison of Predictions with QSA-Values

Since September 1951 according to this method about 20 large-distance predictions with more than 10,000 km have been calculated every month. Most stations which received our predictions sent some QSA-results. Comparing them with our predictions we found a good correspondence in most of the cases.

1. In October 1951 calculations have been made for the *antipodal distance* Dakar-Nouméa which is rather near to the equator. For this nearly 20,000 km distance propagation is possible over quite different great circles and thus with quite different azimuth angles. Therefore we calculated a set of predictions with different azimuths which have finally been combined by superposition. Figure 7 gives the result of the combination. By intercomparison with the different basic

graphs also the variation of the optimum azimuth with hour and frequency is here obtained.

Our predictions have been tested by experience in this case. From 9 October, until 10 November 1951 the frequencies 11·4 have been hourly tested from 03 to 11 TU, QSA values have been noted. The average QSA values are also shown in Fig. 7 (dashed curves). Except for the very beginning of the transmission period the agreement between prediction and exception is highly satisfying.

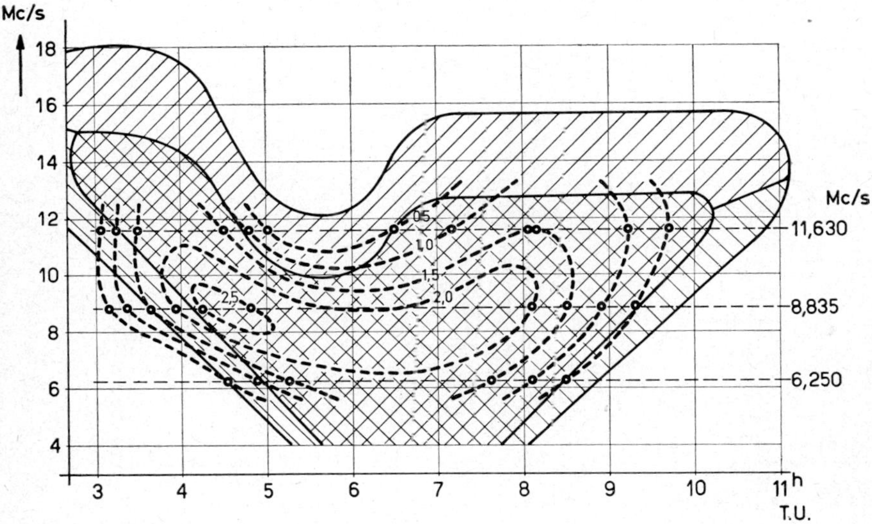

FIG. 7.

2. Another interesting feature of large distance traffic is the echo phenomenon. An echo is obtained when propagation is possible on both the forward and the backward great circle path. For these hours too round the world echoes are probable along the given great circle. For the distance Paris-Buenos Aires, October 1953, Fig. 8 gives a prediction for medium power over both great circle paths; the backward path is indicated with dashed curves.

The occurrence of these echoes has been investigated for a large number of distances by Mögel; he has published his results for the year 1927–29 (maximum of the solar cycle) and 1933–34 (minimum of the solar cycle). Figure 9 gives one example covering the period 1933–34. For the distance Buenos-Aires-Berlin the frequency of echo occurrence has been observed as a function of the hour and season. We see from Fig. 9 that, e.g., in the lower frequency range (5·5–9·8 MHz) echoes begin to appear about at 07.30 T.U., the maximum of occurrence is reached at 08.40 T.U., echoes disappear about 09.30 T.U. Our calculations have been made with the data for the last minimum of the solar cycle 1953–54, Mögels observations have been made 1933–34 also a minimum period. His results (black spots in Fig. 9 representing the reception probability) agree quite well with our calculations (hatched zones in Fig. 9).

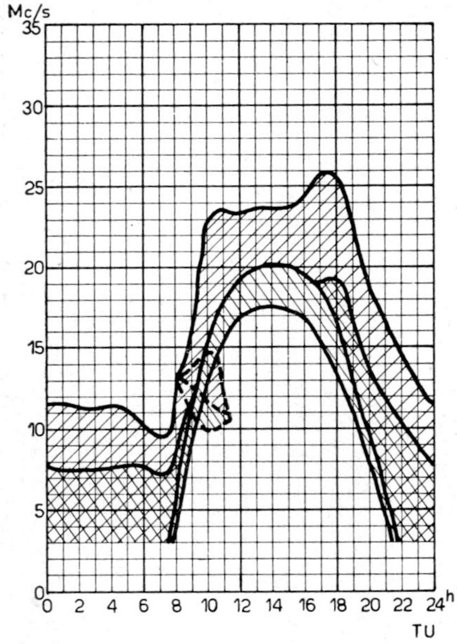

FIG. 8.

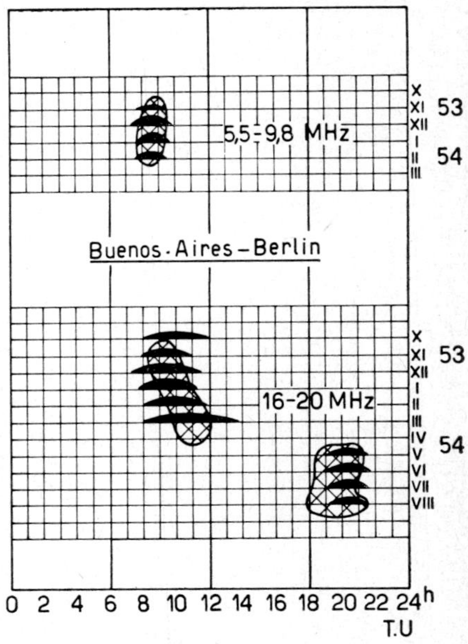

FIG. 9.

3. Some intercomparison with round the world echoes is made in Fig. 10(a) and (b). During his systematic investigation on such phenomena Hess in 1942–43 obtained many records. Figure 10 gives results obtained with a transmitter in

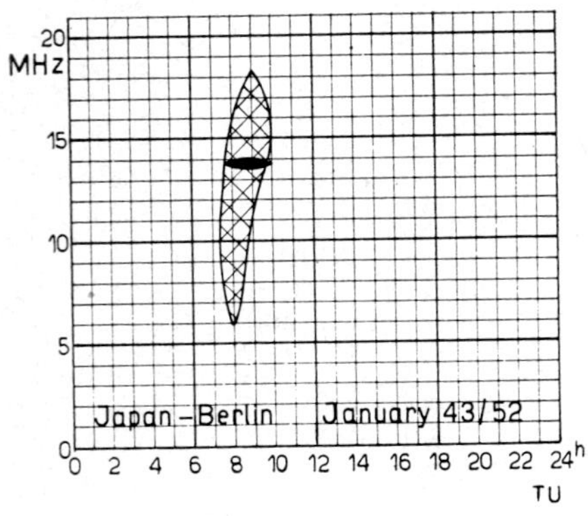

FIG. 10*a*.

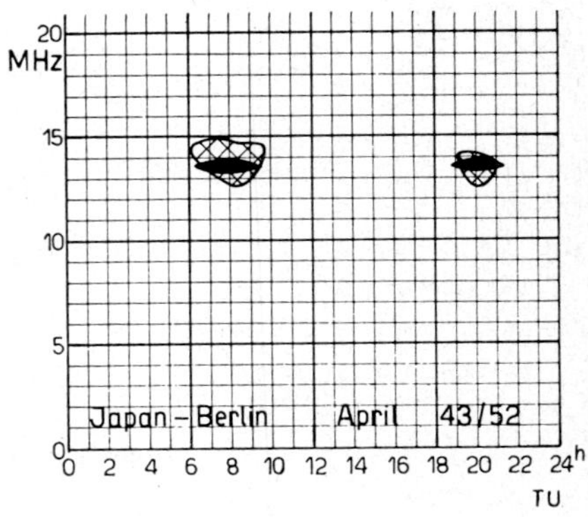

FIG. 10*b*.

Japan, the records have been made in Central Europe. He observed the frequencies between 13·5 and 14 MHz. We compare his results obtained about one year before a minimum of the solar cycle with our calculations for the year 1952, which has a similar position in the solar cycle. One can see from the example of

Fig. 10(a) January, 10(b) April, that the zones of echo occurrence in April are quite different from those in January, nevertheless the agreement between calculations and observations is highly satisfying.

In all these cases the results of our calculations have been verified by the experiments. We are aware that the basic data we have at our disposal have a rather poor accuracy seen all over the world. A more homogeneous distribution of basic ionospheric data all cover the world would greatly improve the calculations. On the other side more test experiments are highly desired. With respect to such tests we must finally make a statement concerning the antennae. Propagation is obtained with different azimuths on very large distances especially on those near 20,000 km. It follows that for propagation tests over such distances directive antennae are not at all desirable. A certain directivity for vertical angles with priority for small elevation angles can yet be admitted. At any case directivity in the horizontal plane should not be used for very large distance propagation tests.

References

1. Harnischmacher, E. (1953). *C.R. Acad. Sci. U.R.S.S.* **237**, 1071–73.
2. Rawer, K. (1952). *Wireless Engr* **29**, 287-301.
3. Rawer, K. (1953). *J. atmos. terr. Phys.* **3**, 123–124.
4. Rawer, K. (1954). Rep. of Phys. Soc. Conference on "The Physics of the Ionosphere", Cavendish Lab., Cambridge, 270–275.
5. Mögel, H. (1935). *Telefunkenztg* No. 67.
6. Hess, H. A. (1952). *Proc. Inst. Radio Engrs, N.Y.* **40**, 1065–1068.

Meteor Activity as a Factor in Ionospheric Scatter Propagation

G. A. Isted

Marconi's Wireless Telegraph Company, Ltd., Research Division, Great Baddow, Essex

Ionospheric scatter links for both experimental and communication purposes have been operating now for some years but, so far, no completely satisfactory theory has been advanced to explain all the observed propagation effects. Of the theories so far advanced, meteor activity and turbulence in the E-region are the most popular, and each is held by some to be self-sufficient to maintain the medium in the state necessary for the scattering of useful amounts of energy in the forward direction.

There can be no doubt that ionization by meteors does contribute to the general phenomenon of ionospheric forward scattering; indeed, much can be said for the theory that meteor activity may be responsible for a greater part of it. From time to time, however, certain characteristics have been observed which make the theory less plausible.

In recent years the author has suggested that weather-cloud discharges may be capable of producing bursts of ionization similar to those produced by meteors. Evidence has been found, for example, of groups of bursts having regular spacing between bursts; furthermore some of these groups are associated with thunder-cloud discharges in a manner which cannot be ignored. Interesting evidence has come to light which shows that the seasonal variation of relatively local lightning activity has clearly defined minima at each of the equinoxes; similar well-defined seasonal minima have been observed in the United Kingdom during independent investigations of ionospheric forward-scatter propagation—this despite the fact that the meteor theory would predict maximum ionization at the autumnal equinox.

One of the natural phenomena associated with ionospheric forward scatter transmission is the large number of audible Doppler beats which are present at most times. It has generally been accepted that they are produced, in association with a radio frequency wave, by meteors entering the earth's denser atmosphere. The careful analysis of these Doppler disturbances discloses the fact that there are as many which ascend in frequency as those which descend. The probability that the ascending type of beat could be caused by meteors has been investigated and the conclusion is reached that out of six Doppler disturbances only one should ascend.

Doppler disturbances have a similar diurnal distribution to that of transient echoes observed at vertical incidence on a frequency of 10 Mc/s, and it is suggested that they are each caused by a common mechanism not of meteoric origin.

Meteor activity may be found to play a relatively small part in ionospheric forward scatter propagation and it is possible that the weather-cloud discharge, or some other mechanism as yet unidentified, may instead prove to be the dominant factor.

I. Introduction

Communication by means of scattering processes in the E-region of the ionosphere is now well established, and invaluable communication circuits, operating in the lower VHF band, now exist between the United States of America and the United Kingdom, via intermediate relay stations situated in Greenland and Iceland; their performance outmodes all previous means of radio telegraph communication over routes which pass near or through the auroral zone. This is particularly apparent when magnetic disturbances render useless the orthodox modes of H.F. communication utilizing the F_2 reflecting processes while, at the same time, the ionospheric scattering processes are observed to be relatively immune to such disturbances.

Although satisfactory means have been found to make the best use of ionospheric scattering processes it cannot be said that agreement is unanimous upon the precise propagation mechanisms involved; of the theories so far advanced meteor activity and turbulence in the E-region are the most popular, and each is held by some to be self-sufficient to maintain the medium in the state necessary for the scattering of useful amounts of energy in the forward direction. Many of the less dogmatic of the theorists are in favour of a combination of both mechanisms in varying proportions.

There can be no doubt that transient echoes, caused by meteor ionization, contribute to the general phenomenon of ionospheric forward scattering; indeed, much can be said for the theory that meteor activity may be responsible for a major proportion of it. The principle characteristics in favour of the theory are the frequent burst-like character of the signal, the pronounced evening minimum and the systematic deviation of the angle of arrival with respect to the great-circle bearing—the latter two characteristics being dependent upon local time. From time to time, however, certain observations have been made which make the meteor theory less plausible and the object of this paper is to examine some of the unique characteristics of transient echoes in relation to the observed characteristics of ionospheric scatter propagation. Certain experiments will be described which have been carried out in an attempt to obtain a better understanding of the mechanisms involved in ionospheric scatter propagation.

II. Characteristics of Scattered Signals

A brief description of the salient characteristics of the scattered signal will enable the discussion which follows to be better understood. The outstanding feature of the signal in general is undoubtedly its extremely variable amplitude; it would appear to consist of several components, not all of which are present together:

(a) The Background Signal. This component varies rapidly in amplitude by some tens of decibels, with fading rates of the order of a few cycles per second. It is always present, but it nevertheless exhibits marked diurnal and seasonal variations.

(b) Transient Echoes or Bursts. Frequent sudden enhancements, with durations of the order of a few seconds, are superimposed upon the background signal. Some of these bursts are accompanied by Doppler frequency shifts which, together with a direct C.W. signal, give rise to a musical note ascending or descending in frequency.

(c) Long Bursts. Occasional enhancements of signal by some 30–40 db which last several minutes are observed on frequent occasions. It is believed that this type of burst is the result of drifting clouds of ionization, but just how these are produced is not clearly understood.

(d) Sporadic-E Activity. On occasions, the median signal level increases by some 60–80 db for periods lasting from several minutes to several hours; this is more likely to occur in the summer months during the hours of daylight, although it has also been observed at other seasons and during darkness. As its name implies it is sporadic both in time of occurrence and geographical situation.

(e) Seasonal and Diurnal Variations. The signal resulting from a combination of the components (a), (b) and (c) above shows a marked minimum of amplitude each day at about 18.00 hours local time. The maximum occurs sometime before noon but its exact location seems to depend upon the horizontal beamwidth of the combined transmitting and receiving aerials. Well defined seasonal minima have been observed [3, 7] at both equinoxes from transmissions along great circle routes running approximately North and South.

(f) Doppler Frequency-shift Effects. As stated in (b) above short enhancements of signals are often accompanied by a Doppler frequency shift which produces a musical note, either ascending or descending in frequency, when a C.W. signal is being observed. It has generally been supposed that this frequency shift is the result of a reflection from the moving head of a column of ionization, possibly produced by a meteor in its rapid flight through the atmosphere. It has been observed [8] that Doppler beat notes of this type have a well defined diurnal variation reaching a maximum at 02.00 hours and a minimum at 18.00 hours local time.

III. Examination of Unique Characteristics of Transient Echoes and Ionospheric Scatter

A. Height of Scattering Region

By general agreement it is considered that the mean height of the effective scattering region is somewhat below the normal E-layer, its mean height being about 86 km above ground. The height of this scattering region has been investigated by Bailey *et al.* [1] and Pineo [14] who, with a one-way pulse transmission, used the observed height of the tropospheric scattering region as their reference; later a go-and-return pulse technique was used.

The results of these oblique incidence experiments indicate that a continuous narrow stratum of ionization is present most of the time, and it is probably this stratum which sustains the continuous background signal. The theory that the background signal is the result of overlapping of many reflections from ionized meteor trails at diverse heights above the ground would therefore seem to have been proved untenable.

However, on the other hand, Eastwood and Mercer [5] as early as 1946 carried out a series of experiments using wartime radar installations operating on frequencies between 22 and 44 Mc/s; in these experiments interest was concentrated on "transient echoes" and from their description it is clear that they were similar to those echoes which investigators today would consider to be reflections from meteors. Eastwood and Mercer established beyond reasonable doubt that the transient echoes which they observed were concentrated in a well-defined stratum at a height of 85–86 km above the ground. Thus it is probable that the transient echoes observed by them constitute the continuous stratum observed obliquely by Bailey *et al.* Eastwood came to the conclusion that in many respects the characteristics of transient echoes were similar to those which would be expected from meteor activity; it is difficult, however, to understand why this activity should be concentrated in a narrow stratum at a height which would seem much too low for exclusive meteor effects.

From the foregoing it would seem there is strong evidence in support of the theory that the scatter background signal results from numerous transient echoes of the type observed by Eastwood; while there is some evidence associating transient echoes with meteor ionization, other evidence in this respect is not altogether convincing.

B. Association of Electrical Discharges with Transient Echoes

Wilson [17] in 1925 suggested that, as the height is increased above the ground, the electric force required to reach the critical break-down value decreases much more rapidly than the electric field due to a charged weather-cloud and, because of this, he predicted that weather-clouds would be sufficient to produce ionization at a height of about 80 km; furthermore, in extreme cases, a sudden discharge of the upper pole of a charge cloud to the upper atmosphere might occur, but this discharge need not be in the form of lightning.

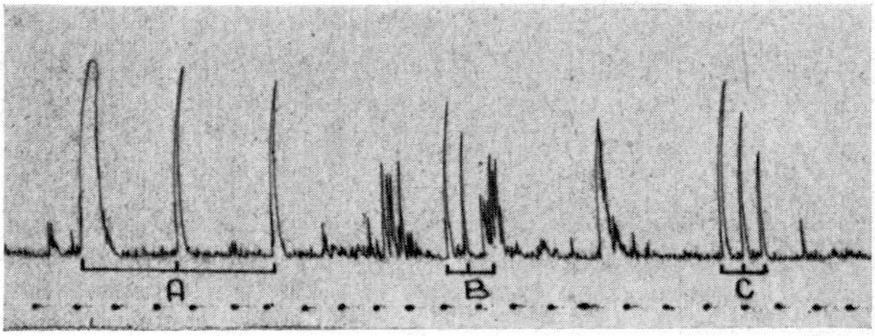

FIG. 1. Examples of regularly spaced bursts of ionization observed at a frequency of 53 Mc/s, 500 km from the transmitter.

Wilson [18] had previously suggested that the earth's vertical electric field was maintained by upwardly directed thunder-cloud discharges, and he considered that the E-region was the means by which the energy so discharged was transferred to areas of clear sky; the energy there leaks down to the ground by conduction through the atmosphere, and gives rise to the observed vertical electric field.

Pawsey [13], Bauer and Flood [2] and Graf [6] have all observed transient radio echoes from thunder-cloud discharges; Rumi [16], who conducted a radar investigation into meteor reflections, stresses the fact that many of the returns which he analysed cannot in fact be explained in terms of known meteor reflection characteristics. He states that they can be attributed to upwardly directed discharges from weather-clouds to the ionosphere.

The author [9, 10] has investigated the signals received from a high-power continuous-wave transmitter radiating at a frequency of 53 Mc/s. The transmitter was situated nearly 500 km from the receiver; at this distance the transient echoes, reputed to be caused by meteor activity, could be observed to the exclusion of the continuous tropospheric signal. When the signal was made to operate a high-speed recorder it was observed that groups of signal bursts frequently occurred in which the individual bursts were equally spaced in time from one another.

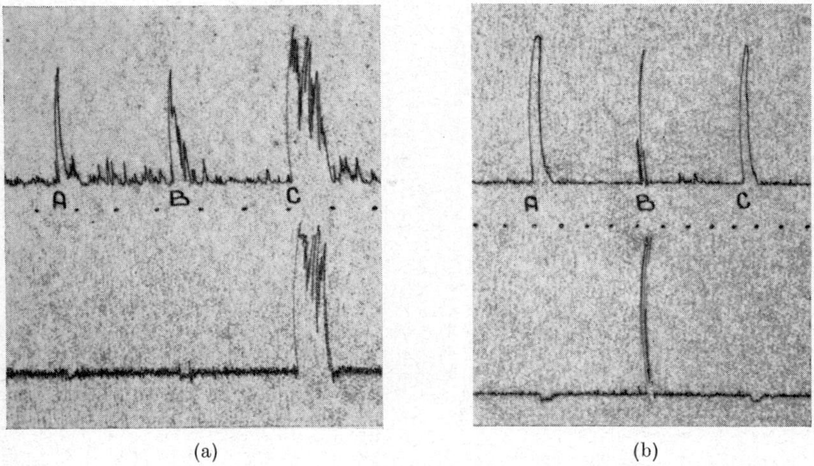

FIG. 2. Examples of the association of lightning with regularly spaced bursts of ionization.

Figure 1 shows such a record where groups of three equally spaced bursts can be seen clearly at A, B and C. In this record time advances from left to right and is marked off at second intervals along the foot of the record; increases of amplitude appear as upward deflections. A unique characteristic to be noted in all three groups illustrated is the progressive decrease in amplitude of the individual bursts in a group.

During relatively local thunderstorm activity the recurrent bursts of the type described appear to be associated with lightning. High-speed records of this effect were made with the aid of a twin channel recorder operated by two separate but similar receivers; one of them was tuned to the signal frequency of 53 Mc/s and recorded on the top trace; the other receiver was slightly detuned to an adjacent free channel and recorded on the bottom trace. Disturbances due to relatively local lightning flashes would record on both traces simultaneously, while the wanted radio signal would appear only on the top trace. Thus, from Fig. 2(a), it can be seen that radio signal bursts A and B precede a lightning flash C, the time spacings between AB and BC being equal. In Fig. 2(b) radio signal bursts A and C are equally disposed in time on either side of the lightning flash B.

It has been shown by Briggs [4] that among a random distribution of signal bursts grouping and regularity in spacing—within given limits—would be expected on a statistical basis, nevertheless the unique characteristics described above, and the growing number of reports which associate transient echoes with weather-cloud discharges, suggest that all meteor-like bursts of ionization may not in fact be due to meteors.

C. Association of Electrical Discharges with Seasonal Variations of E-region Scatter

In the preceding Section the possible instantaneous effects of weather-clouds discharging to the E-region were considered; the long term trends in this respect

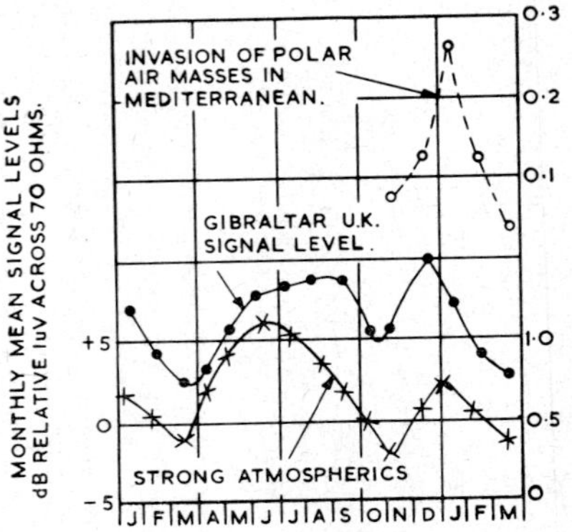

FIG. 3. Seasonal variation of strong atmospherics and polar air masses in the Mediterranean (Reiter) compared with mean signal levels observed from scatter transmissions (Gibraltar–United Kingdom).

—·—·—·—	Gibraltar–U.K. signal level at 37 Mc/s.	
—x—x—x—	Number of strong atmospherics at 4–12 kc/s.	Reiter.
—o—o—o—	Polar air masses in Mediterranean.	

are also worthy of note. In this connection it is interesting to compare the long-term measurement of atmospherics made by Reiter [15] in Munich, with the long-term measurement of signal level carried out by Bray *et al.* [3] and by Isted [7], during separate investigations of ionospheric scatter propagation.

Reiter counted the number of impulses due to lightning occurring within a radius of 2000 km, which he observed in the frequency range 4–12 kc/s. His results showed that over a period of 5 years two well-defined minima and two maxima occur per year; the first maximum occurs in the summer months and can be explained by the frequency of thunderstorms in Central Europe. The second maximum occurs in the winter and is not so easily explained but, in his analysis of the main weather processes, Reiter found that during the winter months violent invasions of polar air-masses into the Mediterranean area take place and form moist unstable gradients over the relatively warm sea which give rise to thunderstorms.

Reiter's results are reproduced in Fig. 3. The curve labelled "Strong Atmospherics" represents the seasonal variation averaged over a period of 5 years; the summer and winter maxima are clearly seen. In the top right-hand part of the figure the number of observed invasions of polar air into the western Mediterranean during the winter months is shown; it would seem that the explanation given by Reiter for the winter maximum of atmospherics is well founded. The centre curve represents the seasonal variation of the mean signal level observed during an investigation carried out into the propagation characteristics of ionospheric forward scatter between Gibraltar and the United Kingdom, at a frequency of 37 Mc/s [7, 11]. The similarity between Reiter's "Atmospheric" curve and the mean signal level of the ionospheric scatter signal is impressive. More impressive still, however, is the comparison shown in Fig. 4 of Reiter's atmospherics observation with the seasonal variation of mean signal level for the same period

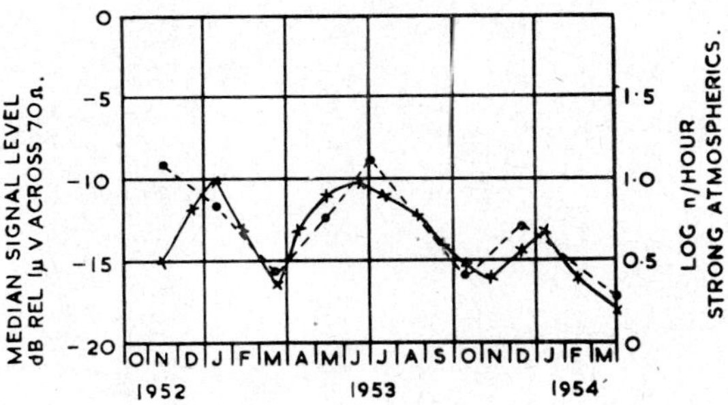

FIG. 4. Seasonal variation of strong atmospherics (Reiter) and median signal level of the Shetlands-Slough scatter transmissions at 41 Mc/s (Bray *et al.*)

—x—x—x— Number of strong atmospherics (Reiter).

.—·—·—·— Scatter median signal (Bray *et al.*)

observed by Bray *et al.* [3], during the Shetlands-Slough ionospheric scatter tests carried out at a frequency of 41 Mc/s.

The similarity in trend between these independent observations indicates the possibility that weather cloud discharges might, in fact, have a major influence on received scatter signal intensities.

It should be noted from these results that, in the United Kingdom at least, the intensity of scatter signals is a minimum at each of the equinoxes while the meteor theory would predict that the highest signal levels should occur at the autumnal equinox.

D. Relation of Transient Echo Activity to Ionospheric Scatter Signal-levels

Many of the signal bursts observed in scatter propagation certainly behave as though they had been produced by meteor activity but, so far as the author is aware, no one has yet related scatter signal levels to an independent measurement of meteors. Probably a test which would inspire a degree of confidence would be one which was largely self-checking. Such an experiment has been attempted. It was assumed that two adjacent and independent routes having similar great circle bearings will be affected equally by sporadic type meteors, and that the influence of general meteor activity observed on one route would be observed simultaneously on the other.

Two suitable test paths were available near Chelmsford, England; one long ionospheric scatter path to Gibraltar in the south, and a short path to the B.B.C. Television Transmitter at Kirk-o-Shotts, Scotland in the north. Bursts of signal from the Kirk-o-Shotts transmitter, reputed to be caused by meteor ionization, could be observed to the exclusion of the continuous background; of these signal bursts, those which exceeded a given amplitude were counted by means of an electro-mechanical device. The Gibraltar ionospheric scatter signal was integrated over a period of 20 sec and recorded; the mean level was then estimated by eye.

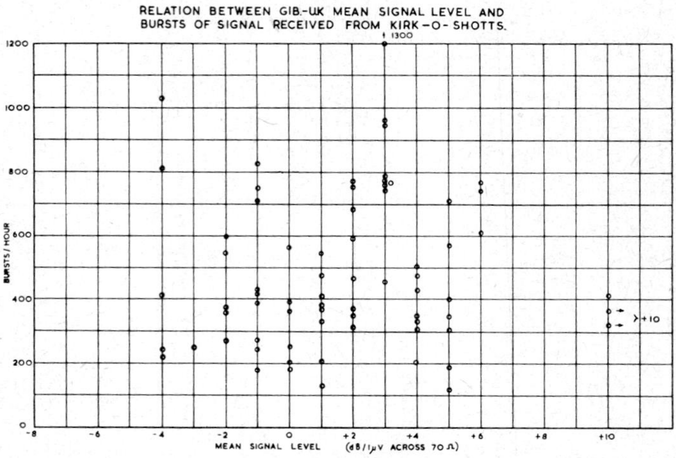

FIG. 5. Relation between the Gibraltar–United Kingdom scatter signal levels and transient echoes received from Kirk-o-Shotts.

Figure 5 shows the number of signal bursts received per hour from the Kirk-o-Shotts transmitter, plotted against the mean signal level received from Gibraltar for the appropriate time and day.

A cursory glance is sufficient to show that the straight-forward correlation sought for was not found. This lack of correlation could imply that the Kirk-o-Shotts signal bursts and the Gibraltar scatter signal are each influenced by a different mechanism or, alternatively, that each may be influenced by a common mechanism which is relatively local in effect—such as weather-cloud discharges. In any event, from this test, it would seem that the contribution of sporadic type meteors to long distance scatter propagation is very small indeed.

E. Transient Echo Activity at Vertical Incidence

When investigating the behaviour of scatter propagation from the E-region at oblique angles of incidence it is appropriate to give some consideration to the equivalent frequency at vertical incidence. Most of the data concerning ionospheric scatter refer to the lower end of the VHF band; therefore the equivalent frequency at vertical incidence is in the region of 10 Mc/s.

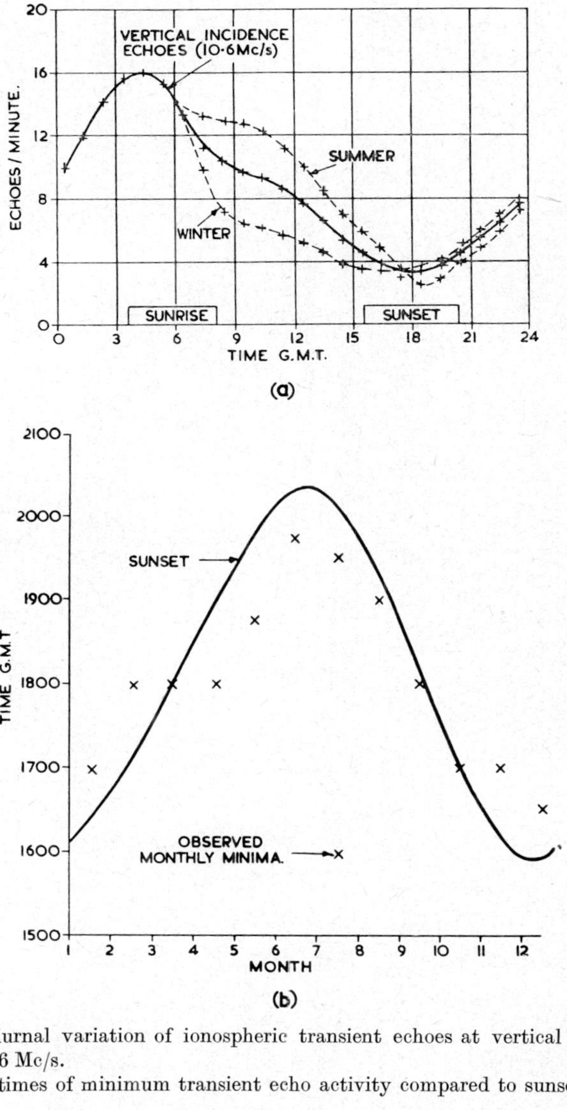

FIG. 6. (a) Diurnal variation of ionospheric transient echoes at vertical incidence on a frequency of 10·6 Mc/s.

(b) Observed times of minimum transient echo activity compared to sunset times.

Observations of transient echo activity were made by the Marconi Company in the years 1944–45 by means of an ionospheric sounder. The transmitter, operating at a fixed frequency of 10·6 Mc/s, was modulated by pulses of 100 μsec duration, at a repetition rate of 25 per sec. The peak-pulse power radiated from a half-wave horizontal dipole situated $\lambda/4$ above ground, was 25 kw. A rhombic aerial, the main lobe of which was directed vertically, energized a suitable receiver the output of which was connected to an "A scan" cathode ray oscilloscope; the transmitted pulse was arranged to trigger the oscilloscope time-base, and a continuously

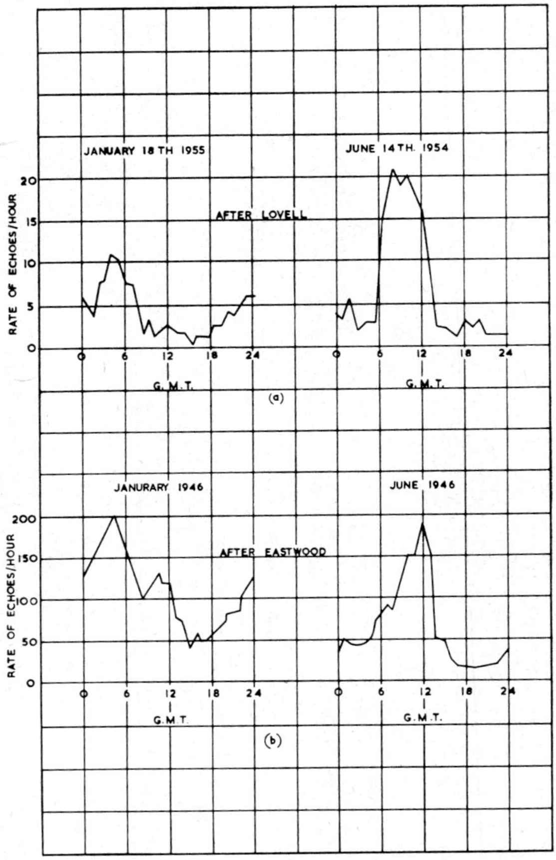

FIG. 7. Seasonal variation of transient echoes as measured. (a) by Lovell and (b) by Eastwood and Mercer.

moving film through a camera recorded the information given by the system. A 5-minute sample was recorded at hourly intervals over a period of nearly 2 years. The complete analysis of the diurnal distribution of the transient echo activity observed under these conditions is presented in a simplified form in Fig. 6(a) where the diurnal distributions are embodied in three curves—winter, summer and yearly mean.

Two outstanding points of importance are illustrated by the curves; firstly, the time of maximum activity throughout the seasons is constant at 04.00 hours G.M.T.

and secondly, the minimum of activity is correlated with sunset time; the latter effect is shown in greater detail in Fig. 6(b), where the calculated sunset times for each month are compared with the observed times of minimum transient echo activity. These observations do not altogether agree with similar observations made by Eastwood and Mercer [5] at frequencies between 22 and 44 Mc/s and by Lovell [12] at a frequency of 70 Mc/s.

Lovell has reported that the analysis of his observations has shown a distinct seasonal variation; this, he claims, is due to the significant effect of day-time meteor streams. He shows that the diurnal distribution for January, observed on an aerial array rotating twice per minute, has a maximum activity at about 04.00 hours G.M.T., while in June the maximum activity was observed at about 10.00 hours G.M.T.; moreover twice the number of echoes were observed during daylight in June than in January during night-time.

The results obtained by Eastwood and Mercer agree very well with those of Lovell except that they observed the same degree of activity throughout the year. The diurnal distributions for the months of January and June observed by Lovell are reproduced in Fig. 7(a) and, for comparison, those observed by Eastwood and Mercer in Fig. 7(b).

Lovell took over a quarter of a million echoes into account in his analysis and the reality of his results must be accepted; however, more than half a million echoes were included in the analysis of the Marconi investigation at 10 Mc/s, and the results obtained must be accepted as equally real. Considering that one set of observations was made at a frequency of 70 Mc/s and the other at 10 Mc/s, it seems likely that two entirely different phenomena, giving similar manifestations, are observed at the two frequencies in varying proportions.

E. Meteor Activity and Doppler Frequency-shift Effects

One of the interesting phenomena of ionospheric scatter transmission over long distances is the large number of audible Doppler beats which can be heard in the form of musical tones descending or ascending in frequency. These can be heard in varying numbers throughout the day and night.

In the past it has been believed that the Doppler beats observed are the result of meteors which are entering the earth's atmosphere from outer space. The audible beat is believed to be caused by a direct radio wave mixing with a wave which is reflected from the moving head of a column of ionization; the ionization is thought to be produced upon combustion of the meteor in its rapid flight into the earth's denser atmosphere.

An analysis has been made of the Doppler beat activity observed during the investigation into ionospheric scatter propagation between Gibraltar and the United Kingdom [7, 8, 11], at a frequency of 37 Mc/s. Samples of the Gibraltar signal were taken by arranging a magnetic tape to record the audio output from the receiver for a period of one minute at half-hourly intervals; the receiver, having a band-width of 1000 c/s, was used without the BFO. The samples were later analysed by playing back the record and counting the number of beats which could be heard above a certain threshold value. The analysis disclosed that there were two main types of beats—those which ascended in frequency and those which descended—and, furthermore, that the two types were present in approximately equal numbers.

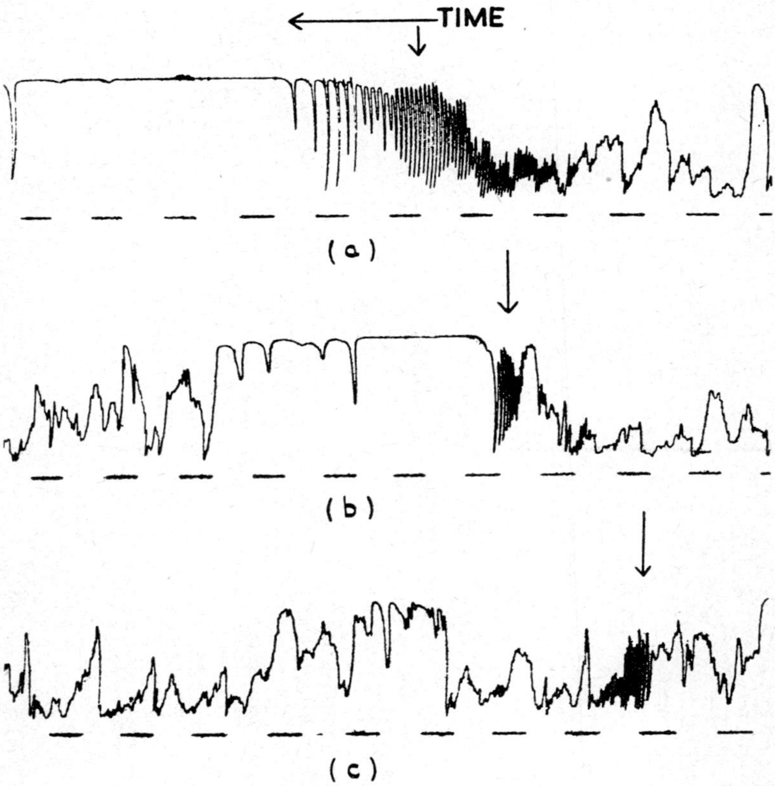

FIG. 8. Examples of Doppler beats.
(a) Long descending beat. (b) Short descending beat. (c) Short ascending beat.

Beats descending in frequency usually became audible between 1000–500 c/s and fell in frequency until they became inaudible below 50 c/s. Ascending beats, on the other hand, became audible at a frequency of about 50 c/s and rose to about 500 c/s. The low frequency limit of audibility of 50 c/s was imposed by the combined frequency response of the electronic equipment and the human ear. To observe frequencies below 50 c/s, however, a high-speed stylo-pen recorder was used; typical records are shown in Fig. 8 where time advances from right to left, and vertical deflections correspond to increases of amplitude. Second time-marks are recorded at the bottom of each record. In Fig. 8(a) an unusually long descending beat is indicated by the arrow; the beat ends within a cycle or two of zero beat, and is followed by a period of greatly enhanced signal. This is typical of the behavior of a meteor. A more common type of descending beat is shown in Fig. 8(b) while an equally common type, but ascending in frequency, is shown in (c). A unique characteristic of the ascending beat is the sudden commencement at, or very near to, zero beat.

The diurnal distribution of both types together shows that the maximum number occur between the hours of 02.00–04.00 G.M.T., and the minimum number at around 18.00 G.M.T. A typical diurnal distribution—that for August—is shown in Fig. 9(a), where it is seen that there is a maximum rate of occurrence of about 35 per min, and a minimum rate of 3 or 4 per min.

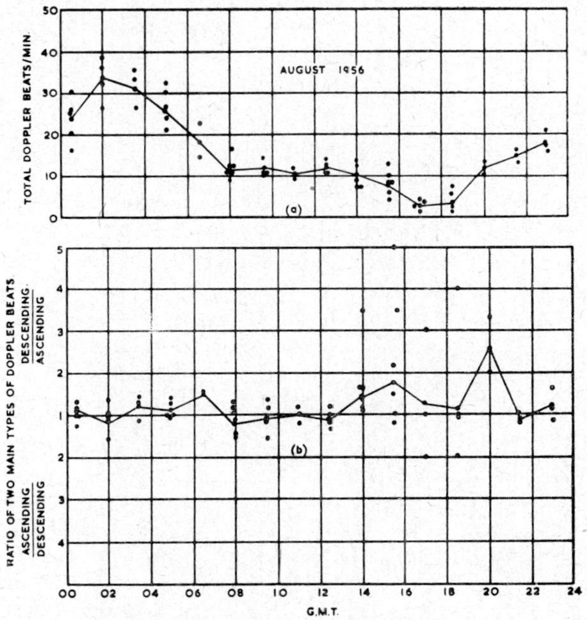

FIG. 9. (a) Diurnal variation of Doppler beats.
(b) Ratio of descending to ascending beats for August 1956. Frequency 37 Mc/s.

The ascending and descending types of Doppler beat were counted separately and the ratio of one to the other is shown in Fig 9(b).

Assuming that beats are caused by sporadic meteors the probability of an ascending beat being produced was investigated, and the number of meteors producing this effect deduced as a fraction F of the total number observed; the assumptions made as regards the average velocity of the meteors, the height at which reflections occur and the lower limit of audibility (50 c/s), were chosen with a view to making the fraction F nearest to unity. Under these conditions the predicted value of F is 0·17, i.e., one ascending beat for every six which descend. Furthermore, the fraction F can be shown to be inversely proportional to the relative velocity of meteors with respect to the scattering region; the velocity of sporadic meteors relative to the earth is determined largely by the earth's orbital velocity, components of this velocity being added to, or subtracted from, the meteor velocity depending upon the earth's rotational position with respect to its orbit. The average relative velocity of sporadic type meteors should therefore reach a maximum at 06.00 hours local time, i.e., when the observing point is on the crown of the leading hemisphere considered in relation to the earth's orbit. Minimum relative velocity should occur at 18.00 hours local time. A diurnal distribution of the fraction F would therefore be expected. In terms of the ratio of ascending to descending Doppler beats, therefore, the diurnal variation should show a maximum at 06.00 hours, and a minimum at 18.00 hours local time.

Reference to Fig. 9(b) will show that this situation is far from being fulfilled, in fact, if any real deviation exists in that direction, it is at 18.00 hours and not at 06.00 hours as predicted; it is believed, however, that the actual deviations observed were caused by errors in counting because, at that time, the signal/noise ratio was low and the number of observable beats was small.

The fact that the analysis showed the ratio of descending to ascending beats to be only slightly greater than unity, and the fact that no systematic diurnal variation in this ratio was present, is probably significant evidence that a mechanism other than meteors alone must be producing the beats. Having made this point it is interesting to compare the diurnal distribution of Doppler beats shown in Fig. 9(a) with the diurnal distribution of transient echoes observed at vertical incidence shown in Fig. 6(a). The similarity of the diurnal distributions shown in the two curves very strongly suggests that they each represent a manifestation of a common mechanism; the evidence does not, however, suggest that the mechanism is of meteoric origin.

IV. Conclusion

From the foregoing examination of the unique characteristics of transient echoes and ionospheric scatter it is clear that further investigation will be needed before it can be said that propagation by the ionospheric forward scatter mode is completely understood.

There would seem to be strong evidence in support of the view held by many investigators that an ionospheric scatter signal is produced, to a large extent, by the overlapping of many transient bursts of ionization which occur at a height of about 86 km above the earth. The origin of these transient bursts of ionization is, however, by no means certain; while there is no doubt that ionization is produced by meteors, some very pertinent questions must be answered by those who hold the view that the origin is wholly meteoric.

Why, for example, is the effective scattering region concentrated in a narrow stratum situated at an average height of 86 km?

Is there, in fact, a strong connection between weather-cloud discharges and transient echoes, or are we to believe that, given suitable conditions, a meteor will itself initiate an electrical discharge? The electrification of clouds is one of the greatest of terrestial natural phenomena. Is it so inconceivable that it should profoundly affect the state of ionization in the E-region through which the resulting discharged energy is known to be dissipated?

It is possible that the ascending Doppler beats which are observed on ionospheric scatter signals provide an important link with upwardly directed weather cloud discharges. If these discharges do take place could they not cause a very intense pocket of ionization which would rapidly expand in the manner of a shell-burst, and give rise to the observed ascending Doppler beat frequency? It is certainly difficult to explain the ascending Doppler beats in terms of meteoric activity, particularly when the beat commences at zero frequency.

In the examination of some of the unique characteristics of transient echoes and ionospheric scatter propagation, considerable evidence is found which is inconsistent with the theory that meteor activity is an important factor in ionospheric scatter propagation. The author has come to the conclusion that meteor activity may in fact be found to be relatively unimportant, but it is possible that the weather-cloud discharge, or some other mechanism as yet unidentified, may instead prove to be the dominant factor.

Acknowledgments

The author wishes to thank the Engineer-in-Chief, Marconi's Wireless Telegraph Company Limited for permission to publish the paper. The author gratefully acknowledges the assistance given by many of his colleagues in making some of the observations discussed in the paper. The ionospheric scatter investigation mentioned in the paper was sponsored by the Admiralty.

References

1. Bailey, D. K., Bateman, R. & Kirby, R. C. (1955) *Proc. Inst. Radio Engrs, N. Y.* **43**, 1181.
2. Bauer, L. H. & Flood, W. A. (1957). *Proc. Inst. Radio Engrs, N.Y.* **45,** 1743.
3. Bray, W. J., Saxton, J. A., White, R. W. & Luscombe, G. W. (1955). *Proc. Instn elect. Engrs* **103**B, 236.
4. Briggs, B. H. (1956). *J. atmos. terr. Phys.* **8,** 171.
5. Eastwood, E. & Mercer, A. K. (1948). *Proc. phys. Soc. Lond.* **61,** 122.
6. Graf, C. R. (1958). *Proc. Inst. Radio Engrs N.Y.* **46,** 915.
7. Isted, G. A. (1958). *Proc. Instn elect. Engrs* **105**B, Supp. no. 8, p. 36.
8. Isted, G. A. (1958). *Proc. Instn elect. Engrs* **105**B, Supp. no. 8, p. 27.
9. Isted, G. A. (1954). *Marconi Rev.* **17,** 37.
10. Isted, G. A. "The Physics of the Ionosphere," Report of 1954 Cambridge Conference, p. 150.
11. Kitchen, F. A. and Millington, G. (1958). *Proc. Instn elect. Engrs,* **105**B, Supp. no. 8, p. 2.
12. Lovell, A. C. B. (1956). *J. atmos. terr. Phys.* **9**, 229.
13. Pawsey, J. L. (1957). *J. atmos. terr. Phys.* **11**, 289.
14. Pineo, V. C. (1956). *J. geophys. Res.* **61,** 165.
15. Reiter, R. (1956). *J. geophys. Res.* **61,** 23.
16. Rumi, G. C. (1957). *J. geophys. Res.* **62,** 547.
17. Wilson, C. T. R. (1925). *Proc. phys. Soc. Lond.* **37,** 32D.
18. Wilson, C. T. R. (1920). *Phil. Trans. A,* **221,** 73.

Diurnal Influences in Tropospheric Propagation

M. W. Gough

Marconi's Wireless Telegraph Company, Ltd., Research Division, Great Baddow, Essex

It has been appreciated since about 1930 that wave propagation in the metric, decimetric and centimetric bands can be profoundly influenced by the troposphere. Spatial and temporal variations in the atmospheric refractive index amounting to a few parts in a million can promote large changes in received signal strength, which because of their scientific and engineering importance have received much study. In the present paper attention is confined to marked diurnal variations in signal strength attributable to the nightly occurrence of tropospheric inversion layers promoted by the cooling by radiation of heated ground when the night sky is clear. While this effect has generally been found to be most clearly defined over land paths well removed from maritime influences, some striking examples have nevertheless been observed over coastal paths in tropical and sub-tropical regions.

After a brief review of observations by various workers of this phenomenon in the metric, decimetric and centimetric bands, the paper appraises 80 Mc/s signal strength recordings maintained continuously for 6 months over a 137 km non-optical path in the Persian Gulf, where the effect persistently occurred to a noteworthy degree. The test path lay mostly over land, between Bahrain Island and Doha on the Qatar Peninsular. The terminal aerial heights were respectively 60 m and 20 m, and the transmitted power was about 20 W C.W.

Throughout the test period, which covered September to February inclusive, 1955–56, there was a marked tendency for the signal strength to be weakest between mid-day and 17.00 hours, local time, and strongest between midnight and 05.00, when the signal level, particularly in September, frequently approached the free space figure. In that month, which exhibited the most pronounced variations of the test period, the night-time level sometimes exceeded the daytime level by as much as 40 db. By contrast in January, when diurnal variations were erratic and least pronounced, the disparity between day and night-time levels virtually disappeared for a time. In February diurnal variations became rapidly intensified once again.

A noteworthy feature of many days was a rapid drop in signal strength from its high nocturnal level at about 09.00–10.00 hours local time. Thereafter a low level, accompanied by some scintillation, usually persisted to about 17.00 hours after which the signal strength regained its nocturnal value fairly rapidly, and except for occasional deep fades maintained this level until once again it dropped abruptly next morning. Samples of signal strength records illustrating these effects are presented, and statistics of diurnal and seasonal variations are touched on.

Modified refractive index profiles, derived from selected meteorological soundings at Bahrain, have shown the association between (a) very weak signals and an approximately standard atmosphere, and (b) very strong signals and pronounced elevated or ground-based inversions.

Theoretical and measured signal strengths are compared on the basis of various idealizations suited to the M profiles investigated. These include curved earth diffraction within a uniformly graded atmosphere, reflection at an idealized elevated layer, and propagation within a duct as treated by the transmission mode theory of Booker.

For the non-optical path under discussion, disparities between theoretical and measured signal strengths indicate that the effects of scattering, and partial or total reflection at grazing incidence from weak elevated layers, are often dominant. Other discrepancies serve to emphasize the limitations inherent in the concept of a uniform M gradient to infinity when surface super-refraction is approaching the "flat earth" value.

Finally, an investigation of 19 ground-based M inversions associated with very high signal strengths indicates 12 cases where the first transmission mode was trapped on the relatively long wavelength of 3·75 m. In two of these cases the second transmission mode was just trapped also. Very high M deficits, reaching 100 M units in one instance, were found in the ducts causing trapping. These observations, made in exceptional climatic conditions, provide evidence of recurrent nocturnal trapping on a wavelength which is perhaps longer than has hitherto been considered possible.

I. Introduction

Since about 1930 it has been realized that wave propagation in the metric, decimetric and centimetric bands can be profoundly influenced by the troposphere, by virtue of changes in its refractive index amounting to only a few parts in a million. In the present discussion attention will be directed to diurnal variations in signal strength which, in favorable circumstances, can be a prominent feature of very short-wave propagation. These variations have an important influence on the engineering and operation of VHF, UHF and SHF radio links.

The salient characteristics of this phenomenon were first observed and satisfactorily explained by Englund, Crawford & Mumford [8] in 1940. In prolonged propagation tests of an optical overland path near New York they observed, on wavelengths of 4 m and 2 m, a marked tendency, particularly in summer, for the signal strength to remain nearly constant during the morning and afternoon, while becoming extremely variable between the hours of 18.00 and 09.00. These nocturnal disturbances, which intermittently gave rise to signal strengths much above the steady daytime level, were attributed to disturbing reflections from elevated inversion layers, in conjunction with variable ray bending near the ground which modified the diffracting effect of the earth's surface.

The diurnal nature of these variations is due to the nightly formation and subsequent dispersal of ground-based or elevated inversion layers promoted by the cooling by radiation of heated ground when the night sky is clear. As will be more fully explained later, the formation of these layers accounts for the nocturnal signal strength behavior often observed. An hour or more after sunrise, atmospheric stratification is usually swept away by convention, after which a well mixed atmosphere normally persists until late afternoon, promoting during this period a relatively steady signal whose magnitude is fairly compatible with a standard atmosphere. Somewhat before sunset atmospheric stratification develops once more, and the nocturnal period of variable propagation is re-established.

The characteristic diurnal behavior outlined above results from a meteorological structure peculiar to heated land, and indeed diurnal variations are virtually absent with oversea paths unless lying near the coast. This is because the sea temperature scarcely varies throughout the 24 hours, so that exclusively nocturnal temperature inversions do not develop.

The rapid development of very short-wave techniques since the early war years has resulted in many observations of diurnal variations on wavelengths ranging between 4 m and 3 cm. To cite a wartime example given by Durst [7], 200 Mc/s radars in Bengal frequently exhibited during the night and early morning a

5-fold increase in range over the daytime figure. Since the war, microwave measurements made over very diverse overland paths have shown in varying degree but with noteworthy similarity of characteristics, the type of diurnal variation described. Measurements in Great Britain have been described by Smith-Rose & Stickland [12] and by Bray & Corke [3], and in the U.S.A. by Durkee [6], Day & Trolese [5] and by Millar & Byam [10]. Vecchiacchi [13] and Carassa [4] have given diurnal statistics for microwave paths in Italy, while Vikramsingh *et al.* [14] have reported comparable influences over a path in the Himalayan foothills. Gough [9] has described diurnal variations over VHF paths in Malaya, Nigeria and Ghana.

Diurnal variations are usually more pronounced in summer than in winter, because higher temperatures increase the humidity lapse rate, thereby increasing the refractive index gradient. For this reason diurnal effects tend to be more marked in tropical regions than in temperate zones. To give an example of the magnitudes likely to be experienced, conventional radio paths selected for VHF radio links in Nigeria, Ghana and Kuwait have been found to give nocturnal signal strengths often 20–25 db above the steady daytime level on a frequency of 180 Mc/s. These occasions are however often interrupted by deep fades.

II. General Description of the Bahrain-Doha Test

To present a clear picture of diurnal effects, the rest of this discussion will be devoted to an analysis of 6 months continuous recording on 80 Mc/s (3·75 m) of a non-optical radio path between Bahrain Island on the Persian Gulf, and the town of Doha 137 km to the south-east. A small fraction of the path crosses the sea and the path remains near the Gulf throughout its length. The proximity of the sea affects the propagation mainly it is thought, by allowing weak inversion layers to persist into the daytime hours which—if the land influence were complete—would be destroyed by daytime convention. In confirmation of this conjecture it has been observed that paths further inland in neighboring Kuwait show remarkably steady daytime signals on 180 Mc/s, testifying to the improved daytime mixing of the atmosphere promoted by the absence of maritime influences.

For the test under review, an 80 Mc/s C.W. transmitter on Bahrain Island supplied 20 W to a 4-element Yagi aerial mounted at a height of 60 m and directed at Doha. A similar aerial at Doha, mounted at a height of 20 m, fed a receiver operating a recorder whose deflection was nearly proportional to the received signal level expressed in decibels above 1μV. The recorder sampled the signal strength once a minute, some of the intervening time being occupied in automatic calibration of the receiver. Except for a few breakdowns, signal strength was continuously recorded from 27 August 1955 to 1 March 1956, so that nearly 4000 hours of effective recording were secured.

III. Features of Signal Behavior

Figure 1 shows, by means of vertical bars, the maximum and minimum signal strength observed in any hour, indicated in local time, while the inter-connecting lines show the hourly median signal strength. The upper chart shows typical diurnal variations observed during September while the lower chart, which is characteristic of January, shows a random pattern with no discernible diurnal variations. September showed the largest variations of the 6-month test period. The difference between mean daytime and night-time levels often amounted to

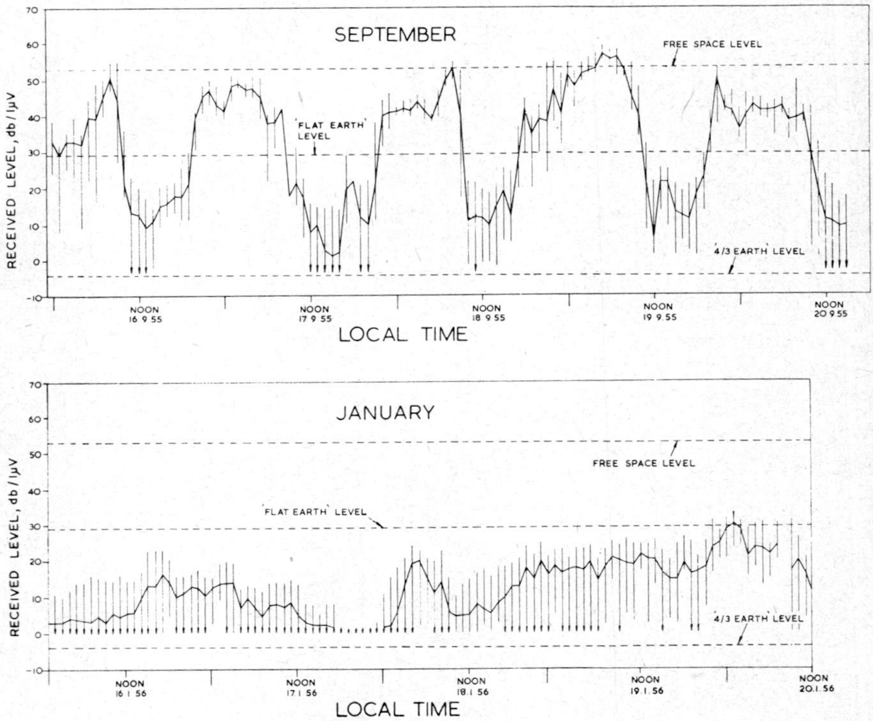

Fig. 1. Hourly signal ranges, 80 Mc/s, Bahrain–Doha.

40 db, and diurnal variations were repeated with remarkable regularity in that month. Well defined variations persisted throughout October, interspersed with occasional erratic periods. During November and December the irregularities increased, but even in the latter month distinct diurnal cycles were apparent for a few successive days. During January diurnal patterns were only occasionally perceptible. February and early March, while very erratic, occasionally showed some very well defined diurnal changes.

Figure 2 shows samples of signal strength records. The upper trace, obtained in September, shows a typical nocturnal disturbance with fading from a high level, resulting probably from interference between strong multiple reflections at grazing incidence from an elevated inversion layer. The decline after 09.00 hours local time from the free space level to that consistent with a standard atmosphere reflects the decay of the inversion due to increasing convection. The lower trace, obtained in October, shows an early morning period with a very high level devoid of fading, followed at about 10.00 hours by a very abrupt drop in signal strength symptomatic of a rapidly dispersing nocturnal inversion. The fluctuations immediately following the fall indicate interference between a partial reflection

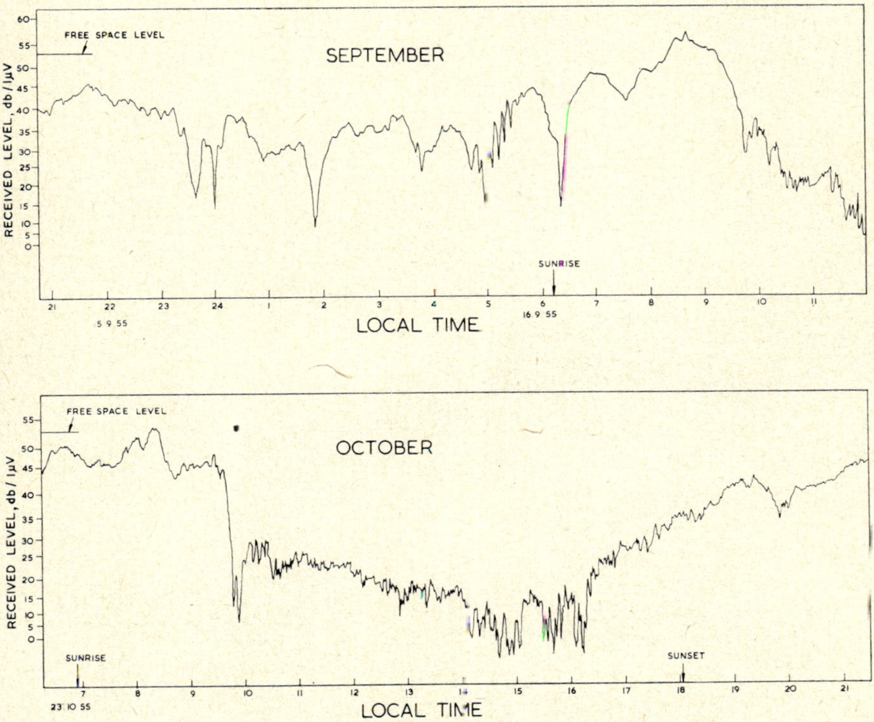

FIG. 2. Illustrative signal records, 80 Mc/s, Bahrain–Doha.

from the dissolving inversion layer and a diffracted wave propagated near the earth's surface. Thereafter until about 14 00 hours, the signal pattern implies a substantially unstratified atmosphere exhibiting steadily declining refraction. After 14.00 hours there is significant interference from a weak reflection at a developing inversion layer which, after 16.00 hours, is providing the dominant signal.

The upper trace on Fig. 3 shows a typical January record. Refraction near the surface appears to be slightly greater than standard in daylight hours, and the resulting weak diffracted signal is very vulnerable to partial reflections from weak inversion layers. There is a small increase in atmospheric stratification and sufače refraction during the evening. Finally, the lower trace shows a diurnal cycle obtained in February. It comprises firstly, a steady early-morning period when the received signal is derived mainly from a near-specular reflection at an elevated inversion. Secondly, there is a mid-day period consistent with a well mixed standard atmosphere and thirdly, there is a steady growth in signal strength after 17.00 hours indicative of the returning nocturnal inversion, which culminates in a period of near-specular reflection after 20.00 hours.

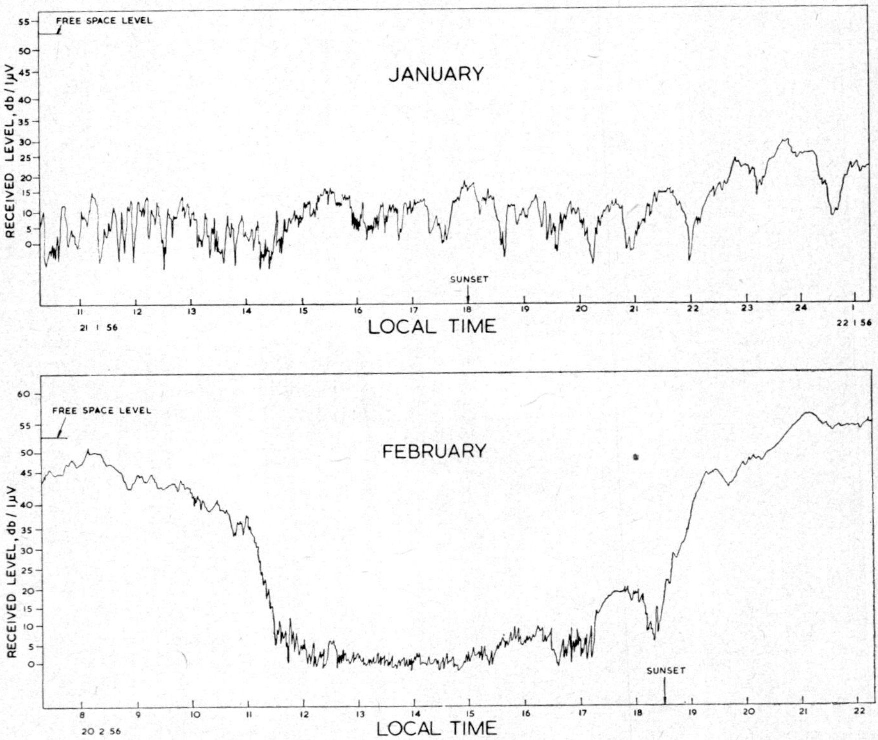

FIG. 3. Illustrative signal records, 80 Mc/s, Bahrain–Doha.

IV. Statistical Features

Figure 4 illustrates the statistical character of the diurnal variations over the 6 months test period. It has been constructed by analysing the variations in the hourly median level for a selected hour of the day during the 160 days for which complete records were obtained. Each curve shows for the specified times of day, the hourly median level exceeded for the indicated percentage of the total number of days tested. It is seen that the weaker hourly median levels, namely those exceeded during a high percentage of days, show progressively reduced diurnal variations, until by the time the 99 % level is reached the diurnal effect is barely perceptible. It is reasonable to conclude that higher percentage levels, which have important engineering significance, will show even less variation, and from this it is inferred that the troughs of fades attain a fairly constant low level which is independent of the time of day.

Attention will now be briefly directed to the behavior during the period 14.00–16.00 hours when the signal strength is least, and the contrasting period 02.00-04.00 hours when the hourly median signal strength is near its maximum. Figure 5

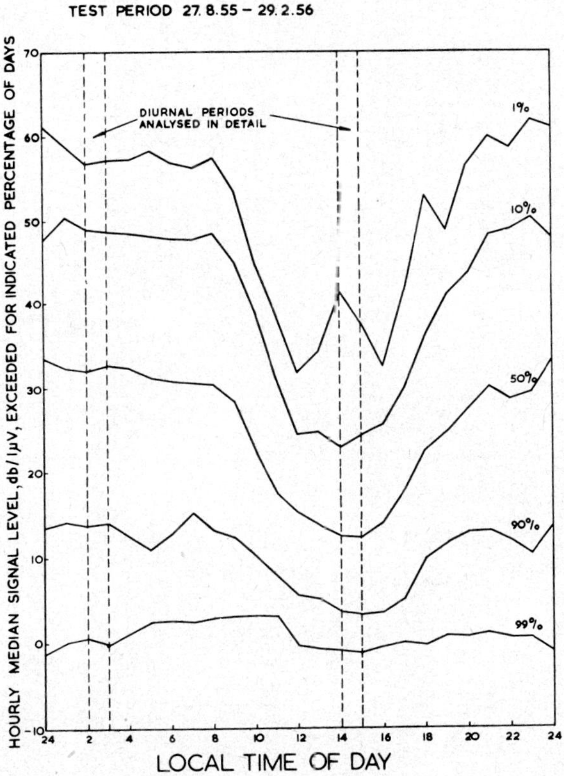

FIG. 4. Statistics of diurnal variations, 80 Mc/s, Bahrain–Doha.

shows the seasonal trends for these extreme periods of the diurnal cycle. To smooth the effects of short-term weather changes the hourly median levels for the respective day and night periods have been averaged over a week, and plotted against the appropriate week of the test. A seasonal trend with a minimum in January is clearly evident, but while the weekly mean night-time levels experience a seasonal variation of perhaps 35 db, the daytime levels experience a variation of only about 20 db. It will be noticed furthermore that by mid-January the difference between daytime and night-time levels has narrowed to insignificance, but that it widens rapidly again thereafter. Extrapolating the trends on the chart to the unrecorded seasons, it seems likely that both daytime and night-time weekly mean levels would show broad maxima in early July.

Figure 6 illustrates the wide divergence between cumulative distributions of hourly median levels applicable respectively to the specified daytime and nighttime

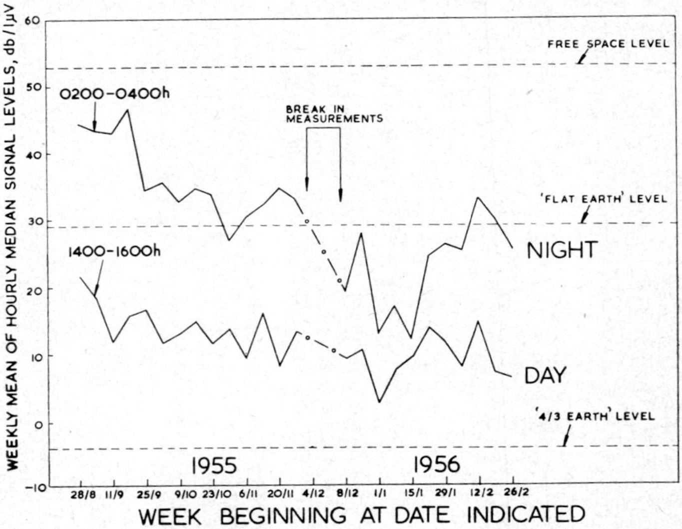

FIG. 5. Seasonal variations for daytime and night-time periods.

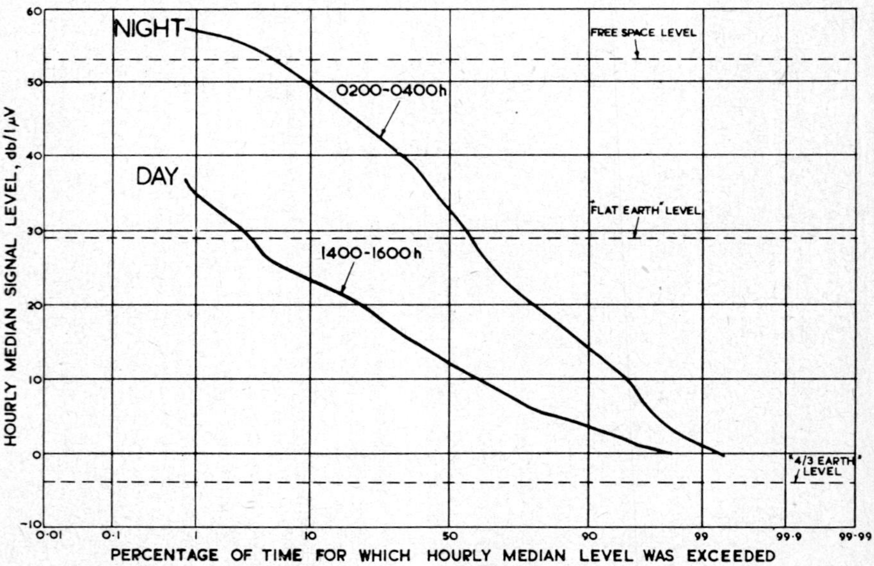

FIG. 6. Cumulative distributions of hourly median levels for daytime and night-time periods.

hours. For the nocturnal hours the free space level was exceeded for as much as 5 % of the time. Similar curves for the period December-January, when signal strengths were weakest, show a much reduced divergence between daytime and nocturnal levels. Moreover the free space level was never reached during that period.

V. Association Between Radio and Meteorological Behavior

Throughout the period of the radio tests, regular meteorological soundings were made by the Royal Air Force near the transmitting terminal at Bahrain. These soundings were made daily at 06.00 and 18.00 hours local time, and modified

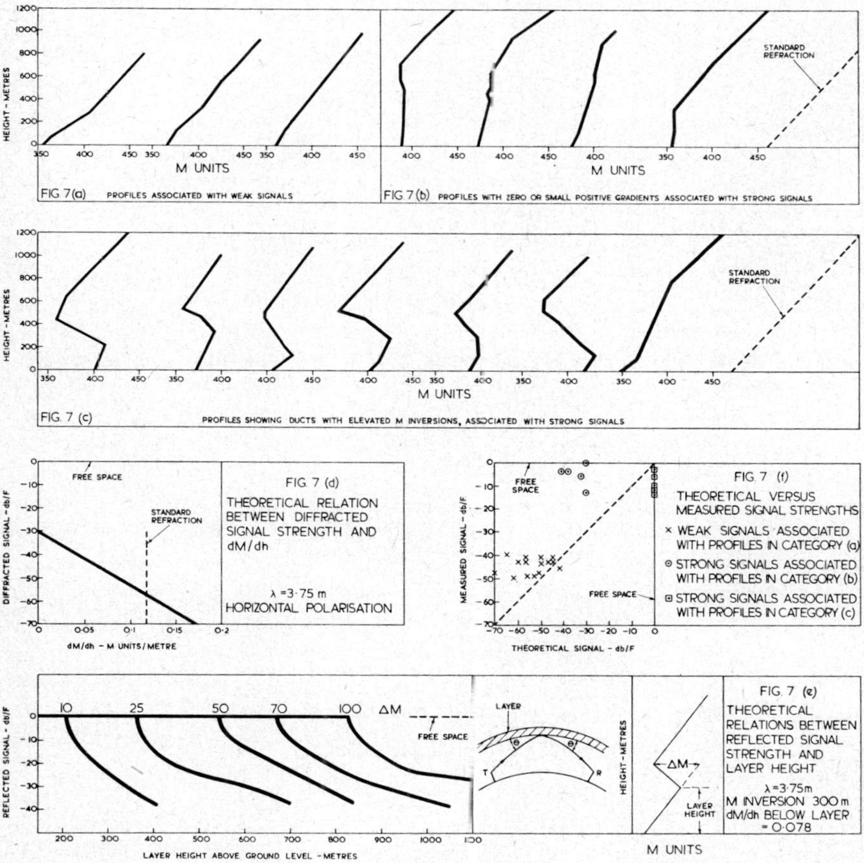

Fig. 7(a). Profiles associated with weak signals.
Fig. 7(b). Profiles with zero or small positive gradients associated with strong signals.
Fig. 7(c). Profiles showing ducts with elevated M inversions, associated with strong signals.
Fig. 7(d). Theoretical relation between diffracted signal strength and dM/dh.
Fig. 7(e). Theoretical relations between reflected signal strength and layer height.
Fig. 7(f). Theoretical versus measured signal strengths.

refractive index profiles derived from them have served to confirm the existence of intense atmospheric stratification inferred from many recordings of the 3·75 m transmission. To limit the labor involved in extracting M profiles, attention has been confined to the synoptic hours associated with very high and very low hourly mean signal strengths.

Inspection of the chart of diurnal variations previously illustrated (Fig. 4), shows the synoptic hour 06.00 to be associated with characteristically high signal levels well suited to this purpose. The synoptic hour 18.00 generally shows intermediate levels rapidly increasing towards the typically high nocturnal values. Thus although of interest as representing a transitional state, this hour has little utility in the investigation of the extreme conditions with which this discussion is concerned. Unfortunately no meteorological soundings exist for the period noon to 16.00 hours associated with low signal strengths, and therefore to obtain examples of low levels it was necessary to search for the few uncharacteristic occasions when weak signals occurred during the synoptic hours available.

Adopting the arbitrary criterion that a weak signal has an hourly mean level at least 40 db below the free space level, 14 examples of such signals were found to occur in the morning synoptic hour 06.00, and two examples in the evening synoptic hour 18.00. Figure 7(a) shows representative M profiles associated with some of these occasions. As expected these exhibit fairly uniformly graded atmospheres with a tendency however for refraction to exceed the standard value.

Attention will now be directed to meteorological soundings associated with very high hourly mean levels, arbitrarily defined as lying within 14 db of the free space level. In this category 31 examples were found for the complete test period of which 5 examples, illustrated in Fig. 7(b), show zero or slightly positive M gradients over the first few hundred metres. The remaining 26 examples show pronounced ducts. Of these, 7 show elevated M inversions as illustrated in Fig. 7(c), while the remaining 19 ducts show ground based M inversions which will be separately considered later.

Apart from one anomaly, where a profile associated with a low signal exhibited a weak surface duct, the character of each profile investigated has been found compatible with the associated signal level. In view of this qualitative correspondence it was considered worth computing, for the test path and the test wavelength, the theoretical signal strengths based on the available M profiles, for comparison with the hourly mean signal strengths measured during the corresponding synoptic hours.

VI. Comparison between Measured and Theoretical Signal Strengths

Taking first the simplest category of profile, illustrated in Fig. 7(a) and having a fairly uniform M gradient, theoretical signal levels have been based mainly on Eckersley's Propagation Curves for a smooth curved earth, whose radius has been suitably adjusted to allow for the required gradient of modified refractive index. Figure 7(d) shows, for horizontal polarization, the theoretical relationship between the positive gradient of modified refractive index dM/dh, and the signal strength over the tested path. It should be noted that the path is non-optical and comprises gently undulating terrain with no significant hills and a short stretch of sea. The relationship of Fig. 7(d) has been restricted to positive values of dM/dh for reasons which will become apparent. It is quite accidental that it is approximately linear over the range considered.

Figure 7(f) shows measured signal strengths plotted against their calculated values for various situations. The crosses refer to the weak signals associated with profiles in category (a). In general the measured levels substantially exceed the theoretical figure due, it is believed, to partial reflections or scattering from atmospheric irregularities. These, while often too small to be revealed by the relatively large-scale meteorological soundings, can nevertheless dominate the weak diffracted signals associated with moderate surface super-refraction.

The circles refer to the high signal levels associated with profiles in category (b), which have zero or small positive values of dM/dh near the surface. One again the measured levels much exceed the theoretical values. While strong reflections at very grazing incidence from small discontinuities may well account for this in some cases, there is another modifying factor worthy of consideration. The condition under discussion is one of incipient duct formation, where the track-width of the first transmission mode, as defined by Booker [2], is very large. In this narrow region it is quite unrealistic to compute signal strength on the basis of a uniform M gradient to infinity, which is the idealization implicit in Fig. 7(d). In practice, uniform M gradients are maintained only over a limited height interval and this shortcoming restricts the valid range of the theoretical relationship. As a rough guide to the magnitudes applicable to the region of the test, M profiles in category (b) show fair uniformity of positive M gradient up to heights of at least 300 m if small-scale irregularities are ignored. Consideration of the track-width of the first transmission mode in relation to the above height leads to the conclusion that Fig 7(d) becomes unusable only when dM/dh becomes less than about 0·01. Although the invalid region is quite small, the situation is worsened in practice because the meteorological measurements are insufficiently accurate to make the small but critical distinction between zero and small positive values of dM/dh. Thus slightly inaccurate surface gradients measured from the M profiles which nominally fall in the valid region of the curve, may actually lie in the unusable region in some cases.

The profiles illustrated in Fig. 7(c) will now be considered. These indicate pronounced elevated M inversions from which radio waves can be reflected. The reflection coefficients of such layers have been calculated by Millington [11], after idealization in a mathematically convenient way. For layers having a negative refractive index gradient and finite thickness Millington has shown that for any thickness of layer, total internal reflection will occur if the grazing angle is less than a critical value depending only on the total change of refractive index across the layer. For grazing angles exceeding the critical value, the reflection coefficient declines in a rapid but complex way governed by the layer thickness expressed in wavelengths. Millington's idealization implies zero refractive index gradient outside the layer, where in practice the gradient is usually negative. The major effect of this departure from the prescribed idealization is to reduce the grazing angle θ on account of downward ray-bending, as indicated diagrammatically on Fig. 7(e). There is also a secondary influence connected with the fact that negative gradients outside the layer necessarily reduce the discontinuity in slope on entering and leaving the layer which is implicit in Millington's idealization. In the limit, when the slopes outside it and inside it become equal, the layer vanishes. A simple modification has been devised which, while conforming to Millington's results for layers suiting his idealization, at the same time gives no reflected signal when the layer vanishes on account of the effect just mentioned. This empirical adjustment, which is not claimed to be rigorously accurate in intermediate cases, involves an M deficit ΔM which is defined on the idealized M curve in Fig. 7(e). The effective total change in refractive index across the layer, which is a fundamental parameter in Millington's analysis, is then taken as $\Delta M \times 10^{-6}$.

Based on these principles, the chart of Fig 7(e) shows theoretical signal strengths over the test path due to reflection at representative elevated layers 300 m thick at various heights above the surface. To conform broadly to M profiles typical of category (c) the chart is based on a uniformly graded atmosphere below the layer corresponding to an effective earth radius factor of 2. The values of ΔM shown on the chart cover a range representative of the M profiles under consideration. Signal strengths are expressed relative to the free space level.

The chart clearly shows the layer heights above which total internal reflection, implying "free space" signal strength, gives way to a rapidly declining level. The seemingly unrelated shapes of the curves in this region result from interference effects characteristic of layers of finite thickness. Being based on various idealizations the curves should not be interpreted too literally, but they serve to illustrate the very high levels obtainable by reflection at even minor discontinuities if the ray incidence is sufficiently grazing. For example when ΔM is only 10, which is probably too small a figure for reliable evaluation from the meteorological soundings, this suffices at a height of 200 m to give a reflected signal of "free space" level. Moreover a much weaker discontinuity or an equal discontinuity at a greater height, is still capable of swamping weak diffracted signals associated with moderate super-refraction near the surface. It is for this reason that the measured signals associated with the crosses on Fig. 7(f) mostly exceed the theoretical figures based on diffraction in a uniformly graded atmosphere.

The reflection chart just considered may be used to estimate the reflected signal strengths from elevated layers in category (c), by applying the layer height and M deficit derived from the appropriate M profile. In this way it has been found that all layers placed in this category give rise to total internal reflection, and hence a "free space" reflected signal. This conforms satisfactorily to the measured signal strengths which, it will be recalled, were selected for their closeness to the "free space" figure. On Fig. 7(f) this association is represented by the line of squares to the right of the chart. In this situation the diffracted signal propagated below the layer is of course negligible in comparison with the reflected signal. The substantial fades sometimes associated with this condition are thought to result from interference between multiple ray paths via the reflecting layer which, as is well known, are geometrically possible with spherical stratification and a concentric reflecting earth.

VII. Propagation in Ducts

We shall now briefly consider some salient features of propagation in ducts having ground-based M inversions, as an aid to assessing particular examples associated with high signal levels over the path under discussion.

The basic features of propagation within a duct have been simply explained by Appleton [1], by considering interference between upgoing ground-reflected waves and downward-refracted waves within an M inversion. In this way Appleton shows that in a simple surface duct having an M deficit ΔM, a wavelength λ will be propagated freely only if the height of the M inversion is $\lambda/2\sqrt{(2\Delta M)}$, where ΔM is here expressed in M units $\times 10^{-6}$. For a smaller duct thickness, upgoing and downcoming waves interfere destructively and therefore do not travel far. Although a wave treatment of the problem as given by Booker [2] is essential for estimating the signal strength in the duct, Appleton's simple treatment shows quite satisfactorily the vital role played by the wavelength in duct propagation. Moreover it is in fair numerical agreement with conclusions based on Booker's principle which states that, when the height of an M inversion equals the track-width of the first transmission mode—subsequently defined—the associated wave is propagated to great distances, because leakage from the top of the track is then suppressed.

When an M inversion is just thick enough to trap the first mode, energy in the higher order modes leaks freely from the top of the duct so that the contributions of these modes to the received signal can usually be neglected. Thus in this case the signal strength depends on the position of the terminals in relation to the track of the first mode, and on the fraction of the transmitted power supplied to that mode—factors which Booker presents in the form of propagation curves for the mode in question. If the M inversion is just thick enough to include the track-width of the second mode, the contributions of the first and second modes, derived from the appropriate propagation curves, must be combined to obtain the requisite signal strength. Once again higher order modes may be neglected in such a case. If many modes are trapped, as might occur for example on centimetric wavelengths in the region under review, the terminals are intervisible from the radio aspect, and the signal strength may be calculated on a ray basis to give a value in the neighborhood of free space.

Without entering deeply into the numerical consequences of Booker's mode theory it may be stated that when the terminals lie within an M inversion extending over the track-widths of the first or higher-order modes, signal strengths within the duct may be expected to be very high at points well beyond the transmitter's geometrical horizon. The non-optical path under discussion has amply confirmed this expectation as will be shown.

The characteristic transmission modes, whose properties are governed by the necessity of satisfying the boundary conditions at the earth's surface, are associated with track-widths whose size increases with the order of the mode. When dM/dh is positive—a condition already considered—the track-width is represented mathematically as a complex quantity, connected with the fact that physically it is somewhat ill-defined due to leakage from the top of the track. We are at present concerned with negative values of dM/dh—the duct condition—where the track-width is precisely defined.

Following Booker, the track-widths of the first three modes under ducting conditions are respectively given in metres by:

$$W_1 = 54{\cdot}1\lambda^{\frac{2}{3}}[-\mathrm{dM/dh}]^{-\frac{1}{3}}$$

$$W_2 = 95{\cdot}2\lambda^{\frac{2}{3}}[-\mathrm{dM/dh}]^{-\frac{1}{3}}$$

$$W_3 = 128{\cdot}6\lambda^{\frac{2}{3}}[-\mathrm{dM/dh}]^{-\frac{1}{3}}$$

where λ is the wavelength in metres and dM/dh—a negative quantity in this case—is the modified refractive index gradient expressed in M units/metre. Fig 8(a) shows the relation between these track-widths and dM/dh for the test wavelength of 3·75 m.

VIII. Analysis of Observed Ducts

M profiles showing ground-based M inversions have been found associated with high signal levels in 19 cases. All but one of them came from the morning meteorological sounding. A selection of the profiles is given in Figs. 8(b) and (c). To determine how many transmission modes, if any, are trapped by a particular duct, the track-widths of the first three modes have been obtained from Fig. 8(a), using initially the surface value of dM/dh derived from the M profile. In 7 cases, of which 6 are illustrated in Fig. 8(b), the track-width of the first mode significantly exceeds the height of the M inversion, indicating absence of trapping. In the remaining 12 cases the M inversion covers the track-width of the first mode, while in two of these cases the second mode is just trapped also.

Figure 8(c) shows examples of this last category. Where a mode is trapped the top of the associated track is indicated with an appropriately marked arrow, adjusted to be consistent with the average refractive index gradient over the whole track. The first two profiles illustrate two noteworthy occasions, both occurring in September, when the second mode was just trapped on the wavelength of 3·75 m. The first profile shows the remarkably large M deficit of 90 M units while the third, which fails by a small margin to trap the second mode, shows a deficit of 100 M units. The author is not aware of M deficits of quite this magnitude having been observed elsewhere, nor can he discover other reliable evidence of trapping on such a long wavelength.

It is of interest that the 19 surface ducts found to be associated with very high signal strengths yielded as many as 7 examples incapable of trapping the test wavelength. It can only be presumed that in this latter category the progress towards trapping was sufficiently advanced to cause a substantial reduction in leakage of the first transmission mode and hence high signal levels. It may well be that a detailed numerical analysis of such cases on the lines described by Booker [2] would yield signal strengths conforming to the high values observed.

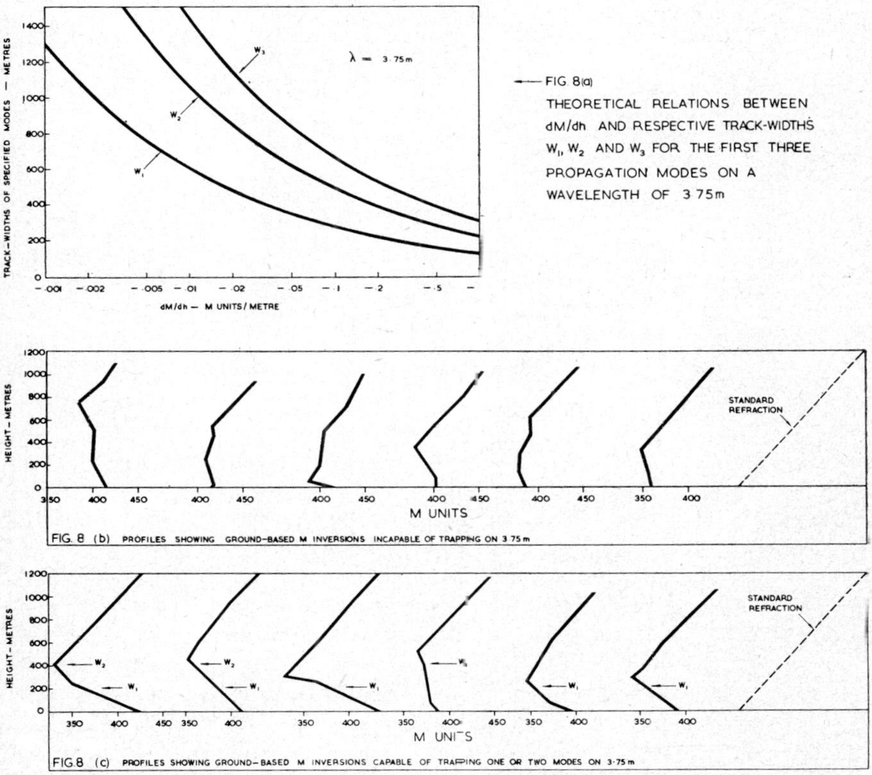

FIG. 8(a). Theoretical relations between dM/dh and respective track-widths W_1, W_2 and W_3 for the first three transmission modes on a wavelength of 3·75 m.

FIG. 8(b). Profiles showing ground-based M inversions incapable of trapping on 3·75 m.

FIG. 8(c). Profiles showing ground-based M inversions capable of trapping one or two modes on 3·75 m.

IX. General Conclusions

It may be of value to review some important features of the investigation just described.

The radio measurements, reinforced by the meteorological soundings, have enabled a clear distinction to be made between three main types of propagation operating over the radio path at various times. Firstly, there is the association between weak signals and an approximately standard atmosphere, which however exhibits signal strengths significantly greater than would be expected from the M gradients observed. This excess is attributable to the influences of scattering and partial reflection from small atmospheric discontinuities which, as is well known, become dominant well in the diffraction region. Secondly, there is the powerful influence of reflections from well defined discontinuities which can easily give rise to "free space" signal levels. Thirdly, many examples have been given of the association between high signal levels and surface ducts, most of which have been found capable of trapping the first mode.

The investigation has shown, furthermore, that high signal levels are also associated with M profiles in the difficult category involving a zero or small positive M gradient maintained over a limited height interval near the ground. This situation, it will be recalled, cannot be assessed by conventional methods because these are invalidated by the great height of the track-width of the first mode in relation to the restricted height of the ground-based linear region of the profile.

In conclusion, a few comments may not be out of place concerning the remarkably intense nocturnal surface ducts encountered. Observation of those ducts was virtually confined to the synoptic hour 06.00. If meteorological soundings had been taken in the period midnight–04.00 hours, when signal strengths were generally at a maximum, it is likely that ducts even more intense than those observed would have been discovered, for on several occasions in this period signal strengths exceeded the free space level by some 6 db. The presence of pronounced ducts certainly conforms broadly to the climatological description of the Persian Gulf given by Durst [7], but his detailed information is unfortunately relevant to the Gulf itself rather than the surrounding terrain, where the situation is profoundly altered as already explained, by the heating and cooling of the land.

A search in the literature for direct evidence of trapping from one-way communication or radar systems yields many examples of very long ranges over sea on wavelengths as long as $1\frac{1}{2}$ m, but any evidence of exceptional ranges over land is very scanty. Durst [7] quotes the examples, already mentioned, of large nocturnal increases in range experienced by $1\frac{1}{2}$ m and 7 m radars in Bengal when looking inland over the Ganges delta. There was, however, an accompanying "skip-distance" effect, indicating near-specular reflection at long ranges from an elevated inversion layer. Although providing an interesting parallel with some of the Persian Gulf observations, this effect is of course quite distinct from trapping and throws no further light on that subject. While it seems that documentary evidence of trapping is confined to wavelengths of 1·5 m and below, it is submitted that the present observations, made in exceptional climatic conditions, provide evidence of recurrent nocturnal trapping on a wavelength of 3·75 m, which is perhaps longer than has hitherto been considered possible.

Acknowledgments

The author wishes to thank the Engineer-in-Chief, Marconi's Wireless Telegraph Company Limited, for permission to publish the paper. He is also indebted to engineers of Cable and Wireless Limited, who made all the radio measurements described. He is further indebted to the Royal Air Force, Bahrain, and the Meteorological Office, London, who provided the upper air data. He also acknowledges the assistance of Mr. F. Immirzi in carrying out many computations.

References

1. Appleton, E. (1946). Phys. Soc. and Royal Met. Soc. Report of a Conference on Meteorological Factors in Radio-Wave Propagation, p. 1.
2. Booker, H. G. & Walkinshaw, W. (1946). Phys. Soc. and Royal Met. Soc. Report of a Conference on Meteorological Factors in Radio-Wave Propagation, p. 80.
3. Bray, W. J. & Corke, R. L. (1952). *Proc. Instn elect. Engrs* **99,** 281.
4. Carassa, F. (1956). *Alta Frequenza,* **25,** 378.
5. Day, J. P. & Trolese, L. G. (1950). *Proc. Inst. Radio Engrs, N.Y.* **38,** 165.
6. Durkee, A. L. (1948). *Proc. Inst. Radio Engrs, N.Y.* **36,** 197.
7. Durst, C. S. (1946). Phys. Soc. and Royal Met. Soc. Report of a Conference on Meteorological Factors in Radio-Wave Propagation, p. 193.
8. Englund, C. R., Crawford, A. B. & Mumford, W. W. (1940). *Proc. Inst. Radio Engrs, N.Y.* **28,** 360.
9. Gough, M. W. (1955). *Proc. Instn elect. Engrs,* Part B, **102,** 43.
10. Millar, J. Z. & Byam, L. A. (1950). *Proc. Inst. Radio Engrs, N.Y.* **38,** 619.
11. Millington, G. (1949). *Marconi Rev.* **12,** 140.
12. Smith-Rose, R. L. & Stickland, A. C. (1946). Phys. Soc. and Royal Met. Soc. Report of a Conference on Meteorological Factors in Radio-Wave Propagation, p. 18.
13. Vecchiacchi, F. (1956). *Alta Frequenza* **25,** 100.
14. Vikramsingh, R., Rao, M. N., Singh, S. & Uda, S. (1958). *J. Inst. Telecom. Engrs,* New Delhi, **4, 147.**

An Interesting Propagation Effect of Sputnik I

EDMOND DEWAN

Electronics Research Directorate, Air Force Cambridge Research Center, Bedford, Mass.

When Sputnik I was announced by the Russians, this laboratory set up four 20-Mc/s interferometers to track it. Pen graphs recorded the intensity of the audio signal versus time and magnetic tapes recorded the signal frequencies. The data revealed that on four consecutive days (18, 19, 20, 21 October), at approximately 18.50 hours, a rather surprising propagation phenomenon occurred.

1. The intensity of the signal received when the satellite was between 730 miles and 2000 miles from the receiver was occasionally as great as the signal received during the direct overhead pass.

2. On 18 October the pattern showed regions of peaked intensity when the satellite was approximately 1380 miles away and again when it was 730 miles away. The effect occurred not only as Sputnik approached but also as it departed.

3. Each day the number of peaked regions reduced successively. There was only on one 21 October and on 22 October there was no evidence of the phenomenon.

4. Great irregularities were noted in the peaked regions.

5. The irregularities in the peaks were found to be due to beat tones produced by the simultaneous reception of up to three closely spaced frequencies in the peaked regions.

6. Near the points where the high intensity suddenly cut off, two of the three frequencies approached the same value.

A mathematical analysis of the shift in received frequency indicates that the signals in the peaked regions arrived from Sputnik (located just below the maximum of the only ionospheric layer present—the F layer) via reflections between the earth and ionosphere. The presence of several frequencies was apparently caused by a superposition of reflection modes as well as the presence of "high" and "low" waves due to electron density variation with altitude. The fact that the phenomenon is possible only when the earth reflections are from a very good conductor, such as the ocean, can explain the daily reduction in the number of peaked regions since the orbit was such that on each successive day there was less possibility for ocean reflections. The merging of two frequencies at the end of a peaked region can be explained by the known merging of high and low waves just before a skip zone is reached. The high intensity at the end of a peaked region and its sudden cut off is for the most part explained by the focusing effect at the edge of a skip zone and the sudden absence of reception when such a zone is entered.

I. INTRODUCTION

Immediately upon the announcement that the Russian Sputnik I was in orbit, the AFCRC Electromagnetic Radiation Laboratory at Bedford, Mass., set up four 20–Mc/s interferometers to track it. Dipole arrays (two each of 6 λ and 1·7 λ) were installed in the north-south (NS) direction and in the east-west (EW) direction. The four signals from these arrays, together with a time signal, were recorded on 5-channel magnetic tape.

Examination of the Brush pen recordings from the tape (see Figs. 1–5) revealed a remarkable phenomenon: when the satellite came up over the horizon and again when it went down toward the horizon in the opposite direction, the signal occasionally had peaks as large as those observed when the satellite was at the zenith of its pass.

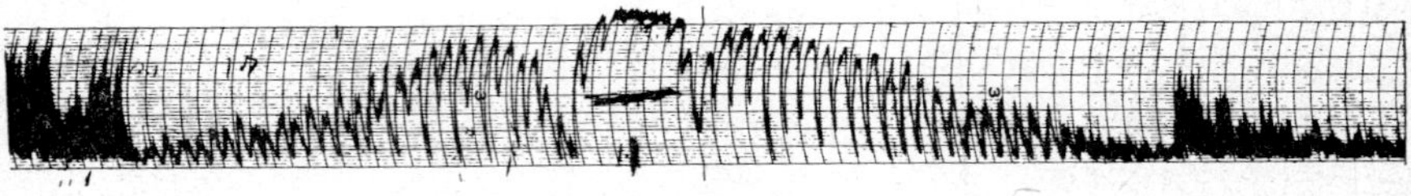

FIG. 1. Brush recording of 1905 hrs pass, October 18, taken from 1.7λ array in EW direction. Each division equals 7.2 sec.

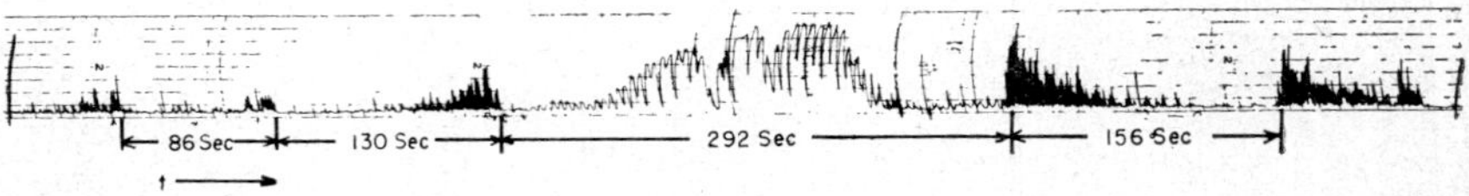

FIG. 2. Brush recording of 1905 hrs pass, October 18, taken from 6λ array in NS direction. Each division equals 7.2 sec; Sputnik velocity ≈ 5 mi/sec.

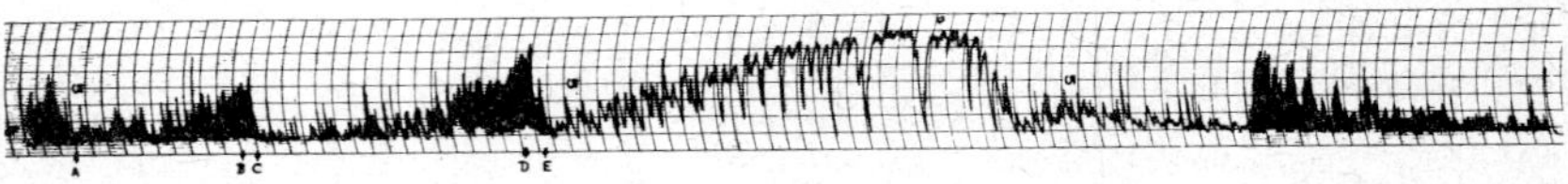

FIG. 3. Brush recording of 1900 hrs pass, October 19. Letters refer to times in Fig. 9. Each division equals 7.2 sec.

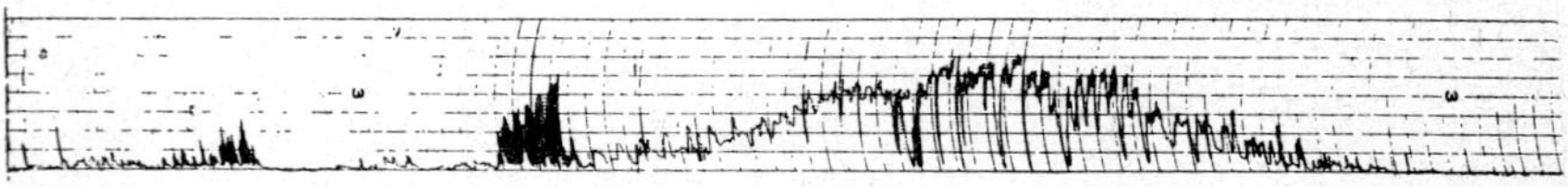

FIG. 4. Brush recording of 1850 hrs pass, October 20. Each division equals 7.2 sec.

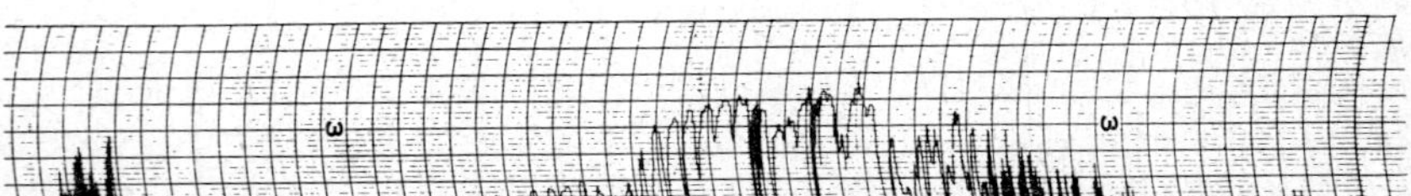

FIG. 5. Brush recording of 1845 hrs pass, October 21. Each division equals 7.2 sec.

II. Observations and Data

During a period of 14 days, 35 Sputnik I passes were recorded over this area (see Appendix A for dates and hours). On four consecutive days, at approximately 18.55 hours, unexpectedly large signals appeared on each side of the main pattern. Because of the suggestive configuration of the pen recording in Fig. 1, these large signal regions are, for convenience of reference, familiarly termed "ears" in the discussion that follows.

A. Brush Pen Recordings

The ordinates of the Brush graphs are proportional to signal strength. Figure 1 (19.05 hr, 18 October 1957) shows large disturbances on each side of the main

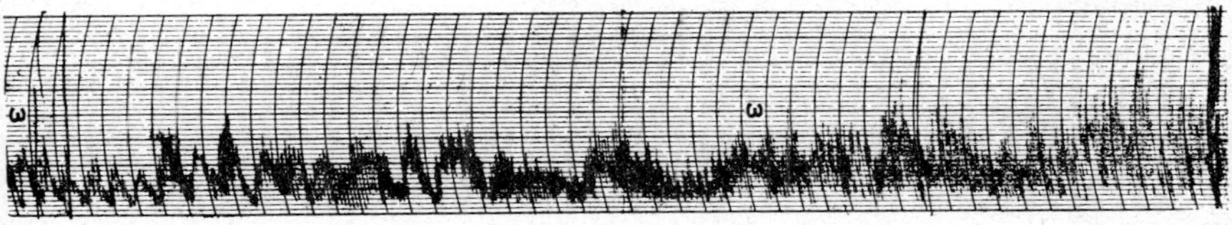

FIG. 6. Brush recording of an ear region. Expanded time scale (each division equals 0.5 sec).

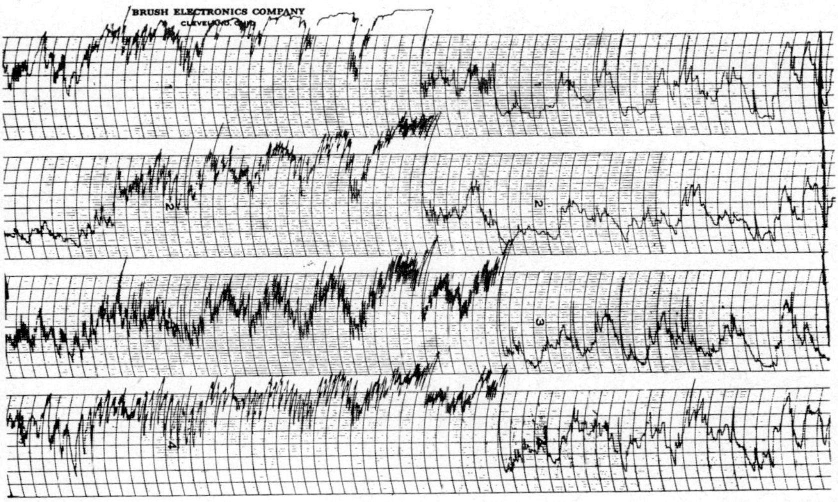

FIG. 7. Part of brush recording of 1850 hr pass, October 20, showing lack of synchronism of cutoffs of ears from NS and EW arrays. Each division equals 1 sec. Channels 1 through 4 are, respectively: NS 1.7λ, NS 6λ, EW 6λ, EW 1.7λ.

pattern. Figure 2 is a full recording of the same pass partially recorded in Fig. 1. The wide difference in the appearance of the two main patterns can be attributed to their being recorded from separate receivers connected to separate interferometers. In all likelihood, the difference is due to the difference in antenna patterns.

Figures 2–7 (evening passes of 18 to 21 October) show the following:

A series of 6-sec dips is superposed on the main pattern, probably because of rotational motion of the satellite.*

The signal intensity in an ear region, recorded when the satellite is about 730 miles from the receiver, is approximately as large as it is at the center of the main pattern.† This seems to be the strangest feature of the phenomenon.

The left ears end in a sharp cutoff; the right ears start with a steep rise.

When multiple ears occur (see Table I), the number of ears decreases daily.

TABLE I

Date	Configurations	
18 Oct 57	3 left ears	2 right ears
19 Oct 57	3 left ears	1 right ear
20 Oct 57	2 left ears	0 right ear
21 Oct 57	1 left ear	0 right ear
22 Oct 57	0 left ear	0 right ear

Figure 2 explicitly shows the time intervals in seconds. During a pass the spacing between ears consistently changes from 86 sec to 156 sec.

In Fig. 7 (20 October) the ear immediately preceding the main pattern cuts off approximately 6 sec earlier on the north-south interferometers than on the east-west interformeters.

Figure 6, which shows a small portion of an ear region on an expanded scale of two divisions per second, clearly brings out a sporadic periodicity. This curious effect prompted a frequency analysis of the magnetic tapes.

B. Magnetic Tape Recordings

Careful auditing of the signal on the tape recordings disclosed that more than one frequency was being received during an ear. These frequencies were close enough to one another to set up low-frequency beat tones. When the frequencies faded in and out the beat tones changed discontinuously. The changing periodicities seen in Fig. 6 are thus explained as nothing more than a visual record of the beat tones.

*Dr. F. Vilbig observed that the light reflected from Sputnik I faded approximately once every 6 sec.

†The Doppler curve discloses that the center of the main pattern corresponds to the time the satellite reaches the point of closest approach (nearly overhead) at an altitude of about 140 miles.

A single frequency, which shifts upward, is audible during the main pattern of all these recordings (Fig. 8). The shift, which is of course due to the Doppler effect, shows that the pass was almost directly overhead. The shift was upward

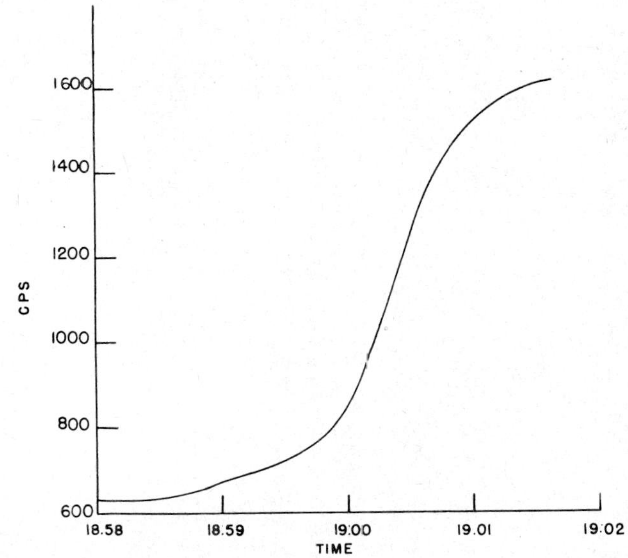

FIG. 8. Doppler curve for 1905 hrs pass, October 18.

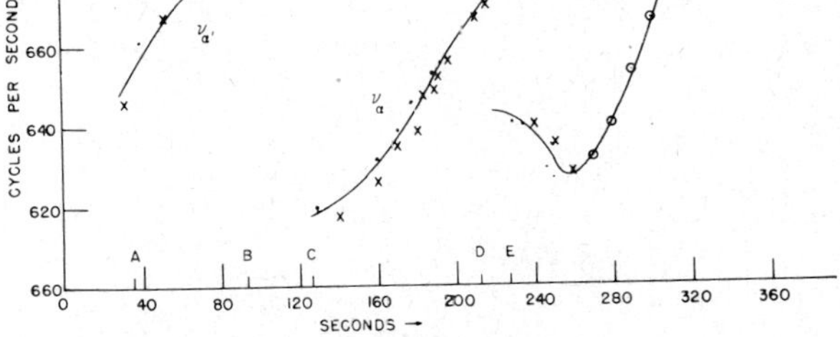

FIG. 9. Frequency analysis of ears shown in Fig. 3.

instead of downward because the beat frequency oscillator (BFO) pitch was set above the frequency of the Sputnik transmitter.

Figure 9 is a frequency analysis of the ear regions of Fig. 3, obtained by means of a counter and a 6-cycle bandwidth filter. Corresponding times in Figs. 3 and 9 are designated by the letters A through E. Figure 9 shows that, within the accuracy of the instruments used, there were at most three closely spaced tones. Figure 3 shows that at point B the signal intensity gradually increased. Figure 9 shows that at point B, two frequencies were present, approaching within 30 to 40 c/s of each other.* Before cut-off, at point C, a third and lower frequency came in. After point C, the pattern more or less repeated itself.

In Fig. 9, the lower frequencies correspond to ν_α and ν_α' and the upper to ν_β. Since the latter faded more than once a second, the direct use of a frequency counter was not possible. Crude measurements were made by direct comparison with an audio oscillator connected to a counter.

Near a cut-off, for example C, E, the ν_β signals became less intermittent and higher in intensity. Just before cut-off, the ν_β and ν_α signals were approximately equal in both intensity and frequency.

The dip in frequency at the beginning of the main pattern in Fig. 9 is apparently due to experimental error, such as a stretching of the tape from uneven pull during the recording process. Figure 9 was not obtained from the original tape but from a recording of it, and neither the careful auditing of the original nor analysis of other recordings of it reveal this dip.

III. Theory

An acceptable theory for the phenomenon observed must explain:

1. The (apparently) impossibly large signal strengths registered on the pen recordings when the satellite was approximately 730 miles from the receiving station.
2. The structure of the "ears" seen in the pen recordings—the gradual buildup of signal intensity followed by sudden cut-off (for right ears, the reverse sequence); the increasing spacing between the ears as a function of time (Fig. 2); and the lack of synchronism on one day (Fig. 7).
3. The disappearance of ears day to day (Table I).
4. The presence of multiple tones and their patterns analyzed in Fig. 9—why the slopes of the ν_α curves increase for each successive ear; why the ν_α and ν_β frequencies approach one another; and why the ν_β frequently is intermittent and feeble during the first part of an ear but becomes stronger and less intermittent until it is approximately as steady and intense as the ν_α tone just before cut-off.

The most important clues to the explanation of this phenomenon are the multiple ears and the multiple frequencies (Fig. 9). They indicate that the signals responsible for an ear region reached the receiver by indirect routes, that is, that the paths were in some way multiply reflected, multiply refracted, or both. More explicitly, the multiple frequencies show that the signal left the satellite at more than one angle (the Doppler shift depends on the angle); and the occurrence

*They may actually have met because the frequency of the audible beat tone went to zero. In other words, the 30 to 40–c/s gap may have been the result of experimental error. A further complication is mentioned in Appendix C.

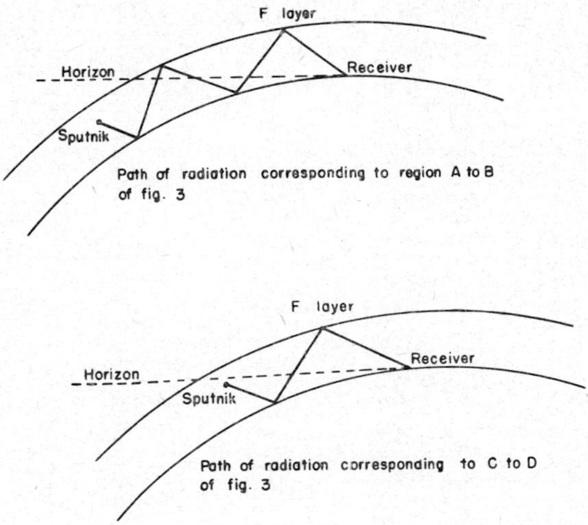

FIG. 10. Theoretical model for ear phenomenon.

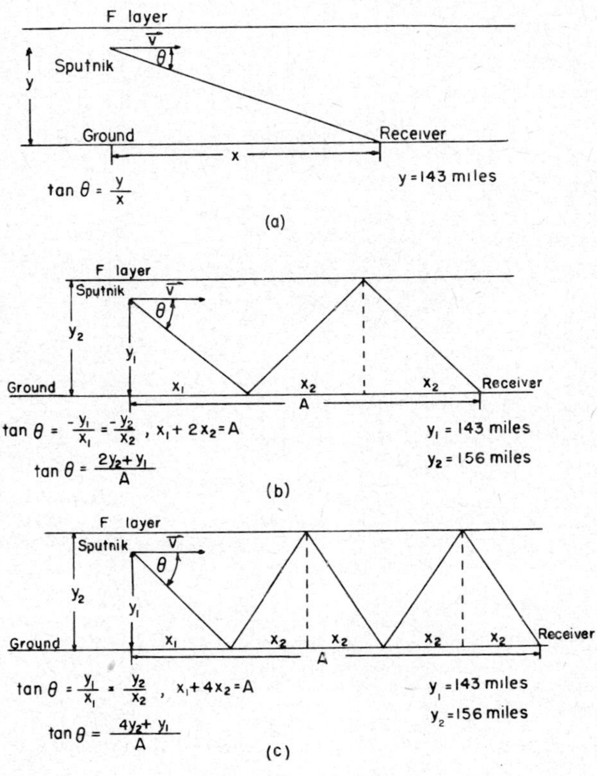

FIG. 11.

of the multiple ears at consistent time-intervals from day to day suggests that the path of the signal described some sort of pattern of multihops when the satellite was over the horizon. The rarity of the phenomenon suggests that its occurrence requires a critical and unusual combination of orbit and ionospheric configurations. The proposed model (Figs. 10 and 11), based on information from the Central Radio Propagation Laboratory (Series D reports) and from the Smithsonian Institute, and corresponding to the reflection patterns causing the two left ears of Fig. 3, is described as follows.*

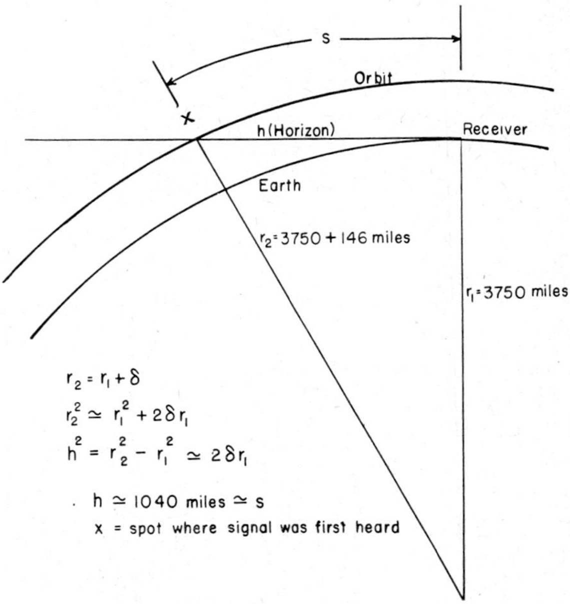

FIG. 12.

Because its altitude was only about 146 miles (Fig. 12), Sputnik remained below the horizon until it came within 1040 miles of our receiving station. The only (known) way for the signal to reach the receiver in most of the ear regions was therefore by earth-ionospheric reflections. When the angle θ (Fig. 11) became too large for reflection, the radiation took a path with smaller θ and hence one with

*Information from CRPL Series D reports: The height of the maximum of the F ionospheric layer was approximately 156 miles, all other layers being absent at this time. The critical frequency was 10 Mc/s, implying (from the "secant law") that a 20-Mc/s wave would have been reflected if the angle of incidence with respect to the normal to the layer had been greater than approximately 60°. This corresponds to an angle of emission from Sputnik of approximately 30° with respect to the velocity of the satellite.

Information from the Smithsonian Institute: The altitude of the satellite for these passes was approximately 146 miles. (See Appendix B for a day-to-day westward drift of orbit for these passes. The listed data refers to 1957 $\alpha - 1$, which did not differ too much from the transmitting 1957 $\alpha - 2$.)

fewer reflections. In general, this process continued until the radiation finally reached the antenna directly. Since the antennas were interferometers with nulls in the horizontal direction, a nonreflected signal could not be detected when the satellite was close to the horizon; and before the satellite was high enough for the direct signal to be detected, it had to come within approximately 900 miles.

The general nonrelativistic Doppler shift for a moving source and a stationery observer is given by

$$\nu' = \frac{\nu}{1 - v/c \cos \theta}$$

where ν is the frequency of the emitter, ν' is the received frequency, v is the velocity of the source, c is the velocity of signal propagation, and θ is the angle between the velocity vector and a vector joining the source and receiver. The 20-Mc/s signal from Sputnik was detected with the help of a beat frequency oscillator. The audio signal frequency is given by:

$$\nu_0' = |\nu_B - \nu'| = \left| \nu_B - \frac{\nu}{1 - \beta \cos \theta} \right|$$

where $\beta = v/c$, ν_B is the BFO frequency, and ν_0' is the audiofrequency. This equation gives the dependence of the audio signal upon θ.

Solving for cos θ, and using $\nu_B > \nu_0'$, we get

$$\cos \theta = \frac{1}{\beta}\left[1 - \frac{\nu}{\nu_B - \nu_0'}\right].$$

Since $\nu_0' \ll \nu_B$,

$$\cos \theta \simeq \frac{1}{\beta}\left[1 - \frac{\nu}{\nu_B} - \frac{\nu\nu_0'}{\nu_B^2}\right].$$

Letting $\delta = \nu_A - \nu$, and using

$$\frac{\delta}{\nu} \ll 1,$$

$$\frac{\nu}{\nu_B} = \frac{\nu}{\nu + \delta} \simeq 1 - \frac{\delta}{\nu},$$

and

$$\frac{\nu\nu_0'}{\nu_B^2} = \frac{\nu\nu_0'}{(\nu + \delta)^2} \simeq \frac{\nu_0'}{\nu}\left(1 - \frac{2\delta}{\nu}\right).$$

Thus,

$$\cos \theta \simeq \frac{1}{\beta}\left[\frac{\delta}{\nu} - \frac{\nu_0'}{\nu} + \frac{2\nu_0'\delta}{\nu^2}\right]$$

or, to a good enough approximation,

$$\cos \theta \simeq \frac{1}{\beta}\left[\frac{\delta}{\nu} - \frac{\nu_0'}{\nu}\right]. \quad (1)$$

Unfortunately, ν_B and hence δ, is not known. Calculations will therefore necessitate more information (to be supplied by geometric considerations).

A simple relation between $d\nu/dt$ and θ is obtained as follows. Assume (a) that the effects of the earth's curvature are not important (maximum error to be expected $\simeq 3°$) for the region of interest, (b) that the proximity of the F layer to Sputnik does not appreciably alter the velocity of the transmitted energy, and (c) that Sputnik passed directly overhead. We have from Fig. 11(a) that $\tan \theta = y/x$, where y is the height of the satellite and x is the horizontal distance between the satellite and the antenna. Because

$$\frac{v}{c} = \frac{5 \text{ mile/sec}}{1{\cdot}86 \times 10^5 \text{ mile/sec}} = 2{\cdot}69 \times 10^{-5} \ll 1,$$

$$\nu' = \frac{\nu}{1 - v/c \cos \theta} \simeq \nu(1 + v/c \cos \theta).$$

Thus,

$$\frac{d\nu'}{dt} = -\frac{\nu v}{c} \sin \theta \frac{d\theta}{dt}.$$

The $d\theta/dt$ is easily related to known parameters as follows: $x = y/\tan \theta$. Therefore,

$$\frac{dx}{dt} = -\frac{y \sec^2 \theta}{\tan^2 \theta} \frac{d\theta}{dt}.$$

and since $v = dx/dt$,

$$\frac{d\theta}{dt} = -\frac{v}{h} \sin^2 \theta.$$

Inserting this into the expression for $d\nu'/dt$, we obtain

$$\frac{d\nu'}{dt} = \frac{\nu v^2}{hc} \sin^3 \theta. \tag{2}$$

Similarly, by making use of Figs. 11(b) and (c), we obtain

$$\frac{d\nu_\alpha}{dt} = \frac{\nu v^2}{c} \frac{\sin^3 \theta}{[2y_2 + y_1]}, \tag{3}$$

and

$$\frac{d\nu_{\alpha'}}{dt} = \frac{\nu v^2}{c} \frac{\sin^3 \theta}{[4y_2 + y_1]}. \tag{4}$$

We can now show that the proposed model is consistent with the data given in Fig. 9. Equations (1) through (4) will be used to give the θs that correspond to the frequencies observed. First, we obtain the value of δ in Equation (1). From Fig. 2, $x = 730$ miles, which implies that $\theta \simeq 11°$ or $12°$ for a direct ray. Using $\theta = 12°$ for $\nu_0' = 620$ c/s in Equation (1), we get $\delta = 1{\cdot}14 \times 10^3$. To check the consistency of using $\theta = 12°$, we can use Equation (2). Because of the spurious dip in the curve

(Fig. 9) when 220 sec $< t <$ 260 sec, we must obtain $d\nu'/dt$ by extrapolation. Taking the estimated $d\nu'/dt$ to be between 0·2 and 0·5, from Equation (2) we obtain $12° < \theta < 18°$, which does not contradict the value $\theta = 12°$. (An independent frequency analysis not shown here gave $d\nu'/dt \simeq 0{\cdot}25$.)

Table II lists the results of applying Equations (1) through (4) to the data (note that the cut-off angle is $30° \pm 3°$).

TABLE II

EXPERIMENT

$$\nu_{\alpha'} = 650 \text{ c/s} \Rightarrow \theta = 26°$$

$$\nu_{\alpha'} = 690 \text{ c/s} \Rightarrow \theta = 33° \text{ (cutoff angle)}$$

$$\left[\frac{d\nu_{\alpha'}}{dt}\right] = 0{\cdot}38 \text{ (from Fig. 9)}$$

$$\nu_{\alpha} = 620 \text{ c/s} \Rightarrow \theta = 14°$$

$$\nu_{\alpha} = 670 \text{ c/s} \Rightarrow \theta = 29° \text{ (cutoff angle)}$$

$$\left[\frac{d\nu_{\alpha}}{dt}\right] = 0{\cdot}68 \text{ (from Fig. 9)}$$

$$\nu_{B} \simeq 715 \text{ c/s} \Rightarrow \theta \simeq 37°$$

TABLE III

THEORY

$$A = 1805 \text{ mi*} \Rightarrow \theta = 23°$$

$$A = 1380 \text{ mi*} \Rightarrow \theta = 29°$$

$$\left[\frac{d\nu_{\alpha''}}{dt}\right] = 0{\cdot}36$$

$$A = 1380 \text{ mi*} \Rightarrow \theta = 18°$$

$$A = 730 \text{ mi*} \Rightarrow \theta = 32°$$

$$\left[\frac{d\nu_{\alpha}}{dt}\right] = 0{\cdot}63$$

*Obtained from Fig. 2

Table III gives theoretical values of θ, obtained from Fig. 11; and gives $d\nu/dt$, obtained from Equations (3) and (4).

The proposed model is thus consistent with the data, and thus the behavior of the ν_α and $\nu_{\alpha'}$ frequencies is completely explained. It should be pointed out that the value of δ used was chosen *before* Table III was calculated. The ν_β frequencies will be explained further on.

The reduction in the number of ears on successive days is easily explained by the fact that large reflected signal intensities are possible only when the earth reflections are from a highly conducting surface such as that of the ocean. If the orbit went from SW to NE over an area of the Atlantic Ocean just east of Boston, the pattern produced would be that of Fig. 3. If, in addition to this, the orbit shifted to the west on each successive day, reception of the signal would have to

FIG. 13. Map showing shift of orbit.

depend on ground reflections. The resulting severe loss in reflecting power would cause the ears to disappear. Figure 13 graphically shows this. From the shape of the land, right ears should be the first to disappear, and this is of course exactly what happens. In appendix B we see that this explanation is supported by the orbit data.*

*In a letter published in *Proc. Inst Radio Engrs, N.Y.* (Mar. 1958) under the title "A Note on Some Signal Characteristics of Sputnik I," Kraus mentioned an unsteadiness of the signal as Sputnik I approached, and advanced an explanation that involved compressed ionized gases in front of the satellite. Inasmuch as his published observation was for a NW to SE (morning) pass, and his antenna was at the appropriate distance south of the Great Lakes, it is quite possible that, contrary to his proposed explanation, he observed some sort of ear phenomenon caused by a reflection from these lakes. This possibility could easily be checked by a frequency analysis of his tape recordings.

The lack of synchronism (6 sec) in the recordings of an ear on 20 October (Fig. 7) could conceivably be due to differing antenna patterns, that is, the signal may have been attenuated by nulls in the NS antenna patterns but not in the EW antenna patterns. An alternative possibility is mentioned in Appendix C.

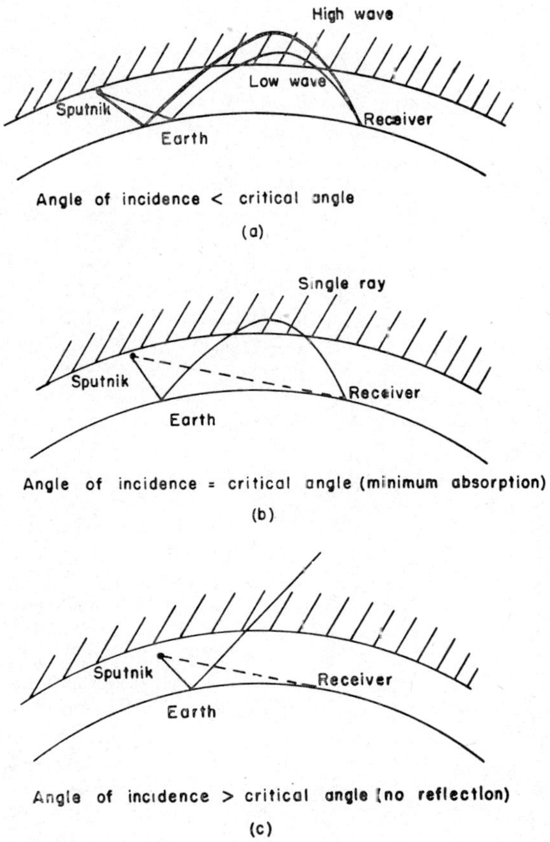

FIG. 14. "High" and "low" waves corresponding to v_β and v_α respectively.

The increase in spacing of the ears as a function of time is most likely due to the changing critical frequency of the ionosphere shortly after sundown—the ear phenomenon occurred approximately one hour after sunset. As the satellite moved north-eastward, it passed through ionospheric regions of lower critical frequency, which changed the critical angle for reflection, and this in turn changed the ear spacings. Details and exact mathematical confirmation of this have not been worked out.

The other aspects of the phenomenon are more easily explained after considering the following case. If a constant-frequency transmitter and a receiver are located on the ground at a distance from each other too large for ground wave reception but small enough for reception due to single ionospheric reflection, then it is possible for the radiation to follow *two* paths to the receiver. These paths correspond to the "high" and "low" waves discussed in the *National Bureau of Standards Circular*, No. 462, pages 67 to 71. If the distance between the transmitter and receiver is then diminished, the two paths approach one another until a critical point—the skip distance—is reached, where both paths unite. Maximum reflection and a sort of focusing occur at this point. Any further diminution of the distance results in a rapid decrease in received signal strength.

Figure 14 shows the sequence of paths proposed for the region C-E of Fig. 3. As the critical angle is approached the signal strength increases, until it reaches a maximum in intensity followed by an abrupt cut-off as the satellite enters the skip zone. The very large peaks seen in an ear region of the pen recording, which are due to both the focusing effect and the exceptionally good reflection from the ocean and the ionosphere, may have been somewhat enhanced by the following properties of the apparatus: (a) It is possible that the pens, which were vibrating at a high speed, could have swung higher than the correct point because of inertial effects, thus registering an erroneously high signal strength. (b) The large difference between the signal of the ear before cut-off, just preceding the main pattern, and the signal of the main pattern was increased because the signals coming along a direct path were attenuated by the null points of the interferometers.

Lastly, the merging of the ν_α and ν_β frequencies is explained by the merging of the high and low rays. Theoretically, the two frequencies *should* meet and the fact that they apparently do not is in all likelihood due to experimental error. In Appendix C this is discussed further on the basis of information obtained by listening directly to the tapes. This proposed explanation for the "ear phenomenon" is convincing not only because of its consistency, but primarily because of its simplicity.

Appendix A

Time of Recorded Passes for Sputnik I ($\pm$ 5 min) EST

1957

9 Oct 06·58	16 Oct 06·55
19·25	
21·00	18 Oct 19·05*
23·00	20·30
10 Oct 21·00	19 Oct 05·05
23.44	19·00*
	20·35
11 Oct 21·00	
23·28	20 Oct 18·50*
	04·55
12 Oct 06·20	20·25
20·60	

	21 Oct 04·50
13 Oct 06·18	17·15
06·50	18·45*
14·25	
20·45	22 Oct 17·09
22·20	18·40
	20·10
14 Oct 20·50	
	23 Oct 17·00
15 Oct 20·55	18·38
22·20	

*Passes exhibiting the ear phenomenon

Appendix B

LONGITUDES OF CROSSINGS OF THE 40th PARALLEL OF $\alpha - 1$ (ROCKET)
(obtained from the Smithsonian Institute)

1957	
18 Oct	61°W.
19 Oct	63°W.
20 Oct	65°W.
21 Oct	67°W.
22 Oct	69°W.

Note: These data are probably good to within 4°. Boston is located at 71°W. The path of the satellite with respect to the moving earth's surface is approximately 33° with the vertical from SW t. NE.

Appendix C

MERGING OF ν_α AND ν_β FREQUENCIES

Repeated direct auditing of the tapes reveals that at the end of an ear the frequency of one of the beat tones went continuously to zero. This suggests that the ν_α and ν_β tones actually met. The situation is, however, more complex because at a time approximately 6 sec later the same thing happened again. The proposed explanation is that there is a further multiplicity of signal paths caused by the Faraday effect—that there are *two* ν_α and *two* ν_β tones present (call them ν_α, ν'_α and ν_β, ν'_β respectively). What probably happens is that ν_α and ν_β meet, then ν'_α andt ν'_β meet about 6 sec later. That the ν_α and ν_β frequencies apparently did not merge (Fig. 9) can be attributed to the confusion arising from frequencies too closely spaced to be separated by the filter, as well as to the crude method of measuring ν_β.

The explanation for the unsynchronized ears in Fig. 7 may involve the Faraday effect because the time differential between NS and EW ears is 6 sec—the same as between the meetings of ν_α and ν_β, and ν'_α and ν'_β. The polarization of the second component of the Faraday split signal (ν'_α and ν'_β) was perhaps such that the NS arrays were less capable of detecting it than the EW arrays were. This would indeed explain Fig. 7 consistently. We feel that this explanation is more likely than the one offered previously.

Further frequency analyses of the tapes will be made to check the auditory observations.

Notes Added in Proof

1. When this report was presented, Professor Dieminger gave the opinion based on the work he has done in back scattering that a large difference between the reflection coefficients of land and sea for 20 Mcps radiation is quite unlikely. A large difference would be needed for explanation of the gradual disappearance of the ears on successive days. (At the meeting, this ear disappearance was called the "Van Gogh Effect".) He suggested that the possibility of unusual sunspot activity during that time should be investigated.

It has subsequently been pointed out by Paul Wild of the University of Bern that on October 18, 1957, exactly the first day of the ear phenomenon was observed, the so-called gegenschein was unusually bright and a comet suddenly came into view. Also, on October 21, the last day of the ear phenomenon, the comet ceased to be visible. These evidences of unusual sunspot activity at exactly the interval of time of interest should not be taken lightly. On the other hand, since the reflections were mainly from wooded areas and since 20 Mcps corresponds to a wavelength very nearly equal to the height of an average tree, there remains the possibility that there was an unusual amount of land absorption for specular reflection at this frequency.

2. Recently, R. Parthasarathy and G. C. Reid of the Geophysical Institute, College, Alaska reported in *Proc. Inst. Radio Engrs, N.Y.*, vol. 47, January 1959, that they observed a phenomenon similar to that of Kraus mentioned above. They said, "On many occasions the entire record is either unbroken or completely broken." Their observation prompts me to mention that on October 19, during the 2035 pass, when the satellite was far out over the ocean, the pen recordings exhibited only ears and no main pattern. This was interpreted to mean that the signals reached the station only by means of multiple ocean-ionosphere reflections.

Acknowledgments

I would like to thank Dr. Werner Gerbes for suggesting this problem and encouraging its solution. In addition, I would like to acknowledge the cooperation and literature references given by certain members of the Electromagnetic Radiation Laboratory and Propagation Sciences Laboratory, especially Noel Stone, Herman Poeverlein, and Hallock Marsh. Thanks are also due Thomas Skinner, for doing a calculation from the Doppler curve, and Mary Regier Hoes, for assistance with the illustrations. In addition, acknowledgment is made of the very painstaking work done by Miss Betty Karasik in editing the paper.

Etude Physique du Feuilletage dans l'Atmosphère

F. DU CASTEL, P. MISME ET J. VOGE

présenté par

P. MISME

Centre National d'Etudes des Télécommunications, Issy-les-Moulineaux (Seine)

Après quelques rappels de mécanique des fluides et des caractéristiques de l'atmosphère, les auteurs analysent le cas d'une atmosphère limpide. Ils en déduisent qu'en dehors de quelques cas particuliers l'atmosphère ne peut pas être le siège de turbulence généralisée. On met ainsi en évidence l'existence de feuillets stables qui coexistent éventuellement avec des feuillets turbulents. On étudie ensuite la loi de répartition de l'indice de réfraction à l'intérieur de chaque feuillet. L'article signale enfin les précautions à prendre pour le dépouillement des mesures faites à bord d'avions.

I. HISTORIQUE ET INTRODUCTION

Dès l'origine des transmissions d'informations par des procédés "sans-fil" il est apparu que la non-constance des propriétés de l'atmosphère influait sur la qualité du signal reçu. De ce fait il est devenu indispensable d'étudier l'atmosphère, considérée comme milieu de propagation, au même titre que le câble dans les liaisons avec "fil." On a successivement simplifié cette étude en considérant que l'atmosphère pouvait s'assimiler au vide, était un milieu homogène, admettait une représentation stratifiée ou au contraire était le siège permanent de mouvement tourbillonnaire. Cette dernière caractéristique elle-même a donné naissance à des théories de turbulence isotropique ou anisotropique.

Dans l'étude ci-dessous on essayera de ne considérer que les propriétés de l'atmosphère considérée comme un mélange gazeux pesant et, par la suite la structure de cette atmosphère servira de base à une théorie de propagation.

II. REGIMES D'ECOULEMENT DANS L'ATMOSPHERE

Pour étudier les propriétés du mélange gazeux qui forme l'atmosphère la mécanique des fluides offre deux conceptions; le point de vue d'Euler et celui de Lagrange.

On dit que l'on se place au point de vue de Lagrange lorsque l'étude est conduite en prenant un point fixe sur une trajectoire comme point de référence. Au point de vue d'Euler on se place en un point fixe par rapport aux parois. Toutes les mesures faites au sol on en avion se ramènent au cas d'Euler. On essayera au contraire dans l'étude qui suivra de se placer au point de vue de Lagrange chaque fois que cela sera possible. D'ailleurs l'utilisation de l'une ou l'autre méthode n'est pas indifférente à l'interprétation des résultats.

On sait en effet que l'écoulement d'un fluide ne peut se faire que par l'un des deux processus: laminaire ou turbulent. La limite entre les deux régimes d'écoulement est nettement marquée et pour des conditions d'expérience choisies, elle dépend du nombre de Reynolds. Lorsque le fluide est compressible, comme l'air, en écoulement laminaire, sa vitesse peut varier d'un point à un autre de l'espace. En se plaçant en un point fixe par rapport au sol on observera des écarts à la vitesse moyenne. Si ces écarts ne semblent pas suivre une loi simple, on suppose qu'ils sont aléatoires, et ce critère est souvent pris comme définition de la turbulence ce qui conduit à une contradiction.

Malheureusement le fait d'admettre une valeur du nombre de Reynolds comme limite des deux types d'écoulement conduit à des calculs pratiquement inextricables dans le cas de l'atmosphère. On va donc essayer d'utiliser une des propriétés de la turbulence, plus accessible à la mesure, pour définir le type de régime.

En atmosphère limpide, si l'on excepte la vapeur d'eau (et éventuellement le gaz carbonique) la composition chimique de l'atmosphère est remarquablement constante à toutes les altitudes, c'est-à-dire contient une masse constante des différents gaz. Par contre la quantité de vapeur d'eau par kilogramme d'air, que l'on appelle l'humidité spécifique, varie continuellement. On constate simplement que cette dernière quantité, en valeur moyenne, diminue avec l'altitude.

On va donc admettre que l'atmosphère est composée d'une gaz homogène qui est l'air sec, dans lequel on peut introduire de la vapeur d'eau.

Les sources de vapeur d'eau, représentées par les océans et la végétation, se trouvent à la surface du globe terrestre. La densité relative de la vapeur d'eau par rapport à l'air est faible. Si l'atmosphère était éternellement au repos il y aurait donc une tendance à ce que cette vapeur d'eau subisse un mouvement ascendant jusqu'à un niveau de condensation. Le coefficient de diffusion de vapeur d'eau vers le haut est très différent suivant le régime d'écoulement du milieu. En régime laminaire sensiblement horizontal il faut considérer le coefficient de diffusion moléculaire. En régime turbulent ce coefficient de diffusion devient environ 10^4 fois plus grand. Pour notre étude on peut donc dire que le mélange air plus vapeur d'eau sera homogène, c'est-à-dire chimiquement constant dans le cas de régime turbulent et ne sera pas homogène dans le cas de régime laminaire. Dans ce dernier cas, en dehors des sources ou puits de vapeur d'eau, l'humidité spécifique décroit avec l'altitude.

III. Quelques Caracteristiques de l'Atmosphere

On a signalé qu'il était très difficile d'effectuer des mesures dans l'atmosphère en se plaçant au point de vue de Lagrange. Cependant quelques résultats ont pu être obtenus en utilisant des ballons ou des motoplaneurs. On en retiendra que la plus forte décroissance de température avec l'altitude est toujours très voisine de la valeur adiabatique, c'est-à-dire environ 10° par 1000 m dans les 2 premiers kilomètres. Cette règle ne souffre d'exception qu'au voisinage d'un sol surchauffé.

Lorsque le gradient vertical de température devient faiblement supérieur à la valeur adiabatique, des mouvements verticaux dirigés de bas en haut prennent naissance. Mais la quantité d'air qui est ainsi élevée doit être remplacée par une quantité égale qui provient donc de mouvement de haut en bas. L'expérience montre que le mouvement ainsi créé est turbulent. On en déduit donc une autre caractéristique du seuil de turbulence.

Enfin l'étude d'un paramètre quelconque met en évidence des irrégularités nombreuses dans les plans verticaux ou horizontaux. Sur les Fig. 1 et 2 on a

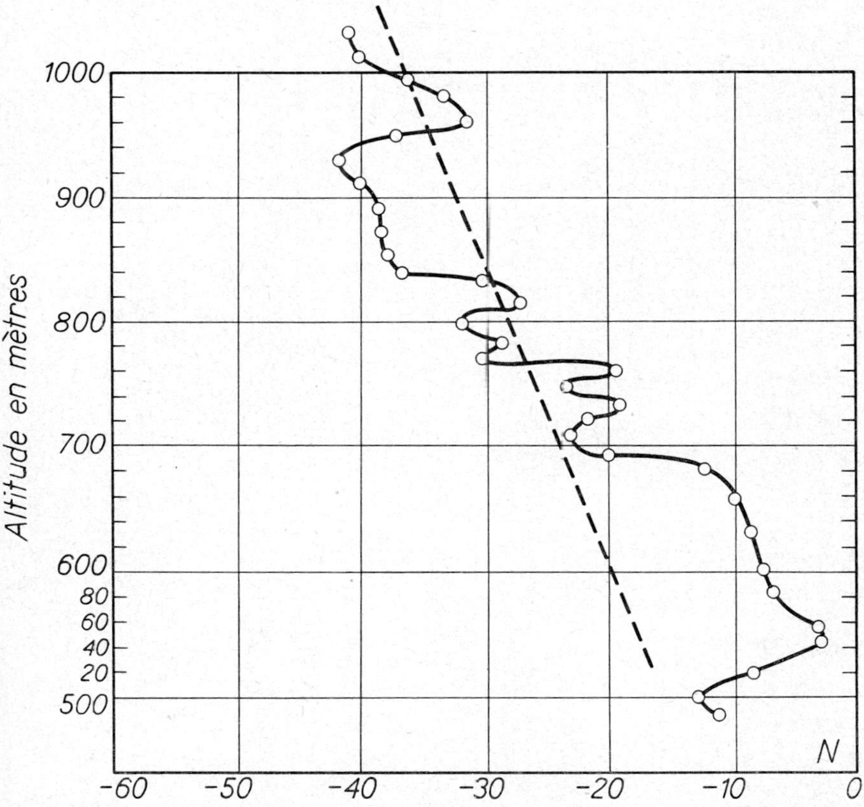

FIG. 1. Exemple obtenu par radiosondage. La précision de l'instrument est de l'ordre de $\pm$ 2 N.

représenté des variations d'indice de réfraction dans un plan vertical et des variations de température dans un plan horizontal. Les constantes de temps des appareils de mesures sont les mêmes dans les deux exemples.

IV. Repartition de l'Indice de Refraction

Nous étudierons la fonction $N = (n - 1)\,10^6$. Les valeurs médianes mensuelles de la variation de N dans le premier kilomètre au-dessus du sol varient suivant la situation géographique entre les limites approximatives de -20 et -80; la valeur -40 représente d'ailleurs assez bien les climats tempérés.

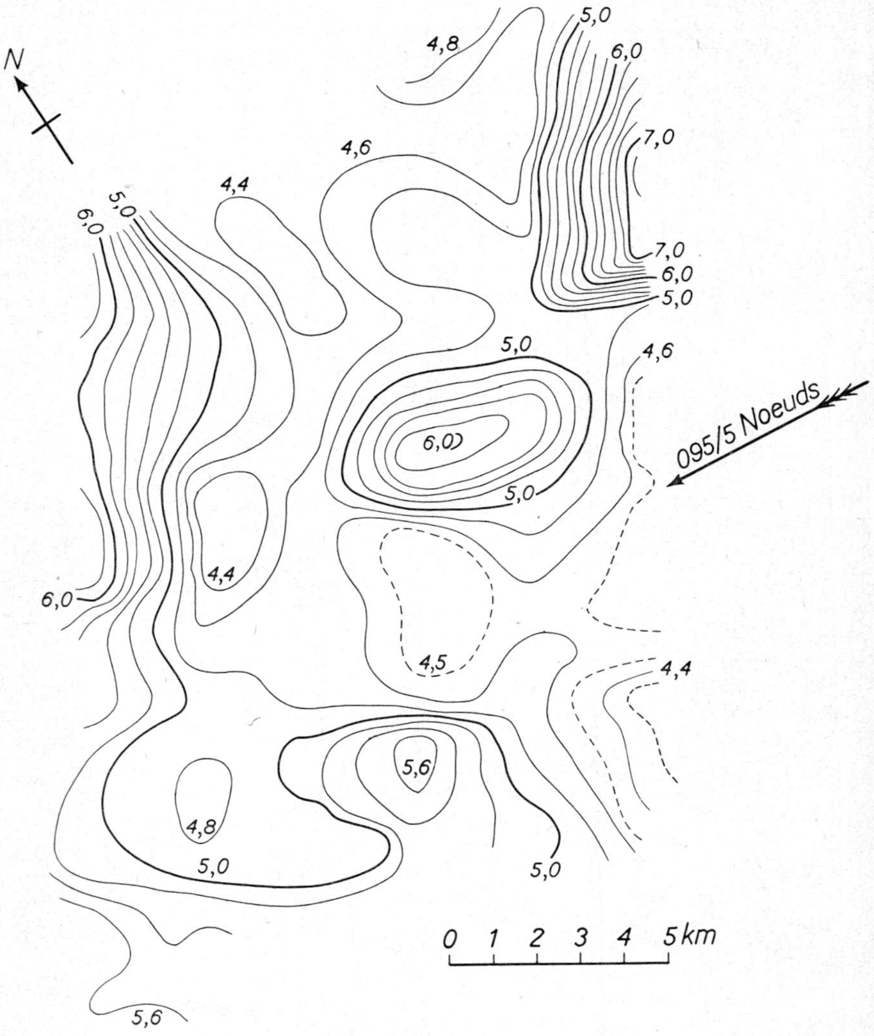

FIG. 2. Structure du champ de température sur une surface de l'ordre de 15×20 km^2. Le vol de l'avion a été effectué à pression constante. L'altitude n'est donc donnée qu'à titre indicatif.

N est fonction de la pression, la température et l'humidité et un ordre de grandeur à 10 % près environ des variations de N dans les basses couches est donné par:

$$\Delta N = 0{\cdot}3 \; \Delta P - \Delta T + 7\Delta s,$$

avec P = pression, T = température, s = humidité spécifique.

P est presque constant dans le premier kilomètre et de l'ordre de 100 mb.

Raisonnons sur une épaisseur de 1 km.

Si l'atmosphère était un mélange chimiquement constant on aurait $\Delta s = 0$. D'autre part T diminue avec l'altitude d'une quantité toujours inférieure à 10°/km. Donc à mélange constant on aurait $20 \leqslant -\Delta N$.

Par suite du gradient vertical de température, la valeur $-\Delta N = 20$ correspond au seuil d'instabilité de l'atmosphère. Etant donné l'importance du coefficient de diffusion turbulente seule la valeur $-\Delta N = 20$ est compatible avec la notion de milieu homogène et turbulent. Donc en dehors de source ou de puits dans l'atmosphère $-\Delta N$ ne peut prendre une valeur inférieure à 20 environ. On a pu vérifier cet ordre de grandeur sur de nombreux sondages, et on sait que la valeur médiane de ΔN/km est de 20 au Sahara et en U.R.S.S. Il est probable que l'éloignement de la mer dans les deux cas conduit à une valeur médiane nulle pour les sources ou puits de vapeur d'eau en altitude.

La turbulence étant nécessaire et suffisante pour rendre rapidement le milieu homogène, on en déduit que lorsque $-\Delta N > 20$ l'atmosphère ne peut pas être le siège d'une turbulence généralisée. Il faut donc envisager l'existence de zones stables et étudier la décroissance de s avec l'altitude.

Considérons une tranche d'atmosphère ne contenant pas de particules liquides en suspension. Nous savons que s décroit lorsque l'altitude h augmente, mais l'expérience prouve que cette décroissance n'est pas monotone.

Il existera des tranches d'atmosphère pour lesquelles $ds/dh < 0$ et éventuellement d'autres caractérisées par $ds/dh = 0$. Nous donnons le nom de feuillet à ces différents éléments. On met ainsi en évidence la possibilité d'existence de deux familles de feuillets différents.

A la suite de ce qui a été dit au sujet de la diffusion de la vapeur d'eau les feuillets tels que $ds/dh = 0$ ne seront pas le siège de mouvements turbulents, on les appelera des feuillets stables. Les autres, quand ils existent, seront turbulents.

V. Etude Verticale des Feuillets Stables

Dans un feuillet stable l'humidité spécifique décroît avec l'altitude. De plus cette décroissance est fonction du temps. En effet s'il n'y a ni "source" ni "puits," c'est-à-dire si on conserve les hypothèses de départ: pas de condensation ni de précipitation, la loi de décroissance de l'humidité sera donnée par l'équation différentielle:

$$\frac{ds}{dt} = -D\frac{\partial^2 s}{\partial h^2},$$

où t est le temps et D le coefficient de diffusion moléculaire.

Lorsque le régime permanent sera atteint, ce qui est probablement possible pour des épaisseurs faibles de l'ordre de la dizaine de mètres, la loi de répartition de s sera une loi linéaire. Cette loi sera sensiblement valable pour $N(h)$. En dehors du régime permanent la solution ne peut pas être explicitée, mais nous remarquons que des lois faisant intervenir des lignes hyperboliques seront solutions. On peut dans ce cas admettre que la variation de $N(h)$ est représentée par les fonctions hypergéométriques adoptées par Epstein.

L'étude du gradient d'indice de réfraction dans une tranche d'atmosphère conduit à un grand nombre de possibilités. En effet, seule la valeur maximum de la décroissance de température avec l'altitude est fixée. Au-delà de cette borne toutes les valeurs sont possible.

Considérons une tranche de 1000 m d'épaisseur présentant une variation de 40 N par exemple (Fig. 3). Soit A le point figuratif de l'indice de réfraction au

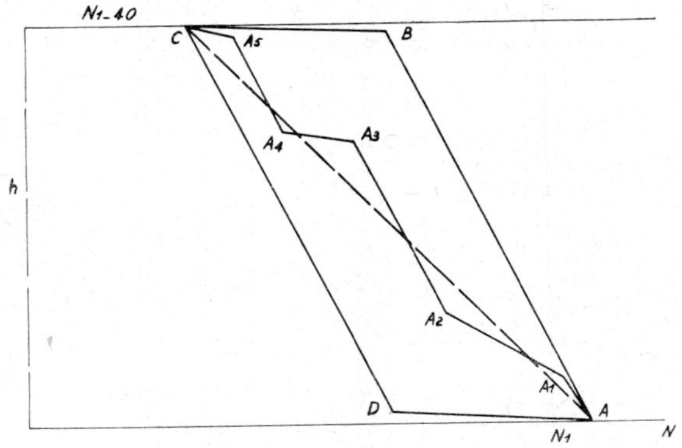

FIG. 3. Limite de variation de $N(h)$ pour une épaisseur de 1 km dans les basses couches. Les gradients des feuillets AA_1, A_2A_3, A_4A_5 sont les mêmes que ceux des feuillets AB et DC.

niveau 0 et C le point figuratif au niveau 0 + 1 km. Les feuillets turbulents quand ils existent seront représentés par des variations parallèle à AB. On peut donc concevoir par exemple les représentations ABC ou ADC telles que BC ou AD aient des gradients presque infinis. De la même façon il peut exister des représentations telles que AA_1 . . . C pour lesquelles A_1A_2, A_3A_4, A_5C, aient des gradients quelconques supérieurs en valeur absolue à 20 N/km. La seule conclusion est que l'épaisseur des feuillets stables peut être quelconque. Dans une étude plus complète il faudra envisager le spectre de ces épaisseurs. Les gradients des feuillets stables sont toujours supérieurs en valeur à 20 N/km et la somme des variations de N dans ces feuillets est l'excès de la variation réelle à la variation adiabatique.

VI. ETUDE VERTICALE DES FEUILLETS TURBULENTS

Le mécanisme qui conduit à l'existence de cellules tourbillonnaires dans l'atmosphère libre est assez obscur. Dans le cadre de l'étude de la mécanique des fluides, d'innombrables études statistiques ont été enterprises par des méthodes théoriques [1] ou expérimentales. Cependant en général les auteurs ont admis que la population ainsi étudiée était homogène ce qui ne semble pas évident *a priori* et on a vu que le point de vue d'Euler que de nombreux expérimentateurs avaient adopté, pouvait conduire à des interprétations erronées.*

*On doit cependant noter que quelques auteurs se sont placés dans les conditions physiques de la turbulence: nombre de Reynolds très élevé.

Le mécanisme physique qui semble le mieux rendre compte de ces mouvements est celui qui donne naissance à des "tourbillons thermoconvectifs." Benard et lord Rayleigh ont les premiers étudié ce phénomène. Depuis, de nombreuses études ont été publiées sur cette question [2, 4].

On peut rapidement résumer le mécanisme des tourbillons thermoconvectifs. Considérons deux surfaces horizontales telles que celle qui occupe la position inférieure soit plus chaude que l'autre. Si certaines conditions de gradient de température et d'épaisseur sont respectées, des tourbillons parfaitement organisés prennent naissance et le régime permanent se forme rapidement. Il semble cependant que les ordres de grandeurs trouvés en laboratoire et dans l'atmosphère ne soient pas en bon accord. Quoiqu'il en soit il est certain que, dès que le gradient de température devient très peu supérieur au gradient adiabatique, il se forme des tourbillons conduisant à une rapide et parfaite homogénéisation du milieu. Les études de laboratoire tendraient à prouver que l'épaisseur de cette turbulence ne peut pas être quelconque compte tenu de fait que le gradient de température est fixé. D'autre part les dimensions des tourbillons élémentaires pourraient être bien calculées et le rapport de l'épaisseur à la longueur est de l'ordre de 2 à 4. Il semble normal d'admettre que certains feuillets turbulents peu épais sont constitués par des tourbillons parfaitement organisés, les feuillets les plus épais au contraire seraient formés de tourbillons élémentaires inorganisés.

Des études de laboratoire [2] permettent d'admettre que la dimension de ces tourbillons élémentaires est sensiblement constante en régime permanent. Dans ce cas les études théoriques et expérimentales de la répartition des différents éléments du point de vue d'Eulcr gardent toute leur valeur. Nous analyserons l'influence de ce type d'atmosphère dans la 2-ème partie de cette étude.

Quel que soit d'ailleurs le mécanisme envisagé, la loi de variation de la température est toujours très voisine de la loi adiabatique qui conduit pour l'indice de réfraction à un gradient de $-20\ N/\mathrm{km}$.

On peut donc sans grave erreur adopter une loi linéaire de l'indice de réfraction en fonction de l'altitude pour caractériser les feuillets turbulents.

En résumé on peut dire que chaque fois que le gradient moyen d'indice de réfraction est supérieur en valeur absolue à $20\ N/\mathrm{km}$ dans l'atmosphère limpide on doit considérer un empilage de feuillets stables et éventuellement de feuillets turbulents.

Le gradient des premiers est quelconque, celui des autres est de $-20\ N/\mathrm{km}$.

Remarquons enfin que l'étude ci-dessus a été conduite pour donner des ordres de grandeur dans les basses couches mais que le raisonnement est valable à n'importe quelle altitude.

VII. Structure Horizontale des Feuillets

Les mesures montrent sans ambiguité que les variations horizontales des paramètres atmosphériques sont beaucoup plus faibles que les variations verticales en dehors de certaines zones particulières. De la même façon le vecteur vitesse du vent est très peu incliné sur l'horizontale. Toutes ces mesures conduisent à donner aux deux familles de feuillets considérés des dimensions horizontales très grandes devant leurs dimensions verticales. On pourra trouver quelques résultats de mesure dans un article déjà publié par les mêmes auteurs [3].

La notion d'horizontale est cependant une approximation.

Une surface caractérisée par la constance d'un paramètre dans l'atmosphère, par exemple une surface isotherme ou iso-indice est déformée par les mouvements de l'atmosphère qui n'est pas au repos dans un plan vertical.

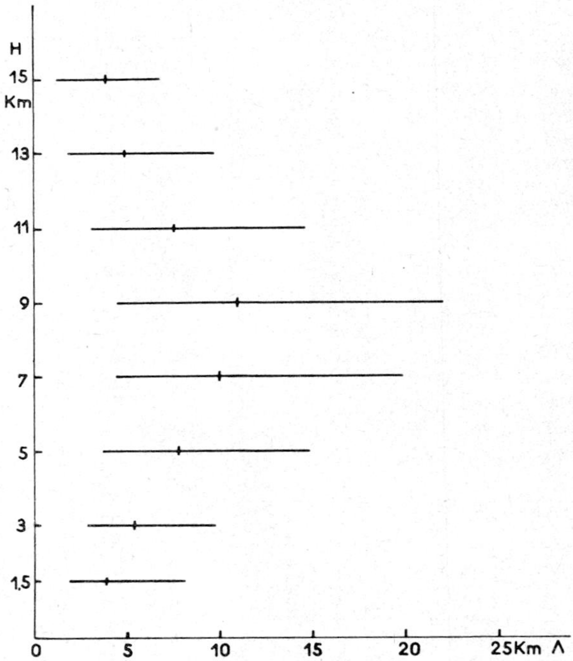

FIG. 4. A chaque niveau on a représenté les valeurs 10 %, 50 %, 90 % des cas calculés pour l'année 1957. Le calcul a été effectué en tenant compte des valeurs moyennes du vent et des valeurs moyennes de températures. On notera que sur la Fig. 2 les longueurs d'onde sont de l'ordre de 7 km.

Si pour une raison quelconque une particule d'air est écartée de sa position d'équilibre, elle sera soumise à une accélération

$$-g\frac{\gamma_a - \gamma}{T}z,$$

où g = accélération de la pesanteur,
γ_a = gradient verticale adiabatique de la température,
γ = gradient vertical réel de la température,
z = distance verticale de la particule à son point d'équilibre.

L'intégration de l'équation du mouvement conduit à un mouvement oscillatoire non amorti si on néglige la viscosité de l'air.* On en déduit que la période est

$$\pi = 2n\sqrt{\left(\frac{T}{g(\gamma-\gamma)_a}\right)},$$

et la longueur d'onde

$$\Lambda = 2nV\sqrt{\left(\frac{T}{g(\gamma_a-\gamma)}\right)},$$

V étant la vitesse d'écoulement, c'est-à-dire la vitesse du vent. Comme T est mesuré en degré absolu, sa variation est faible devant celle de $\gamma_a - \gamma$.

Dans les conditions d'instabilité, c'est-à-dire $\gamma_a = \gamma$, la période et la longueur d'onde deviennent infinies, c'est-à-dire qu'il n'y a plus d'oscillations. Donc seuls les feuillets stables présenteront un écoulement à allure sïnusoidale. On a déjà signalé que fréquemment des feuillets turbulents étaient compris entre des feuillets stables. Si ces derniers subissent un mouvement oscillatoire, la zone turbulente sera limitée par des surfaces ayant ce mouvement. Donc même si on ignore l'organisation de la turbulence on aura une précision sur la forme des surfaces limites des feuillets turbulents.

Les ordres de grandeur des mouvements oscillants signalisés ci-dessus ont été fréquemment utilisés par les planeurs à l'occasion de ce qu'il est convenu d'appeler le vol d'onde. On doit donc voir là une bonne vérification de cette théorie.

A titre indicatif on a donné Fig. 4 la répartition 10 %, 90 % des longueurs d'onde du mouvement oscillatoire des feuillets dans la région parisienne. Ces valeurs ont été calculées à l'aide de mesures de vent effectuées en 1957.

VIII. Irregularite de Surface

Jusqu'à maintenant on a considéré que les surfaces de séparation des feuillets n'étaient soumises qu'à des oscillations de grande longueur d'onde. En réalité dans les couches turbulentes, quelle que soit l'organisation des éléments tourbillonnaires, il existe des mouvements verticaux de l'air. Par suite des inégalités de rayonnement du sol et de l'absorption atmosphérique, ces courants verticaux subissent des fluctuations. De ce fait les surfaces de séparation des feuillets subissent des déformations continuelles que seule une méthode statistique peut permettre d'analyser. Quelques ordres de grandeur ont déjà été cités [3]. Rappelons qu'ils ont été déduits de mesures effectuées du point de vue d'Euler ce qui est normale en cas de régime turbulent.

IX. Utilisation des Mesures Faites par Avion

On a expliqué au début de cet article que la méthode d'analyse physique qui avait été choisie dans cet article conduisait à pouvoir expliquer la structure de

*Les ordres de grandeurs trouvés pour la longueur d'onde montrent que la courbure des filets d'air est très faible, dans ce cas il est normale de négliger les forces de viscosité.

l'atmosphère en une suite de points fixes par rapport à l'écoulement de l'air. L'inconvénient de cette méthode est que la mesure faite à bord d'un mobile est plus difficilement utilisable. En particulier la trajectoire d'un avion en montée est peu inclinée sur l'horizontale: de l'ordre de 1/10 de radian environ. De ce fait l'analyse d'un feuillet de 20 m d'épaisseur sera faite sur une longueur de 200 m et on sera conduit pour l'interprétation des résultats à placer ces deux points sur la même verticale. On voit donc que l'analyse des mesures doit être faite avec prudence. Si on utilise une méthode de dépouillement statistique, sans analyse physique préalable, on comparera les vecteurs vitesse du vent appartenant à des écoulements différents et on en déduira une turbulence généralisée dont on a démontré l'impossibilité ci-dessus.

Par contre si une analyse physique a déjà prouvé que le milieu considéré est turbulent, l'étude statistique est possible puisqu'elle traite le cas d'une population homogène. Ceci doit d'ailleurs être rapproché d'une remarque faite par Crain à l'U.R.S.I. en 1957. Cet auteur signale qu'il semblait illusoire de vouloir définir les propriétés de l'atmosphère par une seule fonction d'autocorrélation.

Sous une autre forme les observations faites en dehors d'un point fixe par rapport à une trajectoire, conduisent à intégrer dans le temps et l'espace et ainsi on croit mettre en évidence une rotation du vecteur vitesse du vent alors que rien ne permet de le supposer. On commet d'ailleurs une erreur supplémentaire en confondant la vitesse de rotation du vecteur vent, c'est-à-dire Rot $\vec{V}$ avec la possibilité de mouvement turbulent. Dans un fluide compressible div $\vec{V} \neq 0$ et une valeur non nulle de Rot $\vec{V}$ est une condition nécessaire mais non suffisante de la turbulence au sens de la mécanique des fluides.

Jusqu'à maintenant on a négligé les possibilités de forts gradients horizontaux des paramètres P, T, S. Des mesures précises ont largement prouvé que les variables T et S peuvant varier très rapidement dans le plan horizontal. Aucune loi autre que la conductibilité de la chaleur dans l'air ne vient limiter la valeur du gradient horizontal de T. Le gradient horizontal de S sera défini par une équation de la même forme dont les constantes diffèreront. L'existence de ces discontinuités horizontales limitent l'aire occupée par les feuillets et leur donnent l'allure d'une répartition aléatoire dans l'espace.

X. Conclusion

On a essayé de montrer par une étude simple et en ne faisant appel qu'à des ordres de grandeur que l'atmospherè ne pouvait pas être un milieu exclusivement turbulent. On en a déduit l'existence de feuillets les uns stables caractérisés par un fort gradient d'indice de réfraction, et les autres turbulents caractérisés par un gradient constant et faible de cet indice. On a signalé que ces feuillets étaient soumis à des ondulations à grande échelle dont les ordres de grandeurs étaient facilement calculables. Ce modèle d'atmosphère qui respecte les lois thermodynamiques et les résultats expérimentaux va servir de baseà l'étude de la propagation radioélectrique au delà de l'horizon.

Remerciements

Une grande partie de cette étude a été réalisée grâce aux nombreux conseils et à la documentation expérimentale de M. Eyraud, Ingénieur en Chef de la Météorologie. Les auteurs sont heureux de l'en remercier.

References

1. Agostini, L. & Bass, J. *Publ. sci. Minist. Air*, No. 237.
2. Avsec, D. *Publ. sci. Minist. Air*, No. 155.
3. Castel, F. du, Misme, P. & Voge, J. *Ann. Télécomm.* No. Juillet-Août, 1958.
4. Vernotte, P. *Publ. sci. Minist. Air*, No. 224.

The Application of the UHF Scatter Mode to Obtain Reliable, Extended Range, Aeronautical Communications

L. A. Ames, E. J. Martin and T. F. Rogers

Electronics Research Directorate, Air Force Cambridge Research Center, Bedford, Mass.

This paper is an attempt to utilize all the presently available body of pertinent knowledge concerning UHF Air-Ground radio wave propagation characteristics to estimate the transmission loss at different altitudes for distances far beyond the radio horizon. Based upon these estimates, equipment parameters such as transmitter power, receiver sensitivity and antenna gains required to achieve a given degree of voice communication reliability at various ranges and altitudes, are indicated.

I. Introduction

The present limited line-of-sight range of UHF air-ground communication systems places an added burden upon flight safety which, coupled with the increased speed and altitude of modern aircraft, serves to increase hazards created by any lack of prompt exchange of information between traffic controllers and aircraft pilots. Probably not more than 75 % of the line-of-sight range as estimated from the expression $D = \sqrt{(2\, h_T)} + \sqrt{(2\, h_R)}$ (D in miles: h_T and h_R in feet) is achieved consistently. This is due in part to anomalous propagation conditions and to equipment limitations and operational techniques. (It seems rather ridiculous that even today an aircraft costing millions of dollars and generating more than the equivalent of 10 million W power on take-off should be restricted to 10 to 100 W radio transmitter power for communication purposes if these purposes are important to the safety of the aircraft.)

In order to extend the reliable communication service radius HF is presently employed. However, atmospheric noise, variability in ionospheric conditions and interference in the already overcrowded HF bands severely limit the usefulness of this mode for reliable communication purposes. The employment of the sideband technique, while reducing the noise and interference problems, still is equally subject to gross variability in ionospheric conditions. Thus, at the present time, reliable communication to distances of say, 350 miles is not achieved.

Considerable information is now available relative to the behavior of tropospherically scattered radio-waves. We shall attempt to summarize the pertinent knowledge and to examine the feasibility of utilizing this mode to provide reliable extended range voice communication to 350 miles. Two general types of operational requirements are considered. One requirement is for reliable communication with piston engine type aircraft operating at altitudes ranging from 8000 to 18,000 ft and courses along certain fixed air traffic control corridors, such as those governing coastal approaches; the other for modern jet aircraft operating at much greater speed and altitude (20,000–40,000 ft) and along courses less restricted than in the first case.

II. Line-of-Sight Aircraft Communication

At the present time, a representative air-ground UHF system has a dynamic range of about 145 db (estimated on the basis of 10 W transmitter power, 3 μV receiver sensitivity, 6 and 2 db antenna gains at ground and airborne terminal respectively. No allowance is made for cable losses or system deterioration.) Figure 1 shows that this system dynamic range will provide, on the average, a

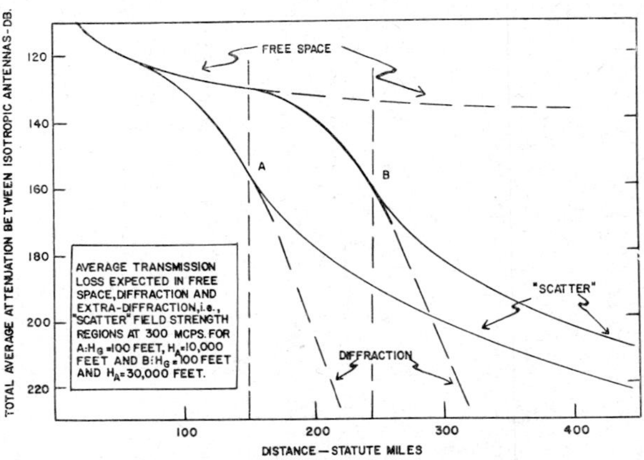

Fig. 1.

range of 140 statute miles when the receiving terminal is at 10,000 ft and the ground terminal at 100 ft. If the airborne terminal is raised to 30,000 ft the average range will be increased to 220 statute miles. (In each case the signal-to-noise ratio is unity.) We also note that, at these ranges, the system dynamic range must be increased by a factor of 10 in order to extend the range only 10 miles. In fact, on the basis of pure diffraction propagation calculations we see that an increase of 70 db in system dynamic range (nearly the ultimate in practical value) would extend the range to 210 statute miles in the case of the aircraft at 10,000 ft and to 305 statute miles in the 30,000 ft case. Of course, variations about these average calculated values will occur because of signal level changes caused by atmospheric refractive index changes. On the average then, the desired 350 mile reliable range cannot be realized and when adequate allowance for a comfortable signal-to-noise ratio and fading range is made, the prospect becomes even less hopeful.

It is now well known that, deep within the diffraction region, the values calculated on the basis of pure diffraction theory are not a true measure of the path loss. A mode of radio wave propagation exists, called "extra-diffraction" or tropospheric "scatter," which extends well beyond the radio horizon and diffraction region. Much data on this mode is now available and our hope for extending the communication range must be based upon the possibility of exploiting this mode. These data have been presented in our companion paper (q.v.).

III. Extra-Diffraction or "Scatter" UHF Air-Ground Propagation

Of particular interest to the air-ground communications problem is the UHF field strength distribution at heights common to aircraft flight. At the Air Force Cambridge Research Center we have been conducting an extensive program of long distance airborne field strength measurements at 220 Mc/s over the past several years and are now in possession of much of the significant propagation data necessary to estimate the possibility of air-ground communication range extension in the UHF extra-diffraction region. The data are unique since they represent the only airborne measurements of tropospherically scattered fields to distances beyond 200 miles at frequencies above 100 Mc/s.

A detailed study of our airborne measurements conducted between 500 and 40,000 ft altitude to distances up to 1000 statute miles gathered in the course of the Cambridge Research Center research program coupled with the ground based data of the FCC, Lincoln Laboratory, Collins Radio Company, Bell Telephone Laboratories and the National Bureau of Standards, permits several general conclusions regarding 220 Mc/s radio wave propagation to be drawn.

(1) The phenomenon of extra-diffraction propagation is *not* simply a surface phenomenon; the fields have been observed to be of the same fundamental character through the range of altitudes measured so far, i.e., from 500 to 40,000 ft, and out to at least 700 miles. Between 700 and 800 miles the character of the signal changes; it then behaves in a manner similar to VHF ionospherically scattered radio waves.

(2) The average rate of signal decay varies with the season. The signal level is higher during summer than in winter, but it's rate of attenuation with distance appears greater.

(3) The average rate of signal decay varies with the altitude of the airborne terminal. At higher altitudes the average signal level is greater but so is the rate of decay with distance, thus the increase in field strength with height lessens as the path length increases up to 600 miles. Near ground level the average received power varies nearly as d^{-9} in the distance range 150–600 miles or a 27 db increase in path loss each time the distance is doubled.

(4) The average short time fading rate is the order of a fraction of a cycle per second.

(5) The short time fading intensity is of the order of 10–20 db and generally follows the Rayleigh distribution law. Also the hourly fading intensity decreases at the greater distances.

These conclusions provide the basis for the modification of the diffraction UHF propagation situation as depicted in Fig. 1 to include the influence of the scatter fields deep within the diffraction region. The scatter curves in this figure provide the best present estimate of the behavior of these fields as related to aircraft altitudes and transmitter heights.

These attenuation values will obtain only over smooth terrain and/or over water. Should the horizon angle be increased only a few tenths of a degree, the values indicated will not be greatly influenced. As the horizon angle is increased, however, the path attenuation will increase 10 db over a 200 mile path for a total horizon angle of 1 degree distributed between terminals.

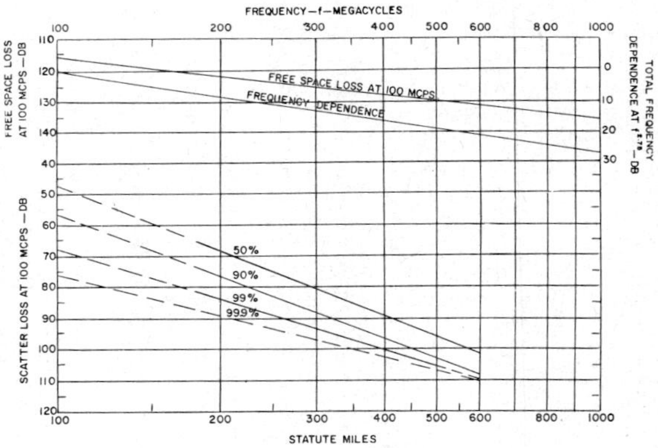

FIG. 2.

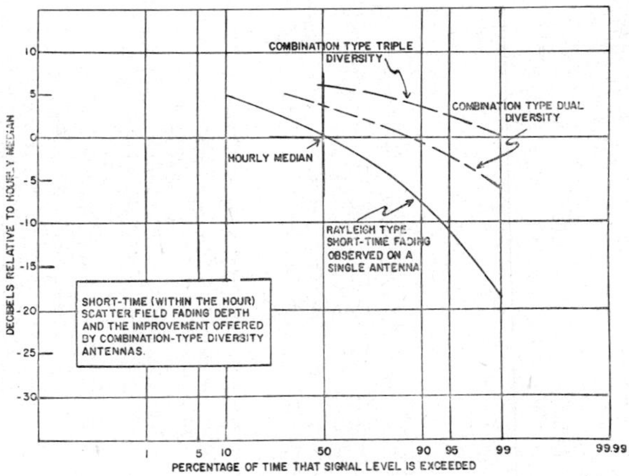

FIG. 3.

It will be noted that considerable height-gain is realized. An important assumption which appears to be valid is that the total path loss depends, to a good first approximation, only upon the free space (d^{-2}) loss plus a scatter field loss which depends upon the distance between the transmitter and receiver radio horizons. The use of this assumption offers a means of extending predictions based upon surface and low altitude measurements up to heights of 40,000 ft.

Figure 2 represents our best estimate of the path loss as related to effective distance. The estimated frequency dependence and the free space loss are also plotted here for convenient reference. We have previously stated that measurements have been made beyond 600 effective miles. These curves do not extend beyond 600 miles because there is now evidence to indicate that the rate of decay is much greater beyond 600 miles although there are not now available sufficient data to evaluate this slope.

We have, as yet, no quantitative information concerning the short-time fading of the scatter signal as received in the air but our airborne observations are not obviously in disagreement with the conclusion that it is Rayleigh distributed. Figure 3 graphs this behavior and also indicates improvement to be expected through the use of diversity techniques.

A remaining question to be answered is one concerning frequency selective fading, i.e., propagation bandwidth, within the scatter region. Point-to-point studies made by several groups including the Cambridge Research Center, have established that for most of the time there is little distortion over a bandwidth of one megacycle at 200 miles, and it is estimated that a relatively distortionless bandwidth of at least 10 kilocycles should be available at 500 miles on wide angle antennas.

We have conducted a few voice tests from 5000 to 30,000 ft to distances of 400 statute miles at speeds ranging from 160 to 270 knots. The results are only qualitative, but in any case it appears that intelligibility is limited only by inadequate signal-to-noise ratio, not by noticeable propagation distortion. Therefore acceptable speech quality should be available over ground-air scatter paths provided an adequate signal-to-noise ratio is maintained.

IV. Required System Parameters

The possibility of good quality, highly reliable voice communication on air ground circuits over 300–400 statute miles is indicated by the foregoing work. The median total transmission loss as noted in Fig. 1, is very large compared to that experienced in present line-of-sight systems, nevertheless, the required system gains can be achieved in an air-ground circuit. Even greater system gains were found necessary for the exploitation of the "scatter" mode for point-to-point multi-channel voice communication.

In addition to the system gain required to overcome the large median transmission loss, a generous allowance must be made for seasonal and second-to-second signal level variation if one is to obtain the desired communication range for 95 % of the time or more, for instance. Figures 2 and 3 indicate these values.

The present low and future high altitude problems are believed to be inherently different in a few important respects; for this reason they will be studied separately.

A. Operation at Low (10,000 *ft*, 20,000 *ft*) *Altitude with Piston Engine Type Aircraft*

Two features characterizing the low altitude operational problems are:

(1) Passenger and cargo aircraft seldom exceed altitudes of say 20,000 ft and much of the time operate at 10,000 ft and lower; (2) the courses flown by these aircraft are usually predetermined by controlled airways.

The altitude limitation of 10,000 ft precludes meeting the 350 mile minimum range in the line-of-sight case even if the ground antenna height is elevated to 1000 ft, thereby extending the line-of-sight and diffraction range. Of course, any terrain features which would increase the ground terminal antenna height should be used in order to minimize the horizon angle and shorten the effective distance thereby decreasing the path loss. Figure 4 illustrates the change in median path loss for various altitudes at 400 miles, as the ground antenna height is varied relative to the assumed antenna height of 100 ft.

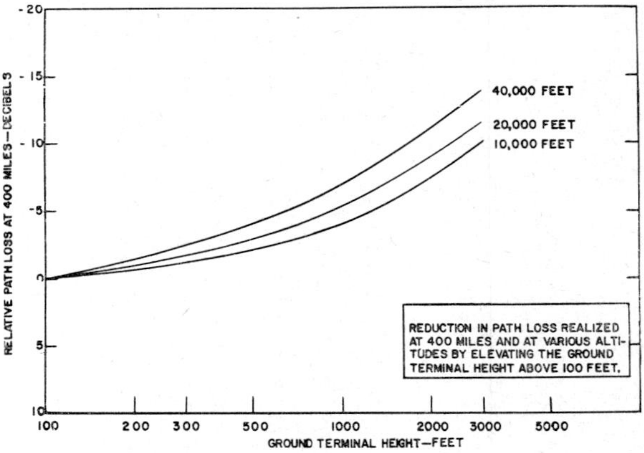

FIG. 4.

The second feature, that of the use of predetermined airways, offers a most important advantage: relatively large ground antenna gains may be realized in the horizontal plane as well as in the vertical plane. Since every decibel of system gain is needed, the additional gain in the horizontal plane should be pressed to the limit permitted by aircraft operational considerations.

It must be emphasized that the system to be discussed here is not intended to represent the best system possible for air-ground communications. Rather it is a system which can be put together today using components which have been air-tested in prototype models and are within the present state-of-the-art.

First, the air-to-ground path. The airborne transmitter average output power should be increased by 20 db. A 500 W average output power 300 Mc/s transmitter should not place undue burden upon either the main power capacity or space in a modern 4-engine aircraft. If the transmitter is adequately representative of the present state of the art in airborne equipment design, a minimum of 500 W average output power seems quite reasonable. At the present time some HF airborne transmitters are rated at 1 kw.

The use of speech processing techniques such as clipping is indicated. Such techniques will increase the effective signal-to-noise ratio in the voice channel, the amount depending upon the input carrier-to-noise level.

The airborne antenna system should realize considerable gain over those presently in use. A simple 5 element Yagi will yield 10 db gain above isotropic and its half power beamwidth will be some 50°. Two such antennas, pointed fore and aft should provide an adequate coverage pattern and provide for changes in aircraft heading under various wind conditions for airways type communication. Our tests have demonstrated that the use of such antennas is feasible although other types may be preferred.

It is possible to improve the 3 μV sensitivity of present ground based receivers by about 26 db. This can be accomplished by the use of low noise front ends and tight frequency control to reduce excessive bandwidths presently in use. The fact that low noise figures can be achieved does not, of course, mean that the receiver necessarily will experience a corresponding increase in sensitivity unless external sources of noise are avoided by appropriate site selection and shielding.

TABLE I

(Air-to-ground path)

P_T (500 watts c.w.; 2 kW peak	= 33 dbw (peak)
G_T (5 element Yagi or equivalent)	= 10 db
G_R (50′ × 25′ dipole array)	= 30 db
S min (0·15 microvolt—50 ohms)	= − 154 dbw
Total dynamic range	= 227 db
Peak Signal to RMS noise ratio	= 10 db plus speech processing gain (5–8 db)
Dual Diversity	

At UHF, cosmic noise and noise from the sun will probably be observed at the ground receiver site because of the large capture area of the ground antennas, as will lighting strokes in or near the main beam. These forms of noise, however, should not exceed the ground receiver noise for more than a small fraction of the time at 300 Mc/s in most locations.

The airborne receiver will probably be affected by lightning for a very small fraction of the time and may be subject to a more serious noise source: precipitation static. Some evidence of this type of noise has been encountered in our airborne tests. This may be a serious problem for high speed aircraft antenna installations. For these and other reasons it appears that the field strength at the airborne terminal should exceed that at the ground terminal.

An antenna gain of some 30 db is required at the ground terminal. This can be achieved with an array size of some 50 ft by 25 ft at 300 Mc/s. Its configuration should be such as to derive the smaller beamwidth in the vertical plane. While such size is certainly novel compared to present ground-air systems, it is much smaller than those used in many point-to-point tropospheric scatter circuits: 60 ft diameter parabolas.

Two or three ground antennas may be used in a horizontally spaced combination type diversity system. At least dual diversity is required to counteract the effect of the Rayleigh distributed short time fading (see Fig. 3). The appropriately

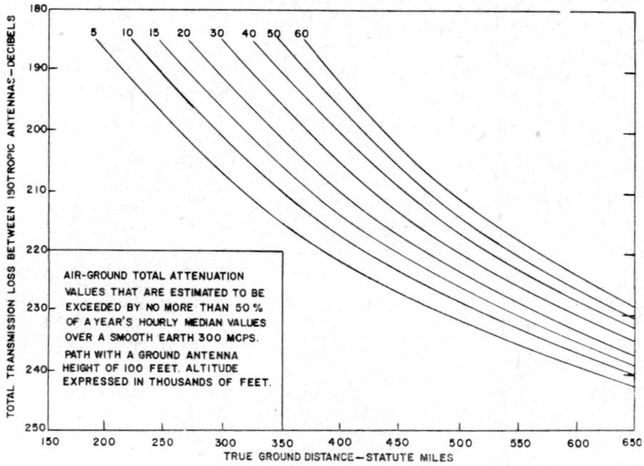

FIG. 5a.

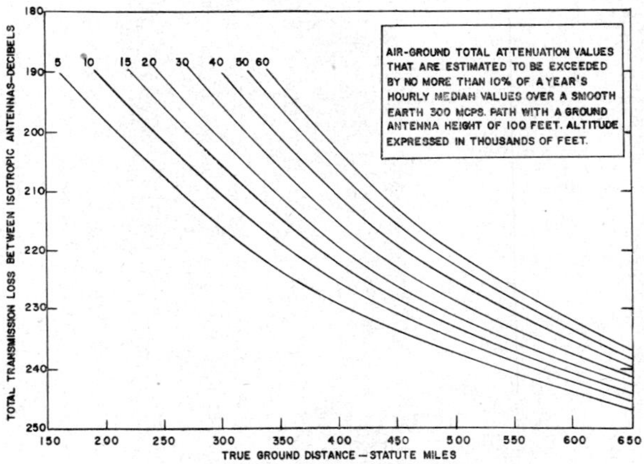

FIG. 5b.

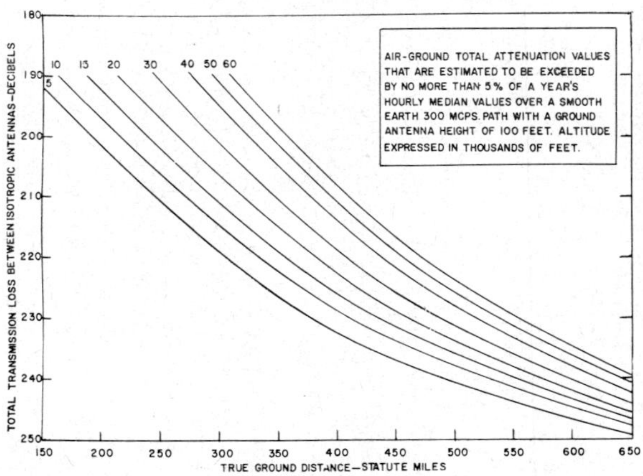

FIG. 5c.

combined outputs should yield a much higher effective signal level for 90 %–99 % of the time within the hour. The azimuth beam width will be nearly 10°; this allows a distance of 35 miles across the radiated cone of energy at a distance of 200 miles and 70 miles across at 400 miles.

Such an array might be supplemented by a dipole for use to distances of say, 100 miles, where the high gain is not required and better vertical coverage may be desired.

Table I summarizes a possible set of basic system parameters that might be used and indicates the total system dynamic range for the air-to-ground path.

Figure 3 shows that we must add to the expected transmission loss 8 db to cope with 90 % of the second-to-second fading or 12 db for the 95 % reliability values unless diversity techniques are employed. We note that dual diversity reduces these margins by 7 and 9 db respectively at the 90 and 95 % short time reliability levels.

Figures 3 and 5(b) show that choice of the indicated system parameters might be expected nearly to eliminate the second-to-second fading as a serious factor for 90 % of the time and provide a communication range of 410 statute miles 90 % of the year's hours for aircraft at altitudes $\geq$ 10,000 ft; 450 miles at 18,000 ft. At altitudes above 10,000 ft, the reliablity will increase to 95 % at 390 miles. Conversely, fixing the reliability at 90 % at 18,000 ft the carrier-to-noise ratio will rise to 15 db at 400 miles and to 25 db at 350 miles. Speech processing may enhance these values by 5–8 db.

The employment of space or frequency diversity reception on the aircraft would materially aid the ground-to-air path but probably would present antenna design problems. Ground transmitter output powers of 10 kW average are now commonplace; the additional 13 db over the air-ground path provided by this power level will compensate for the lack of diversity and provide matched performance on both paths.

It will be noted that no margin has been made for system deterioration under actual operating conditions. No basis for estimating such margin is available to the authors. Our path loss estimates are probably not accurate to better than $\pm$ 5 db due in part to the scarcity of data. It seems that system deterioration in excess of this value would be unreasonably large.

Our discussion has been confined to voice operation only but it seems reasonable to expect that teletype communications could be maintained to distances in excess of 400 miles independent of season for aircraft above 10,000 ft.

B. Operation at above 30,000 *ft with Jet Aircraft*

Aircraft performance characteristics may make the airborne antenna problem more difficult and wider flight paths may be required. Thus, it may be desirable to employ wider azimuth beam widths both in the air and on the ground for this type of operation. Space and weight considerations will probably limit RF power to 500 W average. Let us assume, then, the following system parameters.

One db of this total will be required to compensate for 90 % of the short time fading relative to the hourly median, leaving 218 db to cope with the hourly median path loss. Reference to Fig. 5(b) shows that, at a distance of 440 miles and an altitude of 40,000 ft, this loss (218 db) is expected to be exceeded no more than 10 % of a year's hours.

TABLE II

(Air-to-ground path)

P_T (500 Watts c.w.; 2 kW peak)	= 33 dbw (peak)
G_T	= 6 db/isotropic
G_R ($3° \times 20°$ azimuth)	= 26 db/isotropic
$S_{\min}$[4] (providing 10 db S/N)	= − 154 dbw
Total systems dynamic range	= 219 db
Dual Diversity at ground terminal.	

If the use of diversity at the airborne terminal is not feasible, then a higher power ground transmitter will be required to provide matched system performance. Again, as in the low altitude case, an average output power of some 10 kW C.W. is indicated.

V. Other Considerations

Present line-of-sight UHF communications may, when under the influence of anomalous propagation conditions, encounter signal level reductions of 20 to 30 db due to "radio holes." Systems with parameters as suggested here should

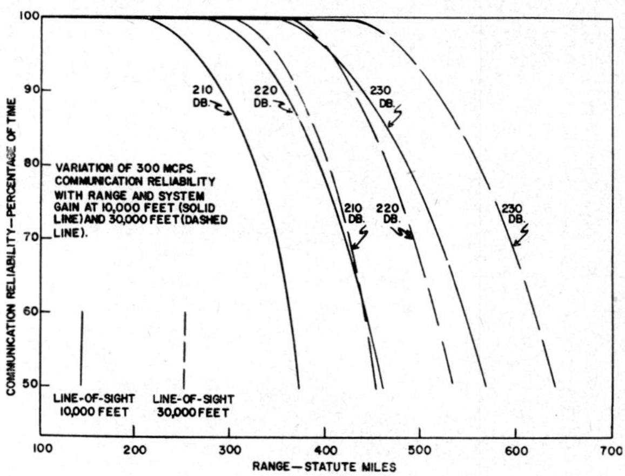

Fig. 6.

maintain their short range reliability during such periods of anomalous propagation and provide highly reliable communications for hundreds of miles.

Another important difference in communication when using the scatter mode is the relative rate of signal decay with distance indicated in Fig. 1. Starting near the radio horizon the rate of attenuation is nearly 1 db per statute mile. Consequently, as the range is increased in this region the loss of communications is very abrupt. At 400 miles the median signal level is decreasing at a rate of some 0·24 db per mile at 25,000 ft and only 0·15 db per mile at 10,000 ft; at the higher reliability levels the signal level will decrease at an even slower rate. Thus the service range will be less well defined and loss in communication should not be as abrupt as in the line-of-sight case.

Figure 6 graphs the expected reliability at various ranges and altitudes for given system dynamic ranges. We see that greater ranges are possible if lower reliabilities are accepted. This fact should be remembered when comparison with a line-of-sight UHF system is made. The range which may be depended upon for 90 % of a year's hours in the present line-of-sight case is probably no more than 70 % of the distance to the radio horizon because of surface lobing and anomalous propagation. Even if we assume this value to be as much as 85 %, then the communication range for an aircraft at 10,000 ft is only about 130 miles assuming the height of the ground terminal to be 100 ft. The median range for the system proposed here using the tropospheric mode and comparable terminal heights, is about 490 miles. Such a method of comparison thus yields an increase in communication range of nearly four times.

VI. Conclusion

Several conclusions may be drawn from this study.

(1) In the light of our present quantitative knowledge of the behavior of "scatter" field strength behavior far beyond the radio horizon in the temperate zone, it seems possible and feasible to extend the range of reliable air-ground UHF voice communications out to distances of at least 400 statute miles.

(2) In order to achieve these ranges ($\geqq$ 400 statute miles) use of the UHF tropospheric "scatter" mode willl be required.

(3) The data at hand appear to be sufficient to permit a reasonably accurate appraisal of most of the propagation loss factors related to this problem.

(4) A *large* increase . . . some 60 to 80 db . . . over present UHF air-to-ground system performance is required to achieve *reliable* voice communication over the long distances required.

(5) Present commercial or laboratory developed equipment and techniques appear to possess the potential of realizing this increased voice range without unduly restricting aircraft performance.

(6) For a given range substantially greater system gain must be employed in the case of aircraft operating at 10,000 ft altitude as compared to those operating at 30,000 ft altitude since at a given distance, field strength decreases with decreasing altitude.

(7) Use of the system gain indicated should eliminate the line-of-sight communication difficulties caused by "radio holes" induced by elevated refraction index discontinuities.

(8) Use of the suggested system parameters coupled with narrow band (100 cycles) receivers should permit teletype communication well beyond 400 statute miles at 10,000 ft and to greater distances at higher altitudes.

Acknowledgments

Most of the airborne field strength data and an appreciable amount of the ground data upon which this study is based were obtained by the writers' co-workers in the Air Force Cambridge Research Center.

Many interesting and helpful discussions have been held through the past few years with individuals from many government, university, and industrial organizations pursuing research programs in radio wave propagation and allied fields. Several have made their private or prepublication reports and data available to us. Their assistance is gratefully acknowledged.

References

1. Ames, L. A., Martin, E. J. & Rogers, T. F. *Air Force Cambridge Research Rep.* TR–55–121 (unpublished). June 1955.
2. Ames, L. A., Martin, E. J. & Rogers, T. F. (1956). *Trans. Inst. Radio Engrs, N.Y.* CS-**4,** no. 1., **102.**
3. Ames, L. A., Martin, E. J. & Rogers, T. F. (1957). *Trans. Inst. Radio Engrs, N.Y.* CS-**5,** 106–121.
4. Ames, L. A., Martin, E. J. & Rogers, T. F. "The Airborne Measurement of 1·36 meter Fields to Ranges in Excess of 900 miles and at Altitudes from the Surface to 40,000 feet." Presented at Liège, Belgium, 6 to 10 October 1958.
5. Ames, L. A., Newman, P. & Rogers, T. F. (1955). *Proc. Inst. Radio Engrs, N.Y.* **43,** 1369.
6. Gerks, I. H. & Svien, A. J. (1956). *Trans. Inst. Radio Engrs, N.Y. Convention Record,* Part I, pp. 3–8.
7. Martin, Edward J. *AFCRC–TR*–58–149, ASTIA, No. AD152415. (Unpublished.)
8. Rice, P. L. & Daniel, F. T. *Trans. Inst. Radio Engrs, N.Y.* AP–3, pp. 50–62.
9. Rogers, T. F. (1955). *Proc. Inst. Radio Engrs, N.Y.* **43,** 623.

Die Ausbreitung von VHF und UHF in unregelmäßigem Gelände

Udo Kühn

Betriebslaboratorium für Rundfunk und Fernsehen, Berlin-Adlershof

Es wurden umfangreiche Feldstärkemessungen in verschiedenen Geländearten im VHF- und UHF-Bereich angestellt. Aus jedem vermessenen Profil wurde ein statistischer Wert der örtlichen Feldstärkestreuung erhalten. Alle Ergebnisse sind in einem Diagramm zusammengefaßt, aus welchem die Abhängigkeit der örtlichen Feldstärkestreuung (1 %-50 % oder 10 %-50 %) von der Frequenz und der Geländeunregelmäßigkeit ermittelt werden kann. Die Abweichungen der praktischen von den theoretischen Feldstärken zeigten eine gute Übereinstimmung mit Messungen, die von RCA und FCC durchgeführt worden sind.

I. Einleitung

Für die Planung von Sendernetzen im VHF- und UHF-Gebiet und zur Ermittlung der Versorgungsgebiete ist die Kenntnis folgender Größen der Wellenausbreitung wichtig:

(1) Feldstärkekurven für 50 % der Zeit und 50 % der Orte zur Bestimmung der Grenze des Versorgungsbereichs.

(2) Angaben über die Feldstärkestreuung zur Bestimmung der Ortswahrscheinlichkeit.

(3) Feldstärkekurven für 50 % der Orte und 1 % bzw. 10 % der Zeit zur Ermittlung der Interferenzstörung.

Da besonders zum Punkt (2) noch nicht sehr zahlreiche Messungen vorliegen und für die Zwecke der Sendernetzplanung meist mit einem einheitlichen Wert der örtlichen Streuung gerechnet wird, bestand die Absicht, die Frequenzabhängigkeit dieses Wertes und die regionale Verteilung der Feldstärkestreuung innerhalb der DDR zu ermitteln. Der in erster Linie aus amerikanischen Messungen hervorgegangene Wert der örtlichen Streuung von 19·3 db gilt für mittleres Gelände im Osten der Vereinigten Staaten, und es wurde vermutet, daß dieser Wert für flaches und leicht hügliges Gelände zumindest im VHF-Gebiet zu groß ist, während im Bergland wahrscheinlich höhere Werte der örtlichen Streuung zu erwarten sind.

Man kann sich grundsätzlich zweierlei Methoden bedienen, um die örtliche Streuung zu erfassen: Das ist einmal die punktförmige Messung und zum andern die Feldstärkemessung vom fahrenden Meßwagen aus. Das erste Verfahren ist zur Erlangung eines ausreichenden statistischen Materials mit einem verhältnismäßig großen Aufwand verbunden. Es wurde daher bei den von uns durchgeführten Messungen ausschließlich von dem Verfahren der Feldstärkeregistrierung vom fahrenden Meßwagen aus Gebrauch gemacht. Die Messungen erfolgten in den Fernsehbändern I und III und im UKW-Rundfunkband II, sowie im Band IV bei einer Frequenz von 480 MHz und im Band V bei einer Frequenz von 780 MHz.

Neben diesen Untersuchungen der örtlichen Feldstärkestreuung wurden die Abweichungen der praktisch gemessenen von den theoretisch berechneten Feldstärken in den verschiedenen Geländearten untersucht und vorläufige praktische Ausbreitungskurven für Band IV und V mittels dieser Methode festgelegt.

Im UKW-Rundfunkband wurden auch stationäre Feldstärkebeobachtungen zur Ermittlung der zeitlichen Feldstärkestreuung und zur Festlegung der für Interferenzrechnungen notwendigen Feldstärkekurven für 10 % bzw. 1 % der Zeit angestellt, über diese Ergebnisse wird jedoch in diesem Zusammenhang nicht berichtet.

II. Meßeinrichtung

Da die Feldstärkeregistrierungen vom fahrenden Meßwagen aus erfolgen sollten, sind die Feldstärkemeßeinrichtungen in ein entsprechendes Spezialfahrzeug eingebaut worden. Dabei wurden für die Messungen im VHF-Bereich entweder industrielle Feldstärkemeßgeräte oder Meßgeräte eigener Entwicklung benutzt. Für die Messungen im Band IV und V kamen ausschließlich Geräte eigener Entwicklung zur Verwendung. Die Stromversorgung erfolgte aus Batterien über einen Umformer, und es bestand die Möglichkeit, die Meßbatterien mit einer zweiten Lichtmaschine laufend zu laden. Teilweise wurden die Registrierungen mit einem einfachen Linienschreiber vorgenommen, später konnte ein Kompensationsbandschreiber mit einer ausnutzbaren Schreibstreifenbreite von 28 cm verwendet werden. Der Antrieb des Papiervorschubs erfolgte dabei von dem Getriebe des Meßwagens aus, so daß eine Synchronisation zwischen abgefahrener Meßstrecke und Papiervorschub vorhanden war. Die Meßantennen waren an einem Teleskopmast montiert. Für die VHF-Messungen wurden ausschließlich Runddipole verwendet. Die Messungen bei höheren Frequenzen erfolgten mit gestockten Kreuzdipolen, die ebenfalls Rundempfangscharakteristik besaßen.

Für die Registrierungen im VHF-Bereich wurden ausschließlich im praktischen Betrieb arbeitende UKW-Rundfunk- und Fernsehsender benutzt. Die Band IV- und V-Messungen erfolgten mit eigenen Versuchssendern. Unter Verwendung von gerichteten Sendeantennen mit einer Halbwertsbreite von etwa 35° ergab sich eine ERP von etwa 1 kW sowohl bei 480 als auch bei 780 MHz. Die Messungen erfolgten ausschließlich bei horizontaler Polarisation.

III. Durchführung der Messungen und Auswertung

Die Registrierungen erfolgten, soweit dies das Straßensystem zuließ, in radialer Richtung vom Sender. Die mittlere Entfernung, die dabei jeweils vermessen wurde, betrug meist 70 bis 80 km vom Sender. Wesentlich größere Entfernungen wurden meist nicht vermessen, um nicht zusätzliche zeitliche Feldstärkestreuungen mit zu registrieren, die von den örtlichen Feldstärkeveränderungen nicht getrennt werden können. In all den Fällen, in denen eine Registrierung noch bis in größere Entfernungen hin erfolgte, war durch stationäre

Messungen über einen kürzeren Zeitraum hinweg nachgeprüft worden, ob zeitliche Feldstärkeveränderungen vorhanden waren. In Abbildung 1 sind alle Meßstrahlen wiedergegeben, die im VHF-Gebiet vermessen worden sind. Man erkennt, daß über das gesamte Land etwa gleichmäßig verteilt die Meßstrahlen liegen. Abbildung 2 gibt die Profile wieder, die im Band IV bei einer Frequenz von 480

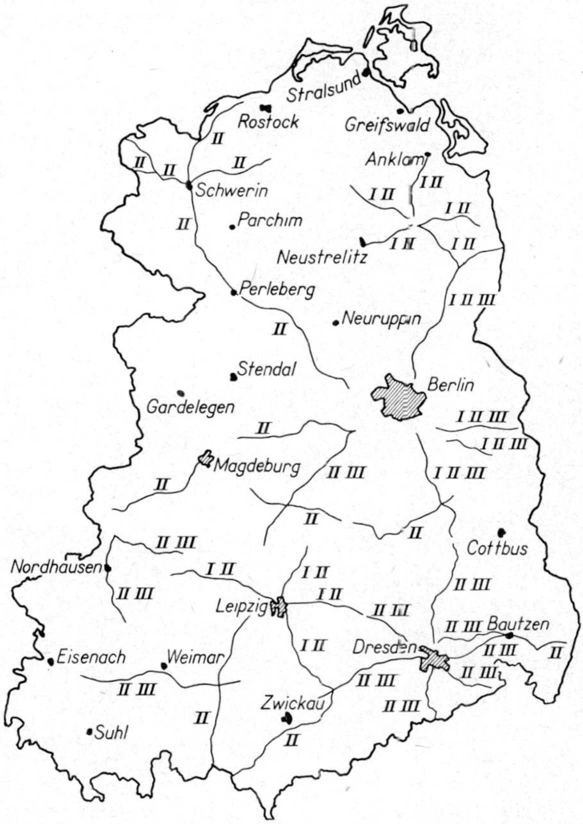

ABB. 1. Strecken, entlang denen Feldstärkeregistrierungen in den Bändern I, II oder III durchgeführt wurden.

MHz vermessen worden sind. Hierbei konzentrierten sich die Messungen in erster Linie auf den gebirgigen südlichen Teil der DDR, da es dort leichter möglich war, gute Aufstellungsorte für den Versuchssender zu finden, um zufriedenstellende Reichweiten zu erzielen. Die Messungen im Band V sind noch nicht abgeschlossen.

Die Auswertung zur Ermittlung der örtlichen Feldstärkestreuung wurde folgendermaßen vorgenommen: Der Feldstärkeverlauf, der während der Fahrt des Meßwagens auf dem Registrierpapier aufgezeichnet wurde, wird über dem Geländeprofil aufgetragen. Danach wird die mittlere Feldstärkeabnahme, das ist der Wert für 50 % der Orte und 50 % der Zeit, ermittelt und parallel zu dieser mittleren Kurve im Abstand von 1 oder 2 db werden gleichmäßige Feldstärkeabnahmekurven gezeichnet. Entlang dieser Kurven wird dann jeweils der Prozentsatz der Orte ermittelt, an denen die Feldstärke diese Kurve überschritten hat. Bei Zusammenstellung dieser Prozentwerte in Wahrscheinlichkeitspapier läßt sich daraus sehr leicht die Differenz der Werte zwischen 50 % und 10 % bzw. 50 %

ABB. 2. Strecken, entlang denen Feldstärkeregistrierungen bei 480 MHz durchgeführt wurden. Die Pfeile bedeuten Aufstellungsorte des Versuchssenders.

und 1 % der Orte entnehmen. Diese Auswertung wurde für jedes Profil getrennt vorgenommen. Bei der Anfertigung der Karten zur regionalen Verteilung der örtlichen Streuung wurde der Wert der örtlichen Streuung jeweils in dem Gebiet eingetragen, für das das gemessene Profil charakteristisch erschien.

Die Ermittlung der Unterschiede zwischen den praktisch gemessenen und den theoretisch berechneten Feldstärken erfolgte in der Weise, daß unter Berücksichtigung des mittleren Sendervorgeländes und der bekannten Senderleistung die Feldstärke für bestimmte Entfernungsbereiche ausgerechnet und diesen Werten die praktisch gemessenen Feldstärken gegenübergestellt wurden.

IV. Beispiele von Feldstärkeregistrierungen in verschiedenen Geländearten und bei verschiedenen Frequenzen

Abbildung 3 bringt eine Feldstärkeregistrierung im Band II in leicht hügligem Gelände. Wie aus der Abbildung hervorgeht, findet sich eine Übereinstimmung zwischen Geländeverlauf und Feldstärke besonders an den Hindernissen in etwa 30 km und 75 km Entfernung. Am erstgenannten Hindernis beträgt der Feldstärkerückgang hinter der beugenden Kante etwa 20 db. Sonst sind jedoch größere Schwankungen des Feldstärkeverlaufs nicht vorhanden. Wesentlich anders sieht eine derartige Registrierung in leicht bergigem Gelände aus. Ein

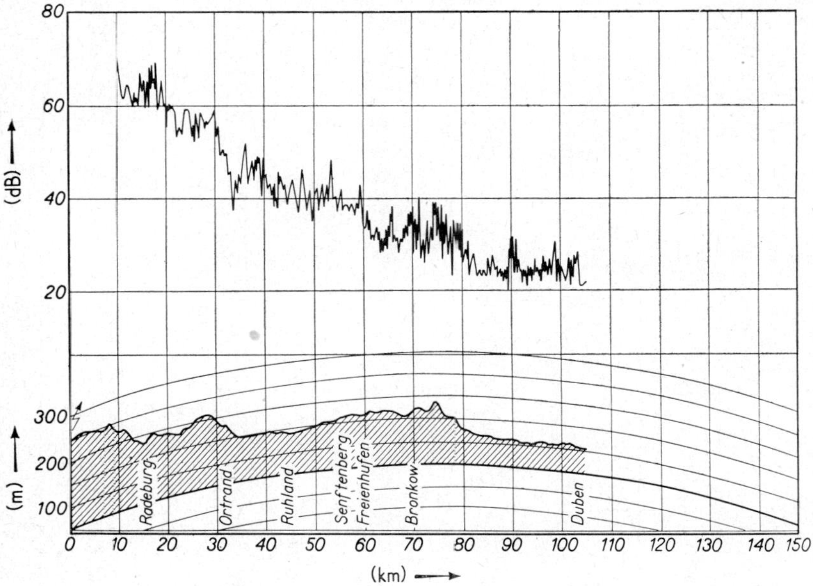

Abb. 3. Feldstärkeregistrierung im Band II in hügligem Gelände.

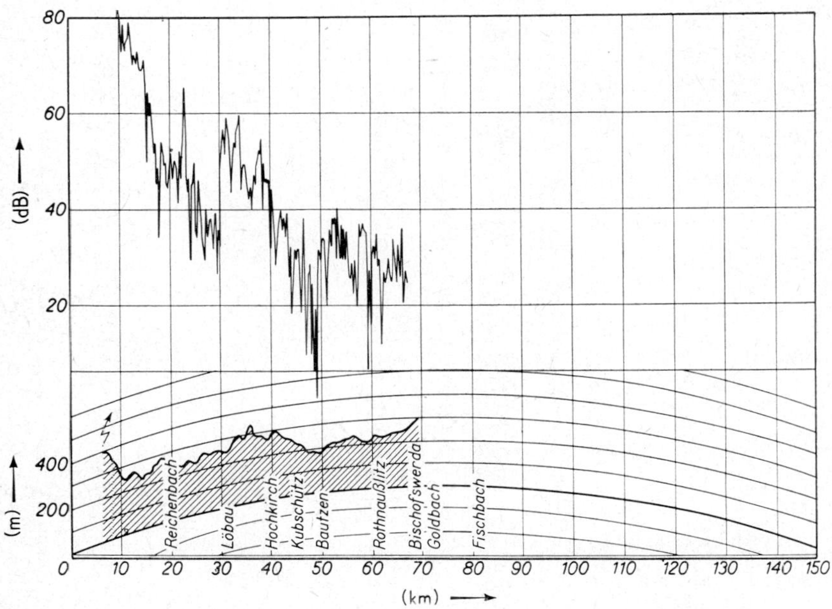

Abb. 4. Feldstärkeregistrierung im Band II in leicht bergigem Gelände.

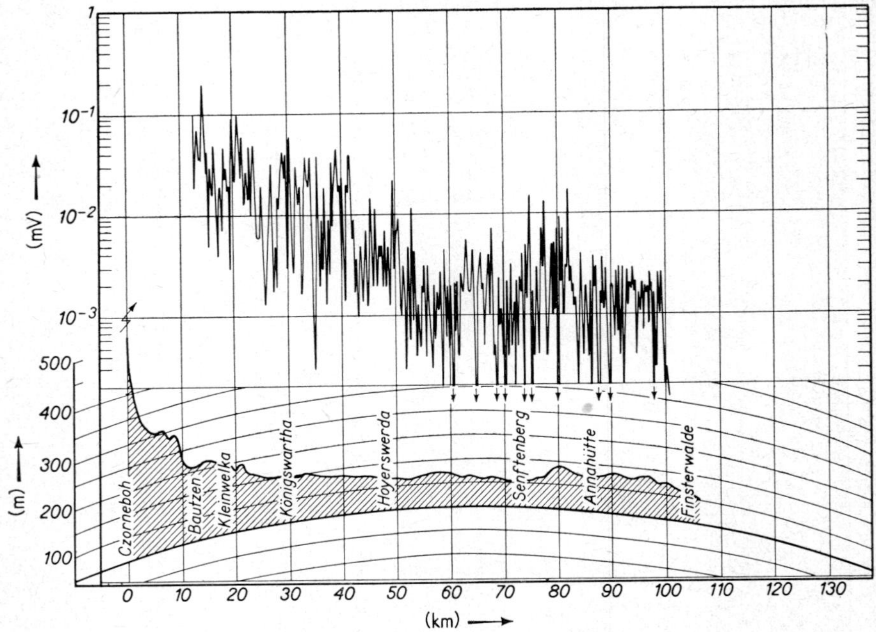

ABB. 5. Feldstärkeregistrierung im Band IV in flachem oder leicht hügligem Gelände.

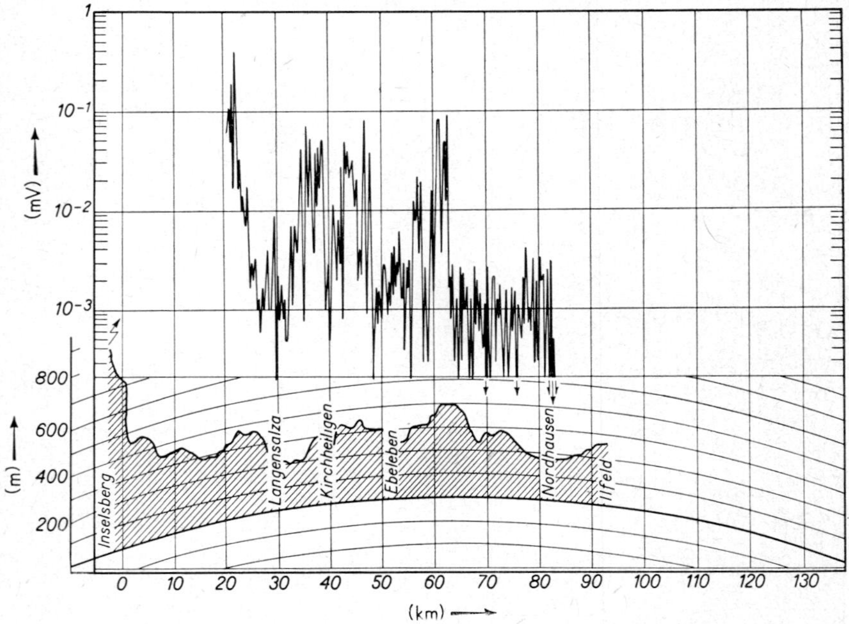

ABB. 6. Feldstärkeregistrierung im Band IV im bergigen Gelände.

Beispiel dafür bringt Abbildung 4. Die Messung erfolgte in Ostsachsen am UKW-Rundfunksender Landeskrone im Band II, und aus der Abbildung geht deutlich hervor, welch bedeutenden Schwankungen die Feldstärke in einem derartigen Gelände bereits unterliegt.

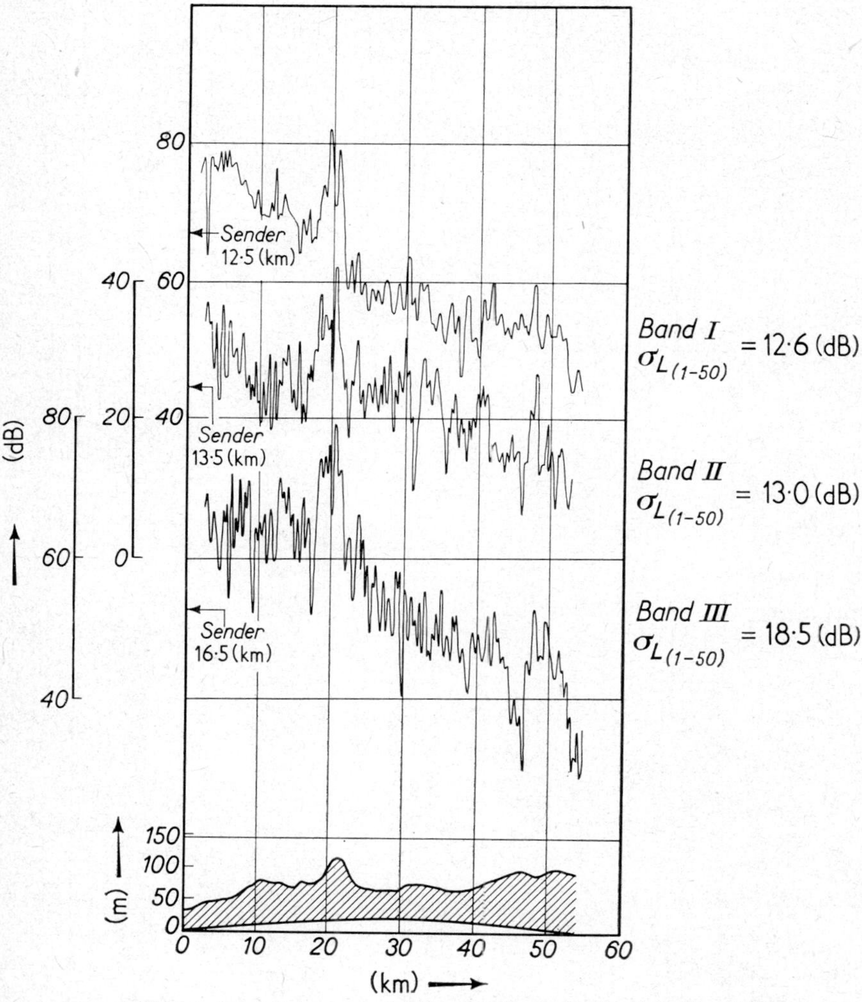

ABB. 7. Feldstärkeregistrierung in hügligem Gelände im Band, I, II und III.

Zum Vergleich dazu zeigt Abbildung 5 die Vermessung des Versuchssenders Czorneboh bei 480 MHz über ebenem oder leicht hügligem Gelände. Wie man erkennt, sind trotz des Fehlens von Unregelmäßigkeiten in der Topographie recht maßgebliche Feldstärkeschwankungen vorhanden, die durch alle möglichen anderen Hindernisse, wie Wälder, Brücken, Häuser u.ä. hervorgebracht werden.

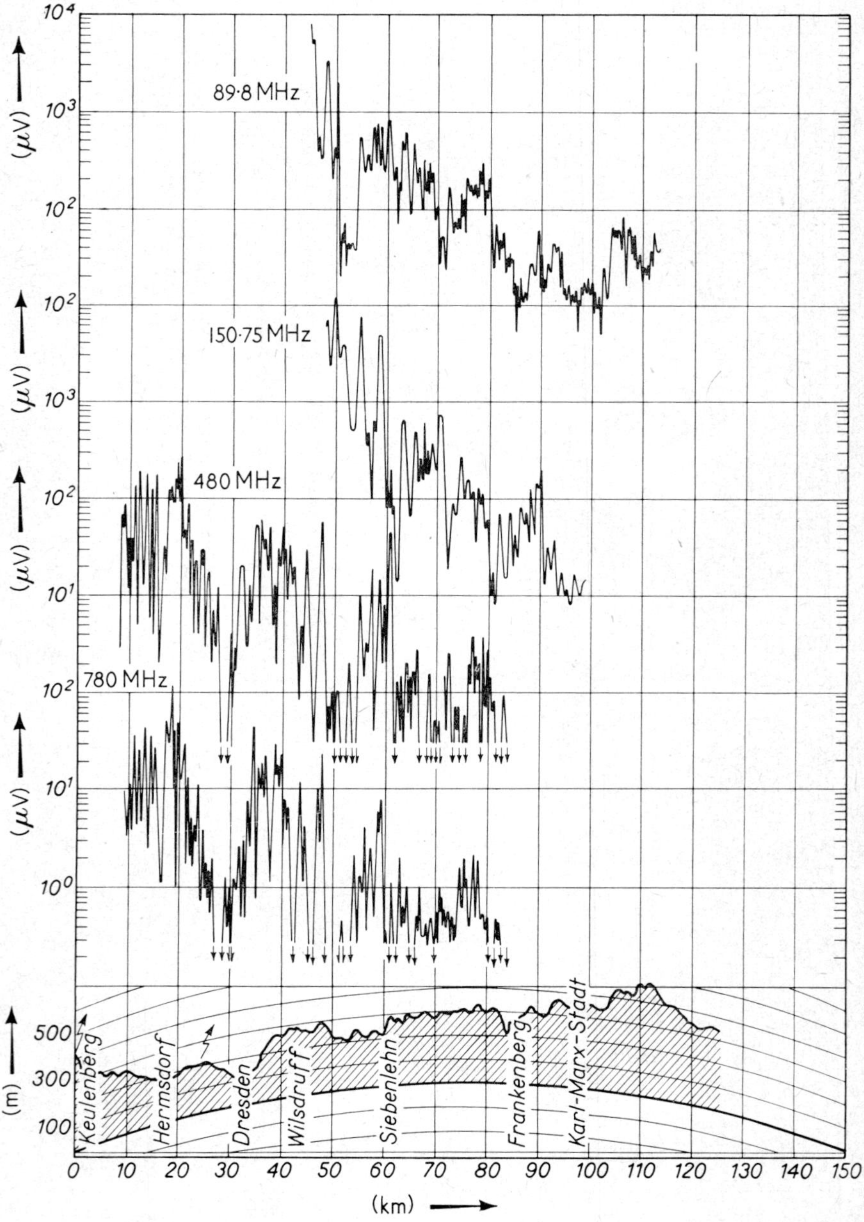

ABB. 8. Feldstärkeregistrierung in leicht bergigem Gelände im Band II, bei 150 MHz und im Band IV und V.

Der Unterschied in der Ausbreitungscharakteristik zwischen dem Band II und dem Band IV wird besonders deutlich bei einem Vergleich dieser Abbildung mit der Abbildung 3.

Abbildung 6 zeigt eine Feldstärkeregistrierung im Raum Thüringen. Der Versuchssender wurde auf dem Inselsberg aufgestellt. Bei der verwendeten Frequenz von 480 MHz sind die Feldstärkeschwankungen außerordentlich groß und überstreichen annähernd 60 db zwischen den höchsten und niedrigsten Werten. Besonders markant sind die Einbrüche hinter dem Hindernis bei etwa 25 km, bei 38 km, 47 km und 63 km.

In Abbildung 7 sind drei Berliner Sender auf der Autobahn in Richtung Frankfurt vermessen worden. Das Gelände ist dort leicht hüglig. Die Messung erfolgte im VHF-Bereich bei zwei Fernseh-und einem UKW-Rundfunksender. Die örtliche Streuung steigt mit zunehmender Frequenz an. Man erkennt dies auch bereits bei einer oberflächlichen Betrachtung des Kurvenverlaufs. Besonders auffallend macht sich das Hindernis am km 22 bemerkbar.

In Abbildung 8 sind einige Vergleichsregistrierungen bei Frequenzen um 90, 150, 480 und 780 MHz mitgeteilt, wobei die Dezimetersender auf dem Keulenberg Aufstellung fanden, wogegen zur Messung im VHF-Gebiet der Fernseh- und UKW-Rundfunksender Dresden benutzt wurde. Bei den Dezimetermessungen fällt besonders der Feldstärkerückgang im Elbtal bei Dresden etwa am km 30 auf.

V. Häufigkeitsstatistische Auswertung

Es wurde bereits eingangs erwähnt, daß von jedem vermessenen Profil eine häufigkeitsstatistische Auswertung vorgenommen wurde. Um einen besseren Überblick geben zu können, wurden jeweils für die einzelnen Frequenzbänder sämtliche gemessenen Profile zusammengefaßt und nach den Geländeunregelmäßigkeiten aufgetragen. Abbildung 9 zeigt die häufigkeitsstatistischen Verteilungen aller gemessenen Profile im Band II für flaches, hügliges, bergiges und gebirgiges Gelände. Aus der verschiedenen Neigung der Kurven geht sehr deutlich die Zunahme der Feldstärkestreuung mit stärker werdender topographischer Gliederung hervor. Abbildung 10 bringt eine Zusammenfassung aller Messungen im Band IV, unterschieden nach hügligem, bergigem und gebirgigem Gelände. Die Zahl der dort im flachen Gelände vorliegenden Messungen reicht für eine statistische Auswertung noch nicht aus. Vergleicht man jedoch diese Verteilungen mit denjenigen im Band II, so fällt sofort die beträchtlich höhere örtliche Streuung auf.

Ausgehend von den örtlichen Streuungen, die auf den einzelnen Profilen gemessen wurden, bestand die Absicht, ein Diagramm zu entwerfen, mit dem es möglich ist, die örtliche Streuung direkt in Abhängigkeit von der Geländegestaltung abzulesen. Als Maß für die Geländegestaltung wurde dabei der Höhenunterschied in Metern Δh zwischen den höchsten und tiefsten Punkten des Übertragungsweges, auf dem die Messung erfolgte, benutzt. Auf der Ordinate wurde dann die örtliche Streuung in db aufgetragen, in Abbildung 11 für die Differenz zwischen 10 % und 50 % der Orte und in Abbildung 12 zwischen 1 % und 50 %

der Orte. Es wird unter Verwendung dieses Diagramms möglich, auch für Geländearten, in denen keine vorherige Messung vorgenommen wurde, lediglich aus der Kenntnis der Topographie eine Abschätzung über die Größe der zu

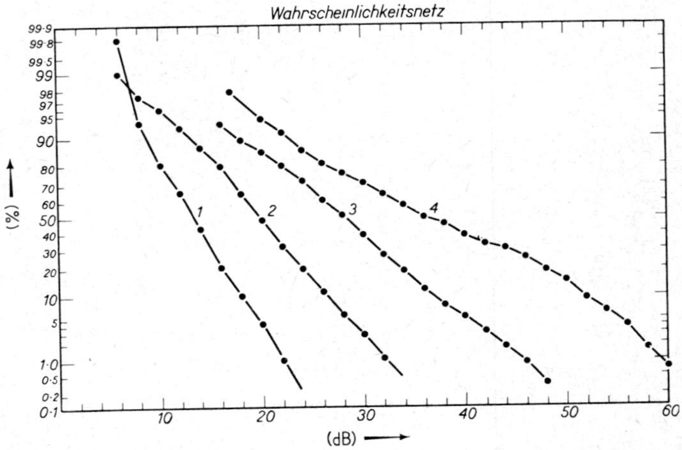

Abb. 9. Häufigkeitsstatistik der örtlichen Feldstärkeveränderlichkeit im Band II für verschiedene Geländearten.

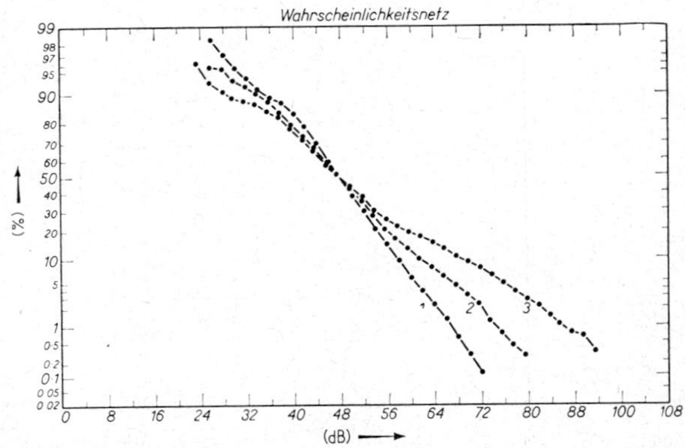

Abb. 10. Häufigkeitsstatistik der örtlichen Feldstärkeveränderlichkeit bei 480 MHz für verschiedene Geländearten.

erwartenden Feldstärkestreuung vorzunehmen. Dabei ist jedoch zu berücksichtigen, daß die Werte im einzelnen recht stark streuen können und die hier angegebenen Kurven mittlere Werte darstellen. Diese Streuung der Einzelwerte ist verhältnismäßig klein im Band I und II, etwas größer im Band III und bereits recht beachtlich im Band IV und V, eine Tatsache, die offensichtlich darauf zurückzuführen ist, daß mit zunehmender Frequenz sich in stärkerem Maße auch kleinere Hindernisse, wie Bäume, Häuser, Wälder u.ä. bemerkbar machen, diese Einflußnahme der Hindernisse aber nicht bei der Festlegung der topographischen Unregelmäßigkeiten berücksichtigt werden kann. Wie man aus den Diagrammen

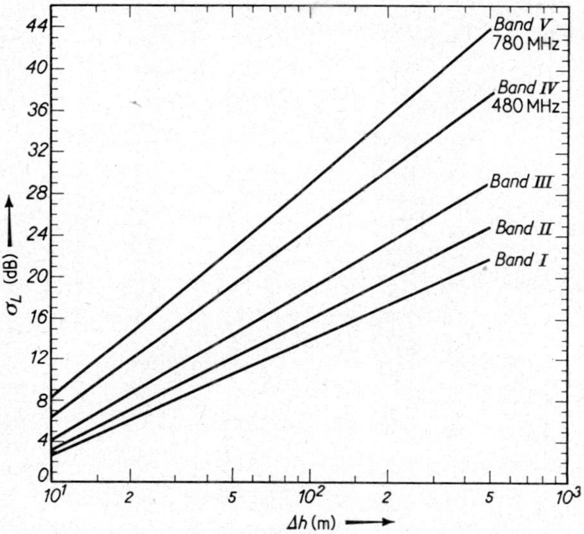

ABB. 11. Feldstärkestreuung zwischen 10 % und 50 % der Orte für verschiedene Geländearten. Δh bedeutet Höhenunterschied zwischen höchsten und tiefsten Punkten des Geländeprofils.

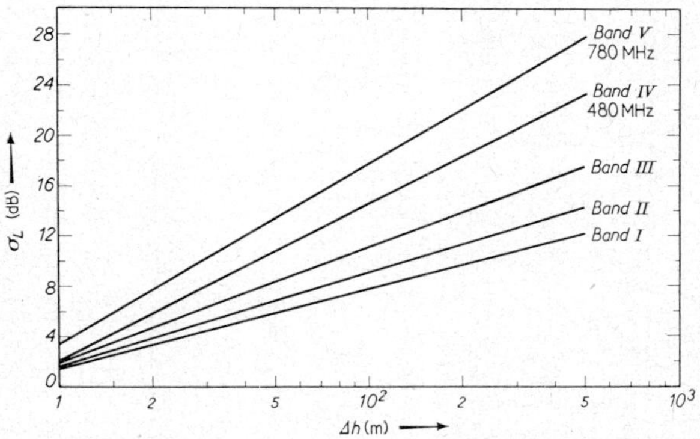

ABB. 12. Feldstärkestreuung zwischen 1 % und 50 % der Orte für verschiedene Geländearten.

ersieht, gilt der Streuwert der Feldstärke, von 19·3 db, wie er aus den amerikanischen Messungen hervorgeht, für das VHF-Gebiet für eine Geländegestaltung mit Höhenunterschieden zwischen 100 und etwa 280 m, wobei bereits innerhalb des VHF-Bereichs recht maßgebliche Unterschiede zwischen den einzelnen Bändern auftreten. Wie aus den Kurven ferner hervorgeht, wird der Frequenzunterschied der örtlichen Streuung mit zunehmender Geländeunregelmäßigkeit größer. Die Messungen wurden jedoch nur bis zu Geländearten durchgeführt, die Höhenunterschiede von 350 m aufwiesen. Die Kurven sind von da bis 500 m extrapoliert.

VI. Regionale Verteilung der Örtlichen Streuung

Jeder Wert der örtlichen Streuung, der auf einem gemessenen Profil ermittelt wurde, ist in eine Übersichtskarte eingetragen worden, und es wurden danach alle die Gebiete sinnvoll miteinander verbunden, die etwa die gleiche örtliche Streuung aufzuweisen haben. Abbildung 13 gibt die Ergebnisse nach Messungen

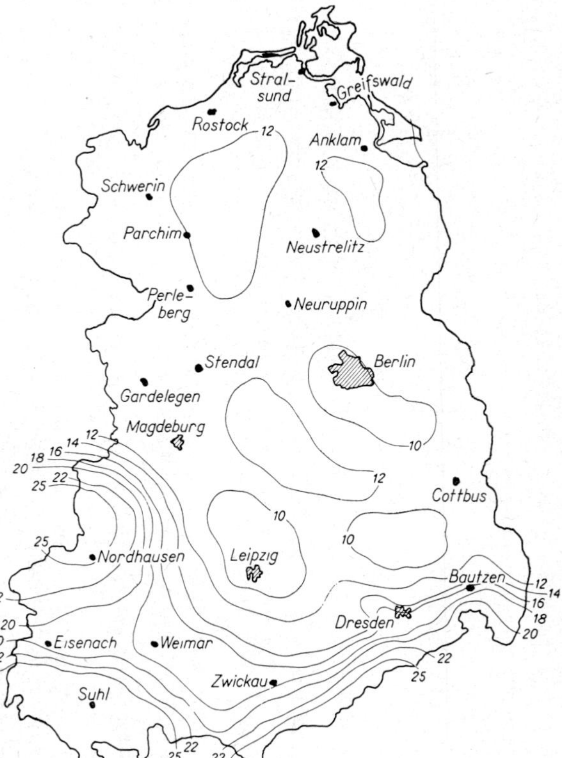

Abb. 13. Regionale Verteilung des Wertes der örtlichen Feldstärkestreuung zwischen 1 % nd 50 % der Orte in Band II.

im Band II wieder, und man erkennt aus dieser Karte, welchen Veränderungen dieser Wert auf dem Gebiet der DDR unterworfen ist. Während im mittleren und nördlichen Teil des Landes meist recht einheitliche örtliche Streuungen zwischen 10 und 12 db vorherrschen, zeigt sich im südlichen Teil eine recht maßgebliche

Veränderung, wobei die mittlere Streuung zwischen 18 und 20 db auf leicht bergiges Gelände beschränkt ist. Es geht aus dieser Karte hervor, daß bei einer Senderplanung mit einem eineitlichen Wert der örtlichen Streuung von 19·3 db im mittleren und nördlichen Teil der DDR zu kleine Versorgungsgebiete errechnet werden, während im leicht bergigen Gelände die theoretisch vorhergesagten Versorgungsgebiete etwa mit den wirklichen Versorgungsgebieten übereinstimmen. Im topographisch noch stärker gegliederten Gelände würden jedoch Rechnungen mit einem solchen Wert zu optimistische Verhältnisse vortäuschen. Derartige Karten lassen sich auch unter Verwendung des Diagramms der Abbildung 12 herstellen. Von diesem Verfahren ist aber hierbei nicht Gebrauch gemacht worden. Abbildung 14 zeigt die regionale Verteilung der örtlichen

Abb. 14. Regionale Verteilung des Wertes der örtlichen Feldstärkestreuung zwischen 1 % und 50 % der Orte für 480 MHz.

Streuung für 480 MHz, und aus der Karte geht hervor, daß im mittleren und nördlichen Teil des Landes die Feldstärkestreuungen zwischen 18 und 20 db liegen, während im gebirgigen Teil die Werte über 30 db ansteigen.

Selbstverständlich sind die hier vorgelegten Karten nicht so zu deuten, daß innerhalb eines kleinen Gebietes, z.B. von einigen Quadratkilometern, die örtliche Feldstärkestreuung diesen Wert erreicht, sondern die örtliche Streuung ist nur im Zusammenhang mit einer Fläche des Landes zu sehen, die etwa in der Größenordnung des Versorgungsgebietes eines Senders liegt. Es kann jedoch der Fall auftreten, daß für Versorgungsrechnungen in verschiedenen Richtungen vom Sender mit verschiedenen Werten der örtlichen Streuung gearbeitet werden muß, da die Geländegestaltung eine derartige Berücksichtigung verlangt.

VII. Abweichungen der praktisch gemessenen von den theoretisch berechneten Feldstärken

Derartige Messungen sind vor allem in U.S.A. vorgenommen worden, und über die Ergebnisse hat Egli zusammenfassend berichtet. Aus dem von uns gemessenen Material, vor allem im VHF-Gebiet, wurden analoge Betrachtungen vorgenommen, wobei jedoch eine Unterscheidung der Abweichung der praktischen von den theoretischen Feldstärken nach verschiedenen Entfernungsbereichen vorgenommen wurde. Abbildung 15 zeigt die Ergebnisse, und wie ersichtlich, ergibt sich

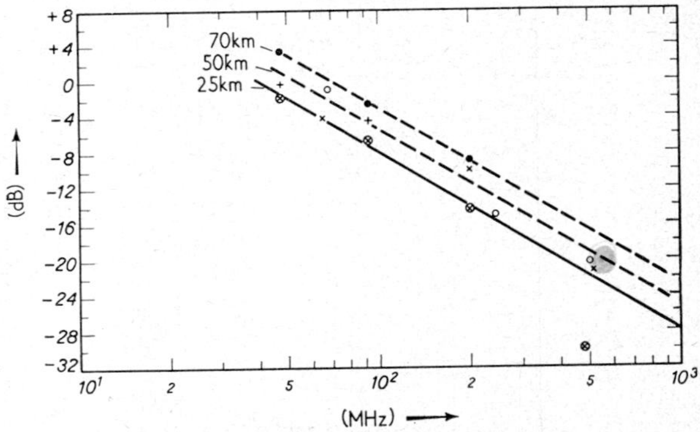

Abb. 15. Abweichung der praktisch gemessenen von den theoretisch berechneten Feldstärken in unregelmäßigem Gelände × Messungen von FCC; ○ Messungen von RCA; ⊗ Messungen des Verfassers für d = 25 km; + Messungen des Verfassers für d = 50 km; ● Messungen des Verfassers für d = 70 km.

eine sehr gute Übereinstimmung zwischen den Messungen der FCC, RCA und unseren Ergebnissen. Es geht außerdem aus der Darstellung hervor, daß die Abweichungen der praktischen von den theoretischen Feldstärken mit zunehmender Entfernung für höhere Frequenzen geringer werden. Der von uns bei 480 MHz eingetragene Wert liegt jedoch wesentlich zu tief. Dies ist darauf zurückzuführen, daß absolute Feldstärkewerte in diesem Frequenzbereich uns zur Zeit der Durchführung der Messungen noch nicht zur Verfügung standen.

Diese Abweichungen der praktischen von den theoretischen Feldstärken sind ihrerseits geländeabhängig, das heißt, bei sehr flachem Gelände wird die Übereinstimmung der theoretischen mit den praktischen Werten besser sein, als dies bei sehr unregelmäßig gestaltetem Gelände der Fall ist. Zur Klärung dieser Frage wurde für die einzelnen Profile der Band IV-Messung diese Abhängigkeit der praktischen von den theoretischen Feldstärken von den Geländeunregelmäßigkeiten aufgetragen. Es zeigt sich in Abbildung 16, daß, wie bereits vermutet, diese Abweichungen mit zunehmenden Geländeunterschieden größer werden. Auf der Abszisse ist dabei der Unterschied zwischen den höchsten und tiefsten Stellen des Profils in Metern aufgetragen. Aus dem gleichen Grunde, der bereits bei dem Wert bei 480 MHz in Abbildung 15 genannt wurde, gibt der Maßstab auf der Abszisse keine Absolutwerte an. In Wirklichkeit liegt nach neueren Messungen der absolute Pegel um etwa 6 db höher als in diesem Maßstab angegeben. Es ist

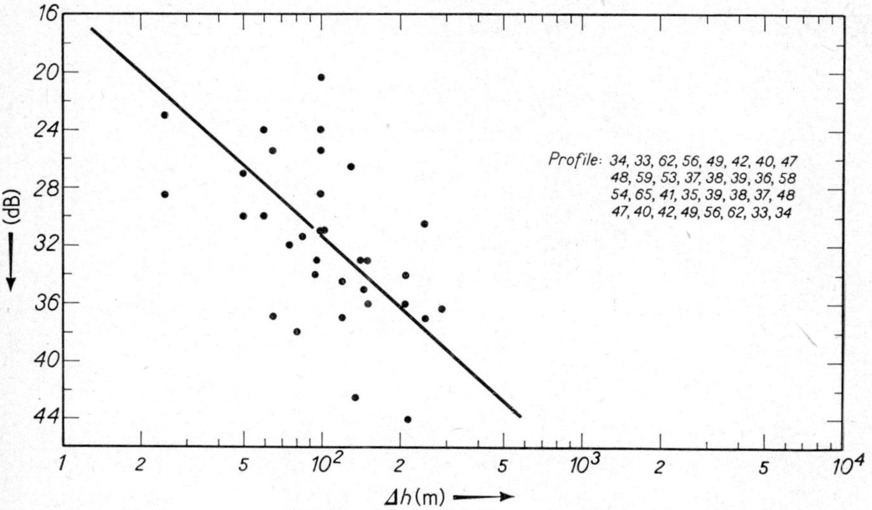

ABB. 16. Abhängigkeit der praktisch gemessenen von den theoretisch berechneten Feldstärken bei verschiedenen Geländearten. Δh bedeutet Höhenunterschied zwischen den höchsten und niedrigsten Punkten der Übertragungsstrecke.

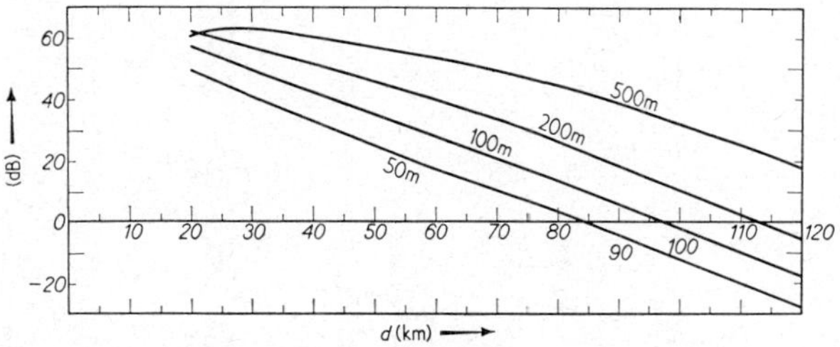

ABB. 17. Feldstärkekurven für 500 MHz für 50 % der Zeit und 50 % der Orte für unregelmäßiges Gelände.

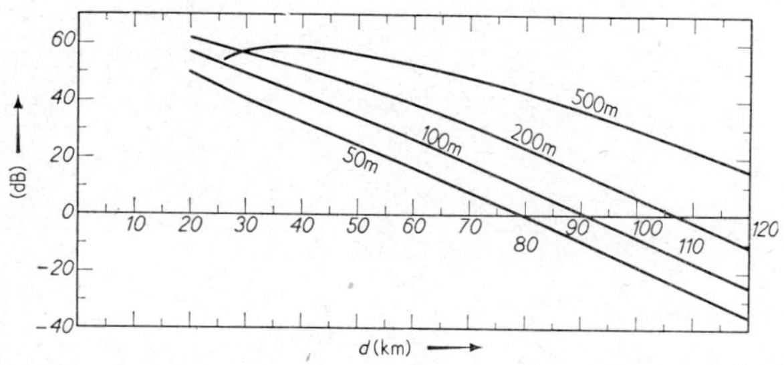

ABB. 18. Feldstärkekurven für 700 MHz für 50 % der Zeit und 50 % der Orte für unregelmäßiges Gelände.

bei den Abbildungen 15 und 16 zu berücksichtigen, daß die Abweichung der praktisch gemessenen von den theoretisch berechneten Feldstärken nicht zurückgeht auf einzelne Punktmessungen, sondern daß jeweils die mittleren Feldstärkewerte aus den vom Meßwagen registrierten Feldstärkekurven entnommen wurden. Diese Werte haben daher eine größere Sicherheit als dies bei Punktmessungen der Fall wäre.

Mit Hilfe dieser letztgenannten Ergebnisse sind praktische Ausbreitungskurven für 50 % der Orte und 50 % der Zeit aufgestellt worden, die aber nur als vorläufige Kurven angesehen werden können. Bei der Ermittlung dieser Kurven wurde zunächst so vorgegangen, daß diese Kurven für glatte Erde $k = 4/3$ und elektrische Bodenbeiwerte von $\epsilon = 10$, $\sigma = 10^{-13}$ EME theoretisch berechnet und die Größe der Abweichungen der praktischen von den theoretischen Kurven nach Abbildung 15 an diesen Kurven, vergl. Abbildungen 17 und 18, angebracht wurden.

Propagation Measurements at 858 Mc/s over Paths up to 585 Km

G. C. Rider

Marconi's Wireless Telegraph Company, Ltd., Research Division, Great Baddow, Essex

The tests described in the paper have been made over five paths in Great Britain during the last $2\frac{1}{2}$ years. The objective was to obtain the "know-how" in the application of the tropospheric scatter mode of propagation. The paths investigated have been progressively increased in length up to a total distance of 585 km.

The transmitting site has remained fixed throughout the tests. Facilities for scanning were provided on the transmitting aerial, so that the beam could be directed towards the various receiving sites. The first test was over a distance of 158 km, and this was followed by 12 months of recording over a 322 km path. During this period height-gain and space-diversity tests were carried out. Finally the range was extended to 575 and 585 km; for these latter paths concurrent signal and fading recordings were made at a site in the mid-path region.

The mean diurnal variations of signal are presented for three paths, and their significance is discussed in relation to meteorological conditions. The effect of mean refractive index gradient upon median signal levels is examined. A good correlation is found between the presence of low signal and a polar air mass over the propagation path.

The fading ranges and rates measured over the 322 km path are presented in some detail, and the correlations—positive between fading rate and wind velocity, and negative between signal level and wind velocity—are discussed. Correlation coefficients between the fading on spaced aerials are shown for aerial spacings varied both horizontally and vertically, and a height gain curve has been derived from the vertical measurements. The height gain effect is of limited practical significance, but the persistence of the vestigial lobe pattern with increasing height appears to favour the reflecting layer mechanism of propagation. The negative correlation between signal strength and wind velocity also seems to point in this direction. Wind velocity has been found to be associated with the time-lag for maximum cross-correlation between the fading on horizontally spaced aerials.

An experimental approach to the problem of assessing the bandwidth potentialities of a tropospheric scatter system has been made by recording the phase variations of a 1 Mc/s modulation, the record being interpreted as a plot of changes in the mean path length with time.

Finally the obtrusive signal flutter attributed to aircraft is discussed and assessed.

I. Introduction

The purpose of this paper is to present the results of some propagation measurements at a frequency of 858 Mc/s made in the United Kingdom over the last $2\frac{1}{2}$ years. The motive has been to obtain practical knowledge and experience in the tropospheric scatter propagation field. The routine measurement of signal strength and fading parameters has been accompanied by tests of polarisation, height-gain and diversity, and an investigation of multipath time-delays. Figure 1 shows the position of the receiving sites, and the path lengths. The first test site, which was at Grantham, was used for three months and that at Catterick for over a year. The remaining sites were used for periods of two or more weeks.

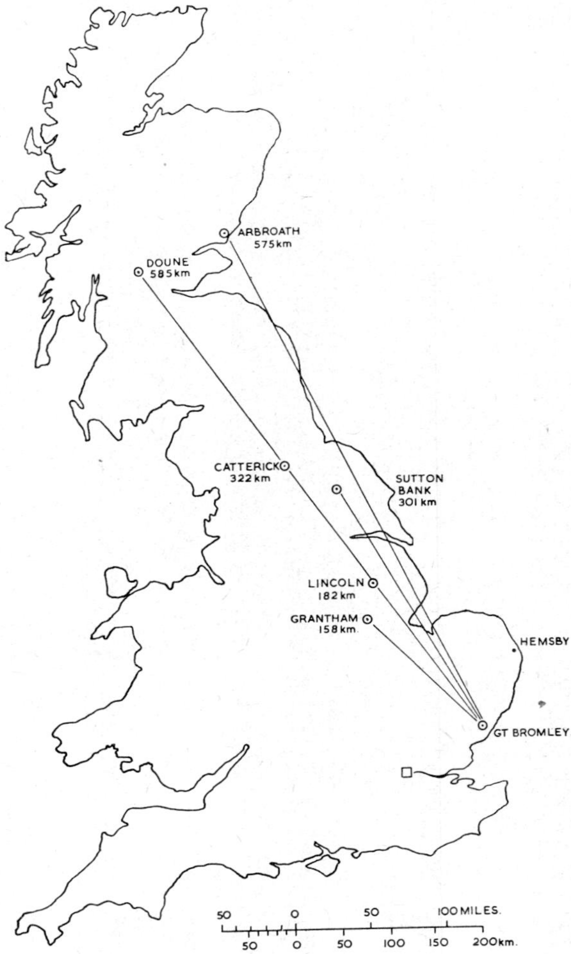

FIG. 1. Map showing propagation paths and distances.

II. EQUIPMENT

The transmitter had a mean output power of 200 W when amplitude modulated with a 1 kc/s sine wave, and it was installed in a trailer (Fig. 2) with the aerial feed horn mounted upon the roof on a steel-lattice boom. The trailer was movable over an arc of railway-line, and the movement served to steer the aerial beam in azimuth. Vertical movement of the beam was achieved by elevating or depressing the boom and feed horn. A 9·1 m (30 ft) diameter spherical reflector having an aperture ratio 1·5 was used as the transmitting aerial in conjunction with the movable feed horn, the long focus permitting the beam to be scanned through $\pm$ 12° in azimuth, and from 0° to 12° in elevation, without appreciable degradation of gain or lobe pattern.

FIG. 2. 858 Mc/s transmitter cabin and aerial—Great Bromley.

In each of the two receivers a crystal-controlled oscillator was used for frequency changing, and a low-noise balanced crystal mixer was followed by a 13·5 Mc/s I.F. amplifier, and a 1 kc/s audio amplifier of bandwidth 50 c/s. This arrangement limited the effective receiver bandwidth, and allowed very low signals to be measured without demanding an unreasonable degree of frequency stability either from the transmitter drive, or from the receiver frequency-change oscillators.

Two pen-recorders were employed, the fading data being taken from the records of a fast response recorder capable of faithfully reproducing frequencies up to 10 c/s, whilst the hourly median levels were taken from a recorder driven by a circuit averaging the receiver output over a period of 20 sec. The receiver had an accurately linear characteristic.

The main receiving aerial was a 3 m (10 ft) diameter paraboloid, mounted on a trailer at a height of 3 m; for the twin-receiver tests, a 1·23 m (4 ft) diameter paraboloid was used together with a horn of similar gain.

III. HOURLY MEDIAN PATH ATTENUATION

The hourly median path attenuation varies with the time of day, the period of the year and the path length, and will be considered first; consideration of the fading parameters will then follow.

Figure 3 presents the mean of the hourly median path attenuation for each hour of the day, measured over the paths to Grantham, Catterick and Arbroath at distances of 158, 322 and 575 km respectively. The shape of these curves should be

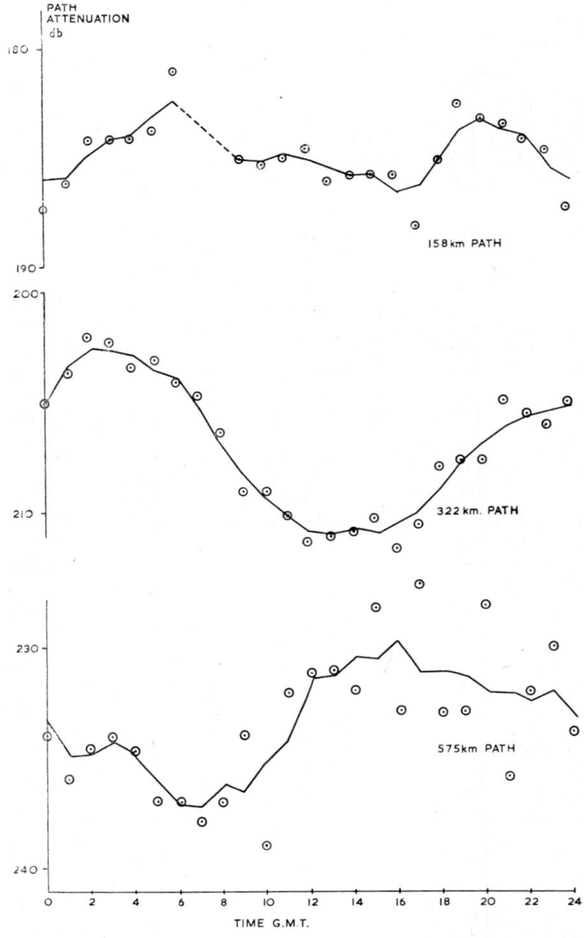

FIG. 3. Diurnal variation of hourly median path attenuations.
⊙ Average of hourly median.
— 3-hour moving average.

accepted with some reserve however, since although they present a true summary of measurements made, they are not necessarily typical of days in general. The curve for the 575 km path, for instance, is based upon far fewer measurements than the curves for the shorter paths; it should be noted, furthermore, that calm radiation nights—which largely contribute to night-time high signal-level—were rarely encountered during the period of observation over this path. It should be pointed out, however, that observations of path attenuations carried out during the daytime are unlikely to give an over-optimistic estimate of the performance of the path for other times. The monthly mean path attenuations shown in Fig. 4 for

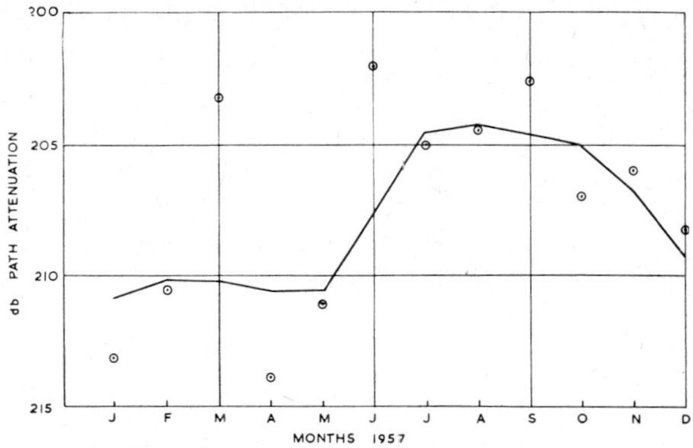

FIG. 4. Seasonal changes in hourly median path attenuation, 322 km path.
⊙ Mean of hourly medians 12.00–16.00 hr. G.M.T.
— 3-month moving average (weighted for number of observations).

the 322 km path are therefore taken for the mid-day hours alone in order to present, as far as possible, only the seasonal changes attributable to variations in the scattering mechanism.

IV. Meteorology and Hourly Median Path Attenuations

Two attempts will now be described to relate these time-variations in the hourly median path attenuation to meteorological conditions. The first approach is summarised in Fig. 5. Here the mean refractive index gradient, expressed as effective earth's-radius, is plotted against the hourly median attenuation for the appropriate hour. The meteorological data used were those published for the upper-air station at Hemsby [9] (see Fig. 1). The dashed line on the graph is calculated on the hypothesis that the signal variation is caused by a change in the scattering angle consequent upon a change in the ray curvature due to refraction. A ($\sin^{-4} \theta/2$) scattering law was used.

The trend of the observed points shows sufficient agreement with the predicted slope to suggest that the postulated mechanism is operating, modified however by the dependence of the scattering coefficient upon height, which has not been considered here.

In the second approach the path attenuations are grouped, for the synoptic hours, in accordance with the air-mass present in the common scattering volume. Figure 6 shows the signal distribution obtained with two such groups and, as more data become available, it is hoped to make a more detailed identification of the air-mass present. This approach appears to offer a useful empirical method for comparing the median path attenuations on similar paths in differing climates.

It would appear that there is more reason to attribute the changes in signal from hour to hour, and indeed from season to season, to the successive advent of

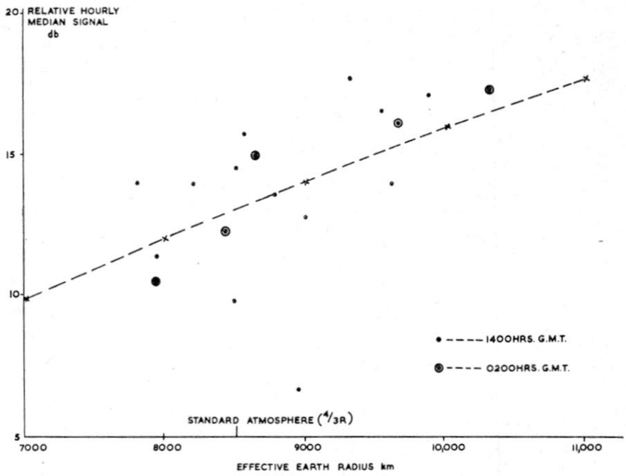

FIG. 5. Hourly median signal versus effective earth's radius.
·– 14.00 hr. G.M.T.
⊙ 02.00 hr. G.M.T.
× Computed points.

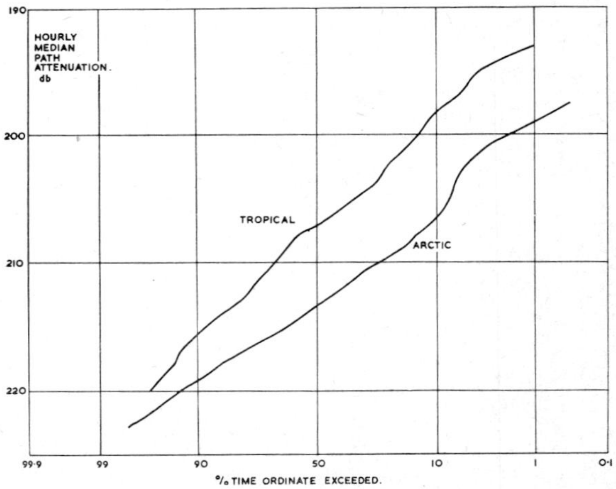

FIG. 6. Distribution of hourly median path attenuation for arctic and tropical air in the common scattering region.

anticyclones and depressions than to the diurnal changes in the meteorological elements, which are relatively of less importance in the middle and upper troposphere.

V. Path Attenuation versus Distance

The hourly median path attenuations are finally displayed, in Fig. 7, according to their dependence upon path length. The average of all the hourly figures for each site has been used for this purpose. Following Norton *et al.* [5] angular

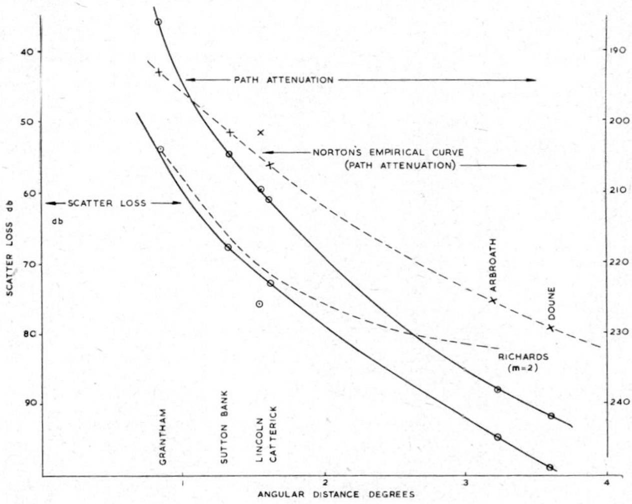

Fig. 7. Path attenuation and scatter loss versus angular distance.
⊙ Measurements.
× Calculated using Norton's [5] empirical curve.

distance is used as abscissa. The rate of attenuation is seen to be larger than that derived from Norton's [5] empirical curve, and larger also than that predicted by Richards [7] for an inverse square decay of $\overline{\Delta_n^2}$ with height.

VI. Fast Fading

Two parameters of the fading signal have been measured from the fast response recordings. The fading rate "F" is defined as the number of positive crossings per minute of the median level of the sample, and the fading range "R" as the difference in the levels exceeded by the sample for 10 % and 90 % of the time. One-minute samples were analyzed for each parameter and a cumulative distribution of the values obtained over the 322 km path, to which the bulk of the data accumulated refers, is given in Fig. 8. The fading rates measured at 575 km are also shown. The dependence on distance mentioned by Norton *et al.* [6] was not observed.

As will be seen later, the signal is quite frequently affected by reflection effects from aircraft. The samples analyzed for Fig. 8 were taken at regular time intervals,

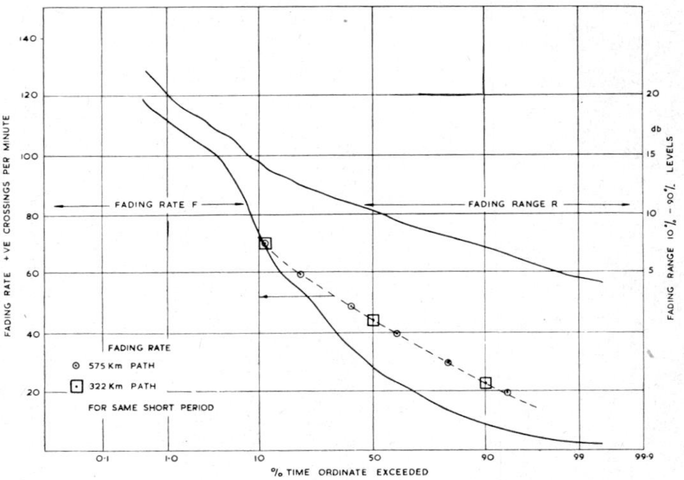

FIG. 8. Distribution of fading rate and fading range.
— 322 km path, 1 year.
□ 322 km path } same three weeks.
⊙ 575 km path }

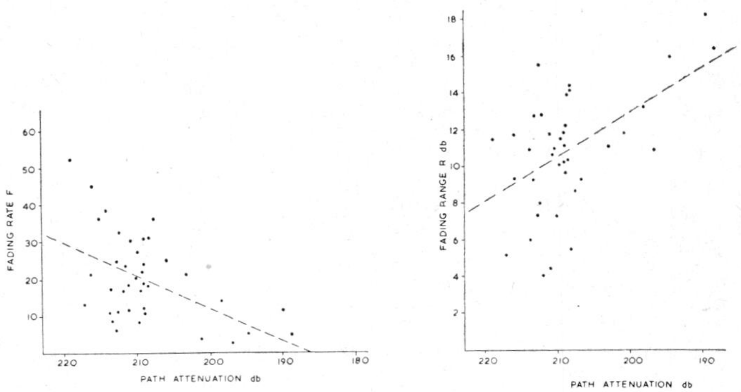

FIG. 9. Fading rate and fading range versus path attenuation for 322 km path. Aircraft effects excluded.

and are therefore representative of the conditions actually prevailing. These recordings were re-examined for a sample period, and those showing any sign of the aircraft effect were discarded. The remaining data, which are representative simply of propagation conditions, are presented in Fig. 9, and show well the association between the lowest fading rates and the greatest fading ranges. These occur at times of highest signal, usually at night, and contribute largely to the form of the diurnal curve already shown.

VII. Fading and Wind

The fading parameters have been considered in relation to the wind velocity in an attempt to produce some evidence distinguishing between the turbulent scatter model and that which invokes portions of reflecting layers. The wind is thought of as promoting propagation in the former case by increasing the turbulence, and as inhibiting the propagation in the latter case by breaking up and dissolving any stratification which existed. The normal presence of some measure of stratification is shown both by published data of refractometer soundings and by observation of the clouds.

The wind velocity data used—the best available—are taken from the regular meteorological soundings at a height of 4200 m over Hemsby. Hemsby is about 160 km from the scattering region, and this fact doubtless increases the element of randomness in the correlation computed.

The correlation between wind and fading is summarized in Table I and is discussed later in conjunction with further evidence arising from the diversity tests.

TABLE I

Correlation Coefficients

	Component of wind		Wind Speed
	Across Path.	Along Path.	
Fading Rate	+ 0·5	− 0·4	—
Fading Range	Not Significant	− 0·3	—
Hourly Medn. P.A.	—	—	− 0·2

The magnitude of the correlation between signal level and wind speed is insufficient to give much support to the reflection hypothesis although the sign of the correlation suggests it.

It may be of interest to note that synchronous fading records over 182 and 322 km paths show, as expected, no sign of correlation. The same is true of the hourly median values, where for the period of the tests—about 50 hours—a coefficient of + 0·04 has been evaluated.

VIII. Spaced Diversity and Height Gain

Spaced diversity and height gain tests were carried out at Catterick, over the 322 km path. The aerial spacings were varied, up to 30 m horizontally, and up to 25 m vertically. Here a 30 m (100 ft) mast was used, having mounted upon it a vertical railway carrying the aerial and head amplifier. A reference aerial, the standard receiving aerial at this site, was located at a height of 3 m above ground; it was an electromagnetic horn, very similar in gain and beamwidth to the 1·2 m (4 ft) diameter movable paraboloid.

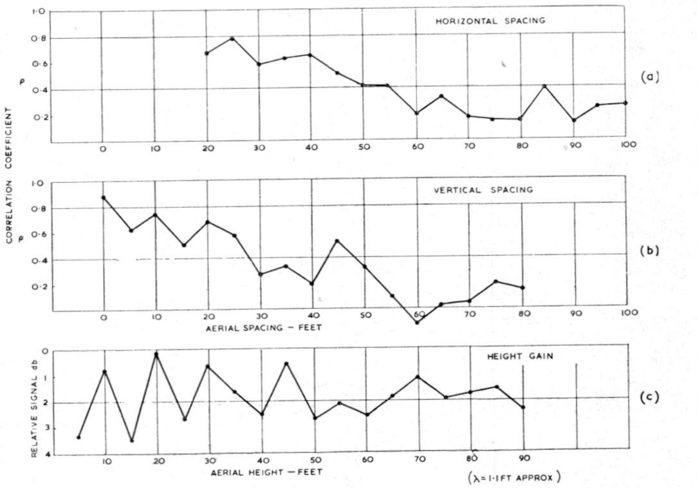

Fig. 10. Diversity fading correlations and height-gain curve for 322 km path.
(a) Horizontal aerial spacing versus fading correlation.
(b) Vertical aerial spacing versus fading correlation.
(c) Height-gain curve.

In each test, simultaneous one-minute samples were taken of levels from the respective spaced aerials, using the twin-channel fast recorder. Levels for the computation of correlations were read at 1·25 sec intervals. About 20 runs were made under various conditions, and Fig. 10 shows the results. Figures 10(a) and (b) are aligned to share a common scale of aerial spacings. Figure 10(c) is a "Height-Gain" curve derived from the observational data of Fig. 10(b). The ordinate in this case is aerial height above ground, and the ordinate scale is aligned with that of Fig. 10(b) so that, for example, an aerial height of 20 ft represents a vertical spacing of 10 ft. The abscissa here is the mean difference in level between the signals on the two aerials. This has been referred to the value at a height of 20 ft, since its actual value is irrelevant, depending on the slight difference in aerial gain. The optium height was found by averaging to be 20 ft, with the ground conditions obtaining, and quite a well marked lobe structure, albeit of small amplitude, may be seen in the curve Fig. 10(c). The pattern repeats itself well as between successive tests up to a height of about 30λ, above which level the coherence between different tests rapidly vanishes.

It is of interest to compare this curve with that of correlation coefficients for varying aerial spacings. The fluctuations in the relative signal level over about the first 30λ are seen to be closely mirrored by the variations in the mean correlation coefficient, which are superimposed upon the steady decay of correlation with increased aerial spacing.

This effect, and the decay of the height-gain pattern, may most easily be understood in terms of the polar diagram produced by the moving aerial and its image in the ground. As the aerial height increases, the lobes become sharper and sweep downward, a maximum signal position occurring as the lobe sweeps over the scattering region. For small aerial heights the width of the lobe is sufficient to embrace the whole scattering region, i.e., there is as yet no diversity effect due to the relative phases from individual scatterers differing for the two aerials (here aerial and image). However, when the lobes become too narrow to embrace the scattering volume, there will no longer be a coherent variation of signal with aerial height.

Attention will now be directed to the correlation plot. When both aerials are at positions of maximum signal—in our case the reference aerial coincides with the first maximum—each has a lobe embracing the scattering region, and positively correlated signals will appear on each. If now one aerial moves to a minimum, so that the trough between its lobes is directed to the centre of the scattering region, its signal will be contributed predominantly by the peripheral scatterers, and will be less correlated with that on the other aerial.

An estimate of the vertical angle subtended by the scattering region can be deduced from this argument.

Since the height above ground of the scattering region is great compared with that of the receiving aerials the ground reflection will take place near to the aerial, and simple plane-earth geometry will suffice.

We consider the lobe pattern of the receiving aerial together with its image in the ground. Since the aperture of the reflector is small, (about 4λ) the lobe pattern in the direction of the scattering region will be unaffected by the directional pattern of the aerial itself.

Let θ be the vertical angle between maxima for low angle lobes, and h the aerial height above the ground reflecting plane. We may simply obtain

$$\theta = \frac{\lambda}{2h}$$

From Fig. 10(b) we may see that the fluctuations in the correlation coefficient superimposed upon its general decay have ceased by the time the value h equals 19 m (60 ft). If this value of h is taken as giving a lobe spacing such that two lobes are directed towards the scattering region, we have, from the above expression, a vertical angle of $32'$ subtended by the scattering region.

IX. Time Lag in Horizontal Diversity Fading

A point of interest which is quite evident on inspection of the horizontal diversity fading records, is the time lag between the pattern of fading upon widely spaced aerials. For a number of tests on aerials at 30 m (100 ft) spacing, correlation coefficients have been evaluated for various time lags.

Referring now to Fig. 11, we see in graph (a) some of these lag correlograms.

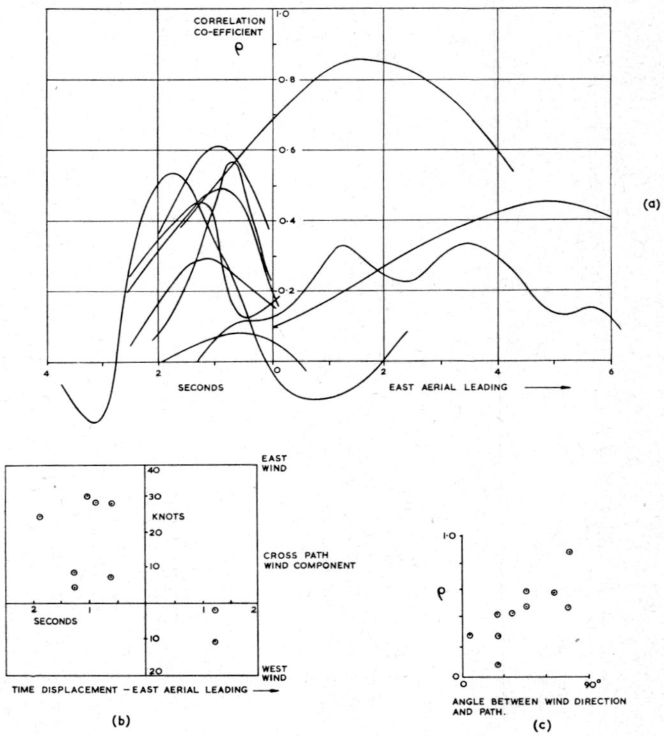

Fig. 11. The effect of cross path wind velocity upon diversity fading correlations—322 km path.

(a) Fading correlation versus time displacement.

(b) Time displacement for maximum correlation versus cross path component of wind speed.

(c) Maximum correlation versus inclination of wind direction to propagation path.

Graph (b) shows the time lag for maximum correlation plotted against the best available figures for the cross wind component in the common scattering region—that for Hemsby at a height of 3000 m. There appears to be little doubt that the time lags are in fact caused by the drift of the scattering sources through the common scattering volume when under the influence of the prevailing wind. Graph (c) also substantiates this, showing the maximum lag correlation to be higher when the wind is blowing more exactly across the path.

If the correlation between fading rate and wind velocity shown in Section VII is also considered, it seems evident that the movement of scattering sources with wind is an important cause of fading; this influence would not vary with path length except insofar as the wind itself is dependent on height. Fading rates measured over the 575 km path were seen (Fig. 8) to accord well with those measured at 322 km.

X. MODULATION PHASE MEASUREMENTS

A pulse system was first tried to assess the time delays involved in the propagation and hence the bandwidth potentialities of the medium. However, the signal/noise ratio was found to be inadequate for too large a part of the time, and the system now described was devised to enable useful time delay measurements to be made whilst retaining the narrow bandwidth used for field strength measurements.

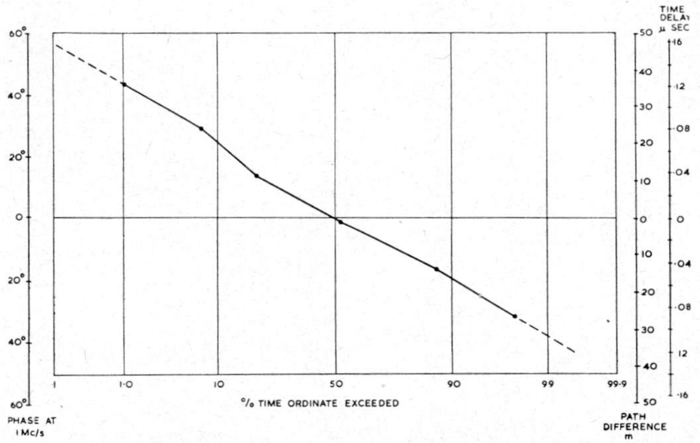

FIG. 12. Cumulative distribution of time delay over 322 km path. (The ordinate origins are placed on the median values.)

High stability crystal oscillators were used at both terminals; at the transmitter, 1 Mc/s sine-wave amplitude modulation was derived from the oscillator. At the receiver, crystal controlled frequencies of 999 kc/s and 1 kc/s were derived, the former being mixed with the incoming signal modulation to give a 1 kc/s tone having phase variations identical with those of the received modulation. This was then passed through the receiver audio-filter, giving an output of adequate S|N ratio to work a phase comparator. The crystal-controlled 1 kc/s source was used as the reference. The phase record is interpreted as a measure of the changes in path length, or time delay, suffered by the scatter signal, and Fig. 12 shows the distribution obtained from the data available. The instantaneous path is seen to differ from its median value by more than 38 m in length or 0·15 μsec in delay for 1 % of the time. It may be calculated that a ray-path, to give this additional path length, would require to be scattered from a point offset by 1·5° from the direction of the centre of scattering. A value for the angle subtended at the receiver by the effective scattering region is thus provided, and may be compared with the figure of 0·5° (32′) found in the last section. This latter figure applies only to the vertical angle subtended at the receiver by the scattering region, and in view of the somewhat arbitrary assumptions made the two measurements seem not entirely incompatible.

XI. Aircraft Reflections

The frequent superposition of a regular beat pattern upon the random scatter signal is immediately noticeable upon examination of the records from most of the sites during daylight hours. Figure 13 shows an example of the effect; a

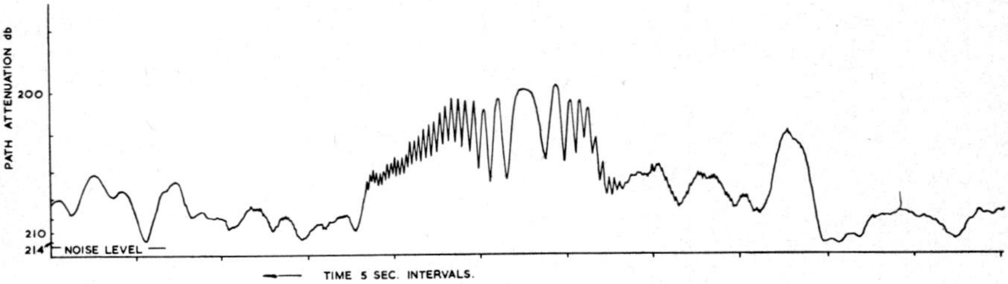

Fig. 13. Sample recording of aircraft effect.

similar effect is described by Mellen *et al.* [4], Altman *et al.* [1], Kiely *et al.* [3] Angell *et al.* [2] and Rider [8]. The latter papers bring forward considerable evidence to ascribe the effect to aircraft, and point out the great increase in effectiveness of aircraft in the vicinity of a site terminal.

XII. Significance of the Aircraft Reflections

Significant amplitude changes (i.e. > 3 db) have not been observed at beat frequencies greater than about 4 c/s so that an A.G.C. system would not be defeated.

Considering space diversity, an examination of the recorded effects at 100 ft horizontal spacing suggests that the slowest beats are often in phase on the two aerials, due to the correlation in the scatter fading sometimes observed at periods of slow fading and high signal. However, with normal fading rates and a negligible cross-correlation between the fading, the aircraft beats are also generally uncorrelated, though a number of successive beats may be in phase.

The most feared effect is of course the time delay in the reflected signal. Aircraft beats have been recorded during the modulation phase tests described above, and the range of time delays encountered is tabulated below. Unfortunately very little data is available, but it is felt that even an idea of the magnitudes involved will be of value.

A more disturbing picture is presented by Mellen *et al.* [4] quoting time delays up to 1·5 μsec, but using 6° beamwidths at both terminals.

TABLE II

TIME DELAYS OBSERVED DUE TO AIRCRAFT REFLECTIONS

No. of cases	14
Greatest Time Delay	0·5 μsec
Median Time Delay	0·07
Least Time Delay	0 (no observable change superimposed on the scatter changes)

XIII. POLARIZATION

Tests of received polarization have been made using a rotatable dipole in the receiving dish aerial, and turning it to a minimum signal position. This was invariably found to be at 90° to the launched polarization, which was varied through these tests. The "crossed" polarization position gave a signal 30 db or more below the normal level. The median signal was also recorded for successive periods of 10 min on Vertical and Horizontal polarization, and no significant difference in path attenuation was observed. For this test the launched polarization was 45° to the vertical.

XIV. CONCLUSION

The general picture presented by the measurements described, accords well with comparable published results.

However, the attenuation rate of 0·103 db/km may be noted, as also the diversity distances which are larger than was expected. Finally, the failure to detect a dependence of fading rate upon distance is worthy of note.

ACKNOWLEDGMENTS

The author is grateful to the Engineer-in-Chief, Marconi's Wireless Telegraph Company Limited, for permission to publish the paper. The assistance given by the author's colleagues in the provision of the equipment and the conducting of the tests, and particularly the help of Mr. F. S. Immirzi in the reduction of data, is gratefully acknowledged.

REFERENCES

1. Altman, F. J., Gray, R. E., Kandoian, A. G. & Sichak, W. (1955). *Trans. Inst. Radio Engrs, N.Y.* **MTT3** No. 6, p. 22.
2. Angell, B. C., Foot, J. B. L., Lucas, W. J .& Thompson, G. T. (1958). *Proc. Instn elect. Engrs* **105B**, Supp. 8, p. 128.

3. Kiely, D. G., Robinson, S. J. & Chesterman, F. C. (1957). *J. Brit. Inst. Radio Engrs* **17,** no. 3, 161.
4. Mellen, G. L., Morrow, W. E., Pote, A. J., Radford, W. L. & Wiesner, J. B. (1955). *Proc. Inst. Radio Engrs, N.Y.* **43,** p. 1269.
5. Norton, K. A., Rice, P. L. & Vogler, L. E. (1955). *Proc. Inst. Radio Engrs, N.Y.* **43,** 1488.
6. Norton, K. A., Rice, P. L., James, H. B. & Barsis, A. P. (1955). *Proc. Inst. Radio Engrs, N.Y.* **43,** 1341.
7. Richards, E. G. (1958). *Proc. Instn elect. Engrs* **105**B, Supp. no. 8, p. 177.
8. Rider, G. C. (1958). *Proc. Instn elect. Engrs* **105**B, Supp. no. 8, p. 143.
9. The Daily Aerological Record, and The Daily Weather Report. The Meteorogical Office, Kingsway, London, W.C.2.

Intercomparison of Different Calculation Methods of the Sky-wave Field-strength

K. Rawer

Ionosphären Institut Breisach im Fernmeldetechnischen Zentralamt der Deutschen Bundespost, Breisach (Rhein)

The four methods of fieldstrength-calculation which we discuss here are all of the analytical type, i.e., they suppose a detailed model of sky-wave propagation and use formulas (or graphs) which are intended to reproduce physical properties of the different phenomena involved in propagation. Of course the model is—like all models in theoretical physics—somewhat simplified as compared with the true phenomena. This simplification is even important as the phenomena are very complicated. It is always the aim of a practical method to come to a compromise between a model which has a maximum of agreement with the facts on one side while introducing a minimum of calculation work on the other. Of course these conditions can be met in different ways and there exists some ambiguity as to where a suitable compromise could be established just now with the rapid progress of automatic calculation methods. Some graphical methods which have been considered as easy for human workers cannot be adapted easily to machines, whilst some calculation methods which seemed too laborious in the past must now be considered as not complicated at all.

I. The Main Influences Effective in Sky-wave Propagation

The model which is introduced for such calculations refers to ray-optics only. We consider the propagation of waves between a more or less mirror-like earth and a refracting and absorbing ionosphere which is composed of several layers. When considering the geometrical optics of this system three groups of questions arise:

1. If absorption and scatter are neglected we have to do with loss-less refracting layers, it results either total reflection or total transmission. We thus obtain the *unabsorbed field-intensity*. The solution is trivial for the case of plane lossless, earth and ionosphere. Two points must be discussed: the influence of the curvature causing eventually focusing and the losses occurring with reflections at the ground (for multihop propagation).

2. Now *absorption* must be taken into account. It is brought about by collisions between electrons and neutral particles or ions. Different approximations are usable when the collision number is high or when it is low. In the first case we obtain appreciable absorption with a pretty low electron density, i.e., even when refraction effects can be neglected. This is called "non-deviative absorption"; it occurs only in lower layers of the ionosphere (D and bottom of E). The frequency-variation of the corresponding attenuation is very smooth.

In the higher layers absorption becomes only important when the electron density is high enough to give appreciable refraction effects. This is called "deviative absorption." It is a highly selective phenomenon with maxima at cusp-frequencies and critical frequencies.

All these are mainly interesting for lower layers which are penetrated by the rays with or without refraction-bending. Finally the absorption losses in the neighborhood of the reflecting level may also sometimes be considered.

3. Analytical methods by definition consider the *propagation modes** as individuals. One problem is to decide which modes must be taken into account. The most important effect in this respect is that of blanketing, i.e., the reflection of rays to rather small distances by the E-region, the same rays being otherwise useful for larger distances via the F-region. This "cut-off" is illustrated by

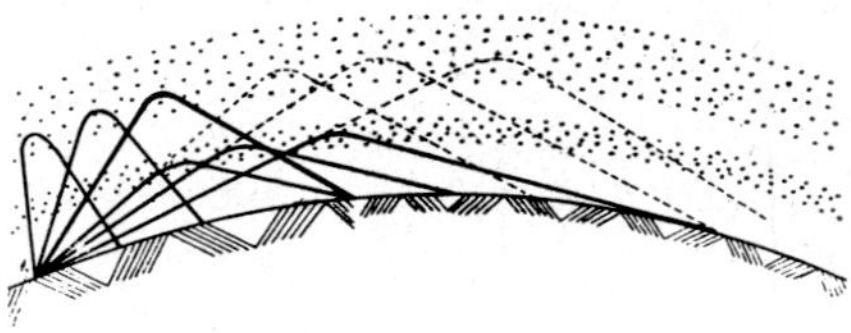

FIG. 1.

Fig. 1. It is a problem of geometrical optics which can easily be resolved for a thin blanketing layer like Es. With a thick blanketing layer like E considerable refraction occurs in the lower layer. Strictly speaking a geometrical ray-path with transmission through the E-region exists for every frequency higher than the E

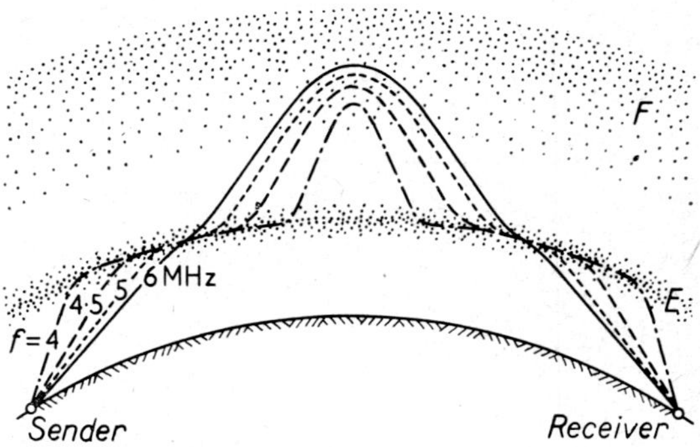

FIG. 2.

critical one (Fig. 2). But transmission with large deflection of rays by refraction suffers from a defocusing effect and from selective absorption (see Section II *C*) too. The problem thus turns out to be one of absorption.

The useful elevation angle is a problem which is closely connected with that of the modes. It is of outstanding importance when the practical use of predictions is under discussion. As things stand a complete solution of the mode problem practically involves a rough determination of that of usable elevation angles.

*(The term "modes" is used here to designate a certain path with a well defined number of ionospheric reflections in the E — or F — region and—normally—intermediate ground reflections. The same term is often used with a different meaning for long wave problems as well as with wave-guides.)

Author Organization	First Report	2·3 Modes		2·4 LUF		
		…istinction	·2 Blanketing	·1 Noise Level		·2 Day to Day Variation
		·2 Long Distance		·1 Day	·2 Night	
N. Smith US Nat. Bur. Stands.	1948	–	E only (med. dist.)	Atmospheric Noise (Median)		–
F. Dickson US Signal Corps	1950	+	+ E only	–		–
A. V. Kazancev Nizmir-USSR	1947	lowest hop number only	–	estimated E_{min}		–
K. Rawer ZfF Germany SPIM France	1943	+	+ E and E_s	Man-made Noise, estimated	Atmosph. Noise, variable with MUF	+

4. A problem which is closely connected with the field strength calculation is that of the determination of the *lowest useful frequency* (LUF). Essentially it turns out to be a simple intercomparison between calculated field strength values and some minimum value. The latter is normally given by some noise level, but it may also be due to the poorness of equipment or to the interference with other radio stations. Thus the essential problem for LUF calculation is that of the minimal field strength which can be accepted. Rather different solutions have been proposed.

II. Intercomparison of Different Fieldstrength Calculation Methods

We consider in the following the four main methods in the order in which they have been presented to international organizations such as CCIR. It is seen from the second column of our table that this order is not identical with the chronological one. We may be allowed to use the names of the chiefs of the different teams to represent the different methods: N. Smith [5], F. Dickson [3], A. Kazancev [4], K. Rawer [6, 9].

The problems are discussed in the order given in Section I, the main results being indicated in the intercomparison table. For this table we have replaced by approximative formulas some relations which are given in a graphical form by the American or Russian authors; these relations are given in bold type in the table. Generally a positive answer to a certain problem is indicated by a cross; that a problem has not been taken into account is shown by a dash.

A. Unabsorbed Field

1. *Focusing* is considered only by Rawer. He obtains three conditions for focusing: (1) at the skip-distance, (2) for the tangential ray, (3) for the antipodes [7].

In 1948 he calculated the complete geometrical optics in the case of a parabolic refracting ionosphere with loss-less ground; as focusing effects are highly variable with ionization he retains only the lower envelope of his theoretical curves. Focusing at the antipodes is nearly suppressed in his curves by the argument that homogeneity of the layers cannot be supposed for such large distances [8].

2. *Ground reflection loss* is not taken into account by Rawer, Kazancev introduces the reflection coefficient of the horizontal polarization, Dickson distinguishes three cases with respect to ground conditions. From the graphs given by Smith we obtain the approximation formula given in the table. It corresponds to a very high ground reflection loss; this formula is obtained from the large distance graphs of Smith, where the different modes are not distinguished.

3. *Empirical adaptation constant.* The American authors introduce an adaptation value which is arbitrarily introduced into the results of the model calculations and which is adjusted by experience.

4. *Discussion.* The problem of focusing as well as that of ground losses is not yet resolved. The practical importance of these effects may be highly variable. In

order to distinguish between the different contributions analytical tests should be made. At the present moment no such investigation is known to the author.

B. Absorption

1. *Non-deviative absorption.* A rather general formula for the absorption decrement is:

$$\delta_0 = \sec \alpha \, . \, P_0 \, . \, (1 + \gamma \bar{R}) \, . \, F(\cos \chi) \, . \, G(f) \tag{1}$$

with the symbols: f = frequency, χ = solar zenith angle, R = smoothed sunspot number, P_0 = standard absorption value (for zero sunspot number, $\chi = 0$, $G = 1$), γ = coefficient of solar activity, α = angle of incidence into the absorbing layer. F is usually normalized to $F(1) = 1$. A formula of this form is effectively used by all four authors. Smith gives G by a graph which can well be approximated in the interesting H.F. range. In Kazancev's formulas there appears always (foE), the critical frequency of the E-layer. We have introduced into his formulas the well-known relations for (foE); we use the empirical formulas of Harnischmacher-Rawer:

$$foE = (3,4 + \cdot 0055\bar{R}) \, . \, (\cos \chi)^{0,3} \text{ MHz} \tag{2}$$

It is generally asumed that the ordinary component gives the most important contribution, at least for frequencies in the neighborhood of the LUF which are of primary interest for absorption effects.

(a) For the *frequency variation* Kazancev and Rawer use Booker's approximation (o-ray). The longitudinal component of the gyrofrequency, f_L, appears in Booker's formula, the power being—2 [2]. The variation of f_L with the direction is not introduced into the practical calculations, this would be too troublesome. As a mean value something between 1 and 1·2 MHz is usually inserted for medium latitudes (lower values must be used in the tropics). Dickson has the full gyrofrequency f_H instead of f_L, he has found by pure experience a power of—1·98, sensibly identical with—2. The graphs given by Smith can similarly be approximated, at least in the H.F. range. (For lower frequencies see Section II *E*.)

(b) Absorption as a function of the *sun's position* has first been found by theoretical calculation to follow a law with $(\cos \chi)^{3/2}$ this means a rather sharp maximum at noon. Observations have shown, however, that this law is too sharp and, furthermore, that rapid fluctuations occur very often. The exponents given by the different authors have been obtained by experience, except for Kazancev.

(c) The sun's *limiting position* (i.e., the zenith angle for which absorption disappears, $\chi(F = 0)$), is assumed between 90° and 102·2°. It seems that some absorption persists after sunset such that the limit should somewhat exceed 90°.

(d) The *standard absorption* value, P_0, has been determined, independently, from experiments by the different authors. The results agree pretty well with each other.

(e) The *solar cycle* influence can well be approximated by a linear function. The coefficients, found also by independent experiments, agree well with each other with only the exception of Smith.

(f) *Discussion.* The laws adopted by the four authors are not widely different. We have stated that numerical values obtained in different countries and with different experiments agreed very well.

C. Deviative Absorption

Deviative absorption of transmitted rays is discussed only by Rawer who in 1952 [1] made a detailed calculation with a model composed of two parabolic layers corresponding to regions E and F, taking account of non-deviative and deviative absorption (see also Section III 2). From this model calculated LUF-factors have been deduced which are applied to the *foE*-value. These factors can be used like the secant-factor obtained with a thin layer model. Blanketing and deviative absorption are considered here as two aspects of the same phenomenon; the limiting frequency of blanketing obtained with the simpler concept is now replaced by a continuous variation of the total attenuation as a function of frequency.

D. Absorption at the Reflecting Level

Smith's graphs contain some deviative E-region absorption for the case of reflection by this region. For the same case Kazancev gives a linear approximation to a formula obtained by theory. He supposes some absorption in the F-region too. His formula has been obtained under the hypothesis that electron-ion collisions are predominant in this region. It follows that this absorption increases rapidly with the frequency, like f^2.

This hypothesis has been confirmed by various experiments and theoretical considerations during recent years. However the numerical value, which is assumed by Kazancev, is very high; if we take vertical incidence and a frequency of 10 MHz we obtain a decrement of 17 db which is by far too high. It seems thus that this influence is over-estimated by Kazancev. Also a better theory of the effect should give an important influence of the ion-density in the F-region which is not obtained with Kazancev's formulas.

E. Different Modes

1. The *distinction* of the different propagation modes is an essential feature of analytical methods. The past experience with global formulae has shown that the complex structure of sky wave propagation cannot be accounted for without considering this most important feature (see, for example, Fig. 10). Smith who formerly had used a global formula introduces the mode concept only for medium distances (400–3200 km); thus important discrepancies with the global formula appear at the limiting distances [9]. Kazancev also uses the mode concept in a strict form for small and medium distances only; for long distances his method considers the mode with the lowest number of hops, i.e., that of lowest elevation angle.

2. *Blanketing* is not taken into account for large distances, either by Smith or by Kazancev. Dickson fully takes account of blanketing brought about by the normal E-layer. Rawer considers blanketing by normal E as well as by Es; by his statistical representation of prediction results the consideration of a highly fluctuating phenomenon like Es is facilitated (see Section II *F* 2). Blanketing by a thick layer like normal E is presented in his more recent publication as a problem of deviative absorption (see Section II *C*) [10].

3. *Discussion.* There is little doubt that the mode concept and blanketing are very important for large distances too. It seems to the author that the LUF problem for high power large distance transmission is essentially one of blanketing.

Intermediate reflections of the type $(nE + mF)$ as introduced by Harnischmacher, multiple M paths as proposed by Lepechinsky and multiple rays without ground reflection as proposed by Stein and Choikova should also be considered for large distance propagation (this problem is discussed in the contribution to this volume by Harnischmacher).

F. Lowest Useful Frequency (*LUF*)

1. If the field strength calculation has been carried out, the essential parameter which must still be introduced is the noise level. This problem is not discussed in Dickson's report. Smith gives a detailed discussion using empirical (median) levels of atmospheric noise. The daily variation of this noise as well as the variation with the frequency is taken into account; the LUF value is found by graphical intersection of field strength and noise level plotted both as a function of frequency. Kazancev supposes that estimated minimum field strength levels may be compared with the calculated value. Rawer uses different methods for daytime and at night. He thinks that for modest equipment the very low daytime atmospheric noise level is not the effective limitation, but that industrial (man-made) noise is decisive; this leads to the insertion, in daytime of fixed minimum values for practical calculations. At night Rawer uses an approximative formula for the frequency variation of atmospheric noise which is deduced from the American curves. He introduces an important correction in the case where the LUF approaches to the MUF; he thinks that in these cases the increase of the skip zone causes a decrease of the effective atmospheric noise level.

2. The day to day variation of the ionospheric parameters is only considered in Rawer's method: two different predictions are normally made with an optimistic and a pessimistic set of basic data in order to obtain different results corresponding to two different probability values; this gives the possible range of fluctuations.

3. *Discussion.* Like most problems of adaptation of calculated results to practical traffic data the problem of the effective noise level is far from being resolved. The question of the atmospheric noise level is still under consideration by the C.C.I.R., it seems that not enough measurements have yet been made in enough parts of the world. Details as the variation of noise level as a function of propagation conditions have not been seriously studied. Very little is known on man-made noise.

There can be no doubt that the fluctuations of ionospheric conditions from which the day-to-day variation results are important for all practical applications; for certain types of service a solution of the field strength problem given in terms of median values only is not very useful.

III. Control by Theoretical Considerations

1. Calculation methods for practical use cannot take account of all physically existing facts. They must use simplified models, sometimes even oversimplified ones. But detailed model calculations can be carried out for a limited number of examples; by intercomparison with the result of the simplified calculation one can find out where the approximation is not valid.

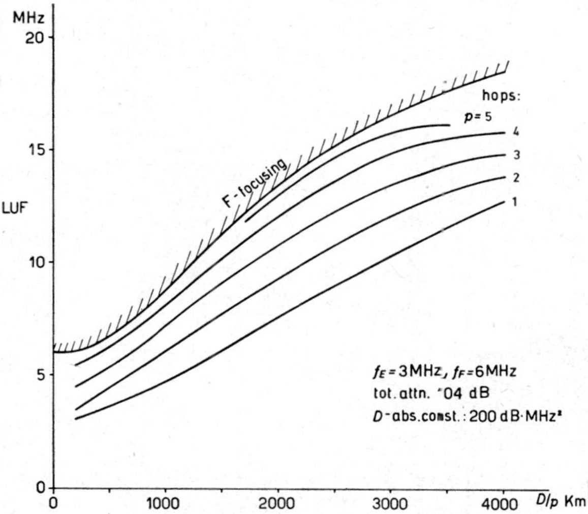

Fig. 3.

2. We have made a complete model calculation with two parabolic layers representing the E- and F-region respectively [1]. This calculation is in the frame of geometrical optics taking account of a non-selective absorption (like that in the D- and E-layer) and selective absorption in the E-layer itself. With Booker's quasi-longitudinal approximation [2] the effect of the earth's magnetic field has been taken into account in the cases of non-deviative absorption. No magnetic effect has been used in the selective absorption formula, but the influence of the collision number decreasing with hight is accounted for. Between two points at the earth's surface with only one reflection the normal (lower) as well as the Pedersen (upper) ray can be observed. The total attenuation by dilution and absorption is calculated separately as a function of frequency taking into account the frequency dependent refractive deviation (Fig. 2).

We give some results in Figs. 3–9. Figure 3 shows the calculated LUF value in the case of a rather powerful transmitter for different modes. As our model takes account of focusing at the skip distance the LUF cannot exceed the MUF value. On the abscissa we have plotted the distance of *one* hop. For intercomparison it is preferable to have the total distance as abscissa as in Figs. 4–6. In these figures the usable frequency range has been plotted as a function of distance for different modes reflected by the F-region. Figure 4 corresponds to rather normal conditions

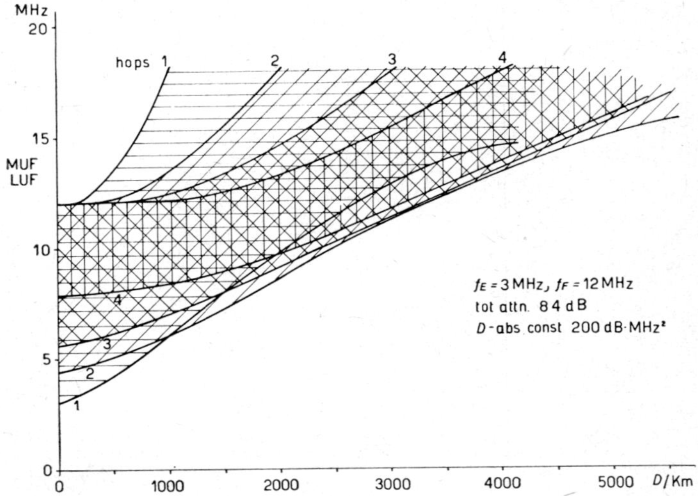

FIG. 4.

with medium power; the ranges corresponding to different modes overlap; the highest frequencies are transmitted with the lowest number of hops. The situation is more complicated for the LUF where blanketing appears beginning with about 1000 km; for higher distances the first mode has a higher LUF than the second one.

Let us now ask: what is the influence of the refracting properties of the F-region on the frequency range? Figure 5 has been obtained with the same transmitted power and the same absorption data. Only *foE* has been slightly increased and it has been supposed that the critical frequency of the F-region was rather near to that of the E-region. In this case we obtain very small usable frequency bands which follow the focusing of the skip distance for the different modes. In this case the lowest usable frequency is obtained with the highest number of hops and we have large gaps in the usable frequency range.

We may now repeat the calculation under the same ionospheric conditions but with higher power. In this case we obtain Fig. 6 where the frequency ranges are enlarged but the gaps have not yet completely disappeared. It is seen by intercomparison of Figs. 4 and 6 that an important influence of refraction effects must be admitted in cases where the ionization of the F-region is rather small. This influence is completely neglected in usual field strength calculations.

Figure 7 shows an example of the variation of the LUF with non-deviative absorption; it is rather small as blanketing is the most important factor in this case.

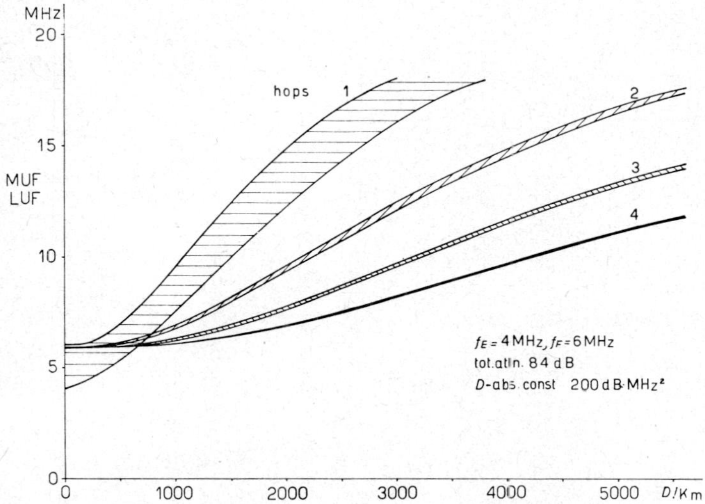

FIG. 5.

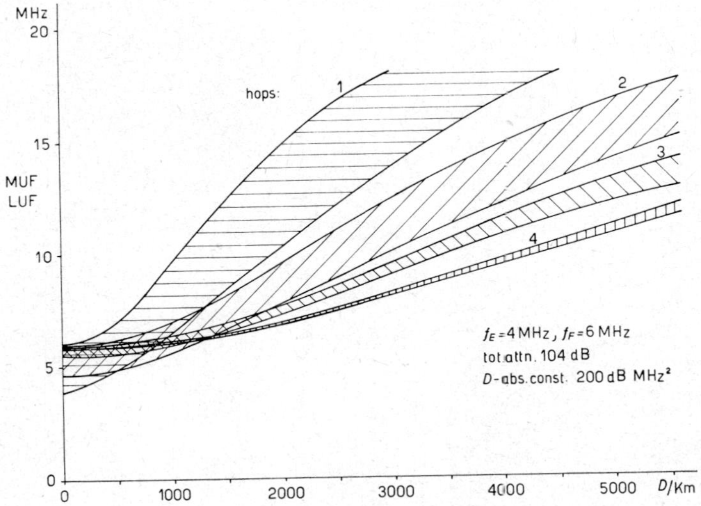

FIG. 6.

The field strength is given by the total attenuation which is plotted in Figs. 8 and 9 as a function of frequency. Figure 8 shows the intensity for the normal ray with two different values of the critical frequency of the F-region. In this special example the difference is only small. For one of these curves we have

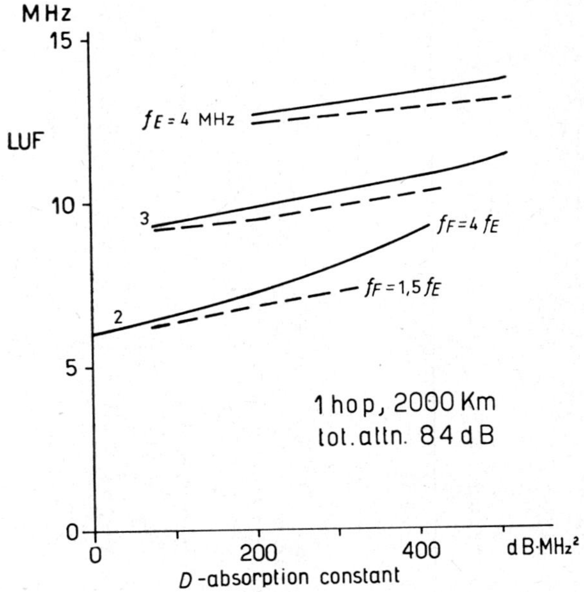

FIG. 7.

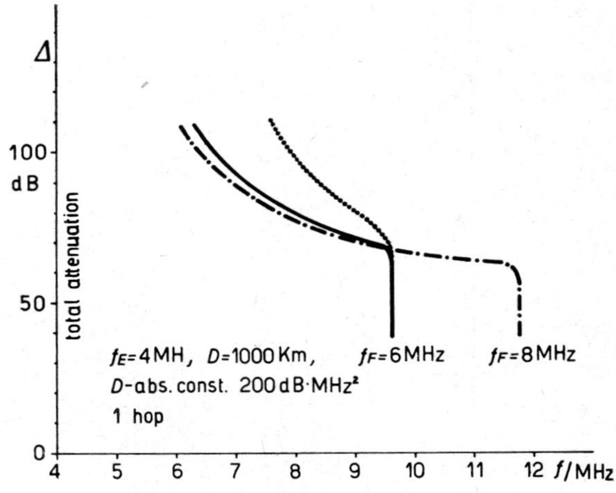

FIG. 8.

introduced the attenuation of the Pedersen ray (dotted curve); it is higher than that of the normal ray, the difference varying between 5 and 25 db.

Considering again the case of a low ionization in the F-region we see that in such cases the Pedersen ray can have nearly the same attenuation as the normal

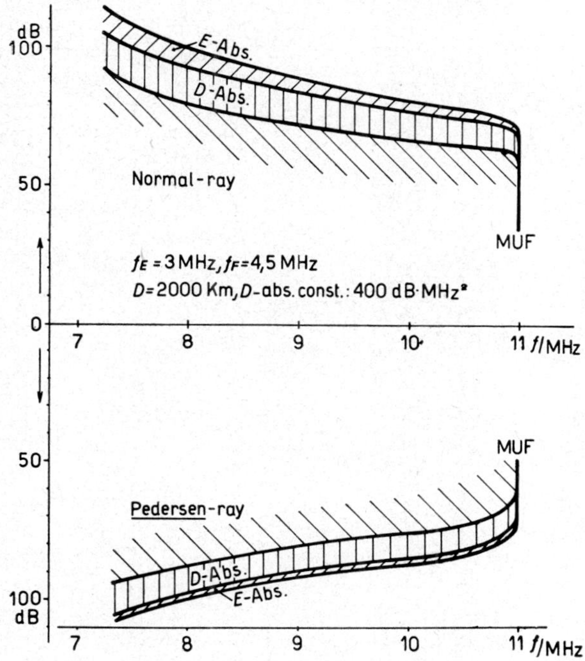

FIG. 9.

FIG. 10. Freiburg-Dakar.

one; this is due to the fact that the Pedersen ray attacks the D-layer more steeply than does the normal ray and thus suffers less from D-absorption. This can be seen from Fig. 9 and such conditions have been obtained occasionally on one of our oblique incidence lines (Freiburg-Dakar, Fig. 10). We may still note that with the mostly used rhombic antenna the Pedersen ray is also privileged by the antenna radiation pattern.

3. Other influences which should be investigated by a more complete theory will only be summarized:

The influence of the magnetic field on the propagation phenomena has not yet been investigated thoroughly.

We have already mentioned the possibility of complex propagation modes under Section II *E* 3.

A large field for future investigations is given by the fact that the layers are not homogeneous in the horizontal direction. We here have the problems of sporadic E, of scatter, of deviations from the great circle and the corresponding MUF correction [11].

IV. Control by Experiments

1. Most comparisons which have been made used C.W. transmission. With these observations it is very difficult to have a detailed control of any calculation method; the contributions of different modes are completely mixed and it is impossible in most cases to come to a reasonable analysis.

2. The only way to have a scientific test of field-strength calculation methods is pulse transmission on variable or fixed frequency. By such experiments the mechanisms and parameters can be distinguished and determined separately. It would be necessary of course to record the amplitude of the different echoes. Figure 10 gives an example of oblique incidence experiments over a 5000 km path. It can be seen that different modes are predominant in daytime and at night and that only an analytic evaluation can be useful. If in this case only the total field had been introduced we would come to erroneous conclusions neglecting the interchange of modes which appears from the pulse record.

3. For test experiments the problem of the antenna diagram is very difficult. A serious test can only be made if the diagrams have been measured on both sides. Many experiments are known where a calculated antenna diagram has been consulted and the maximum gain has been introduced into the calculation; this cannot be admitted as the elevation angle of the mode in question generally does not coincide with the main lobe direction.

We feel that a better solution of the problem could be obtained if people interested in propagation data would not indicate the received field intensity with a given transmitting station but if they would try to evaluate their observations in terms of the total transmission attenuation.*

*The total transmission attenuation is the ratio (expressed in decibels) between the fieldstrength at a distance of 1 km from the transmitter in the direction of propagation to the fieldstrength at the receiving place.

References

1. Bibl, K., Rawer, K. & Theissen, E. Service de Prévision Ionosphérique Militaire, rapport R 11 (1951); (1952). *Nature, Lond.* **169**, 147–148.
2. Booker, H. G. (1935). *Proc. roy. Soc. A* 150, 267.
3. Dickson, F., Laitinen, P. O. & Haydon, G. W. (1950). *Radio Propagation Unit, Tech. Rep.* No. 9.
4. Kazancev, A. N. *Bull. Acad. Sci. U.R.S.S., O.T.N.* 1947, no. 9; *O.T.N.* 1956, no. 9.
5. *Nat. Bur. Stand. Circ.* no. 462 (1948).
6. Rawer, K. (1947). *Rev. Sci.* **85**, 361; (1951). *Arch. elektr. Übertr.* **5**, 154–167.
7. Rawer, K. (1948). *Rev. Sci.* **86**, 481 and 585.
8. Rawer, K. (1949). Service de Prévision Ionosphérique Militaire (France), rapport R 6.
9. Rawer, K. (1952). *Wireless Engr* **29**, 287–301.
10. Rawer, K. (1956). *Nuovo Cim.* Suppl. al vol. iv. ser. x, **4**, 1460–1476.
11. Rawer, K. (1957). *Ann. Télécomm.* **12**, 35–38.

On the Computation of Diffraction Fields for Grazing Angles

JAMES R. WAIT and ALYCE M. CONDA

National Bureau of Standards, Boulder, Colo.

The diffraction of electromagnetic waves by a convex cylindrical surface is considered. Attention is confined primarily to the region near the light-shadow boundary. The complex-integral representation for the field is utilized to obtain a correction to the Kirchoff theory. Numerical results are presented which illustrate the influence of surface curvature and polarization on the diffraction pattern. Good agreement with the experimental results of Bachynski and Neugebauer is obtained.

I. INTRODUCTION

The Kirchoff theory of diffraction has been quite successful in the prediction of field strength when the transmission path is obstructed by obstacles such as mountainous ridges [4, 6, 9, 10, 11, 14, 16, 17]. In its usual form, however, this theory disregards the difference between vertical and horizontal polarization. Furthermore, the electrical properties of the diffracting obstacle are not considered since in effect, the obstacle is regarded as a "black screen" or absorbing knife-edge. A brief mention of the various developments (both old and new), will be made in order to properly orient the reader.

Schelleng, Burrows & Ferrell [16] in a now classic paper, applied knife-edge Kirchoff theory to the computation of the fields behind mountainous obstacles and other pronounced terrain features. The influence of the terrain in the immediate vicinity of the terminals was accounted for by the introduction of ground reflected rays which are also diffracted by the obstacles. A similar approach wa smade by Wickizer, Herbstreit, Dickson & Egli [6], who demonstrated that the presence of an obstacle can, in some cases, produce an increase of field strength beyond that expected in the absence of the obstacle. This phenomenon has been described rather aptly as "obstacle gain" and has been used to advantage in the siting of the terminal and repeater stations in microwave relay links.

The Japanese workers [10, 11] have also been very active in this field. Presumably, the mountainous features of their country have motivated much of this work. Matsuo [10] in particular has applied Kirchoff theory to many geometrical forms including conical shaped obstacles. Some fundamental theoretical work has been carried out by Furutsu [8] who has solved, at least in a formal way, the diffraction by very general forms of obstacles.

S. O. Rice [15] has presented a comprehensive solution for the diffraction of a plane wave by a perfectly-conducting parabolic cylinder which was considered to be a more realistic model of a terrain obstacle. Particular attention was paid to the fields near the horizon plane on the boundary region between light and shadow. It was shown that to a first approximation the diffraction pattern behind the obstacle was very similar to that for a knife-edge. The main effect of the finite radius of curvature of the crest was to increase the field strength for vertical polarization and to decrease it for horizontal polarization. A similar observation had been made by Artmann [1] sometime earlier who employed a circular cylindrical model.

A still further approach to this problem was adopted by Norton, Rice & Vogler [13]. They represented the diffracting obstacle first by a sphere for the case of a very large radius of curvature and then by a knife-edge for vanishing small radii of curvature. A curve of diffraction loss as a function of radius of curvature was then drawn by interpolating between these two limiting cases. According to their results, the diminishing of the radii of curvature is accompanied by enormous increases in the field strength even though the separation between terminals and the angular distance is unchanged.

Refinements of Kirchoff diffraction theory when applied to the diffraction by smooth cylindrical models were made by Neugebauer & Bachynski [12]. The basic idea of their analysis is that the equivalent radiating aperture is illuminated not only by the incident rays, but from rays reflected from the illuminated side of the cylindrical crest. The subsequent radiation from this aperture plane, which extends from the apex of the crest upward to infinity, is again reflected from the shadow side of the crest. Neugebauer & Bachynski are quite willing to admit that this is not a rigorous procedure, but nevertheless, it leads to results which are in good agreement with experiment.

It is the purpose of the present paper to present a method for the computation of diffracted fields by convex surfaces which is both rigorous and easy to apply. The mathematical basis of this method was described in a previous paper [18]. This was an extension of the fundamental work of Van der Pol and Bremmer, and Fock. The reader is referred to the above quoted paper for an account of the early theoretical work.

Unfortunately, the existing solutions for the diffraction by spheres and cylinders are poorly convergent in the line-of-sight region. Consequently, these formal solutions require a great deal of numerical work despite the fact that they are mathematically elegant. A significant break-through was made when it was suggested that the complex-integral representation of the field was easier to evaluate by numerical means than attempting to sum the residues of the poles of the integrand. While this point was probably recognized by Van der Pol & Bremmer [3], in the 1930's, it was not until 1946 that Academician V. A. Fock actually demonstrated this in what is now regarded as a classic paper [7]. Fock's method was applied by the present authors to the computation of the radiation pattern of an antenna mounted on or near a convex lossy surface [18]. The present problem is really an extension of this work. Only a statement of the working formulas will be given at this time. In a subsequent paper, the detailed mathematical derivations will be presented along with additional numerical results [19].

II. Statement of Theoretical Results

The situation is illustrated in Fig. 1(a). The transmitting antenna is located at A_1 and a receiving antenna at A_2. The diffracting obstacle in this case is a smooth convex surface S of radius of curvature a in the principal plane. The tangent ray from A_1 meets the surface S at the point T_1. The length of this ray is s_1. Correspondingly, the tangent ray from A_2 meets S at the point T_2 and the length of this ray is s_2. The angle subtended at the center of curvature of S is also θ. (This angle is often called the angular distance and is usually expressed in milli-radians.) The illustration in Fig. 1(a) corresponds to positive values of θ when the terminals are "obstructed" by the surface. The other case of interest is when θ is negative. This is illustrated in Fig. 1(*b*) where now the rays s_1 and s_2 are drawn from A_1 and A_2 respectively, to the surface S, meeting it at the point T_0 which can be described rather loosely as a reflection point. Note that the rays from A_1 and A_2 subtend an angle $-\theta/2$ (which is positive) with the tangent plane of the surface S drawn through T_0.

In the cases considered here, θ will also be a small angle since attention is being confined to the region near the light-shadow boundary (i.e., $\theta = 0$). More specifically, s_1 and s_2 are both assumed to be large compared to the quantity $|a\theta|$. (The vertical scale in Figs. 1(a) and (b) is somewhat exaggerated.)

The basic transmission loss, L_{bf} for propagation from A_1 and A_2 in the absence of the obstacle, is given by the wellknown relation [13]

$$L_{bf} = 20 \log (4\pi d/\lambda) \tag{1}$$

where d is the distance between A_1 and A_2 and λ is the wavelength.* Since $|\theta|$ is a small number, d can be replaced by $s_1 + s_2$ for convenience of calculation. The resultant basic transmission loss, L_b in the presence of the diffraction obstacle, is now written

$$L_b = L_{bf} + L_d \tag{2}$$

where L_d is the loss due to diffraction. If the obstacle is idealized by a knife edge, the diffraction loss has the following form†

$$L_d = -20 \log |F_0(\alpha)| \tag{3}$$

where

$$F_0(\alpha) = \frac{1}{\sqrt{2}} e^{+i\frac{\pi}{4}} \left\{ \left[\frac{1}{2} - \overline{C}(\alpha)\right] - i \left[\frac{1}{2} - \overline{S}(\alpha)\right] \right\} \tag{4}$$

for $\alpha > 0$, and

$$F_0(\alpha) = 1 - F_0(-\alpha)$$

for $\alpha < 0$. In the above

$$\alpha \simeq \left[\frac{2ks_1s_2}{s_1 + s_2}\right]^{\frac{1}{2}} \theta/2, \; k = 2\pi/\text{wavelength},$$

$$\overline{C}(\alpha) = \left(\frac{2}{\pi}\right)^{\frac{1}{2}} \int_0^\alpha \cos^2 \alpha \, d\alpha \text{ and } \overline{S}(\alpha) = \left(\frac{2}{\pi}\right)^{\frac{1}{2}} \int_0^\alpha \sin^2 \alpha \, d\alpha.$$

The function $F_0(\alpha)$, or its equivalent, has been discussed on numerous occasions and is relatively easy to compute since the Fresnel integrals $\overline{C}(\alpha)$ and $\overline{S}(\alpha)$ are well tabulated.

*The logarithms are to base 10.

†Ground reflected rays are not considered here. They can be accounted for by the "four-ray technique" described by Schelleng, Burrows and Ferrell [16].

It is to be noted that when $(-\alpha) \gg 1$ corresponding to points sufficiently above the shadow boundary, L_d is zero. On the other hand, if $\alpha = 0$, (i.e. $\theta = 0$) the rays just graze the crest of the obstacle and L_d is 6·02 db.

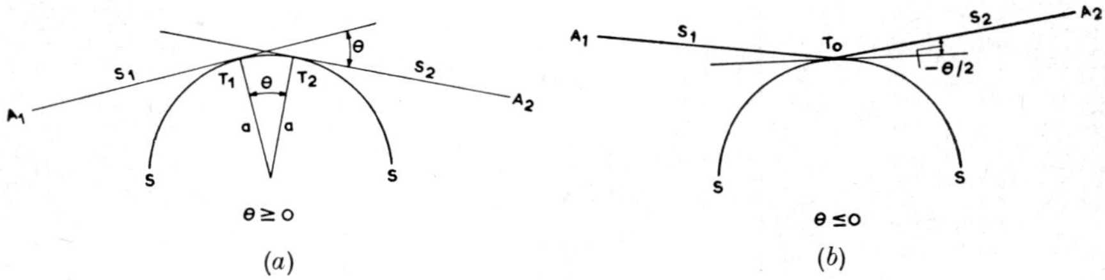

FIG. 1. Geometry for diffraction by a cylindrical surface.

Unfortunately, the knife-edge representation is not adequate, at least for VHF propagation over terrain obstacles. It is possible, however, to recast the results from rigorous diffraction theory in a form which shows that the knife-edge formulas are indeed the leading or zeroth approximation to diffraction by smooth convex surfaces. In this case,

$$L_d = -20 \log | F(\alpha) | \tag{5}$$

where

$$F(\alpha) = F_0(\alpha) + F_1(\alpha) e^{-i\alpha^2} \tag{6}$$

and $F_1(\alpha)$ is the correction term which is a function of the radius of curvature of the surface S and the electrical constants of this surface. Using a cylindrical model for S with the idealizations used previously, it can be demonstrated that

$$F_1(\alpha) = -\frac{G(X)}{u} \tag{7a}$$

where

$$u = \frac{\alpha}{X} = \left(\frac{2ks_1s_2}{s_1 + s_2}\right)^{\frac{1}{2}} \frac{(2/ka)^{\frac{1}{3}}}{2}$$

and

$$G(X) = \frac{e^{-i\pi/4}}{(\pi)^{\frac{1}{2}}} \int_0^\infty e^{-iXt} \frac{v'(t) - qv(t)}{w_1'(t) - qw_1(t)} dt + \frac{e^{i\pi/12}}{(\pi)^{\frac{1}{2}}} \int_0^\infty e^{-\frac{Xt}{2}(\sqrt{3} - i)} \left[\frac{v'(t) - qe^{-i2\pi/3}v(t)}{w_2'(t) - qe^{-i2\pi/3}w_2(t)}\right] dt \tag{7b}$$

with $X = \left(\frac{ka}{2}\right)^{\frac{1}{3}} \theta$. In the above integrals for $G(X)$, the functions $v(t)$, $w_1(t)$ and $w_2(t)$ are Airy integrals [7, 18] and the corresponding primes indicate derivatives with respect to t. The complex quantity q involves the electrical properties of the medium. For vertical polarization

$$q = q_v = -i \left(\frac{ka}{2}\right)^{\frac{1}{3}} Z/\eta_0, \; \eta_0 = 120\pi$$

where Z is the surface impedance of the boundary S. For horizontal polarization.

$$q = q_h = -i\left(\frac{ka}{2}\right)^{\frac{1}{3}} Y\eta_0$$

where Y is the surface admittance of the boundary S. η_0 is, of course, the characteristic impedance of free space and is approximately 120π ohms. For grazing incidence

$$Z/\eta_0 = \eta/\eta_0[1 - (\eta/\eta_0)^2]^{\frac{1}{2}} \tag{8a}$$

and

$$Y\eta_0 = \eta_0/\eta[1 - (\eta/\eta_0)^2]^{\frac{1}{2}} \tag{8b}$$

where η is the intrinsic impedance of the material (assumed homogeneous) making up the obstacle (note that Z and Y are not quite reciprocal). In terms of the conductivity σ, the dielectric constant ϵ, and the magnetic permeability μ,

$$\eta = [i\mu\omega/(\sigma + i\omega\epsilon)]^{\frac{1}{2}}.$$

For perfect conductivity, it is seen that $q = q_v = 0$ for vertical polarization, but for horizontal polarization $q = q_h = \infty$. For metallic surfaces these values are still quite appropriate since $Z/\eta_0 \ll (2/ka)^{\frac{1}{3}}$ and $Y\eta_0 \gg (2/ka)^{\frac{1}{3}}$ for nearly all cases of practical interest. For lossy surfaces, however, q_v can have almost any value. In a previous paper [18], considerable attention was devoted to a situation where $|q_v|$ was of the order of unity. In connection with VHF diffraction by mountains with smooth crests, however, the effective value of q_v will be very large and can be replaced by infinity in the integral for $G(X)$. With horizontal polarization, $|q_h|$ may be replaced by infinity in this case even for low frequencies as long as $(ka/2)^{\frac{1}{3}} \gg 1$.

In the present work two cases will be considered only, namely, $q = 0$ and $q = \infty$. For metallic surfaces they are referred to as vertical and horizontal polarization, respectively. In a subsequent paper results for other values of g are given [19].

III. Numerical Results

Basic to the present work is the function $G(X)$. In the region near $X = 0$ the revelant integrals cannot apparently be evaluated in closed form. The integrands are quite well behaved however, and it is not too difficult to evaluate them by numerical means. In Figs. 2(a) and 2(b) the real and imaginary parts of $G(X)$ are plotted as a function of X. The two cases considered are $q = 0$ and $q = \infty$.

Before discussing the diffraction pattern it is of interest to first consider the behavior of the field along the horizon plane. Clearly, this corresponds to $\theta = 0$, $X = 0$ and $\alpha = 0$. The function $F(\alpha)$ then becomes

$$F(0) = \frac{1}{2} - \frac{G(0)}{u}. \tag{9}$$

where

$$u = \left(\frac{2ks_1 s_2}{s_1 + s_2}\right)^{\frac{1}{2}} \frac{(2/ka)^{\frac{1}{3}}}{2} \text{ as before.}$$

Numerical integration yields the following values:

$$\text{for } q = 0, \quad G(0) = -0{\cdot}295 + i\,0{\cdot}0811 \text{ and}$$

$$\text{for } q = \infty, \; G(0) = 0{\cdot}344 - i\,0{\cdot}0918.$$

This immediately indicates that for vertical polarization with a metallic convex surface, the field along the shadow boundary is increased above that for a knife-edge. The converse is true for horizontal polarization. The increase or reduction

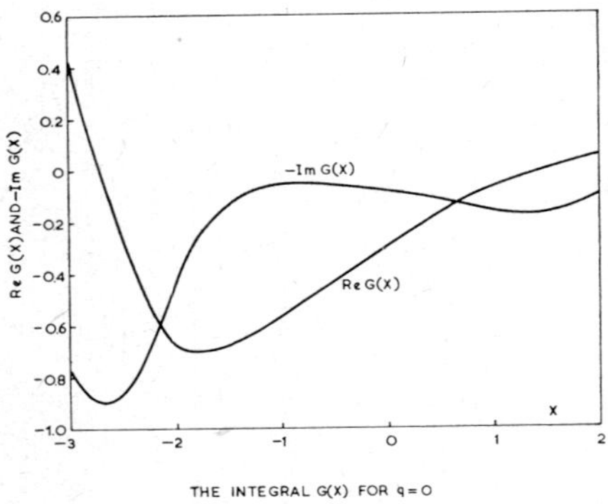

FIG. 2a.

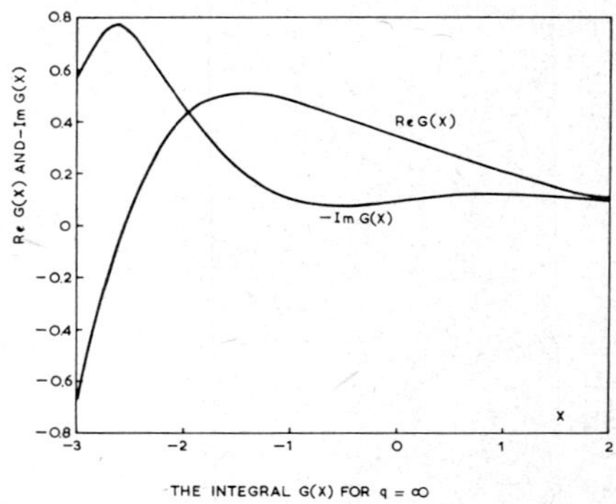

FIG. 2b.

from 1/2 is approximately the real part of the function $[-G(0)/u]$. Therefore for $q = 0$, $|F(0)| \simeq 0{\cdot}500 + 0{\cdot}295/u$ and for $q = \infty$, $|F(0)| \simeq 0{\cdot}500 - 0{\cdot}344/u$. It is immediately of interest to compare this departure from 1/2 with Rice's prediction using a parabolic cylinder model. The agreement is good if we equate the curvature of the apex of his parabolic cylinder with our a. It is also noted that his assumed plane-wave incidence is equivalent to moving our source to infinity (i.e., $d_1 = \infty$). Then on identifying Rice's h with $ka/2$ and his Λ with

$$\left(\frac{s_1 + s_2}{2\pi k s_1 s_2}\right)^{\frac{1}{2}}$$

we find according to Rice for $q = 0$, $|F(0)| \simeq 0{\cdot}500 + 0{\cdot}296/u$ and for $q = \infty$, $|F(0)| \simeq 0{\cdot}500 - 0{\cdot}342/u$ being in close agreement with our values quoted above. On the other hand, Artmann, using a circular cylindrical model, gives for $q = 0$, $|F(0)| \simeq 0{\cdot}500 + 0{\cdot}165/u$ and for $q = \infty$, $|F(0)| \simeq 0{\cdot}500 - 0{\cdot}33/u$. The disagreement in the case $q = 0$ with our result should be noted. It is suggested that the asymptotic approximation for the Hankel functions used by Artmann is not fully justified.

Neugebauer and Bachyinski have also computed a value on the shadow boundary using a semi-empirical modification to Kirchoff theory. Again transferring to our notation they would give*

$$\text{for } q = 0, \quad |F(0)| \simeq 0{\cdot}500 + 0{\cdot}271/u, \text{ and}$$

$$\text{for } q = \infty, \; |F(0)| \simeq 0{\cdot}500 - 0{\cdot}282/u.$$

*Their quantity $A = -8\pi^{\frac{1}{3}}\,2^{-\frac{1}{3}}$ times the real part of our $G(0)$.

The agreement of these values with ours (and those of Rice) is reasonable in view of the entirely different theoretical approaches employed. It is probable that at least part of the discrepancies can be attributed to the non-validity of some of the geometrical-optical arguments used by Neugebauer and Bachynski.

IV. The Diffraction Patterns

The diffraction pattern behind the obstacle is specified by the behavior of the function $F(\alpha)$ as α varies. The amplitude $F(\alpha)$ itself and the resulting diffraction loss L_d are plotted in Figs. 3(a) and (b) respectively, for various values of u with $q = 0$ and ∞. As u tends to infinity, the patterns approach the Kirchoff form described by the function $F_0(\alpha)$.

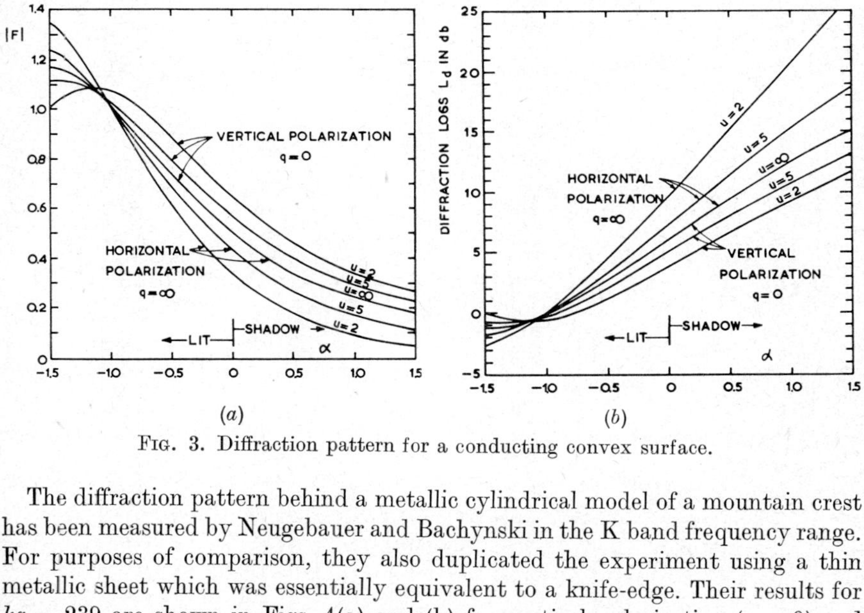

FIG. 3. Diffraction pattern for a conducting convex surface.

The diffraction pattern behind a metallic cylindrical model of a mountain crest has been measured by Neugebauer and Bachynski in the K band frequency range. For purposes of comparison, they also duplicated the experiment using a thin metallic sheet which was essentially equivalent to a knife-edge. Their results for $ka = 239$ are shown in Figs. 4(a) and (b) for vertical polarization ($q = 0$) and horizontal polarization ($q = \infty$) respectively. The ordinates are relative power and the abscissa are distance d_2 measured from the crest to the observed. The quantity h is the vertical displacement of the receiver above or below the horizon plane. In both Figs. 4(a) and (b) the source which is a transmitting horn is at a fixed location, being a distance d_1 ($= 150\,\lambda$) from the crest. Values calculated from Equation (5) are also shown on the curves. The agreement is reasonably good.

Neugebauer and Bachynski also measured the fields between the crest for various values of the curvature parameter ka. This is shown in Fig. 5 where the ordinate is relative power and the abscissa is the deflection angle θ. The corresponding calculated curves are also shown in Fig. 5 taking $d_1 = 150\,\lambda$ and $d_2 = 113\,\lambda$. The agreement is still fairly good and probably is within experimental error.

These curves illustrate in a striking way the influence of curvature on the diffraction fields. In the case of horizontal polarization and deep in the shadow, the finite radius of curvature greatly reduces the fields below the Kirchoff or knife-edge value. This is in accord with the results of Norton, Rice & Vogler [13].

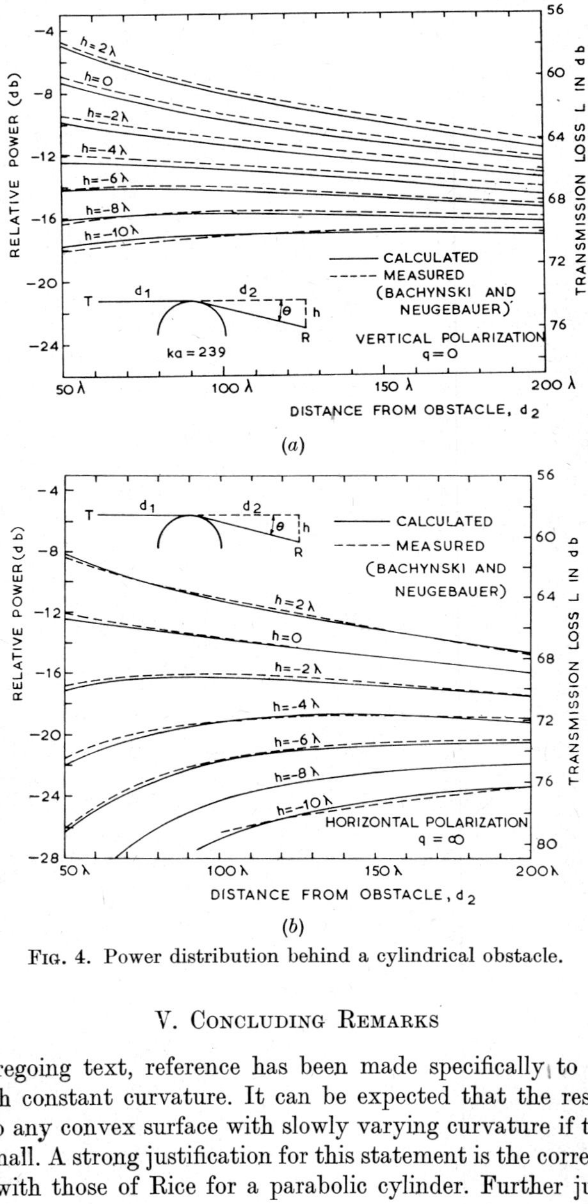

FIG. 4. Power distribution behind a cylindrical obstacle.

V. CONCLUDING REMARKS

In the foregoing text, reference has been made specifically to a cylindrical obstacle with constant curvature. It can be expected that the results are also applicable to any convex surface with slowly varying curvature if the deflection angles are small. A strong justification for this statement is the correspondence of our results with those of Rice for a parabolic cylinder. Further justification is obtained by applying our formulas for diffraction by a cylindrical obstacle to the calculated field intensity over a spherical earth in the case of high antennas. An example is chosen which corresponds to antenna heights of 500 m over a spherical earth of effective radius 8480 km. The frequency is 300 Mc ($\lambda = 1$ m) and the

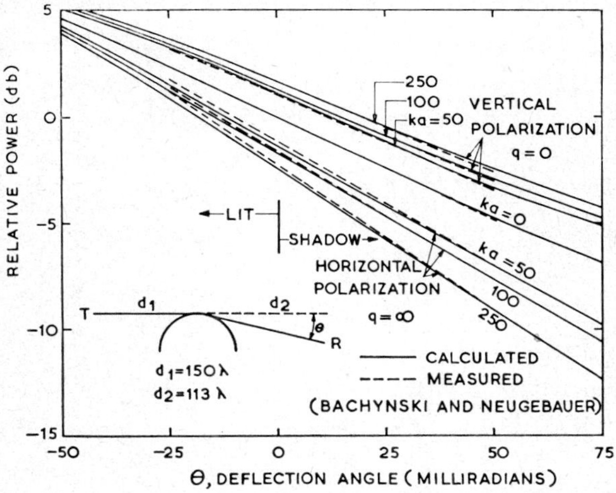

FIG. 5. Effect of curvature on diffraction by a convex surface.

polarization is taken to be horizontal. Our computed curve for this case (taking $q = \infty$) is shown in Fig. 6 where the ordinate is the field intensity in decibels above 1 μV for 1 kW radiated. The abscissa is the distance s between the transmitting and receiving antennas. The other curve is taken from the published C.C.I.R.

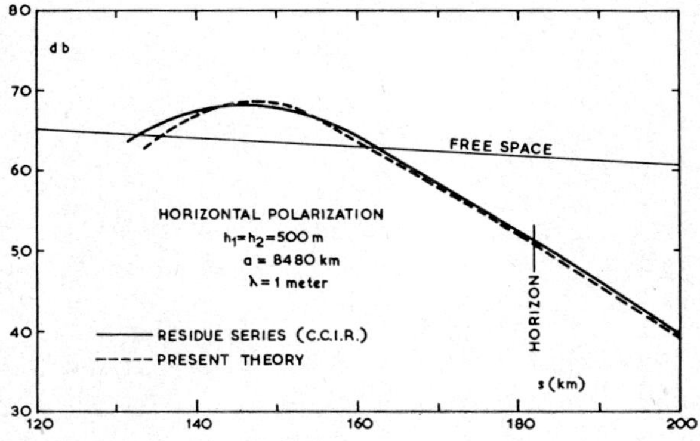

FIG. 6.

curves [2] and is based on the residue series representation in the form developed by Van der Pol and Bremmer. For the range of distances shown, it can be seen by inspection that the C.C.I.R. curves are practically independent of ground conductivity and polarization. In this case, the effective value of q is infinity. The agreement between the two sets of curves is good. The residue series is very poorly convergent in this penumbral region, and requires much computation whereas the universal diffraction formulas given in this paper are relatively simple to use. It is suggested that they could be used advantageously in VHF ground wave calculations for high antenna heights as a bridge between the residue-series deep in the shadow and geometrical-optics for the lit region.

References

1. Artmann, K. (1950). *Z. Phys.* **127,** 468–494.
2. "Atlas des courbes de propagation de l'onde de sol pour les frequences comprises entre 30 et 300 Mc/s." Publié par l'Union Internationale de Télécommunications, Genève (1955).
3. Bremmer, H. "Terrestrial Radio Waves." Elsevier Publishing Company, Amsterdam (1949). (A comprehensive review of the early work of Watson, Van der Pol and Bremmer is given.)
4. Bullington, K. (1947). *Proc. Inst. Radio Engrs, N.Y.* **35,** 1122–1136.
5. Crysdale, J. H., Day, J. W. B., Cook, W. S., Psutka, M. E. & Robillard, P. E. (1957). *Trans. Inst. Radio Engrs, N.Y.* AP–**5,** 203–208.
6. Dickson, F. H., Egli, J. J., Herbstreit, J. W. & Wickizer, G. S. (1953). *Proc. Inst. Radio Engrs, N.Y.* **41,** 967–969. (See comments on this paper by J. H. Crysdale, *Proc. Inst. Radio Engrs, N.Y.* **43,** 1955.)
7. Fock, V. A. (1946). *J. Phys., Moscow* **10,** 399–409. and *C.R. Acad. Sci. U.R.S.S.* (1946).
8. Furutsu, K. (1956). *J. Radio Res. Labs, Tokyo,* **3,** 331.
9. Kirby, R. S., Dougherty, H. T. & McQuate, P. L. (1955). *Proc. Inst. Radio Engrs, N.Y.* **43,** 1467–1472.
10. Kono, T., Uesugi, Y., Hirai, M., Niwa, S. & Irie, H. (1954). *J. Radio Res. Labs., Tokyo,* **1,** 1–16.
11. Matsuo, S. (1950). Report of the Electrical Communications Laboratory 621.30.001.6 (047.3). Tokyo, Japan.
12. Neugebauer, H. E. J. & Bachynski, M. P. Commission II Document, URSI General Assembly, Boulder, Colorado, September, 1957. See also Res. Rep. No. 7–100–2, August 1957, RCA Victor Co. Ltd. Montreal, Canada.
13. Norton, K. A., Rice, P. L. & Vogler, L. E. (1955). *Proc. Inst. Radio Engrs, N.Y.* **43,** 1480–1526.
14. Norton, K. A., Schulkin, M. & Kirby, R. S. Reference C to the report of the *ad hoc* Committee of the Federal Communications Commission for the evaluation of the radio propagation factors concerning the television and frequency modulation broadcasting services in the frequency range between 50 and 250 mc. 6 June, 1949.
15. Rice, S. O. (1954). *Bell Syst. tech. J.* **33,** 417–504.
16. Schelleng, J. C., Burrows, C. R. & Ferrell, E. B. (1933). *Proc. Inst. Radio Engrs, N.Y.* **21,** 427–463.
17. Selvidge, H. (1941). *Proc. Inst. Radio Engrs, N.Y.* **29,** 10–16.
18. Wait, J. R. & Conda, A. M. (1958). *Trans. Inst. Radio Engrs, AP*-**6,** 348–353.
19. Wait, J. R. & Conda, A. M. (1959). *J. Res. nat. Bur. Stand.* **63**D, 181–198.

Sur le Rôle des Phénomènes de Réflexion dans la Propagation Lointaine des Ondes Ultracourtes

F. du Castel, P. Misme et J. Voge

présenté par

F. du Castel

Centre National d'Etudes des Télécommunications, Issy-les-Moulineaux (Seine)

Le phénomène à l'origine de la propagation lointaine des ondes ultracourtes fait l'objet depuis plusieurs années de discussions théoriques, sans qu'une analyse entièrement satisfaisante des processus de propagation ait été obtenue. Nous nous proposons de montrer qu'en faisant appel à un phénomène de réflexion en atmosphère feuilletée, il est possible de rendre compte des principaux résultats expérimentaux [1].

Nous traiterons successivement de la réflexion par un feuillet et de la puissance réfléchie par les types de feuilletage qui peuvent exister dans l'atmosphère. Nous montrerons en conclusion les conséquences qui découlent de la théorie et étudierons brièvement les résultats expérimentaux qui permettent de conclure à l'existence d'un processus de réflexion. Cette conclusion ne signifie d'ailleurs pas que ce processus soit exclusif d'autres mécanismes de propagation, les notions de diffusion ou de réfraction gardant leur valeur dans certains cas.

I. Reflexion par un Feuillet

Si l'atmosphère présente une petite discontinuité δn de l'indice de réfraction ($\delta n \ll 1$) à une altitude Z, le coefficient de réflexion r, en champ, d'une onde attaquant la couche sous un angle rasant α ($\alpha \ll 1$)—ce qui correspond au cas de la propagation lointaine des ondes ultracourtes—est donné par la formule de Fresnel, qui s'écrit alors:

$$r = \frac{\alpha - \sqrt{(\alpha^2 + 2\delta n)}}{\alpha + \sqrt{(\alpha^2 + 2\delta n)}}.$$

Dans le cas ou la discontinuité est faible devant l'angle d'attaque, le coefficient r peut s'écrire:

$$r = \frac{\delta n}{2\alpha^2}; \qquad 2\delta n \ll \alpha^2.$$

Considérons (Fig. 1) un feuillet à l'altitude Z, d'épaisseur h suffisamment faible pour que l'angle d'attaque α puisse être considéré comme constant dans le feuillet. Soit $n(z)$ la variation d'indice dans le feuillet et $g(z)$ la variation du gradient d'indice.

$$g(z) = \frac{dn(z)}{dz}.$$

Le coefficient de réflexion élémentaire à l'intérieur du feuillet s'écrit, en introduisant la phase:

$$dr = \frac{g(z)}{2\alpha^2}\, e^{-jKz}$$

en appelant K le nombre d'onde d'espace:*

$$K = 2\,\frac{2\pi}{\lambda}\,\alpha.$$

Pour tout le feuillet le coefficient de réflexion en champ peut donc s'écrire:

$$r = \int_Z^{Z+h} \frac{g(z)}{2\alpha^2}\, e^{-jKz}\, dz. \tag{1}$$

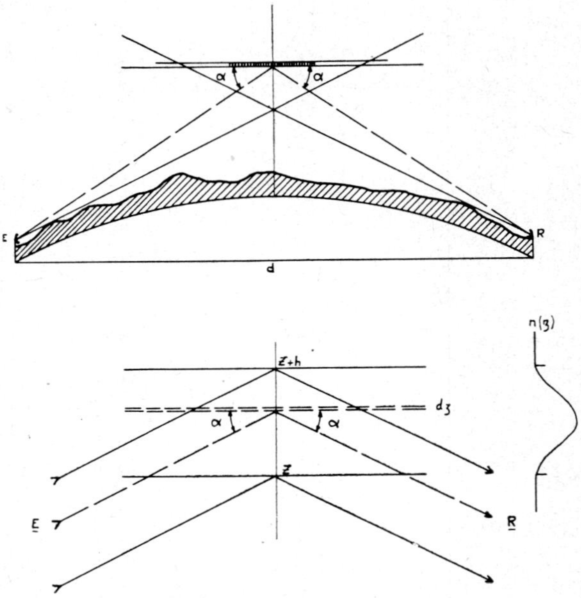

FIG. 1. Phénomène de réflexion.

Il est possible de donner à (1) une expression développée, par une intégration par partie, en introduisant les dérivées successives du gradient d'indice $g'(z)$, $g''(z)$ . . .

$$r = \frac{1}{2\alpha^2}\left\{\left[j\frac{g(z)}{K} + \frac{g'(z)}{K^2} - j\frac{g''(z)}{K^3} + \ldots\right]e^{-jKz}\right\}_Z^{Z+h}. \tag{2}$$

*Certains auteurs appellent nombre d'onde d'espace la quantité $\frac{2\pi}{\lambda}\alpha$ mais une autre définition peut être acceptable.

A. Feuillet Analytique

Si la variation d'indice dans le feuillet peut être représentée par une courbe explicite $n(z)$, la formule (1) montre que le coefficient de réflexion est la transformée de Fourrier $G(K)$ du gradient d'indice dans la couche [3].

$$G(K) = 2\alpha^2 r = \int_{-\infty}^{\infty} g(z) e^{-jKz}\, dz.$$

Le Tableau I donne l'expression du coefficient de réflexion en champ pour un feuillet à variation linéaire, parabolique, cosinusoïdale ou continue.

TABLEAU I

COEFFICIENT DE REFLEXION EN AMPLITUDE D'UN FEUILLET A VARIATION ANALYTIQUE

Discontinuité élementaire δn. Coefficient de réflexion $r = \dfrac{\delta n}{2\alpha^2}$

Feuillet analytique $g(z) = \dfrac{dn(z)}{dz}$. Coefficient de réflexion $r = \dfrac{1}{2\alpha^2}\displaystyle\int_{-\infty}^{\infty} g(z)\, e^{-jKz}\, dz$

	$n(z)$	$g(z)$
Feuillet linéaire		
$g(z) = g \quad Z < z < Z + h$	$r = \dfrac{gh}{2\alpha^2}$	$\dfrac{\sin (Kh/2)}{Kh/2}$
Feuillet parabolique		
$g(z) = g\left(1 - \dfrac{z}{h}\right) \quad Z < z < Z+h$	$r = \dfrac{gh}{2\alpha^2}$	$\left(\dfrac{\sin (Kh/2)}{Kh/2}\right)^2$
Feuillet sinusoïdal		
$g(z) = g \cos \dfrac{\pi}{2} \dfrac{z}{h}. \quad Z < z < Z + \mathrm{h}$	$r = \dfrac{gh}{2\alpha^2}$	$\dfrac{\cos (Kh/2)}{1 - (Kh/2)}$
Feuillet continu		
$g(z) = g \dfrac{1}{ch^2 \dfrac{z}{h}}$	$r = \dfrac{gh}{2\alpha^2}$	$\dfrac{Kh/2}{sh(Kh/2)}$

On a représenté sur la Fig. 2 les variations comparées des coefficients de réflexion d'un feuillet linéaire et d'un feuillet continu. On voit que des résultats peuvent être très différents suivant la loi de variation d'indice à l'intérieur du feuillet.

Il n'est donc possible de faire un choix sur la loi de variation de $n(z)$ qu'à partir de considérations physiques sur la nature du feuilletage atmosphérique. De telles considérations [6] conduisent à considérer la validité d'un modèle à variation linéaire, ou dans certains cas d'un modèle à variation continue faisant intervenir des fonctions hyperboliques.

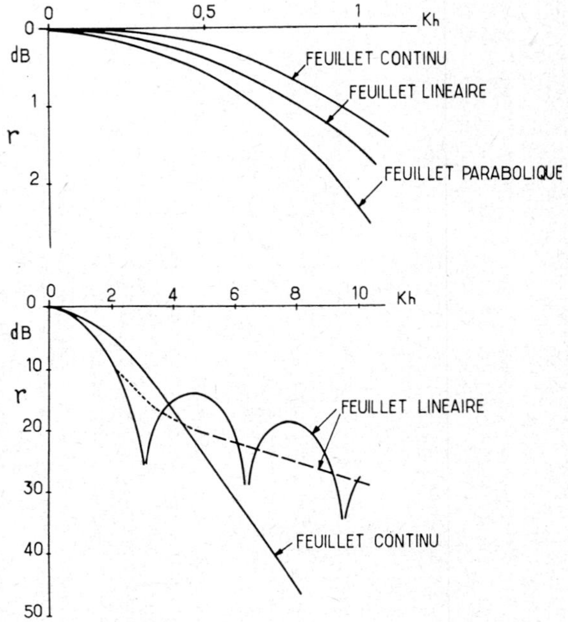

FIG. 2. Coefficient de réflexion d'un feuillet en fonction de son épaisseur.

B. Feuilletage Réel de l'Atmosphère

Les études météorologiques du feuilletage de l'atmosphère [6] montrent qu'il est possible de distinguers assez nettement deux types de feuilletage qui peuvent d'ailleurs coexister.

Type I, correspondant à une atmosphère instable (turbulente), où le feuilletage est permanent jusqu'à haute altitude et constitué par un empilage de petits feuillets. Ceux-ci ont une variation d'amplitude faible, sur une faible épaisseur, et des dimensions horizontales petites.

Ils sont en perpétuel mouvement et leur position peut correspondre à des orientations très différentes autour de l'horizontale.

L'ordre de grandeur des dimensions correspondant à ce type est:

amplitude δn de l'ordre de quelques unités N dans les basses couches,
épaisseur h de l'ordre de quelques mètres,
dimensions l de l'ordre de quelques mètres ou dizaines de mètres.

Type II, correspondant à des tranches d'atmosphère stable; type non permanent oú le feuilletage, généralement superposé au type précédent, est constitué de feuillets en nombre limité. Ceux-ci ont une amplitude plus importante, une épaisseur plus grande, et leurs dimensions horizontales peuvent être considérables. La loi de variation interne de l'indice, ou du gradient, est difficile à préciser, mais nous avons signalé plus haut qu'une hypothèse de variation linéaire ou dans certains cas, en régime permanent, une loi de variation continue, sont des approximations physiquement plausibles.

La durée de vie de ces feuillets peut être assez longue. Une ondulation de leur position dans l'espace et le temps est probablement générale [6].

L'ordre de grandeur des dimensions correspondants à ce type est:

amplitude δn de l'ordre de la dizaine d'unités N,
épaisseur h de l'ordre de la dizaine de mètres,
dimensions l de l'ordre du kilomètre,
espacement en hauteur H de l'ordre de la centaine de mètres,

Ces deux types conduisent à un traitement mathématique différent de la puissance réfléchie par l'atmosphère.

II. Puissance Reflechie par l'Atmosphere Feuilletee

La puissance totale réfléchie par l'atmosphère s'obtient par sommation des contributions de chaque feuillet. Or cette sommation doit s'entendre aussi bien à la verticale que dans le plan horizontal.

A. Sommation Verticale

Les deux types de feuilletage distingués plus haut doivent faire l'objet d'un traitement différent. On peut prendre comme paramètre le terme de phase Kh.

1. *Type I*: $Kh \ll 1$

L'empilage des couches conduit dans ce cas à un traitement aléatoire de la puissance totale réfléchie R. La contribution de deux feuillets voisins à l'intérieur de la hauteur totale H qui intervient et pour laquelle on a $KH \gg 1$ donnera [5]:

$$R = \int_Z^{Z+H} \frac{g(z)}{2\alpha^2} \mathrm{e}^{-jKz} \, dz \int_Z^{Z+H} \frac{g(z')}{2\alpha^2} \mathrm{e}^{+jKz'} \, dz'$$

ce qui peut s'écrire, en posant $z = Z + \zeta$ et $z' = z + h$, et en étendant les limites d'intégration à l'infini (puisque l'atmosphère n'est supposée intervenir qu'entre les altitudes Z et $Z + H$),

$$R = \frac{1}{4\alpha^4} \int_{-\infty}^{+\infty} \mathrm{e}^{-jKz} \left[\int_{-\infty}^{\infty} g(\zeta) g(\zeta + h) d\zeta \right] dh.$$

Appelons $\rho(h)$ la fonction d'autocorrélation des variations du gradient d'indice, et $\bar{g}^2$ la valeur quadratique moyenne des variations du gradient, on a:

$$\rho(h) = \frac{1}{H} \int_0^{H-h} \frac{g(\zeta) g(\zeta + h)}{\bar{g}^2} d\zeta$$

et appelons $S(K)$ la transformée de Fourrier de $\rho(h)$, c'est-à-dire le spectre de puissance des variations du gradient, on a:

$$S(K) = \frac{1}{H} \mathscr{R} \left[\int_0^H \rho(h)\, \mathrm{e}^{-jKh}\, dh \right]$$

$$\rho(h) = \frac{1}{H} \frac{1}{\pi} \int_0^H S(K)\, \mathrm{e}^{jKh}\, dK.$$

L'expression de la valeur moyenne de la puissance réfléchie devient:

$$\overline{R} = \frac{H^2 \overline{g}^2}{4\alpha^4} S(K). \tag{3}$$

Ce traitement mathématique est analogue à celui utilisé dans la théorie de la diffusion turbulentielle [8]. Nous ne retiendrons pour l'instant que cette différence: il s'agit d'un spectre unidimensionnel (en altitude) et non tridimensionnel. Nous utiliserons alors les travaux sur la turbulence pour chercher une expression spectrale acceptable également en théorie des réflexions; en soulignant cependant les limites d'une telle comparaison. En particulier l'anisotropie, considérée là comme un phénomène secondaire, est ici un aspect fondamental en conséquence de l'étude physique de milieu.

On sait que deux théories sont actuellement en présence:

La théorie d'Oboukhof-Kolmogorof, où la redistribution de l'énergie incidente entre tourbillons de la zone turbulentielle se fait par un phénomène de convection, conduit à une expression spectrale proportionnelle à $K^{-\frac{5}{3}}$

La théorie du gradient de mélange, où le mouvement des tourbillons introduit une variation sensiblement adiabatique du gradient, conduit à une expression spectrale proportionnelle à K^{-3}.

Considérons donc le cas où, dans un certain domaine, on a:

$$S(K) \simeq \left(\frac{K}{K_0}\right)^{-\nu} \tag{4}$$

avec $K_0 = 1/H_0$, H_0 caractérisant l'échelle de cohérence des variations du gradient dans l'empilage des feuillets.

L'expression de la puissance réfléchie s'écrit alors:

$$R = \frac{1}{4} \frac{1}{(4\pi)^\nu} \overline{g}^2 \frac{H^2}{H^\nu} \frac{\lambda^\nu}{\alpha^{\nu+4}}. \tag{5}$$

Les travaux sur le spectre de turbulence conduisent à des valeurs decomprises entre 1 et 3. Pour des valeurs deégales à 1, 2 ou 3, par exemple, on aurait pour expression de $\overline{R}$.

$$R = \frac{1}{16\pi} g^2 \frac{H^2}{H_0} \frac{\lambda}{\alpha^5} \qquad (\nu = 1) \tag{6}$$

$$R = \frac{1}{64\pi^2} g^2 \frac{H^2}{H_0^2} \frac{\lambda^2}{\alpha^6} \qquad (\nu = 2) \tag{7}$$

$$R = \frac{1}{256\pi^3} g^2 \frac{H^2}{H_0^3} \frac{\lambda^3}{\alpha^7} \qquad (\nu = 3). \tag{8}$$

Notons que des expressions spectrales $S(K)$ telles que (4) ne peuvent être admissibles que dans un domaine limité. Elles conduisaient en effet à des expressions de la fonction de corrélation (h) de la forme:

$$\rho(h) \simeq \left(\frac{h}{H_0}\right)^{\nu-1}$$

qui sont seulement acceptables comme approximation au voisinage de $h = H_0$.

2. *Type II*: $Kh > 1$

L'expression du coefficient de réflexion sous la forme (2) montre que, à partir du moment où on peut considérer les termes de phase aux deux extrémités d'un feuillet comme indépendants, la puissance réfléchie peut s'exprimer en fonction de la discontinuité g_p de la dérivée p-ième du gradient d'indice $g^{(p)}(z)$ [3].

$$R = \frac{1}{4}\frac{1}{(4\pi)^{2p+2}} g_p^2 \frac{\lambda^{2p+2}}{\alpha^{2p+6}}. \tag{9}$$

Nous avons indiqué qu'il semblait plausible de considérer le cas d'un feuilletage linéaire pour lequel l'expression (9) conduit à:

$$R = \frac{1}{64\pi^2} g^2 \frac{\lambda^2}{\alpha^6}. \tag{10}$$

Notons que dans le cas d'un feuilletage continu, en revenant à l'expression (1) du coefficient de réflexion, on trouverait:

$$R = g^2 h^2 \frac{1}{\alpha^4} e^{-4\pi \frac{\alpha K}{\lambda}}. \tag{11}$$

Si plusieurs feuillets interviennent, avec un déphasage qui peut être quelconque, la puissance réfléchie correspondant aux expressions précédentes pourra varier par suite de l'interférence de plusieurs termes. Par suite de la décroissance rapide avec α, seuls interviendront les termes provenant de réflexions sur des couches assez voisines. A titre d'exemple, si α peut être considéré comme ayant même valeur pour plusieurs feuillets succesifs en altitude, la puissance moyenne réfléchie dépendra du nombre n de couches intervenant, et prendra une valeur [7] comprise entre R pour $n = 2$, 0·76 R pour $n = 3$ et 0·69 R pour $n = 4$.

Remarquons que des feuillets de ce type peuvent avoir des dimensions très variables et que, dans certains cas de grands feuillets, les approximations précédentes peuvent cesser d'être valables. Il est possible d'obtenir alors des puissances réflechies R beaucoup plus élevées. On doit noter ici que le même mécanisme physique peut permettre d'interpréter des observations différentes, telles que le guidage atmosphérique.

B. Sommation Horizontale

La sommation des contributions des feuillets dans un plan sensiblement horizontal fait intervenir d'une part les dimensions horizontales des feuillets, d'autre part leur inclinaison, leur courbure ou leurs irrégularités.

1. *Dimensions Horizontales des Feuillets*

On peut considérer que tous les points du plan horizontal, tels que le déphasage entre deux d'entre eux, soit inférieur à une valeur de l'ordre de $\pi/4$ donnent une contribution en phase. Il faut donc comparer les dimensions l des feuillets avec l'ellipse de Fresnel correspondante, dont les dimensions d'axes sont transversalement $L_1 = \sqrt{\lambda d}$ et longitudinalement $L_2 = \frac{\sqrt{\lambda d}}{\alpha}$ (Fig. 3).

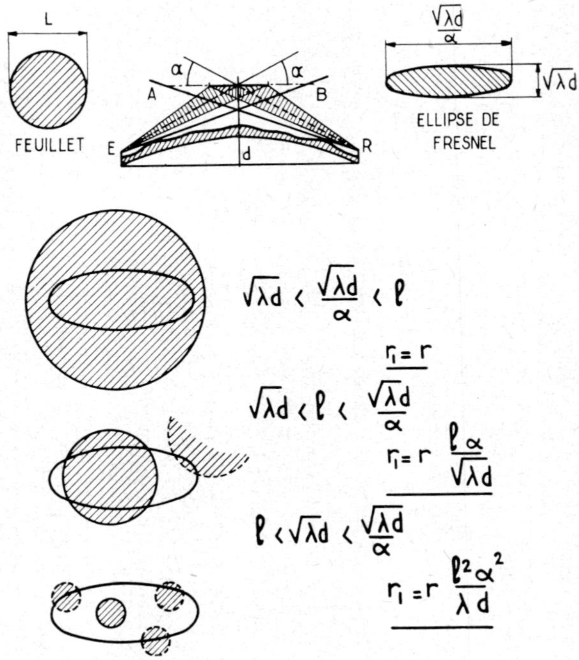

FIG. 3. Coefficient de réflexion en fonction des dimensions horizontales du feuillets.

Pour donner des ordres de grandeur, dans une liaison de $d = 300$ km avec la longueur d'onde $\lambda = 1$ m, et pour un angle d'attaque au voisinage des horizontales des extrémités $\alpha = 17{\cdot}5$ mR, on aurait $L_1 = 300$ m, $L_2 = 27$ km.

Si donc il s'agit de feuillets de grandes dimensions $l > L_2$ la puissance réfléchie sera inchangée $R_1 = R$.

S'il s'agit de feuillets de dimensions moyennes $L_2 > l > L_1$, ce qui sera le cas général du type II et pourra être celui du type I, la puissance réfléchie deviendra:

$$R_1 = R\frac{l^2\alpha^2}{\lambda d}. \tag{12}$$

S'il s'agit de feuillets de dimensions petites $L_1 > l$ ce qui sera le cas fréquent du type I, la puissance réfléchie deviendra:

$$R_1 = R\,\frac{l^4\alpha^4}{\lambda^2 d^2}. \tag{13}$$

Notons qu'il faudrait en outre faire intervenir le nombre de feuillets pouvant se trouver simultanément dans la même ellipse de Fresnel, ce qui réduit l'importance du coefficient en l.

2. *Irrégularités de Surface**

Une irrégularité de la surface intervient pour les feuillets de type II. Elle a pour conséquence d'introduire une réduction de la puissance réfléchie. Ce coefficient de réduction peut d'ailleurs dépendre de la longueur d'onde, ainsi que l'a montré un travail préliminaire [2].

3. *Inclinaison des Feuillets*

Il convient en outre de tenir compte que les réflexions pourront provenir de feuillets inclinés sur la verticale, ce qui augmente les possibilités d'interférences entre réflexions multiples, ou se traduit par une augmentation du nombre de contributions possibles de réflexions en phase.

Cette inclinaison augmente en outre la zone d'atmosphère qui peut être à l'origine du phénomène de réflexion.

On a signalé plus haut qu'une certaine ondulation des feuillets du type II était systématique [6]; ce phénomène peut évidemment introduire également des réflexions sur des zones du feuillet plus ou moins inclinés sur l'horizontale.

4. *Courbure des Feuillets*

D'autre part, une certaine courbure de la surface horizontale des feuillets peut conduire à une convergence du rayonnement. Nous nous contenterons de signaler ce fait ici, sans l'étudier plus avant, mais en soulignant toutefois que ceci pourra se traduire, dans le cas de feuillets de dimensions assez grandes (types II), par une importante augmentation de la puissance réfléchie.

III. Consequences de la Theorie des Reflexions

Tous ces phénomènes montrent que la puissance réfléchie peut varier considérablement et qu'il est difficile (ou trop facile) de donner des résultats numériques absolus, ou correspondant à des observations expérimentales, en raison du nombre de paramètres qui interviennent dans le calcul.

Il est intéressant cependant d'essayer de distinguer des types de propagation, susceptibles d'être rattachés aux types de feuilletage de l'atmosphère.

Si l'on observe des enregistrements à vitesse rapide de champs lointains en ondes ultracourtes, on peut assez nettement distinguer 2 types de propagation, qu'il est possible de rattacher aux 2 types d'atmosphère précédents.

*Dans ce paragraphe, comme dans les deux suivants, nous nous contenterons d'indications générales sans approfondir les calculs. Ceux-ci font l'object actuellement de travaux dont nous comptons publier les résultats ultérieurement.

La figure 4 met en évidence ces types observés sur une liaison de 250 km à 470 MHz. Le premier correspond à un champ très fluctuant, de niveau moyen variable. Il peut s'interpréter comme provenant de réflexions sur les petites fluctuations aléatoires correspondant au feuilletage du type I.

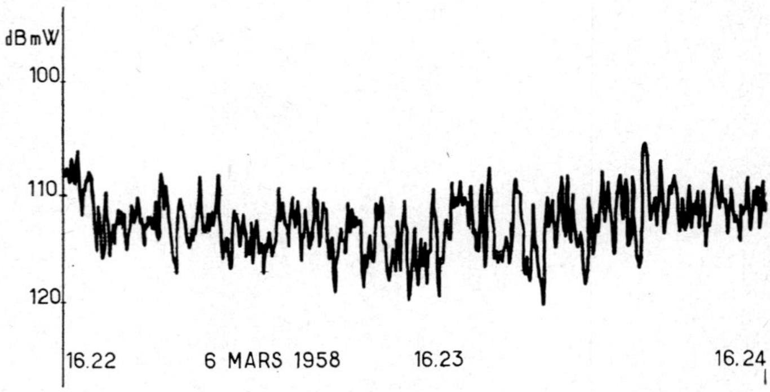

FIG. 4*a*

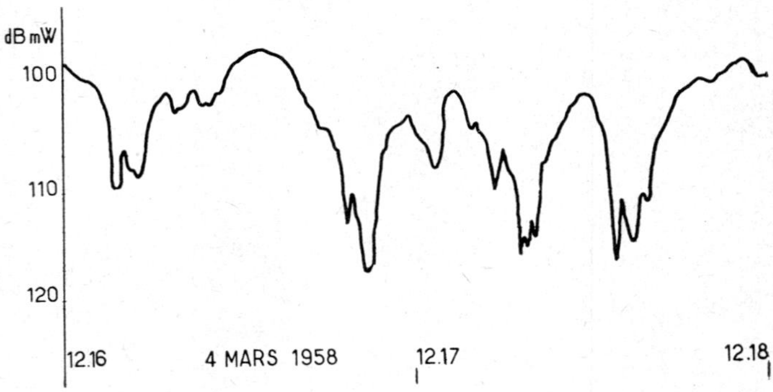

FIG. 4*b*

FIG. 4. Types de fluctuations de champs lointains.

Le second correspond à un champ affecté d'évanouissements relativement lents, et de niveau moyen variable. Il peut s'interpréter comme provenant de l'interférence de quelques réflexions sur des feuillets de type II. Notons que les niveaux des deux types I et II pouvant être voisins, le champ pourra être plus complexe et donner un résultat intermédiaire entre les deux types.

On peut également étudier les lois de variation auxquelles conduit la théorie, pour les comparer aux données expérimentales. Notons d'ailleurs que l'étude statistique des données expérimentales est en général globale, et qu'il conviendrait mieux de faire une étude statistique par type de propagation, qui permettrait de considérer chaque type comme une population homogène [4] donc susceptible de considérations physiques plus valables à partir de ses propriétés statistiques.

A. Variation de la Puissance Réfléchie avec la Fréquence

On sait que, dans une gamme limitée tout au moins, telle que 100 à 1000 MHz, une loi proportionnelle à la longueur d'onde, semble rendre compte assez bien des résultats expérimentaux.

Une telle loi peut par exemple être obtenue avec le type I, pour un exposant spectral $\nu = 3$ (formule 8) et des dimensions l petites (formule 13), on a alors un affaiblissement de propagation, exprimé par rapport à l'espace libre, tel que

$$A_{\mathrm{I}} \sim \frac{\lambda}{\alpha^3 d^2}. \tag{14}$$

Une telle loi peut également être obtenue avec le type II, dans une hypothèse de feuilletage linéaire (formule 10) et des dimensions l moyennes (formule 12), on a alors pour l'affaiblissement:

$$A_{\mathrm{II}} \sim \frac{\lambda}{\alpha^4 d}. \tag{15}$$

Remarquons qu'une loi en λ ne peut cependant être considérée comme certaine, ni surtout comme valable dans une gamme étendue de fréquence. Les limites qui interviennent dans la théorie des réflexions sont d'une part le terme de phase Kh, c'est-à-dire l'épaisseur du feuillet comparée à la longueur d'onde d'espace $\Lambda = \lambda/2\alpha$, et d'autre part les dimensions de l'ellipse de Fresnel, c'est-à-dire l comparé à $\sqrt{\lambda d}$ et $\sqrt{\alpha\lambda d}/\alpha$. La nature différente de ces deux limites offre une grande souplesse dans les possibilités d'expliquer des lois en fréquences expérimentales.

Nous avons déjà signalé ailleurs [1] qu'il s'agit là d'une différence importante avec la théorie de la diffusion turbulentielle.

Notons en outre que les applications numériques montrent que, pour les fréquences élevées, le type I aura tendance à dominer le type II, c'est-à-dire à donner des champs plus fluctuants.

Ajoutons que le choix des modèles, conduisant aux expressions (14) et (15) des affaiblissements, est quelque peu arbitraire puisqu'on pourrait aussi bien trouver une loi analogue en λ avec un autre choix.

Enfin, on doit signaler que le même phénomène permet d'interpréter la propagation à grande distance depuis l'horizon, puisque au voisinage de celui-ci le coefficient de réflexion tend vers l'unité et permet de retrouver les résultats de la théorie de la diffraction sphérique.

B. Variation de la Puissance Réfléchie avec la Distance

La distance intervient dans l'expression de la puissance réfléchie par le terme d dans les formules telles que (12) et (13) et par le terme α dans les formules telles que (6) à (8) ou (10), puisque α, ou tout au moins α minimum, est proportionnel à d (Fig. 1).

Les modèles précédents conduisent aux formules (14) et (15) qui indiquent une loi en d^{-5} de l'affaiblissement par rapport à l'espace libre—ce qui peut être acceptable dans un certain domaine de distance.

Remarquons que le rôle de α, par rapport au rôle propre de d, donne une importance moins grande à la notion de distance angulaire que dans la théorie de la diffusion.

Pour les très grandes distances, où la zone d'atmosphère en visibilité des extrémités peut être située à très haute altitude, il subsiste une difficulté. En l'absence de données suffisantes sur les conditions de feuilletage du gradient dans la haute atmosphère, ou sur les observations expérimentales de loi de décroissance du champ, il n'est pas possible d'envisager encore la réponse à donner à ce problème. Il semble cependant qu'il faille donner plus d'importance aux phénomènes à basse altitude, y compris dans la zone située en-dessous des horizons.

Notons d'autre part que les applications numériques montrent, à grande distance, que le type I aura tendance à dominer le type II, ce qui correspond à des champs plus fluctuants.

Les conséquences de la théorie des réflexions, dégagées jusqu'ici, mettent en évidence la possibilité d'interpréter les résultats d'observation dans la propagation à grande distance des ondes ultracourtes, à partir d'un phénomène de réflexion. Ces conséquences ne démontrent pas que le processus de réflexion soit nécessairement le phénomène fondamental.

Cette dernière démonstration pourrait résulter de deux ordres de considération.

(*a*) La première a trait aux conséquences expérimentales découlant de la théorie, et de leur comparaison aux autres théories. On a déjà signalé les conséquences de l'étude des paramètres fréquence et distance. Mais il en est trois autres qui paraissent importants.

En premier lieu la quantification du phénomène de propagation, introduit par la théorie des réflexions, permet d'interpréter les observations expérimentales sur certains cas de propagation hors du plan de grand cercle de la liaison.

En deuxième lieu, les importantes variations de niveau moyen et de caractéristiques de fluctuations, observées sur les champs lointains, peuvent être interprétées comme résultant d'un unique phénomène de réflexion, mais pouvant se produire sur des feuillets atmosphériques de caractéristiques différentes. Cette unité de phénomène donne sans doute une vue plus cohérente des observations.

En troisième lieu, les observations expérimentales sur le renforcement du champ en atmosphère stable, trouvent une interprétation aisée par le passage d'un phénomène de réflexion sur une atmosphère à feuilletage de type I en atmosphère instable, à un phénomène de réflexion sur une atmosphère à feuilletage de type II en atmosphère stable.

(*b*) Le second ordre de considérations, sur le rôle du phénomène de réflexion, peut résulter d'études sur la structure fine du champ.

Une analyse du champ instantané permettrait de saisir le phénomène élémentaire. S'il s'agit d'un processus de diffusion, le champ instantané doit résulter d'un grand nombre de composantes aléatoires. C'est ce que traduit le traitement mathématique de la théorie de la diffusion, dans lequel la sommation des composantes a lieu au sein de la puissance diffusée élémentaire. S'il s'agit au contraire d'un processus de réflexion, le champ instantané doit résulter d'un petit nombre de points brillants de position localisée. Le traitement mathématique de la théorie des réflexions traduit ce fait dans une sommation *a posteriori* des puissances réfléchies élémentaires.*

Si l'intervalle de temps considéré n'est pas suffisamment faible, ces points brillants se composent entre eux et les deux phénomènes se traduisent par des effets semblables.

On pourrait dire qu'à un instant donné, la distribution spatiale de l'indice de réfraction possède une structure feuilletée qui conduit à des phénomènes de réflexion. Mais les variations de cette distribution au cours du temps se font très souvent suivant des processus aléatoires, ayant le caractère d'une turbulence, de sorte que le phénomène, considéré sur un certain temps, peut se présenter avec des caractéristiques statistiques analogues à celles d'un champ de diffusion turbulente. De ces remarques résulte le grand intérêt d'expériences séparant la variable temps et la variable espace.

A cet égard, les expériences effectuées par A. T. Waterman [9], de rotation rapide d'un lobe d'antenne étroit, semblent démontrer le bien fondé de ces hypothèses.

REFERENCES

1. Castel, F. du (1958). *Rapport d'étude C.N.E.T.* no. 483 T.
2. Castel, F. du & Misme, P. (1958). *C.R. Acad. Sci., Paris*, **246**, 82–84.
3. Castel, F. du, Misme, P. & Voge, J. (1958). *C.R. Acad. Sci., Paris*, **246**, 1838–1840.
4. Castel, F. du & Rivet, P. (1957). *Ann. Télécomm.* **12**, 246–252.
5. Feinstein, J. (1952). *Trans. Inst. Radio Engrs, N.Y.*, PGAP4.
6. Misme, P. (1958). *Ann. Télécomm.* **13**, 7–8 et 9–10, et avec Castel, F. du & Voge, J. Communication au Congrès de Propagation de Liège, octobre 1958.
7. Simon, J. C. (1955). *Ann. Radio*, **10**, no. 39.
8. Staras, H. & Wheelon, A. D. (1959). *Trans. Inst. Radio Engrs A-P7.*
9. Waterman, A. T. Communication au Congrès de Propagation de Liège, octobre 1958.

*A titre de comparaison on pourrait trouver un exemple de la différence de nature de ces deux phénomènes dans le cas d'ondes lumineuses. Un pinceau lumineux dans le brouillard diffuse de la lumière de façon isotrope. Au contraire, une lumière réfléchie sur une mer légèrement agitée apparait comme décomposée en points brillants en déplacement perpétuel.

Tropospheric Scatter Developments

C. L. MACKEY

Rome Air Development Center, Griffiss Air Force Base, Rome, N.Y.

Developments in the field of tropospheric scatter equipment are reviewed and advances in the state of the art in power klystrons, diversity, solid state devices, antennas, modulators, and miniaturization techniques for mobile application are discussed and areas of future investigation are indicated.

I. GENERAL

Tropospheric scatter has grown within a few years from a propagation phenomena into one of the accepted methods of communication along with wire, radio, radio relay and other modes of transmission. This is rather remarkable considering that while a propagation phenomena was demonstrated by Marconi in 1932, the theory of tropospheric scattering was not rediscovered until 1945.

From 1945 to 1955, many major contributions were made by Bullington, Gerks, McGraw and others to the knowledge of this propagation phenomena. However, no practical use was made of this transmission medium. Several experimental circuits were established purely for test purposes primarily to obtain propagation, phase, fade rate and other basic information; however, no operational circuits were constructed.

In 1955, the United States Air Force on the basis of very limited data regarding actual transmission characteristics decided to gamble on installing an operational chain of stations in an inaccessible region of Labrador. This system consisted primarily of modified commercial equipment plus a power amplifier fabricated by using available klystron tubes designed for commercial television service. This circuit, while experiencing many difficulties as expected in any new system of this magnitude, is, with some modifications, in operation today with a very high degree of reliability.

Since this time the Unites States Air Force has installed and in operation approximately eight thousand miles of tropospheric scatter circuits. Each circuit carries from 12 to 72 voice channels of nominal 4 kc bandwidth.

The state of the art has advanced tremendously since this first circuit was installed. New developments have resulted in improvements in every phase of equipment employed in tropospheric scatter operation.

Major areas in which developments have been made to achieve the present state of the art are klystrons, diversity, solid state devices, mobile applications, antennas frequency application and modulation methods.

II. POWER KLYSTRONS

New tubes which have been developed exclusively for tropospheric application such as the four cavity klystron shown in Fig. 1 have improved the efficiency, linearity and modulation characteristics to a marked degree. This tube employs internal cavities which are an integral part of the tube and are sealed and evacuated. This makes for very stable operation as no R.F. leakage can occur to cause unstable or erratic conditions. Also tuning is greatly simplified since all the cavities can be tuned to the same frequency and the output coupling is fixed.

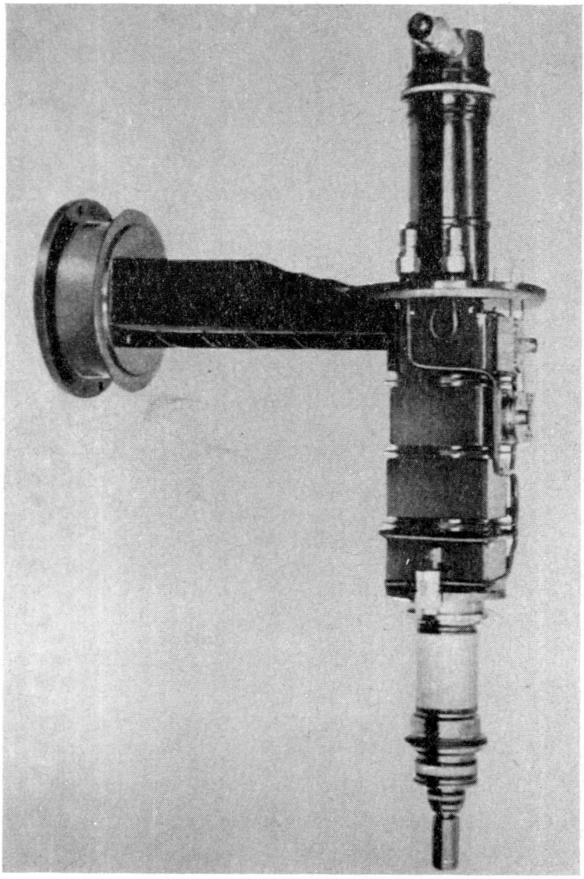

FIG. 1.

This, together with temperature and voltage compensation of the magnetic circuits, requires only periodic tuning of the tube even if the equipment is turned on and off repeatedly.

Development is now underway to provide permanent magnets for the klystron to replace the electro-magnetic circuits now employed. This will also improve the stability and efficiency of the output circuit since variation in voltage will not effect power output and tuning.

Previous four cavity tubes as shown in Fig. 2 employed external cavities which were placed around the ceramic windows. This permitted wide tuning range but radiation of R.F. energy is difficult to control and the third cavity must be detuned slightly to prevent oscillation which will wreck the tube. This tube also had an adjustable output coupling which was very critical as to tuning. Since no

FIG. 2.

compensation was used in the magnetic circuits, the tube required retuning each time the transmitter was turned on in order to focus the beam directly on the collector to prevent the body currents from exceeding the maximum limit. Short tube life results especially in field operations where inexperienced personnel are used to operate the equipment.

Three cavity klystrons were initially used. These were designed primarily for commercial television application and had poor efficiency and require very high beam voltage.

III. DIVERSITY AND RELIABILITY

The greatest contribution to increased reliability of tropospheric scatter has been in the application of diversity. Some type of diversity is required on all tropospheric circuits if any degree of reliability is to be achieved. Today all circuits are engineered to provide at least 99·9 % reliability. Reliability is a function of a number of parameters, namely, long and short term fading, path length, take-off angle, frequency, antenna gain, bandwidth, order of diversity and type of

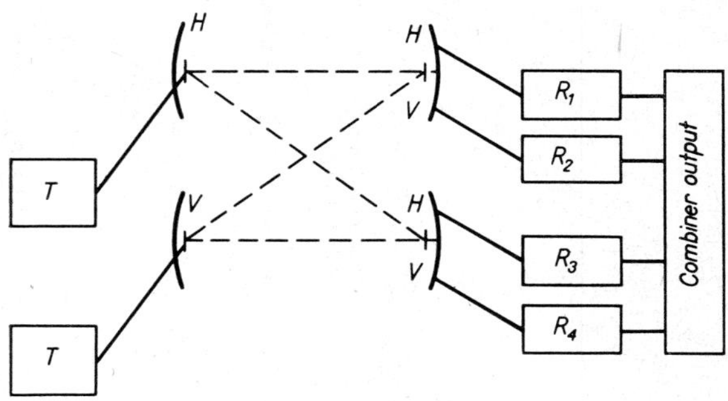

Fig. 3.

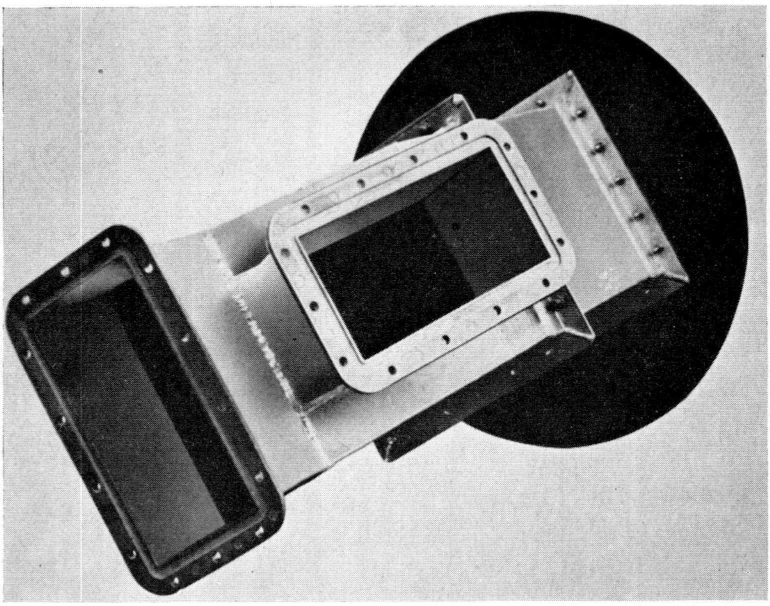

Fig. 4.

combining. All of these must be considered in the design of each link of any circuit. Fade rates are a function of the medium and beyond our control. Path length and take-off angle can be controlled but for economic and site location reasons, these parameters are generally limited. Antenna gain is controlled by the aperture over which correlation of the signal can be maintained. Bandwidth and frequency are a function of the information to be transmitted and frequency limitations.

In the application of diversity, there is a wide choice. At present the type in most general use is shown in Fig. 3. This is referred to as space and polarization. This is a misnomer as polarization does not provide diversity but does provide an

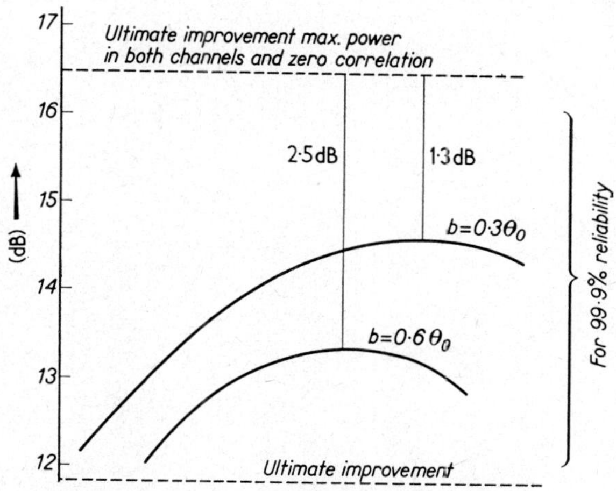

FIG. 5.

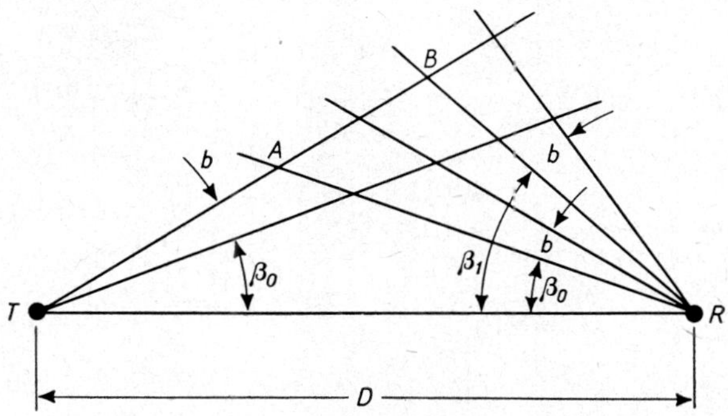

FIG. 6.

additional degree of space diversity by utilizing space separation of both the horizontal and vertical polarized waves. Figure 4 illustrates the type of feed horn used in this system.

Dual and triple diversity was originally used primarily with switching type combiners. Ratio squarer and linear adder type combiners are employed in all circuits at present.

Rome Air Development Center is at present experimenting with several forms of angular diversity. Dr. Gordon of Cornell University has theoretically shown that angular diversity can be achieved if two beams from an antenna are separated by one beamwidth. The gain is only 1·3 db below that obtained by space as shown in Fig. 5 but eliminates the need of an additional antenna. Or we can employ this with the present antenna configuration and obtain 8 fold diversity by additional feed horns. The geometry of angular diversity is shown in Fig. 6. Test of this type of diversity is being conducted at Rome Air Development Centre at the present time.

IV. Solid State Devices

Solid state devices are being utilized wherever possible in all tropospheric scatter equipment to provide more reliable equipment. Transistors and diodes have replaced all tubes in power supplies. These require no replacement or maintenance during the life of the equipment. In addition, solid state devices such

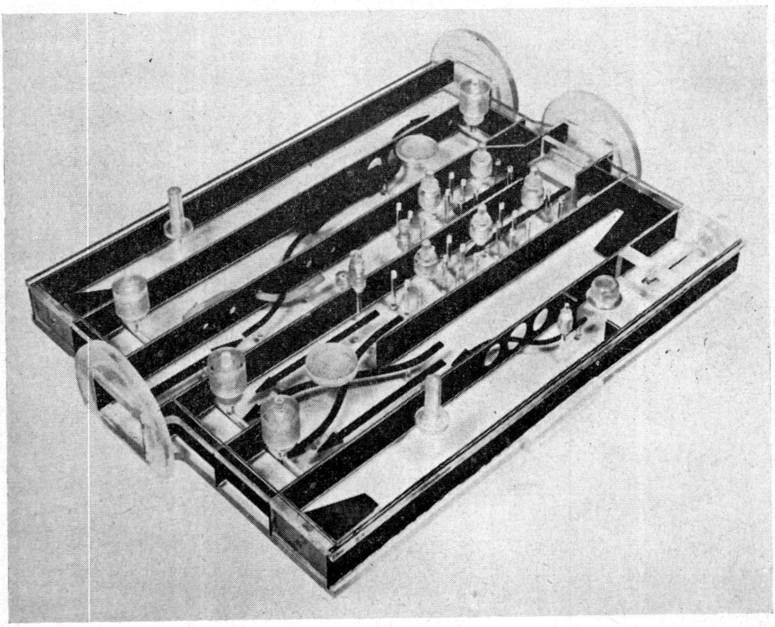

Fig. 7.

as transistors, diodes and ferromagnetic devices are being employed in the construction of immediate and audio amplifiers and in some application of the radio frequency components of the sets.

Rome Air Development Centre now has under development low noise front ends using parametric amplifiers. These are lossless non-linear reactances which exhibit a negative resistance characteristic under certain conditions or can perform as an upconverter with gain under other conditions. Because the reactance is lossless, no thermal noise is generated. An R.F. energy source or pump is required but no cooling is required as in a mazer.

Noise figures of receivers in production using conventional techniques are 7 db. By the use of parametric amplifiers, overall noise figures in the order of approximately 2 db have been obtained in laboratory models. While this is far from the figure that can be realized by the use of mazers, these are practical for use in equipment at the present time. Additional studies being conducted should materially improve the quality of these devices.

V. Mobile Scatter

The majority of applications of tropospheric scatter is for fixed station application where size is not of prime importance but solid state devices are used to

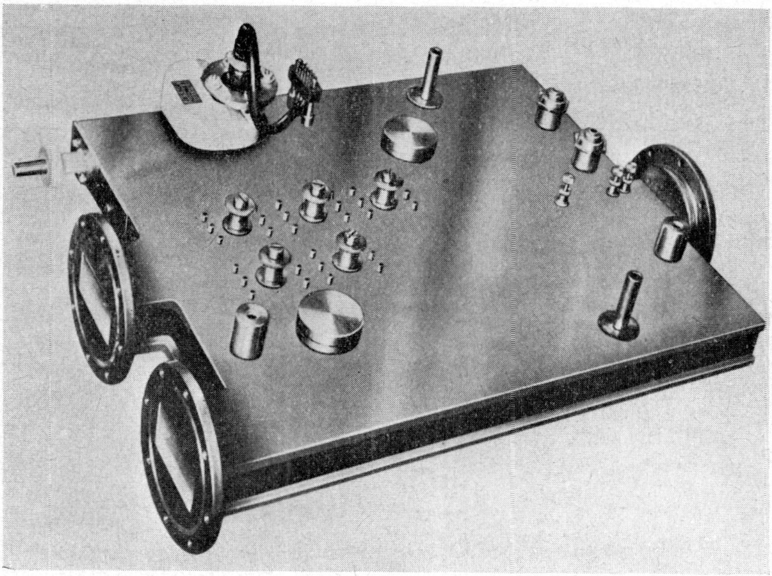

Fig. 8.

improve reliability. There are, however, some requirements for highly mobile scatter equipment. Under development is a transportable unit known as Radio Set AN/TRC–66. In this unit, every effort is being made to reduce size and weight. The development of a radio frequency block as shown in Figs. 7 and 8 has greatly facilitated the packaging into one rack of equipment. This block incorporates all of the R.F. circuits into one unit. It contains the receiver local oscillator, mixer, antenna input, transmitter input, power monitor and provides filtering for the receiver and transmitter.

This together with the development of an air-cooled klystron which will use permanent magnets for focusing in place of electro-magnets, and the use of transistors and miniaturized components has resulted in the development of a complete transmitter and dual diversity receiver which is mounted in a single rack as shown in Fig. 9. By mounting this rack in a shelter together with an antenna and other ancillary equipment, a complete mobile station is provided as shown in Fig. 10. Two of these shelters provided a complete quadruple diversity terminal. The shelter can be readily transported by truck or helicopter to inaccessible locations.

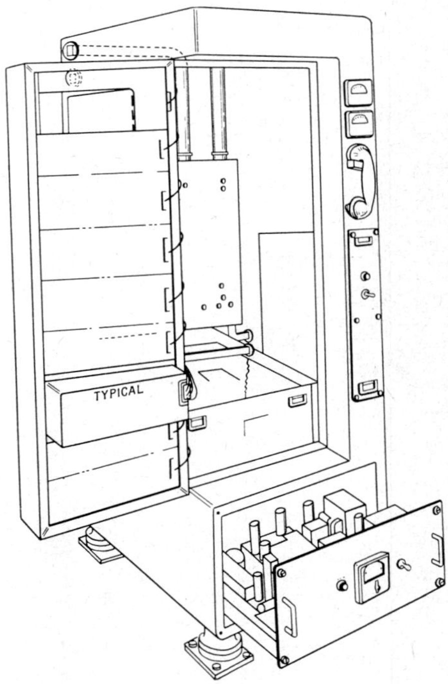

FIG. 9.

All necessary components are available so that the station can be operated either as a radio relay station for line of sight or as a tropospheric scatter station. The only changes being in the antenna configuration and in operating the high power amplifier. In radio relay application, the center section only of the antenna is employed. In scatter application, additional wings are added to provide the full size 14 ft parabolic dish.

VI. ANTENNAS

In the field of antennas and transmission lines, developments have progressed along with other components in tropospheric scatter. The major improvements have been in mechanical design, especially in maintaining tolerances required for large size wave guide for the lower frequencies and in the parabolic dish for higher frequency application.

The use of light weight material and structural techniques has resulted in light weight structures that can withstand all but the most severe climatic conditions.

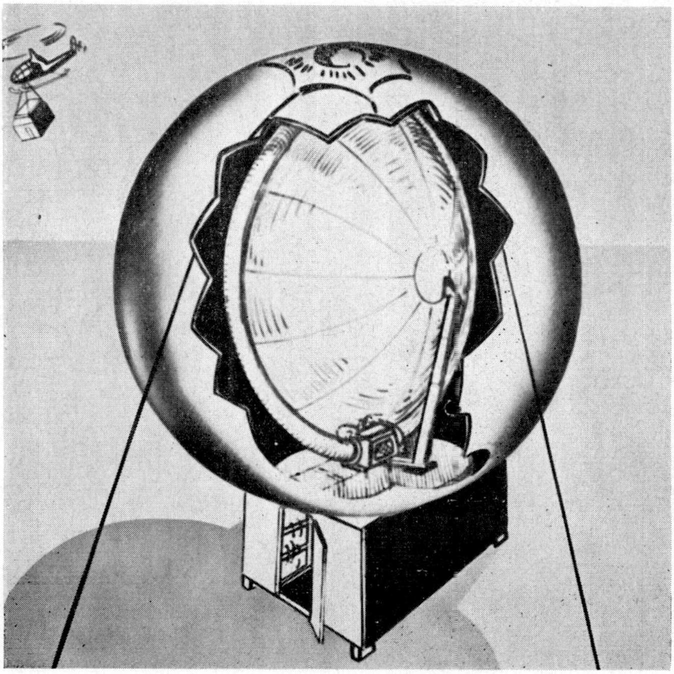

FIG. 10.

For very severe conditions, a structure such as is shown in Fig. 11 is employed. This is a 60 ft parabola which is fabricated from beam and covered with solid or strips of metal sheets. This type construction is used in 60 and 120 ft diameter parabolas. For all but the severest conditions, parabolas of the type shown in Fig. 12 are used. This type is much lighter and more easily erected. This type is employed for 60, 30 and 14 ft diameter parabolas.

A typical station employing a number of antennas at one terminal is shown in Fig. 13.

VII. FREQUENCY APPLICATION

The first tropospheric scatter equipment operated in the frequency range of approximately 1000 Mc. At present operational and test circuits are being operated on frequencies from 400 Mc to 8000 Mc. Typical equipment design for 1000 Mc equipment is illustrated in Fig. 14 which is a composite photograph of Radio Set AN/FRC–39 which is a standard Air Force set.

Fig. 11.

VIII. Modulation

Up until very recently, frequency modulation has been used almost exclusively in tropospheric scatter circuits. Recently single sideband modulation has been employed in a few applications.

The choice of which type modulation is best depends entirely upon the application of the equipment. Under heavy loading conditions or where a wide bandwidth of several megacycles is required, frequency modulation is required and very little is to be gained by the use of single sideband. In circuits which are primarily for voice application and where long distance is required in which the bandwidth must be as narrow as possible, single sideband is definitely to be preferred and is a necessity.

FIG. 12.

FIG. 13.

At RADC a circuit is in operation which transmits one way on frequency modulation and the other on single sideband. This will be used to obtain comparison data on all types of information to be transmitted under various propagation conditions. The circuit has been in operation for only a relatively short period and insufficient data is available at present to provide any definite conclusions. A simplified block diagram of the single sideband transmitter is shown in Fig. 15

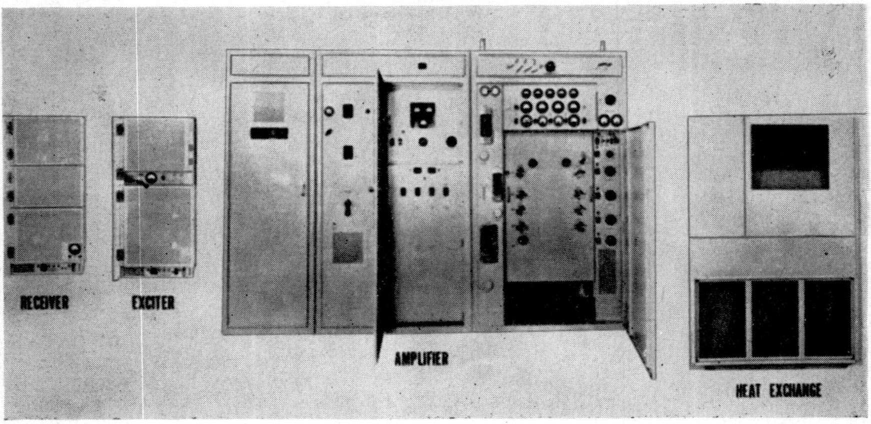

FIG. 14.

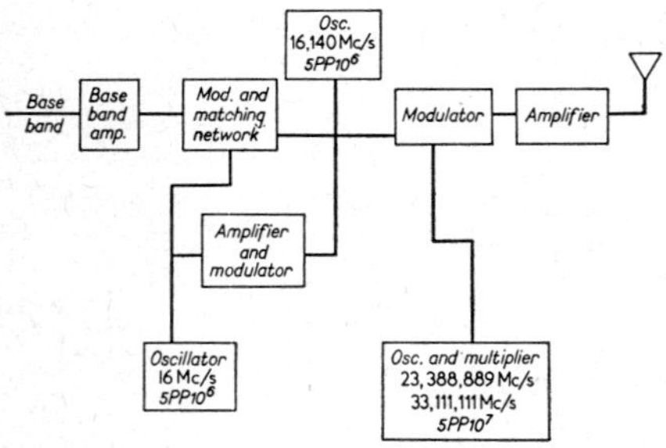

FIG. 15.

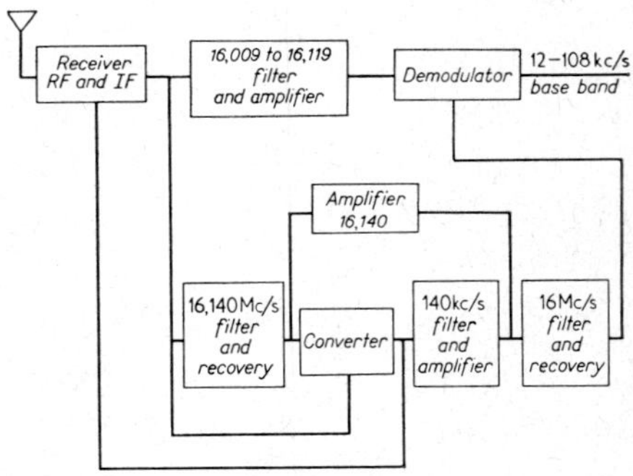

FIG. 16.

and receiver in Fig. 16. By utilizing two pilot frequencies at 16 and 16·140 Mc and placing the information between them, it is only necessary to maintain the frequencies reference to a high degree of accuracy in order to recover the intelligence. The R.F. frequency then does not require an extremely rigid control to be maintained.

IX. Future Investigations and Developments

Tropospheric scatter has grown within a few years from a propagation phenomena into one of the accepted methods of communication along with wire, radio relay, and other means of transmission. There remains, however, much work which can be accomplished to improve reliability, efficiency and circuit capacity.

New power amplifier tubes such as carcinotrons, traveling wave tubes and other devices are being investigated and tested to improve the output efficiency.

Improved antenna patterns are needed to reduce fading and improve reliability. Test on the pincushion antenna may solve some of these problems.

Wideband linear exciters are required for transmission of several hundred voice channels or for wideband video transmission. Additional investigations are needed on single sideband modulators to improve efficiency and permit wider band transmission.

And finally, additional work is required on parametric amplifiers and mazers for providing low noise front ends for receivers.

Ergebnisse von kombinierten Backscatter- und Impulsfernübertragungsversuchen

WALTER DIEMINGER, HANS-GEORG MÖLLER UND GERHARD ROSE*

Max-Planck-Institut für Aeronomie, Lindau (Harz)

Es wird über kombinierte Senkrechtlotungs-, Fernübertragungs- und Backscatter-Versuche berichtet, die auf der Strecke Lindau-Sodankylä (2000 km) und Lindau-Tsumeb (8000 km) durchgeführt werden. Insbesondere werden die Differenzen diskutiert, die sich aus der rechnerischen Ableitung der MUF aus Senkrechtlotungen und den beobachteten Werten bei der Fernübertragung ergeben, sowie über die Ubereinstimmung zwischen Fernübertragung und Backscatter. Die Bedeutung des high-angle ray und die Feinstruktur in der Nähe der MUF wird diskutiert und über die Schwierigkeiten der Deutung der Vielfachsprünge bei Backscatter-Beobachtungen berichtet.

I. UNTERSUCHUNGSMETHODEN

Im folgenden soll berichtet werden über experimentelle Untersuchungen der Ausbreitung elektrischer Wellen nach zwei verschiedenen Methoden. Die eine Methode ist die sog. Rückstreu- oder Backscattermethode, die andere die Impulsfernübertragung.

Die Backscattermethode beruht auf der Tatsache [1, 2, 5, 7], daß der Erdboden elektrische Wellen nicht ideal spiegelt, sondern einen Teil der Energie nach allen Seiten streut (Bild 1). Ein gewisser Bruchteil wird in Richtung des einfallenden

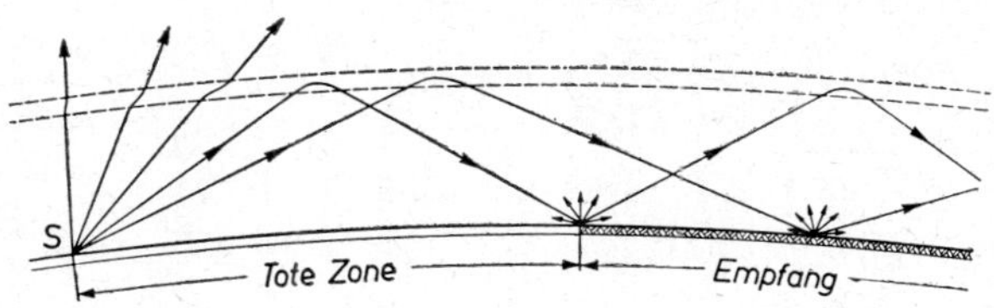

ABB. 1. Entstehung des Backscatter.

Strahles zurückgestreut und kommt auf dem gleichen Wege, den die Welle vom Sender zum Streugebiet zurückgelegt hat, wieder zum Empfänger zurück. Bei Verwendung von Impulsen kann man aus der Laufzeit dieser Streuechos auf den Abstand des Streugebietes vom Sender schließen. Verwendet man noch eine Richtantenne, so kann man auch die Richtung der einfallenden Wellen festlegen und damit das rückstreuende Gebiet orten [8, 11].

Die Impulsfernübertragung ist eine Modifikation der Echolotung mit senkrechter Inzidenz. Bei der Echolotung stehen Sender und Empfänger unmittelbar nebeneinander und aus der Laufzeit der Echos kann die scheinbare Höhe der reflektierten Schicht bestimmt werden. Bei der Impulsfernübertragung steht der

*Vorgetragen von W. Dieminger.

Empfänger in einem gewissen Abstand vom Sender, wobei der Abstand veile tausend km betragen kann. Aus den einzelnen am Empfänger ankommenden Zeichen kann auf die verschiedenen Ausbreitungswege und ihre Länge geschlossen werden. Außerdem gibt die Stärke der einzelnen Zeichen ein Maß für die Absorption der Wellen auf den einzelnen Wegen. Der Vorteil dieser Methode gegenüber den üblichen Feldstärkeregistrierungen ist der folgende: Bei der Registrierung von Dauerstrichsendern erhält man lediglich das resultierende Feld am Empfangsort, das sich zusammensetzt aus den Einzelbeiträgen der verschiedenen Ausbreitungswege. Ein Rückschluß auf die einzelnen Ausbreitungswege ist dabei in komplizierteren Fällen nicht mehr möglich. Bei der Impulsfernübertragung werden dagegen die einzelnen Wege, soweit sie verschiedene Laufzeiten haben, deutlich getrennt, so daß es ohne weiteres möglich ist, auf die Zahl und Art der verschiedenen Ausbreitungswege Schlüsse zu ziehen.

Das Besondere an unseren Versuchen ist nun, daß wir diese beiden Methoden nicht nur unabhängig nebeneinander, sondern gleichzeitig angewendet haben. Damit kann man zum Beispiel feststellen, inwieweit die Backscattermethode ein richtiges Bild der Ausbreitungsbedingungen gibt.

II. Versuchsstrecken

Die Versuche werden auf zwei Strecken durchgeführt. Die eine Versuchslinie führt von Sodankylä in Finnland 200 km nördl. von Rovaniemi nach Lindau

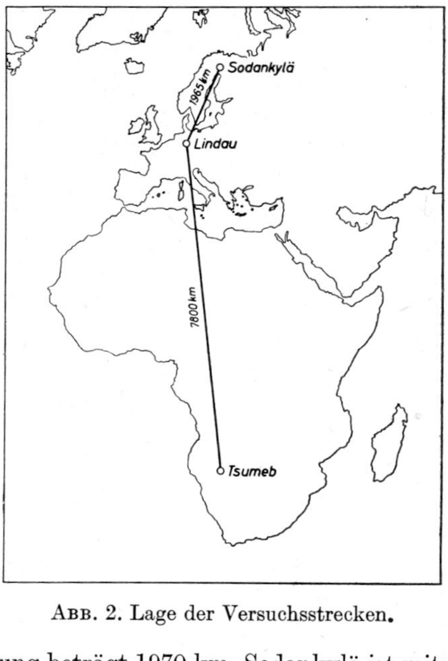

Abb. 2. Lage der Versuchsstrecken.

(Bild 2); die Entfernung beträgt 1970 km. Sodankylä ist mit einem Impulssender von etwa 30 kW Leistung ausgerüstet, der den Frequenzbereich von 1·4 bis

22·6 MHz in 8 Min durchläuft. In Lindau befindet sich eine Empfangsanlage, die den gleichen Frequenzbereich durchläuft. Dabei ist die Frequenzvariation des Empfängers und die Zeitbasis des Oszillographenrohres genau mit der Trägerfrequenz bzw. der Pulsfrequenz des Senders in Sodankylä synchronisiert. Die Synchronisierung erfolgt durch ein Paar von Quarzuhren. Auf Einzelheiten der Technik soll hier nicht näher eingegangen werden, da hierüber an anderer Stelle bereits ausführlich berichtet wurde [3]. An beiden Enden der Strecke, nämlich in Sodankylä und in Lindau, werden Senkrechtlotungen durchgeführt, außerdem stehen uns die Messungen zur Verfügung, die etwa in der Mitte der Strecke vom Observatorium Uppsala aus gemacht werden.

Die zweite Strecke führt von Lindau nach Tsumeb in Südwestafrika. Sie hat eine Länge von rd. 7800 km. Hier werden sehr viel höhere Frequenzen übertragen, als auf der kürzeren Strecke Finnland—Deutschland. Leider steht im Augenblick noch kein Gerät mit kontinuierlich veränderlicher Frequenz für diese Strecke zur Verfügung. Statt dessen wird in Lindau ein Sender mit 100 kW Leistung betrieben, der auf 7 feste Frequenzen zwischen 10 und 40 MHz eingestellt werden kann. Der Empfänger in Tsumeb wird von Hand auf die gleichen Frequenzen abgestimmt. Die Synchronisierung der Impulsfolgefrequenz in Lindau und der Zeitbasis des Oszillographen in Tsumeb erfolgt wiederum durch ein Paar von Quarzuhren. Die Rückstreuechos dieser Anlage werden in Lindau ebenfalls registriert. Als Antennen werden in allen Fällen sowohl beim Sender als auch beim Empfänger Rhombusantennen verwendet, die so dimensioniert sind, daß sie den verwendeten Frequenzbereich möglichst gleichmäßig ausstrahlen bzw. empfangen.

III. Ergebnisse der Strecke Sodankylä—Lindau

1. *Ungestörte Verhältnisse*

Zunächst sollen einige Ergebnisse der Versuchsstrecke Sodankylä—Lindau besprochen werden. Bild 3 zeigt einen Querschnitt durch die Strecke mit einigen

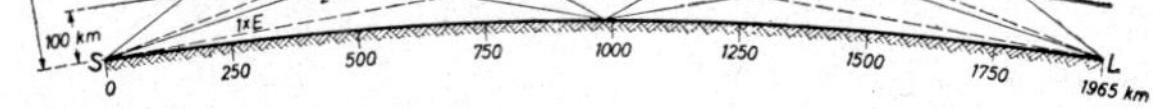

Abb. 3. Querschnitt durch die Versuchsstrecke Sodankylä—Lindau.

Übertragungswegen. Bei einer Schichthöhe von 300 km entspricht der 1 × F-Weg einem Erhebungswinkel von 12°. Für den 2 × F-Weg ergibt sich ein Erhebungswinkel von 28°. Der 1 × E-Weg verläuft bereits außerordentlich flach. Der Erhebungswinkel beträgt bei 120 km Schichthöhe nur 2·3°. Für den 2 × E-Weg ergibt sich dagegen ein Erhebungswinkel von 9°. Bild 4 zeigt ein typisches Ionogramm, das auf der Strecke Sodankylä—Lindau gewonnen wurde. Als Abszisse ist die Frequenz f und als Ordinate die Laufwegdifferenz zwischen den einzelnen

Übertragungswegen $\Delta p'$ aufgetragen. Eine Nullmarke ist auf den Registrierungen nicht vorhanden, da die Bodenwelle auf die Entfernung von nahezu 2000 km nicht mehr übertragen wird. Ihre Lage kann jedoch aus der E-Reflexion bestimmt werden, die ja eine ziemlich definierte Höhe hat. Die Lücken, die an einigen Stellen der Echospuren auftreten, sind nicht durch mangelhafte Synchronisierung bedingt, sondern durch sehr starke Sender, die den Empfänger in bestimmten

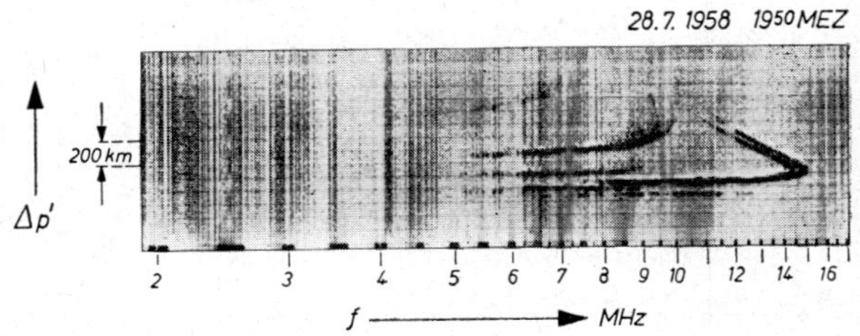

Abb. 4. Fernübertragungs-Ionogramm Sodankylä—Lindau.

Frequenzbereichen übersteuern. Es handelt sich dabei in erster Linie um die sogenannten Kurzwellenrundfunkbänder. Im einzelnen zeigt die Registrierung folgende Spuren: Die unterste Spur entspricht der 1 × E-Übertragung. Sie erstreckt sich bis etwa 13·5 MHz. Die darüber liegende Spur der 2 × E-Übertragung endet bei etwa 9 MHz. Die nächste sehr kräftige Spur entspricht der 1 × F-Übertragung. Auffallend ist die starke Ausbildung des sog. Übertragungssporns (im angelsächsischen Schrifttum "Pedersen Ray"). Er ist von der Maximal Usable Frequency (MUF) von etwa 15 MHz bis 11 MHz deutlich zu verfolgen,

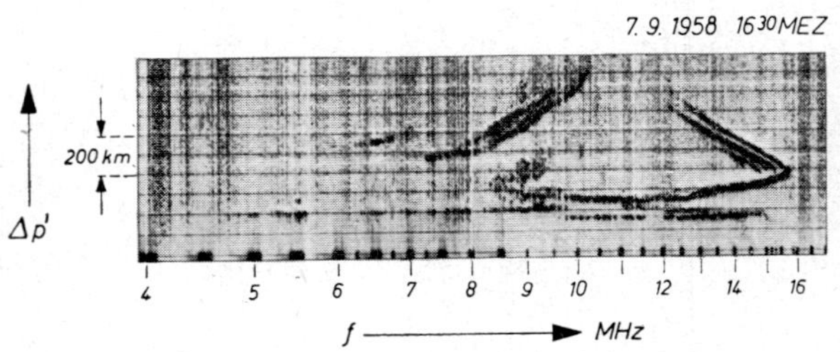

Abb. 5. Fernübertragungs-Ionogramm Sodankylä—Lindau mit E-Sporn.

wobei seine Laufzeitverzögerung dauernd ansteigt. Auch die magnetische Doppelbrechung ist klar zu erkennen. Die nächste Spur, die bis etwa 9 MHz reicht, entspricht einem sog. kombinierten E-F-Weg. Die folgende Spur entsteht durch die 2 × F-Übertragung. Auch hier ist die Ausbildung des Sporns und die magnetische Aufspaltung deutlich zu erkennen. Schließlich folgen noch in größerer Höhe Andeutungen des 3 × F-Weges. Nach niederen Frequenzen sind die einzelnen Spuren durch Absorption bzw. Abdeckung begrenzt. Es ist deutlich zu erkennen, daß wegen des flachen Abstrahlwinkels der $1 \times$ E-Weg bei wesentlich höheren Frequenzen abbricht als der $2 \times$ E-Weg. Das Abbrechen des $1 \times$ F-Weges ist durch Abdeckung durch die darunterliegende E-Schicht auf niederen Frequenzen bedingt. Erstaunlich ist die Tatsache, daß bei einer Entfernung von nahezu 2000 km und einem entsprechenden Einfallswinkel von weniger als 3° überhaupt eine $1 \times$ E-Übertragung, und zwar regelmäßig, auftritt. Wahrscheinlich spielt dabei eine zusätzliche Brechung in der unteren Atmosphäre eine Rolle,

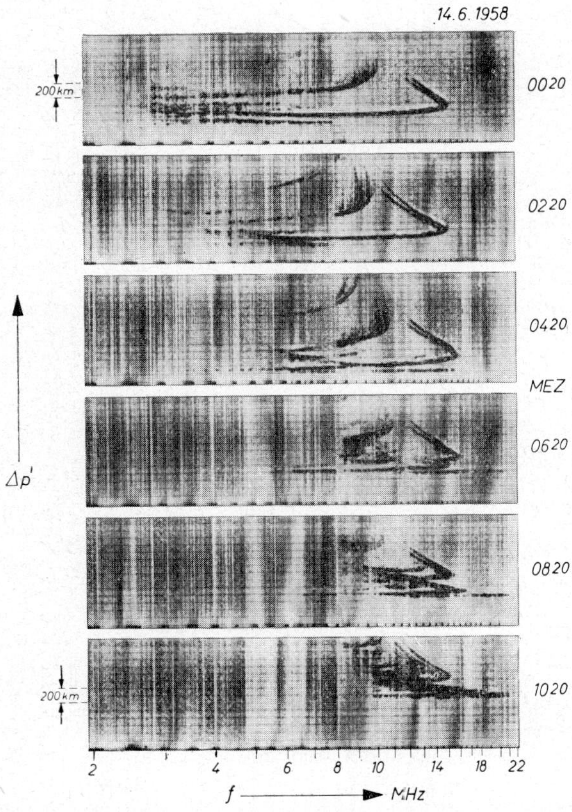

Abb. 6. Fernübertragungs-Ionogramm Sodankylä—Lindau, Übergang von Nacht auf Tag.

so daß der tatsächliche Erhebungswinkel, unter dem die Welle die Erde verläßt, etwas größer ist.

Eine weitere Überraschung ist die Tatsache, daß gelegentlich in der E-Übertragung ein deutlicher Sporn auftritt (Bild 5). Man hatte bisher angenommen, daß wegen der hohen Dämpfung, die die Welle bei einem längeren Weg in der E-Schicht erleidet, eine Spornbildung nicht möglich ist. Eine Verwechslung mit der $2 \times$ E-Übertragung ist unwahrscheinlich, da diese gleichzeitig in etwas größerer Höhe über dem oberen Sporn sichtbar ist.

Eine Serie von Aufnahmen, die im Abstand von je 2 Stunden gemacht wurden, sind in Bild 6 dargestellt. Die Bildfolge zeigt den Übergang von Nachtbedingungen auf Tagbedingungen in einem Sommermonat. Auf der obersten Registrierung reichen die Echospuren herunter bis etwa 2·7 MHz. Bereits auf der folgenden Aufnahme um 02·20 h ist der Einfluß der Dämpfung auf den niederen Frequenzen zu sehen. Dabei ist zu beachten, daß im Juni die Sonne in Sodankylä um 02·20 h

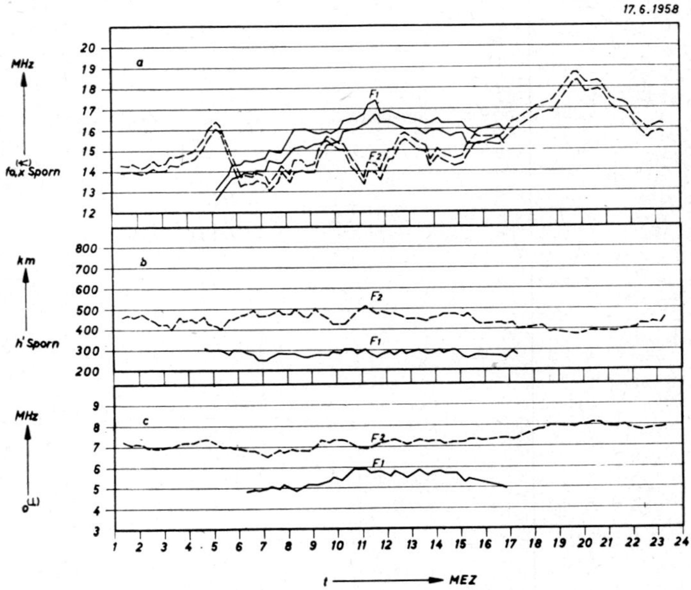

ABB. 7. Fernübertragung Sodankylä—Lindau.
(*a*) Beobachtete F1- und F2-MUF.
(*b*) Spornhöhe.
(*c*) Abgeleitete Senkrechtgrenzfrequenz.

schon merklich über dem Horizont steht. Um 04·20 h hat sich bereits eine Übertragung über die normale E-Schicht ausgebildet. Die 1 × F-Übertragung reicht bis etwa 15·5 MHz. Auf der folgenden Registrierung 06·20 h hat sich die E-MUF auf 20 MHz vergrößert, außerdem macht sich bei der F-Übertragung der Einfluß der F1-Schicht durch die Ausbildung einer weiteren Nase unter der F2-Nase bemerkbar. Auf der folgenden Registrierung um 08·20 h ist dieser Prozeß weiter fortgeschritten, so daß jetzt die F1-MUF wesentlich höher als die F2-MUF liegt. Um 10·20 h hat sich dieses Bild noch deutlicher ausgeprägt. Außerdem hat die Dämpfung soweit zugenommen, daß Echospuren unter 8·5 MHz nicht mehr beobachtet werden. Bemerkenswert ist noch die Tatsache, daß die Laufzeit der Echos, die als erste am Empfänger ankommen, nach niederen Frequenzen nicht ab-sondern zunimmt. Die Echos mit der kleinsten absoluten Laufzeit treten in der Nähe der MUF auf.

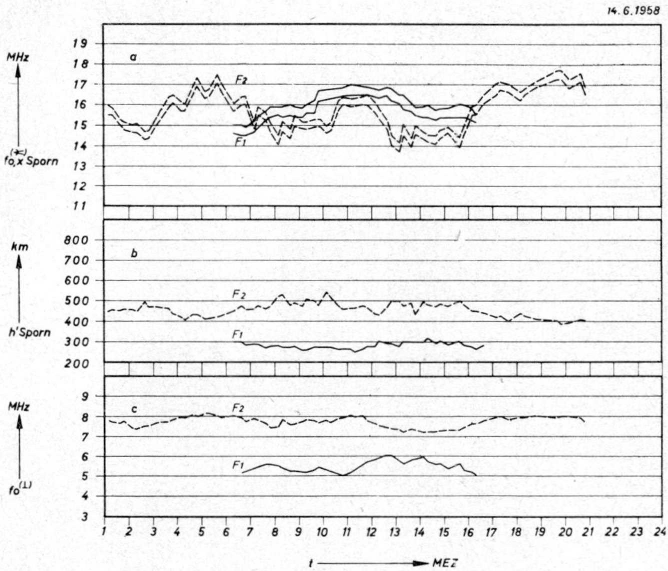

Abb. 8. Fernübertragung Sodankylä—Lindau.
(*a*) Beobachtete F1- und F2-MUF.
(*b*) Spornhöhe.
(*c*) Abgeleitete Senkrechtgrenzfrequenz.

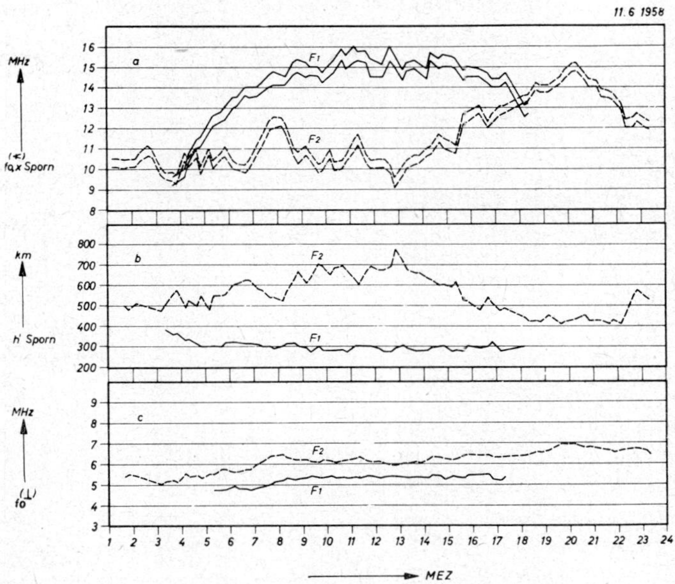

Abb. 9. Fernübertragung Sodankylä-Lindau bei mittelstarker Ionosphären-Störung.
(*a*) Beobachtete F1- und F2-MUF.
(*b*) Spornhöhe.
(*c*) Abgeleitete Senkrechtgrenzfrequenz.

Die routinemäßige Auswertung der Fernübertragungsversuche ergab, daß im Sommer tagsüber die MUF der F1-Schicht regelmäßig größer als die MUF der F2-Schicht ist. In Bild 7 und 8 ist der beobachtete Verlauf der F1- und F2-MUF im Verlaufe von 2 ungestörten Tagen aufgetragen. Man erkennt deutlich, daß etwa von Sonnenaufgang bis Sonnenuntergang in unseren Breiten die F1-MUF

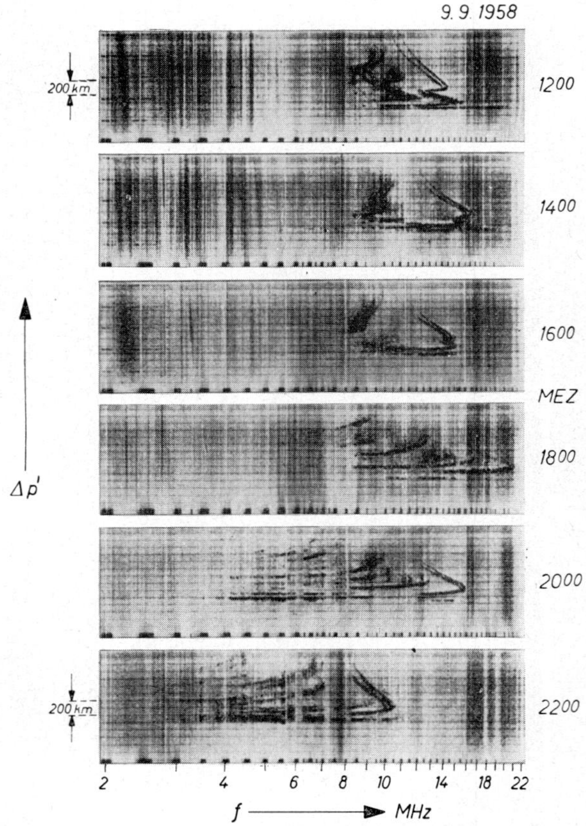

ABB. 10. Fernübertragungs-Ionogramme Sodankylä—Lindau. Zeitliche Änderung bei leichter Ionosphären-Störung.

höher liegt als die F2-MUF. Der Abstand beträgt gegen Mittag etwa 1·5 MHz. An gestörten Tagen (Bild 9), an denen die F2-Grenzfrequenz vermindert ist, wird der Abstand noch größer. Aus dieser Tatsache ergibt sich eine wichtige Konsequenz: Bei Ionosphärenstörungen ist im allgemeinen die F1-Grenzfrequenz kaum beeinflußt. Nachdem aber die F1-MUF über der F2-MUF liegt, macht sich die Verminderung der F2-MUF überhaupt nicht bemerkbar. Außerdem ist gleichzeitig noch eine E-Übertragung vorhanden, die, wenn auch mit geringeren Feldstärken, bis zu noch höheren Frequenzen reicht. Dieses Ergebnis stimmt gut mit den Erfahrungen der Deutschen Bundespost überein, die festgestellt hat, daß auf der Strecke Deutschland—Finnland an Sommertagen der Einfluß von Ionosphärenstörungen auf die MUF viel geringer ist, als erwartet. Dieses Ergebnis darf allerdings nicht verallgemeinert werden, da sich für andere Entfernungen eine andere Geometrie der Ausbreitungswege ergibt.

2. *Der Einfluß von Störungen*

Im folgenden Abschnitt werden einige Beispiele von Übertragungsstörungen behandelt, die auf der Strecke Sodankylä—Lindau auftraten. Bild 10 stellt zunächst eine Serie von Ionogrammen dar, die am 9.9.58 zwischen 12 und 22 h gewonnen wurden. Die 3stündigen erdmagnetischen Kennziffern lagen in diesem

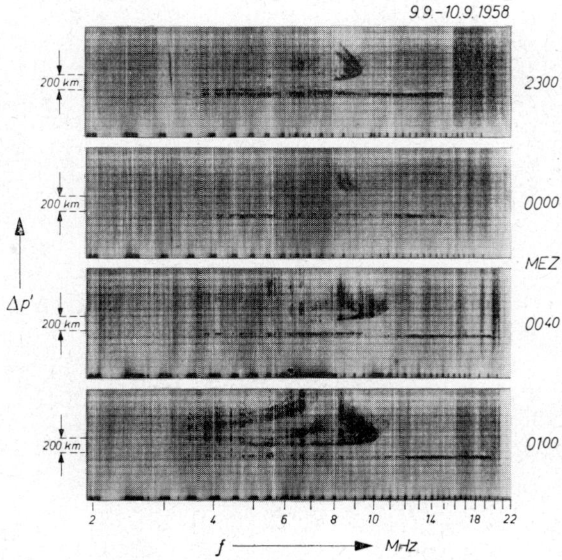

Abb. 11. Fernübertragungs-Ionogramme Sodankylä—Lindau. Wirkung der Absorption bei mittelstarker Ionosphären-Störung.

Zeitraum zwischen 4 und 5. Die Aufnahme um 12 h zeigt zunächst ein normales Bild. Die F1-MUF ist merklich größer als die F2-MUF. Auf der nächsten Aufnahme ist die F2-MUF so stark angewachsen, daß sie größer als die F1-MUF geworden ist. Um 16 h ist dieser Prozeß wieder rückläufig geworden. Sehr auffallend ist die folgende Aufnahme um 18 h. Die F1-Übertragung ist vollständig verschwunden. Die F2-MUF reicht nunmehr bis zu 21 MHz, hat also sehr stark zugenommen. Gleichzeitig tritt ein ausgeprägter kombinierter E-F-Weg auf, der bis etwa 17 MHz reicht. Auf den Senkrechtlotungen in Lindau und Uppsala wurde gleichzeitig ein merkliches Ansteigen der F2-Grenzfrequenz und eine Abnahme der Schichthöhe beobachtet. Beides bewirkt eine Erhöhung der MUF. Außerdem trat in Lindau eine starke sporadische E-Schicht mit einer Top-Frequenz von 9 MHz auf. Diese erklärt den gleichzeitigen E-F-Weg. Anschließend fällt die F2-MUF stark ab und erreicht um 22 h einen Wert von 10·5 MHz. Der weitere Verlauf ist in Bild 11 dargestellt. Die F2-MUF bleibt im weiteren Verlauf der Nacht fast konstant, nimmt sogar gegen 01 h wieder etwas zu. Gleichzeitig setzt jedoch Übertragung über die Nordlicht-E-Schicht und starke Absorption der F-Übertragung ein. Um 01 h haben sich die Verhältnisse etwa wieder normalisiert. Ein ähnlicher Fall wurde in den frühen Morgenstunden des 3.8.58 beobachtet (Bild 12). Auch zu

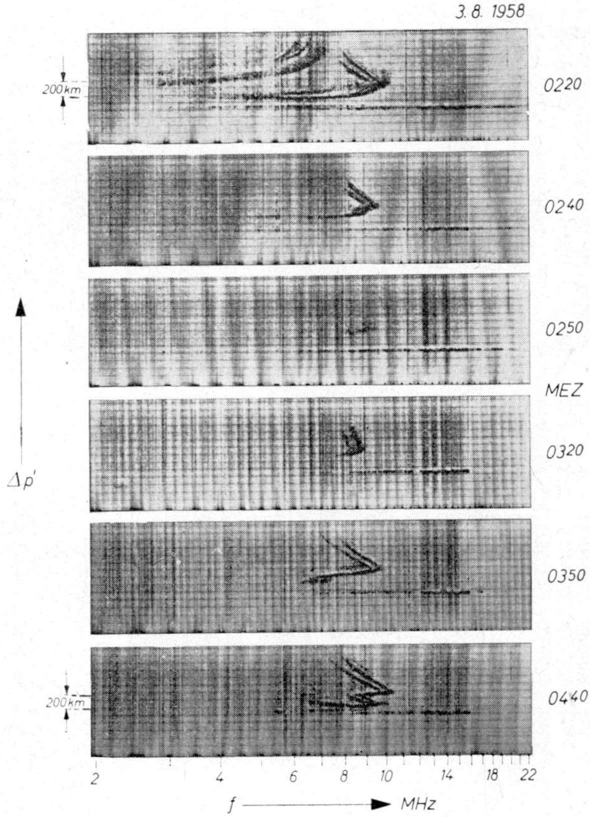

Abb. 12. Fernübertragungs-Ionogramme Sodankylä—Lindau. Wirkung der Absorption bei mittelstarker Ionosphären-Störung.

dieser Zeit lagen die 3stündigen erdmagnetischen Kennziffern zwischen 4 und 5. Die erste Aufnahme um 02·20 ist noch ungestört. Zwischen 02·40 und 03·20 h tritt wieder starke Absorption der F2-Übertragung und gleichzeitige Reflexion an Nordlicht-E auf. Ab 03·50 h normalisieren sich die Verhältnisse wieder.

Der geschilderte Verlauf ist charakteristisch für leichte Ionosphärenstörungen während der Nacht. Die alte Faustregel, daß bei Korpuskularstörungen auf niedrige Verkehrsfrequenzen übergegangen werden soll, erweist sich auf der

Strecke Lindau—Sodankylä somit als irrig. Tatsächlich ist eher ein Ausweichen gegen höhere Frequenzen anzuraten.

Nimmt die Intensität der Störung weiter zu, so setzt die Übertragung auf sämtlichen Frequenzen durch Absorption aus. Hieran ist offensichtlich in erster

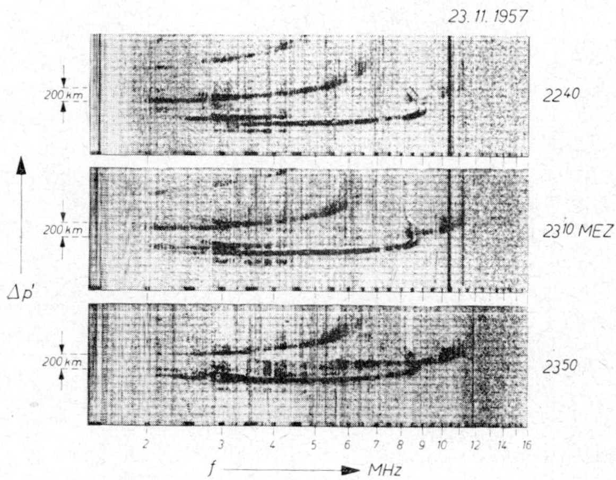

Abb. 13. Fernübertragungs-Ionogramme Sodankylä—Lindau. Ausbildung eines zusätzlichen Übertragungsweges.

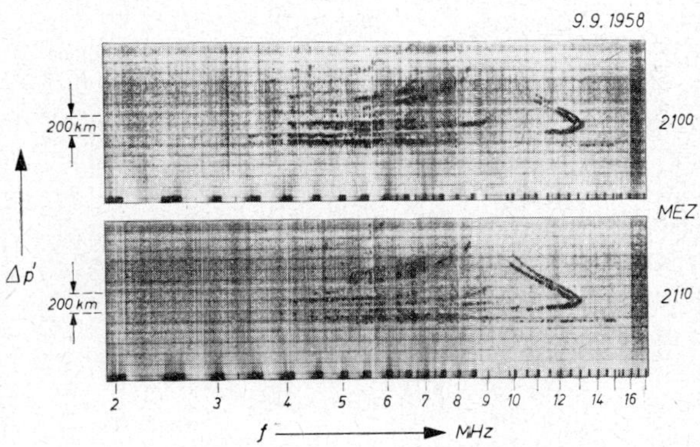

Abb. 14. Fernübertragungs-Ionogramme Sodankylä—Lindau. Ausfall eines Frequenzbereiches.

Linie das nördliche Ende der Übertragungsstrecke beteiligt, da dort im Verlaufe eines polar-blackout die Absorption unterhalb der E-Schicht viel stärker zunimmt als in mittleren Breiten. Auch in diesem Falle hilft ein Ausweichen gegen niedere Frequenzen nicht.

Ein Fall, in dem während der Nacht ein zusätzlicher Ausbreitungsweg entsteht, ist in Bild 13 dargestellt. Bereits auf der ersten Aufnahme um 22·40 h lassen sich Spuren von Echos oberhalb der F2-MUF erkennen. Im Laufe der folgenden 70 Min. bilden sich diese Echos zu einem zusammenhängenden Kurvenzug aus, der etwas oberhalb der 1 × F-Übertragung liegt. Die MUF dieses Übertragungsweges liegt 2·5 MHz über der F2-MUF. Die Vermutung, daß es sich bei dieser Übertragung um eine Zwischenreflexion an einer Nordlichtschicht handelt, konnte noch nicht bestätigt werden.

Einen anderen interessanten Fall stellt Bild 14 dar, das ebenfalls aus der leicht gestörten Nacht des 9.9.58 stammt. Auf der Aufnahme um 21 h tritt eine sog. Inselbildung auf, d.h. der übertragene Frequenzbereich ist nicht zusammenhängend, sondern weist zwischen 9 und 11 MHz eine Lücke auf. Gleichzeitig tritt eine deutliche kombinierte E-F-Spur auf, die bei etwa 9 MHz endet. Die Erklärung für diese Erscheinung ist folgende: Wie aus den Senkrechtlotungen in Lindau hervorgeht, trat um die fragliche Zeit eine sporadische E-Schicht mit einer Top-Frequenz von etwa 9 MHz auf. Diese sporadische E-Schicht ermöglichte das Auftreten eines E-F-Weges, schirmte aber gleichzeitig die 1 × F-Übertragung bis zu 11 MHz hinauf ab. Das Abklingen dieser Erscheinung ist in der folgenden Aufnahme um 21·10 h deutlich zu erkennen. Die kombinierte Spur ist schwächer geworden, gleichzeitig hat die abschirmende Wirkung für den F2-Sporn, der steiler durch die E-Schicht hindurchgeht, abgenommen, während die Flachstrahlung des F2-Weges noch abgeschirmt ist. Gerade solche Sonderfälle zeigen die Überlegenheit der Impulsfernübertragungsmethode mit veränderlicher Frequenz gegenüber anderen Beobachtungsmethoden.

3. *Vergleich zwischen Backscatter, Fernübertragung und Senkrechtlotung*

Bei den Backscatterbeobachtungen steht im Vordergrund des Interesses, inwieweit die aus der Senkrechtlotung in der Streckenmitte berechnete MUF, die aus den Backscatterbeobachtungen abgeleitete MUF und die in der Fernübertragung tatsächlich beobachtete MUF übereinstimmten. Das Verfahren, das für den Vergleich der Fernübertragung und der Backscatterbeobachtungen angewendet wurde, ist in Bild 15 schematisch dargestellt. Aus der Fernübertragung (untere Hälfte) wird zunächst die MUF abgelesen. Für diese Frequenz muß der Radius der Toten Zone gerade gleich der Entfernung Sodankylä—Lindau sein. Es müßte somit die Front der auf dieser Frequenz in Sodankylä beobachteten Backscatterechos gerade aus der Entfernung Sodankylä—Lindau, gemessen über die Ionosphäre, kommen. Tatsächlich wurde dies in ungestörten Fällen mit erstaunlicher Genauigkeit bestätigt. Die beobachteten Abweichungen im Laufweg wurden auf äquivalente Abweichungen der MUF umgerechnet. Das Ergebnis der Auswertung ist in Bild 16 (untere Hälfte) dargestellt, und zwar ist die Summenhäufigkeitskurve der relativen Differenz zwischen der beobachteten MUF und der aus der Backscatteraufnahme abgeleiteten MUF dargestellt. Die Abweichung des Medianwertes ist kleiner als 1 % und die Streuung ist erstaunlich gering. Wesentlich größer sind die Abweichungen zwischen der beobachteten MUF und der aus der Senkrechtlotung in der Mitte der Strecke berechneten MUF (obere Hälfte).

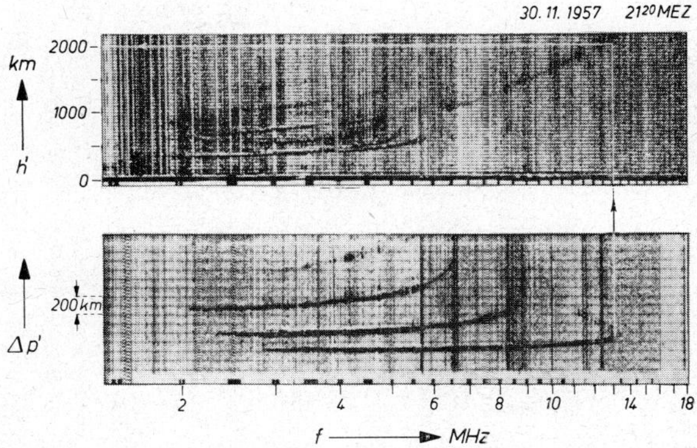

ABB. 15. Backscatteraufnahme (oben) und gleichzeitiges Fernübertragungs-Ionogramm Sodankylä—Lindau bei ungestörter Ausbreitung.

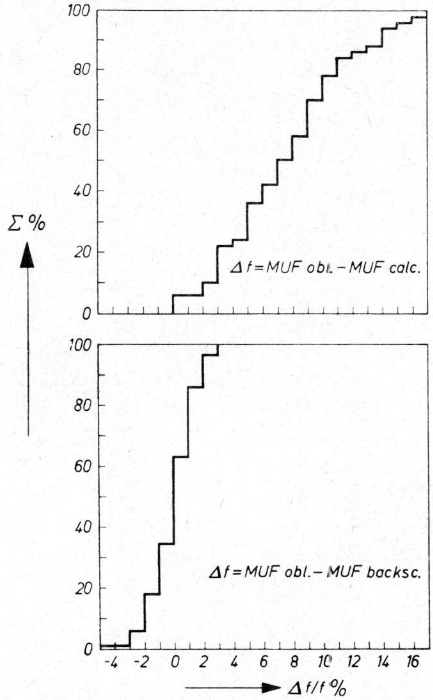

ABB. 16. Summenhäufigkeitsverteilung der relativen Differenz zwischen der beobachteten MUF und der aus der Senkrechtlotung berechneten MUF (oben); desgl. für die relative Differenz zwischen der beobachteten MUF und der aus Backscatter-Beobachtungen abgeleiteten MUF (unten).

Hier liegt der Medianwert bei + 7 %, d.h. die beobachtete MUF war im Mittel 7 % höher als die berechnete. Außerdem ist die Streuung wesentlich größer. Die größere Streuung ist verständlich, da die Senkrechtlotung eben nur die Verhältnisse an einem Punkt der Übertragungsstrecke wiedergibt, während die Backscatterbeobachtungen den ganzen Ausbreitungsweg erfassen. Bei der Differenz der Medianwerte ist zu berücksichtigen, daß bei den Berechnungen der MUF die

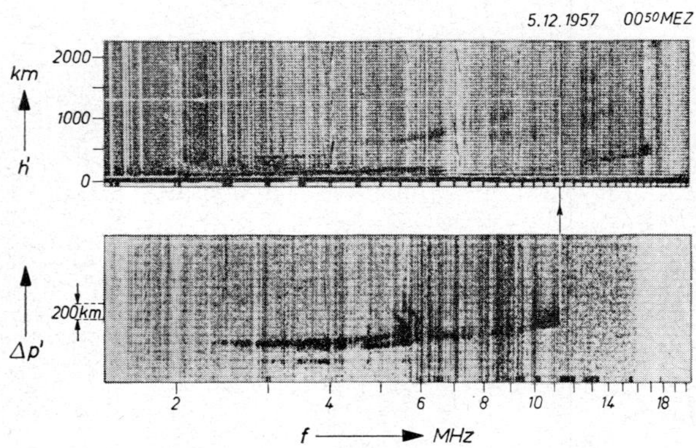

Abb. 17. Backscatteraufnahme (oben) und gleichzeitiges Fernübertragungs-Ionogramm Sodankylä—Lindau bei gestörter Ausbreitung.

Erde als gekrümmt, die Ionosphäre dagegen als eben angenommen wurde. Außerdem wurde der Einfluß des erdmagnetischen Feldes auf die ordentliche Komponente vernachlässigt. Eine frühere Überschlagsrechnung für eine Strecke von 1300 km zeigte, daß die Ionosphärenkrümmung für diese Entfernung etwa nur 2 % ausmacht. Selbst wenn man für 2000 km den doppelten Wert annimmt, so bleibt eine mittlere Abweichung von 3 % bestehen, deren Aufklärung bis jetzt noch nicht gelungen ist. Insgesamt stimmen die Ergebnisse überein mit den Resultaten früherer Versuche auf einer Strecke von 1320 km [4, 6] und mit den Beobachtungen von Silberstein [9] über eine Strecke von 2370 km in U.S.A. Auch hierbei ergab sich, daß die beobachtete MUF im Mittel größer ist als die aus der Senkrechtlotung berechnete.

Diese Ergebnisse gelten für ungestörte Ausbreitungsbedingungen, d.h. für Zeiten, in denen das Übertragungsionogramm den normalen Charakter zeigt. Bei Ionosphärenstörungen konnte bis jetzt kein klarer Zusammenhang zwischen den Fernübertragungskurven und den Backscatterechos festgestellt werden (Bild 17). An der Aufklärung der Zusammenhänge wird z.Z. gearbeitet, indem von einfachen Fällen zu komplizierteren übergegangen wird.

IV. Ergebnisse der Strecke Lindau—Tsumeb

Ein Querschnitt durch die Strecke Lindau—Tsumeb ist in Bild 18 wiedergegeben. Die Entfernung beträgt 7800 km. Ein 1 × F-Weg ist nicht möglich, für den

2 × F-Weg ergibt sich tangentiale Ausstrahlung bei einer Schichthöhe von 300 km. Für den 3 × F-Weg beträgt der Abstrahlwinkel 7°. Die Backscatterversuche in Lindau sind seit dem 1 März 1958 in Gang, die Fernübertragungsbeobachtungen in Tsumeb seit dem 1 August d.J. Eine Auswertung konnte bisher nur stichprobenweise durchgeführt werden, die Ergebnisse sind daher noch spärlich. Ein Beispiel einer Backscatterregistrierung auf 34 MHz zeigt Bild 19. Das erste Echo

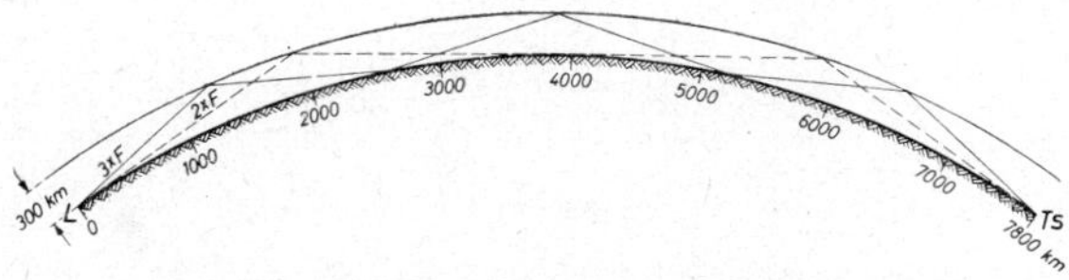

ABB. 18. Querschnitt durch die Versuchsstrecke Lindau—Tsumeb.

beginnt etwa bei 2000 km und erstreckt sich bis rd. 4000 km. Backscatterechos kommen also bei dieser Apparatur nicht nur vom Rande der Toten Zone, sondern bis zur Grenzreichweite des Tangentialstrahles. Im Gegensatz zu den klaren Verhältnissen beim ersten Echo ist die Deutung des 2. Echos schwierig. Gegen 09 h beginnt es wohl, wie zu erwarten, in einer Entfernung, die deg.

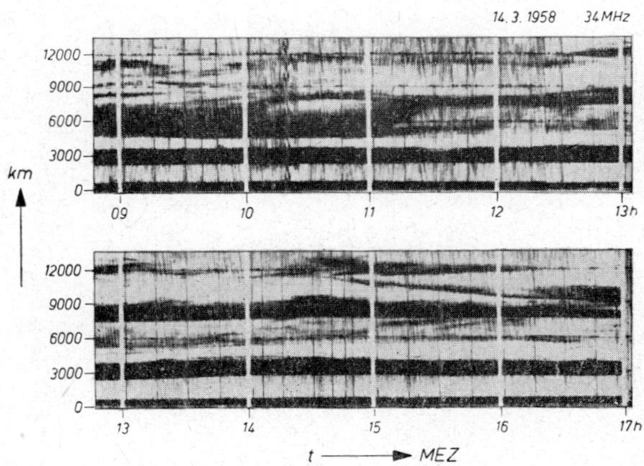

ABB. 19. Backscatterregistrierung in Lindau auf 34 MHz.

Doppelten des einfachen Übertragungsweges entspricht und erstreckt sich bis zur Grenzreichweite des doppelten Sprunges, nämlich bis zu etwa 8000 km. Zwischen 11 und 12 h ändert sich jedoch der Charakter erheblich. Das zweite Hauptecho rückt in eine Entfernung von etwa 7500 km und bleibt dort für den Rest des Nachmittags. Aus der theoretischen Entfernung des doppelten Sprunges kommt dagegen nur ein relativ schwaches Echo an. Noch komplizierter sind die Verhältnisse bei den Echos höherer Laufzeit.

Das Verhalten der Echos beim Aussetzen der Übertragungsmöglichkeiten entspricht der Erwartung (Bild 20). Das Echo wird dabei immer schmäler und verschwindet schließlich in einer Entfernung von 4500 km. Auch die z.B. von Villard, Stein und Yeh [12] beobachteten Echos sehr langer Laufzeit traten bei den Versuchen auf. Bild 21 zeigt eine Registrierung, die auf 18 MHz bei Sonnenaufgang gewonnen wurde. Vor dem Einsetzen der 1 × F-Übertragung treten bereits Echos aus 5000 und später auch sehr stark aus 6000 km Entfernung auf. Die Entstehung dieser Echos wird üblicherweise durch Schichtneigung erklärt [10].

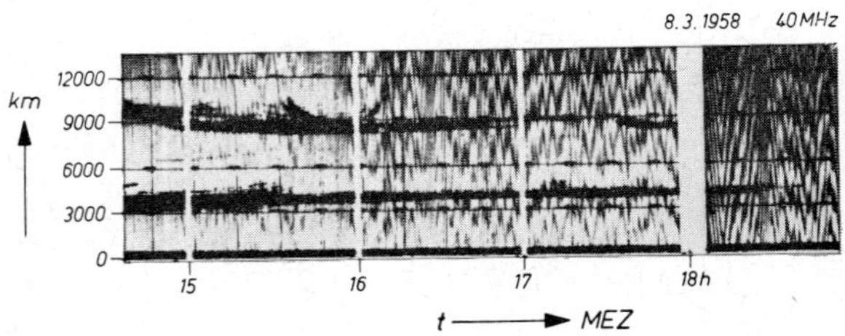

ABB. 20. Backscatterregistrierung in Lindau auf 40 MHz bei Aussetzen der Echos.

Die Möglichkeit, die Backscatterbeobachtungen auf verschiedenen Frequenzen durchzuführen, gestattet es, die Größe der Toten Zone als Funktion der Frequenz wenigstens punktweise zu bestimmen. Bild 22 zeigt zwei solche Backscatter-Ionogramme. Man kann nun versuchen, die gemessene Größe der Toten Zone mit derjenigen zu vergleichen, die sich theoretisch aus der Senkrechtgrenzfrequenz und den MUF-Faktoren für verschiedene Entfernungen ergibt. Man muß dabei

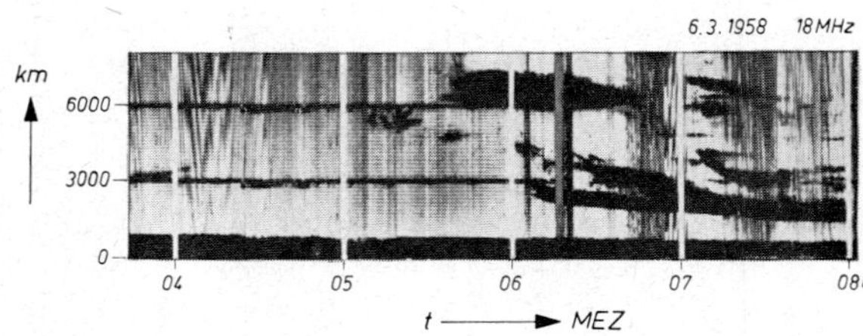

ABB. 21. Backscatterregistrierung in Lindau auf 18 MHz mit Echos langer Laufzeit.

allerdings berücksichtigen, daß im Backscatter-Ionogramm nicht direkt die Größe der Toten Zone, sondern die Laufzeit der Echos, die vom Rande der Toten Zone kommen, gemessen wird. Man kann jedoch bei bekannter Schichthöhe den Umweg der Echos rechnerisch berücksichtigen. In Abb. 23 ist der theoretische Zusammenhang zwischen dem MUF-Faktor und der Größe der Toten Zone als gestrichelte Kurve dargestellt. Die ausgezogene Kurve gibt den Zusammenhang zwischen den MUF-Faktoren und dem minimalen Laufweg der Backscatter-Echos. Die eingetragenen Punkte sind die experimentellen MUF-Faktoren, die sich aus dem Verhältnis Betriebsfrequenz zu Senkrechtgrenzfrequenz für die verschiedenen Entfernungen ergeben. Sie sollten alle auf der ausgezogenen Kurve liegen. Tatsächlich ist die Übereinstimmung bis zu einer Entfernung von 3000 km recht gut, für 4000 km liegen allerdings die MUF-Faktoren höher als die theoretischen. Außerdem ist die Abweichung für die Messungen des 19.3.58 für sämtliche Entfernungen über 1500 km beträchtlich. Es wurde vermutet, daß die Abweichung damit zusammenhängt,

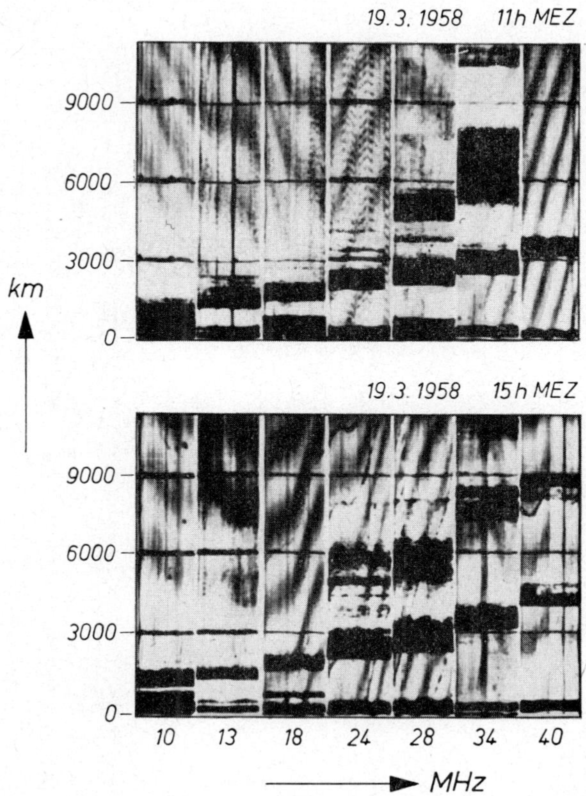

Abb. 22. Backscatter-Ionogramme Lindau 10–40 MHz.

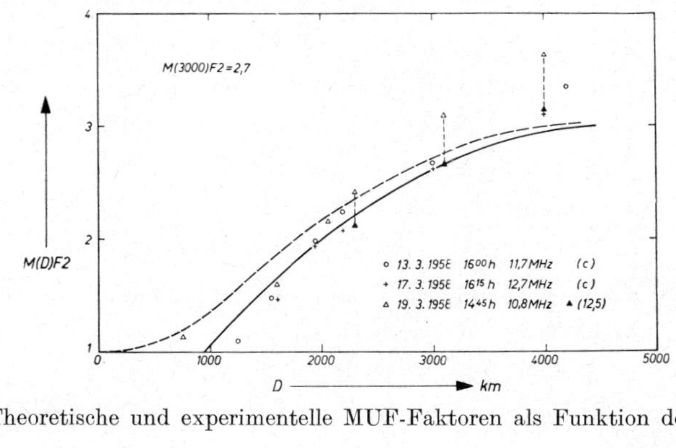

Abb. 23. Theoretische und experimentelle MUF-Faktoren als Funktion der Entfernung.

daß als Senkrechtgrenzfrequenz die in Lindau gemessene zugrunde gelegt wird. Es ist aber bekannt, daß die Grenzfrequenz nach Süden hin zunimmt, wodurch dann eine Zunahme des MUF-Faktors vorgetäuscht wird. Dieser Effekt muß sich am stärksten auswirken für die größten Entfernungen, weil dann der Reflexionspunkt am weitesten nach Süden rückt. Diese Erklärung scheint zuzutreffen. Jedenfalls ergab sich für den 19.3.58 folgendes: Die Grenzfrequenz war an diesem Tag in Lindau besonders niedrig. Werden für die Entfernungen ab 2000 km die gleichzeitig gemessenen Grenzfrequenzwerte von San Alessio bei Rom herangezogen, so rücken die Punkte wesentlich näher an die theoretische Kurve heran. Für die beiden anderen Tage konnte leider eine derartige Korrektur nicht durchgeführt werden, da die Messungen in San Alessio zu den fraglichen Terminen ausgefallen sind. Besonders deutlich wird dieser Effekt in dem folgenden Bild Nr. 24. Es sind hier zwei Tage gegenübergestellt. Am 14.3.58 war die Differenz

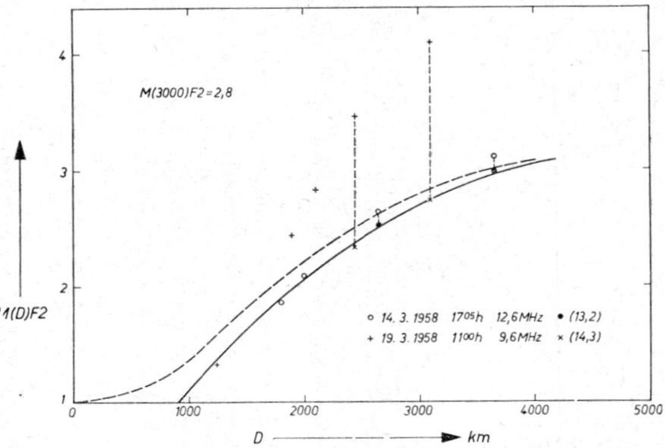

ABB. 24. Theoretische und experimentelle MUF-Faktoren als Funktion der Entfernung.

zwischen der Grenzfrequenz in Lindau und in San Alessio nur 0·6 MHz. Die für Lindauer Grenzfrequenz berechneten Kurven stimmen schon relativ gut mit den theoretischen Werten überein. Der Einschluß der Werte von San Alessio für große Entfernungen gibt eine weitere Verbesserung der Übereinstimmung. Ganz anders sind die Verhältnisse am 19.3.58, an dem ein mittlerer Ionosphärensturm herrschte. An diesem Tage war die Grenzfrequenz in Lindau besonders niedrig, in San Alessio dagegen höher als normal. Die Abweichung der MUF-Faktoren nach oben ist dementsprechend besonders groß. Eine Umrechnung mit Hilfe der Werte von San Alessio für größere Entfernungen ergibt wieder eine ausgezeichnete Übereinstimmung zwischen Theorie und Messung.

Die Ergebnisse zeigen umgekehrt, daß es möglich ist, aus den Backscatter-Ionogrammen mit recht guter Genauigkeit auf den Verlauf der Grenzfrequenz längs der Ausbreitungsstrecke zu schließen. Voraussetzung dafür ist allerdings, daß Antennen mit guter Bündelung verwendet werden, damit die Rückstreuung aus einer definierten Richtung kommt [13].

Über den Zusammenhang zwischen den Backscatter-Erscheinungen und den Ausbreitungsbedingungen auf der Strecke Lindau—Tsumeb kann z.Z. noch nichts gesagt werden, da die Auswertung gerade erst begonnen hat. Da bei dieser langen Strecke die Ausbreitung in Mehrfachsprüngen erfolgt, sind die Zusammenhänge sicher komplizierter als auf der Strecke Sodankylä—Lindau.

Einen Eindruck von Fernübertragungs-Registrierungen vermittelt die Abb. 25. Sie zeigt die zeitliche Änderung der Übertragungswege auf 27 MHz vom 20.9.58 15 h bis 21.9.58 11 h. Zunächst sind 3 gut definierte Ausbreitungswege und daran anschließend eine streuende Echospur sichtbar. Die beiden Spuren mit der kürzesten Laufzeit setzen nacheinander aus (17·00 h bzw. 18·45 h). Gegen Sonnenuntergang treten mehrfache Ausbreitungswege sehr langer Laufzeit auf, etwas

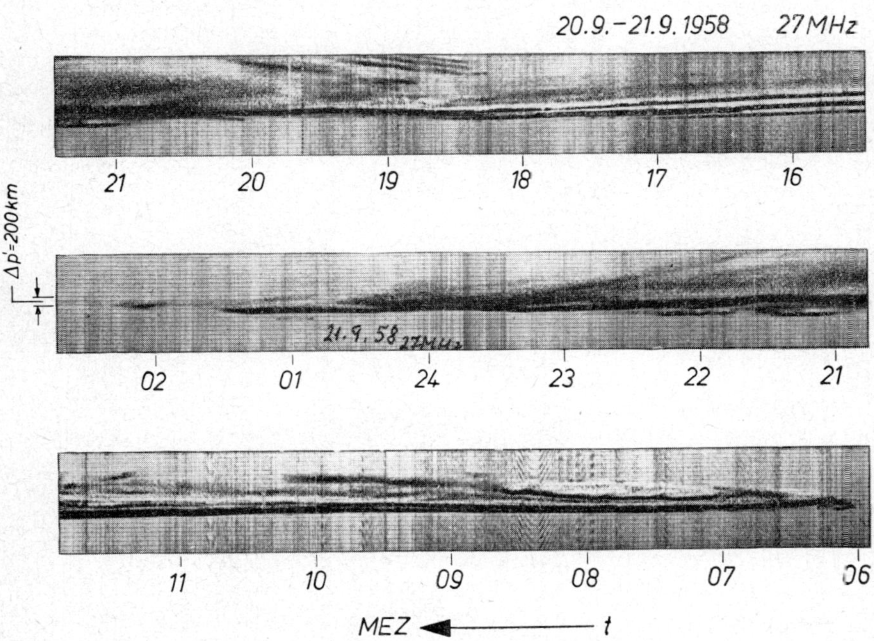

ABB. 25. Fernübertragung auf 27 MHz Lindau—Tsumeb.

später zieht sich die streuende Spur über einen großen Laufzeitbereich auseinander. Nach Mitternacht vermindert sich die Streuung, und um 01·30 h setzt die Hauptübertragungsspur aus. Eine weitere schwache Spur verschwindet gegen 02·15 h. Kurz nach 06 h setzen die verschiedenen Übertragungswege nacheinander wieder ein und gegen 10 h geht die 3. Spur in Streuung über. Eine Deutung im einzelnen muß einer späteren Veröffentlichung vorbehalten bleiben, ebenso ein Vergleich mit den gleichzeitigen Rückstreubeobachtungen. Vermutlich entspricht das Zeichen mit der kürzesten Laufzeit dem 2 × F-Weg, das 2. Zeichen dem 3 × F-Weg und das 3. Zeichen dem 4 × F-Weg. Die starke Streuung fällt zusammen mit dem Auftreten eines Flutterfadings, das sämtlichen Signalen auf der Strecke Europa—Südafrika in der 1. Nachthälfte anhaftet. Vermutlich steht es mit dem Auftreten des äquatorialen spread-F in ursächlichem Zusammenhang.

Zusammenfassung: Die bisherigen Ergebnisse zeigen, daß im Bereich des ersten Sprunges ein enger Zusammenhang zwischen Backscatter und Fernübertragung besteht. Bei stärkeren Störungen treten im Gebiet der Polarlichtzone allerdings Verhältnisse auf, die noch nicht voll zu übersehen sind. Über die Verhältnisse im Gebiet der Mehrsprünge können noch keine Aussagen gemacht werden, da die Auswertungsarbeiten erst begonnen haben.

References

1. Abel, W. G. & Edwards, L. C. (1951). *Proc. Inst. Radio Engrs, N.Y.* **39**, 1539.
2. Dieminger, W. (1951). *Proc. phys. Soc. Lond.* **64**, 142.
3. Dieminger, W., Geisweid, K. H. & Möller, H.-G. (1955). *NTZ* **8**, 578.
4. Dieminger, W. & Möller, H.-G. (1956). *Nuovo Cim.* no. 4, del Supp. al **4**, serie x, 1532.
5. Kono, T. (1950). *Rep. Ionospheric Res., Japan*, **4**, 127, 189.
6. Möller, H.-G. (1956). *J.A.T.P.* **9**, 155.
7. Peterson, A. M. (1951). *J. geophys. Res.* **56**, 221.
8. Shearman, E. D. R. (1956). *Instn elect. Engrs*, **103**, 121.
9. Silberstein, R. (1954). *Trans. Inst. Radio Engrs, N.Y.* AP–**2**.
10. Stein, S. (1958). *J. geophys. Res.* **63**, 217.
11. Villard, O. G. Jr. & Peterson, A. M. (1952). *Trans. Inst. Radio Engrs, N.Y.* PGAP–3.
12. Villard, O. G. Jr., Stein, S. & Yeh, K. C. (1957). *Inst. Radio Engrs, Nat. Convention, N.Y.*
13. Widdel, H. U. (1957). *Arch. elektr. Übertr.* **11**, 429.

Geomagnetic Control of Ionospheric Scatter Signals

H. S. Marsh

Air Force Cambridge Research Center, Bedford, Mass.

An interesting anomaly appears in studying E layer characteristics as indicated by the behavior of signals operating below and above the so-called "Maximum Useful Frequency."

Nelson [6] and others have shown that in the former case, the semi-specular reflection or refraction of signals below the MUF, from the E layer, develops characteristics that correlate closely with diurnal and seasonal variations in sunlight intensity. This intensity of photoionization is governed largely by the sun's zenith angle (X) from the point of reflection [2]. This zenith distance can be used to predict quite accurately propagation conditions for frequencies below the E layer MUF [7].

Now, studying the effects of what appears to be the same E layer region on frequencies above the MUF, we find a vast difference in behavior of this same layer. One of the most striking differences is that, instead of a smooth annual curve, following the sunshine pattern of a maximum in summer and a minimum in winter, the scatter signals, at frequencies above the MUF, show a semi-annual amplitude curve, with maxima in both winter and summer, the strongest and weakest sunshine seasons, and two minima at the two equinoxes. This appears clearly in our Figs. 1(a), 4(b), and with the June maximum partly masked by absorption, in Fig. 5. Evidently other agencies, not having visible effects on reflected signals, are overriding photoionization as the controlling factor for the scatter signals, for at least a large part of the year.

The question that arises now is, what agencies are causing this anomaly; or, what mechanisms in addition to mid-point photoionization scatter are active in VHF scatter propagation, and under what conditions?

This is not an easy question to answer. Several possible solutions have been proposed, such as meteor ionization and auroral reflection. Before considering the data and possible inferences from it in detail, let us consider a few more general ionospheric concepts.

As previously stated, Nelson and others have shown that the E layer, at frequencies below the MUF, follows solar control very closely. The F2 layer, on the other hand, is shown by them to exhibit characteristics correlating poorly with sunlight or zenith distance factors, but showing strong correlation with indices of geomagnetic activity. In other words, the F2 layer displays strong geomagnetic control. One way of loosely visualizing this is to picture an ionized plasma, whose numeric electronic content might vary quite directly with the sun's zenith distance, but whose form, thickness, and other parameters effecting propagation might vary as the results of modulation of this plasma by such possible agents as winds, tides, variations in magnetic fields, or trapped ion fields due to possible corpuscular bombardment during magnetic storms.

The mechanism connecting F2 layer behavior with magnetic instability is not well understood at this stage.

Another layer formation showing strong correlation with geomagnetic indices, and even less with photoionization, is the sporadic E, or Es. This layer occurs at unpredictable times, at heights slightly above the regular E layer, has several recognized forms, occurring consistently at different latitudes. Again, most of these forms show strong geomagnetic controls. Several generating agencies have been suggested for various forms of sporadic E, but very little appears to be definitively supportable concerning its generation at present.

To recapitulate—the E layer, at frequencies low enough to be reflected or refracted, shows direct sunlight control. Conversely, the F2 and Es layers, and the E layer at frequencies high enough to penetrate it, and be scattered en route, all show strong symptoms of geomagnetic control, possibly with features of the controlling mechanism common to all of them.

What this may mean is that a closer understanding of the control mechanism of any one of these latter modes may well prove to be applicable to the others. On the other hand, some knowledge of the variations of the "reflective" layers, as outlined above, is basic to a satisfactory understanding of the inferences that may be drawn from the ionospheric E layer scatter data available.

The main goal of this paper will be to select some features from the AFCRC ionospheric scatter link data that point up this departure from the photoionization, or Booker-Gordon type of control [8], and to hazard some qualitative speculation as to what inferences they might justify concerning possible mechanisms.

Geomagnetic A_k indexes, since made available, were not at hand at the time of this study, and so were not used.

A change in emphasis since the summary of this paper was first proposed has prevented some of the analytic work that was then anticipated. However, this change was caused by the initiation of an exploratory study of HF arctic signal blackouts, and abnormal enhancements. Some of the preliminary data from this new study may appropriately shed some information on the objectives to be included in the present work.

It is a little difficult initially to visualize a mechanism in which the E layer plasma shows widely different characteristics, varying under totally different controls, to waves in two different wavelength or frequency groupings.

One crude picture that can be used to visualize such a process is as follows:—consider the E layer to be replaced by a series of layers of wire screens or nettings, vertically stacked, with possible variability in the mean sizes and separations of the upper layers.

Consider now a stream of ping pong balls shot or blown at this layer of screens. If the size of the balls is large compared to the mesh of the bottom screen, corresponding to electrically long waves, at frequencies below the MUF, none of the balls will penetrate through it, but all will be reflected or bounced back downwards at angles and energy distributions governed by the laws defining the structure of this bottom screen only.

Now consider the results with a second stream of balls of diameter small compared with the mesh of the bottom screen, corresponding to wavelengths of frequencies well above the MUF.

Dependent on the wavelength factor and angle of incidence, a few of these small balls will strike the wires of the bottom screen, and be scattered back downward, but most of them will penetrate the bottom screening and encounter the upper layers of screening. It can be seen that a large or small number of them can be scattered back downward and through the bottom screen, depending on mesh sizes and vertical separations of the upper layers.

To stretch the analogy, if we could assume the number and mesh sizes of the screens to be analogous to the total electronic population, directly generated and governed by solar ionization and assume the vertical distribution of the upper screens, and possibly their mesh sizes, or mean free paths, to be controlled by geomagnetic, tidal or other independent effects, it can be seen that two totally different sets of laws might seem to be presented to the two different sets of ball sizes, or wavelengths.

With these simplified pictures in mind, let us take a quick look at some of the data accumulated on the AFCRC ionospheric scatter circuits, and try to draw some qualitative inferences from the anomalies appearing.

The illustration presented, the references and some of the descriptive material, have been taken from a forthcoming "Summary Report on AFCRC-Scituate VHF Ionospheric Scatter Circuits at High Sunspot Numbers." Certain slides and illustrations have been selected without changing the numbering from this report, so our illustration or reference numbers will not all be internally consistent.

In brief, amplitude and other observations have been recorded on a 38 Mc/s 20 Kw signal from Cedar Rapids, Iowa, to Scituate, Massachusetts, about 1730 km distance, for over a year beginning in 1956, around the top of the sunspot cycle. Good scale correspondence with the classic NBS FPIS experiments around 1951–52 [1] was found but several interesting anomalies partly connected with the present high sunspot number have been noticed.

Of particular interest to our current inquiry is the sensitivity of the scatter signal to approach to Es MUF resulting in enhancement, sometimes of the scatter signal, sometimes of absorption. Variations of apparent antenna and frequency sensitivity with these phenomena, as evidenced by marked amplitude sensitivity, are strikingly evident, and are under further study.

Commencing October 1957, 50 Mc/s signals have also been available permitting some frequency gain, as well as antenna gain, comparisons. Through an inadvertence the same antenna was used for transmitting both frequencies alternately, making its 50 Mc scale height far above "optimum," and depressing its beam's vertical take-off angle nearly one degree below that for 38 Mc. This causes different adjacent patches of the ionosphere to be illuminated by the two frequencies.

A rather startling result of the antenna mis-scaling noted is that minor peaks and valleys in the diurnal curves appear to be repeated perhaps one-half hour later on 38 Mc than on 50 Mc. A tentative explanation for this strange delay has been outlined, but the effect, if real, is not yet understood.

This anomalous effect, together with other results of the multipath and antenna frequency sensitivity criteria emerging, show a need for critical examination of, and possible addition to, the simple mid-path scattering model adapted from tropospheric scatter theory.

Diurnal Median Values

Monthly averages of hourly median values show quite strong correspondence with the NBS low SSN variations (sunspot number) [1], scaled upwards as a result of the higher solar activity. Major and minor *high signal* strength periods appear around the summer and winter solstices, and *low signal* periods around the spring and fall equinoxes as before. However, the June values, which by the classical

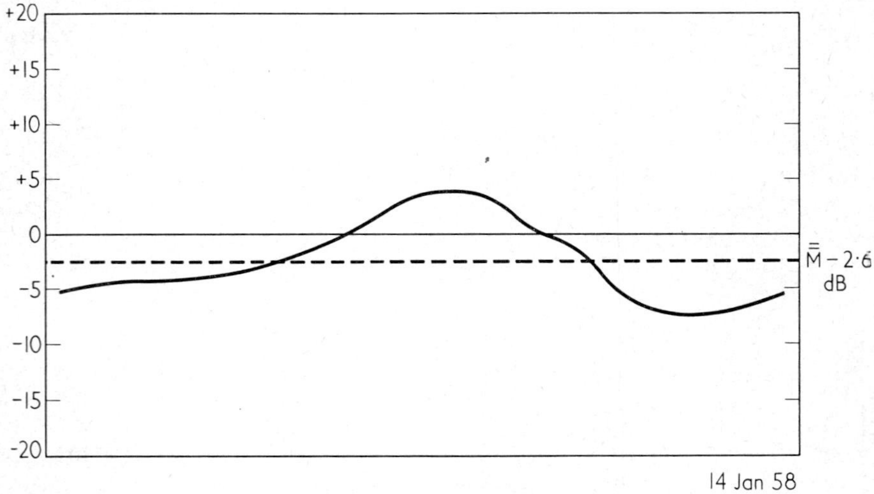

Fig. 1a. Average of median values for year 1957 (excluding Aug.) (only one run for Sept.) 38 mc beam. Cedar Rapids-Scituate.

photoionization and mid-point scatter theory, would be expected to be the greatest, are actually several decibels below the May and July values, particularly at the upper percentile levels. This is believed due to increased absorption during these high Es MUF periods. Figure 1(a) shows the yearly average of hourly medians and Fig. 1(b) shows the average diurnal median field curves for March, September, June and December, 1957, illustrating the seasonal variations. Since only one or two 24 hour runs per week were possible, variations due to individual disturbed days have introduced distortions more noticeable than with larger statistical accumulations of data. Major enhancements due to below-MUF layer propagation and other sources have been largely eliminated by arbitrarily interpolating for values where an abnormal enhancement, defined as rising over 4 db between half hourly readings, occurred.

Close correlation between Es predicted MUF and variation of observed scatter median levels for various months will be described later, in connection with cumulative distribution variations.

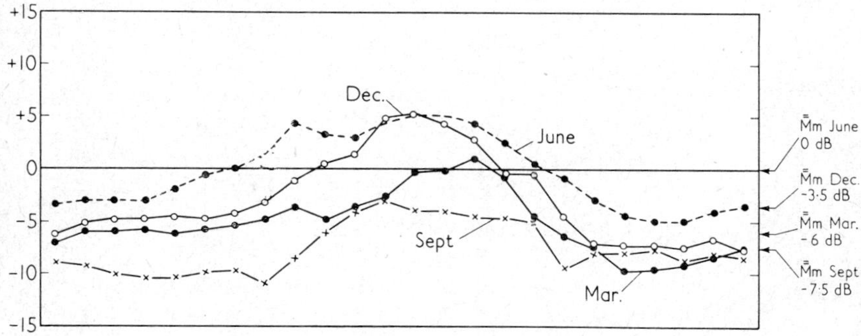

Fig. 1B. Quarterly diurnal median levels, 1957.

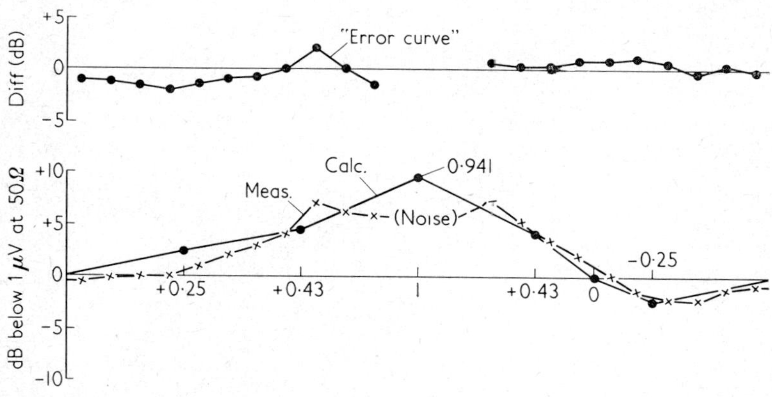

Fig. 2. Cedar Rapids-Scituate calculated and observed median signals. June '57'. 38.48 mc beam.

Predicted values based on photoionization agree quite closely with the June observed values at these fairly low latitudes. The predictions are based on semi-empiric calculations from the Chapmann-Bartels instantaneous photoionization expression (Equation 43, p. 507, ref. 2). This is shown by Fig. 2 for June 1958 and follows less closely for other seasons having lower photoionization densities, where other factors than the direct photo-illumination might govern.

Cumulative Amplitude Probability Distributions

Cumulative distributions of intermittent observations during 1956 and for the year 1957 are shown in Figs. 3 and 4(a). Distributions for the four key months of 1957 are shown in Fig. 4(b). It is noted that the distribution curves slide laterally from month to month without too great a change in slope appearing. A factor appearing obvious in Figs. 3 and 4(a), and noticeable in the monthly curves of Fig. 4(b) is the apparent presence of two distinct distributions. It is seen that the

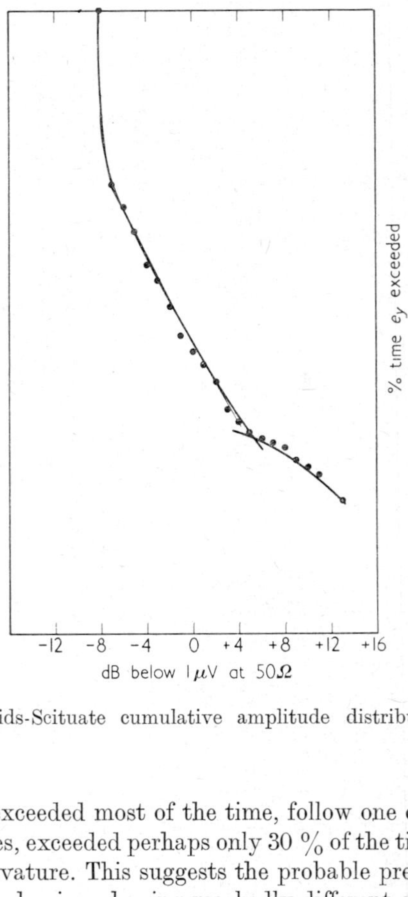

Fig. 3. Cedar Rapids-Scituate cumulative amplitude distribution. 38.6 mc. Beam-year 1956.

lower signal values, exceeded most of the time, follow one curve quite smoothly while the higher values, exceeded perhaps only 30 % of the time, follow a different curve of reversed curvature. This suggests the probable presence of at least two separate preferred mechanisms having markedly different characteristics.

For closer scrutiny the values exceeded each month for 2 %, 50 % (median) and 98 % of the time are plotted as bar charts for each month in the center of Fig. 5. Below are plotted the predicted Es and F layer MUFs for the path for each month and above are shown the number of abnormal enhancements (defined as rises greater than 10 db above running median) for each month. Clearly, the amount of

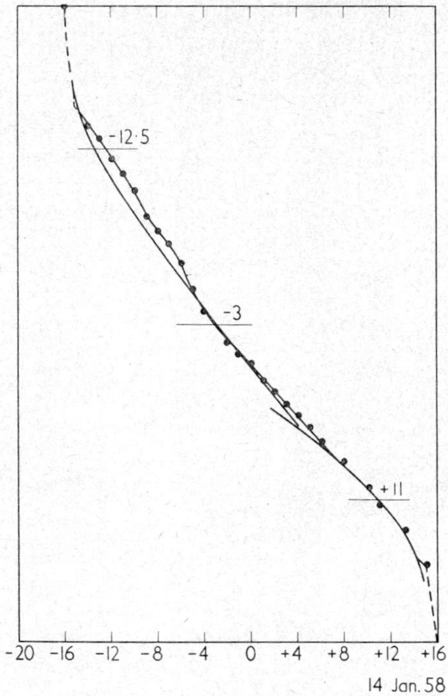

FIG. 4A. Cedar Rapids-Scituate cumulative distribution for 1957. 38-48 mc beam.

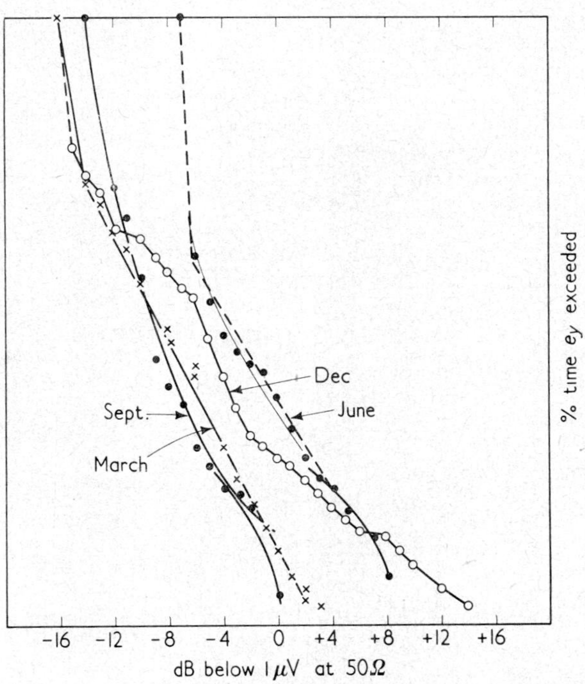

FIG. 4B. Cedar Rapids-Scituate cumulative amplitude distributions by quarters. 38-48 mc beam 1957.

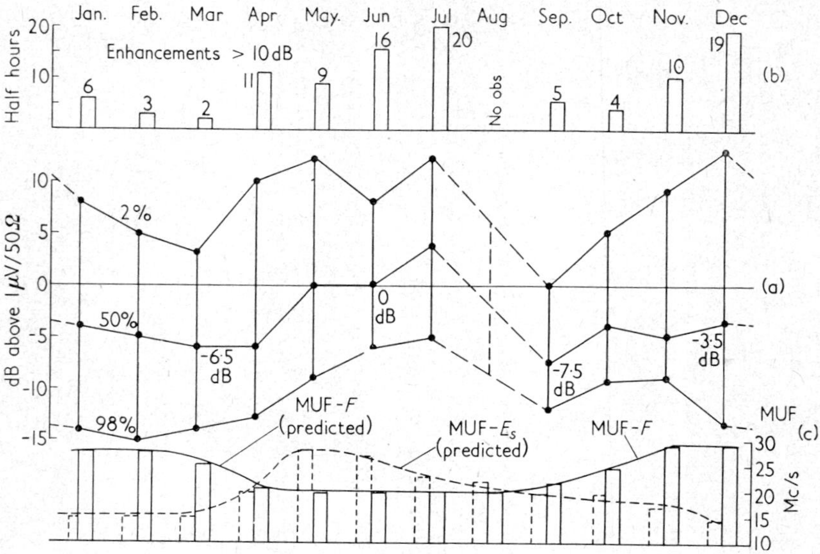

FIG. 5.

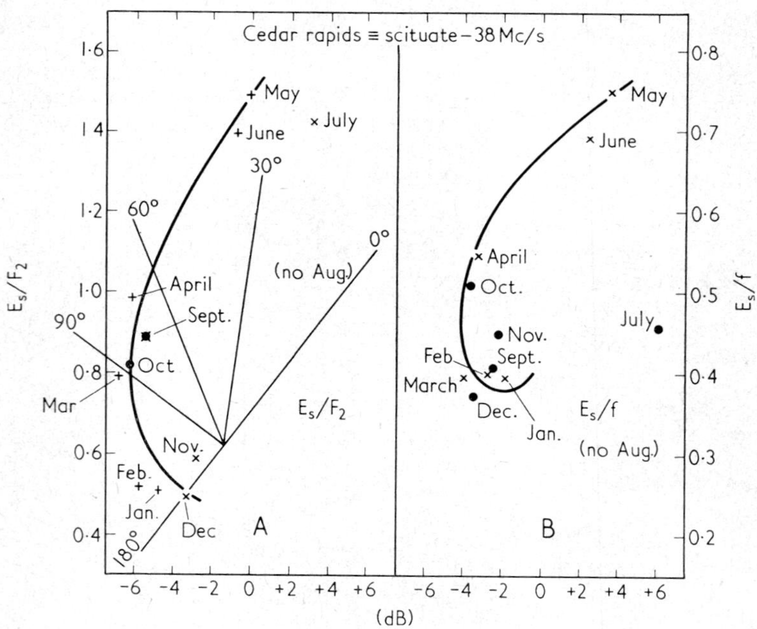

FIG. 6. Cedar Rapids-Scituate—38 mc. Variation of monthly median with: (*a*) predicted E_s muf/F muf; (*b*) predicted E_s muf/operating frequency.

abnormal enhancement to the scatter signal follows the rise of the Es MUF fairly well, drops at the equinoxes, then rises again with the winter rise of F2 MUF.

Variation of the scatter monthly medians with predicted Es for the month is shown in a different manner in Fig. 6. In Fig. 6(a) the number of decibels the monthly median is above or below the yearly median is plotted against the ratio of the predicted Es MUF to the predicted F MUF for the month. The points on this curve lie quite close to a hyperbolic spiral except for the June high absorption anamoly noted above.

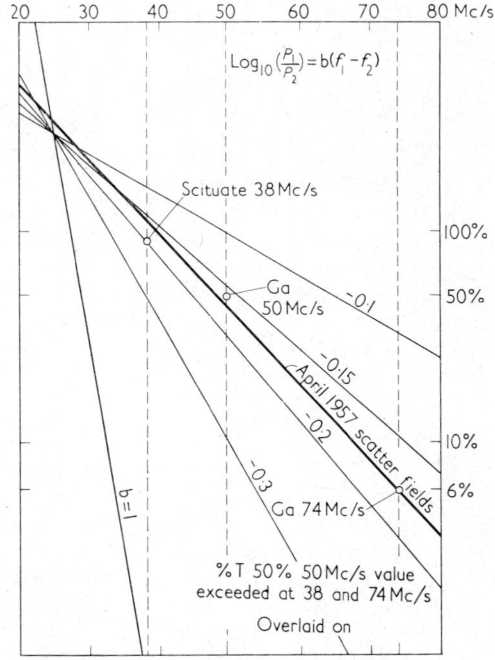

FIG. 7. (L-P fig. II G-4): Smith; e_s frequency dependence curve.

In Fig. 6(b) the same monthly median decibel variations are plotted against the ratio of predicted Es MUF to actual operating frequency, resulting in a tighter spiral curve of similar form

It can be seen that a strong connection is indicated between the monthly median field strengths and a MUF factor comprising the ratio of the Es and the F2 predicted MUFs, along the locus of the hyperbolic spiral. Equivalent seasons seem to fall into closely adjacent batches, progressing from 180° through a batch of spring values centered around 80°, and rising to the summer batch centered around 30°. (Here the June anomaly, attributed to absorption, drops the June value below the expectedly lower values for May and July.)

From the summer values the points again progress regularly back down the curve through autumn to winter values. A physical explanation for the hyperbolic form of this apparent connection has not yet been established. The value of the figure at this stage is that it seems to indicate a close connection between the yearly scatter field variation, and the relative variations of the two MUF factors, both magnetically controlled, instead of the close solar control that mid-point photo-illumination theory would lead us to expect.

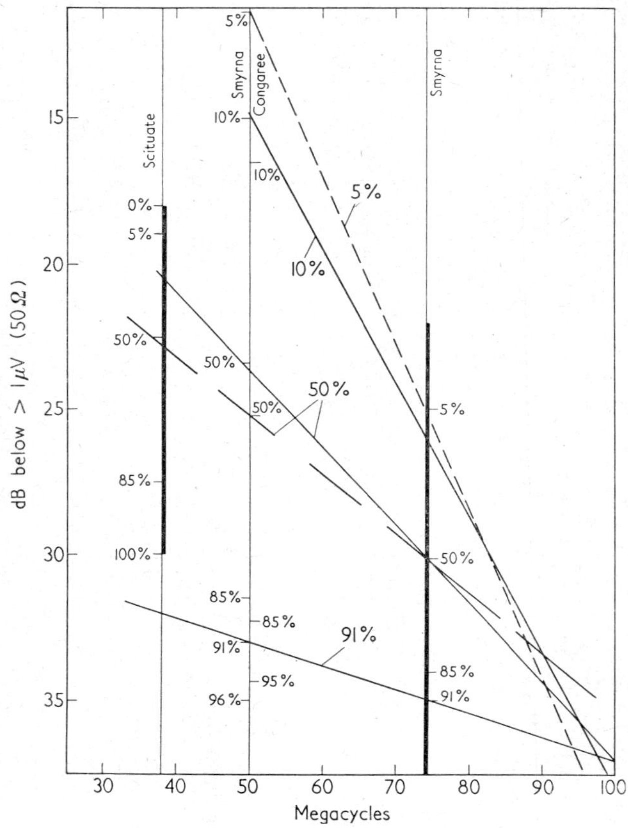

Fig. 8. Variation of scatter field distribution with frequency and antenna pattern.

Our Fig. 7 is based on Smith's study of sporadic E [3]. He quotes Mrs. Lindemann-Phillips's expression for Es frequency dependence, of the form "$\log_{10} (P1/P2) = b(f_1 - f_2)$": he shows a graph of this for various "b" factors (relatable to the frequency dependence exponent of ionospheric scatter) as his Fig. II G 4. Our Fig. 7 shows a portion of this curve with our April 1957 scatter signal probability values for given levels and frequencies on 3 circuits superimposed on it.

The three points shown by crosses are for Cedar Rapids to Scituate, Mass., 38 Mc and from Walpole, Mass., to the neighborhood of Atlanta, Ga., on 50 and

74 Mc, on very roughly comparable circuits. The crosses show that the same normalized value which was exceeded 50 % of the time on 50 Mc was exceeded around 99 % of the time on 38 Mc but only 6 % of the time on 74 Mc. All three points liequite close on Mrs. Philips's chart to her ($b = -0{\cdot}18$) line. This indicates close consistency between her Es frequency-dependence law and our scatter frequency dependence exponent, comparable to her "*b*" factor.

Evidently the mechanism which controls Es generation and its frequency dependence and that which controls ionospheric scatter propagation frequency dependence are closely allied, at least for the lower latitudes.

To explore this apparent relation more directly, Fig. 8 was set up using data available through courtesy of Georgia Tech. Research Institute. This figure is plotted on linear coordinate paper with frequencies as abscissas as before, but with measured field strengths in decibels above 1 μV/50 Ω as ordinates, in place of probability values. On this the cumulative probabilities of occurrence of various field strengths for three circuits above are marked, plus a Walpole, Mass. to Congree (Columbia, S.C.) 50 Mc signal. These values are all taken from April 1957 distributions. Contour lines connecting the 5, 10 %, 50 % and 91 % probability points for 50 Mc to Congaree and 74 Mc to Smyrna (Atlanta) produce an approximately confocal cone of rays focusing roughly around 95 Mc and $-$ 36 db, and in fairly close agreement with the 50 Mc Smyrna distribution. (The 50 Mc Congaree signal was chosen as representing assertedly more reliable measurements than the 50 Mc Smyrna signal.)

(By using the linear display of logarithmic decibel values in place of the logarithmic scale of probability values, a rough correspondence with Mrs. Phillips's Frequency Dependence Chart for sporadic E is maintained.)

Apparently if the scatter signals are governed by the same or similar agencies to those generating sporadic E, strong internal consistence should be expected between the contours of various percentiles at the different frequencies.

Examination shows strong correspondence between the percentiles of the three higher frequency signals. However, the 38 Mc Scituate signal distribution is highly inconsistent with the other three. Its median value is reasonably aligned with the 50 % contour describing the frequency dependence of the other signals, but its range of variation is vastly shrunken.

Part of this could be accounted for by enhanced absorption at the higher electronic densities accompanying the higher signal levels, but the shrinkage appears to be somewhat symmetrical.

Two other environmental differences might be considered as possibly accountable for the decreased range of variation. The first is that the Scituate path is roughly East-West while the others are North-South. The other, intuitively more attractive, is that the Scituate circuit used high gain, narrow beam antennas, focused sharply on the path mid-point, while the other circuits used broad, single Yagi antennas, with a half-power horizontal beamwidth of about 60°.

Now, if we considered the Scituate circuit antennas as responsive only to the single, mid-point path, while the broad-beam Yagis on the other circuits were also responsive to several other possible side-paths, and with their mid-point path contribution lessened by the lower forward gain of the Yagis, it is credible that their received fields might fall far below or far above the restricted variational range of the mid-point path alone, dependent on the activity and transmission characteristics of the several off-beam paths possible. Long delay multipath measurements on these circuits have indicated the probable existence of such auxiliary paths, and possible mechanisms, such as F layer propagation plus back-scatter, and auroral reflection, found compatible with the delays measured.

Acknowledgment

In conclusion, thanks are due to Mr. R. W. Miner and Sgt. W. H. Cameron, who were responsible for operation and adjustment of the equipment and reduction of the data. Airmen J. Temples and F. Reiner also gave valuable assistance both in maintenance of the equipment and manipulation of the data.

References

1. *Proc. Inst. Radio Engrs, N.Y.* **43**, 10.
2. Chapmann & Bartels, "Geomagnetism".
3. Smith, E. "World wide Occurrence of Sporadic-E", *Nat. Bur. Stand. Circ. Rep.* 3575.
4. Reports Raytheon AFCRC Contract AF 19(604)2422.
5. Bailey, etc. (1952). *Phys. Rev.* **86**, 141 ff.
6. Miller, *Trans. Inst. Radio Engrs, N.Y. PGCS* CS-**2**, 2. (A). Nelson, "Radio Weather Forecasting Techniques", *Trans. Inst. Radio Engrs, N.Y.* (B).
7. "Ionospheric Radio Propagation", Nat. Bur. Stand. Circ. 462.
8. Booker (1956). *J. atmos. terr. Phys.* **8**, 204.